BIOLOGY

BIOLOGY
Second Edition

KNUT NORSTOG

Fairchild Tropical Garden, Miami, Florida

ANDREW J. MEYERRIECKS

University of South Florida, Tampa, Florida

Charles E. Merrill Publishing Company
A Bell & Howell Company
Columbus Toronto London Sydney

Published by
Charles E. Merrill Publishing Company
A Bell & Howell Company
Columbus, Ohio 43216

This book was set in Novarese
Administrative Editor: Robert Lakemacher
Developmental Editor: Jennifer Knerr
Text Designer: Cynthia Brunk
Production Coordination: Martha Morss
Cover Design: Cathy Watterson
Cover Art: Charles Harper. Used by permission of Cincinnati
Nature Center.

Library of Congress Catalog Card Number: 84–61747
International Standard Book Number: 0–675–20419–4
Printed in the United States of America
1 2 3 4 5 6 7 8 9 10—89 88 87 86 85

Preface

We have attempted in the pages of this book to achieve a balance between several levels of biological organization. One of our guiding principles has been our observation, during many years of college teaching, that while a textbook usually has a short life, the concepts students acquire in their undergraduate courses in biology often must last a lifetime. Because the first course in biology, for most students, is the only formal exposure to such concepts they will have, a balanced presentation of biological principles is vital.

NEW TO THIS EDITION

■ The text is specifically redirected to the **nonscience major** market. It is shorter than the first edition; it has fewer chapters (26 instead of 36); and the art-to-text ratio has increased.

■ **Color** has been added to the prologue to enhance the introduction of the nonmajor student to the biological world, the scientific method, and the concepts and practice of biology. Other color photographs enliven the later presentation of plant and animal diversity as well as ecology.

■ Although the text is directed to nonscience majors, it maintains a solid approach to the principles of biology presented in a traditional format. The **opening unit**—including biochemistry, molecular and cellular biology, and genetics—has been strengthened and updated. Chapters on mitosis, meiosis, and cellular and molecular genetics have been completely rewritten. Genetic engineering topics reflect current research. A section on nitrogen fixation in the Photosynthesis chapter sets the coverage of this text off from its competitors.

■ The presentation of the **diversity** of life has been significantly **streamlined** and moved back in the text (Part III of four parts). The material on plants has been significantly condensed, but the botany in the text continues to be a major strength and a distinguishing feature from other introductory texts which typically slight the Plant Kingdom.

■ A chapter entitled "**Health**" has been added which covers topics including nutrition, stress, drugs, disease, immunity, and cancer. It also features a number of health-related aspects of cell and molecular biology. This chapter is unique among introductory texts of this kind; reviewers indicate it is a welcome and long-overdue departure.

- Approximately 80 **new illustrations**, 50 **new photographs** (including some quality electron micrographs), and 20 **new color photographs** have been added. Artwork from the First Edition has been corrected, when necessary, and enhanced by a new color treatment.

- **New boxed essays** have been added covering subjects as diverse as split genes, hermaphroditic hyenas, and earthworm excavation.

- The unit on organismal biology has been reorganized to create unusual and **consolidated presentations** of topics including inheritance, multicellularity, and hormones. **New material** on the nervous system and on muscles has also been incorporated into this unit.

- The presentation of **evolution** has been **concentrated** in the last unit, "Ecology and Evolution". **New material** on genetic drift, ecology, and the green revolution has been added to this unit.

- Reviewers acclaim the **readability** of the text and the value of the **in-text aids**, including chapter summaries, key words, review questions, and suggested readings.

- The **Instructor's Manual** and **Student Study Guide** have been revised for the new text by Albert A. Latina at the University of South Florida. A **test bank** is available in both written and computerized formats, and two-color overhead **transparencies** will be provided to adopters.

Organization The 26 chapters of this text are for the most part manageable as single reading assignments. Those that are not are divisible into well-circumscribed subject-matter areas. Although the information is presented in a "graded-sequence," some flexibility in chapter sequencing is possible. Certain chapters are readily interchangeable, and some may be omitted. Although only one chapter is specifically titled "Evolution," a strong evolutionary theme is maintained throughout. The early chapters deal primarily with molecular and cellular phenomena. In these, examples are drawn from all of the kingdoms (5 or 6, depending upon inclusion or exclusion of the Archaebacteria). In subsequent chapters, however, we have separated matter pertaining to structure and functions of protistans, fungi, plants, and animals. We do not think it appropriate, for example, to equate circulation of blood with transport in xylem and translocation in phloem.

Pedagogy Several "convenience" features of this book will prove useful for instructors and students. Chapter summaries succinctly recapitulate each chapter's content so that students have ongoing content reviews available as they read. Other texts omit these valuable chapter summaries or relegate them to the study guide. At the end of each chapter is a listing of Key Words, analogous to the "key words" listed in many scientific papers. These can be used as a rapid review of the main points covered in the text. Also found there are Questions for Review and Discussion, containing items for direct testing of concepts learned, along with more thought-provoking problems. Chapters end with a Suggested Reading section, which we have updated to the time of publication. Inclusion was based on easy-to-find sources. At the back of the book, in the Appendix, is a complete listing of the classification system used throughout, followed by a Glossary, which contains definitions for Key Words and other important terms. The glossary is, we think, very complete, containing nearly double the terms and definitions found in certain other books. In addition, interesting vignettes are interspersed throughout chapters to stimulate and maintain student awareness.

ACKNOWLEDGEMENTS

We have again been encouraged by the kind and helpful suggestions of many colleagues, students, and other friends. We have consulted with and/or received advice from the following:

Betty D. Allamong, Ball State University
Marvin Anderson, Spokane Community College
Michael Bell, Richland Community College
Dan Benjamin, Central Michigan University
Herman Brockman, Illinois State University
Thomas A. Cole, Wabash College
Donald Collins, Orange Coast Community College
Louis F. DeWein, Capital University
Roger del Moral, University of Washington
Richard Firenze, Broome Community College
Jack Fisher, Fairchild Tropical Garden
David J. Fox, University of Tennessee
Elon Frampton, Northern Illinois University
Paul K. Glasoe, Wittenberg University
William Grey, Northern Illinois University
James Grosklags, Northern Illinois University
Arnold Hampell, Northern Illinois University
Laszlo Hanzeley, Northern Illinois University
Bruce Harris, M.D., Chicago, Illinois
David T. Jenkins, University of Alabama in Birmingham
Malcolm Jollie, Northern Illinois University

William H. Leonard, Louisiana State University
Joseph Linton, University of South Florida
James McCleary, Northern Illinois University
Thomas R. Mertens, Ball State University
Norma Meyerriecks, Tampa Veterans' Administration
   Hospital
Sidney Mittler, Northern Illinois University
Dian Molsen, Northern Illinois University
William O'Dell, University of Nebraska
Janet Piwoworski, Parke-Davis, Chicago
Kathryn Podwall, Nassau Community College
Kadaba Pralad, Northern Illinois University
Brian Rigby, Northern Illinois University
Arne Schjeide, Northern Illinois University
Joseph Simon, University of South Florida
John Skok, Northern Illinois University
Dennis Stevenson, Barnard College of Columbia
   University
Gerald Summers, University of Missouri
Paul Umbeck, University of Tennessee
Leonard S. Vincent, Georgia Southern College
Paul Voth, University of Chicago
E. W. Wickersham, Penn State University
C. C. Wolfe, Northern Virginia Community College
John L. Zimmerman, Kansas State University

We also are indebted to those who provided illustrative material, often going to considerable trouble to do so. We have acknowledged these sources in the legends of the illustrations, thinking this a more positive expression of our appreciation than a listing in an appendix. We also are indebted to Richard Lampert for early encouragement in the development of the manuscript, to Robert Lakemacher for his seemingly tireless work in behalf of the two editions, and to Jennifer Knerr for similar devotion to the development of this edition. We also thank Martha Morss, Jean Simmons Brown, April Nelson, Stuart Weibel, Lorraine Woost, Ronald G. Boisvert, Jim Hubbard, Connie Geldis, Ann Mirels, and Cleo Eddie for their combined efforts on development and production. In particular, we thank Priscilla Fawcett for her preparation of a number of original drawings, Babs Klein and Betty Hallinger for preparing what we think is a very complete and useful index, and to James Watson for his help with the aforementioned glossary.

Knut Norstog
Andrew J. Meyerriecks

Brief Contents

Contents

Prologue

The male of a certain firefly species sends his typical coded sequence of flashes, to which a nearby female firefly will flash a coded response. The male then moves to the female in order to copulate with her. This time, however, the returning flashes lure him to his death; a female of a second species has intercepted his signal and mimicked the answering flashing pattern, and, when he approaches, this "femme fatale" kills and devours him. How, in so simple an animal as an insect, does one explain this bizarre behavior, reminiscent of a spy drama on the late-late movie?

A Florida family is proud of a clump of poinsettias growing alongside their carport. Each December in the past, the plants have produced lovely blooms, but this year they fail to flower. A neighbor explains that the new yard light installed in midsummer and timed to turn on at sunset is at fault. The explanation for this puzzling situation is that flowering in poinsettias is induced by a series of exposures to dark periods that are slightly longer than the alternating light periods. The poinsettia is, therefore, a short-day plant (actually a misnomer, as the long dark period is what brings on the flowers). If the dark period is broken by light or shortened, as in the present case, flowering will not occur. In fact, some short-day plants are so sensitive that a brief exposure during the night to light from a flashlight, or even from the striking of a match, can delay or prevent flowering. Understanding the way in which plants sense light and dark and measure their durations is one of the many remarkable advances in modern science.

Dramatic advances have been made on many other fronts. About 60 years ago a pair of Canadians, F. G. Banting and C. H. Best, discovered a way to extract and purify the hormone insulin from the pancreases of dogs. The Nobel Prize was awarded for this work, because it provided for the first time a means of saving the lives of millions of diabetics. In spite of many attempts to find economical alternatives, until recently the only practical source of insulin has been the pancreases of slaughtered animals. Many diabetics, however, have uncomfortable, sometimes severe, allergic reactions to such insulins, which differ in subtle ways from human insulin. Now, however, by procedures referred to under the general heading "recombinant DNA," a common intestinal bacterium, *Escherichia coli*, has been "taught" to make human-type insulin. Commercial quantities of such insulin are currently in production. This is another example of genetic engineering, a field of human endeavor with far-reaching implications. What is the biological ba-

1

sis of genetic engineering, and how may it affect human life? This subject will be discussed in coming chapters.

Many other intriguing short stories could be extracted from the pages of any current biology textbook, or for that matter from a good newspaper. All such stories exemplify some of the excitement that modern biology holds for the informed person, as well as the impact it can and will have on the lives of all of us.

BIOLOGY AND THE WORLD OF LIFE

It would be difficult to imagine the earth without its living populations of plants, animals, and various microorganisms (color plates 1–4). Just as we find it difficult to visualize **life** on earth without its human element, so too we know we could not exist without the many kinds of nonhuman life with which we share our planet. It is a rare individual indeed who does not see beauty in his or her natural surroundings, or does not marvel at the interlocking relationships of different forms of life to each other. In the pages of this book, we will explore many of these relationships, beginning with the atoms and molecules of which life is composed, then considering the chemical and physiological actions that make the assemblages of atoms behave as true living organisms. We hope to show not only that the basic units of life, the cells, are the source of all the functions we think of as "life," but also that such functions are integrated at higher levels in whole organisms, often composed of thousands, millions, and even billions of cells. Then we will seek to understand the ways in which individuals of all kinds act collectively in their own societies and how these societies interact with other societies and with the environments in which they live. The study of all these relationships is biology, which, like other sciences, attempts to arrive at certain operating principles, not by reason alone, but by reason based on observation and experimentation.

SCIENCE AND THE SCIENTIFIC METHOD
What is Science?

The effects of science are many. Science has led to antibiotics, jet airplanes, computers, and hybrid corn, as well as toxic wastes, radiation, and possibly the means of the eventual extinction of the human race; those, however, are products of science, not science itself. What is there about the workings of science that has enabled humans to rise above and dominate the environment in so spectacular a fashion?

Science, which literally means "to know," generally is considered to refer to a series of thought processes by means of which new concepts are formulated. These concepts of science relate to the natural world and, although often quite theoretical and abstract, deal with forces or objects that have a physical existence and can be measured: the shape of the earth, the force of gravity, or the structure and function of cells. Thus science is not directly concerned with the study of such philosophical matters as ethics, morals, or religion, even though all of these become involved in making decisions about the uses of scientific discoveries. Science is, instead, a systematic way of thinking about things and analyzing them as objectively as possible and, in so doing, minimizing wrong decisions and bad explanations.

The speculative nature of science is its greatest strength. Were it a rigid, dogmatic system, little progress would have resulted. Instead, the nature of science and those who practice it is such that concepts constantly are tested, and either confirmed, revised, or rejected. When a concept such as a chemical, nutritional, or evolutionary principle is found to be faulty, it is evidence of the strength rather than the weakness of science.

How Science Works

All through history, individuals have employed methodologies of science in their attempts to discover new facts about nature. It is generally accepted that science is based upon a productive kind of methodology called the **scientific method.** Although not a rigid procedure, this methodology has as its starting point an **observation** of some particular natural phenomenon (e.g., the falling of an object through space; the development of maggots in rotting meat; the imprint of a leaf in a piece of rock). The second step in the scientific method usually consists of the formulation of an explanation, or **hypothesis** (gravity attracts all falling bodies equally; maggots arise from eggs that flies have deposited in spoiling meat; rocks were once soft mud and therefore may retain the impression of leaves and other objects). Then, most importantly, the hypothesis is put to test by further **experimentation** and observation (large and small objects are dropped from the same height and their rate of fall noted; meat is placed in a vessel from which flies are excluded by a screen; the rate of deposition and consolidation of mud in present times is carefully measured). Depending on the results of experimentation, the hypothesis is accepted, rejected, or modified.

Many fields of study are considered science. These include the **social sciences** (political science, sociology, anthropology, some aspects of psychology, etc.), in which observation rather than experimentation tends to prevail, and the **natural sciences** (physics, chemistry,

PLATE 1

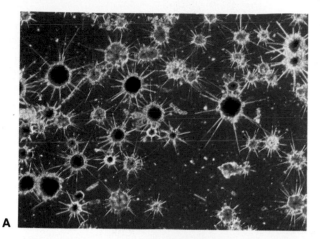

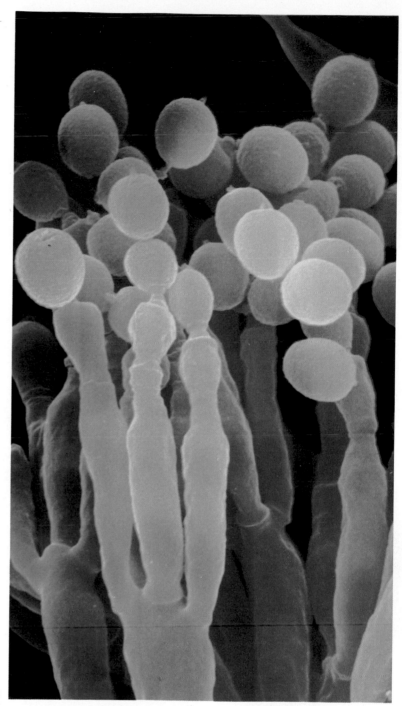

A. Radiolarians. Kingdom Protista, Phylum Protozoa, Class Sarcodina, (© Manfred Kage/Peter Arnold, Inc.) **B.** Stinkhorn fungus, *Clathrum* sp. Kingdom Fungi, Subdivision Basidiomycotina. **C.** Spores, asexual reproduction of a fungus, *Penicillium* sp. Kingdom Fungi, Subdivision Ascomycotina (sac fungi). (© Manfred Kage/Peter Arnold, Inc.)

PLATE 2

A

B

A. Red tube sponge in foreground, hard and soft corals in background. (James A. Bohnsack) B. Hybrid angelfish. A natural hybrid of *Holocentrus ciliaris* × *H. bermudensis.* (James A. Bohnsack) C. Pacific tide pool plant and animal life. Starfish in foreground, barnacles in background, sea palm at center. Phylum Echinodermata, Phylum Arthropoda, Division Phaeophyta (brown algae).

C

PLATE 3

A

B

C

A. Students examining life in Pacific tide pools, Vancouver Island, B.C., Canada. **B.** Floating mat of green algae and water fern in a cypress pond, Florida. **C.** Colorful bract of *Anthurium* is not a flower but a modified leaf that functions as a pollinator attractant. Very small flowers make up the flowering stalk (inflorescence) extending from the bract.

PLATE 4

A. Rainforest scene, Malaysia. A row of caterpillars (Phylum Arthropoda) feed on the leaf of a tropical shrub. (David Lee) **B.** Female cone of a cycad, *Encephalartos ferox,* a primitive seed plant native to South Africa. Kingdom Plantae, Division Pinophyta. **C.** Rainforest tree. Buttressed trunks are common in larger rainforest trees. They probably aid in support of large trees in shallow rainforest soils. (David Lee) **D.** Tropical savanna, Africa. A savanna is a tropical or subtropical open grassland interspersed with shrubs and copses of trees. (Warren Garst/Tom Stack)

mathematics, geology, astronomy, biology, etc), in which measurements and experimentation commonly are employed to verify hypotheses.

Biology is at the center of the natural sciences, drawing on all the other sciences for support. No other science has more ramifications, uses a greater variety of tools and techniques, or studies such widely ranging phenomena as does biology in studying the processes known collectively as life.

WHAT IS LIFE?

As a result of centuries of scientific progress, a great deal has been learned about the organization and operation of those things referred to as living. Now it is known that plants and animals are complex entities, or **organisms,** composed of **cells;** that cells are made up of smaller parts called **organelles;** and that organelles in turn consist of **molecules** and **atoms.** In addition, many of the chemical processes that occur in the living cell are understood. Unfortunately, when it comes to the question of what exactly life is, there is a problem, because scientists really do not know how to measure life. When they take a living organism apart, there remains no specific residue that could be called life. Life cannot be defined in such terms as weight, volume, or wavelength, nor can it be said that life is a liquid, a solid, or a gas. No doubt this baffling nature of life contributes in large measure to the fascination biology holds for many individuals.

In the absence of a specific definition of life, generalizations are often resorted to. Life may be defined as that unique property of organisms that makes them living; that is, it is the sum of all the activities of living organisms. Among these are the ability to react to stimuli (**irritability**); the ability to transform energy from one form into another by photosynthesis and respiration (**metabolism**); the ability to use this energy to make new chemicals (**biosynthesis**); and, finally, the ability to manufacture more living matter and produce a new generation of organisms (**reproduction**).

MAJOR CONCEPTS OF BIOLOGY

Every science is based on a cluster of fundamental ideas that have been verified by observation and experimentation. Without an understanding of these basic principles, the study of a science becomes a boring recital of seemingly unrelated facts. However, if one approaches the learning of a science with a grounding in the underlying ideas of that science, then factual matter becomes comprehensible. Generations of scientists have pro-

duced a number of basic concepts to which biologists adhere and which constitute the framework of biology. Among these are the following:

1 *The earth is ancient.* The earth is about 4.5 billion years old. Life, which was not present during the first 2 billion years of earth history, originated spontaneously from nonliving matter by processes referred to as **chemical evolution.**

2 *Life comes from preexisting life.* In the present, all forms of life on earth result from the reproduction of parental life, not spontaneously from nonliving matter. (This concept appears to be in conflict with the preceding statement, a paradox that is discussed in Chapter 2).

3 *The basic unit of life is the cell.* Living things are cellular in nature. The structural and functional unit of life is a discrete entity, the cell. Cells of various forms of life are remarkably similar. Some simple lifeforms are single cells; others, including plants and animals, are composed of many cells. Both one-celled and many-celled forms of life constitute organisms (organized life).

4 *The cell is an energy conversion machine.* Cells absorb, convert, and use several forms of energy in their life processes. The initial energy source is the sun, whose **radiant energy** (light) is **converted** into **chemical energy** (food) by **photosynthesis** occurring in green cells of plants (Fig. P-1). Animals and other nonphotosynthetic organisms directly or indirectly utilize food produced by plants. All cells, green and nongreen, use food energy in their life processes.

5 *Organisms are interdependent.* All life is interlocked in

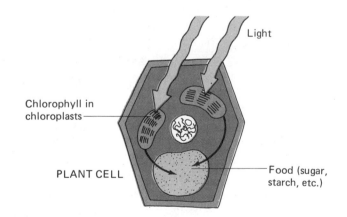

FIGURE P-1
Energy relationships in a plant cell. Light energy is absorbed by chlorophyll in chloroplasts (a type of organelle) and converted into food energy (chemical energy) in the form of sugars, starch, and other substances.

energy relationships such as exist between plants, as the primary producers of food, and animals and other food-consuming organisms that feed on plants and on one another. Life also is interdependent in many often-subtle ways, including cooperative relationships in which two organisms coexist with mutual benefit and other arrangements in which some organisms act as hosts and others as parasites. Such interrelationships among organisms, as well as interactions with the physical environment, are the subject of the branch of biology known as **ecology.**

6   *Organisms are self-regulating.* Organisms are able to regulate and modify their life processes in relation to changes in both external and internal conditions. This capacity is called **homeostasis.**

7   *Organisms grow from within.* Organisms have the unique ability to **grow** from within themselves. This is accomplished by internal additions to the substance of living cells and also by reproduction of cells. (The growth sometimes observed in nonliving objects such as icicles, stalagmites, stalactites, and many crystalline substances occurs by addition of matter to the outside.)

8   *Organisms are self-reproducing.* Organisms reproduce themselves, either by simple division of the whole organism or, in the case of multicellular organisms, by forming specialized sex cells which produce a new organism. In either case, the new organism, by processes of growth and specialization of its parts, attains the form of the parent organism. This is **development.**

9   *Development is controlled by genes.* The development of an organism is directed by subcellular entities, each of which determines a specific structural or functional characteristic of an organism. These entities are called **genes** (Fig. P-2); the structural and functional characteristics of both a developing organism and a mature organism are regulated by the collective actions of all the genes. The study of the actions of genes is called **genetics.**

10 *Genes are comprised of* DNA. A chemical, **deoxyribonucleic acid (DNA),** composes the genetic information (all the genes) of the cell and organism. When an organism reproduces, its DNA also is predictably and accurately reproduced, so each new generation possesses essentially the same genetic information as preceding generations.

11 *Genes are capable of change.* On rare occasions, a gene may change into a new form. Such changes, called **mutations,** are a result of modification in the struc-

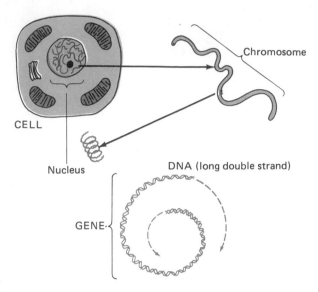

FIGURE P-2
Structural relationships of genes in a cell. Genes are located in larger, elongated bodies called **chromosomes,** which in turn are found within the **nucleus** of the cell.

ture of the DNA of the gene. A mutation of a gene produces a modification of a particular characteristic of an organism.

12 *Mutant genes may be inherited.* If mutations occurring in an organism are not so harmful as to cause death, they may be passed on to succeeding generations. Over long periods of time, **inheritance** and other processes may lead to the evolution of new forms of life.

13 *Nature selects optimal gene combinations.* Organisms may be viewed as being in competition with one another. In any given circumstance, organisms best adapted structurally and functionally to their environment are more apt to survive and reproduce. Such adaptations originate from gene mutations. This process is known as **evolution by natural selection.**

14 *Mutations of genes may lead to new species.* An accumulation of several or many mutations in the line of descent of a group of similar organisms, acted upon by natural selection, may produce a new and unique form of life, known as a **species.** The accumulation of genetic changes over millions of years has produced the great diversity among organisms known today.

Because much of the remainder of this book is based upon the concepts just outlined, keeping them in mind will prove a great aid to interpretation and understanding of modern biology.

SUMMARY

Biology, because it deals with forces, processes, and objects that can be measured and studied experimentally, is one of the natural sciences. In seeking to discover new information about life and its processes, biologists employ the scientific method, which has as its starting point an observation of some phenomenon. This observation leads to the development of a hypothesis, an explanation or interpretation of the phenomenon. The hypothesis is tested by observation and experimentation, and broad generalizations called concepts are developed. Many such concepts make up the body of knowledge that constitutes biology.

The Biology
of Cells

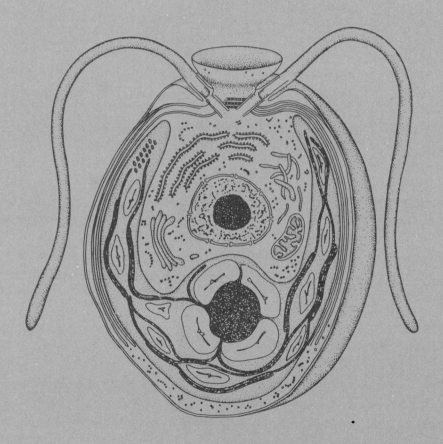

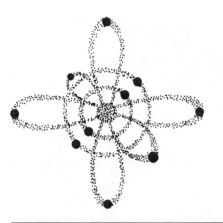

1

The Chemistry of Life

Key atoms of carbon, hydrogen, and oxygen make up the wood of our dining room table, compose the cooking oil in the kitchen cupboard, and are combined in the rubbing alcohol stored in the medicine cabinet. A relative handful of other kinds of atoms, arranged in a great variety of combinations, constitute all the material things we are likely to encounter in our daily lives. That a few basic elementary units compose everything substantial in the universe is now a widely accepted concept, even though not one of us has ever seen a solitary atom of anything. The concept of elementary units, however, is not new! The ancient Greeks reduced the composition of all things, both living and nonliving, to just four elements (earth, air, fire, and water), assuming that when substances are burned, air is needed, flames and water are produced, and ashes (earth) remain. Later, the Greek philosopher Democritus (460?–370 B.C.), proposed that the elements might consist of a few kinds of indivisible particles, or **atoms,** and, in addition, that such atoms might differ from substance to substance. Atoms of water perhaps were smooth and slippery; other atoms possibly were rough and heavy, and so on, but this was pure conjecture—the Greeks did no experiments to verify their suppositions.

The ancient Egyptians, who tended to be more matter-of-fact and less philosophical than the Greeks, recognized seven **elements**—gold, silver, mercury, iron, copper, tin, and lead—and "electrum," which was actually a mixture of gold and silver. With the exception of electrum, their elements were substances that were elemental in nature; that is, they were substances that could not be broken down further to yield new or different elements. To this day, with certain reservations, this remains the definition of an element.

As time went on, arsenic and antimony were added to the list of elements, and more was learned of element characteristics and of ways in which the substances could be used. Progress was slow and halting, however, and in the Western World a long and dismal period called the Dark Ages (about A.D. 500 to 1,000) inhibited the pursuit of knowledge.

Despite a general lack of scientific progress, an interesting group, the **alchemists** (Fig. 1-1), during the latter part of the Dark Ages and into the Middle Ages (A.D. 1100–1500), pursued a vocation which might be loosely defined as science. Although alchemists did not contribute much in the way of organized chemical information, they did recognize seven metallic elements, each associated with a heavenly body. Gold was paired with the

9

FIGURE 1-1
The alchemist. (Photo from The Bettmann Archive, Inc.)

sun, silver with the moon, iron with Mars, tin with Jupiter, lead with Saturn, copper with Venus, and mercury with Mercury. Because all these elements can be extracted from ores through the use of fire, fire itself was sometimes considered to be an element. Elemental sulfur also was known, as were a few more complex substances, including hydrochloric acid, sulfuric acid, nitric acid, vinegar (acetic acid), salt (sodium chloride), lye (sodium hydroxide), ammonia, vitriol (copper sulfate), and saltpeter (potassium nitrate). The alchemists were not aware that these more complex materials are composed of combinations of elements, and they had neither the means nor the knowledge for discovering this fact.

Eventually methods of chemistry became more experimental, and alchemists were replaced by true chemists. By 1783 the great French chemist Antoine Lavoisier (1743–1794) (who was to die under the guillotine during the French Revolution) had identified 33 elements. He also recognized that some substances such as salt, silica, alumina, and potash were not elements, and predicted that they would be found to yield further elements. Currently, some 90 elements are known to exist in nature, and some 16 others, not known to occur naturally, have been produced in atomic energy reactors and atomic particle accelerators. The elements usually are presented in a periodic table in ascending order of atomic numbers and in groups of elements having similar characteristics (see Fig. 1-8).

Why did it take so many centuries to learn the identities of the elements? Simply because most of them do not exist naturally as single elements but instead occur in specific combinations with other elements as substances we call **compounds**. We define a compound as a substance composed of two or more different elements in a specific atomic ratio; for example, table salt, sodium chloride, is composed of sodium and chlorine, both elements, in a 1:1 atomic ratio.

THE ORGANIZATION OF MATTER

The smallest particle in which a compound can exist is a **molecule**. The water molecule (written H_2O) contains one atom of oxygen and two atoms of hydrogen. Although most molecules are composed of two or more different atoms, some contain only one kind of atom (e.g., oxygen, nitrogen, and hydrogen are gaseous molecules, each composed of two atoms: O_2, N_2, H_2).

Compounds should not be confused with **mixtures**. Mixtures are aggregates composed of variable proportions of two or more substances, for example, salt and water, sugar and water, air (mostly nitrogen and oxygen plus some carbon dioxide), and soil (clay, sand, humus, etc.). Both compounds and mixtures are important in life processes because elements in pure form are generally not utilized by living organisms. Instead, substances are ingested as compounds, processed, used, and finally eliminated, still as compounds. The compounds themselves are almost always part of mixtures, especially a kind of mixture known as a **solution**. Solutions are mixtures in which molecules of two or more kinds are evenly distributed and can move about freely with respect to one another. Many solutions consist of solid matter dissolved in a liquid (salt in water); others are gases dissolved in a liquid (carbon dioxide in a soft drink) or in other gases (oxygen, nitrogen, and carbon dioxide in the air we breathe). In solutions, the **solute** is the dissolved substance, the dissolving medium the **solvent**. Air, seawater, soil water, blood serum, urine, plant saps, and many other natural substances are solutions.

Atomic Structure

When atoms were first recognized to be invariable components of molecules, nearly two centuries ago, they were thought to be tiny solid particles which somehow

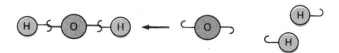

FIGURE 1-2
An early, naive concept of the relationships of atoms in a molecule of water. Atoms at one time were thought to be solid objects.

had such strong affinities for each other that in most cases it was very difficult to "pry" them loose and identify them as elements. The methods employed in isolating atoms of elements need not concern us here except to say that considerable energy is required to do so. Quite pertinent, however, are the characteristics of the attractive forces that hold atoms together in molecules.

Some early chemists proposed that tiny hooks held atoms together in a molecule (Fig. 1-2). Thus an oxygen atom would have two hooks, and hydrogen atoms would have single hooks; in that way one could account for the association of hydrogen and oxygen in the precise 2:1 atomic ratio found in a water molecule (H_2O). Now it is known that the "hooks" that hold atoms together are forces called **chemical bonds,** which exist between one atom and another in a molecule, but understanding the characteristics of these forces requires a further exploration of the structure of atoms.

The Divisible Atom Early chemists thought that atoms differed from each other in size and weight, but did not visualize them as being composed of even smaller units. Until 1902 scientists generally believed that atoms were not further divisible.

Studies performed as early as 1879 demonstrated that when metals are heated in a vacuum and subjected to an electrical potential (by being connected to the positive and negative poles of a battery), they give off a kind of energy. That energy was difficult to detect except when directed against a substance such as phosphorus, producing visible light (phosphorus, for this reason, is used to coat television screens; the same principle applies). This energy is now known to be composed of a stream of **electrons,** the smallest primary components of the atom. Additional experiments demonstrated that some elements, such as radium, disintegrate (decay) spontaneously, forming new elements. As a result of such information, the English physicist Ernest Rutherford (1871–1937) deduced that the atom is divisible after all and is composed of subunits.

Shortly after Rutherford explained that the atom is composed of subunits, a Danish physicist, Niels Bohr (1885–1962), developed a theory of the atom in which the hydrogen atom was pictured (Fig. 1-3) as consisting of a nucleus composed of one **proton,** about which an

electron moved in an orbit, much as the earth orbits the sun.

The Hydrogen Atom Hydrogen is the smallest and simplest atom known. Its mass relative to other atoms is 1, and its diameter is such that, if one could somehow line up 50 million hydrogen atoms in a row, the length of the row would be about 1 centimeter (cm) long. (However, all atomic diameters are small, ranging only from about 2 to about 5 Ångstrom units; $1Å = 1 \times 10^{-8}$ cm.) The hydrogen atom has a nucleus consisting of one positively charged particle, the **proton,** which composes nearly all the weight or mass of the atom and is about 1836 times as heavy as the electron.

More Complex Atoms Atoms larger than hydrogen have nuclei containing one or more **neutrons** in addition to protons. Neutrons have the same mass as protons but are electrically neutral. Although elements differ markedly in their physical and chemical properties, all are composed of electrons, protons, and neutrons, differing only in the number and arrangement of these subunits. Atoms of helium gas, for example, have two neutrons, two protons, and two electrons; those of oxygen, another gas, have eight protons, eight neutrons, and eight electrons (Fig. 1-4). Atoms of five other elements are intermediate in complexity between helium and oxygen. The remainder of the 106 known elements have atoms with more subunits.

In all atoms, regardless of the numbers of protons, neutrons, and electrons, the protons in the atom's nucleus are always numerically equal to the electrons outside the nucleus. Because each proton carries one positive electrical charge and each electron carries one negative charge, the atom as a whole is electrically neutral. The number of neutrons in the nucleus, on the other hand, may vary in number in relation to proton and electron numbers, but they are electrically neutral particles and do not change the neutrality of the atom.

Isotopes About one out of 5000 hydrogen atoms is twice as heavy as the others. Such atoms are called **deuterium,** to distinguish them from hydrogen. Their addi-

FIGURE 1-3
Bohr's model of the hydrogen atom.

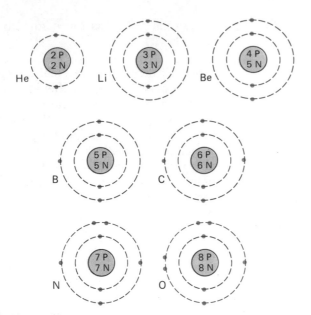

FIGURE 1-4
Bohr diagrams of atoms of helium (He), lithium (Li), beryllium (Be), boron (B), carbon (C), nitrogen (N), and oxygen (O).

tional mass is due to the presence in the nucleus of a neutron as well as a proton. Thus, a deuterium atom is composed of one proton, one neutron, and one electron. Chemically, deuterium behaves in the same way as does hydrogen. Two deuterium atoms react with one oxygen atom to form "heavy water," or "D_2O," which is almost indistinguishable from water except that it is 10 percent more dense. Another kind of hydrogen has two neutrons and is called **tritium.**

A number of other elements exhibit the relationship illustrated by tritium, deuterium, and hydrogen. All such forms of an element are called **isotopes.** Isotopes of a given element differ from each other only in the number of neutrons they possess. Many isotopes are stable, but a number, including tritium, are rather unstable. Such unstable isotopes, or **radioisotopes,** tend to break down spontaneously, giving off radiation in the form of subatomic particles (protons, neutrons, electrons) and X rays. Radioisotopes are among the most useful, as well as the most destructive, atoms known; those of uranium and plutonium are the explosive components of atom bombs.

Some radioisotopes are useful in biological experiments because their radiation can be tracked with detection instruments called scintillation counters. Tritium, for example, has been used to form compounds that are incorporated into living cells and then take part in various reactions of cellular chemistry. The pathways of these tritium-containing compounds can be followed by checking samples for radioactivity.

The Orbital Model of the Atom

Although the Bohr model is acceptable as a generalized concept, it does not agree well with more recent information about the energy relationships and behavior of electrons. For example, electrons are no longer envisioned as following predictable, planetary orbits, but instead are considered to occupy unpredictable positions within three-dimensional volumes called **orbitals,** each of which is capable of containing a maximum of two electrons (Fig. 1-5). The hydrogen atom thus has one electron in one orbital.

In larger and heavier atoms additional orbitals are present, and the orientations of these orbitals are such that the distances between their respective electron pairs are maximal. As a result, the orbitals are often pictured as projecting three-dimensionally in space, and the atoms of these three-dimensional molecules are attached at specific angles (Fig. 1-6).

Electrons are not permanently fixed in their orbitals but can absorb energy from light or heat or by transfer from another electron and move temporarily into higher energy positions. In its higher energy state, the electron can exist for a time as an **excited electron.** Its added energy is transitory, however, and is either emitted as light or employed in a chemical reaction. With the loss of its extra energy, the electron "drops" back into its original orbital position and is said to be at its **ground state.** It should be noted, however, that only the outermost electrons of an atom behave in this fashion. Because only the outermost orbitals are likely to have room for additional electrons, only outermost electrons can move into higher energy shells.

Chemical Reactions

Perhaps you are wondering what the foregoing has to do with the chemistry of life, even though it is common knowledge that all the material things of this world, living and nonliving, are composed of atoms and molecules. If that were the sum of our knowledge we could stop at this point. To do so, however, would be to ignore the fact that all the functions of things like motor cars,

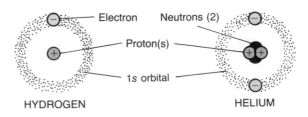

FIGURE 1-5
The orbitals of hydrogen and helium atoms. The innermost orbital is termed 1s.

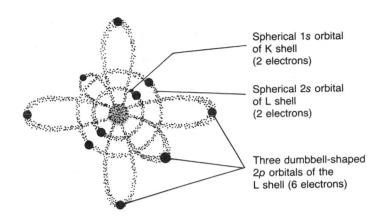

FIGURE 1-6
A diagram showing the orbitals of the neon atom. Its ten electrons completely fill five orbitals. Neon is an inert element, a characteristic accounted for by its lack of any unfilled orbital. Here, in addition to the innermost orbital (called the 1s), we see four orbitals: a spherical orbital (the 2s) and three ellipitical or dumbbell-shaped orbitals (called 2p orbitals). The axes of the 2p orbitals are positioned so that the distances between their elecron pairs are maximal and the orbitals project three-dimensionally in space.

Spherical 1s orbital of K shell (2 electrons)

Spherical 2s orbital of L shell (2 electrons)

Three dumbbell-shaped 2p orbitals of the L shell (6 electrons)

refrigerators, and televisions, to say nothing of the activities of running, jumping, eating, shouting, drinking, growing, and reproducing of living organisms, result from interactions between atoms and molecules. That being the case, we need to know more about these interactions and how they come about. First of all, they are **chemical reactions,** in which energy is used to convert relatively low-energy molecules into those having higher levels of energy or in which energy is liberated when molecules of relatively high energy are converted into molecules having lower energy levels. This process results in conversion of a substance of one kind into a substance of another kind by rearrangement and recombination of atoms. The original substance that is modified in a chemical reaction is the **reactant;** the new substance produced is the **product.** Although single substances sometimes break down spontaneously or as a result of heating or exposure to radiation, reactions usually occur as a result of recombinations of atoms of two or more substances in contact with one another. Such contact commonly occurs when the reacting substances are dissolved in an **aqueous** (water) solution or on a surface in contact with water.

How Atoms and Molecules Act and React

Molecular Action Molecules of any substance are continually in motion—freely moving in gases and liquids, less active in solids. The speed of this motion is relatively high, about 1280 km/h (800 mi/h) at room temperature, and is a reflection of the **kinetic energy** (energy of motion) of atoms and molecules. The movement of any one molecule is entirely random and its direction unpredictable. As you can imagine, collisions between such fast-moving and erratic particles must be very frequent; as they bump around they become more and more dispersed. This is the explanation for **diffusion,** which is the tendency of molecules to disperse. Diffusion

is readily visualized by picturing the spreading of a drop of ink in water or sensed as a rapidly penetrating fragrance when a heavily perfumed person enters a room.

The knowledge that molecules move rapidly about and often collide with each other has significance to the understanding of chemical reactions. First, a chemical reaction between two molecules will not occur unless they are in contact with each other—as when colliding. Second, the collision must be sufficiently forceful—the molecules must be going fast enough to generate an effective impact. Third, when the molecules collide, they must be oriented in a certain way if they are to react with each other—the parts of each molecule that will become joined to the other molecule must make contact. Fourth, the molecules must each possess sufficient inherent energy so that they will be able to react—some kinds of molecules react more (or less) readily than others.

The rapidity of a chemical reaction depends on all of the factors just described. Because one of the conditions for a chemical reaction to occur is the frequency of collisions, it follows that any factor that increases the probability of intermolecular collisions will increase the rate at which the molecules will undergo a chemical reaction. One rather obvious means to that end is simply to add more molecules, increasing their concentration (Fig. 1-7B). Another way to increase the reaction rate is to make the molecules move faster (Fig. 1-7C). This can be done by raising the temperature of the mixture of molecules—for every rise of 10°C, the rate of a reaction approximately doubles. We should note, however, that for many mixtures of molecules, the reaction between them can be very slow, even nonexistent, because the requirements outlined above have not been satisfied.

For every kind of chemical reaction, there is a minimum energy required if colliding molecules are to interact with each other. This amount of energy is called the **activation energy** (see Fig. 1-12). Then, when the first

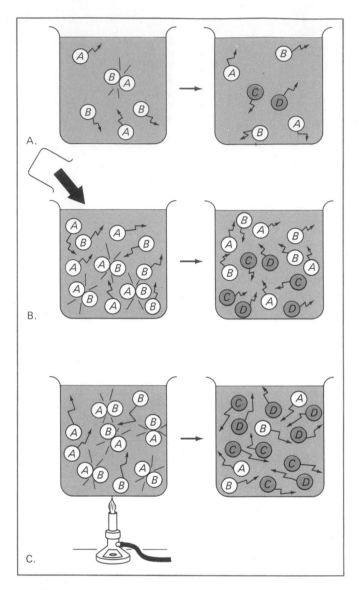

FIGURE 1-7
Effects of increasing concentrations and temperatures upon a chemical reaction within an unspecified time. Molecules (reactants) are labeled *A* and *B*; products are designated *C* and *D*. The reaction described is $A + B \rightarrow C + D$. **A.** Reaction at low temperature and low reactant concentration. **B.** Effect of increasing the concentration of reactants. **C.** Effect of increasing temperature as compared to B.

The result of this chemical reaction is the formation of water molecules and the release of a comparatively large amount of energy in the form of heat and light. We can write the chemical equation for this reaction as follows:

$$2H_2 + O_2 \longrightarrow 2H_2O + \text{energy}$$

Note that two molecules of hydrogen react with one of oxygen to produce two molecules of water. This, as we have written it, is a balanced **equation**—the same number and kinds of atoms are listed on each side of the arrow. What can you infer regarding the requirement that atoms be specifically oriented upon colliding if a chemical reaction is to occur? Simply that two hydrogen molecules must simultaneously impact with one oxygen molecule in a certain way to make a molecule of water.

Catalysts   The concept of **catalysts** is one of the most important, perhaps *the* most important in the chemistry of life. You perhaps have known individuals who are said to act as "catalysts" at meetings and parties. Imagine a room full of strangers standing awkwardly around wondering what to do or say. Into this room comes a lively, personable individual who takes charge, introducing the shy ones to each other, setting things into motion. Couples form, chat, perhaps dance a bit. The "catalyst" has transformed the entire evening! In test tubes and, more importantly, in living cells, there are catalysts that start and speed up chemical reactions, not by increasing the numbers of molecules or raising the temperature but simply by "performing introductions." Living things cannot withstand high temperatures, and we cannot, therefore, start a reaction within a living cell by touching a match to it. Catalysts do the job instead.

How do catalysts actually work? The concept of catalysis is that the catalyst temporarily attaches certain molecules to itself in such a way that they can readily undergo a chemical reaction. When that is accomplished, the catalyst releases the new kind (or kinds) of molecule and is once again free to "perform a new introduction." One definition of the action of catalysts is that they lower the minimum energy of activation required to start a chemical reaction. We will explore catalysis again later in this chapter in connection with **enzymes,** the catalysts of living cells.

Most reactions are complicated, and analysis of their progress is difficult. The reacting substance becomes used up, the concentration of new product increases, and temperature change occurs. Some reactions proceed until all or nearly all the reactant is used up; others slow down before much of the reactant is used up. In a **reversible reaction,** a back reaction occurs between molecules of the product to reconstitute molecules of the reactant. When the rates of the forward and

molecules enter into a chemical reaction, enough additional energy is released to keep the reaction going. An often-cited example is a mixture of hydrogen and oxygen molecules (H_2 and O_2) enclosed within a balloon. Normally nothing happens, although oxygen and hydrogen molecules are continually colliding with each other, but if we touch a match to the mixture a violent reaction occurs. All that is necessary is to raise the temperature to the point where the molecules are moving rapidly enough to interact with each other when they collide.

PERIOD 1	1 1.0080 H Hydrogen						2 4.0026 He Helium

	3 6.941 Li Lithium	4 9.0122 Be Beryllium	5 10.811 B Boron	6 12.011 C Carbon	7 14.0067 N Nitrogen	8 15.9994 O Oxygen	9 15.9994 F Fluorine	10 20.179 Ne Neon
PERIOD 2								

FIGURE 1-8
The periodic table. Only the first two periods are shown here. Note that the first period, corresponding to the first electron shell, contains only two elements. The second period corresponds to the second set of orbitals, or energy shell (2s + 2p orbitals). The number in the upper left of each box is the atomic number and equals the number of protons. The second number is the atomic mass. The chemical symbol follows, then the name of the element.

back reactions are stabilized, a **chemical equilibrium** exists. This state is indicated by arrows pointing in both directions between reactants and products (e.g., A + B $\leftrightarrows$ C + D). Later on, we will see that in some reactions occurring in cells a reaction is made to go in one direction in one set of circumstances, the reverse direction in other circumstances. Often enzymes will participate in determining the directions of such reactions.

Atomic Interaction Some atoms enter into chemical reactions much more readily than others. The **inert gases** such as **helium** and **neon** are extremely nonreactive because their completely filled outer orbitals have no accommodations for additional electrons (Fig. 1-6). We might expect then that an atom of an element having but one unfilled orbital would exhibit a strong tendency to fill that orbital by gaining an electron from some other atom. In this way it would attain a stable state similar to that of helium or neon. Following this logic, it would seem also that an atom of an element having only one electron in its outermost shell would very readily give up an electron and also attain the stability associated with having a complete outer shell (in this case, the next innermost). (The exception would appear to be hydrogen, which has just one orbital and one electron; removal of the electron would leave a naked proton.)

Activities of elements having 2 to 6 outer orbital electrons are somewhat lower in comparison with those having 1 and 7 orbital electrons; those having 2 or 3 orbital electrons tend generally to give them up to other atoms, while those having 5 or 6 outer orbital electrons tend to gain electrons. As you might suspect, elements with 4 outer orbital electrons are somewhat ambivalent, tending neither to give or gain but to **share** electrons. In Fig. 1-8, a portion of the **periodic table of elements** is shown. The first ten elements are listed, together with

numerical order and atomic weight. Interestingly, these include the elements most frequently encountered in compounds of living cells (**hydrogen, carbon, nitrogen, oxygen**).

Atoms of even the simplest molecules known are involved in electron-sharing of the kind described. Two hydrogen atoms share their two electrons (as shown in Fig. 1-9A); similarly, four electrons of two oxygen atoms are shared in molecular oxygen. Then, when atoms of hydrogen and oxygen react to form water, electron shar-

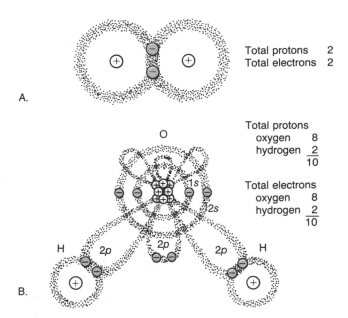

FIGURE 1-9
Electron sharing. **A.** The hydrogen molecule. **B.** The water molecule. Note that all of the orbitals of hydrogen and oxygen atoms are filled. Note also that the shape of the water molecule resembles that of a boomerang and is determined by the angles of the 2p orbitals which share electrons with hydrogen atoms. Neutrons are not included.

ing again occurs, but this time between two hydrogen atoms and one oxygen atom (Fig. 1-9B).

Chemical reactions often involve restructuring of atomic combinations by the sharing or even complete transfer of electrons between one kind of atom and another. In other reactions, parts of molecules, called **ions,** simply exchange places without any further electron sharing or electron transfer. These reactions are known as **displacement reactions.**

Oxidation and Reduction   Reactions involving either electron transfer or electron sharing between atoms are called **oxidation-reduction reactions.** Often these reactions are accompanied by the release of energy. Sometimes this energy release is violent, as when hydrogen and oxygen react explosively. An often-cited example is the disaster that befell the German zeppelin *Hindenburg,* which exploded and burned during its landing at Lakehurst, New Jersey, on May 6, 1937 (Fig. 1-10). Zeppelins were filled with hydrogen gas, and great care was taken to avoid leaks and sparks that might set off a holocaust. When contained within a balloon, hydrogen cannot be ignited; only if oxygen (or air) is present, as in the case of a leak, may an explosion occur.

Many other reactions, unless they are artificially accelerated, occur much more slowly. An example of slow reaction is the combination of atoms of iron and oxygen to form iron oxides, or rust. The explosion of hydrogen and the rusting of iron have one thing in common: both involve a chemical reaction with oxygen. Traditionally, such reactions are called **oxidations,** because they involve the acquisition by oxygen of electrons from some other substance. In the two examples given, hydrogen and iron are **oxidized,** and at the same time oxygen is **reduced.** However, some other reactions have been noted that involve either the sharing or the transfer of electrons between atoms other than oxygen. For example, the reaction between sodium and chlorine involves the actual transfer of an electron from the sodium atom to the chlorine atom. This reaction also is an oxidation, in this case of sodium, even though oxygen is not present. The atom that has given up its electron has been oxidized, the atom that received electrons into its orbit has been reduced; therefore the action is an oxidation-reduction reaction (Fig. 1-11).

Energy and Work   Not all oxidation-reduction reactions involve the release of energy, but some of those that do are important in the lives of organisms. **Energy** can be defined as the capacity to do **work.** Obviously, in the rapid burning of hydrogen in the *Hindenburg* disaster, much energy was expended, but was any work accomplished? Certainly the explosion did not accomplish any useful work, but the potential to do useful work was

FIGURE 1-10
Explosion of the Hindenburg. (Photo from The Bettmann Archive, Inc.)

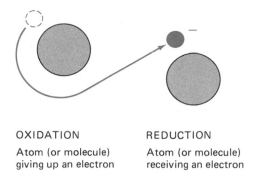

OXIDATION
Atom (or molecule)
giving up an electron

REDUCTION
Atom (or molecule)
receiving an electron

FIGURE 1-11
Oxidation and reduction.

present, and under other circumstances might have been used productively. In fact, much the same kind of explosive energy is harnessed in automobile engines. Work is done also in many chemical reactions that take place more slowly, such as the burning of food (oxidation) in cells of the human body.

The **first law of thermodynamics** states that energy is neither created nor destroyed, merely changed from one form into another. The energy given off as heat and light (classified as **radiant energy**) by the burning hydrogen in the *Hindenburg* disaster was originally present as **potential energy** in molecules of hydrogen and oxygen

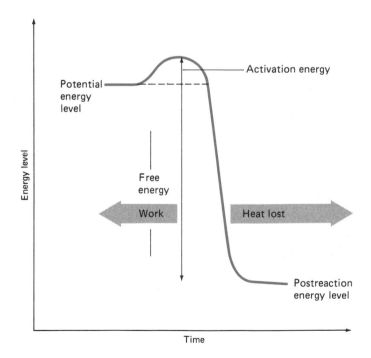

FIGURE 1-12
Relationship between activation energy and chemical reaction. In many reactions the input of a relatively small amount of energy, the activation energy, is required to raise the temperature of the reactants to the point at which the reaction will proceed.

(Fig. 1-12). Potential energy may take several forms. Water in a reservoir at the top of a hill has potential energy of position. If the water is allowed to run downhill, its momentum may be used to do useful work. This energy, which can be harnessed in several ways, is called **free energy**. For example, the free energy in the reservoir example could be used to turn an electrical generator, and to produce a form of free energy known as **electrical energy**. Free energy is the maximum capacity to do work obtainable from a given system.

When water from the reservoir comes to rest in a pond at the bottom of the hill, most of its potential energy will be gone. Its free energy will have been used to produce other useful forms of energy, but some energy will have been dissipated through friction as **heat energy** and lost. If the water is to be used again, much energy must be expended in raising it to the top of the hill. In nature, this work is done by the sun, which causes the water to evaporate and rise into the clouds, from which it descends as precipitation to fill the reservoir (Fig. 1-13).

In the case of hydrogen and oxygen mentioned earlier, the potential energy exists as **chemical energy**. This form of potential energy resides in the structure of molecules and, in particular, in the nature of interatomic bonds. When bonds are broken, their potential energy may be converted into electrical or radiant energy. Conversely, when bonds are reconstituted, these forms of energy may be converted back into chemical energy.

Whether potential energy is made available for work by a reaction, or work is done to transform some other kind of energy into potential energy, some energy is always lost in the process, just as in the earlier example of water power. No system for the conversion and utilization of energy is completely efficient. The general trend for highly useful and organized forms of energy to become degraded to disorganized and essentially nonuseful energy is given by the **second law of thermodynamics**: energy always tends to become dispersed and disorganized in the universe. In its useless state, energy is in the form of random movements of molecules in space. Because of this dissipation of energy into space, the amount of useful energy in any closed system, whether a living cell or a millstream, always tends to decrease.

Chemical reactions are of two general types. Those in which free energy is utilized to do work, with an overall net decrease in potential energy, are called **exergonic** ("energy-liberating") reactions. The explosion of the *Hindenburg* was an exergonic reaction. On the other hand, those reactions that require an input of free energy and that result in a net increase of potential energy are **endergonic** ("energy-storing") reactions. Photosynthesis is

FIGURE 1-13
Energy relationships in nature and industry.

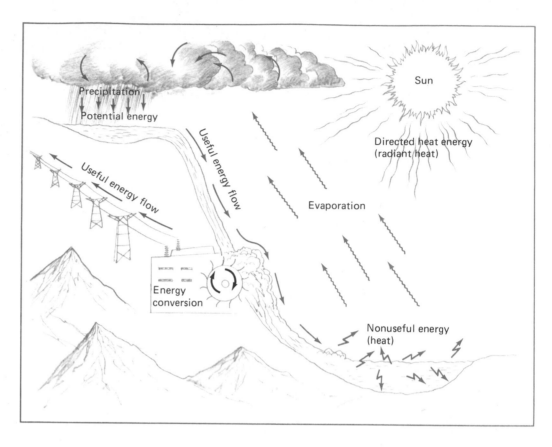

an example of an endergonic reaction. Light energy is used to produce complex food molecules that are high in potential energy. Biologists refer to such reactions as **anabolic** (*anabolism*, "all building processes of cells"). When food is eaten, digested, and oxidized in the cells of the body, its potential energy is converted into other forms of energy in a series of exergonic reactions. This is an example of **catabolism**, and includes all processes in which substances are broken down and their energy utilized to do work. In all the various reactions occurring in living cells, energy is converted from one form to another, and work is performed in both the manufacture and the breaking down of complex chemical substances.

How Atoms are Held Together—Chemical Bonds

Bond formation accounts for the ability of elements to form molecules. Most atoms do not occur as independent units but exist as molecules in which each atom has achieved a stable complete-outer-shell configuration, either by sharing electrons or by giving up or accepting electrons from other atoms. Molecules may consist of atoms of the same elements, as in hydrogen gas (H_2), oxygen gas (O_2), and nitrogen gas (N_2), or of

different elements, such as water (H_2O), sodium chloride (NaCl), and so forth.

Covalent Bonds When atoms share electrons, the atoms involved often are held tightly together by their mutual attachment for the shared electrons; however one kind of atom in a molecule may have a stronger affinity for the shared electron (or electrons). In the water molecule, the oxygen atom has the stronger affinity for the shared electrons; the hydrogen atoms have the weaker affinity. Because neither the oxygen atom of water or its two hydrogen atoms give up or accept electrons completely, but rather share them, the force, or chemical bond, that holds them together in the water molecule is called a **covalent bond**. Customarily, covalent bonds are depicted by a single line joining the symbols of the atoms involved or by pairs of dots representing shared electrons. Water may be shown diagrammatically as H—O—H or as H:O:H.

Sharing of electrons occurs also between atoms of many other kinds of molecules. The carbon atom, which has four electrons in its outermost shell, can share electrons with four hydrogen atoms. The resulting molecule (CH_4), which is methane (Fig. 1-14), is called a **hydrocarbon** because it contains only hydrogen and carbon. Car-

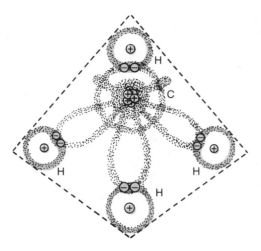

FIGURE 1-14
The methane molecule. Four electrons of four hydrogen atoms share with carbon electrons the four orbitals of the L shell. Because of the three-dimensional orientation of these orbitals, the molecule has a four-angled or tetrahedral shape.

bon may also react with atoms other than hydrogen—for example, with oxygen to produce carbon dioxide or with both hydrogen and oxygen to form carbohydrates and many other kinds of molecules. Such molecules are collectively called **organic compounds.** In them, the bonds between carbon and other atoms, or between adjacent carbon atoms, are the consequence of electron sharing and therefore are covalent.

Ionic Bonds   We have noted that collisions between atoms may produce transfers of electrons from orbitals of one atom to another. This occurs most readily when the atoms involved are metals such as sodium and potassium, which have several energy shells and a single outer-orbital electron. Because such an electron is some distance from its nucleus it tends to be more easily and completely removed from its orbital. If then, an atom of chlorine, which has but one unfilled orbital, is a partner in the collision, there is a high probability that it will accept the sodium electron, forming a molecule of sodium chloride (Fig. 1-15). The energy required to accomplish this transfer is a type of activation energy called **ionization energy,** and the products of this transfer are a pair of ions held together by an **ionic bond.** The sodium ion has one more proton than it has electrons and so carries a positive charge. Positively charged ions are also called **cations** because they are attracted to the negative pole or **cathode** of a battery. The chlorine ion, on the other hand, has one more electron than it has protons, hence carries a negative charge, and is called an **anion** because it is attracted to the positive pole, the

anode, of a battery. Anions and cations are attracted to each other much as are the opposite poles of a pair of magnets (Fig. 1-16).

When dry, anions and cations are closely associated, forming crystalline substances such as table salt. When hydrated (dissolved in water) the ions are more loosely associated with each other. This separation of ions in solution is known as **dissociation.**

Elements with between two and six electrons in their outer shells may form ions and ionic bonds (Ca^{2+}, Fe^{2+}, S^{2-}, etc.) in much the same way as sodium and chlorine, or they may form covalent bonds. Other compounds may have more complex ions composed of two or more atoms in which both electron sharing and electron transfer occur. Among these are the sulfate ion (SO_4^{2-}) and the phosphate ion (PO_4^{3-}). In each case, the kind of bonding and the reactivity of the atoms will depend on the number of electrons in the outer shells, as well as on other factors such as the size of the atom and the nature of its orbitals.

Many organic molecules contain both covalent and ionic bonds. Although these molecules are composed of carbon atoms held together with covalent bonds, one or more atoms, such as an oxygen or a nitrogen, may, when bonded to a carbon atom of the molecule, retain the ability to form one or more ionic bonds with some other atoms or groups of atoms.

The number of electrons that an atom of a particular element uses to form bonds with another atom is the **valence** of the atom. Valence is indicated by a su-

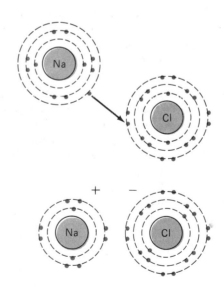

FIGURE 1-15
Reaction between sodium and chlorine atoms, an example of ionic bond formation. Note that this is a two-dimensional model and does not depict the three-dimensional shape of the orbitals.

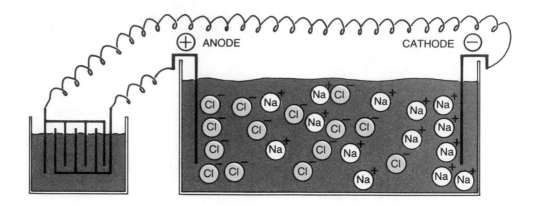

FIGURE 1-16
Ionic dissociation. The dissociation of sodium and chlorine (chloride) ions in water is shown, as well as their attraction to opposite poles of a battery. Note, however, that an electric current is not necessary for dissociation of sodium chloride to occur.

perscript$^+$ (e.g., H^+, Na^+, Ca^{2+}) for those atoms that donate or tend to donate electrons to other atoms and a superscript$^-$ (e.g., Cl^-, S^{2-}) for those that gain or tend to gain electrons from other atoms. In writing the formula for a compound, care must be taken to see that the total positive and negative charges are equal. For example: Na^+ and Cl^- make NaCl, but note that Mg^{2+} and Cl^- do not make MgCl; they make $MgCl_2$.

Acids and Bases—The pH Scale Dissociation and formation of hydrogen ions is characteristic of all **acids**. The hydrogen ion (H^+) is responsible for their corrosive nature as well as for other chemical characteristics. Not all acids dissociate to the same degree to form H^+ ions; some are weak acids and do not dissociate as freely as stronger acids. As a result, they do not release as many H^+ ions into solutions. The degree of dissociation is related to the **pH** of the solution. The pH scale extends from 0 to 14 (Fig. 1-17). Solutions having a pH of less than 7 are acidic, with the lower numbers indicating the stronger acids and higher concentrations of H^+ ions.

Compounds known as **bases** are said to be **alkaline,** and act as opposites to acids, in that addition of a base to an acid solution will result in **neutralization** of the solution. Bases also dissociate into ions when added to water; they have pH values greater than 7 (Fig. 1-17). Household lye, or sodium hydroxide (NaOH), is a common base. When it dissociates it forms sodium ions (Na^+) and **hydroxyl ions** (OH^-). The OH^- ion is responsible for the corrosive nature of a base.

When a base such as sodium hydroxide is added to a solution containing an equivalent amount of an acid such as hydrochloric acid, neutralization occurs. In this process the respective OH^- and H^+ ions combine by a displacement reaction to form water, and a new compound, sodium chloride (NaCl), is formed (NaOH + HCl → NaCl + H_2O). Sodium chloride and other similarly formed compounds generally are called **salts.**

Potassium chloride (KCl) is formed by combining hydrochloric acid (HCl) and potassium hydroxide (KOH); calcium chloride ($CaCl_2$) may be formed from calcium hydroxide (Ca[OH]$_2$) and two molecules of hydrochloric acid (2 HCl); and so on. Salts are biologically important because many elements taken up by living things are in the form of salts, commonly called **mineral salts**, or simply **minerals.**

Buffers Acids as well as bases are present in all cells and sometimes are secreted by them. Hydrochloric acid, for example, is produced by certain stomach cells and secreted into the stomach cavity when food is present,

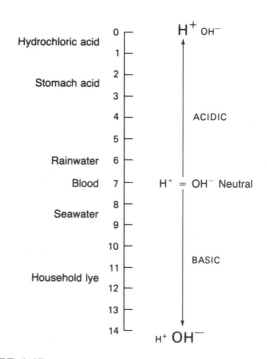

FIGURE 1-17
The pH scale. The size of the symbols OH^- and H^+ denotes the relative proportion of each.

ACID RAIN

Natural, unpolluted rain has a pH of about 5.6, slightly acid due to absorbed carbon dioxide (in the form of carbonic acid). More acidic rains generally are attributable to the noxious gas sulfur dioxide (SO_2), which is produced by the burning of fossil fuels (natural gas, petroleum, and coal) in home heaters, motor vehicles, and industry. The sulfur dioxide given off into the air is further oxidized to sulfur trioxide (SO_3) by reaction with atmospheric oxygen ($2 SO_2 + O_2 \rightarrow 2 SO_3$); the sulfur trioxide then may combine with water vapor to produce sulfuric acid ($SO_3 + H_2O \rightarrow H_2SO_4$).

Although there is little historical information upon which to base calculations of trends in acid rainfall, it is generally thought that acid rains are increasing. An acid rainfall study recently completed by the Norwegian government indicates that acidity in the environment has increased during the past 30 years. In another study, ice samples taken at several depths in the Greenland ice cap showed a steady increase in acidity beginning about 1860, when coal burning increased in the early years of the Industrial Revolution. In nonindustrialized areas, such as rural Norway, Greenland, and northeastern North America, acid rains usually are attributed to long-distance transport of air pollutants from industrialized regions. In actuality, tracing the sources of acid rains has proven very difficult. Most such pollution has a variety of origins, some local, some distant.

The most acidic rainfall recorded in the United States fell in Wheeling, West Virginia, and measured 1.4 in pH (vinegar is 3.5 and battery acid is 1.0). In 1979 the average February rainfall in Toronto, Canada, had a pH of 3.5, and the average pH of rainfall over the northeastern region of North America in that year was 4.3. High acidity is found also in northern Europe and in scattered areas of Asia.

In many places, acid rain becomes neutralized by limestone and other alkaline components of soil and water. However, in certain mountainous regions, where limestone is scarce, acid rains tend not to be neutralized, and lakes and streams therefore become quite acidic. Most fish species die at a pH of 4.5–5.0, and those in mountain lakes are particularly vulnerable because the lake bottoms are composed of granitic sands that have little ability to neutralize acidity. In 1975 a survey of 217 Adirondack lakes at elevations above 600 m (2000 ft) showed 51 percent to be acidified to a pH of 5.0 or lower; 90 percent of these contained no fish. Acidification of lakes not only directly affects animal life, but as water becomes more and more acid, it loses its ability to hold certain essential nutrients. This causes decreased yields of plant life and disturbs the entire food chain.

Smaller acidic lakes can sometimes be neutralized by adding crushed limestone, but this remedy is expensive and usually needs to be repeated from time to time. The long-range solution to the problem of acid rainfall will require not only additional research on causes and effects but also establishment and enforcement of more rigorous air-pollution standards.

WATER

An ancient joke describes the plight of a scientist who has invented "the universal solvent"; only problem is: what does one keep it in? Water comes near to being such a universal solvent, and insofar as life is concerned, water is indeed *the* universal solvent. Because cells, tissues, and, yes, organisms are mostly water (70 to 98 percent), the properties of water are a vital concern. All polar substances dissolve in water; this includes proteins, carbohydrates, alcohols, fatty acids, and others, but not lipids. The last named are nonpolar; however, they can form thin films in water and this too has significance for life. The enclosing membrane of the living cell, which surrounds the aqueous interior, is just such a thin lipid film.

Water molecules are cohesive. They adhere to each other and consequently form a kind of surface skin; one can float a needle in a glass of water or see a water strider run over the water surface. This same adherence of water molecule to water molecule enables a "thread" of water to be pulled up from the roots to the tops of the tallest trees by the evaporative forces occurring in leaves.

Water has a high specific heat; that is, it absorbs a lot of heat without exhibiting great changes in temperature. Water acts therefore as a sort of temperature buffer, preventing sudden excessive changes in body temperatures of organisms. Another remarkable and important characteristic is seen when water freezes. Water, unlike other liquids, is less dense as a solid than as a liquid. Lakes therefore freeze from the top down and not from the bottom up. A layer of floating surface ice effectively insulates the underlying water, protects its inhabitants from freezing, and, of course, makes ice-fishing possible.

Science fiction speculates about which substances might substitute for water in the life of an alien planet. It is difficult to imagine this, but if some other liquid were to be the equivalent of water, then might not water itself, the "universal solvent" of our world, be a deadly poison? In this scenario, the earthling spaceperson, armed with a trusty water pistol, wipes out a horde of menacing extraterrestrials!

as an aid to digestion. Although such secretions generally have a rather low pH (quite acidic, pH 4), pH levels within cells normally are near neutrality. Extremes of pH in cells are suppressed by stabilizing molecules called **buffers**. Buffers are salts of weak acids; sodium bicarbonate ($NaHCO_3$) is one. When H^+ ions of a strong acid react with bicarbonate ions (HCO_3^-), carbonic acid (H_2CO_3) is formed. Being a weak acid, carbonic acid does not dissociate much. As a result, free H^+ ions tend to be removed from solution and the pH tends not to be lowered further. A number of compounds, including salts of organic acids, may function as buffers in living cells.

Weak Bonds Molecules of many substances are attracted to and held in association with each other by intermolecular forces known as **weak bonds**. Such bonds depend on unequal electron distributions in molecules so that one part of a molecule is electrically more negative than the rest of the molecule. Molecules of this type, called **polar molecules**, have an affinity for other polar molecules and for ions. Reference to the structure of the water molecule in Fig. 1-9 shows it to have an asymmetrical shape because of the angles of its $2p$ orbitals. The molecule has a "boomerang" form and its two "arms" are electronegative with respect to its center,

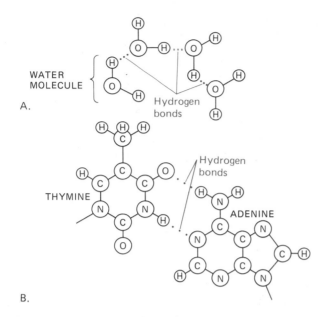

FIGURE 1-18
Hydrogen bonding. **A.** Hydrogen bonds of water molecules. Note that the bonds between oxygen and hydrogen atoms are angled. This accentuates the polarity of the water molecule. **B.** Hydrogen bonds between the adenine and thymine bases of the DNA molecule. Refer to Figs. 5-4 and 5-5 to see how these relate to the structure of DNA.

which has the eight protons of the oxygen atom. The water molecule, therefore, is a polar molecule.

Hydrogen Bonds When hydrogen atoms are attached to certain larger atoms such as oxygen or nitrogen, the electrons of the hydrogen atom tend to be attracted into closer association with the large nuclei of oxygen or nitrogen. The result is that the hydrogen portion of the molecule tends to be more positively charged than the oxygen or nitrogen part. In addition, as previously noted, the water molecule has a bent configuration (see Figs. 1-9B and 1-18A). Its bonding angles contribute to the electrical asymmetry (polarity) of the molecule. A consequence of polarity is that when, for example, the hydrogens of a water molecule come near the oxygen of an adjacent water molecule, an attractive force, or **hydrogen bond**, occurs between the two molecules (Fig. 1-18A).

Hydrogen bonds between water molecules account for several of water's characteristics: its surface tension; its boiling and freezing points; and its ability to dissolve a great many other kinds of molecules, including many biologically important organic molecules (Fig. 1-18B). In some large molecules, which may be bent back upon themselves, hydrogen bonds can also occur between different parts of the same molecule.

THE CHEMICALS OF LIFE

Although about 90 elements occur naturally, and a few others have been created in atomic reactors and accelerators, fewer than one-fourth are deemed essential for life processes. Eighteen elements are required by all organisms, and about 11 others (some are questionable) have been found to be necessary for some but not all organisms (Table 1-1).

Many organisms obtain their **essential elements** directly from soil and the atmosphere and their energy from sunlight; others obtain such elements by consuming other organisms or their products. In ecological terminology, the former are **producers** and the latter are **consumers**. Because most producers are plants, plants

TABLE 1-1
Biologically important elements

Element	Chemical Symbol	Atomic Number
Aluminum*†	Al	13
Barium*†	Ba	56
Boron*†	B	5
Cadmium*†	Cd	48
Calcium	Ca	20
Carbon	C	6
Chlorine	Cl	17
Cobalt†	Co	27
Copper†	Cu	29
Fluorine†	F	9
Hydrogen	H	1
Iodine†	I	53
Iron	Fe	26
Lithium*†	Li	3
Magnesium	Mg	12
Manganese†	Mn	25
Molybdenum†	Mo	42
Nickel*†	Ni	28
Nitrogen	N	7
Oxygen	O	8
Phosphorus	P	15
Potassium	K	19
Selenium*†	Se	34
Silicon*	Si	14
Sodium	Na	11
Sulfur	S	16
Tin*†	Sn	50
Vanadium*†	V	23
Zinc*†	Zn	30

Note: The atomic number of an element refers to the number of protons in its nucleus.
Boldface elements: Occur in all living things and are indispensible to life as we know it.
*Required by some but not all organisms.
†Trace element.

are the direct link between elements of the environment and the elements that constitute the chemicals of living matter—the **chemicals of life**. This fundamental relationship has long been recognized.

Because water composes somewhere between 40 and 98 percent of the mass of cells (depending on the type), one could say that cells are mostly hydrogen and oxygen. Several other elements, including carbon, nitrogen, and sulfur, also are relatively abundant in cells. Carbon atoms form the "backbone" of many of the compounds of which all cells are made (the organic compounds), and nitrogen and sulfur are essential atoms in proteins. This does not mean that some of the other elements found in cells are unimportant. Certain elements known as **trace elements** are absolutely essential to life, even though they are present in almost infinitesimal concentrations (refer to Table 1-1).

Organic Compounds

With the exception of water, a majority of the molecules composing the living cell are themselves composed principally of carbon, hydrogen, and oxygen. These organic molecules are vital parts of organisms or else are products of living organisms. The list of such substances is almost endless and includes **fats, proteins, carbohydrates, vitamins, alcohols, aldehydes, organic acids, amino acids,** and many others.

At one time it was thought that only organisms could make organic compounds, hence their name. Now, however, we know that some organic molecules are spontaneously produced in nature, and, of course, many others are the products of chemistry and the industrial laboratory. Of these, the ones that living organisms are able to utilize and to break down are referred to as **biodegradable;** those that cannot be utilized or broken down in this fashion are **nonbiodegradable.** Among the latter are many synthetic organic molecules, including plastics, solvents, and certain pesticides. They can be very persistent environmental pollutants.

Structure of Organic Molecules   Organic molecules are usually composed of many atoms, even though the kinds of atoms involved are few in number (most commonly **carbon, hydrogen, oxygen, nitrogen,** and **sulfur**). We can visualize them in the form of the "tinker toy" models used in biology and chemistry classes and, with increasing frequency, even in magazine illustrations. The brightly colored "atoms" are held together with pins that represent covalent bonds, and our diagrams usually use dashes to represent these same bonds. The simple organic molecule **methane** is portrayed diagrammatically as

Because organic molecules always contain carbon atoms, it is helpful to remember that each carbon can form four covalent bonds, as in methane. However, these bonds need not be single covalent bonds, as in methane. **Double covalent bonds** are common in many organic molecules, as in the gas **ethylene:**

$$H—C=C—H$$

Another name for a double bond of this kind is **unsaturated,** implying that the carbons involved in the double bond are not carrying as many hydrogens as they could. Unsaturated fats, so often mentioned in connection with cooking oils and butter substitutes, have more double bonds between certain carbon atoms in their molecules than do saturated fats. In the latter, double bonds between adjacent carbons are few or absent (refer to Fig. 15-4C, D).

Occasionally, in diagrams of organic molecules, one sees triple covalent bonds, as in **acetylene** gas:

$$H—C≡C—H$$

While many organic molecules are depicted as being relatively straight chains of carbon atoms, as in the case of **hexane:**

rings of carbon atoms also are quite common, as in **cyclohexane:**

Looking at these diagrams of organic molecules, one gets the impression that they are flat two-dimensional objects, like pressed flowers. Instead, as three-dimensional models of important organic molecules reveal, they have fronts and backs as well as tops and sides. In Fig. 1-14, a diagram illustrates the three-dimensionality of the methane molecule, showing that, because its covalent bonds project at angles to each other, the molecule has a tetrahedral (four-sided) shape. Molecules more complex than methane (composed of chains or rings of carbon atoms, together with hydrogens and perhaps other atoms such as oxygen, nitrogen, and sulfur) have unique and characteristic three-dimensional shapes due to the angles of their covalent bonds. These are impossible to portray on the printed page but approximations are possible in some cases. Cyclohexane, for instance, could be shown as having a folded shape something like this (note: this is the same molecule diagrammed above, but the carbons and hydrogens are represented only by corner dots):

Chair

This three-dimensional shape is called the "chair," and studies show it to be quite rigid and stable. In some cases, however, if considerable force is applied, the molecule can attain a different conformation, known as the "boat":

Boat

Changes of this sort are called **conformational changes**. The observation that organic molecules can undergo conformational changes with application or release of energy is important in comprehending certain biological phenomena, including the action of the large protein molecules we know as enzymes (Fig. 1-28).

You will have noted that all the organic molecules mentioned are composed solely of carbon and hydrogen; they therefore are called **hydrocarbons**. The majority of organic molecules in living cells are not hydrocarbons, however, but contain other atoms in addition to carbon and hydrogen, most commonly *oxygen* and *nitrogen*, and, less frequently, *sulfur*.

In addition, many organic molecules are large and composed of many atoms, sometimes hundreds and even thousands, but these dimensions are achieved by the repetition of a few basic kinds of subunits. Among these subunits are alcohols, aldehydes, organic acids, and amino acids (Fig. 1-19).

FIGURE 1-19
Some simple organic molecules and their active groups. **A.** Methane, a hydrocarbon. **B.** Methyl alcohol (methanol). **C.** Formaldehyde. **D.** Formic acid. **E.** Methylamine, an amine. **F.** Glycine, an amino acid.

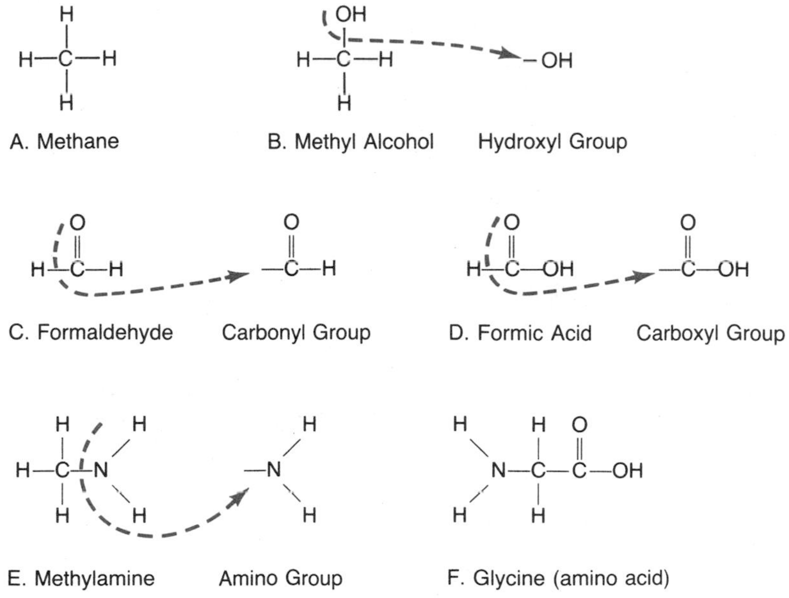

A. Methane

B. Methyl Alcohol   Hydroxyl Group

C. Formaldehyde   Carbonyl Group

D. Formic Acid   Carboxyl Group

E. Methylamine   Amino Group

F. Glycine (amino acid)

Alcohols An alcohol is formed when one (or more) hydrogen of a hydrocarbon is replaced by a **hydroxyl group (—OH)**. This simplest alcohol, as one might suspect, is based on the methane molecule and is **methanol** (also called **methyl alcohol**):

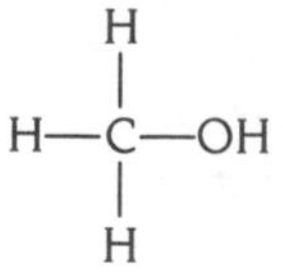

$$\begin{array}{c} H \\ | \\ H—C—OH \\ | \\ H \end{array}$$

Ethanol (ethyl alcohol) is:

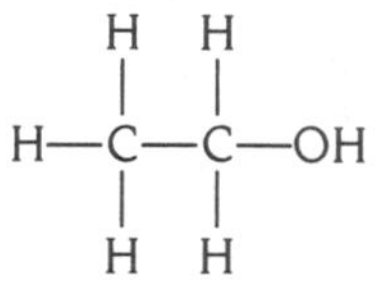

$$\begin{array}{c} H \quad H \\ | \quad | \\ H—C—C—OH \\ | \quad | \\ H \quad H \end{array}$$

and **propanol (propyl alcohol)** is:

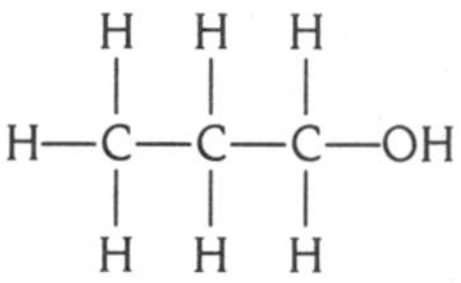

$$\begin{array}{c} H \quad H \quad H \\ | \quad | \quad | \\ H—C—C—C—OH \\ | \quad | \quad | \\ H \quad H \quad H \end{array}$$

An important alcohol in living systems is **glycerol**, which has three carbons and three —OH groups:

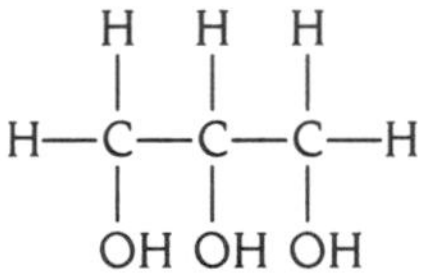

$$\begin{array}{c} H \quad H \quad H \\ | \quad | \quad | \\ H—C—C—C—H \\ | \quad | \quad | \\ OH \ OH \ OH \end{array}$$

Glycerol is a component of fats (Fig. 1-24).

Many alcohols in the chemistry laboratory are synthetic products and poisonous to cells, but ethanol is a biological product of the brewing industry and is the alcohol found in beers, wines, and other alcoholic beverages.

Aldehydes If one removes one of the three remaining hydrogens of the methanol molecule and the hydrogen of the OH group as well, what remains is **formaldehyde, CH_2O**, diagrammatically:

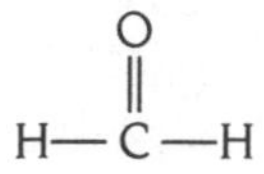

$$\begin{array}{c} O \\ \| \\ H—C—H \end{array}$$

Formaldehyde is the simplest of a group of organic compounds called aldehydes, among which are **sugars**, the basic units of carbohydrates. There are a number of different sugars in cells; these commonly have carbon atoms ranging from 3 to 7, although more are present in some cases. The generalized names for these sugars are based on the number of carbon atoms per molecule:

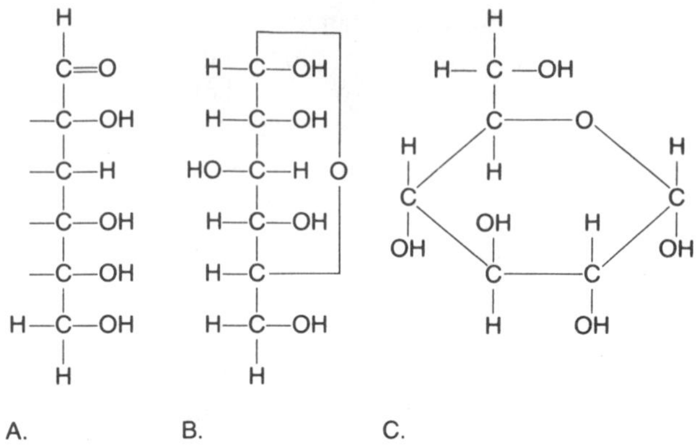

FIGURE 1-20
Glucose molecule, a simple hexose sugar. **A.** Glucose depicted as a straight-chain aldehyde. **B.** Same molecule, showing how the ring configuration is attained. **C.** Glucose as a commonly depicted six-membered ring.

triose (3-C), **tetrose** (4-C), **pentose** (5-C), **hexose** (6-C), and **heptose** (7-C). Sugars with five or more carbons can exist either as straight chains or as rings (by formation of an **oxygen bridge;** see Figs. 1-20, 1-22).

Organic Acids If one of the remaining two hydrogens of a formaldehyde molecule is replaced with an —OH, the new molecule is an **organic acid**, specifically **formic acid:**

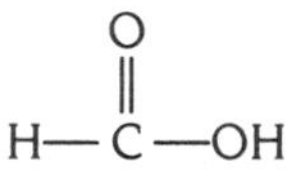

$$\begin{array}{c} O \\ \| \\ H—C—OH \end{array}$$

Formic acid is employed as a chemical warfare weapon by some species of ants (*formica*, "ant"). Another organic acid, **acetic acid,**

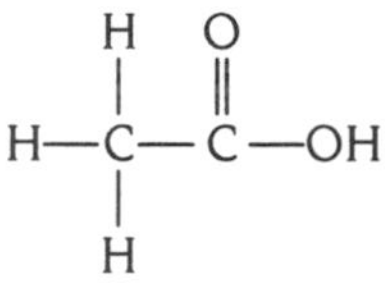

$$\begin{array}{c} H \quad O \\ | \quad \| \\ H—C—C—OH \\ | \\ H \end{array}$$

also is naturally occurring (in vinegar and in cells) and is an important molecule in the energy relationships in cells. Long-chain organic acids, known as **fatty acids,** are a common component of cells. Fatty acids commonly have from 12 to 24 linearly arranged carbons with a **carboxyl group** $\begin{array}{c} OH \\ | \\ (—C=O) \end{array}$ at one end. They are usually attached to glycerol to form fats.

The carboxyl group, found in formic acid, acetic acid, fatty acids, and all other organic acids, can dissociate in water to form H^+ ions characteristic of acids in general. Organic acids tend to be weak acids, not dissociating to the same extent as do such strong mineral acids as HCl (hydrochloric acid), H_2SO_4 (sulfuric acid), and HNO_3 (nitric acid).

Amino Acids and Amines A group of organic molecules known as **amines** are characterized by the attachment of an amino group ($-N \begin{smallmatrix} H \\ \\ H \end{smallmatrix}$) to a carbon somewhere in the molecule (Fig. 1-19E). The resulting amine is water soluble and will have basic properties, owing to the ability of the amino group to combine with water to produce OH^- ions. When an amino group is attached to an organic acid, the entire molecule is an amino acid (Fig. 1-19F). Depending on conditions, an amino acid may act either as an acid or as a base.

Macromolecules and Life

A macromolecule is a large and complex organic molecule formed by the bonding together of simpler molecules (Fig. 1-22A). Many macromolecules are known; for the present discussion they are classified into four major categories: **polysaccharides, lipids, nucleic acids,** and **polypeptides** (proteins). Some of these molecules are composed of several or many different kinds of molecules; others, called **polymers,** consist of many repeating units **(monomers)** of the same kind of molecule.

Each kind of macromolecule has a characteristic type of **linkage** or bond holding its monomers together; these linkages are produced between adjacent monomers by enzyme-directed actions which remove an OH^- and an H^+, respectively, from the pair of participating monomers. The result is the creation of a linkage and the liberation of a water molecule. An example is the formation of an **oxygen bridge** between two monomers of a carbohydrate macromolecule such as starch:

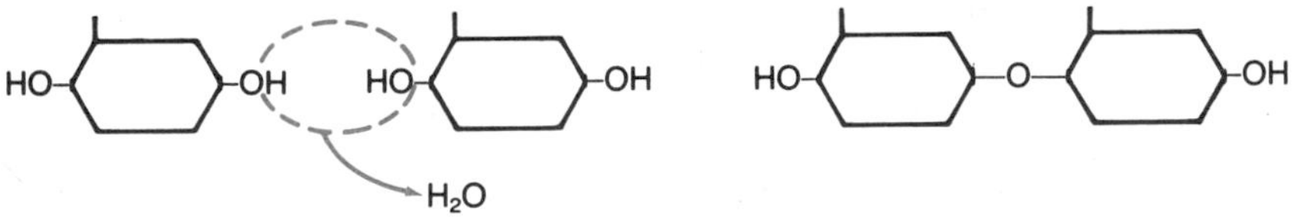

Repetition of this process between many adjacent monomers forms a long polymer molecule. Because the formation of each linkage is accompanied by the production of a molecule of water (in effect, removal of water from the polymer), this kind of reaction is designated a **dehydration** or a **condensation.**

In addition to a supply of monomers and an enzyme capable of directing the assembly of the polymer,

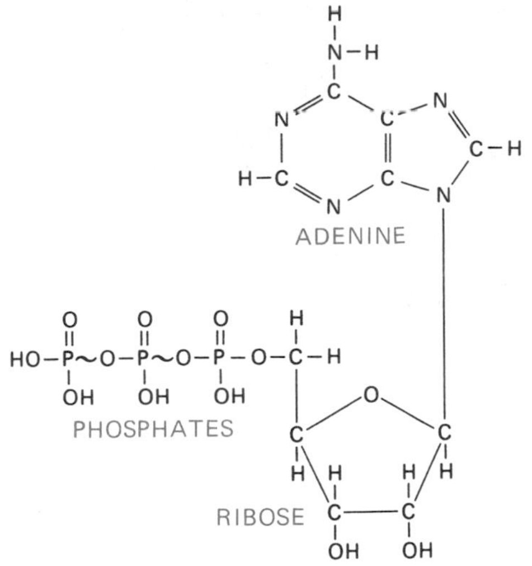

FIGURE 1-21

Structure of ATP. Note that the components of this molecule are adenine, a sugar (ribose), and three phosphates, and that the bonds between two of the phosphates (the terminal two) are represented by "squiggles." The squiggles represent potential energy, which can be released in reactions with other molecules. Normally, only the terminal phosphate and its energy-rich bond (squiggle) is involved in such reactions.

condensation reactions also require the input of chemical energy. This energy commonly is derived from energy-rich molecules of ATP (**adenosine triphosphate,** Fig. 1-21); these molecules occur in all living cells and are the driving force of many chemical reactions in cells. The production and utilization of ATP in cells are discussed in later chapters so perhaps it will suffice to say here that the ATP molecule is a ready source of energy in cells. The molecule has been likened to a tiny power source, such as a flashlight battery, that can be "plugged in" where needed to do a particular bit of the cell's work. Such analogies are all right if not taken too literally; consider them symbolically, not factually.

Carbohydrates and Polysaccharides Among the many carbohydrates (Fig. 1-22) occurring in cells are simple sugars (trioses, tetroses, pentoses, hexoses, etc.) which participate in the functions of cells either as single entities (e.g., **monosaccharides,** "single sugar") or as subunits of polymers. For example, the monosaccharide glucose (Fig. 1-20) functions as a monomer in three important polymers: **starch, cellulose,** and **glycogen** (Fig.

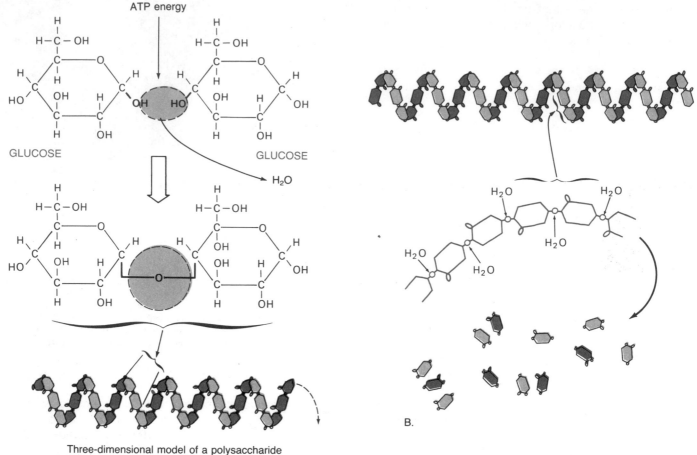

FIGURE 1-22
Carbohydrates. **A.** The formation of an oxygen bridge between two monosaccharide molecules (in this example, glucose) is shown. Further condensation reactions produce a long-chain polysaccharide. **B.** Hydrolysis. A macromolecule such as a polysaccharide (e.g., starch or cellulose) is broken down into subunits (monomers) by addition of one water molecule at each linkage.

1-23). Polymers composed of sugars only are called **polysaccharides**. In addition to polysaccharides, there are double sugars or **disaccharides** composed either of two molecules of the same sugar (as shown in Fig. 1-22) or of two different sugars. Common table sugar **(sucrose)** is, for example, composed of two different hexose sugars, glucose and **fructose**. Sugars also can be combined with other nonsugar monomers to form polymers. Examples that will be discussed extensively in future chapters are **DNA** and **RNA (deoxyribonucleic acid** and **ribonucleic acid)**. The "D" ("deoxyribo-") and "R" ("ribo-") in the names of these two polymers stand for two similar pentose sugars, **deoxyribose** and **ribose**. (This combination of subunits occurs also in ATP; Fig. 1-21.)

You may have wondered how three different polysaccharides—starch, cellulose, and glycogen—could differ in their characteristics yet be composed of the same subunits, for example, glucose molecules. The explanation is that in cellulose, glycogen, and starch, the oxygen bridges between atoms of the glucose molecule differ in their bonding angles and in which carbon atoms are involved in bond formation. In addition, cellulose is a straight-chain polymer whereas starch and glycogen are branching polymers, glycogen being the more highly branched (Fig. 1-23).

Lipids Lipids are molecules which are insoluble in water. They contain carbon, hydrogen, and oxygen, but in different proportions than found in carbohydrates. Cer-

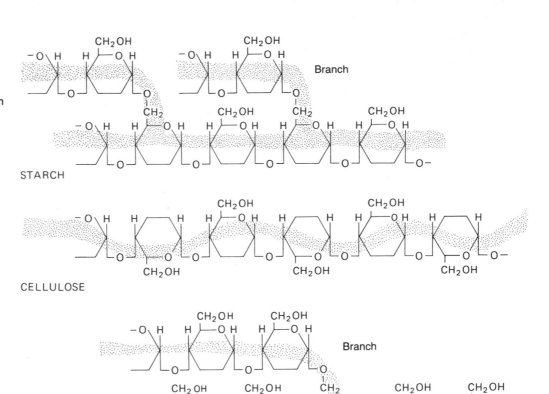

FIGURE 1-23
Starch, cellulose, and glycogen. Note the similarity of starch and glycogen; cellulose is depicted as an unbranched chain having different bonding angles between alternate glucose molecules.

STARCH

CELLULOSE

GLYCOGEN

tain lipids may also contain phosphorus or nitrogen. **Fatty acids** are the simplest lipids. They are organic acids, those in fats most commonly of 16 or 18 carbon atoms, and may be further characterized as **saturated** (without double bonds between any adjacent carbon atoms) or **unsaturated** (with one or more such double bonds—refer to Figs. 1-24 and 15-4D).

Waxes and fats are lipids composed of conjoined fatty acids and alcohols. Waxes are composed of an alcohol and a fatty acid joined together by an ester linkage formed between the carboxyl group of the fatty acid and the —OH of the alcohol. The reaction is a dehydration:

$$-\overset{\overset{\displaystyle O}{\|}}{C}-O\!\!\fbox{H}\ \ \fbox{$H\ O$}-\overset{\overset{\displaystyle H}{|}}{\underset{\underset{\displaystyle H}{|}}{C}}-\ \longrightarrow\ -\overset{\overset{\displaystyle O}{\|}}{C}-\underbrace{O-}_{\text{ester linkage}}\overset{\overset{\displaystyle H}{|}}{\underset{\underset{\displaystyle H}{|}}{C}}-\ +\fbox{H_2O}$$

There are many kinds of lipid molecules of importance in biological systems. Some familiar ones are **triglycerides,** represented in the kitchen and on grocery shelves by vegetable oils and animal fats. Triglycerides are composed of one molecule of the alcohol glycerol and three molecules of fatty acids (Fig. 1-24).

Phospholipids resemble triglycerides, but they bear a phosphate-containing part in a position occupied by one of the three fatty acids of a triglyceride. This phosphate-containing part makes the molecule partially soluble in water, a characteristic of great importance in the organization of biological systems. The outer and inner membranes of cells are constructed of a double layer of phospholipid molecules, arranged with the phosphates facing outward, in contact with external and internal aqueous solutions (Fig. 1-25C).

Steroids are lipids that differ from fatty acids, triglycerides, and phospholipids in not being composed of, or containing, fatty acids. Nevertheless, they are classified with lipids because of their general fatlike properties: insolubility in water, solubility in fats, oils, and solvents of fats and oils. They are complex molecules composed of four interlocking carbon-containing rings plus hydrocarbon side-chains. A number of biologically important lipids are steroids, including cholesterol (Fig. 15-4A), vitamin D, cortisol, and certain sex hormones. All of these will be discussed in later chapters.

Nucleic Acids Nucleic acids are macromolecules found in all living cells. They are composed of three

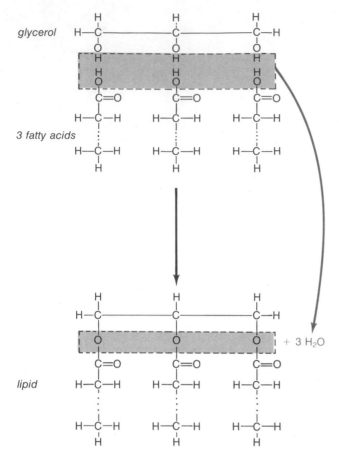

FIGURE 1-24
The formation of ester linkages between glycerol (an alcohol with three —OH groups) and three fatty acid molecules to make a triglyceride. The total lengths of fatty acids are not shown. Intermediate carbons in each fatty acid are indicated by dots.

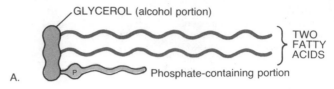

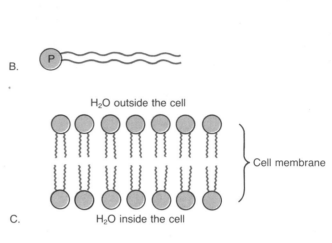

FIGURE 1-25
Structural relationships of a phospholipid. **A.** The general arrangement of a phospholipid. **B.** A conventional way of depicting a phospholipid, with a phosphate-containing "head" and two parallel fatty acid portions. **C.** The arrangement of phospholipids in a cell membrane.

classes of subunits: pentose sugars (ribose or deoxyribose), phosphate ions, and amines. The last are of two basic types, **purines** and **pyrimidines,** and there are subtypes of each. The relationships of the subunits of a nucleic acid are as follows:

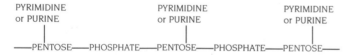

Because of their great importance in the lives of cells, nucleic acids will be discussed extensively in Chapter 5, but you may wish to obtain a preview of their structure by referring to Fig. 5-4. One kind of nucleic acid, deoxyribonucleic acid (DNA), contains the genetic information of the cell; another kind, ribonucleic acid (RNA), is employed in reactions that translate the information encoded in DNA into cellular structures and functional molecules such as enzymes.

Polypeptides (Proteins) and Enzymes Polypeptides are long-chain molecules composed of amino acids linked together by **peptide bonds** between adjacent carboxyl and amino groups (see also Fig. 1-26):

Simple polypeptides may be composed of repetitions of a single or only a few different kinds of amino acids; more complex ones are comprised of a greater variety of amino acids. Twenty different amino acids have been identified as components of various polypeptides, all resembling one another in having an amino group attached to the carbon atom next to the carboxyl group. (That carbon is an **alpha [α] carbon,** so polypeptides are said to be composed of **alpha amino acids.**) Proteins are made up of one or more polypeptides; insulin, one of the smallest protein molecules and the first in which the amino acid sequence was determined, contains two polypeptides, one with 21 amino acids, the other with 30 (Fig. 1-26). The two polypeptides of insulin are held together by a covalent bond, called the **disulfide bridge,** between adjacent sulfur atoms. In addition to bonding

FIGURE 1-26

Formation of a peptide linkage between two amino acids, and the relationship of the resulting dipeptide to a polypeptide. The peptide linkage also is called an amide bond.

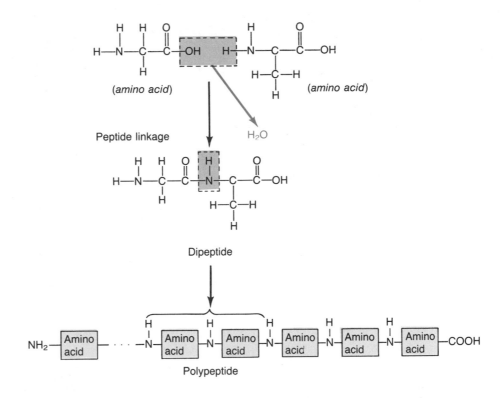

polypeptides, disulfide bridges are important in maintaining the three-dimensional form of protein molecules, which often are folded back upon themselves. Hydrogen bonds also may be present between some amino acids, further contributing to the maintenance of the three-dimensional architecture of proteins.

Enzymes are proteins with a highly organized three-dimensional structure (Figs. 1-27 and 1-28). This organization is responsible for a region of the molecule called the **active site.** The active site serves as a kind of molecular "jig" or "clamp" that holds reacting molecules in such a way (sometimes referred to as the **key-in-lock principle**) that activation energy is lowered and the react-

ants are modified (reactants are called **substrate molecules,** or simply the **substrate**). Once the reaction has occurred, newly altered molecules are released and the enzyme is free to catalyze another reaction.

In addition to combining molecules into new arrangements, enzymes can also act to split larger substrate molecules into smaller subunits. Macromolecules such as starches and proteins often are broken down into their respective monomers during digestive processes. Enzymes catalyze most such reactions, which are called **hydrolyses** (sing. **hydrolysis,** "breaking with water"): a water molecule (actually an H^+ ion and an OH^- ion) is introduced in place of an existing linkage

FIGURE 1-27

Insulin molecule.

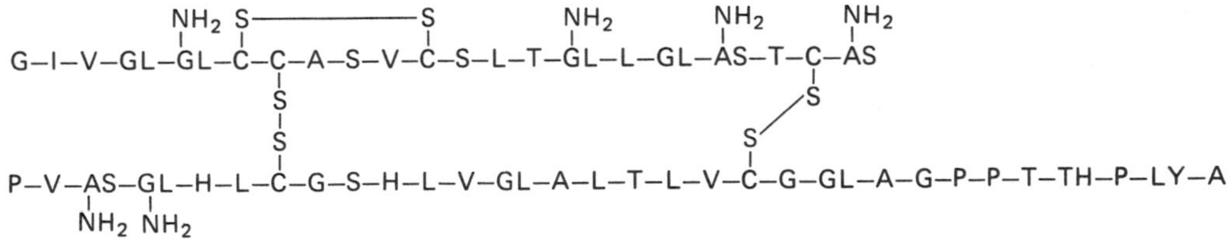

$$NH_2 \quad S—————S \qquad NH_2 \quad NH_2 \quad NH_2$$
$$G–I–V–GL–GL–C–C–A–S–V–C–S–L–T–GL–L–GL–AS–T–C–AS$$
$$\qquad\qquad\qquad\qquad\qquad S \qquad\qquad\qquad\qquad\qquad\qquad\qquad S$$
$$\qquad\qquad\qquad\qquad\qquad S \qquad\qquad\qquad\qquad\qquad\qquad\qquad S$$
$$P–V–AS–GL–H–L–C–G–S–H–L–V–GL–A–L–T–L–V–C–G–GL–A–G–P–P–T–TH–P–LY–A$$
$$\qquad\quad NH_2 \ NH_2$$

A—alanine	L—leucine
AR—arginine	LY—lysine
AS—aspartic acid	P—proline
C—cysteine	S—serine
G—glycine	T—tyrosine
GL—glutamic acid	TH—threonine
H—histidine	V—valine
I—isoleucine	

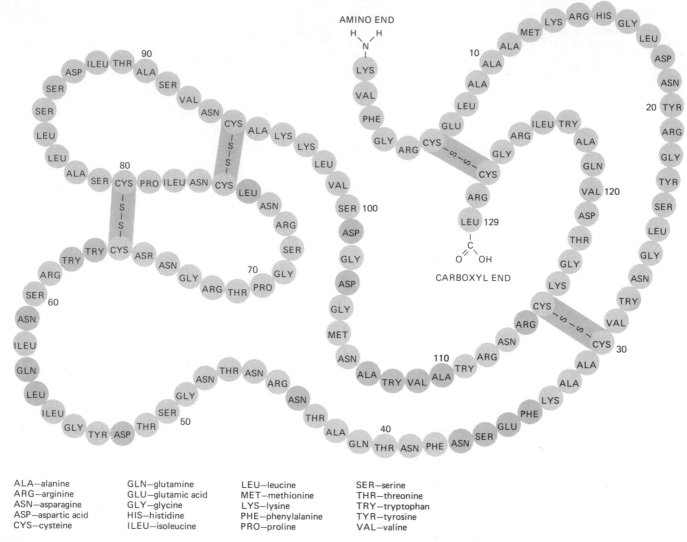

FIGURE 1-28
Molecular model of the enzyme lysozyme. (Redrawn from D. C. Phillips, The three-dimensional structure of an enzyme molecule. Copyright © by Scientific American, Inc. All rights reserved.)

ALA—alanine
ARG—arginine
ASN—asparagine
ASP—aspartic acid
CYS—cysteine

GLN—glutamine
GLU—glutamic acid
GLY—glycine
HIS—histidine
ILEU—isoleucine

LEU—leucine
MET—methionine
LYS—lysine
PHE—phenylalanine
PRO—proline

SER—serine
THR—threonine
TRY—tryptophan
TYR—tyrosine
VAL—valine

between the monomers (Fig. 1-22B). An example is the enzyme ptyalin (salivary amylase) in human saliva. Ptyalin catalyzes the breakdown of starch in food, producing sugar molecules.

The Key-in-Lock Principle of Specificity Molecular recognition is an important strategy in cellular processes. This recognition depends on molecular size and shape, which vary considerably, as can be seen in diagrams of molecules presented in this chapter. Some molecules "recognize" and transport other molecules through cell membranes; others are activated only by specifically shaped molecules. Our sense of smell works because we have specifically shaped odor-receptor sites in our nostrils which "recognize" odor molecules by their individual shapes and send messages to the brain when a particular odor molecule fits a specific receptor site and generates a nerve impulse. Enzymes also operate according to this kind of recognition process, which is called the "key-in-lock" principle because it resembles the opening of a lock only when the properly shaped key is inserted and turned (Fig. 1-29). An enzyme interacts with substrate molecules, binding them in such a way that the parts of the substrate molecules in which bonds will be made or broken impinge upon the active site of the enzyme. During this process, the enzyme may undergo a conformational change (as shown in Fig. 1-29), which facilitates either bond formation or bond breaking, as the case may be, and (or) subsequent release of the product(s).

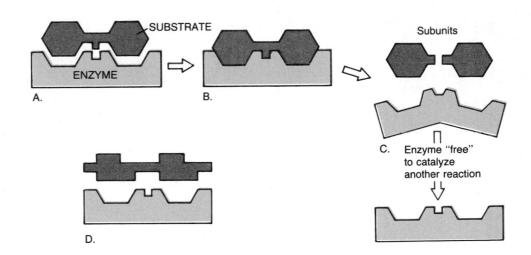

FIGURE 1-29

The key-in-lock principle of enzyme action. An enzyme is visualized as having an active site **(A)** that matches only the shape of certain substrate molecules **(B)**. The same enzyme may assemble two molecules into a more complex one, or disassemble a complex one into subunits. When substrate bonds are broken (or made) we can visualize them as no longer exactly fitting the active site, which may in addition have undergone a conformational change, causing the product to be released from the enzyme **(C and D).**

Some enzyme-catalyzed reactions are spontaneous; others require the activation of the enzyme by accessory molecules or ions called **cofactors.** ATP is a cofactor in many enzyme actions, supplying energy for making substrate bonds. Certain metal ions, such as Cu^{2+}, Zn^{2+}, Mg^{2+}, Mn^{2+}, also function as cofactors, aiding the attachment of substrate molecules to the active site of an enzyme. This is the explanation for the importance of trace elements to life (see Table 1-1). Other cofactors, known specifically as **coenzymes,** act in concert with enzymes by accepting hydrogen ions and electrons from, or giving them up to, substrate molecules in oxidation-reduction reactions.

We have noted that an essential aspect of enzyme catalysis is the fit, or complimentarity, of enzyme and substrate. However, certain nonsubstrate molecules are sufficiently like the substrate that they can "compete" with the substrate for the active site of the enzyme. This results, as one might expect, in lowering the efficiency of that particular chemical process. This effect is called **competitive inhibition,** and we will see examples in later chapters.

SUMMARY

Today about 90 natural elements are recognized, and a few others have been produced in the laboratory. Of these, only about a quarter appear to be essential for the life of plant and animal cells. Present knowledge of the essential elements comes from experiments in which the elements that plant and animal cells take from the environment are carefully recorded.

Elements, including those essential to life, exist in the form of molecules comprised of atoms that always combine in definite proportions. All atoms have a similar construction, being composed of three kinds of particles: the electrically negative electron, the electrically positive proton, and the electrically neutral neutron.

All atoms have a nucleus composed of closely packed and relatively heavy protons and neutrons, about which the relatively small and light electrons move in orbitals. Atoms are joined together by forces established when electrons either are transferred from one atom to another (ionic bonds) or are shared between atoms (covalent bonds). Covalent bonds between carbon, hydrogen, oxygen, and a few other elements form organic compounds, the compounds most prevalent in living organisms.

When substances of one kind are converted into other kinds of substances by the rearrangement of atoms, they are said to have undergone a chemical reaction. In some reactions, only a rearrangement of ions occurs. For example, in a neutralization reaction between HCl and NaOH (an acid and a base), NaCl (a salt) is formed, and H+ and OH$^-$ ions combine to form water. Such ionic reactions usually start spontaneously and are quite rapid. In other cases the reaction may proceed either slowly or rapidly, according to the nature of the molecules. Many chemical reactions require an initial input of outside energy, the activation energy, but thereafter are self-sustaining. Certain substances, called catalysts, have the ability to lower the activation energy requirement, and thereby to control the speed of a reaction. In living cells, catalytic functions are performed by enzymes.

When electrons are shared between atoms in certain chemical reactions, oxidation and reduction of the participating atoms occurs; the electron donor becomes oxidized, and the electron acceptor becomes reduced. Although oxygen often is an electron acceptor in such reactions, other atoms may also serve this function. All such actions, whether or not oxygen is involved, are oxidation-reduction reactions.

Many reactions (especially those involving oxidation and reduction) are accompanied by changes in energy possessed by the reacting substances. Reactions that make energy available for performing useful work are exergonic. The oxidation of food in the human body is an example. Reactions that require the input of energy and form products high in potential energy are endergonic. Photosynthesis, a natural food-producing process, is an example. Thus it is clear that atoms of elements are the fundamental units that carry out life processes, particularly in the energy exchanges which operate the life machinery in the cell.

KEY WORDS

element	ionic bond	organic compound
atom	cation	amino acid
compound	anion	macromolecule
molecule	hydrogen bond	hydrolysis
electron	acid	polysaccharide
proton	pH	lipid
neutron	base	protein
orbital	salt	polypeptide
isotope	buffer	enzyme
chemical bond	oxidation-reduction reaction	substrate
valence	potential energy	cofactor
chemical reaction	free energy	coenzyme
activation energy	work	competitive inhibition
catalyst	heat	
covalent bond	essential element	

QUESTIONS FOR REVIEW AND DISCUSSION

1 What is meant by the term *element*? Name some elemental substances. Are they the same as compounds? Explain.

2 List some of the important biological elements. Discuss their importance to organisms.

3 Define *molecule*. Describe the relationship between atoms and molecules. Give some examples of different kinds of molecules.

4 Discuss the bonds that hold atoms together in molecules. What are ions? Anions and cations? Give examples of each.

5 Why is hydrogen bonding particularly important in biological processes?

6 Define *acid*, *base*, and *salt*. What is meant by the term *neutralization*? What is the pH scale? Define *chemical equilibrium*.

7 In simple terms, describe an oxidation-reduction reaction. Is oxygen required?

8 Give an example from your own experience that illustrates various states of energy. Give a good illustration of activation energy.

9 Describe the key-in-lock principle of enzyme action.

10 Give a specific example of an enzyme action upon a macromolecule.

11 What kinds of bonds hold the subunits of macromolecules together? Specifically describe the bonds in a polysaccharide, a protein, and a lipid.

12 What is the biological significance of hydrolysis?

SUGGESTED READINGS

FRIEDEN, E. 1972. The chemical elements of life. *Sci. Amer.* 227(1): 52–60. (Lists the essential elements, including a few "new" ones. Also describes composition of earth, sea water, and human body in terms of elemental content.)

GROBSTEIN, C. 1965. *The strategy of life.* San Francisco: W. H. Freeman and Co. (The first chapters are a good introduction to the molecular basis of life.)

HILL, J. W. 1984. *Chemistry for changing times*, 4th ed. Minneapolis: Burgess. (A contemporary approach to chemistry written for the novice.)

KOSHLAND, D. E., JR. 1973. Protein shape and biological control. *Sci. Amer.* 229(4):52–64. (Enzyme actions are described in terms of molecular structure and symmetry.)

MASTERTON, W. L., E. J. SLOWINSKI, and C. L. STANITSKI. 1981. *Chemical principles*, 5th ed. Philadelphia: Saunders College Publishing. (This modern chemistry textbook covers basic principles in a lucid, easily comprehensible fashion. Includes a valuable chapter introducing biochemistry.)

SOLOMONS, T. W. G. 1984. *Organic chemistry*, 3rd ed. New York: John Wiley and Sons. (This understandable text emphasizes bio-organic chemistry.)

SUMMERLIN, L. R. 1981. *Chemistry for the life sciences.* New York: Random House. (Facilitates understanding of the concepts covered in this chapter and the next.)

WHITE, E. T. 1964. *Chemical background of the biological sciences.* Englewood Cliffs, N.J.: Prentice-Hall. (Presents a comprehensive explanation of biologically important chemistry.)

WOLFE, S. L. 1981. *Biology of the cell*, 2nd ed. Belmont, Calif.: Wadsworth Publishing Co. (This excellent cytology text contains a well-written chapter on the origins of cellular life.)

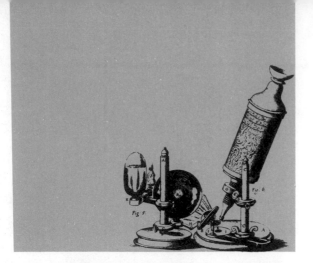

Cells 2

During the Dark Ages and the centuries of the Renaissance (the fourteenth through the seventeenth), it was widely believed that animals could arise from muck and mud, even be transformed into other animals, and finally sink back into the lifeless matter from which they came. Frogs were supposed to be formed from rainwater, or from leaves in ponds. It was believed that in winter they burrowed into the bottoms of ponds and swamps, underwent drastic transformation, and in the spring reappeared in the form of swallows and swifts. Geese were said to have been transformed from barnacles, barnacles to have grown from inanimate wood and rock, horsehairs in watering troughs to have turned into slender horsehair worms (nematodes), and maggots to have arisen spontaneously from decaying meat. Up until about 100 years ago, most people, including some eminent scientists, believed in **spontaneous generation** of life, or **abiogenesis.**

The spontaneous generation of life hypothesis has had a curious history, including periods of wide acceptance as well as times of scorn. Although many of the theories now seem like fairy tales, they were at one time matters of serious controversy. In fact, one of the earliest controlled experiments known was done by the Italian nobleman Francesco Redi (1626–1698). (Note: In a con-

trolled experiment, the experiment is performed in duplicate. One experiment, the **control,** represents the normal or natural state of affairs; the second, or **variable** experiment, differs from the control in one aspect, that which is being tested.)

Redi set out to explain the occurrence of maggots in decaying flesh. Many of his contemporaries believed that the maggots were formed spontaneously by the process of putrefaction. Redi used three sets of pots, all containing animal flesh (snake, fish, eel, or veal) (Fig. 2-1). One set consisted of jars with tight paper coverings, another of jars covered by screening, and the third of open jars (the control). Maggots grew only in the flesh in the open jars, where it was observed that flies had free access; Redi stated that maggots therefore developed only from eggs previously deposited in dead animal flesh by flies. He used both closed and screened jars to answer the objections of those persons who argued that some property, perhaps air, was required in addition to rotting flesh for spontaneous generation to occur and had been excluded by the impervious lids. In this way, at least one theory of spontaneous generation was disposed of, at least from the scientists' point of view. Nevertheless, the concept lingered on for two more centuries.

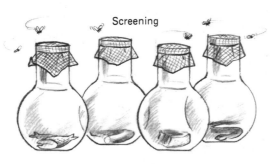

FIGURE 2-1
Redi's experiments with maggots and spontaneous generation.

During the early 1800s science had attained sufficient sophistication that gross examples of spontaneous generation were known to be nonsense. However, the discovery of bacteria in 1683 by van Leeuwenhoek introduced a new source of controversy. Because bacteria grew in broths of various kinds, even after long cooking, it was thought that they developed spontaneously out of the nutrients present.

A number of distinguished scientists supported the concept of spontaneous generation of bacteria and attempted to prove it. Others devised ingenious experiments designed to show that life did not arise from inert matter. Eventually, the controversy was settled by Louis Pasteur (1822–1895) in a series of simple experiments in which glass flasks having recurved goosenecks were used (Fig. 2-2). Broth boiled in such flasks remained uncontaminated even though air could enter through the necks of the flasks. The down-curved portion of the neck acted as a bacterial trap, and Pasteur's flasks remained uncontaminated for years. In fact, one of his original flasks, in the Pasteur Institute in Paris, is still bacteria free more than a hundred years later.

Scientists do not believe that life develops spontaneously in the present world. Millions of experiments that would have discovered any tendency for microorganisms to arise from nonliving matter have been conducted. Every time a bacteriologist prepares a sterile (bacteria-free) culture in a test tube, the hypothesis of spontaneous generation once again is tested. Each time a surgeon disinfects the skin of a patient preparatory to doing an operation and all the instruments, the bandages, the clothing worn by the surgical team, and even the air of the operating theater are sterilized, the theory of spontaneous generation receives another test. To date, all scientific experience leads to the conclusion that living organisms arise only by the reproduction of previous generations of living organisms. Paradoxically, a widely accepted theory of the origin of life is based on spontaneous generation. Basically such a theory de-

FIGURE 2-2
Pasteur flask.

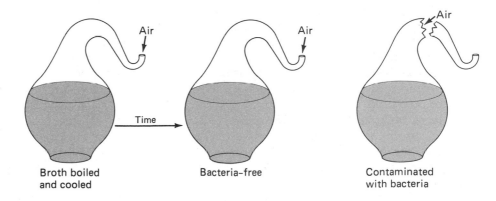

Broth boiled and cooled   Bacteria-free   Contaminated with bacteria

pends on three points: (1) an immense length of time was available for evolution to operate; (2) past environments differed from that of the present; and (3) cells were not present to consume accumulating molecules.

THE ORIGIN OF LIFE AND THE EVOLUTION OF CELLS

Age of the Earth

During the past 200 years, concepts of the age of the earth have been revolutionized. Prior to that time, most Europeans and Americans reckoned the age of the earth on the basis of biblical chronologies. For instance, one Irish bishop calculated that the earth was created in 4004 B.C.

In 1830 a Scottish geologist, Charles Lyell (1797–1875), suggested that the hypothesis of **uniformitarianism** be used to estimate the ages of deeply buried fossils and rock strata. This hypothesis proposed that the deposition of muds, sands, and soils in ages past must have occurred at the same rate as in the present. Thus the age of ancient deposits found in the faces of cliffs, in canyons, or in deep well borings may be estimated by measuring the thicknesses of overlying deposits. The application of uniformitarianism by Lyell and others resulted in estimates of the age of the earth very many times greater than previous figures. Today, although the thickness of rock strata still is used to determine the age of fossils, there are other, more sophisticated, methods at hand. Present knowledge of the age of the earth has been derived through the use of many approaches.

As noted in Chapter 1, certain isotopes of elements are radioactive as a consequence of their decay (disintegration). Uranium is one of the elements for which several radioactive isotopes are known. Among these is uranium 238 (^{238}U), which as it decays undergoes a series of transformations to form lead 206 (^{206}Pb) and helium. The rate of this decay has been calculated as sufficient to convert half of the original amount of ^{238}U to lead and helium in about 4.5 billion years. This rate is called the **half-life** of ^{238}U. The half-lives of other radioactive isotopes have been calculated; for example, that of carbon 14 (^{14}C) is about 5700 years, and that of bismuth 214 (^{214}Bi) is 19.7 minutes. Measurements made with radiation counters on rocks containing radioactive isotopes can be used to estimate the age of the rocks. Calculations based on the present ratios of ^{238}U, residual helium, and lead in the most ancient rocks known have suggested that the earth is about 4.5 billion years old.

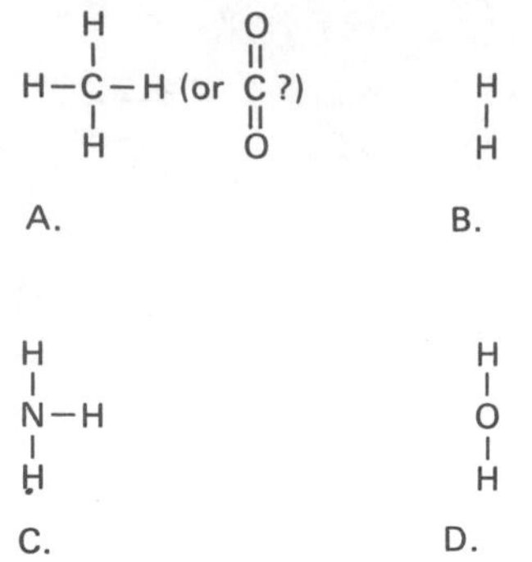

FIGURE 2-3
Gases of the ancient atmosphere. **A.** Methane (CH_4). (Recently, it has been proposed that carbon dioxide in addition to, or rather than, methane was present.) **B.** Hydrogen (H_2). **C.** Ammonia (NH_3). **D.** Water vapor (H_2O).

The Ancient Atmosphere

Other studies of ancient rocks that have been deeply buried beneath the earth's crust and exposed by erosion, glaciation, or oil-well drill cores, indicate that for the first 2 billion years the earth was much different than it is today. The atmosphere apparently contained no oxygen; it may have been composed of hydrogen, ammonia, methane, and water vapor (Fig. 2-3). So far as is known, there were no living organisms. Could such an atmosphere actually have existed in our solar system? In recent years space probes have been able to analyze the atmospheres of Saturn and Jupiter. At present, both of these planets, and Titan, a moon of Saturn, have atmospheres like that postulated for the earth in its early existence. Although these discoveries are not proof that the earth's early atmosphere was composed of similar gases, they do indicate that a similar atmosphere could have existed on earth in some past time.

Chemical Evolution

Molecules A number of scientists have wondered whether life originated in ancient bodies of water by a slow process in which simple chemicals evolved into more and more complex ones. It has been postulated that the molecules in the primitive atmosphere might have reacted with each other to form more complex molecules, a process called **chemical evolution**. It is believed that chemical evolution was accelerated by solar radiation because present-day experiments show that exposure to certain kinds of radiation can speed up reactions between molecules. Energy from ultraviolet light, a component of sunlight, may be absorbed by atoms in such a way that they briefly become more reactive. Today much of the ultraviolet radiation reaching the earth is absorbed in the upper atmosphere by oxygen and **ozone** (a blue gas found in the atmosphere and composed of three oxygen atoms, O_3). However, the primordial atmosphere had no oxygen or ozone layer; hence much more ultraviolet radiation reached the earth's surface than does now. This radiation probably caused the simpler molecules in the ancient atmosphere to react with others to form larger and more complex molecules. These larger molecules of various kinds are thought to have accumulated in the ancient seas during an interval of many millions of years.

In 1953 Stanley Miller, a graduate student at the University of Chicago, together with his professor, Dr. Harold C. Urey, described the results of an ingenious experiment designed to test the concept of chemical evolution. They constructed an apparatus in which water vapor from a boiling flask rose in a column filled with hydrogen, methane, and ammonia to a smaller chamber where electrodes continually produced electrical sparks (Fig. 2–4). These sparks produced ultraviolet radiation as well as visible light and heat. Under this bombardment of radiation, the molecules in the synthetic primordial atmosphere reacted with each other and formed 19 different kinds of organic molecules. These condensed in another part of the apparatus and were collected and analyzed. Among the molecules produced during the week the experiment was run were five amino acids known to occur in the proteins of living cells.

Subsequently, other scientists have run elaborate experiments in which more complex organic molecules were formed, including sugars, **polypeptides** (chains of amino acids), and even nucleic acid bases (portions of DNA, the genetic material of the cell) (see Table 2–1).

TABLE 2-1
Partial listing of organic molecules formed in primitive-atmosphere experiments

Amino acids	Aldehydes and sugars
Alanine	Formaldehyde
Aspartic acid	Ribose
Glutamic acid	Deoxyribose
Glycine	Etc.
Isoleucine	Organic acids
Leucine	Acetic acid
Phenylalanine	Formic acid
Proline	Glycolic acid
Serine	Lactic acid
Threonine	Propionic acid
Valine	Purines and pyrimidines
Hydrocarbons	Adenine
Dimethylbutane	Cytosine
Methylpentane	
Dimethylhexane	

FIGURE 2-4
Miller-Urey experiment.

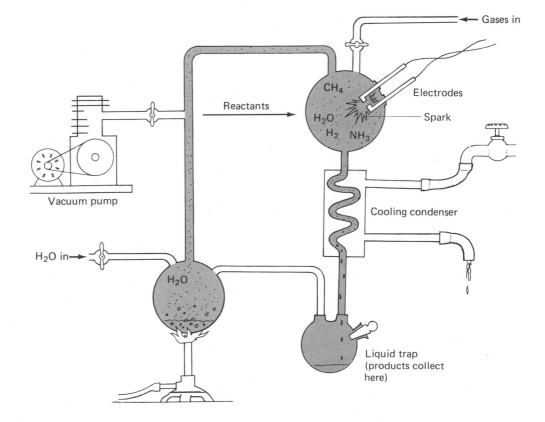

Scientists speculate that many kinds of organic molecules may have been formed in the earth's early atmosphere by processes similar to those described in the experiments just mentioned (see Fig. 2-5 for the structures of some simple organic molecules). Such molecules might have accumulated for thousands and perhaps millions of years, condensing in rainfall and concentrating in ponds, lakes, and seas. Possibly because of evaporation, some bodies of water may have achieved very concentrated solutions. The term "primordial soup" has been coined to describe this condition. Then, the further association of molecules into even larger **macromolecules** may have occurred, and these in turn might have formed aggregates called **colloids,** containing enough large reactive molecules to resemble cells. (A colloid is a mixture of suspended particles too large to dissolve but too small to settle out of suspension.)

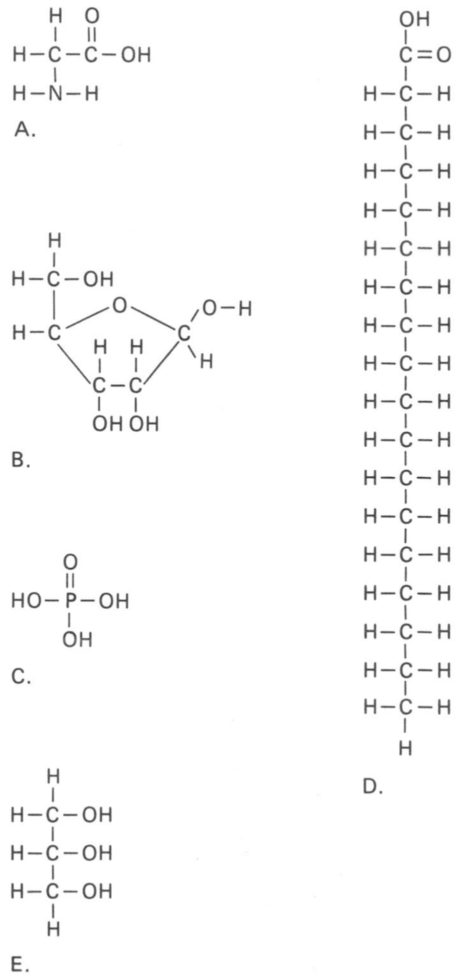

Protenoids One theory of the organization of matter into cell-like structures has been proposed by Sidney Fox. Fox was dissatisfied with proposals that aggregates of amino acids and peptides (short chains of amino acids) in a primordial soup could have produced membranes and cellular structures. However, when such solutions were dried, then rehydrated, Fox discovered that cell-like globules were produced. He refers to these structures as **protenoids** and suggests that cells perhaps originated similarly in temporary pools subjected to alternate drying and wetting. Other scientists, however, think that synthesis of DNA and RNA preceded evolution of cells and cell-like structures.

The First Cells If, in fact, cells evolved by chemical evolution, they at first may have been simple baglike structures containing DNA and perhaps a few hundred other large molecules, including enzymes. That being the case, an essential step in cell evolution was the development of the enclosing "bag." Life is manifested by an amazing number of complex chemical processes working in coordination with one another. This would be impossible if cell contents were merely floating about in water. Because membranes similar to the limiting membranes of cells can assemble themselves in solutions of macromolecules of proteins and **lipids** (fats and oils), it is thought that such self-assembly may also have occurred in the primordial soup. Moreover, artificial lipid-protein membranes have been examined by electron microscopy

FIGURE 2-5
Structural formulas of some biologically important molecules. **A.** Glycine, an amino acid ($C_2H_5NO_2$). **B.** Ribose, a simple sugar ($C_5H_{10}O_5$). **C.** Phosphoric acid (H_3PO_4). **D.** Stearic acid, a long-chain fatty acid and component of animal fats ($C_{18}H_{36}O_2$). **E.** Glycerol ($C_3H_8O_3$).

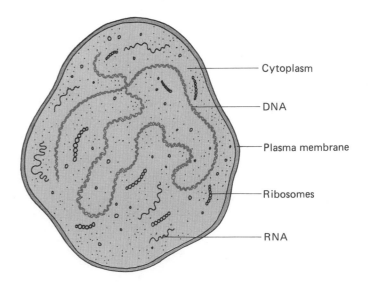

FIGURE 2-6
Essential components of a mycoplasma organism.

and appear nearly indistinguishable from real bounding membranes of cells.

What might these first cells have looked like? Possibly they resembled a group of small, irregularly shaped, baglike cells called **mycoplasmas,** which are considered to be primitive forms of bacterial life and are smaller and less complex than a typical bacterium (compare Fig. 2-6 with Fig. 19-2; also refer to Fig. 2-7). A mycoplasma consists of a delicate external membrane, the **plasma membrane,** enclosing a length of DNA, smaller molecules of several kinds of RNA, several hundred enzymes, and numerous **ribosomes.** The last are very small organelles that function as protein assembly points. Mycoplasmas also resemble the hypothetical first cells in their nutrition, because they are unable to process their own food. Normally they exist as parasites in other organisms, living on predigested food, but they can be free-living if a complex culture medium of life-supporting chemicals is provided. The primordial soup, in theory, would resemble such a culture medium.

The first cells probably reproduced by simple breakage into smaller cells, which then grew and also became fragmented. Eventually the primordial soup came to support a rather dense population of simple cells, and, as a result, a new problem arose: overpopulation and diminishing resources. At that point, the evolution of life might have ended before it had fairly well begun, except that some cells developed the capacity for harvesting energy from sunlight and using it to manufacture food. With the evolution of such **photosynthetic** systems, life became self-sustaining, and further diversification became possible.

Evidence from the study of ancient fossilized cells suggests that cellular life was established more than 3 billion years ago. The oldest fossilized cells are at least 3 billion years old and resemble bacteria (Fig. 19-2) and photosynthetic **cyanobacteria** (blue-green algae) (Fig. 19-5). Because these organisms are more complex than the hypothetical first cells, it is assumed that cells arose even earlier in the earth's history.

As has been noted, the primitive atmosphere contained no oxygen. Primitive cells, therefore, lived by using energy-releasing processes that used electron acceptors other than oxygen in their oxidation reactions. However, beginning about 3 billion years ago, oxygen from photosynthesis began to accumulate in the atmosphere (photosynthesis produces oxygen by splitting water into hydrogen and oxygen). This early production of atmospheric oxygen by photosynthesis is deduced by dating the oldest rocks containing iron oxides, whose history coincides roughly with the fossil history of the cyanobacteria. The earth's present atmosphere, in which iron oxidizes (rusts) very rapidly, contains 20 percent oxygen and supports many oxygen-requiring forms of life. It is the product of nearly 3 billion years of photosynthesis.

The Cell Theory During the last century and a half, biologists have come to realize that all organisms and their parts are basically cellular in nature, and that cells are not empty spaces but are functional complexes of fluids, semifluids, and discernable internal structures. This concept is dignified with the name of **cell theory,** a sweeping generalization first proposed in 1838 by two Germans, Mathias Schleiden (1804–1881) and Theodor Schwann (1810–1882), and has since undergone both amplification and some qualification. Today, the cell theory states that:

1 All living things are composed of cells.

2 All cells are similar in structure and function, but are subdivisible into two basic types: prokaryotes and eukaryotes (those without a nucleus and those having a nucleus).

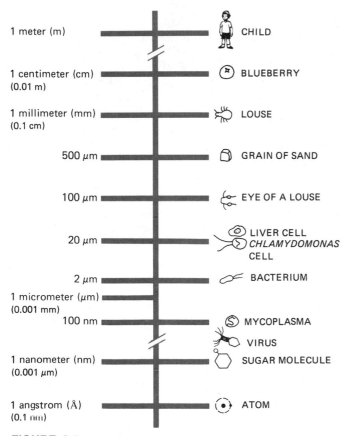

FIGURE 2-7
Metric scale and its application to the dimensions of natural objects.

THE STUDY OF CELLS

Many scientists working during the past 300 years have contributed to a growing appreciation of the roles of cells in the living world. An Englishman, Robert Hooke, is credited with the discovery of cells in 1665; that date may be considered the starting point of the science of **cytology** (*kytos,* "container"), the study of cells. Hooke was probably the foremost scientist of his day and had an unusual occupation for his times in that he was employed solely to do scientific research. Today, to be paid to carry on scientific research is the usual thing, and amateur scientists do not often play an important role in science. In Hooke's day, nearly all scientists were amateurs. Hooke, however, had demonstrated such a remarkable ability for scientific innovation that he was hired by the Royal Society of London, the foremost scientific society of his day, to demonstrate one new scientific discovery to the members each week—a very tall order in any era! Hooke often must have felt hard-pressed to find something new to catch the attention of his audience. He invented the balance spring which made pocket watches possible and developed a powerful vacuum pump for use in studies on the behavior of gases. However, biologists are most familiar with Hooke's work with microscopes; he invented a form of microscope that is surprisingly similar to those used today (Fig. 1). With his microscope he studied many biological phenomena, and eventually wrote a beautifully illustrated book, *Micrographia,* about them. Hooke's microscope undoubtedly gave him many new demonstrations for his weekly exhibits. In one such study he cut bottle cork into very thin slices and saw that it was composed of small, empty chambers (Fig. 2). These chambers, which Hooke called "cells" (after the rooms in the abbey where monks studied and prayed), account for the lightness of cork.

Although Hooke saw only empty cells, it is now known that cells of cork are living only during their development and become empty cell walls when they mature. At present, the term *cell* is applied to living cells, so mature cork would be said to be composed of dead cell walls.

Hooke (1632–1703) was contemporary with another self-made microscopist, Anton van Leeuwenhoek (1632–1723); undoubtedly the two were familiar with each other's work because the latter corresponded regularly with the Royal Society. Where Hooke was a scientific generalist, van Leeuwenhoek was

3 All cells originate by cell division from preexisting cells.

4 The structure and functioning of an organism is produced by the organization and actions of all of its cells.

The term **prokaryote** ("before nucleus") currently is used in reference to bacteria and to a group of simple photosynthetic cells called cyanobacteria. Cells of all other organisms contain nuclei and are termed **eukary**otes ("true nucleus"). (Structural and other aspects of prokaryotes are described in Chapter 19.)

Microscopy

Robert Hooke often is credited with the first observations of cells as well as their naming, but another microscopist was the first to observe individual living cells. He was Anton van Leeuwenhoek, a prosperous Dutch cloth merchant whose hobby was microscopy. During the

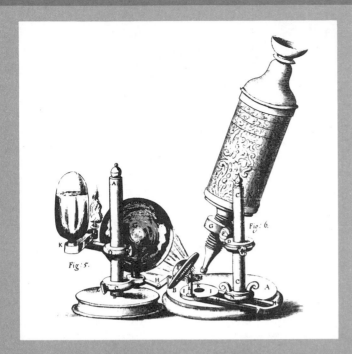

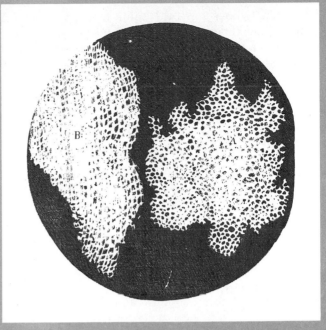

FIGURE 1
Microscope invented by Robert Hooke. Note the basic similarity in arrangement of the instrument and its accessories to present-day microscopes (see Fig. 2-8A). (From *Micrographia*, 1665)

FIGURE 2
Hooke's drawing of cork cells. (From *Micrographia*, 1665)

something of a specialist and apparently very much the "gentleman scholar." He constructed for himself some exquisitely simple but high-powered microscopes with which he observed bacteria as well as other microorganisms. Recently, a set of very thin sections of cork and the pith of elder were discovered in some envelopes in London and were reexamined using modern electron microscopes. It is known that van Leeuwenhoek used a "sharp shaving razor" to cut his specimens. In a recently published account, Brian J. Ford describes finding red and white blood cells as well as staphylococcus bacteria adhering to van Leeuwenhoek's thin-sections of elder pith. Were these cells that adhered to his specimens from van Leeuwenhoek himself? Brian Ford concludes that they may well have been.*

*From a condensation by Brian J. Ford. 1984. Bacteria and cells of human origin on van Leeuwenhoek's sections of 1674. *BioScience* 34(2):106–107.

latter part of the seventeenth century, Leeuwenhoek made a number of simple microscopes with which he was able to see bacteria, simple one-celled organisms called protozoans, and human sperm. He sometimes is referred to as the father of **microbiology,** as the study of microorganisms is called.

Since the pioneering days of microscopy, a great deal more has been learned about the nature of cells. During the last 20 years there has been almost an explosion of new knowledge, much of which has come from the use of the electron microscope and new biochemical tools.

There are two types of microscopes available to the biologist: the **light microscope** and the **electron microscope.** If a bacterial cell is examined with the best light microscope available today, a few more details than Leeuwenhoek was able to see will be distinguished; an electron microscope is needed to see small details of cellular structure. With the electron microscope, bacterial cells can be seen to be simpler in construction than the cells of most other organisms in that they lack nuclei and organelles.

All microscopes have the same function: to produce an enlarged image of the object under observation. This image is formed upon the retina of the eye, or can be projected onto a screen or the film in a camera. In the eye, the size of the retinal image depends on the angle at which light rays are bent by the lens system of the eye and, before they enter the eye, by the additional bending of light rays in the microscope.

In the light microscope, glass lenses bend, or **refract,** rays of light; in the electron microscope, magnetic lenses bend rays of electrons (Fig. 2-8). Because electrons cannot be seen by the eye, the image is first produced on a fluorescent screen like that of a television tube, or on photographic film. This image may be magnified further by a photographic enlarger or a viewing microscope.

The ability of any optical system to distinguish detail is called **resolution.** The resolution of the human eye from a distance of about 25 cm (10 in) is about 0.1–0.2 millimeter (mm); that is, objects as small as 0.1 mm can be distinguished, although only as dots or points of light. The resolution of the best light microscope is on the order of 0.1 micrometer (μm), or 1000 times that of the eye alone. The resolution of the electron microscope is about 0.1 nanometer (nm), or 10 million times that of the eye alone. Although still better resolutions theoretically are achievable, 0.1 nm seems to be about the practical maximum obtainable with the electron microscope. Some idea of the dimensions involved may be gained by referring to Fig. 2-7. See also Table 2-2.

CELL STRUCTURES

Organelles of Eukaryotic Cells

Eukaryotic cells contain in addition to one or more nuclei per cell, a variety of organelles which carry on different aspects of the work of the cytoplasm and enhance the efficiency of life processes. Organelles also enable

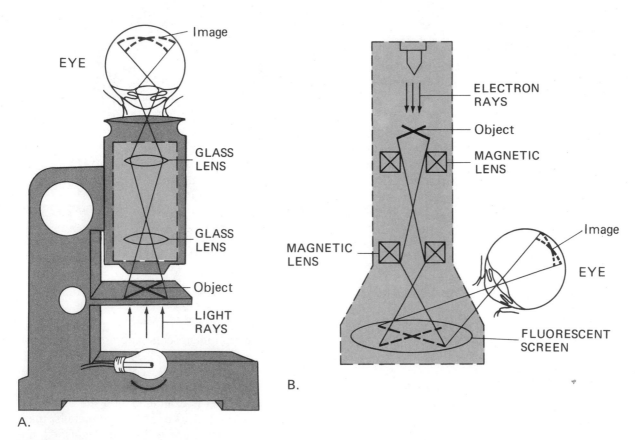

FIGURE 2-8
Light **(A)** and electron **(B)** microscopes. The maximum direct magnification attained with light microscopes is about 2000×; that of electron microscopes is approximately 200,000×. Further magnifications are obtained by photographic enlargements.

TABLE 2-2
Some metric units and English equivalents

Measurements	Metric	English
Linear	kilometer (km) 1000 m)	0.62 mile (mi)
	meter (m)	39.4 inches (in)
	centimeter (cm) (0.01 m)	0.39 in
	millimeter (mm) (0.001 m)	0.039 in
	micrometer (μm) (0.001 mm)	0.000039 in
	nanometer (nm) (0.001 μm)	
	angstrom (Å) (0.1 nm)	
Area	hectare (ha)	2.47 acres
Mass	kilogram (kg) (1000 g)	2.2 pounds (lb)
	gram (g)	0.035 ounce (oz)
	milligram (mg) (0.001 g)	
	microgram (μg) (0.001 mg)	
Volume	*Solids*	
	cubic meter (m^3)	35.3 cubic feet (ft^3)
	cubic centimeter (cm^3 or cc)	0.06 cubic inch (in^3)
	Liquids	
	liter (l)	1.1 quart
	milliliter (ml) (0.001 l)	0.034 fluid ounce
	microliter (μl) (0.001 ml)	

eukaryotic cells to do more different kinds of things and to become more highly specialized than prokaryotes.

It would be difficult indeed to select a cell to represent all the different types classified as eukaryotic. The best approach may be to look at the single cell of a simple green water plant known as a green **alga** (pl. **algae**) or at any of its near-relatives (Fig. 2-9).

The major functions performed by the organelles of green algae and similar cells are:

1   Support and protection.

2   Transport of molecules.

3   Photosynthesis.

4   Respiration.

5   Cellular control and reproduction.

6   Locomotion.

The organelles that carry out these functions are discussed in the sections that follow (refer to Fig. 2-9 throughout).

Cell Membranes   The plasma membrane is the outermost living component of the cell. It is not simply a wrapping around the cytoplasm, but a complex and dynamic structure involved in the movement of many substances into and out of the cell. Although cells differ greatly among species of organisms and, in most cases, among the various parts of individual organisms, the plasma membranes of all cells are thought to be structurally similar. Electron microscope studies, as well as

chemical and physical analyses, lead scientists to think that the plasma membrane and internal cellular membranes are composed of a double layer of phospholipid molecules. In this model, the phosphate part of each lipid molecule is directed outward, and the fatty acid components have an inward orientation (see Fig. 2-10). Associated with the phospholipids of the membrane of an animal cell are protein molecules and protein-carbohydrate molecules called **mucopolysaccharides** (Fig. 2-10). These constitute a surface layer called the **cell coat**. Among the proteins are enzymes and also substrate-transporting molecules that actively move substances through the plasma membrane (their actions are described in the next chapter). Some mucopolysaccharides of the cell coat function as cell recognition factors important in the binding of certain kinds of particles, including other cells, to the plasma membrane.

The Cell Wall   The outer part of the living cytoplasm of cells is the delicate plasma membrane (see Fig. 2-10). In cells of bacteria, fungi, algae, and all more complex green plants, the plasma membrane is covered by an inert and nonliving **cell wall** that is secreted by the living cell. The cell walls of bacteria and some fungi are composed of a complex association of proteins and noncellulosic carbohydrates; those of some fungi, most algae, and all higher green plants are composed of cellulose. Cell membranes of protistans and animals are not enclosed in cellulose walls, but may be supported and strengthened by calcium salts, silicon, or other materials.

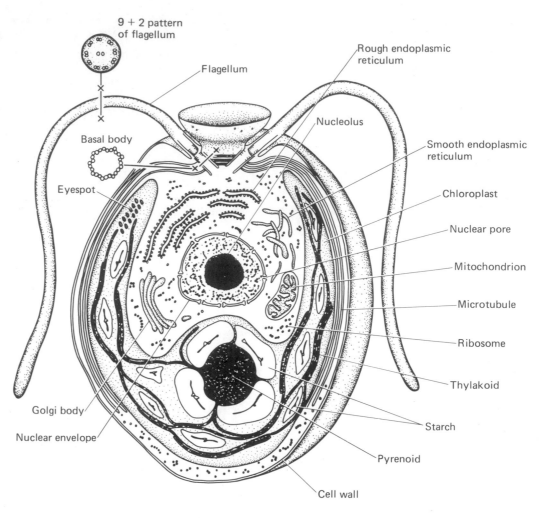

FIGURE 2-9
Cell of the eukaryotic green alga *Chlamydomonas.*

9 + 2 pattern of flagellum

Flagellum

Basal body

Eyespot

Golgi body

Nuclear envelope

Rough endoplasmic reticulum

Nucleolus

Smooth endoplasmic reticulum

Chloroplast

Nuclear pore

Mitochondrion

Microtubule

Ribosome

Thylakoid

Starch

Pyrenoid

Cell wall

An important function of the cell wall is to protect the bacterial, fungal, algal, or plant cell from the disruptive effects of excessive fluid pressure within the cell. In this respect the wall may be compared to the leather covering of a football or basketball, which keeps the air pressure within the inner rubber bladder from blowing out. The cell wall also protects the delicate plasma membrane within it from breaking or tearing when the cell makes contact with sharp objects or experiences rough treatment.

Cytoplasm and Cytoplasmic Organelles Once thought to be a frothy mixture of droplets, gels, food particles, and water, the **cytoplasm** of the cell now is known to be a highly organized network of at least three kinds of filaments—microtubules, microfilaments, and intermediate filaments, which form a lattice about the nucleus (not shown in Fig. 2-10). In addition, there is an interconnecting system, the **microtrabecular lattice** (Fig. 2-10), of very thin filaments. Rather than being a sort of biochemical mishmash, or bag of enzymes—as once vi-

sualized by cell chemists—the cytoplasm is apparently highly organized. The state of this organization is just beginning to be revealed, but it is clear that the architecture of the cytoplasm is very complex. In addition to its filaments and various soluble substances, the cytoplasm contains numerous small (about 20-nm diam) spherical bodies called ribosomes. They are composed of about 100 to 150 proteins and 3 or 4 strands of RNA. The importance of ribosomes in cell processes cannot be overstressed and will be discussed again in Chapter 5.

Also prominent in the cytoplasm are membranous sheets, tubules, and sacs associated with synthesis and transport of intracellular products. These membranous elements are components of the endoplasmic reticulum and the Golgi bodies (Fig. 2-11).

There are two kinds of endoplasmic reticulum: **rough endoplasmic reticulum** and **smooth endoplasmic reticulum.** Both are convoluted tubular or flattened sac-like structures composed of lipoprotein membranes (phospholipid and protein membranes similar to the membrane in Fig. 2-10 but lacking mucopolysaccha-

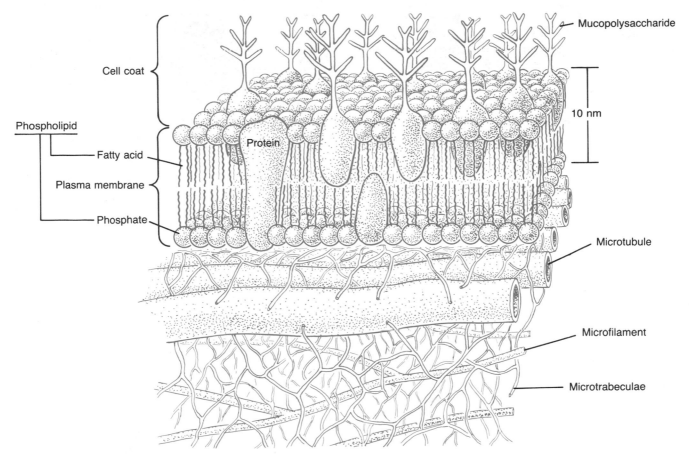

FIGURE 2-10
Three-dimensional model of the plasma membrane and adjacent cytoplasm.

rides). In sectional electron microscope views, they often look like meandering, closely spaced pairs of parallel membranes (Fig. 2-11). The rough endoplasmic reticulum has on its outer membrane surface a coating of ribosomes from which it gets its name. The smooth en-

doplasmic reticulum lacks ribosomes. Both types of endoplasmic reticulum are associated with the synthesis and transport of macromolecules.

The **Golgi body**, in electron microscope views, resembles a pile of flattened sacs; in three-dimensional

FIGURE 2-11
Relationship between the endoplasmic reticulum and the Golgi body.

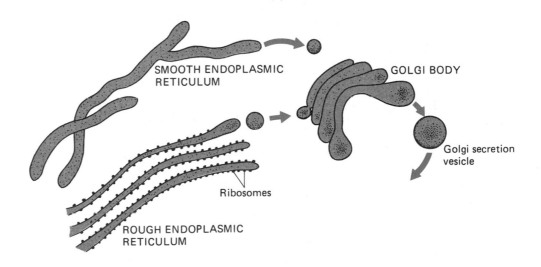

aspect it has been likened to a stack of pancakes from the edges of which **vesicles** ("small sacs") appear to be budding off (Fig. 2-11). It may be compared to a packaging factory in which the products of the endoplasmic reticulum, with which it is closely associated, are enclosed within vesicles and released into the cytoplasm.

Vacuoles Vacuoles have several functions. One type, found in cells that actively take in large food particles, is the **food vacuole**; another type, the **contractile vacuole,** is present in some one-celled organisms. Contractile vacuoles are primarily connected with the regulation of fluid concentrations in cells, and function as pumping structures which keep fluid from accumulating. Food vacuoles are food storage and digestion organelles in those cells that ingest particulate food. Algae and plant cells do not have food vacuoles because they do not ingest food particles. However, large vacuoles which are neither contractile nor digestive but are involved in water uptake and storage do occur in most plant cells.

Chloroplasts The **chloroplast** (see Fig. 2-12) is the organelle in eukaryotes that is specialized to carry out photosynthesis. The actual conversion of light energy into chemical energy is accomplished by a team of pigment molecules, including several kinds of **chlorophyll.** These photosynthetic molecules are located in the internal membrane system of the chloroplast.

Chloroplast membranes are arranged in parallel sheets called **thylakoids.** In many but not all chloroplasts, the thylakoids are cylindrical stacks of disklike **grana** (sing. **granum**); in others, thylakoids are not

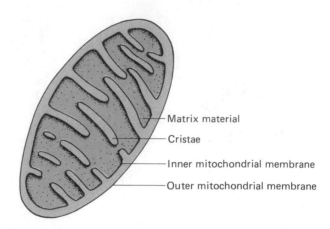

FIGURE 2-13
Mitochondrion.

stacked, and therefore the chloroplast is said to be **agranal**. (These terms are important to later discussions of photosynthesis and should be remembered.) The nonmembranous part of the chloroplast is the **stroma,** a complex association of ribosomes, DNA, RNA, starch, and enzymes, as well as smaller molecules and ions.

Starch is the major macromolecule synthesized by the chloroplast, and during times of active photosynthesis chloroplasts often become filled with starch grains. In cells of algae, chloroplasts also contain **pyrenoids,** which consist of a dense central part containing protein surrounded by a zone of starch grains. Pyrenoids are thought to regulate the formation as well as the breakdown of starch. Chloroplasts of higher plants do not have pyrenoids.

Mitochondria The chief oxidative organelle of eukaryotic cells is the **mitochondrion** (pl. **mitochondria,** "threadlike body") (Fig. 2-13). Most kinds of cells have many mitochondria; a few highly specialized cell types have just one very complex mitochondrion. Mitochondria resemble chloroplasts in having an envelope of two membranes, but differ from them in internal structure. The inner mitochondrial membrane is folded in such a way as to produce incomplete partitions, called **cristae,** which are the sites of many of the enzymes of the respiratory processes of the cell. Other enzymes, DNA, RNA, other macromolecules, and ribosomes are found in the **matrix** of the mitochondrion, a region comparable to the stroma of chloroplasts. ("Matrix" is a general term referring to filling or background matter.) A major function of the mitochondrion is the production of ATP molecules by the oxidation of food (**oxidative respiration**).

The Nucleus and Nucleolus In the eukaryotic cell, control of function and structure is under the direction

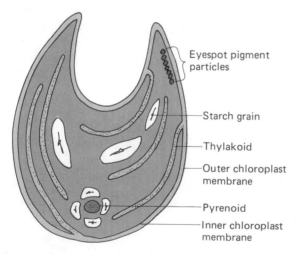

FIGURE 2-12
Chloroplast. This chloroplast is cup-shaped, but many other forms of chloroplasts are found in the different plant groups. Those of higher plants (green land plants) usually are "football"-shaped.

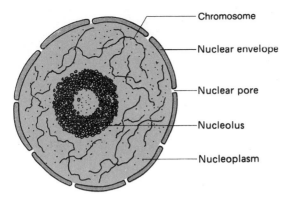

FIGURE 2-14
Nucleus.

Labels (top to bottom): Chromosome, Nuclear envelope, Nuclear pore, Nucleolus, Nucleoplasm

of chromosomes and their DNA. During cell reproduction **(cell division)** the chromosomes and their DNA are duplicated, and each new cell thus receives a complete set of genetic instructions for the maintenance of the cell.

The **nucleus** (see Fig. 2-14) consists of the **nuclear envelope,** the **nucleoplasm,** and the **nucleolus.** The nuclear envelope is a double membrane equipped with many small openings, the **nuclear pores.** Cytoplasm and nucleoplasm both consist in part of a slightly fibrous-appearing matrix. During cell division, when the nuclear membrane is absent, there is no real demarcation between nucleoplasm and cytoplasm.

Nucleoplasm contains **chromosomes,** which are diffuse and threadlike in a nondividing cell but become condensed and rodlike during cell division. Their function is the retention and replication of the genetic code (the DNA) and the copying, or **transcription,** of the code into a message (in the form of messenger RNA) containing information for the manufacturing of proteins by ribosomes. The function of the nucleolus is the manufacture of ribosomes, which subsequently migrate through the nuclear pores out into the cytoplasm. In the cytoplasm, ribosomes may exist as free single units, in linear arrays as **polyribosomes (polysomes)** or attached to the rough endoplasmic reticulum (see Figs. 2-9 and 2-11).

Microtubules Microtubules are protein tubules about 25 nm in diameter and up to 1.5 μm long. A major function is support; they are partly responsible for the shape of the cell (see Fig. 2-15). This internal skeleton of microtubules is properly named the **cytoskeleton.** Experiments have shown that when chemicals are used to suppress the development of microtubules in some kinds of cells, those cells become deformed.

Locomotion is another function of microtubules. The tubules of cilia and flagella are modified microtubules, as are those of basal bodies. In some motile cells

basal bodies are attached to the cytoskeleton (Fig. 2-9); in this way, the thrust forces of the flagella are transferred to the cell in much the same way as the oarlocks of a boat permit a boat to be rowed through the water. Microtubules also are involved in cell division, forming a structure known as the **spindle,** which is involved in several aspects of nuclear and cell division.

Microfilaments Also present in the cytoplasm of the living cell are **microfilaments** (Fig. 2-10). Microfilaments are considerably narrower than microtubules, are composed of a contractile protein called **actin,** and are considered to be responsible for cytoplasmic contractions and other internal movements in cells. Contractile filaments are especially abundant and highly active in muscle cells (see Fig. 14-8).

Flagella and Cilia The locomotory organelles of a green alga cell are constructed on an almost universal pattern among eukaryotes. Locomotion is accomplished by the beating or stroking in oarlike fashion of one or more **flagella** (sing **flagellum**). Each flagellum is a long, flexible rod (Fig. 2-9). The outer covering is an extension of the plasma membrane (see Figs. 2-9 and 2-10). Inside is a circle of nine paired microtubules (commonly called **doublets**) running the length of the flagellum. In the center of the flagellum are two additional longitudinal tubules. This 9 + 2 pattern is characteristic of all flagella.

Each flagellum is attached to a **basal body,** which is a short cylinder made up of a circle of nine elements, each composed of three parallel microtubules. The basal body has no central tubules. The new flagella developed after cell division are seen to grow out from the basal bodies; therefore a function of basal bodies is to serve

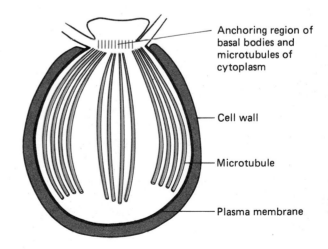

FIGURE 2-15
Cytoskeleton and cell wall of an algal cell.

Labels (top to bottom): Anchoring region of basal bodies and microtubules of cytoplasm, Cell wall, Microtubule, Plasma membrane

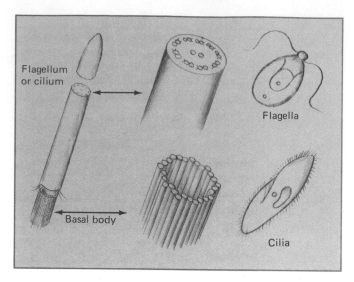

FIGURE 2-16
Cilia and flagella.

Cilia are found in some other types of motile cell. They have much the same function as flagella, with which they share the 9 + 2 structure, but cilia are considerably shorter than flagella and also more numerous. Some cells have hundreds of cilia, whereas flagella usually occur singly or in pairs (Fig. 2-16). For example, cells lining human respiratory passages have numerous cilia, whereas human male sex cells have a single flagellum.

This concludes our survey of the major structures and functions of a generalized cell. Some kinds of cells will have organelles other than those mentioned, and most cell types will differ from each other in the number and arrangement of their organelles. Nevertheless, there is a basic similarity in the structure of eukaryotic cells, which will become apparent as various types are studied.

as a starting point for the production of flagella. Basal bodies also anchor the flagella in the cytoplasm of the cell.

A small light receptor, or **eyespot,** is present near the basal bodies in motile cells of green algae. The function of this organelle is to coordinate the action of flagella with respect to the orientation of cells in the region of optimum light intensity.

Animal Cells

The cells of the human body, as shown by studies using the latest techniques of microscopy and chemistry, are remarkably similar in their basic organization to cells of other organisms, including those of green algae just considered. Yet there are obvious differences as well, not only those to be expected in cells of two unique kingdoms of life, but also those to be observed in cells from different organs. Because it is difficult—perhaps impossible—to find a single, completely representative cell, a

FIGURE 2-17
Composite animal cell.

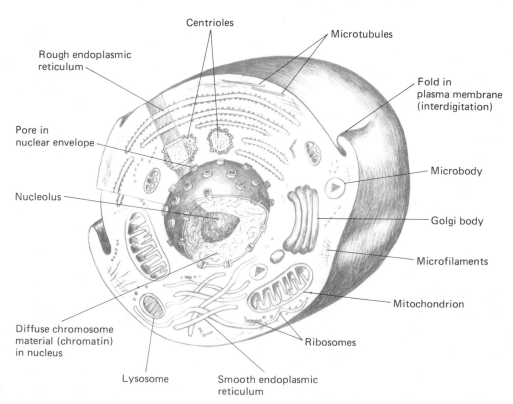

composite animal cell usually is constructed to represent animal cells in general (Fig. 2-17).

In comparing the **animal cell** with a green algal cell, as well as with cells of other plants, it is noted in particular that the animal cell does not have a wall, and of course it lacks chloroplasts. In addition, near the nucleus of the animal cell, there is a pair of minute bodies, the **centrioles** (Fig. 2-17), which are seldom found in plants. Centrioles are involved in cell replication, where they function as centers for the organization of the spindle. They also serve as templates for the organization and development of cilia and flagella, should the cell become equipped with these locomotory organelles. In that event, the centrioles migrate to the edge of the cell and become basal bodies (Fig. 2-16). If numerous cilia are present, the centrioles reproduce many times before becoming basal bodies. It is noteworthy that flagellated plant cells, such as motile algae and motile sex cells of some primitive land plants, also have centrioles.

Many animal cells have active secretory functions. Correlated with this is a pronounced development of both rough endoplasmic reticulum and Golgi bodies. The secretion vesicles of the Golgi bodies may develop into small bags of enzymes called **lysosomes,** which may either be secreted to the outside through the plasma membrane or function in internal digestion. One kind of lysosome is active in forming **autophagic vesicles,** in which worn-out cell components (e.g., mitochondria) are digested and their molecules released into the cytoplasm for reprocessing into new components. Autophagic vesicles have been called "suicide bags," because when a cell is injured, the vesicles release their digestive enzymes into the cytoplasm, resulting in death and digestion of the cytoplasm and cell. This action is one reason that animal flesh decomposes rapidly and is an explanation for the importance of keeping meat under refrigeration.

In addition to lysosomes, there are small enzyme-containing vesicles known as **microbodies,** which have oxidative functions. Some types are involved in the elimination of hydrogen peroxide, a toxic by-product of certain metabolic reactions. Because such microbodies contain the enzyme peroxidase, they also are called **peroxisomes.**

Folds and indentations are evident in many animal cells. In solitary cells, folds and pockets of the plasma membrane may participate in the taking in of food particles by **phagocytosis** ("cell eating"). In multicellular organisms, the free surface of the cell may engulf food in this way, but in surfaces in contact with adjacent cells these folds serve to increase the area of membrane available for transport between cells. Folds in adjacent cells are interlocked like pieces of a jigsaw puzzle and

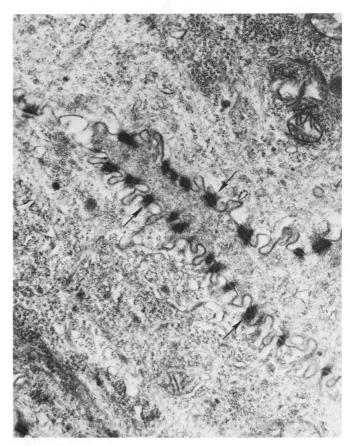

FIGURE 2-18
Electron micrograph of cell-to-cell contact region in epithelium. Note the interdigitations (membrane folds) and desmosomes (dark, fibrous bodies at arrows) which aid in binding cell to cell. (Photo courtesy David Mason)

are called **interdigitations** (Figs. 2-17 and 2-18). Where covering cells (epidermal cells) of animals adjoin, their adjacent membranes often develop thickened areas called **desmosomes** that aid in holding the cells together. Desmosomes are particularly evident in cells subject to stretching (Fig. 2-18).

In most other respects, including the structure of the nucleus, the basic organization of cytoplasm, and the distribution and functions of mitochondria, ribosomes, microtubules, microfilaments, and other cellular structures, animals cells conform to the basic eukaryotic plan previously described.

Plant Cells

Although motile algae are classified among the plants, their cells differ somewhat from typical cells of multicellular green land plants (compare Figs. 2-9 and 2-19). The latter cells lack flagella, because of course they are not

FIGURE 2-19
Generalized plant cell.

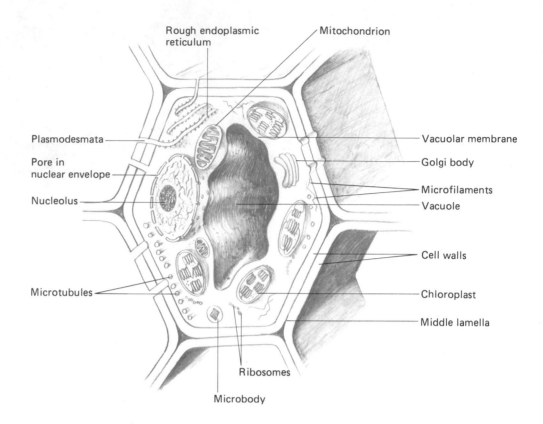

motile cells. In fact, few land plants have any cells equipped for locomotion, and those few are the male sex cells of such plants as mosses, ferns, cycads, and ginkgoes. Such motile cells also are the only plant cells having centrioles; all other cells of multicellular green plants lack them.

In multicellular plants, cell walls of adjacent cells are cemented together by the **middle lamella,** a thin layer of pectin, a polysaccharide. Adjacent cells are interconnected by cytoplasmic strands, the **plasmodesmata** (sing. **plasmodesma,** "strand"), which extend from cell to cell through tiny pores in cell walls.

In most other respects the cells of green land plants bear a marked similarity to the green algal cell. Their chloroplasts are similar in structure to those of green algae, but typically are more numerous. There is no contractile vacuole in the land plant cell, but in most cells a large noncontractile vacuole having a storage and transport function is present. Otherwise, the cytoplasmic organization and organelles generally are alike in both groups. Cells of the multicellular plant are commonly rather angular, due in part to their contact with adjacent cells, and also to the fact that higher plant cells divide by formation of a new cell wall in a rather direct plane across the dividing cell.

Origin of Eukaryotic Cells

Prokaryotes evidently preceded eukaryotes on the earth by as much as one billion years. During this billion years, some prokaryotes became specialized as photosynthetic cells, and oxygen, a by-product of photosynthesis, began to accumulate. This accumulation of oxygen led to further elaboration of cellular life and very likely to the evolution of eukaryotes. Because prokaryotes came first, scientists speculate about how they might have given rise to eukaryotic cells.

Invagination Theory   One theory of the origin of eukaryotes, proposed by J. D. Robertson of Harvard University, suggests that prokaryotic cells evolved into eukaryotes by **invagination** (infolding growth) of the plasma membrane. As a result, an internal membrane system was developed, and gave rise to cell organelles. An argument favoring this concept is that the endoplasmic reticulum, nuclear membrane, mitochondria, and chloroplasts have double membranes, as would be expected if they originated by the invagination of the plasma membrane. Another point favoring the theory is that certain membranous organelles are interconnected: the nuclear envelope with the endoplasmic reticulum and sometimes the endoplasmic reticulum with the Golgi body. Mitochondria and chloroplasts, however, are not connected to other membrane systems, and this seems a weakness in the Robertson hypothesis.

Symbiotic Theory Another proposal for the origin of eukaryotes is the **symbiotic theory** (Fig. 2-20). Symbiosis is an intimate association of two unlike organisms that benefits one or both organisms. Essentially, the symbiotic theory suggests that primitive motile eukaryotes feeding on photosynthetic prokaryotes (and possibly nonphotosynthetic bacteria) incorporated some of the prokaryotes permanently rather than digesting them for their immediate food value. The incorporated prokaryotes are assumed to have continued to function as photosynthetic and aerobic respiratory organelles. (**Aerobic respiration** at the cellular level refers to the use of oxygen to "burn," or oxidize, food in the cell.)

Earlier it was noted that competition for nutrients in the "primordial soup" may have resulted in the evolution of more efficient methods of nutrient use by primitive cells. Photosynthesis may have originated early in the earth's history as a consequence of overpopulation and food shortages, and then aerobic respiration, an energy-efficient process, evolved as oxygen from photosynthesis accumulated. Overpopulation and food shortages

still are evolutionary forces today, even though life is far more complex.

It is presumed that the primitive motile cell was incapable, in its original state, of either photosynthesis or aerobic respiration. Instead of the latter, it employed **fermentation,** a life process of low energy yield that does not require oxygen. Its mode of life may have been to move about and absorb particles of food. Possibly it swam by means of flagella; if so, they probably were of the 9 + 2 type, because that is the basic pattern among all present-day eukaryotes. Subsequently, adoption of prokaryotic cells, and their gradual conversion into photosynthetic and respiratory bodies (chloroplasts and mitochondria, respectively), produced primitive motile cells that were more efficient in energy-conversion processes. Many one-celled motile organisms living today appear to have evolved from primitive ancestors of the kind described. (*Note:* A large number of unicellular organisms are classified in the kingdom Protista [see Chapter 19]; others are classified with fungi and algae [see Chapters 19 and 20].)

There is no way to prove the symbiotic hypothesis, but there is evidence favoring it. First, modern eukaryotes have no other basic pattern of cilia or flagella than the 9 + 2 type; this certainly suggests that they all have a common ancestor. Second, it has been discovered in recent years that mitochondria and chloroplasts are partially independent bodies and reproduce themselves within the cytoplasm of the cell in which they occur; therefore they once might have been completely independent cells. A further similarity is that chloroplasts and mitochondria respond to antibiotics in much the same way as do bacteria; the same antibiotics that inactivate them also inactivate bacteria. Moreover, all have similar ribosomes, smaller than the cytoplasmic ribosomes that are present in the eukaryotic cell. Another striking similarity is that prokaryotes lack nuclei and chromosomes, having their genetic information encoded in "naked" DNA. Both chloroplasts and mitochondria possess similarly "naked" DNA, but their genetic information is far less extensive.

As noted in connection with the Robertson hypothesis, mitochondria and chloroplasts are enclosed within a pair of membranes. The outer of these may represent a portion of the original membrane of a host cell, enfolding the prokaryote; the inner one may represent the original plasma membrane of the bacterial or algal cell. Several living eukaryotic organisms have identifiable cyanobacteria living symbiotically within their cytoplasm (Fig. 2-21). This at least indicates that the initial stages of cell evolution proposed in the symbiotic theory can occur.

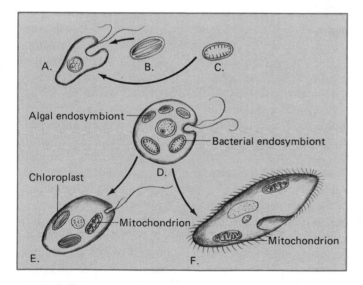

FIGURE 2-20
Symbiotic theory of evolution. **A.** Primitive predatory cell. **B.** Photosynthetic prokaryote. **C.** Bacterial cell. **D.** Primitive symbiotic cell. The bacteria and algae within the cell are called symbionts or endosymbionts ("internal symbionts"). **E.** Present-day photosynthetic alga. **F.** Present-day protozoan (protozoans are one-celled foodgathering eukaryotes classified in the kingdom Protista).

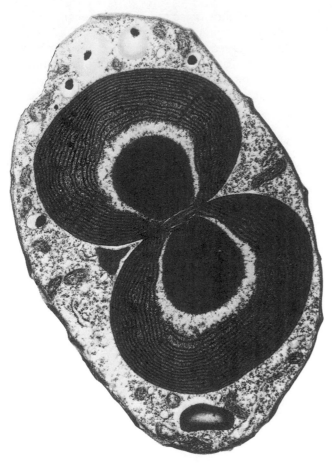

FIGURE 2-21
Cyanophora paradoxa, a flagellated eukaryote containing a cyanobacterium symbiont. The cyanobacterium was dividing at the time
this preparation was made. (Photo from J. Pickett-Heaps, 1972,
New Phytol. 71:561)

CELL PROCESSES

There are several ways substances can move into and
out of cells. The exact mechanism depends on the size
of particles, whether or not they are soluble, whether
they are electrically charged (ionic), and their relative
concentrations inside and outside the cell.

Movement of substances into and within cells can
be evaluated on the basis of whether or not the cell
must do work in the particular process. If the absorption
and transport of substances require very little work, the
processes are said to be passive (**passive absorption,
passive transport**). All passive absorption processes involve **diffusion,** which may be defined as the movement
of substances from a region where they are concentrated
into regions where they are relatively less abundant. Another way of stating this is to say that substances diffuse
down a **concentration gradient.** On the other hand, if a

cell must do work in the transport of substances, **active
absorption,** also called **active transport,** is occurring.
Any movement of substances into, out of, and within
cells against a concentration gradient requires active
transport. In other words, work must be done to concentrate materials inside a cell when those materials are
scarce outside the cell (i.e., energy must be expended to
move substances from regions of low concentrations to
regions of higher concentrations).

A living cell constantly exchanges substances with
its environment. That is to say, the cell takes in substances from outside itself and secretes and excretes
substances into its environment. The process of taking
in and giving off substances is vital to the cell, for in this
way the cell acquires the raw materials necessary for its
continuing existence and gets rid of waste matter.

Passive Absorption and Transport

It is incorrect to think there is no energy involved in passive absorption and transport, for the random movements of atoms and molecules constitute the energy of
heat. Only at absolute zero ($-273°C$) are atoms and
molecules completely free of motion (i.e., free of heat).
At temperatures above absolute zero, atoms and molecules are in motion, and their temperature is a measure
of that motion. Such motion may be put to use if, as a
result of work, substances become concentrated in one
region as opposed to other regions and diffusion results.
(In a steam engine work must be done to concentrate
molecules of water vapor under pressure in a boiler;
their diffusion into the cylinders and against the pistons
of the engine translates the initial work into mechanical
energy. There are some similar relationships in organisms where diffusion occurs after work has been done to
concentrate molecules of liquids and gases in certain regions and structures: lungs and blood vessels, water flow
in stems and leaves, etc.).

Diffusion If a particle of soluble dye, such as methy-
lene blue, is placed in a beaker of water, it will dissolve
in the water and its molecules gradually will become dispersed until the water in the beaker is colored a uniform
blue (Fig. 2-22). The explanation of this movement of
solute molecules has its basis in an observation by Robert Brown, discoverer of the cell nucleus. Brown noted
that small particles, visible only with a microscope, were
in constant, random motion. This motion, called **Brownian movement** (Fig. 2-23), is caused by the constant jos-
tling of the particles by randomly moving water and solute molecules and is evidence of the kinetic energy (energy of motion) of molecules, as a result of which diffusion occurs.

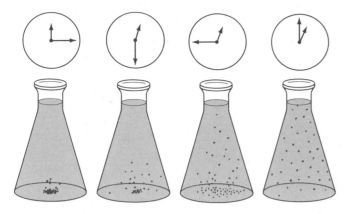

FIGURE 2-22
Diffusion of dye molecules in water.

In this example, it is important to note that molecules disperse along a decreasing concentration gradient. Also important is the fact that two kinds of molecules and two concentration gradients are involved: those of dye molecules and those of water molecules. Where the dye was added, dye molecules are in relatively high concentration and water molecules are in relatively low concentration. This relative concentration of particles in solutions is of great importance in describing and explaining diffusion. The concern is not so much with the total number of molecules in any region of the solution as with their relative numbers with respect to those in some other region and with respect to molecules of other kinds.

Osmosis   The preceding example can be turned into a demonstration of **osmosis** by making a few alterations in the experimental methodology. Osmosis is the diffusion of molecules of water through holes in a membrane so

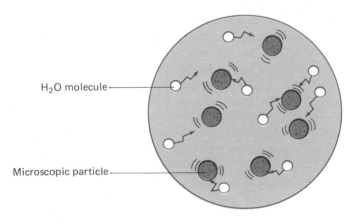

H₂O molecule

Microscopic particle

FIGURE 2-23
Brownian movement.

small as to exclude molecules of a solute. Membranes capable of excluding only solute molecules, or in some cases large but not small solute molecules, are **differentially permeable.**

A short length of cellulose tubing of the type now used as sausage casing, or sold by scientific suppliers as dialysis tubing, can serve as a differentially permeable membrane. This material is a thin, porous film of cellulose acetate, a type of cellophane. The holes in the cellulose acetate film are large enough to allow passage of water molecules, but so small that they exclude larger dye particles or sugar molecules. One end of a short length of tubing is tied tightly to make a bag; a concentrated solution of dark molasses and water is added, until the bag is about half full (molasses is used instead of refined sugar for visibility). The other end of the bag is tied tightly, and the bag is immersed in a beaker of distilled water. Within a few hours the bag has become distended with fluid and the color is less intense, but the water in the beaker remains uncolored. If the experiment continues, the bag may become even more distended and, in time, may break because of internal pressure.

Osmotic pressure can be measured by connecting an upright tube to a bag of the type just described (Fig. 2-24), making an instrument called an **osmometer.** (Osmotic pressure is the tendency of water molecules to enter the bag in response to their concentration gradient.) In this case, the pressure of water molecules diffusing into the bag will result in the rising of its contents within the tube. The distance the fluid rises will be proportional to the osmotic pressure generated by the system.

The explanation of what has just been described is based on diffusion of water molecules through a differentially permeable membrane in which the pores are so small that they do not permit any sugar molecules to pass through. Note that diffusion always occurs in the direction of a lower molecular concentration from a higher one, and that water molecules are in high concentration outside the bag, where no solute molecules are present, and in relatively lower concentration inside the bag, where sugar molecules are present. The result is that more water molecules move inward than outward and the tendency will be for the water molecules to diffuse down the concentration gradient into the bag. This result is closely paralleled in experiments with living cells. The plasma membrane of the cell is a differentially permeable membrane. Although pores have not been seen in plasma membranes, even under the electron microscope, the existence of pores or channels is inferred from osmotic experiments (Fig. 2-25).

Red blood cells often have been used in osmotic experiments. A red blood cell is immersed in blood plasma, which contains molecules of about the same to-

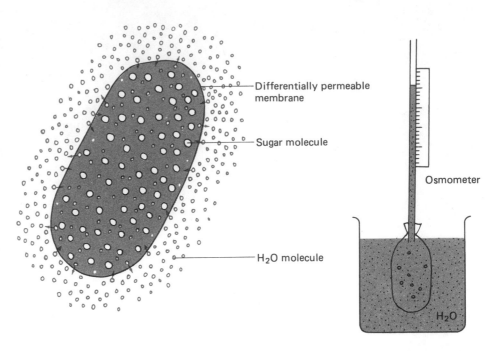

FIGURE 2-24
Osmosis.

Differentially permeable membrane

Sugar molecule

Osmometer

H₂O molecule

H₂O

tal concentration as the solutes within the cell. Such a solution is said to be **isotonic; tonicity** is a term given to the concentration of solutions outside cells with respect to cytoplasmic fluids.

Hemolysis If blood plasma in which red cells are present is diluted with water, the cells will burst. This is called **hemolysis,** a condition attributable to an osmotic pressure greater than that which the red cell membrane can sustain. A **hypotonic** (less than isotonic) solution is one capable of producing such excess osmotic pressure. On the other hand, if sugar is added to blood serum so that the solute concentration outside the cell is greatly increased (a **hypertonic** solution—greater than isotonic),

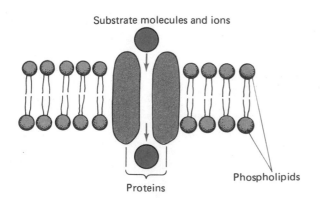

Substrate molecules and ions

Proteins

Phospholipids

FIGURE 2-25
Model of the plasma membrane illustrating the possible structure of pores or channels. Such pores or channels would permit the passage of water molecules, small ions, and perhaps other small molecules.

the cells will lose water and shrivel. Precisely the same effect could be produced by replacing the water in the beaker in an osmosis experiment with a highly concentrated sugar solution. The explanation for the loss of water from the bag under these circumstances is that now the concentration of water within the bag is relatively greater than that outside the bag. More water molecules move out than move in.

Plasmolysis Diffusion of water out of a cell is called **plasmolysis** (Fig. 2-26). One of the more readily observed examples of plasmolysis is seen in simple experiments using cells of the green water plant *Elodea*. Plant cells commonly contain large vacuoles surrounded by cytoplasm and the plasma membrane (see Fig. 2-19). The vacuoles contain a "sap" of water, sugar, and other solutes, and therefore the cell resembles a bag and its contents in an osmosis experiment. If an *Elodea* leaf is placed in a drop of distilled water on a microscope slide, the cytoplasm of leaf cells will be seen to fill out completely the spaces inside the cell walls. However, if *Elodea* cells are immersed in a concentrated sugar solution instead of water alone, the cells will lose water, contract, and pull away from the cell walls. Then, if the sugar solution outside *Elodea* cells is diluted by adding water, the cell will gradually expand until it once again touches the walls. At that point, the outside solution will be isotonic with respect to the solution within the cell, and the cells are said to have become **deplasmolyzed** (Fig. 2-26). If the original concentration of sugar outside the cell and the percentage of its dilution have been determined, the solute concentration within the cell can be calculated: it

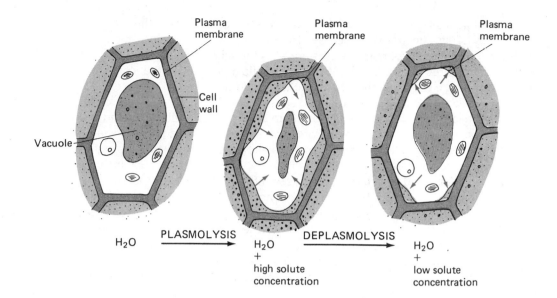

FIGURE 2-26
Plasmolysis and deplasmolysis
of a plant cell.

Plasma membrane

Plasma membrane

Plasma membrane

Cell wall

Vacuole

H_2O → **PLASMOLYSIS** → H_2O + high solute concentration → **DEPLASMOLYSIS** → H_2O + low solute concentration

will be equivalent to that sugar solution outside the cell that neither plasmolyzes nor deplasmolyzes the cell. Solute concentrations in many kinds of plant cells have been estimated by this procedure. In some plant cells these concentrations are relatively high, being equivalent to a sugar solution of about 12 percent; in others they are considerably lower. The average is equivalent to a solution of about 6 percent glucose.

When an *Elodea* leaf is removed from a sugar solution, rinsed, and placed in distilled water on a microscope slide, a slight swelling of the leaf and its cells may be apparent, in which case the leaf and cells are said to be **turgid**. The movement of water into each cell exerts osmotic pressure, often referred to as **turgor pressure**. Under some circumstances this pressure is greater than 10 atmospheres (1 atmosphere of pressure is sufficient to raise a column of water nearly 10 m [33 ft]). Osmotic pressure in plants is resisted by the cell walls, but if a tube could be connected to a leaf cell, as in an osmometer, the osmotic pressure developed might be sufficient to raise the internal solution as much as 100 m (330 ft). The significance of this will be apparent later, when water movement in plants is considered.

Dialysis   In the osmosis experiment described earlier, only water molecules moved through the membrane. However, if table salt (NaCl) were to be included in the solution within the bag and the external solution tested from time to time, sodium and chloride ions would be seen eventually to attain about the same concentration on both sides of the membrane. This movement of solute particles (in this case ions) through a membrane, with retention of other molecules, is called **dialysis**. The movement of ions through a cell membrane is a more

complicated process because of electrical charges of both ions and membrane.

In the kidney, dialysis is involved in the movements of solutes into and out of the urine. In cases of kidney disease, where kidney cells have lost their ability to excrete waste or reabsorb water, artificial dialysis is becoming a commonplace treatment. The blood of the patient is allowed to run through coiled dialysis tubes of cellulose acetate. These tubes are immersed in an isotonic solution into which blood impurities diffuse and are rinsed away. The procedure is more complicated than outlined here, but the principle is the same.

Active Transport

There are many instances in which solvents and solutes are transported into cells against concentration gradients (in direct opposition to diffusion); for example, kidney cells resorb solutes as well as water from the urine, and root cells of plants absorb mineral solutes that are at a lower concentration outside the cells that within them. In both cases, work is done, and hence active absorption and transport occur.

Several mechanisms have been proposed to explain active transport, and probably no one mechanism would apply to all examples. In general, in active transport, ions and molecules appear to be transported through the plasma membrane by proteins functioning as carriers.

Because protein molecules are three-dimensional, they can exhibit specificity with respect to their association with ions or other molecules. Here, specificity means that certain proteins are able to "recognize" specific kinds of ions or other molecules. For example, enzymes exhibit high specificity and have active sites capable of binding substrate molecules. Carrier proteins involved in active transport molecules are also thought to have active sites and to be able to bind specific molecules. But unlike enzymes, the carrier molecules do not make or break bonds within the attached substrate molecules. Instead, the carrier molecule picks up an ion or molecule on one side of the membrane, transports it through the membrane, and discharges it on the other side in an unaltered state.

The binding and discharge processes of active transport require energy in the form of adenosine triphosphate (ATP). Chemical energy in the high-energy phosphate bond of ATP is transferred to the carrier and its substrate. Upon giving up a high-energy bond, ATP is converted to **adenosine diphosphate (ADP)** and a phosphate ion. The process is illustrated diagrammatically in Fig. 2-27.

Among the substances transported by active transport carriers are sugars, amino acids, and potassium and sodium ions. The transport of the potassium and sodium ions is important in maintaining salt balance in cells and also for the electrical properties of cell membranes.

Ingestion   The capture and **ingestion** of food poses particular problems for single-celled organisms and, indeed, for individual cells of multicellular organisms. The integrity of the cell is maintained by the plasma membrane, which not only contains the cytoplasmic cell contents but also functions as a barrier to the entrance of toxic substances and microorganisms. While barring such entry, the cell at the same time must take in food from and expel wastes into the environment. Cells solve this problem in several ways: by forming intake pockets used to ingest food particles, by forming minute channels through which liquids can be taken in, and by producing small membranous sacs (vesicles) in which food and secretions, including wastes, can be collected, digested, converted, absorbed, or secreted.

Many cells are capable of engulfing particles, sometimes of relatively large size. Phagocytosis is the simplest kind of ingestion process, for it appears not to require specialized regions of the cell (Fig. 2-28). The *Amoeba*, a one-celled organism, lives by taking in food particles, including bacteria and algal cells, by phagocytosis (refer to Fig. 19-15), using extensions of the cytoplasm and its plasma membrane to engulf the food. These extensions, called **pseudopodia** ("false feet"), encircle the food particle and form a pocket in the cell surface. Further deepening of the pocket, followed by constriction of the open end, forms a food vacuole. Although this process may be thought of as taking food into the cell, the food particles actually remain outside the cytoplasm and plasma membrane.

The concept of "insideness" or "outsideness" of animals and cells often is misunderstood. Food lying in the human digestive tract remains outside the cells of the body. Only when the food is digested into soluble molecules is it absorbed by cells by transport through plasma membranes, and therefore truly inside the body. In a like manner, when a food particle is taken into a cell as a food vacuole by phagocytosis, it is outside the bit of plasma membrane that forms the boundary of the vacuole. The food does not become incorporated into the cell's cytoplasm until digestion is completed and its molecules move through the vacuolar membrane, whether by diffusion or by active transport.

The process of phagocytosis and digestion incorporates most of the activities previously described in the discussion of transport and biosynthesis by cell organ-

FIGURE 2-27
Active transport of substrate molecules by a carrier molecule in the plasma membrane.

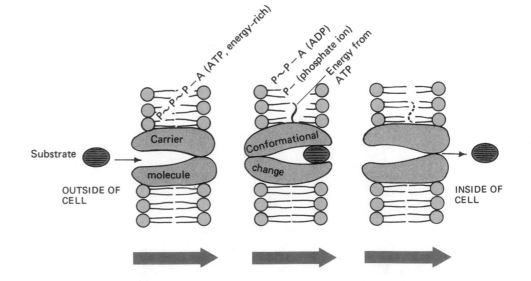

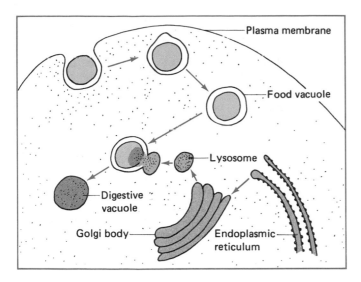

FIGURE 2-28
Phagocytosis. Formation of a food vacuole, and involvement of the endoplasmic reticulum, Golgi body, and lysosome in digestion, are shown.

elles, including the rough and smooth endoplasmic reticulum, Golgi bodies, and lysosomes.

The *Amoeba* has been used as an example of phagocytosis. Other cells ingest comparatively large food particles in much the same fashion. The digestive cells of simple multicellular animals such as sponges and corals are phagocytotic, as are the **white cells (leukocytes)** present in human blood and lymph (see Fig. 15-10). Leukocytes are important because they ingest dead and disrupted (broken-down) cells of the body as well as cells of foreign origin (bacteria and other invading microorganisms). Sometimes, however, the body loses control of their reproduction and an overpopulation of leukocytes begins to destroy the tissues of the body itself. This is the disease leukemia, a type of cancer.

Whereas the *Amoeba* and leukocytes apparently can form phagocytic pockets anywhere on their surfaces, other kinds of cells have localized structures in which phagocytosis takes place. Another microscopic predator, *Paramecium*, has a channel or groove on its side into which food particles are swept by ciliary action (see Fig. 19-13). At the inner end of the groove, food vacuoles are formed in exactly the same way as in the *Amoeba* and leukocytes. *Didinium*, another ciliated microorganism, has the remarkable ability to ingest paramecia as large as or even larger than itself (see Fig. 19-6). Even in this organism, the actual ingestion of the food is accomplished by formation of a food vacuole as described previously.

Phagocytosis involves the ingestion of food particles and the later dissolution of the food by digestive enzymes in a food vacuole. Another process, **pinocytosis**

("cell drinking"), is performed in much the same way but involves the intake of liquids. Very small invaginations, called **pinocytotic channels,** form at specific sites on the plasma membrane, fill with fluid, and produce **pinocytotic vesicles** by pinching off a small pocket of membrane and fluid at the inner end of the channel (Fig. 2-29). The channels and vesicles are so small that they are visible only in electron micrographs.

Autolysis Earlier it was stated that the digestion of food in a food vacuole involves the fusion of the vacuole with membrane-bounded products of the Golgi body, the **lysosomes** ("dissolving bodies"). Lysosomes contain digestive enzymes capable of catalyzing the hydrolysis of macromolecules present in food. Under normal conditions, the lysosomes and their enzymes break down food only in food vacuoles (Fig. 2-28). However, if the cell dies, the lysosomal enzymes are released into the cytoplasm and quickly go to work on the macromolecules of the cell itself. The result is **autolysis** ("self-dissolving"), a part of the death process in cells. It is an explanation, for example, for the deterioration of meat not kept refrigerated. Even in healthy cells, some lysosomes, called **autophagic vesicles** ("self-feeding sacs"), perform somewhat the same function in the cell as white cells do in the body: acting as scavengers of worn-out matter. The autophagic vesicle, in some way not yet understood, "recognizes" a damaged organelle such as a mitochondrion and digests it. The result is that molecules of damaged organelles can be used again by the cell to make new structures.

When tadpoles undergo metamorphosis into adult frogs or toads, their tails are absorbed and disappear.

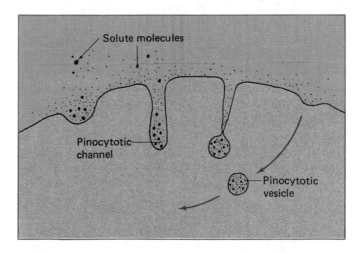

FIGURE 2-29
Pinocytosis. It is thought that solute molecules become attached to specific binding sites on the membrane and that the sites move into the pinocytotic channels.

This absorption is due to the action of autophagic vesicles, which make the substances of the tail cells available for the metabolism of the remaining body cells.

Waste Removal When an *Amoeba* ingests a bacterium or other food particle and digests its macromolecules, not all parts of the bacterial cell or food particle are digested and absorbed; **waste** particles remain in the food vacuole. The vacuole does not simply disappear, leaving the waste behind; instead, it moves back toward the cell membrane. Eventually, it fuses with the plasma membrane and discharges the wastes to the outside of the cell. In essence, the membrane of the food vacuole becomes incorporated into the cell membrane. In an active cell, as many vacuoles and vesicles rejoin the cell membrane as are produced from it. Because the cell does not have an endless supply of new membranes for forming food vacuoles, this recycling process maintains the membrane supply in the cell.

Regulation of Cell Processes

In the present and preceding chapter a number of the activities of cells have been described, including movement, temperature relationships, and the synthesis and breaking down of macromolecules. Normally, these and many other actions are balanced so that cells and organisms do not overdo any one activity to the detriment of others. Exactly how a cell manages to adjust to the demands made upon it by changes in environmental factors is only now beginning to be understood; however, in general terms, cells have a remarkable capacity for self-adjustment. Unlike an intelligent organism, a cell does not consciously assess a situation and then decide on a proper reaction. Rather, the cell may be thought of as a complex system of checks and balances which tend to maintain the status quo. A simple example or two will help in understanding the principle of cellular and organismal self-adjustment, called **homeostasis** ("same state"). Homeostasis involves a regulatory operation termed **feedback**. Feedback **control** uses the product of a particular action to control the action itself. Involved is some kind of product-sensing device, coupled to a shut-down or start-up regulation of product production.

A mechanical model of homeostasis and feedback action is the familiar flush toilet (Fig. 2-30). When the toilet is flushed, the tank automatically begins to fill with water, but always stops before it overflows (or nearly always!). One of the requirements in homeostasis is a sensing device that will recognize a change in the status of whatever is being regulated. In the flush toilet,

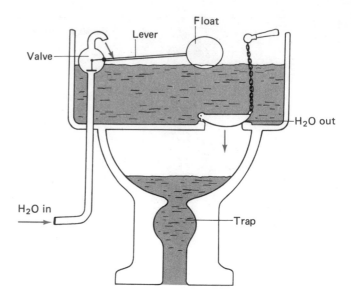

FIGURE 2-30
Flush toilet, a simple example of feedback regulation.

a float located in the tank is the sensor. The float is connected to one end of a lever, the other end of which is attached to a shut-off valve. When the float "senses" that the water has reached the proper level in the tank, it "tells" the lever to shut off the water-intake valve. Should the water level drop even slightly, the float "senses" this and the valve is reopened for a few moments. In this simple example, the float, lever, and shut-off valve constitute a feedback system that controls the process of water transport. Much more complicated feedback devices are in daily use in the home and workplace, but the feedback systems that maintain homeostasis in cells and organisms are the most complex of all.

Among the many examples of feedback regulation of biological processes is the opening and closing of pores in the leaves of plants. Green land plants regulate the diffusion of carbon dioxide, oxygen, and water vapor between the atmosphere and the plant by opening and closing many microscopic pores, the **stomata** (sing. **stoma**, "mouth"), located in surfaces of leaves and other green parts. Each stoma is bordered by a pair of liplike **guard cells** (Fig. 2-31), which open and close in response to environmental factors (amount of moisture; amount of light) and internal conditions (photosynthesis; amount of water, carbon dioxide, oxygen, and sugar).

The opening of stomata is produced by swelling of the guard cells, which causes them to assume a ringlike shape, and stomatal closing occurs when guard cells shrink and are pressed together. Swelling results when guard cells take up water, and shrinking occurs when they lose water.

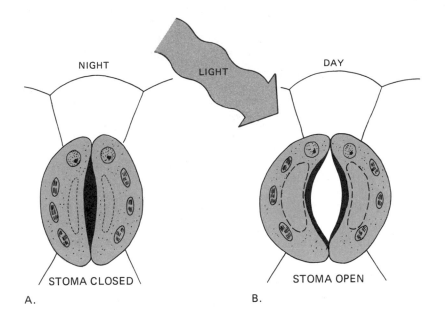

FIGURE 2-31
The action of stomata in plants.
A. At night, when carbon dioxide
concentrations are high within
the leaf, stomata are closed. **B.**
During the day, when carbon
dioxide levels within the leaf are
lower, stomata are open.

NIGHT

LIGHT

DAY

STOMA CLOSED

STOMA OPEN

A.

B.

Normally, carbon dioxide accumulates in leaves at night, when photosynthesis, which requires both light and carbon dioxide, is not taking place. During the day, photosynthesis occurs, and because it is a process in which carbon dioxide is converted by light energy via chlorophyll into carbohydrates, the result is the lowering of carbon dioxide concentrations inside leaves. It has been shown experimentally that introducing carbon dioxide over open stomata causes them to close, and removing carbon dioxide from the vicinity of guard cells results in open stomata. Carbon dioxide thus is equivalent to the water level in the toilet tank example; it is the reactant or product whose presence or absence is sensed by the guard cell. The result is permitting water to escape (when carbon dioxide levels are high; Fig. 2-31A) and letting water into the guard cell (when carbon dioxide levels are low; Fig. 2-31B). Carbon dioxide itself does not appear to produce the actual water movement but rather seems to act as a signal for guard cells to take up water by osmosis in light and lose water by plasmolysis in darkness; the result is that the stomata are normally open during the day, when photosynthesis with its requirement for carbon dioxide is taking place, and closed at night, when carbon dioxide is not required (respiration produces carbon dioxide and water from carbohydrates; hence carbon dioxide within the cells increases at night).

Many complex kinds of regulation occur in organisms; the larger and more advanced the organism, the more involved are the regulatory mechanisms. Nevertheless, all regulation ultimately can be traced to events occurring in cells, between cells, and between cells and the environment.

SUMMARY

It is estimated that the earth is almost 4.5 billion years old. There is no record of living cells for the first 2 billion years. During that period, scientists think that the process of chemical evolution occurred in an atmosphere quite different from that of the present. That early atmosphere may have consisted of hydrogen, methane, ammonia, and water vapor, but no oxygen. Experiments in which these gases have been combined in chemical retorts and exposed to radiation have shown that organic molecules can be formed. These molecules are identical to certain molecules manufactured in living cells.

Among the molecules formed in chemical evolution experiments are sugars, amino acids, fatty acids, purines, and pyrimidines. Possibly these and others existing in the "primordial soup" formed macromolecules, and these aggregated further to form simple cells.

A scientific model should enable the researcher to correlate known scientific information and make predictions about future events and discoveries. The cell theory is such a model. In essence, the cell theory states that all living things are composed of cells, and that the cell is the basic unit of structure and function in an organism. Organisms may be one-celled (unicellular) or many-celled (multicellular). They are divided further into two basic types: prokaryotes and eukaryotes. Prokaryotes are relatively simple cells lacking a nucleus and many other internal structures found in eukaryotes. The eukaryotes are all other cells and include the higher plants and animals.

The eukaryotic cell contains a number of bodies, the organelles, which divide the work of the cell. Among the important organelles are cilia, flagella, the plasma membrane, the nucleus, rough endoplasmic reticulum, smooth endoplasmic reticulum, ribosomes, Golgi bodies, chloroplasts, lysosomes, vacuoles, microtubules, and mitochondria. Each of these organelles performs one or more functions such as locomotion (flagella); transport, biosynthesis, and digestion (plasma membrane, endoplasmic reticulum, ribosomes, Golgi bodies, lysosomes, vacuoles); photosynthesis (chloroplasts); respiration (mitochondria); support (microtubules); pro-

tection (plasma membrane); and cellular control and reproduction (nucleus).

Many kinds of molecules are transported through the plasma membrane, including sugars, amino acids, and certain ions such as sodium and potassium. Larger particles are taken into or excreted from cells by formation of membranous pockets, channels, and vesicles in the cytoplasm. Examples are phagocytosis, the ingestion of large particles including other cells, and pinocytosis, or cell drinking.

Many processes occur in cells. All are regulated by the cell so that integration of cell functions is sustained. This is accomplished by many kinds of self-adjustment reactions which result in the maintenance of a steady-state condition, or homeostasis.

KEY WORDS

abiogenesis	cell membrane	diffusion
controlled experiment	cytoplasm	osmosis
chemical evolution	endoplasmic reticulum	hemolysis
protenoid	vacuole	plasmolysis
mycoplasma	Golgi body	dialysis
macromolecule	chloroplast	active transport
hydrolysis	mitochondrion	ingestion
polysaccharide	flagellum	phagocytosis
lipid	cilia	pinocytosis
DNA	cell wall	autolysis
RNA	microtubule	wastes
cell theory	microfilament	regulation
nucleus	animal cell	homeostasis
nucleoplasm	plant cell	feedback control
prokaryotic cell	symbiotic theory	
eukaryotic cell	passive transport	

QUESTIONS FOR REVIEW AND DISCUSSION

1 What is a controlled experiment? Describe a controlled experiment designed to test the variables in the concept of abiogenesis.

2 Describe the primitive atmosphere in terms of its approximate duration, its components, and the nature of its reactions.

3 What is the relationship between ultraviolet radiation, ozone in the atmosphere, and early life? Why are ecologists concerned about the effects of aerosols and ultrasonic aircraft on the ozone layer?

4   Explain how some of the molecules produced in primitive atmosphere experiments relate to the macromolecules important in life processes.

5   Name six organelles, and give at least one function for each.

6   Contrast the structures of plant cells and animal cells. What are some specific similarities and differences? How do these relate to the lives and functions of the respective forms of life?

7   Recently, a photosynthetic prokaryote was discovered as a symbiont living among the epidermal cells of some primitive marine animals. The prokaryote was green, rather than blue green, and had pigments very similar to those of higher plant chloroplasts. What significance might be attached to this discovery?

8   What importance may be attached to the almost universal occurrence of 9 + 2 cilia and flagella among eukaryotes? What does this indicate about the chronology of origins of the major groups of organisms?

9   Explain the observation that, in osmosis, water molecules diffuse in both directions through a differentially permeable membrane but increase their numbers on the inside of the membrane system.

10   Why is it said that a bacterium or other food particle taken into an *Amoeba* by phagocytosis remains outside the cell until digestion has occurred?

11   What is the relationship between phagocytosis, lysosome formation, food vacuoles, and active transport in a cell?

12   Describe a simple mechanical feedback system in your home, other than the one described in this chapter.

13   What are the relationships between carbon dioxide, photosynthesis, and stomatal opening and closing?

SUGGESTED READING

CALVIN, M. 1969. *Chemical evolution.* New York: Oxford University Press. (A technical exploration of many aspects of the chemical evolution of life hypothesis.)

DICKERSON, R. E. 1978. Chemical evolution and the origin of life. *Sci. Amer.* 239(3):70–86. (Presents a concise account of evolutionary life as well as an excellent overview of structure and synthesis of macromolecules.)

DUSTIN, P. 1980. Microtubules. *Sci. Amer.* 242(2):67–76. (The marvelous intricacies of motile cells and the roles of microtubules in movement and structure of cells are discussed.)

GOODENOUGH, U. W., and R. P. LEVINE. 1970. The genetic identity of mitochondria and chloroplasts. *Sci. Amer.* 223(5):22–31. (Cites similarities between these organelles and prokaryotes.)

HOLTER, H. 1961. How things get into cells. *Sci. Amer.* 205(3):167–80. (Explains passive and active transport, phagocytosis, and pinocytosis by means of clear diagrams.)

HOOKE, R. 1665. Of the schematisme or texture of cork, and of the cells and pores of some other such frothy bodies. In *Micrographia.* London. Reprinted in *Great experiments in biology.* M. L. Gabriel and S. Fogel, eds. Englewood Cliffs, N.J.: Prentice-Hall, 1955, pp. 3–6. (The reprint does not include Hooke's drawings of cells and his microscope.)

LAZARIDES, E., and J. P. REVEL. 1979. The molecular basis of cell movement. *Sci. Amer.* 240(3):100–113. (Describes the nature of microfilaments and their role in cell movement and maintaining cell shape.)

LODISH, H. F., and J. E. ROTHMAN. 1979. The assembly of cell membranes. *Sci. Amer.* 240(1):48–63. (Discusses the growth of cell membranes in relation to other cell components.)

MAYR, O. 1970. The origins of feedback control *Sci. Amer.* 223:110–18. (Discusses mechanical principles of feedback regulation.)

PORTER, K. R., and J. B. TUCKER. 1981. The ground substance of the living cell. *Sci. Amer.* 244(3):57–67. (The high-voltage electron microscope reveals a complex meshwork of cytoplasmic filaments.)

ROBERTSON, J. D. 1962. The membrane of the living cell. *Sci. Amer.* 206(4):65–72. (Presents an interesting account of early and modern studies of cell membranes, in addition to a theory of the evolution of eukaryotic cells.)

SOLOMON, A. K. 1960. Pores in the cell membrane. *Sci. Amer.* 203(6):145–56. (Describes experimental but indirect evidence relating to dimensions of plasma membrane pores—they are calculated to be about 0.8 nm in diameter.)

STAEHELIN, L. A., and B. E. HALL. 1978. Junctions between living cells. *Sci. Amer.* 238(5):140–52. (Describes how animal cells are held together. Magnificent electron micrographs.)

STRYER, L. 1980. *Biochemistry*, 2nd ed. San Francisco: W. H. Freeman and Co. (Renowned for its discussion of cellular respiration.)

UNWIN, N., and R. HENDERSON. 1984. The structure of proteins in biological membranes. *Sci. Amer.* 250(2):78–94. (The three-dimensional structure of several proteins of bacterial cell membranes is described and illustrated.)

VIDAL, G. 1984. The oldest eukaryotic cells. *Sci. Amer.* 250(2):48–57. (More than 1.5 billion years old, they seem to resemble some living eukaryotic cells.)

WESSELLS, N. K. 1971. How living cells change shape. *Sci. Amer.* 225(4):76–82. (This brief article serves to introduce the reader to structural and mechanically active proteins—microtubules and microfilaments.)

3 Photosynthesis and Nitrogen Fixation

Two hundred years ago it was thought that many substances, including food, contained something called **phlogiston,** which was thought to be liberated from matter by animal respiration or by burning. When an animal was confined in an airtight container, it lived only as long as the air could absorb the phlogiston the animal produced when it consumed its food. When the air became filled to capacity with phlogiston, the animal died. In the same manner, a candle continued to burn only until the air became saturated with phlogiston, then it went out.

Joseph Priestley (1733–1804), an English clergyman and scientist, became interested in the phlogiston concept, and this led him to do some experiments with plants and animals. In an experiment conducted in 1771, he placed a leafy sprig of a mint plant in a closed container. Contrary to expectation, the plant did not die. Moreover, Priestly discovered that the mint plant could restore the air in which a mouse had died of suffocation, or in which a candle had burned out. His interpretation was that a green plant had the power to remove phlogiston from the air and to make the air habitable for animals and capable of supporting combustion.

Scientists now know that Priestley had the expla-

nation for his results turned around. Plants do not remove from the air some substance that inhibits combustion, but rather they put a necessary something back in. This fact was recognized a few years later when Lavoisier, the great French chemist, gave the correct interpretation of Priestley's discovery and also coined the name **oxygen** for the substance put into the air by plants. Conversely, it is oxygen that is removed from air by animal **respiration** or by combustion; this gas-exchange principle is of fundamental importance.

Priestley had another problem with his phlogiston experiments. He could not always get plants to restore the air. Sometimes the experiments would work, other times not. Then in 1779 an Austrian scientist, Jan Ingenhousz (1730–1799), discovered that plants restored air only during daylight hours and that only the green parts of plants were effective. Not only was the requirement for light discovered, but also the fact that only the green parts of plants produce oxygen in light. Subsequently, other scientists found that the product of green plants exposed to light is sugar, and that oxygen is a by-product rather than the main product of the process. Both water and carbon dioxide are required in addition to light; **photosynthesis** is the sum of these processes.

PHOTOSYNTHESIS

By the twentieth century, biology students could write the following general equation for photosynthesis:

Carbon dioxide + water + light energy →
$$\text{sugar (glucose)} + \text{oxygen}$$

and this one for respiration (also called **biological oxidation**):

Sugar (glucose) + oxygen →
$$\text{energy} + \text{water} + \text{carbon dioxide}$$

Each of these two equations is the apparent reverse of the other, and they are fundamentally correct in indicating the interdependence of the two life processes. However, photosynthesis and respiration are highly complex processes occurring in many steps and are not reversible except in the broadest ecological sense.

A strategy that can be used when not all of the steps in a complicated process are known is the "black box" approach. In this approach, one conceives of substances going into a box together with energy to operate the box and sees products and by-products come out of the box. Then one attempts to imagine what is going on inside the box. Naturally, any clues will be helpful, so if the box can be penetrated in some way to give a glimpse of a step here and there, the entire concept can be refined.

Imagine that the chloroplast is the "black box" of photosynthesis, with light energy, water, and carbon dioxide going in, and oxygen, sugar (CH_2O, a generic symbol for carbohydrate), and some water coming out. How does it operate?

Since 1900, biologists and chemists have been able to look inside the photosynthetic "box" and have found that they could separate the overall process into two different sets of reactions. Reactions of one set require light and water, and are called **light reactions**. Those of the other set require carbon dioxide but not light and therefore are called **dark reactions**. However, dark reactions are not restricted to darkness, but take place both in light and in darkness; "dark reactions" mean only that light is not a requirement.

The immediate products of the light reactions are hydrogen ions (H^+), electrons (e^-), and oxygen, which are derived by **photolysis**, the "splitting" of water molecules by energy from light. Hydrogen ions and electrons are used in further photosynthetic reactions, but not as free and independent particles; instead, they are attached to molecules known as **electron carriers**, or **electron transport molecules**. When the electron carriers receive H^+ ions and electrons, they become chemically reduced; when they give up H^+ ions and electrons, they become oxidized.

In addition to the release of hydrogen ions and electrons, energy is also stored in molecules of ATP (**adenosine triphosphate**). In chloroplasts ATP is made by channeling light energy, via excited electrons, into chemical-bond energy between phosphate ions (here symbolized as —P) and **ADP** (**adenosine diphosphate**). The phosphate bonds thus produced are said to be **energy-rich bonds** and are depicted by ~; ATP thus is written as A—P~P~P, where each ~P is equivalent to about 7000 calories of energy. Usually only the last ~P is involved in actions of ATP in its energy transactions (A—P~P + —P + energy ⇆ A—P~P~P). The attachment of phosphate is called **phosphorylation** (**photophosphorylation** when light energy is involved).

Now the two interrelated sets of reactions may be depicted as shown below, and the following equations may be written to represent the reactions going on in each box:

Light reactions:
$$2\ H_2O + ADP + {-P} + \text{light energy} \rightarrow$$
$$O_2 + 4\ H^+ + 4\ e^- + ATP$$

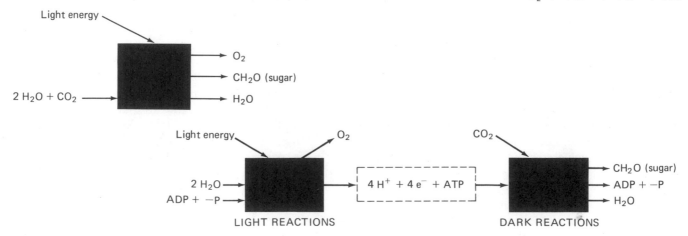

Dark reactions:
$$4 H^+ + 4 e^- + ATP + CO_2 \rightarrow$$
$$CH_2O + ADP + \text{—}P + H_2O$$

In the light reactions, the energy input is light; the reactants are water, ADP, and phosphate (—P); the products are electrons, H^+ ions, and ATP; the by-product is oxygen. The products of the light reactions are used in the dark reactions to produce sugar (CH_2O) by the chemical reduction, or **fixation,** of carbon dioxide.

Water is both a raw material in the light reactions and a by-product of the dark reactions. In the light reactions, it is the source of oxygen, as well as of hydrogen ions and electrons, which are added to carbon dioxide in the dark reactions (carbon dioxide fixation). During fixation, the oxygen of carbon dioxide is removed by combination with some of the H^+ ions and is liberated as water. Hence water appears on both sides of the photosynthesis equation. The addition of electrons and H^+ ions to carbon dioxide constitutes chemical reduction of carbon, which by a series of further reactions culminates in the production of sugar.

In interpreting and explaining the experiments and results of experiments such as those of Priestley and Ingenhousz, an important fact may be overlooked. It is not only animals that respire and render the atmosphere unfit for life. Plant cells also respire, and although their oxidative metabolism usually is not as active as that in animal cells, plants kept in the dark in a closed container will die as surely as will a mouse or other animal. Plants, like animals, have an absolute requirement for oxygen and produce carbon dioxide. Priestley thus could as well have run his whole experiment with mint plants. In darkness, net concentrations of carbon dioxide would have increased while oxygen levels would have decreased. In light, respiration would have continued unabated, but would have been "masked" by the consumption of carbon dioxide and production of oxygen by photosynthetic reactions.

Light Reactions

Light energy is absorbed by chloroplasts and converted by photosynthesis into the energy of foods. Photosynthesis encompasses two complex sets of oxidation-reduction reactions which occur in separate compartments of the chloroplast: light reactions occur in the compartment called the **granum** (pl. **grana**), and dark reactions take place in the **stroma**, the fluid material outside the grana (Fig. 3-1). Arrays of chlorophyll molecules within the thylakoids (layers) of a granum are the initial absorbers of light energy.

Chlorophyll The **chlorophyll** molecule (Fig. 3-1) belongs to a class of organic molecules named **porphyrins,** all of which have four interconnected nitrogen-containing rings. The porphyrins include several kinds of chlorophyll as well as other important pigments such as **heme** (a component of hemoglobin (Fig. 9-21) found in blood and in legume root nodules) and **cytochromes.** The latter are important energy transfer molecules.

A characteristic of pigments in general, which is shared by chlorophyll, is the capacity to selectively absorb light rays present in the **visible spectrum** (Fig. 3-2). The visible spectrum is composed of a mixture of light rays with different wavelengths, each corresponding to a specific color; short wavelengths are blue, longer one range from green to yellow to orange, and the longest visible wavelengths are red. Those wavelengths that are somewhat shorter than blue are termed **ultraviolet** and are invisible to the human eye. Wavelengths slightly longer than red are called **far red** and also are invisible.

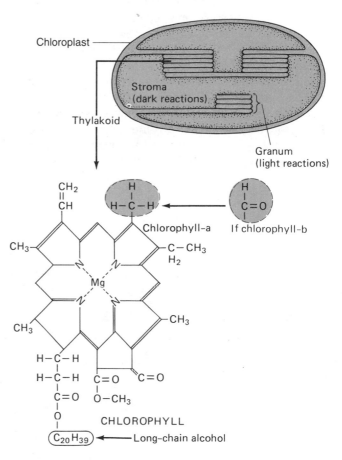

FIGURE 3-1
Chloroplast and chlorophyll. The two most common types of chlorophyll, chlorophyll-a and chlorophyll-b, are illustrated. These are present in all green land plants.

Those beyond far red are **infrared** and generally are referred to as **heat waves;** they, too, are invisible.

All these wavelengths are present in sunlight. Normally, only a blend of them, referred to as **white light,** is seen; however, if sunlight is passed through a **prism,** which has the ability to bend each wavelength at a different angle, the colors of all the wavelengths are seen separated in the color bands of the visible spectrum (Fig. 3-2). (Atmospheric water may also act as a prism, producing a visible spectrum in the form of the familiar rainbow.)

When light is passed through a chlorophyll solution, the solution appears green. This results because chlorophyll molecules absorb light in the red and blue bands of the spectrum; green light is the main wavelength transmitted (Fig. 3-2). It is the absorbed red and blue light, or **absorption spectrum,** that is most important in photosynthesis. Other pigments, however, are associated with chlorophyll in chloroplasts and are involved in photosynthesis as components of photosynthetic systems. Such **accessory pigments** have their own distinctive absorption spectra.

There are several ways to show which wavelengths of light are active in photosynthesis. A simple method is to expose a green water plant such as *Elodea* to different wavelengths and measure the amount of oxygen given off. This is easily done in the case of *Elodea,* because it is an underwater plant; one need only count the oxygen bubbles given off per unit of time. When this is accomplished, a graph can be plotted to show the photosynthetic efficiency of spectral color bands. Such a graph, or

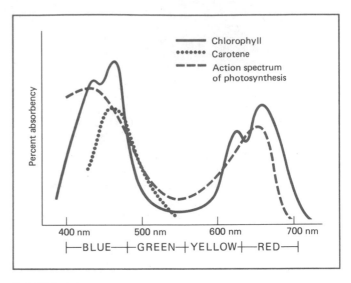

FIGURE 3-3
Absorption spectra of chlorophyll and carotene superimposed on the action spectrum of photosynthesis.

action spectrum, approximates the absorption spectrum for chlorophyll, but with some differences (Fig. 3-3). The reason for the discrepancies is that accessory pigments, principally yellow to red **carotenoids,** also absorb some light energy. The carotenoids pass on this energy to chlorophyll, with the result that the action spectrum of photosynthesis does not quite agree with the absorption spectrum of pure chlorophyll.

Light Energy   There are two theories about the nature of light. One theory states that light is in the form of energy waves. The other states that light is composed of particles. Explanations of photosynthesis make use of both theories. Chlorophyll and accessory pigments selectively absorb wavelengths of light energy, but viewing this energy as a stream of particles, or **photons,** is useful in explaining its conversion into chemical energy. Photons are considered to be particles of light energy having some of the characteristics of electrons, except that they are not electrically charged. The path of a beam of electrons can be bent by a magnet, as in the electron microscope. A beam of photons, on the other hand, cannot be influenced by a magnet, but can be bent by a prism or a lens, as in the light microscope (photons will pass through glass whereas electrons will not). When a molecule such as chlorophyll absorbs a photon, the energy of that photon is transferred to an electron, which attains a higher quantum level and is described as being in the **excited state,** or as an **excited electron.** This **electron transfer** is a fundamentally important part of photosynthesis. (You may at this point wish to refer to the discussion of excited electrons in Chapter 1, p. 5.)

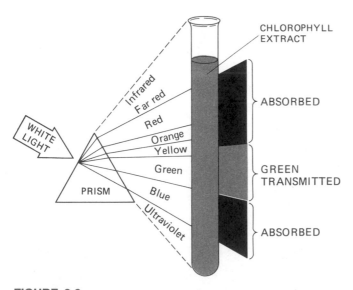

FIGURE 3-2
Visible spectrum contrasted with the absorption spectrum of chlorophyll. An experiment resembling the one in this illustration can be done using an instrument called a **spectroscope,** which projects a light spectrum through a chlorophyll solution.

Photon Trapping   The system of pigment molecules in a chloroplast may be called a **photon trap**. The photon trap is located in the thylakoids of a chloroplast, possibly in particles called **quantasomes** (Fig. 3-4), a name based on the observation that a photon consists of a definite amount, or quantum, of energy. Each photon trap contains about 300 molecules associated with the light reactions. Among these are "antenna" pigments that absorb photons; these pigments transfer their quanta of energy from molecule to molecule and finally to a **reaction center** made up of one of two kinds of chlorophyll-a. These two forms of chlorophyll-a, together with their associated molecules, constitute the centers of two very similar and intimately associated photon traps, known as **photosystem I (PS-I)** and **photosystem II (PS-II)**. In addition to chlorophyll-a, which is blue green, and chlorophyll-b, which is yellow green, the photosystems contain carotenoid pigments ranging from yellow to red. In some kinds of plants other chlorophylls and other pigments may be present. The pigment system of a photosystem may be envisioned as an imaginary funnel in which energy is directed toward the small end, where the active centers are located.

Photosystems I and II   When a photon of light is absorbed by the pigment system of a quantasome, the energy of the photon is transferred to a specific molecule of chlorophyll-a in PS-I or PS-II. In that chlorophyll molecule, the energy is transferred to a single electron in the molecule, which thereafter exists in the excited state. The excited electron is not attached to any single atom of the molecule but moves continuously among the atoms of the molecule. Two things can happen to the energy of the excited electron: it can leave the electron and escape as light energy, a process known as **fluorescence,**

or it and the electron can be transferred to another molecule.

Chlorophyll in solution fluoresces when exposed to light; no matter what wavelength was initially absorbed, the chlorophyll always emits light in the red range of the spectrum. Thus, if chlorophyll is exposed to white light and one looks indirectly rather than directly at the light coming through the solution, the chlorophyll will look dark red. This shows that chlorophyll can emit energy, in this case light energy. However, in chloroplasts of living plants chlorophyll always is associated with electron carrier molecules. These molecules have the capability of spontaneously accepting energy, in the form of an excited electron from chlorophyll-a, and fluorescence does not occur. The acceptance of an electron by an atom or molecule constitutes a chemical reduction; therefore the carrier molecule becomes reduced and the chlorophyll molecule becomes oxidized.

In PS-I, excited electrons are channeled through a series of electron carriers, ending in the carrier molecule **nicotinamide adenine dinucleotide phosphate (NADP)**, which occurs in all cells and also figures in reactions other than those of photosynthesis.

A rule in spontaneous oxidation-reduction reactions, of which this is an example, is that electron transfers tend to run in a "downhill" direction. Each transfer of a high-energy electron is to a molecule having less free energy than the last (remember that free energy refers to the capacity to do work). The difference between the free-energy levels of the reaction molecules may simply be lost as heat or it may be harnessed to do work.

In PS-II, excited electrons pass through a series of stairstep transfers involving several unique electron carriers, including cytochrome (see Fig. 3-5). Finally, having

FIGURE 3-4
Associated pigment molecules involved in photon trapping in a quantasome.

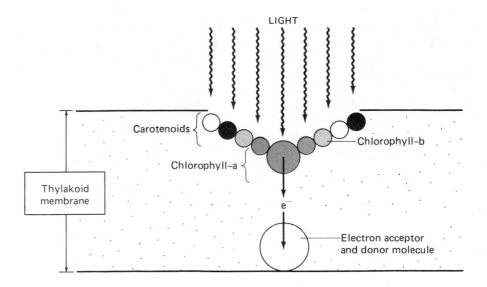

FIGURE 3-5
Summary of events occurring in the light reactions. In this diagram, X represents an uncharacterized molecule sometimes called **FRS (ferredoxin-reducing substance)**; F is **ferredoxin**; NADP is nicotinamide adenine dinucleotide phosphate; Q is an uncharacterized molecule called a **fluorescence quencher**; and C represents a cytochrome associated with the light reactions. The term ½ O_2 is a conventional way of indicating a reaction involving one atom of oxygen. Several steps have been omitted here.

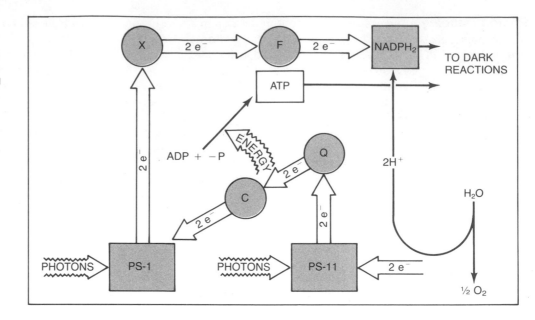

lost much of their energy, the excited electrons are given over to the chlorophyll molecules of PS-I to replace those electrons given up to NADP. At each electron transfer in PS-II, some energy is made available for producing molecules of ATP. This apparently is accomplished by channeling H^+ ions to specific carrier or "coupling" molecules in the thylakoid membranes. There are no oxidations or reductions in this process, but a local change in the electrical charge and pH occurs across the membrane, and the energy differential is sufficient to drive the synthesis of a molecule of ATP from ADP + —P. This explanation of photophosphorylation, called the **chemiosmotic theory,** was first proposed in the early 1960s by Peter Mitchell, an English biochemist, and gained acceptance after an initial period of skepticism.

The production of ATP and the donation of electrons to PS-I are important functions of PS-II. Of equal importance, PS-II has the ability to split water molecules into H^+ ions, electrons, and oxygen molecules. These are disposed of as follows:

1 The electrons derived from water are absorbed by the chlorophylls of PS-II, in place of the excited electrons that are given up to the series of ATP-producing electron carriers and ultimately to PS-I.

2 The H^+ ions derived from water accompany NADP and its newly acquired electrons during its ensuing trip to the dark reactions (the abbreviation $NADPH_2$ designates NADP with its load of two electrons and two H^+ ions, but sometimes it is written in a different form, as for example $NADPH + H^+$, to indicate a loose association; see second footnote, p. 84).

3 The oxygen derived from water either is used by the cells in their oxidations or escapes from the leaf as a free gas.

The relationship between PS-I and PS-II resembles a game of musical chairs in which electrons move through an intricate system of coordinated oxidation-reduction reactions. The overall result of these reactions is that light energy enters the quantasome, and chemical energy in the form of ATP and $NADPH_2$ is produced (Fig. 3-5). In addition, free oxygen is liberated.

Dark Reactions

Although carbon dioxide fixation is the principal chemical event in the dark reactions, about as many steps are involved as occur in the light reactions. These steps consititute a series of reactions, or **pathways,** known as the **Calvin-Benson cycle.** In addition, several other pathways commonly are associated with the Calvin-Benson cycle. They include the **Hatch-Slack pathway** and the **CAM pathway.** Another pathway, called **photorespiration,** is not photosynthetic but is closely associated with the Calvin-Benson cycle. The Calvin-Benson cycle occurs in all green plants, either as the sole photosynthetic system or in cooperation with the Hatch-Slack pathway or the CAM pathway.

Calvin-Benson Cycle (C-3 Pathway) The Calvin-Benson cycle is also called the C-3 **pathway,** because a 3-carbon sugar phosphate, **phosphoglyceraldehyde (PGAL),** is a key product (Fig. 3-6). The C-3 pathway was discovered by Melvin Calvin and colleagues at the University of California shortly after World War II, when car-

FIGURE 3-6

The Calvin-Benson, or C-3, carbon dioxide fixation cycle. Three "turns" of the cycle (indicated as X3) produce a net gain of one molecule of a 3-carbon sugar (triose). The cycle begins with the attachment of CO_2 to a 5-carbon (pentose) sugar to form an unstable intermediate molecule. This key step is illustrated with chemical structures of the molecules involved. The unstable intermediate molecule breaks down spontaneously in the presence of H_2O to produce two molecules of organic acid. These are reduced by $NADPH_2$ and ATP during the light reactions of photosynthesis, and two molecules of a 3-carbon triose sugar are produced. Three turns of the cycle produce six trioses, five of which combine in a series of intermediate steps to generate three pentoses to support the continuation of CO_2 fixation. The sixth triose is the net gain from the fixation of three molecules of CO_2. RuBP = ribulose bisphosphate, PGA = phosphoglyceric acid, PGAL = phosphoglyceraldehyde.

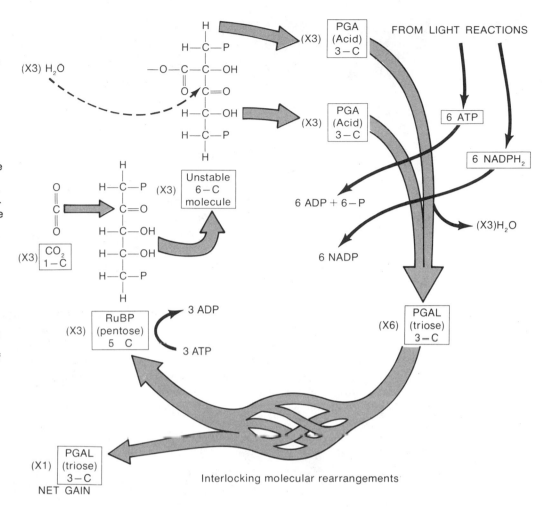

Interlocking molecular rearrangements

bon 14 (^{14}C), the radioactive isotope of common atmospheric carbon (carbon 12 [^{12}C]), became available for scientific use. A simple unicellular green alga named *Chlorella* was used in experiments designed to identify the steps of the dark reactions. *Chlorella* was ideal for this because it could be grown in test tube solutions, and these could be instantly and uniformly exposed to a flash of light and given $^{14}CO_2$ (radioactive carbon dioxide). Then the algae could be killed immediately by strong chemicals, stopping all further reactions. When the dark reactions were stopped in this manner at different intervals of time, and the compounds present isolated and identified using a Geiger counter to trace ^{14}C atoms, it was possible to know the intermediate steps in production of sugar by chloroplasts. Dr. Calvin subsequently received the Nobel Prize for this work.

The first product of the C-3 cycle to be identified is an unstable 6-carbon molecule formed by adding carbon dioxide to one end of a 5-carbon sugar named **ribulose bisphosphate (RuBP)**. This unstable molecule breaks down quickly by hydrolysis into two molecules of a 3-

carbon organic acid, **phosphoglyceric acid (PGA).** Then PGA becomes reduced to the 3-carbon (triose) sugar PGAL by electrons and H^+ ions from $NADPH_2$. The energy of ATP also is used in this conversion, and a molecule of water is derived from each reduction of PGA to PGAL. Although this seems a very roundabout way of making sugar from carbon dioxide, biochemical processes often are complicated, for the cell performs its chemical reactions one step at a time, building complex molecules from simpler ones. One reason for this is that each reaction is catalyzed by a different enzyme: most enzymes are very specific. Another reason is that, in the cell, great changes in energy do not take place in single reactions; instead energy transactions are the cumulative results of many smaller energy transfers. Explosions of energy do not occur in cells.

Looking further into the C-3 cycle, the trioses are involved in a series of molecular rearrangements which restore RuBP used in the initial CO_2 fixation. The end result, provided there is RuBP to start with, is that for every three carbon dioxide molecules used, three new

ARTIFICIAL PHOTOSYNTHESIS

The light reactions, the mechanism of energy trapping in photosynthesis, are of great interest to many scientists. The organization of the pigments in the quantasomes or reaction centers of the thylakoid membranes is such that the excited electron (e^-) and the chlorophyll site (Ch^+) are separated efficiently. If an artificial system of photosynthesis could be developed, perhaps it could be used to drive electrical or chemical reactions in a solar cell and contribute to the solution of the current energy problems of the planet earth. So far, the model systems have not been able to mimic the natural system in rapidly and efficiently separating the charges: the model systems have a back reaction that occurs too quickly to allow a good energy yield.

In 1984 a team of ten workers associated with four laboratories in the United States, France, and the United Kingdom succeeded in synthesizing a complex molecule which mimics aspects of the photosynthetic separation of charge. The molecule consists of three parts: P, pigment; D, electron donor; and A, electron acceptor. The pigment portion sits in the middle and has a structure similar to chlorophyll; it is a good light absorber. The parts may be represented as D—P—A and the sequence of events would be:

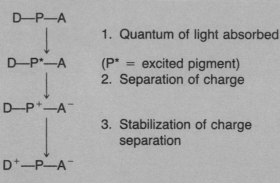

In natural systems Step 1 occurs in 100 picoseconds (1 picosecond = 10^{-12} seconds; 100 picoseconds = 10^{-10} seconds) and the back reaction is 100 times slower thereby giving time for D—P*—A to be converted to D—P$^+$—A$^-$, rather than falling back to D—P⤳A (fluorescence). In the model system, Step 1 occurs in 200 picoseconds and the back reaction is more than a million times slower, thus enhancing the occurrence of Steps 2 and 3.

The energy trapped in the system can reach 25 percent of the incident light. Perhaps this system is a major advance in overcoming the obstacles to an artificial photosynthetic system.

molecules of RuBP are formed, with a net gain of just one triose. Without RuBP as a starting point, the photosynthetic production of trioses would not take place. In Fig. 3-6 the steps in the fixation of carbon dioxide are shown according to the pathways worked out by Calvin and Benson.

Hatch-Slack (C-4) Pathway Certain plants have been found to be more efficient photosynthesizers than others. Among them is sugarcane, the most productive plant in the world; corn (more properly called maize or Indian corn); and crabgrass. These plants have two photosynthetic pathways, the Calvin-Benson or C-3 cycle and the Hatch-Slack or **C-4 cycle**, but are referred to as **C-4 plants**. Credit for discovery of the C-4 pathway is shared principally by the Australian plant physiologists M. D. Hatch and C. R. Slack. C-4 plants have two kinds of chloroplasts: **mesophyll chloroplasts** and **bundle-sheath chloroplasts** (see Figs. 3-7, 3-8, and 8-15 for locations of these two types). The two forms of chloro-

FIGURE 3-7
Leaves of C-3 and C-4 plants in cross section. Locations of mesophyll cells in each and bundle-sheath cells and chloroplasts in a C-4 plant are shown.

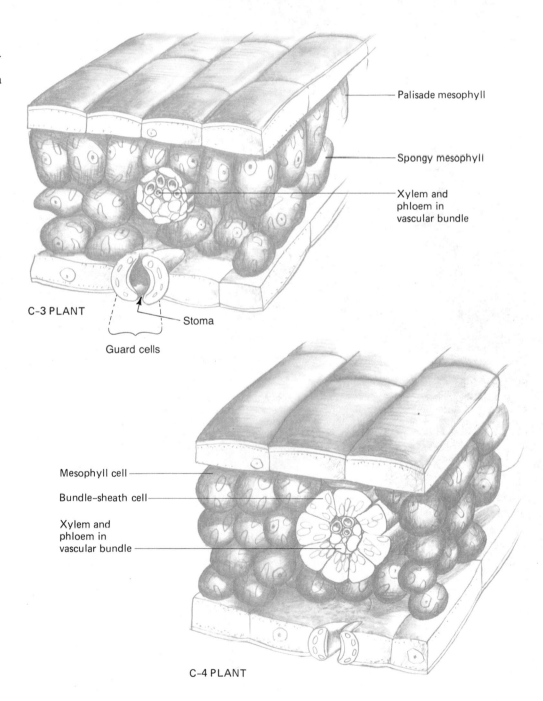

Palisade mesophyll

Spongy mesophyll

Xylem and phloem in vascular bundle

C-3 PLANT

Stoma

Guard cells

Mesophyll cell

Bundle-sheath cell

Xylem and phloem in vascular bundle

C-4 PLANT

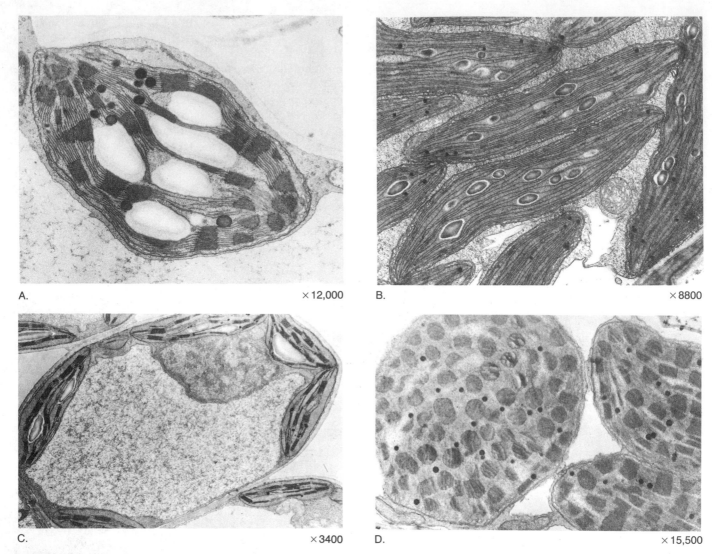

A. ×12,000

B. ×8800

C. ×3400

D. ×15,500

FIGURE 3-8

A. and **B.** Chloroplasts of C-4 plants. **A.** Mesophyll of *Muhlenbergia racemosa* (a grass) showing grana composed of stacks of thylakoids. **B.** Bundle-sheath chloroplasts of *Zea mays* (maize) with parallel thylakoids but no grana. **C.** and **D.** Chloroplasts of C-3 plants. **C.** Cell of *Helianthus* (sunflower) with well-developed grana in chloroplasts. **D.** Unusual view of the grana in a section cut parallel to the surface of the grana and showing their cylindrical form. (A, photo courtesy Watson Laetsch; B, photo courtesy Barbara Breisch Peterson; C and D, photos courtesy Joe Baba)

plasts differ in structure as well as in location. Bundle-sheath chloroplasts lack grana; mesophyll chloroplasts have grana.

The C-4 pathway occurs in the mesophyll cells. In this pathway, carbon dioxide is not attached to RuBP but instead to the 3-carbon molecule **phosphoenol pyruvic acid (PEP)** (Fig. 3-9). The resulting 4-carbon molecule, which is an organic acid **(malic acid)**, not a sugar, is transferred to bundle-sheath cells. Here carbon dioxide is removed and the remaining 3-carbon molecule (PEP) moves back to the mesophyll chloroplasts, where it is again used to fix more carbon dioxide. Carbon dioxide

then is fixed in the C-3 cycle, which occurs in the bundle-sheath chloroplasts. Although this operation seems inefficient because carbon dioxide is processed twice, it actually is highly efficient, for it concentrates carbon dioxide in the bundle sheath, where it may be utilized to produce sugars. C-4 plants also are more efficient than **C-3 plants** in carbon dioxide fixation, because PEP has a greater affinity for carbon dioxide than does RuBP, the initial carbon dioxide receptor in the C-3 cycle (Fig. 3-6).

C-4 plants are especially adapted to the tropics, where the intensity of light is great. However, maize, a highly productive temperate plant, employs the C-4 pathway, as does crabgrass, a late summer pest. Other lawn grasses are C-3 plants, and they do not compete very well with crabgrass under dry conditions. Indeed, in a closed system with limited carbon dioxide, a C-4 plant can kill a C-3 plant. However, C-3 plants are more competitive than C-4 plants under low-light conditions because they require fewer ATPs to fix CO_2.

Cam Plants Plants in which the CAM pathway occurs resemble C-4 plants in their photosynthetic reactions. CAM is the acronym for **crassulacean acid metabolism,** in reference to malic acid and other organic acids active in carbon dioxide fixation and transport in desert plants such as *Crassula*. **CAM plants** are quite efficient under desert conditions where severe problems of water conservation prevail. Their stomata are open at night, when they fix carbon dioxide by combining it with PEP derived from starch. The product, malic acid, is broken down in light, releasing carbon dioxide, which then is fixed via the C-3 cycle.

Photorespiration An organelle called the **microbody** is involved with the photosynthesis in C-3 plants (Fig. 3-10). The microbody, usually found close to the chloroplast, takes in **glycolic acid,** a 2-carbon compound produced in the chloroplast by oxidation of RuBP and fur-

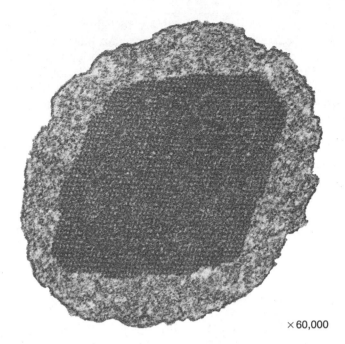

×60,000

FIGURE 3-10
Electron micrograph of a microbody. The crystal within the microbody is composed of the enzyme catalase, which is active in photorespiration. (Photo courtesy Eugene Vigil)

ther oxidizes it to **glyoxylic acid** and **hydrogen peroxide (H_2O_2).** Oxygen is split off from hydrogen peroxide by the enzyme **catalase,** while glyoxylic acid is combined with an amino group to make the simple amino acid **glycine.** Glycine may be transported to mitochondria where it can be converted into other amino acids or into metabolic intermediates.

The overall result of photorespiration is a loss in photosynthetic efficiency (because RuBP is used up), making the function of the process something of a puzzle. Some biologists think photorespiration is an ancient process left over from primitive cells that were highly sensitive to excess oxygen (the C-3 cycle evolved at a time when the atmosphere contained considerably less oxygen than now, and possibly oxygen subsequently acted as a poison until plants evolved photorespiration to deal with it).

Factors Affecting Photosynthesis

Light Intensity The nature of light and the relationships between the light spectrum and chlorophyll have been discussed. Light intensity also is important in photosynthetic reactions. At low light intensities, the rate of photosynthesis is directly proportional to the supply of photons (i.e., the intensity of the light); however, at high light intensities, the supply of carbon dioxide is the factor most limiting to the rate of photosynthesis. What

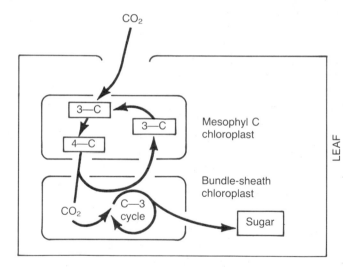

FIGURE 3-9
The Hatch-Slack, C-4 pathway of carbon dioxide fixation. 3—C and 4—C are organic acids which function as intermediary molecules in the transport and fixation of CO_2 (3—C = pyruvic acid, 4—C = malic acid). Malic acid is produced by the fixation of CO_2 in mesophyll chloroplasts, then transported into bundle-sheath chloroplasts, where it releases CO_2 to RuBP of the Calvin-Benson, C-3 cycle (refer to Fig. 3-6, also see Figs. 3-7 and 3-8 for the structures of mesophyll and bundle-sheath chloroplasts). The CAM pathway resembles the Hatch-Slack pathway, but, in CAM, malic acid is produced by day and gives up CO_2 to the C-3 cycle by night; CAM plants have only mesophyll chloroplasts.

this means is that under ordinary daylight conditions in nature, sufficient light is present for maximum photosynthesis to occur. However, if the concentration of carbon dioxide, which normally is about 0.03 percent of the atmosphere, is increased experimentally, then supplemental light may be required for maximum photosynthesis to take place.

Carbon Dioxide If plants are grown in a chamber where the concentration of carbon dioxide in the air can be adjusted, the productivity of photosynthesis can be increased by adding carbon dioxide to the air, provided sufficient light is present. Studies have shown that at a carbon dioxide concentration of about 6 percent, the rate of photosynthesis is approximately five times greater than that at atmospheric carbon dioxide levels.

Is the concentration of atmospheric carbon dioxide presently increasing in the earth's atmosphere? A great deal of carbon dioxide is given off by the combustion of fuels such as coal, natural gas, gasoline, and heating oil. Scientists have predicted that this will lead to a gradual increase in atmospheric carbon dioxide (since the beginning of the Industrial Revolution about 200 years ago, atmospheric carbon dioxide has increased about 10 percent). It is unlikely that this will be great enough to bring about large increases in photosynthesis. However, atmospheric carbon dioxide also contributes to the retention of heat in the atmosphere. This is called the **greenhouse effect**, because the carbon dioxide in the atmosphere transmits heat in the same way as greenhouse glass or an automobile window on a sunny day. In the long run, the greenhouse effect might have a greater impact on plant growth by prolonging growing seasons than by direct effect on photosynthesis caused by higher levels of atmospheric carbon dioxide.

Water Because water is a raw material in photosynthetic reactions, its supply to the green tissues of the plant is critical. With the exception of those living in wet places, most plants maintain a fine balance between evaporation and water conservation. Plants require open stomata during periods of photosynthesis because carbon dioxide is in demand, but this results in the evaporation of water from within the leaf through the stomata (Fig. 3-11). The amount of water consumed by photosynthesis is very small compared to that lost by evaporation through leaf surfaces.

Normally, the stomata are effective in maintaining a balance between the requirement for atmospheric carbon dioxide and the loss of water, so water loss does not become a critical factor. However, when a plant becomes dry, the stomata will close, and this cuts off the

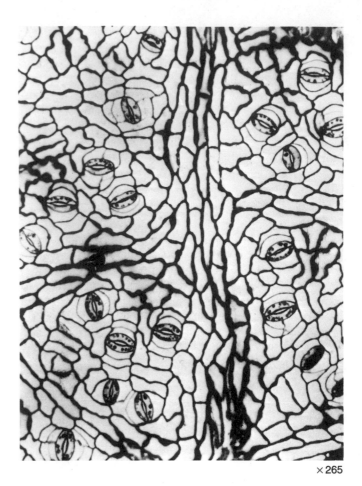

×265

FIGURE 3-11
Epidermal cells of a cottonwood tree leaf *(Populus deltoides)*. Note that only guard cells of stomata contain chloroplasts (dark-staining bodies). The pathways of underlying conductive strands (xylem and phloem) are indicated by the bands of narrower, epidermal cells. (From Russin, W. D., and R. F. Evert, 1984. *Amer. J. Bot.* 71:1368.)

supply of carbon dioxide. Under these circumstances, water is a limiting factor in photosynthesis.

Plant Biosynthesis

The production of compounds by the living cell is called **biosynthesis.** The sugar produced by photosynthesis may be used by a plant in many biosynthetic reactions. It may be converted to starch for later use or into many other kinds of molecules. In this short presentation, mention can be made of only a few of the substances produced in plants that have their origin in photosynthesis. Starch is a common stored plant food and is formed by the polymerization of glucose, with linkages that form branching chains. Cellulose is also a glucose polymer, but is unbranched and is a structural material.

In addition, plants can transform glucose into organic acids, and, by adding nitrogen, make amino acids and proteins. The list is almost limitless; added to the preceding examples are fats and oils, vitamins, nucleic acids, ATP, cytochromes, and chlorophylls and other pigments, as well as drugs, dyes, resins, and many plant products used in commerce.

The production of amino acids is one of the most important contributions of plants in their relationships with other organisms in the environment. Although animals can transfer amino groups from amino acids to organic acids and thus make other amino acids, they cannot synthesize amino groups in the first place. Photosynthesis receives a great deal of emphasis in biology and deservedly so, because transformation of radiant energy into the chemical energy of food is a key reaction in the living world. Nevertheless, another vital process, nitrogen fixation, is equally important. Without it we should have no proteins and therefore no life as we know it.

NITROGEN FIXATION

Nitrogen fixation is the conversion of nitrogen gas (N_2) into a combined form of nitrogen: nitrogen dioxide (NO_2), nitrous acid (HNO_2), nitric acid (HNO_3), or ammonia (NH_3). About 10 percent of the annual production of combined nitrogen occurs as a result of lightning and combustion, which produce oxidized nitrogen. Automobile exhausts contain nitrogen dioxide (NO_2), which can combine further with water to make nitrous and nitric acids—one of the reasons that emission controls have been imposed on autos and other sources of oxidized nitrogen. Lightning also produces these acids, which, when washed into the soil, react with bases and salts of sodium, potassium, and calcium to produce nitrates, an important source of nitrogen for plants. In fact, nitrates are the preferred form of nitrogen for most plants.

Man-made ammonia and nitrates also are an important source of nitrogen for agriculture and industry, accounting for about 25 percent of the combined nitrogen introduced into the world each year. The Haber process, which requires huge quantities of hydrocarbons, is an important commercial source of nitrogen fertilizers. Fertilizers are obtained also by the mining of nitrogen-containing rocks and the harvesting of guano from seabirds and bats.

Biological nitrogen fixation (dinitrogen fixation) accounts for about 60 percent of the world's annual yield of fixed nitrogen. (Note: dinitrogen fixation is the term often used for biological nitrogen fixation because organisms fix (reduce) both atoms of the nitrogen molecule (N_2) simultaneously, rather than separately as in industrial processes.) It is estimated that the atmosphere contains about 10^{15} metric tons (1 metric ton [tonne] = about 2,200 lbs) of N_2; about ten times that much is present in seawater and incorporated in rocks. The nitrogen cycle (all the reactions involving nitrogen in the ecosystem) involves about 3×10^9 tonnes per year, an annual turnover of but a small fraction of the total nitrogen of the world ecosystem.

Biological Nitrogen Fixation (Dinitrogen Fixation)

Dinitrogen fixation (N_2 fixation) is the conversion by bacteria of N_2 into NH_3. The reactions of N_2 fixation have been difficult to work out because nitrogenase, the enzyme system that catalyzes the conversion of N_2 to NH_3, is poisoned by air, or, more specifically, destroyed by oxygen. Not only does oxygen poison the reactions of N_2 fixation, but it also effectively degrades the enzymes of the nitrogenase system so that their purification and characterization require an absolutely oxygen-free environment—not an easy attainment. The reactions catalyzed by the nitrogenase system are oxidation-reduction reactions requiring a reducing agent (i.e., a donor, such as NADP, of electrons and hydrogen ions) and energy from ATP. Both are provided by metabolic processes occurring in the cytoplasm of N_2-fixing bacteria. The overall reaction is summarized as follows:

$$N_2 + 6H^+ + 6e + 12\ ATP \xrightarrow{Nitrogenase} 2\ NH_3 + 12\ ADP + 12 - P$$

N_2 fixation is energy-expensive, requiring at least 2 ATP for each electron transferred to nitrogen, a total of 12 ATP for each N_2 molecule converted into ammonia.

Nitrogen-fixing Bacteria N_2 fixation occurs exclusively in bacteria; no eukaryotic nitrogen-fixing organism has yet been discovered. The popular conception of N_2 fixation by plants of the clover family (i.e., legumes) is correct only to the extent that the actual fixation of N_2 is accomplished by bacteria living in symbiosis within root nodules of leguminous plants (clovers, soybeans, peas, beans, vetches, alfalfa, etc.; see Fig. 3-12).

N_2-fixing bacteria must be able to maintain their nitrogenase enzyme system in an oxygen-free compartment of their cell, or else must live in an oxygen-free environment. Free-living bacteria capable of meeting these requirements account for only about 2 percent of N_2 fixation. Some of these species carry on N_2 fixation by living under anaerobic conditions in such places as

FIGURE 3-12
A red clover plant showing its root nodules. (Photo by Carolina Biological Supply Company)

ules of a number of nonlegumes, including elders, Australian pines (*Casuarina*), some blackberry species, arctic roses, and several others, but in no case is the process as efficient as in the *Rhizobium*-legume symbiosis.

Because symbiotic N_2 fixation far exceeds that of free-living bacteria, the question that may occur to one is why this relationship is so productive. What do the two partners gain from their association? Obviously the clover plant gains a source of nitrogen, which is in short supply in many soils. This benefit, however, is not without cost to the clover plant, which supplies the bacterium with two of its necessities: an oxygen-free environment, required for protection of the nitrogenase system, and food from which to derive both ATP and hydrogen ions and electrons. This is no small contribution; the drain on the clover plant amounts to about 16 percent of its total photosynthetic food production. Nevertheless, in nitrogen-poor soils the exchange is a bargain.

Root Nodules The key to the success of N_2-fixing symbiosis is the structure of the **root nodule** (Fig. 3-12). Bacteria first enter the root of a legume via a channel referred to as an **infection thread** (Fig. 3-13) and work their way into the interior of the young root. There the bacteria become enfolded by the walls, membranes, and cytoplasm of root cells, which at the same time grow and multiply, forming a compact ball of cells—the root nodule (Figs. 3-14 and 3-15)—with the bacteria in the interior. The relationship guarantees a relatively oxygen-free dwelling place for the bacteria. The site is further cleared of O_2 by requirements for oxygen in the cellular respiration of both the host cells and the bacteria. In addition, because roots are underground structures, photosynthesis does not occur in them and that source of oxygen is absent. Still another mechanism has evolved to ensure an oxygen-free environment for the symbiont. Many years ago, attention was drawn to the pinkish to red color of legume nodules, and the pigment was identified as a form of hemoglobin called **leghemoglobin**. At that time, however, the significance of leghemoglobin was not known, and it remained a curiosity without explana-

swamps and polluted waters, where oxygen is scarce and is rapidly depleted by decay processes. Other free-living bacteria have developed internal compartments from which oxygen is scavenged by oxidative processes such as those of cellular respiration (covered in the next chapter). Ninety-eight percent of biological nitrogen fixation is symbiotic, occurring within root nodules of a wide variety of plants. Among the more efficient N_2-fixing bacteria are species of **Rhizobium,** which live in root nodules of legumes. An outstanding example of this symbiosis is seen in red clover, reported to produce as much as 300 kilograms of nitrogen per hundred acres per year. This is more nitrogen than is supplied in most chemical fertilizer treatments. Other genera and species of N_2-fixing symbionts are known to occur in root nod-

FIGURE 3-13
The invasion of a young clover root by *Rhizobium* bacteria.

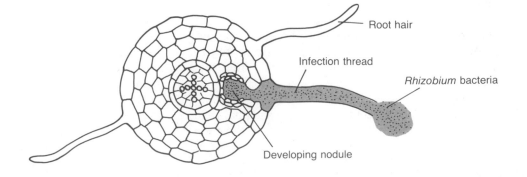

Root hair

Infection thread

Rhizobium bacteria

Developing nodule

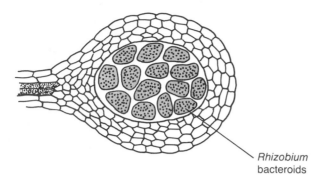

FIGURE 3-14
Cross-section of a red clover root nodule.

Rhizobium bacteroids

tion. When the sensitivity of the nitrogenase enzyme system to oxygen was discovered, the function of leghemoglobin became apparent. It has about the same affinity for oxygen as does animal hemoglobin, and it is thus able to capture and store oxygen. The bacterial cells in the nodule, now commonly referred to as bacteroids because of certain cellular modifications, are literally bathed in leghemoglobin, which traps any free oxygen that might be present. Leghemoglobin protects the nitrogenase system of the bacteroids from oxygen poisoning. It also helps the bacteroids meet their oxygen requirement for respiration by transferring oxygen to oxidase enzymes in the bacteroid cells.

Although the first product of N_2 fixation in legume root nodules is ammonia, plants do not usually take it up directly from bacteroids. Enzymes in bacteroids attach ammonia, as an amino group (—NH_2), to **glutaric acid**, producing **glumatic acid**, an amino acid. A second NH_3 added to glutamic acid produces **glutamine**, a key amino acid in the synthesis of other amino acids in organisms. The host plant acquires its fixed nitrogen from bacteroids in the form of glutamic acid, glutamine, or another amino acid, **asparagine**, which is made by transferring an amino group from glutamine to aspartic acid. Using these amino acids, the plant is able to synthesize the whole array of amino acids required for its protein synthesis.

Nitrogen Fixation by Cyanobacteria (Blue-green Algae) One group of bacteria, the **cyanobacteria**, formerly known as **blue-green algae**, are both photosynthetic and N_2 fixing. In general, N_2 fixation by free-living cyanobacteria is probably of only local importance. However, that performed by an interesting symbiotic relationship known as the *Azolla* system is of considerable importance in the Orient, where rice is an important crop. *Azolla* is a tiny floating water fern which harbors cyanobacteria within its mucus-filled internal cavities. The practice in growing rice in the Orient is to flood a paddy prior to planting time, thereby encouraging the growth of a dense crop of *Azolla*. Later on the paddy is drained; most of the *Azolla* dies and decays, releasing appreciable quantities of fixed nitrogen. Rice then is planted in the nitrogen-rich muck of the paddy. Paddies

FIGURE 3-15
Rhizobium phaseoli, an N_2-fixing symbiont in bean root nodules. Each bacterial cell (bacteroid) is enclosed within host-cell membrane. (Photo courtesy L. M. Baird and B. D. Webster, *Bot. Gaz.* 143(1):41–51, by permission of the University of Chicago Press. © 1982 by the University of Chicago, all rights reserved)

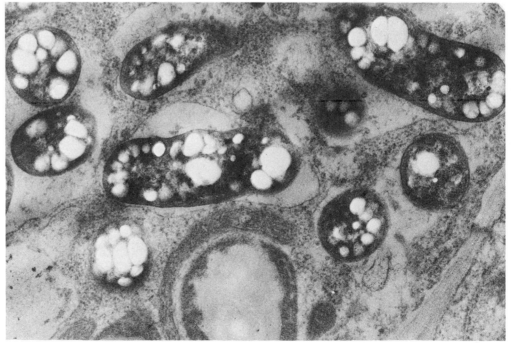

×31,000

in use hundreds of years have maintained their original fertility as a result of *Azolla* culture.

Biological Nitrogen Fixation and Agriculture

In the past few years considerable attention has been focused on N_2 fixation because of the energy demands of synthetic fertilizers. In underdeveloped regions the cost of fertilizers now is prohibitive, and food shortages are worsening. In developed countries also costs of fertilizers are a drain on economies. Biological N_2 fixation is more and more attractive as an alternative to chemical fertilizers.

Crop rotation, in which legumes such as red clover are grown alternately with non-N_2-fixing crops such as maize or wheat, is an old practice. It is not always an attractive option, however, because the cash yield of clover or alfalfa generally is considerably lower than that of a cereal crop. The introduction of soybeans from China was a great benefit to American farmers, because soybeans are not only N_2-fixing legumes, but also a valuable cash crop.

Consideration has been given to developing a technology for inducing N_2-fixing root nodules on non-legume crop plants such as maize, wheat, and cotton. Genetic engineering thus far has not succeeded in doing this in any plant, and it is questionable whether the benefits derived would outweigh the reduction in crop yield due to the drain on photosynthetic productivity by the nitrogenase system. A more realistic approach may be the cultivation of certain soil bacteria which live in close association with roots of some tropical and temperate grasses. One of these, the N_2-fixing bacterium *Azospirillum*, discovered in 1976, can colonize roots of cereals such as maize and sorghum. Fixed nitrogen is transferred to the cereal plant by *Azospirillum*, but apparently not in any appreciable quantity. Thus far there have been no dramatic breakthroughs with this system, although the potential for future agricultural benefit is encouraging.

SUMMARY

Two reactions characterize the energy relations between plants and animals. Photosynthesis is given by

$$CO_2 + 2\ H_2O + \text{light energy} \rightarrow CH_2O + O_2 + H_2O$$

and respiration by

$$CH_2O + O_2 \rightarrow \text{energy} + H_2O + CO_2$$

These two equations demonstrate the fundamental interrelationships of food production (sugar formation), food utilization (sugar oxidation), and food-derived energy.

Photosynthesis consists of a number of coordinated reactions occurring in the chloroplasts of plants (or in the cellular membranes of blue-green algae and some bacteria). Basically, two processes are involved. One requires light; the other is independent of light. In the first of these processes, the light reactions, photons (units of light energy) are absorbed by chlorophyll and associated pigments constituting a photon trap, the energy of the photons resulting in the release of high-energy (excited) electrons. These electrons move through a complex series of reactions occurring in two interlocking sets of electron carrier molecules called photosystems I and II (PS-I and PS-II).

In a second series of reactions, the dark reactions, carbon dioxide is reduced, using electrons and H^+ ions from $NADPH_2$ and chemical energy from ATP. Essentially, the dark reactions consist of the addition of carbon dioxide to one end of a 5-carbon sugar, ribulose bisphosphate (RuBP), and the immediate splitting of this 6-carbon molecule into two 3-carbon molecules of phosphoglyceric acid (PGA), which then is reduced by $NADPH_2$ and becomes a 3-carbon sugar, phosphoglyceraldehyde (PGAL). Some of the PGAL remains as the product of photosynthesis, but most is used in a series of splitting and recombination reactions to replace the RuBP initially used in carbon dioxide fixation. This constitutes a biochemical pathway called the Calvin-Benson (C-3) cycle. Other pathways are also known, including the Hatch-Slack (C-4) pathway and the CAM pathway, which occurs in some desert plants.

Light usually is not a limiting factor in photosynthesis, but if carbon dioxide concentrations are increased, the rate of photosynthesis also increases. Stomata regulate the exchange of carbon dioxide between the atmospheres outside and within the leaf.

Although photosynthesis is the primary source of food for all living organisms, a second vital process, biological nitrogen fixation, is the primary source of proteins. Occurring only in bacteria, N_2 fixation produces ammonia, which then is converted into amino groups of amino acids. The amino acids are transferred to plants, usually in an important symbiotic relationship called the root nodule–bacterium association, which produces about 98 percent of the world's total of biologically fixed nitrogen.

KEY WORDS

phlogiston	photosystem II	CAM plant
oxygen	photon trap	photorespiration
respiration	electron transfer	dinitrogen fixation
biological oxidation	electron carrier	nitrogenase
photosynthesis	ATP	legume
light reactions	NADPH$_2$	*Rhizobium*
dark reactions	chemiosmotic theory	symbiosis
chlorophyll	Calvin-Benson cycle	infection thread
visible spectrum	RuBP	root nodule
absorption spectrum	PGAL	leghemoglobin
action spectrum	Hatch-Slack pathway	cyanobacteria
photon	C-4 plant	*Azolla* system
photosystem I	C-3 plant	*Azospirillum*

QUESTIONS FOR REVIEW AND DISCUSSION

1 Explain why an earlier version of the overall photosynthetic reaction was written $6\ CO_2 + 6\ H_2O + \text{energy} \rightarrow C_6H_{12}O_6 + 6\ O_2$, whereas the current formula is $6\ CO_2 + 12\ H_2O + \text{energy} \rightarrow C_6H_{12}O_6 + 6\ H_2O + 6\ O_2$.

2 What are the products and by-products of the light and dark reactions? How are they interrelated?

3 Differentiate between photosystems I and II in the light reactions. How are the two systems interrelated?

4 Briefly describe the process of carbon dioxide fixation in the Calvin-Benson cycle. Why is it called a cycle? What is the C-4 pathway, and why is it important in the human economy?

5 What evidence is there for stating that the carbon dioxide concentrations in the atmosphere are more apt to limit the rate of photosynthesis than is light intensity?

6 Give the overall equation for dinitrogen fixation.

7 What role does leghemoglobin play in the legume root nodule?

8 It is said that the N_2-fixing symbiosis considerably reduces soy bean productivity. Explain.

SUGGESTED READING

BASSHAM, J. A. 1962. The path of carbon in photosynthesis. *Sci. Amer.* 206(6):88–100. (The C-3 cycle is discussed together with tracer experiments used to isolate and identify the dark reactions.)

BRILL, W. J. 1977. Biological nitrogen fixation. *Sci. Amer.* 236(6):68–81. (Explains the roles of bacteria and blue-green algae in nitrogen fixation and surveys the biochemistry of the process.)

GOVINDJEE, D., and R. GOVINDJEE. 1974. The absorption of light in photosynthesis. *Sci. Amer.* 231(6):68–82. (Presents recent information on electron flow in photosynthesis, in particular the nature of the photon trap.)

MILLER, K. B. 1979. The photosynthetic membrane. *Sci. Amer.* 241(4):102–13. (The importance of thylakoid ultrastructure in relation to the light reactions is discussed; in particular, coupling of ADP phosphorylation is described.)

PETERS, G. A. 1978. Blue-green algae and algal associations. *BioScience* 28(9):580–585. (Describes N_2 fixation by cyanobacteria, including the *Azolla* symbiosis. Other articles on N_2 fixation also are in this issue of *BioScience*.)

Postgate, J. R. 1982. *The fundamentals of nitrogen fixation*. Cambridge, U. K.: The Press Syndicate of Cambridge University. (This is a readable and not difficult treatment of N_2 fixation by a variety of microorganisms.)

TORREY, J. G. 1978. Nitrogen fixation by actinomycete-nodulated angiosperms. *BioScience* 28(9):586–592. (This interesting article discusses an N_2-fixation symbiosis not known until recently.)

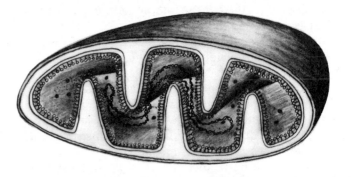

4 Respiration

good deal of the excitement in the science of biology lies in the way that discoveries by many individuals seem to come together as do pieces of a jigsaw puzzle, finally giving a complete picture of processes which at first were quite mysterious. One of the more fascinating of these puzzles is the manner in which chemical energy of food is used to do the work of cells, tissues, organs, and organisms, a process known as **cellular respiration**, or **biological oxidation**.

In past ages it was thought that vitality, or the vital aspect of life seen in living beings and their activities, was not to be compared with ordinary mechanical devices and their actions. Instead life was equated with a quality of the soul or spirit, of which respiration, or breathing, was the visible part. In fact, the word "spirit" comes from the Latin *spiro*, meaning "I breathe."

This failure to recognize a relationship between the vital functions of organisms and mechanical action is understandable. After all, nearly every ancient device, unless powered by wind or water, was driven by animal power, about which almost nothing was known. Then, during the eighteenth century, the steam engine was invented; this completely new power source stimulated new interpretations of the energy sources of animals and eventually of plants.

A steam engine is a **heat engine**. Coal, oil, gas, or wood is burned, using oxygen, and this combustion increases the kinetic energy of water vapor (steam), causing it to diffuse more rapidly. The increased diffusion pressure of the steam then is used to push a piston connected to a set of gears. In this way, the potential energy of the fuel becomes converted into mechanical energy, and work is accomplished. The plunging piston of the steam engine and its connecting shaft resemble, at least superficially, a human arm turning a crank.

Analysis by early 19th century scientists of the energy relations of a steam engine showed that fuel (wood or coal), upon burning, produced heat and released water and carbon dioxide as by-products. Because it was known that animals also took in fuel (food), produced heat, gave off water and carbon dioxide, and did work, a complete analogy between combustion in the animal body and in a steam engine was possible. The animal body could be considered a kind of heat engine. These relationships may be summarized as follows:

food + oxygen → energy + water + carbon dioxide

This is the general equation for the process of biological oxidation.

BIOLOGICAL OXIDATION AND ATP

When fuel is burned in an automobile engine, terrific power is unleashed in a tiny fraction of a second. There is an instantaneous transfer of electrons and hydrogen ions (H^+ + e) from fuel molecules to oxygen atoms, and the potential energy stored in electrons of covalent bonds of the hydrocarbon fuel is given off as heat. This raises the kinetic energy of gas molecules in the cylinders of the engine; the result is an explosive expansion of gases which pushes the pistons in much the same way steam does in a steam engine. Compared with the quiet way a cell burns fuel and does work, the auto engine is a controlled disaster! A cell is not constructed to tolerate such violence and such high temperatures. However, there is one very significant similarity between the automobile engine and the cellular "engine"—in both, energy is released when hydrogen ions and electrons are transferred from fuel to oxygen, and, in both, this energy is used to do work.

If a lump of sugar is burned in a dish, enough heat may result to run a small steam engine. If the sugar is finely powdered and ignited in a puff of air, it can burn very rapidly. When burned in a cell, however, comparatively little heat is released, and the utilization of energy occurs in a series of enzyme-controlled steps. Heat is the energy currency of auto and steam engines, but **adenosine triphosphate (ATP)** is the coinage of energy transactions of cells. This key molecule can store energy in tidy little packages—36 of them made from each molecule of glucose—to be "plugged in" wherever energy is required to perform some cellular task. As will be seen, ATP energy can be transferred to other molecules, producing conformational changes (shape changes), causing intermolecular bonds to be broken or formed anew, and even resulting in the emission of light from certain substances (firefly-tail extract lights up when ATP is added!). The story of biological oxidation is really the story of how ATP is made; not all at once in an energy explosion, but gently and in steps.

As an aid in analyzing respiration in cells, the "black box" method may again be used. In overall terms, oxygen is used to burn food in the form of glucose, here represented simply as CH_2O; the product is energy in the form of ATP, and the by-products are carbon dioxide and water:

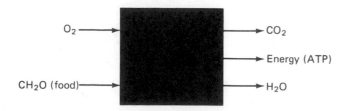

As is the case in photosynthesis, biological respiration is separable into subprocesses. The first of these is glycolysis, the essential steps of which were worked out in the early years of this century by the German chemists Gustav Embden and Otto Meyerhof (the glycolytic reaction therefore is called the **Embden-Meyerhof pathway**). The second of these is the Krebs cycle, which utilizes a product (pyruvic acid) of glycolysis as a further source of energy.

Glycolysis

In **glycolysis** a molecule of glucose is combined with two phosphates derived from ATP, then split into two molecules of the 3-carbon sugar **phosphoglyceraldehyde (PGAL)**. Two additional phosphates, this time derived from phosphoric acid, are attached to the two PGAL molecules, forming two **bisphosphoglyceraldehyde (BPGAL)** molecules. These then are oxidized to **bisphosphoglyceric acid (BPGA)** by the removal of two H^+ ions and two electrons from each DPGAL molecule.* Coincidentally, some of the energy of this oxidation is used in creating energy-rich bonds (~P) in the attached phosphates. The energy-rich phosphates then are removed in reactions with ADP that yield four ATP molecules (a process called **substrate phosphorylation**) and leave two 3-carbon molecules of pyruvic acid as a residue. The 12 reactions of glycolysis, each catalyzed by a different enzyme, are summarized in Fig. 4-1, and shown in greater detail in Fig. 1, p. 87.

ATP is the common energy-rich molecule that cells use to do work. Thus, in analyzing the efficiency of cellular processes, attention should be focused on production of ATP. Of the four molecules of ATP derived during glycolysis, two go to replace those used in the initial phosphorylation; the net production is two ATP molecules. However, the four H^+ ions and four electrons removed from the two 3-carbon sugar phosphates (2 DPGAL) are taken up by an electron-carrier molecule named **nicotinamide adenine dinucleotide (NAD)**, which is similar in action and structure to NADP, discussed in connection with photosynthesis. The two molecules of NAD, when reduced to NADH$_2$** (2 NAD + 4 H^+ + 4 e^- → 2 NADH$_2$), represent considerable potential energy and are capable of energizing the production of an additional four to six ATP molecules (two ATP fewer in eukaryotes than in prokaryotes), provided oxygen is present as a final H^+ ion and electron acceptor

---

*Hydrogen ions (H^+) and electrons (e^-) may participate in further reactions as dissociated ions and electrons or as whole hydrogen atoms.
**This symbol is a simplification. Oxidized NAD is an ion, NAD^+; it is reduced by one hydrogen atom and one electron from a second hydrogen, giving NADH + H^+.

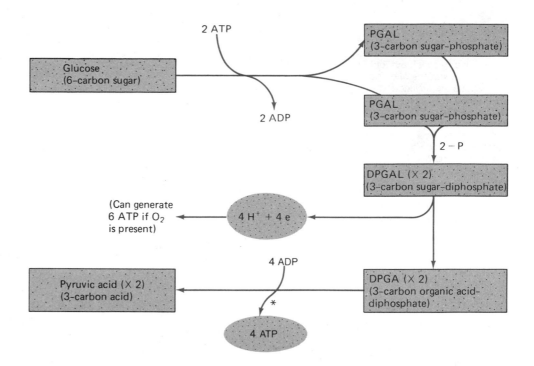

FIGURE 4-1
Summary of the reactions of glycolysis. The products are NADH₂, ATP, and pyruvic acid. Substrate phosphorylation is indicated by an asterisk. Under aerobic conditions, the pyruvic acid is the substrate for the next step, a pathway called the Krebs cycle (see Fig. 4-2). Under anaerobic conditions, pyruvic acid may be converted to alcohol or some other product of fermentation (see Fig. 4-8).

(but see discussions of fermentation and anaerobic respiration, later in this chapter). If oxygen is present, the net production of ATP in glycolysis may be six to eight molecules per glucose molecule, a total energy equivalent of about 40,000 calories per **mole** (1 molecular weight in grams). That is substantial, but not compared to the total free energy of the glucose molecule, which is around 686,000 calories per mole.

The Krebs Cycle

For a time it was not clear how the remaining energy of sugar was converted into additional available energy, but during the 1920s and 1930s, through the work of many individuals and culminating in the discoveries of H. A. Krebs, an English biochemist, the remaining pathways became clear. In overall terms, the pathways of biological oxidation can be pictured as two boxes, the first of which is glycolysis and produces molecules of pyruvic acid, the principal reactant for the second box, which is the **Krebs cycle:**

As indicated, the immediate products of the Krebs cycle include H⁺ ions and electrons; carbon dioxide is a by-product. (It should be noted that more oxygen and hydrogen atoms are produced than were present in pyruvic acid, and that water also is a reactant.) In comparing the energy output of glycolysis and of the Krebs cycle, it is seen that the yield of H⁺ ions and electrons from the Krebs cycle is five times that of glycolysis. In the presence of oxygen, the energy in these H⁺ ions and electrons can be used in the prokaryotic cell, where glycolysis and the Krebs cycle proceed in the cytoplasm, to produce a total of 38 ATP molecules, 8 of which can be attributed to glycolysis. (In eukaryotes, 2 ATP molecules are used to transport NADH₂ into the mitochondrion, the site of the Krebs cycle. Therefore the net yield here is 36 ATP molecules.)

The reactions of the Krebs cycle consist of a series of interlocking steps preceded by the **decarboxylation** of pyruvic acid, a product of glycolysis. (Decarboxylation is

the removal of a carboxyl group, $-\overset{\displaystyle O}{\overset{\|}{C}}-OH$). The re-

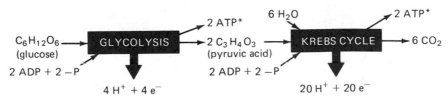

*From substrate phosphorylation

FIGURE 4-2
Simplified and condensed version of the Krebs cycle. Several steps have been left out, but fundamental oxidations and decarboxylations are included. It will be noted that for each pyruvic acid molecule produced by glycolysis and oxidized in the Krebs cycle, five reduced carrier molecules are formed (four $NADH_2$ molecules, and one $FADH_2$). These electron carriers serve to generate ATP via the electron transport system of the cell (see Fig. 4-3). Energy-rich products are in color. Substrate phosphorylation is indicated by an asterisk.

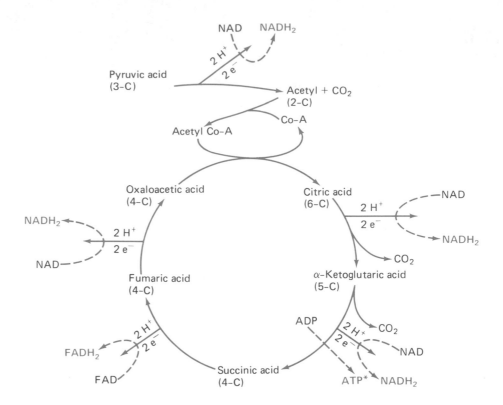

maining 2-carbon fragment (**acetyl**) is combined with a molecule of **coenzyme-A (Co-A)** to form **acetyl Co-A**. (A coenzyme is a molecule that functions as an accessory to an enzyme in certain reactions.)

The decarboxylation of pyruvic acid is an oxidation-reduction reaction and produces H^+ ions and electrons together with acetyl Co-A. Acetyl Co-A then combines with a 4-carbon organic acid, **oxaloacetic acid**. Co-A is released in this reaction, and the resulting 6-carbon molecule, **citric acid**, is progressively oxidized by two decar-

boxylations and the removal of electrons and H^+ ions until all that is left is again a molecule of oxaloacetic acid. The latter then may react with another molecule of acetyl Co-A to start the cycle again. This process, though commonly called the Krebs cycle, also is known as the **citric acid cycle** because citric acid is the starting point for the subsequent oxidations. The reactions of the Krebs cycle are summarized in Fig. 4-2, and presented in greater detail in Fig. 2, p. 88.

INTERMEDIATE MOLECULES, ENZYMES, AND PRODUCTS OF GLYCOLYSIS AND THE KREBS CYCLE

The diagrams of Figs. 4-1 and 4-2 have shown glycolysis and the Krebs cycle in abbreviated and symbolic fashion. If you wish to compare these diagrams with the more complete summary of chemical steps in these two processes, the diagrams of Figs. 1 and 2 may be of interest.

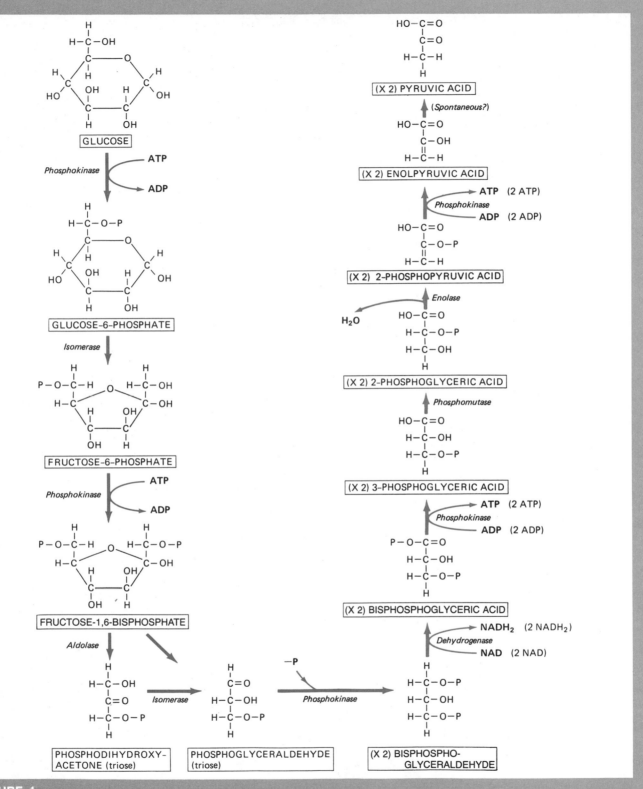

FIGURE 1
Reactions, intermediate molecules, and enzymes in glycolysis. Reactions and products from trioses to pyruvic acid are in duplicate; —P = phosphate ion. (Compare with Fig. 4-1.)

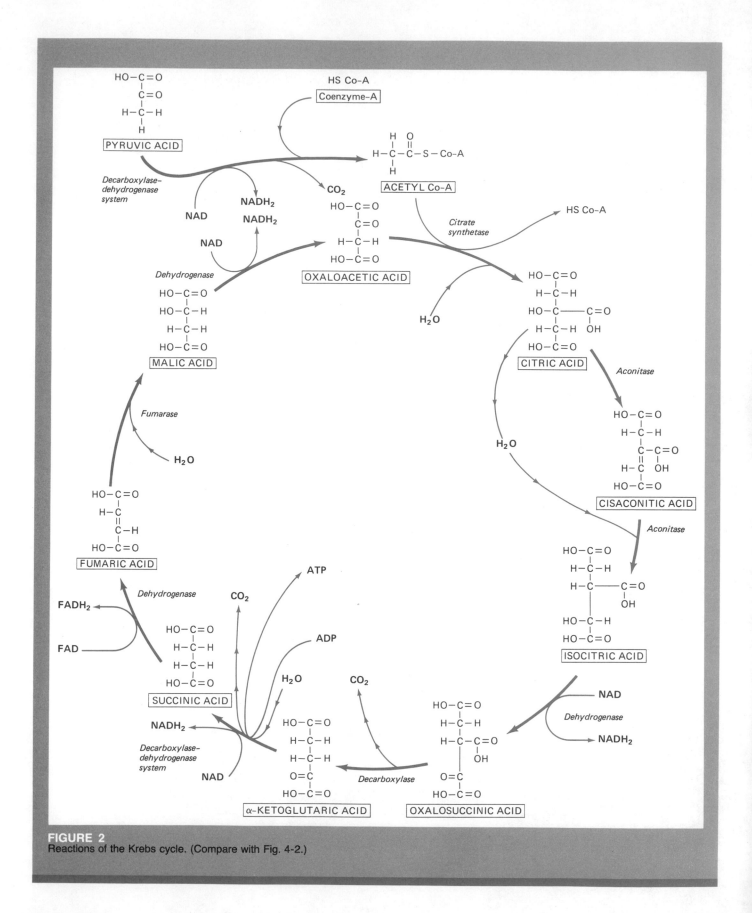

FIGURE 2
Reactions of the Krebs cycle. (Compare with Fig. 4-2.)

The Electron Transport System

In the preceding analysis of glycolysis, it can be seen that H^+ ions and electrons do not react directly with oxygen in the production of ATP, but first participate in the reduction of an acceptor molecule, NAD. Similarly, H^+ ions and electrons, derived from the Krebs cycle, are transferred to the acceptor and carrier molecules, NAD and FAD. These molecules then further transfer their H^+ ions and electrons to a system of electron acceptor and transfer molecules known as the **electron transport system (ETS)** of the cell. Here electrons move down a descending "stairway" of energy-yielding reactions, some of which supply sufficient energy to form molecules of ATP (see Fig. 4-3). Finally, at the end of the chain of electron transport molecules, the electrons and accompanying H^+ ions are combined with oxygen to form water. Oxygen therefore functions as the terminal electron acceptor for the electron transport system.

It should be noted that each $NADH_2$ molecule is capable of generating 3 ATP molecules, but $FADH_2$, which gives up electrons to **coenzyme Q (Co-Q)** of the ETS, generates 2 ATP molecules. The eight pairs of H^+ ions and electrons derived from two turns of the Krebs cycle are therefore equivalent to 22 ATP molecules.

Summary of ATP Yield   In a prokaryotic cell, 4 ATP molecules are derived from substrate phosphorylations occurring in glycolysis and the Krebs cycle. The 10 molecules of $NADH_2$ produced by oxidation of PGAL, pyruvic

acid, and Krebs cycle intermediates can give rise to 30 ATP molecules, and 4 ATP molecules are generated from 2 $FADH_2$ molecules, for a grand total of 38 ATP molecules. In a eukaryotic cell, 2 ATP molecules are required for transport of $NADH_2$ into the mitochondrion; hence the yield is 36 ATP molecules per glucose molecule.

Cytochromes   The electron transport system receives electrons, but not H^+ ions, from $NADH_2$, passing them through a series of carrier molecules including several different **cytochromes** (cell pigments), the last of which transfers them to oxygen at the end of the electron carrier chain. Hydrogen ions from $NADH_2$ and $FADH_2$ are channeled through the inner mitochondrial membrane to generate ATP. Both electrons and H^+ ions eventually combine with oxygen in a terminal oxidation, forming water.

Cytochromes were first observed in 1886, but it was not until 1925 that their role in respiration was discovered by the English biochemist David Keilin. Cytochromes are iron-containing substances composed of protein and **porphyrin,** the latter a ring-shaped molecule also present in hemoglobin and chlorophyll (compare Fig. 4-4 with Fig. 3-1 and Fig. 9-20).

The universal occurrence in cells of cytochromes, of which more than 30 kinds are known, is evidence not only of their vital importance but also of a common origin. Cytochromes seem to have appeared quite early in chemical evolution, and comparisons of their structures are helpful in deducing evolutionary pathways. For ex-

FIGURE 4-3

Simplified electron transport system. Enzymes (E) are specific dehydrogenases. S is a substrate (food) molecule. The electron carriers are NAD (or NADP), FMN (flavin mononucleotide), and several cytochromes (C). According to the chemiosmotic theory, flow of H^+ ions from outside the mitochondrial membrane to the inside provides the energy for the production of ATP.

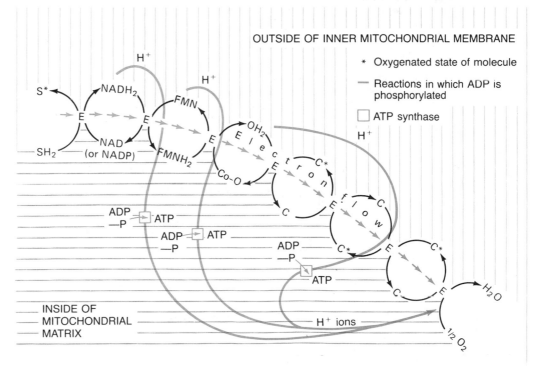

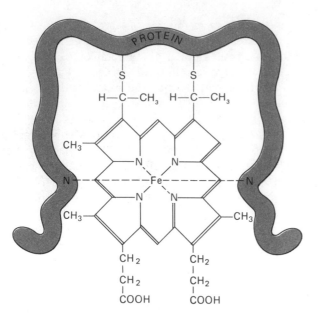

FIGURE 4-4
Cytochrome C. The protein portion of the molecule is composed of 104 amino acids and is folded such that it forms an association with the iron atom at the center of the porphyrin ring.

ample, there are more than 20 varieties of cytochrome C, a member of electron transport chains in all aerobic organisms, and each differs from the others to a greater or lesser extent in the arrangement of the amino acids of the protein component. The cytochrome C proteins of closely related organisms have very similar arrangements of amino acids; those more distantly related show greater discrepancies in amino acid sequences. Humans and chimpanzees have identical cytochrome Cs; those of humans and fish differ in about 18 amino acids, whereas humans and prokaryotes differ in the arrangement of about 40 of the amino acids in their respective cytochrome Cs. Despite these differences, all varieties of cytochrome C have a basic similarity of structure and function.

Generation of ATP For some years there was speculation about the way ATP was produced by the electron transport system, but the process remained mysterious. It was thought that unknown energy transfer molecules acted as intermediates between the electron transfer molecules of the system and ADP, but searches for such molecules have been fruitless. However, as related in Chapter 3, an Englishman named Peter Mitchell proposed that the generation of ATP resulted from channeling of H^+ ions by membranes such as those of chloroplasts and mitochondria. It has been shown that synthesis of ATP by respiration occurs only when mitochondrial membranes are intact, and that separation of positively and negatively charged particles (H^+ ions and electrons) produces differences in electrical potential and pH sufficient to create high-energy phosphate bonds. This theory, the **chemiosmotic theory,** has gained wide acceptance.

The chemiosmotic theory proposes that the inner mitochondrial membrane acts as a hydrogen-ion (proton) barrier between the intermembranal region of the mitochondrion and the mitochondrial matrix (Fig. 4-5). As electrons and protons pass through the electron transport system, protons are channeled out through the inner mitochondrial membrane. That creates an electrical potential difference (capacity to do work) between the outside (more positive) and the inside (more negative) of the inner membrane. Then, at certain points occupied by molecules of the enzyme known as **ATP synthase,** protons flow back through the inner membrane; this flow provides sufficient energy for the enzyme to make ATP (Fig. 4-3).

Summary of Biological Oxidation

The relationship between glycolysis, the Krebs cycle, and the electron transport system in a eukaryotic cell may be diagrammed as follows:

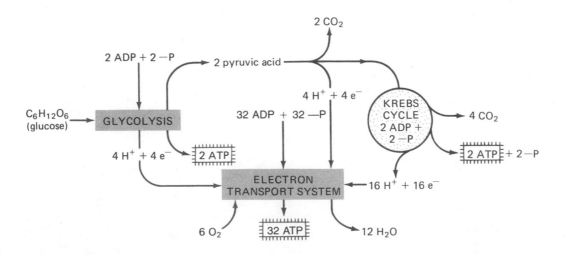

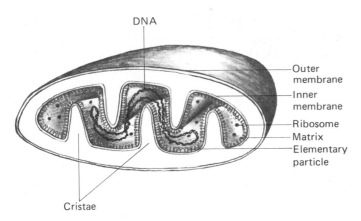

DNA

Outer membrane

Inner membrane

Ribosome

Matrix

Elementary particle

Cristae

FIGURE 4-5

Relationship between the components of a "typical" mitochondrion. (Components are not drawn to scale.)

Summing up,

$$C_6H_{12}O_6 + 6\ H_2O + 6\ O_2 \rightarrow$$
$$\text{energy (32 ATP)} + 6\ CO_2 + 12\ H_2O$$

If 2 ATP molecules produced directly in glycolysis and 2 ATP molecules provided by substrate phosphorylation in the Krebs cycle are added in, the total productivity of biological oxidation is 36 ATP molecules per glucose molecule oxidized. This may be translated into about 260,000 calories, or approximately 40 percent of the free energy of a mole of glucose. Much of the energy of glucose, therefore, is liberated as heat. Even so, the efficiency of biological oxidation is about the same as that of a diesel engine, the most efficient heat engine invented so far.

SITE OF RESPIRATION

In 1850 a then well-known German microscopist named Rudolf Albert Kölliker (1817–1905) discovered granules in **striated muscle cells.** (Striated muscle, also known as **skeletal muscle,** is composed of long cells containing transverse bands easily seen with a microscope, even at low magnification.) Quite independently, granules very much like those found by Kölliker in striated muscle cells were later observed in other kinds of cells, and in 1890 the name **mitochondria** (sing. **mitochondrion**) was given to them.

For years scientists were intrigued by the nature and possible function of mitochondria. In 1894 Richard Altmann (1852–1901), a German physiologist, discovered a stain specific for mitochondria and came to two interesting conclusions about them: (1) that they were involved in some way with oxidation-reduction reactions (because of their staining reactions), and (2) that they might, in fact, be very small organisms in their own right that had become adapted to living in cells. This was not

so far-fetched as it might seem. Mitochondria tend to be about the size and shape of bacteria, around 1 μm in diameter and 3–10 μm long; however, mitochondria are rather variable in size and shape, and some are quite unlike any known bacteria. Even so, Altmann's suggestion that mitochondria might be cellular symbionts is intriguing in terms of the current symbiotic theory of eukaryotic evolution. It will be recalled that this theory, discussed in Chapter 2, suggests that mitochondria and chloroplasts may have evolved from prokaryotic symbionts, possibly bacteria or photosynthetic prokaryotes.

The processes of biological oxidation occur in prokaryotes as well as in eukaryotes. In prokaryotes the entire cell is the site of all the reactions occurring in biological oxidation, including glycolysis, the Krebs cycle, and the electron transport system. As noted earlier, in eukaryotes the mitochondrion is the site of the Krebs cycle and the electron transport system, but glycolysis occurs in the cytoplasm.

That the sites of the eukaryotic cell's aerobic respiration are the mitochondria was proved conclusively in 1948 by A. L. Lehninger and E. P. Kennedy. Lehninger and Kennedy isolated intact mitochondria from broken-up rat liver cells, and discovered that the isolated mitochondrion can utilize pyruvic acid as a substrate and form ATP in the presence of oxygen. They also found that mitochondria contain all the enzymes and electron carriers for the Krebs cycle and the electron transport system. In addition, Lehninger and Kennedy were able to say that some of the Krebs cycle enzymes occur in the inner matrix of the mitochondrion, and others are attached to the inner membrane. Some components of the electron transport system occur as distinct units of the inner membrane surface, possibly associated with small, stalked particles. These particles, called **elementary particles,** are about one-fourth as large as the quantasomes of the chloroplast membrane. A few years ago it was suggested that the entire electron transport system was located in the elementary particle, but now it is thought more likely that the complicated system of cytochromes and other electron carriers may be found at the base of the particle, in the inner mitochondrial membrane.

The mitochondrion has both an outer and an inner membrane, and the inner membrane is pleated in shelf-like **cristae** (Fig. 4-5). The surface area of the cristae provides sites for certain of the reactions of the Krebs cycle (Fig. 14-2)

Within the inner mitochondrial membrane lies the **matrix,** which contains some of the Krebs cycle enzymes as well as ribosomes, messenger RNA, transfer RNA, and one or more loops or circles of DNA. The latter resemble circles of DNA found in the bacterial cell, except for being quite a bit smaller. Their function, as in the case of bacterial DNA, is control of protein synthesis, because mitochondria, like chloroplasts, are semi-independent of the nucleus of the cell. In that respect, mitochondria and chloroplasts are capable of growing, dividing, and replicating their DNA quite independently of the cell's own division.

Although mitochondria differ considerably in size and shape from cell to cell, they have the same function in every cell, namely, aerobic respiration and ATP production. Because the mitochondrion uses pyruvic acid as its principal substrate, it is dependent on glycolysis. As noted earlier, glycolysis does not occur in the mitochondrion itself, but in the cytoplasm of the eukaryotic cell. The outer membrane of the mitochondrion conveys pyruvic acid, ATP, and other substances by active transport processes such as those described in Chapter 2 (Fig. 4-6).

The oxidative processes of the cell utilize many other substances besides glucose as substrates. Fats are broken down into 2-carbon fragments of acetyl, entering the Krebs cycle as acetyl Co-A. Proteins first are broken down into amino acids; then the amino groups are removed in the liver and kidney. The fates of amino groups are various; they may be transferred to other molecules to make new amino acids, or they may be converted into urea and excreted. The organic acids remaining after **deamination** (removal of an amino group) also have various fates, but many of them can be converted into pyruvic acid and oxidized in the Krebs cycle.

It is difficult to generalize about the numbers of mitochondria per cell and about their physical dimensions. Cells that are highly active, such as heart muscle cells, have numerous large mitochondria. Other cells, including many plant cells, are less active and may have fewer and smaller mitochondria. Mitochondrial numbers per cell may range from only one or two up to a thousand or more. In addition, the more active mitochondria commonly have more cristae than do less active ones (Fig. 4-7).

FERMENTATION

Fermentation, as explained earlier, is a respiratory function with a low energy yield. Basically, in fermentation the pathway of glycolysis is followed through to the pro-

FIGURE 4-6
Respiratory pathways in relation to the cytoplasm and mitochondrion.

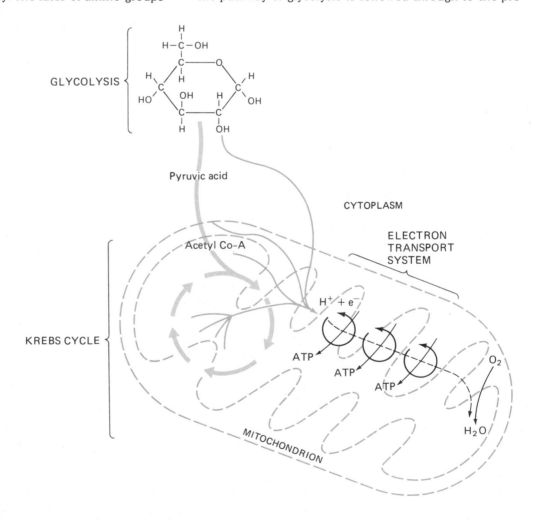

FIGURE 4-7
Electron micrograph of mitochondria (M) in heart muscle. Heart mitochondria are very active, as evidenced by their numerous cristae. (Photo courtesy David Mason)

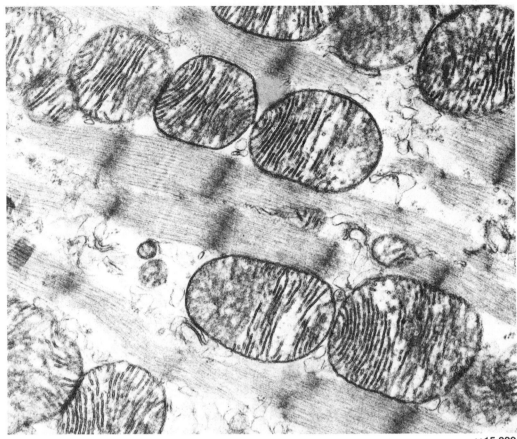

×15,000

duction of pyruvic acid. However, in the absence of free oxygen, $NADH_2$ produced in glycolysis cannot be oxidized by the electron transport system. Without some means of removing electrons and H^+ ions from $NADH_2$ the supply of NAD would soon be exhausted. In fermentation the anaerobically living cell oxidizes $NADH_2$ by a sidestep reaction that forms NAD from $NADH_2$ In this reaction, pyruvic acid itself is the electron- and H^+-ion acceptor from $NADH_2$. In some cells, pyruvic acid is decarboxylated, then reduced, forming carbon dioxide and ethyl alcohol (Fig. 4-8). Still other kinds of fermentation reactions are known. Most of these occur in bacteria and result in a variety of products such as acetic acid, acetone, or various kinds of alcohol.

The net production of ATP in fermentation is lower than that in glycolysis when the products of the latter are coupled with the products of the Krebs cycle. The reason for this is that in glycolysis, the $NADH_2$ produced is directed to the electron transport system, and there its oxidation results in the formation of additional ATP. However, because in fermentation the $NADH_2$ is used in the reduction of pyruvic acid, additional ATP is not produced. As noted previously, this step frees NAD to act again as an electron carrier, but the result is that fermentation of glucose produces only 2 ATP molecules. In contrast, the yield from glycolysis coupled with the electron transport system is 6–8 of the 36–38 ATP molecules (6 in eukaryotes, 8 in prokaryotes) produced per molecule of glucose.

In animal cells where the supply of oxygen is inadequate for completion of aerobic respiration, a fermentation occurs in which the $NADH_2$ formed in glycolysis is used to reduce pyruvic acid to **lactic acid**:

$$
\begin{array}{ccc}
OH & & OH \\
| & & | \\
C=O & & C=O \\
| & & | \\
C=O + NADH_2 \rightarrow NAD + H & H-C-OH \\
| & & | \\
H-C-H & & H-C-H \\
| & & | \\
H & & H \\
\text{Pyruvic Acid} & & \text{Lactic Acid}
\end{array}
$$

NAD then is free and can function again as an electron acceptor for the oxidation of another glucose molecule. Two ATP molecules are produced for each lactic acid

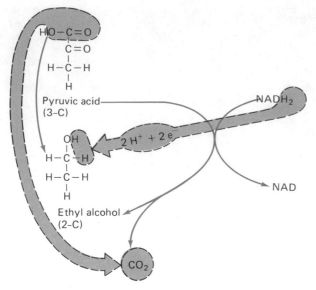

FIGURE 4-8
Steps in a fermentation resulting in the production of carbon dioxide and ethyl alcohol.

molecule produced. Under conditions of oxygen stress (e.g., running), lactic acid accumulates in muscle tissue. During rest, when more oxygen is available, lactic acid is converted back to pyruvic acid, which then may be either further oxidized or converted to glycogen and stored.

Muscle fatigue and cramps have been attributed to lactic acid accumulation, but may be due also to ATP depletion and secretion of pain-inducing molecules called **kinins**.

ANAEROBIC RESPIRATION

Anaerobic respiration and fermentation often are confused with each other because neither requires oxygen. However, fermentation is a low-ATP-yielding oxidation, whereas the ATP yield of anaerobic respiration is about the same as that of aerobic respiration. Anaerobic respiration occurs in certain bacteria, including some that can live only anaerobically **(obligate anaerobes)**. For them, oxygen may act as a poison, preventing their respiration. Anaerobic bacteria function by using some inorganic molecule other than oxygen as the final electron and H^+-ion acceptor in the electron transport system. Nitrate (NO_3^-), for example, is used as an electron acceptor by some kinds of bacteria called **denitrifying bacteria**. A typical reaction is shown diagrammatically in Fig. 4-9.

Other anaerobic pathways among microorganisms are known, and many of them are ecologically important. In general, they function in about the same way as does denitrification.

FIGURE 4-9
Summary of anaerobic respiration. (Compare with Fig. 4-6.)

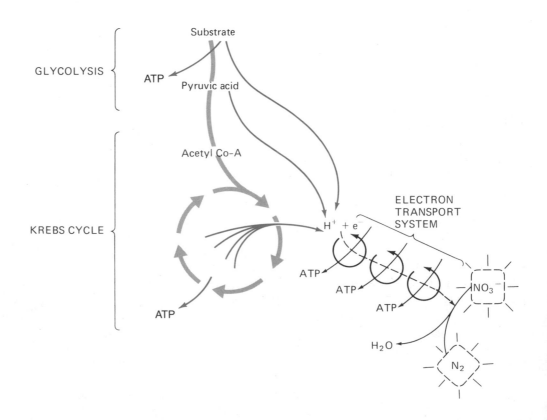

ATP AND MUSCLE

The action of muscles in doing work is an excellent example of use of the energy of ATP in the body. The conversion of ATP energy into the mechanical energy of muscle contractions occurs in the three kinds of muscle cells: striated or skeletal muscle, **visceral** or **smooth muscle,** and **heart** or **cardiac muscle.** Although differing in some aspects of cellular structure and function, all three function on the same principle, that of the sliding of internal filaments past each other.

It has been shown that part of the protein present in muscle cells can be dissolved by washing the cells in a concentrated salt solution. When this solution is diluted, two kinds of protein are precipitated: **actin** and **myosin.** Mixtures containing actin and myosin can be made to produce fine threads of combined protein, which can contract to about two-thirds of their original length when ATP is added. During this contraction, ATP is broken down to ADP and phosphate ions (—P); high-energy phosophate bonds supply the power for the process. Thus the connection between food, its oxidation via glycolysis and the Krebs cycle, production of ATP, and the use of ATP to do work is made clearer.

One problem encountered in studying muscle action is that the amounts of ATP required during work often exceed the total ATP present in the muscle cells. It turns out, however, that ATP is replaced rapidly from a store of high-energy phosphates ($\sim$P) present in molecules of **creatine phosphate** (C$\sim$P):

$$C\sim P + ADP \xrightarrow{enzyme} ATP + C \text{ (creatine)}$$

Additional ATP also is formed from two ADP molecules in the following reaction:

$$2 \text{ ADP (A—P}\sim\text{P)} \rightarrow \text{ATP (A—P}\sim\text{P}\sim\text{P)} + \textbf{AMP}$$
(AMP is adenosine monophosphate, A—P)

Later, when muscle cells are resting, supplies of creatine phosphate can be restored and AMP can be converted to ATP using additional ATP molecules from the Krebs cycle. The use of creatine phosphate and ADP to supply additional energy during rapid muscular activity is an adaptation to stress situations.

SUMMARY

Early scientists and philosophers found respiration to be unexplainable either in mechanical or in chemical terms; later on, as understanding of the principles of oxidation increased, it was recognized that respiration involves more than just the breathing of air. Basically, respiration was seen to be a type of combustion in which the energy of food became converted into another form which was capable of carrying on the work of the body. From this understanding grew an appreciation that respiration is a cellular phenomenon that occurs in all kinds of cells, those of plants as well as those of animals.

Oxidation now is understood to involve three important biochemical pathways:

1 Glycolysis, in which sugar is converted to pyruvic acid, with the production of some ATP and $NADH_2$

2 The Krebs cycle, which produces about four times as much $NADH_2$ as does glycolysis

3 The electron transport system (ETS)

The pathways are intimately related, glycolysis furnishing pyruvic acid as the substrate for the Krebs cycle, and the electron transport system oxidizing $NADH_2$ (as well as $FADH_2$) and producing ATP.

In the electron transport system, $NADH_2$ and $FADH_2$ transfer their electrons and H^+ ions to a series of carrier molecules, including several kinds of cytochromes. At intervals in this system, phosphorylations of ADP occur, so that for each pair of electrons transported through the system, three molecules of ATP are produced. The last step in the electron transport system is the transfer of electrons and H^+ ions to oxygen. This is the only point at which oxygen is directly involved in biological oxidation.

The site of glycolysis in both prokaryotic and eukaryotic cells is the cytoplasm. In prokaryotes, the Krebs cycle, if present, also occurs in the cytoplasm; in eukaryotes, the site of the Krebs cycle is the mitochondrion. The mitochondrion also is the site of the electron transport system.

If free oxygen is unavailable for biological oxidation, the cell may form ATP by fermentation, a low-energy-yielding process in which the by-products are carbon dioxide and alcohol or other organic molecules. Some kinds of anaerobic bacteria use the high-energy pathway of anaerobic respiration.

QUESTIONS FOR REVIEW AND DISCUSSION

1 In what sense is the animal body a "heat engine"? Substituting "energy" for "heat," can the analogy be made more meaningful? If so, explain.

2 According to the "black box" approach to describe biological oxidation, what are the subcompartments of the "box"? What intermediate substances are conveyed from subcompartment to subcompartment?

3 It is stated that oxygen is used in a cell to react with the products of respiration. What is meant by this, and specifically what are the products of respiration with which oxygen will react?

4 Fermentation is said to have a relatively low yield of ATP. Why? What useful by-products are produced?

5 All of the carbon in a molecule of sugar is converted to carbon dioxide in aerobic respiration. Where do the decarboxylations occur? Twelve hydrogens (electrons plus H^+ ions) also are removed from the sugar molecule during respiration. What is their fate?

6 What is meant by "oxidative phosphorylation"? What is its product? Give an example demonstrating its use in the cell.

7 What is the role of hydrogen ions (protons) in ATP production in the mitochondrion?

8 Compare fermentation, anaerobic respiration, and aerobic respiration in terms of energy yield. Why might ethyl alcohol be considered a food as well as a drug? In what organisms does anaerobic respiration occur? What is meant by "denitrification"?

9 What is the relationship between ATP and work done by muscles?

SUGGESTED READING

DeROBERTIS, E. D. P., and E. M. F. DeROBERTIS, Jr. 1981. *Essentials of cell and molecular biology.* Philadelphia: Saunders College Publishing. (A somewhat abbreviated treatment of the essentials of cell biology. Nice balance between cell structure and function.)

DYSON, R. D. 1975. *Essentials of cell biology.* Boston: Allyn and Bacon. (A comparatively difficult cell biology text with an excellent presentation of structure and functioning of chloroplasts and mitochondria. Contains a brief historical survey of respiration.)

HINKLE, P. C., and R. E. McCARTY. 1978. How cells make ATP. *Sci. Amer.* 238(3):104–23. (Presents an introductory explanation of the role of electron carriers in biological oxidations and goes on to discuss the chemiosmotic theory of ATP generation.)

LEHNINGER, A. L. 1961. How cells transform energy. *Sci. Amer.* 205(3):63–73. (In this short article, photosynthesis and respiration are compared. Quite general.)

MARGARIA, R. 1972. The sources of muscular energy. *Sci. Amer.* 226(3):84–91. (Likens muscle to an engine and discusses muscle action in terms of glycolysis and ATP.)

RACKER, E. 1968. The membrane of the mitochondrion. *Sci. Amer.* 218(2):32–39. (Relates the structure of the mitochondrion, particularly the inner membrane and the elementary particles, to oxidative phosphorylation. Nice experimental approach.)

TRIBE, M., and P. WHITTAKER. 1972. *Chloroplasts and mitochondria*. London: Edward Arnold Publishers. Available through Crane, Russak and Co., New York. (This small paperback presents a short historical survey of photosynthesis and respiration, and generally emphasizes the structural features of these organelles.)

Nucleic Acids and Genetic Control

5

How does a cell or an organism, through its reproductive processes, turn out virtually identical copies of itself? The answer is complex, but in essence, growth and reproduction have a universal basis among all living things. That basis is the **storage** of **information** in the genetic code of deoxyribonucelic acid (DNA).

The information encoded in the DNA of a cell can be likened to a central file of computer programs in a purely imaginary factory where many kinds of products are made. Whenever a particular product or part of a product is to be manufactured, that information is **retrieved,** and the program is used to make an instruction tape; the tape is then fed into a machine that stamps, welds, and polishes that particular product according to the specifications of the tape. Suppose further that the output of products is connected by computer both to the warehouse inventory stock and to the sales and distribution network of the company. Then if sales demand is high and the inventory is low, the plant will be "told" by the computer to make more of that product. Conversely, when the sales demand is off and the inventory is high, a feedback system will be in effect and production will be shut down for a time.

Imagine, in addition, that the factory also contains computerized information not only for the manufacture and distribution of products but also for the repair of breakdowns of machinery in the assembly lines. Then if a machine breaks, a pipe bursts, or a fuse blows, the correct information will be sent out for the immediate repair of the damage.

Every one of the factory processes just mentioned has its parallel in the cell. However, the analogy between a factory and a cell is useful only as long as it is remembered that control of cellular processes is vastly more complicated than any set of industrial processes.

DNA AND THE GENE

A cell is composed of many kinds of molecules, including carbohydrates, proteins, nucleic acids, and lipids. How are these molecules made? DNA is the master code that controls the biosynthesis of proteins, one class of which contains the enzymes. Because enzymes catalyze reactions in which many macromolecules other than proteins are made, it follows that enzymes are key molecules in the development and maintenance of cells. For example, cellulose, the main constituent of plant cell walls, is synthesized from glucose by the enzyme cellulase. The phospholipids, which are key elements in all cell membranes, also are synthesized by enzymes. However, other structural elements of cells, such as microtubules and microfilaments, are themselves proteins, and like enzymes are assembled under the direction of DNA and ribonucleic acid (RNA). It should be remembered, therefore, that although enzyme proteins can catalyze the synthesis of all other nonprotein molecules found in cells, only nucleic acids (DNA and RNA) can direct the synthesis of proteins.

DNA was identified in chromosomes years ago, but its function was not understood for some time. Although chromosomes contain both proteins and DNA, for a number of years geneticists believed that proteins composed the **genes** (the units of genetic information). It was not until 1944 that the true role of DNA was revealed. In that year three American geneticists, O. T. Avery, C. M. MacLeod, and M. McCarty, reported their use of two forms of the pneumonia bacterium *Streptococcus pneumoniae*, in a series of classic experiments (Fig. 5-1). One bacterial form produces smooth, glistening colonies on surfaces of culture media and therefore is called the **smooth (S) form**. Its glistening appearance is due to a

slimy coat, called the **capsule**, surrounding each cell. Cells of the second form, the **rough (R) form**, are not encapsulated, and produce rough, nonglistening colonies. It is the presence of a capsule that gives the pneumonia bacterium its ability to produce pneumonia. This capacity to infect another organism and to produce a disease is called **pathogenicity**. The S form of *Streptococcus pneumoniae* is therefore **pathogenic**. The R form does not cause pneumonia and is **nonpathogenic**.

Avery, MacLeod, and McCarty extracted and purified DNA from S-form bacteria (see Fig. 5-1). No trace of living cells was associated with this DNA, but when it was mixed with a culture of R-form bacteria and the bacteria were then injected into mice, the mice died of pneumonia. In the next step, pneumonia bacteria were isolated from the dead mice and cultured. The bacteria recovered from the dead mice turned out to be of the S form. The conclusion was that DNA extracted from S-form cells of *Streptococcus pneumoniae* carried the information for the production of a capsule by a transformed R-form cell.

The introduction of purified DNA into bacteria with a heritable change in the genetic nature of the recipient

FIGURE 5-1

Summary of the work of Avery, MacLeod, and McCarty with the pneumonia bacterium *Streptococcus pneumoniae*. The pathogenic, encapsulated smooth (S) form yields a purified DNA, which when mixed with cells of the nonpathogenic, nonencapsulated rough (R) form, transforms the latter into the pathogenic smooth (S) form.

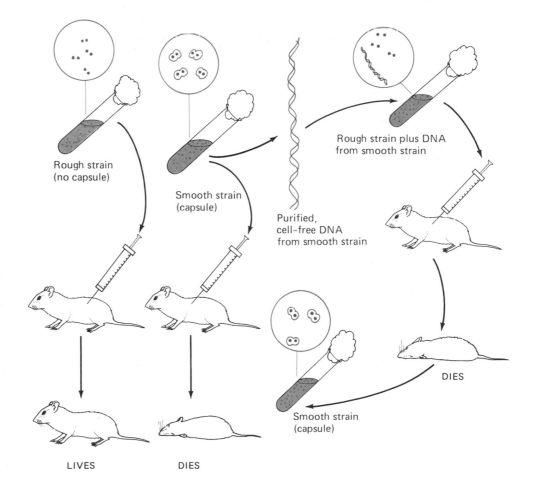

Rough strain (no capsule)

Smooth strain (capsule)

Purified, cell-free DNA from smooth strain

Rough strain plus DNA from smooth strain

Smooth strain (capsule)

LIVES

DIES

DIES

cells is called **transformation**. Therefore, on the basis of transformation experiments, it was recognized that genes are composed of DNA. This conclusion has been verified many times by other kinds of experiments.

What Do Genes Do?

The first experimental evidence of the correspondence of genes to specific molecular characteristics of cells came as a result of experiments with the pink bread mold *Neurospora crassa*. In 1941 George Beadle and Edward Tatum, two Americans at the University of Chicago, announced that certain mutant forms of *Neurospora* could not live and reproduce on simple nutrient media alone, but required additional amino acids or vitamins. These nutritionally dependent types reproduced themselves without reverting to the **wild type,** and consequently were called **biochemical mutants.** (A mutant is any individual or group of individuals differing genetically from the common or natural form, which is called the wild type.) Although they looked like ordinary *Neurospora*, they differed in their biochemical requirements.

When Beadle and Tatum analyzed their biochemical mutants, they discovered that in each case a particular enzyme of a "team" of enzymes required for the synthesis of a particular amino acid was missing (Fig. 5-2). (Although it is now known that DNA and RNA, as well as amino acids and enzymes, are required to make a specific protein, amino acid synthesis requires enzymes but not DNA and RNA. Beadle and Tatum's work did not therefore require exact knowledge of the way DNA functions.) For example, ten reactions, each catalyzed by a different enzyme, are required to make the amino acid histidine. If one of the enzymes is missing in the cytoplasm of *Neurospora*, histidine will not be made, and the mold will die unless histidine is supplied in the culture medium. A corollary is that the missing enzyme can be identified by feeding the mutant mold the intermediate compounds of the histidine pathway. Those compounds, coming after the missing step in the pathway, can substitute for histidine and enable the mold to survive. Further screening of the intermediate compounds can be done to identify the specific enzyme responsible for the failure to make histidine.

A number of amino-acid-requiring mutants of *Neurospora* have been discovered, and in each a correlation has been found between the ability to make a particular amino acid and the absence of a specific enzyme. The phrase "one gene, one enzyme" describes this relationship between genes and proteins. More recently, it has

FIGURE 5-2
Summary of the relationships between genes and enzymes in biochemical mutants of *Neurospora crassa*. The sequence of steps in demonstrating gene-to-enzyme equivalence is indicated at **A, B, C,** and **D.**

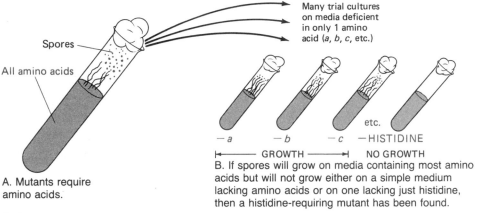

A. Mutants require amino acids.

B. If spores will grow on media containing most amino acids but will not grow either on a simple medium lacking amino acids or on one lacking just histidine, then a histidine-requiring mutant has been found.

C. Synthesis of histidine in the cell occurs in 10 steps, and there are 9 histidine precursor molecules. Each step is catalyzed by a different enzyme.

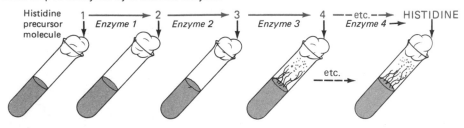

D. If the mutant strain will grow on histidine precursor 4 and all precursors that follow in histidine synthesis, but not on precursor 3 or its predecessors, then the mutant gene is that gene normally controlling production of a single enzyme (Enz. 3), the one that makes precursor 4 from precursor 3.

been shown that some proteins, such as antibodies, are composed of two or more **polypeptides** (long chains of amino acids), each dependent upon a different gene for its synthesis. Therefore the phrase should be modified to "one gene, one polypeptide." In the ensuing discussion we will use polypeptide in preference to protein, but it should be understood that in many, perhaps most, cases the polypeptides are proteins.

Nucleic Acids

Nucleic acids are of vital importance in living organisms; these acids include DNA and several kinds of RNA.

Nucleotides, the constituents of DNA and RNA, are composed of one or more phosphates, one of two 5-carbon sugars (ribose or deoxyribose), and either a purine or a pyrimidine (Fig. 5-3). Although nucleotides are linked together in chains in RNA and DNA, they also occur singly in cells, either as reserve materials for the construction of nucleic acids or as functional entities in their own right. One of the most important of the latter is ATP, which serves as an energy storage and transfer molecule in cells (Fig. 1-21).

DNA DNA constitutes the genetic instructions for determining production in a cell of polypeptides, which in turn are the important structural and functional molecules of the cell. These genetic instructions are in the

PURINES

PYRIMIDINES

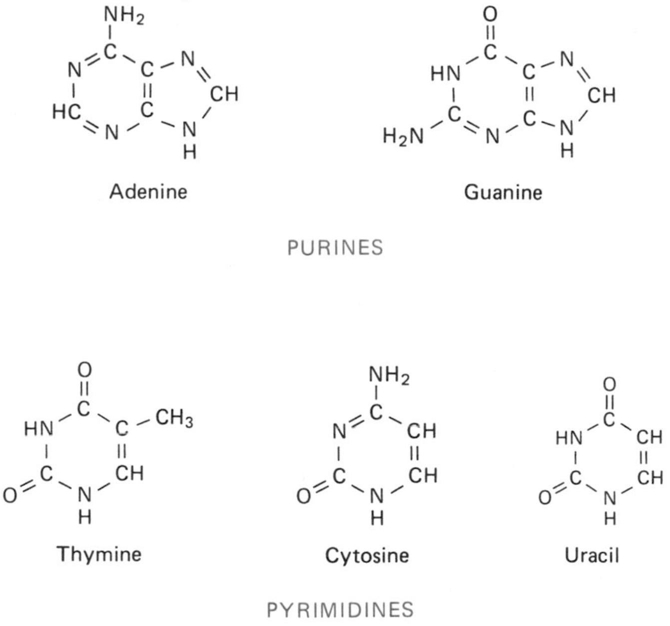

FIGURE 5-3
The five organic bases composing DNA and ribonucleic acid (RNA). Adenine, guanine, cytosine, and thymine are present in DNA; adenine, guanine, cytosine, and uracil are present in RNA.

form of a **genetic code**. The code consists of specific groupings of three subunits called **nucleotides**, with each such **triplet** corresponding to the code for one amino acid. Thus a chain of a hundred or more nucleotide triplets would comprise the genetic code for the arrangement of the many amino acids in a polypeptide. DNA is composed of two intertwined chains of nucleotides, the **double helix**. The two chains of the double helix are held together in zipperlike fashion by hydrogen bonds between opposing pairs of nucleotides (as shown in Fig. 5-4). The double helix model of DNA, announced in 1953, was the result of the work of a number of scientists, with most credit being given to Nobel Prize winners James Watson, Francis Crick, and Maurice Wilkins, who assembled the evidence and devised the model.

RNA There are several kinds of RNA in cells. Each has a specific function relating to carrying out the instructions in the genetic code of DNA. Each consists of chains of nucleotides but, unlike DNA, is single stranded (Fig. 5-4). One RNA, called **messenger RNA (mRNA)**, is a copy of a portion of a strand of DNA corresponding to the code for one polypeptide. Another RNA, **transfer RNA (tRNA)**, "recognizes" individual amino acids in the cell and conveys them to the correct place along a strand of messenger RNA, where they are joined to adjacent amino acids to make a polypeptide. The site of this last operation is the **ribosome**.

THE GENETIC CODE

Many significant advances in understanding the nature and function of DNA have been made since the discovery of transformation and the nature of the gene. The most sensational of these is the concept of the double helix structure of the DNA macromolecule, which is important for at least three reasons: (1) it allowed an explanation of replication, (2) it allowed determination of codes, and (3) it allowed insight into transcription.

DNA Replication

The double nature of DNA provided an explanation for the exact replication of each molecule of DNA by **base pairing**. Base pairing occurs after each DNA strand of the double helix separates from its partner (see Fig. 5-5). This permits a new strand to be formed opposite each original strand by the matching of the original **nucleotide bases** with new, complementary nucleotide bases. (Remember that each nucleotide of DNA contains one of four bases: **adenine** (A), **guanine** (G), **cytosine** (C), or **thymine** (T).) Adenine always pairs with thymine, and guanine always pairs with cytosine; this constitutes base

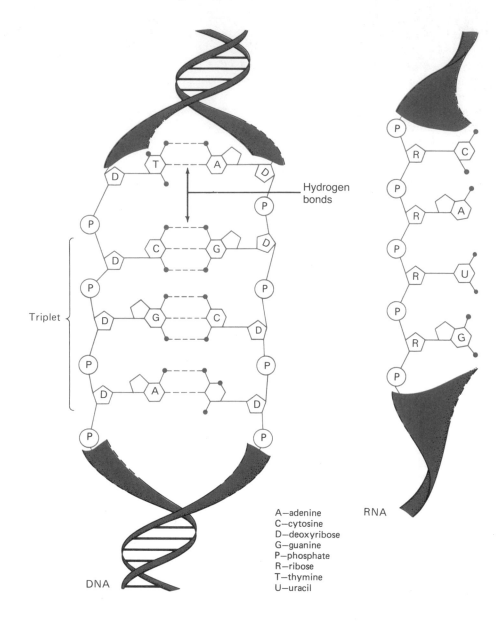

FIGURE 5-4
Structure of DNA and RNA. RNA resembles one strand of DNA except that uracil replaces thymine. Notice that the DNA "code" of base sequences runs in reverse order in the two strands (e.g., T–C–G–A vs. A–G–C–T).

Hydrogen bonds

Triplet

DNA

RNA

A—adenine
C—cytosine
D—deoxyribose
G—guanine
P—phosphate
R—ribose
T—thymine
U—uracil

pairing and is the secret of DNA replication. The four nucleotide bases are derived from nucleotide triphosphates: **deoxyadenosine triphosphate, deoxyguanosine triphosphate, deoxycytosine triphosphate,** and **deoxythymidine triphosphate.** In the presence of the enzyme **DNA polymerase,** nucleotide portions (nucleotide monophosphates) **dAMP, dGMP, dCMP,** and **dTMP** are joined together, leaving a residue of phosphate ions (Fig. 5-5). Although the exact relationship between DNA structure and chromosome structure is unknown, the DNA helix is thought to extend the length of the chromosome. As a result, new chromosome strands can be produced by separation of DNA strands of a double helix and subse-

quent base pairing to produce two new chromosome strands.

Code "Words"

The DNA molecule enabled other workers to devise a "dictionary of code words" which the cell uses in its genetic code. The genetic coding required to produce one polypeptide is a **gene.** The part of a DNA double helix corresponding to one gene consists of a sequence of many nucleotides. Each letter in the alphabet of the code consists of a nucleotide, and three nucleotides, or a triplet, corresponds to three complementary nucleo-

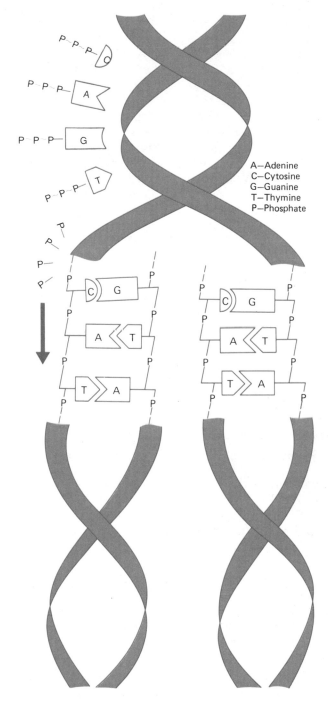

FIGURE 5-5
Replication of DNA by base pairing. (Compare with Fig. 5-4.)

A—Adenine
C—Cytosine
G—Guanine
T—Thymine
P—Phosphate

TABLE 5-1
Genetic code of messenger RNA

First Two Bases	Third Base	Amino Acid Specified
GC	A, C, G, or U	Alanine
AG	A or G	Arginine
CG	A, C, G, or U	Arginine
AA	C or U	Asparagine
GA	C or U	Aspartic acid
UG	C or U	Cysteine
CA	A or G	Glutamine
GA	A or G	Glutamic acid
GG	A, C, G, or U	Glycine
CA	C or U	Histidine
AU	A, C, or U	Isoleucine
CU	A, C, G, or U	Leucine
UU	A or G	Leucine
AA	A or G	Lysine
AU	G	Methionine ("start")
UU	C or U	Phenylalanine
CC	A, C, G, or U	Proline
AG	C or U	Serine
UC	A, C, G, or U	Serine
AC	A, C, G, or U	Threonine
UG	G	Tryptophan
UA	C or U	Tyrosine
GU	A, C, G, or U	Valine

Note: With four different nucleotides (A, C, G, and U), 64 different triplets of nucleotides are possible, coding for 64 different "words," or codons. However, there are only 20 amino acids, and some codons specify the same amino acid. This eliminates the problem of some triplets (codons) coding for nothing and also tends to reduce the effects of codon mutation. One codon, AUG, is the "start word" initiating the genetic message (and is also the codon for methionine); three others, UAA, UAG, and UGA, are "stop words," ending a message.

Transcription—Messenger DNA

Understanding the nature of DNA provided insight into the nature of the "genetic message" required for the production of a protein on a ribosome. Messenger RNA has essentially the same structure as a single strand of DNA, except that **uracil** (U) is substituted for thymine and ribose for deoxyribose (see Figs. 5-4 and 5-6).

An mRNA message is generated when the strands of the DNA helix become separated, and one strand functions as a pattern, or **template;** then the enzyme **RNA polymerase** copies the nucleotide sequence by base pairing (see Fig. 5-6). The copying of DNA into mRNA is called **transcription.**

Each triplet of nucleotides in mRNA corresponds to a specific amino acid of the 20 found in polypeptides. There are more than enough combinations of nucleotides to code for these 20 amino acids; in fact, there is **synonymy** in the code (i.e., some amino acids are coded

tides of messenger RNA, which in turn corresponds to one amino acid of the polypeptide. The actual situation is a bit more complicated than this, because more than one triplet of nucleotides may code for an amino acid. However, this does not detract from the basic concept that nucleotide triplets code for amino acids (Table 5-1).

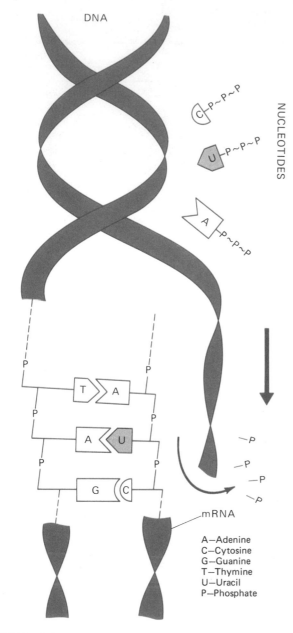

DNA

NUCLEOTIDES

C—P~P~P

U—P~P~P

A—P~P~P

P

P

T > A

P

P

A < U

P

P

G C

~P

~P

~P

~P

mRNA

A—Adenine
C—Cytosine
G—Guanine
T—Thymine
U—Uracil
P—Phosphate

FIGURE 5-6
Transcription of DNA into mRNA.

for by more than one triplet of nucleotides). In mRNA, for example, phenylalanine is coded for by UUU (uridine–uridine–uridine) and UUC (uridine–uridine–cytidine), and histidine by CAU (cytidine–adenine–uridine) and CAC (cytidine–adenine–cytidine), whereas methionine is coded for only by AUG (adenine–uridine–guanidine). The important thing, however, is that no specific triplet codes for more than one amino acid (see Table 5-1).

Prior to transcription of a gene into mRNA, RNA polymerase binds to a **promotor** region of DNA near the beginning of a coding sequence of a gene. The promotor enables RNA polymerase to move along the DNA to an **initiation site** and transcribe the code into mRNA. In addition, four ribonucleotide monophosphates are required: **adenosine triphosphate (ATP)**, **guanosine triphosphate (GTP)**, **cytosine triphosphate (CTP)**, and **uridine triphosphate (GTP)**. They are assembled by the enzyme RNA polymerase (Fig. 5-6). Transcription proceeds to a **terminator site** at the end of the gene, at which point the mRNA is released from the complex of RNA polymerase and DNA. Subsequently, mRNA becomes attached to a ribosome, where its coding sequence becomes translated into the structure of a polypeptide.

Translation

The copying of the coding sequence of nucleotides of a molecule of mRNA into the amino acid sequence in a polypeptide molecule is called **translation**. Translation depends on interactions between several cytoplasmic components, including ribosomes, upon which the assembly of polypeptides takes place by formation of peptide bonds between amino acids.

Ribosomes and Ribosomal RNA Ribosomes are cytoplasmic granules about 25 nm in their largest dimension (a bit more than twice the width of the plasma membrane of the cell). Although quite small, they contain 100–150 proteins, as well as three strands of **ribosomal RNA (rRNA)**. (In eukaryotes, ribosomal RNA is synthesized in the nucleolus, where the ribosome also is assembled; in prokaryotes, ribosomes are assembled in the cytoplasm.)

A ribosome is composed of two subunits, one twice as large as the other and each containing rRNA (Fig. 5-8). At the start of translation, mRNA binds to the smaller ribosomal subunit. Energy from guanosine triphosphate (GTP) is required for this and for the subsequent binding of amino acids.

Translation of the message of mRNA into the amino acid sequence of a polypeptide is accomplished with the aid of ribosomes, and it is noteworthy that *any* ribosome may act in the translation of *any* mRNA into *any* polypeptide. It should be remembered, then, that the program for the individuality of each polypeptide resides in the sequence of nucleotides of an mRNA and not in the specificity of a ribosome.

Lack of ribosomal individuality is a very efficient characteristic. Returning to the factory analogy, the ribosome could be compared to a single machine which could be fed a computer tape on Monday to make electric coffee percolators and fed a different tape on Tuesday to make toasters.

Transfer RNA   The amino acids, which are assembled according to the coding sequence of mRNA, are produced enzymatically in the cytoplasm of a cell by processes like those described earlier for histidine. Up to 20 different kinds of amino acids are required for production of a polypeptide; these must be identified and collected for transport to mRNA prior to or during polypeptide synthesis. This is accomplished by the transfer RNAs of the cell (Fig. 5-7). Each tRNA is a comparatively short single strand in which one nucleotide triplet, the **anticodon**, functions to bind temporarily to a complementary nucleotide triplet, the **codon**, of mRNA. At another site on a tRNA is the amino acid recognition and binding site. Because some amino acids are coded for by more than one nucleotide triplet (see Table 5-1), there are more than 20 kinds of tRNAs involved in the transport of the 20 kinds of protein amino acids in the cell.

The specificity of a molecule of tRNA for an amino acid does not reside in either the amino acid or the tRNA, but in the structure of a specific activating enzyme which selectively binds them together. A covalent bond is formed between the carboxyl group of the amino acid and a terminal group at one end of the tRNA. The energy required for forming this bond is obtained from ATP. When tRNAs acquire their amino acids, they move to ribosomes, where they are bonded temporarily to codons of mRNA. The bond between the tRNA and its amino acid then is broken, and a peptide bond is formed between that amino acid and an adjacent one.

Polypeptide Synthesis   A ribosome has two tRNA receptor sites, and each amino acid–tRNA (AA–tRNA) in turn moves into site 1 and binds to the codon of the mRNA attached to the ribosome (Fig. 5-8). Therefore, at the beginning of translation of an mRNA, the codon of the mRNA corresponding to the first amino acid of the polypeptide occupies site 1 of the ribosome. (Transcription always starts with the amino group–terminal end of a polypeptide and ends with the carboxyl group–terminal end. Refer to Fig. 1-26 for the orientation).

The binding of the anticodon of the first AA–tRNA to the corresponding codon of mRNA (and of all following AA–tRNAs to subsequent codons) requires the presence of a complex of binding agents, including guanosine triphosphate (GTP), whose energy relationships in bond formation resemble those of ATP, and a peptide-bond-forming enzyme, which is one of the proteins of the ribosome itself. Following its attachment to the ribosome in site 1, the first AA–tRNA is shifted to site 2 of the ribosome, and the next AA–tRNA in the sequence moves into the now-vacant site 1. A peptide bond then is made between the two adjacent amino acids, and the first tRNA is released from its amino acid and the codon of mRNA, leaving the tRNA free to transport another amino acid. Other amino acids are added to the growing polypeptide in the same manner, the mRNA being pushed or pulled through the ribosome until a release signal in the form of a terminal nucleotide triplet is reached. The ribosome then separates from the mRNA and from the complete polypeptide. This sequence of events occurs rapidly; it takes only about 10 seconds for a polypeptide of average length (about 300 amino acids) to form. Moreover, several ribosomes may translate one mRNA at the same time, thus forming a kind of assembly line capable of producing an equivalent number of identical polypeptides. Such a strand of mRNA with its several attached ribosomes is called a **polyribosome**, or polysome (Fig. 5-8B).

Gene Mutation

The discovery of Beadle and Tatum that gene mutation could affect a characteristic by modifying a single enzyme in a team of enzymes and making an entire biochemical pathway inoperative was of great importance in defining the nature of mutations. Although this work related specifically to enzyme synthesis, there are many examples in which the synthesis of proteins other than enzymes is controlled by genes.

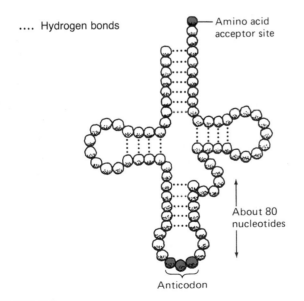

.... Hydrogen bonds

Amino acid acceptor site

About 80 nucleotides

Anticodon

FIGURE 5-7
Transfer RNA (tRNA). Transfer RNAs of the cell are composed of approximately 80 nucleotides and have a "cloverleaf" shape.

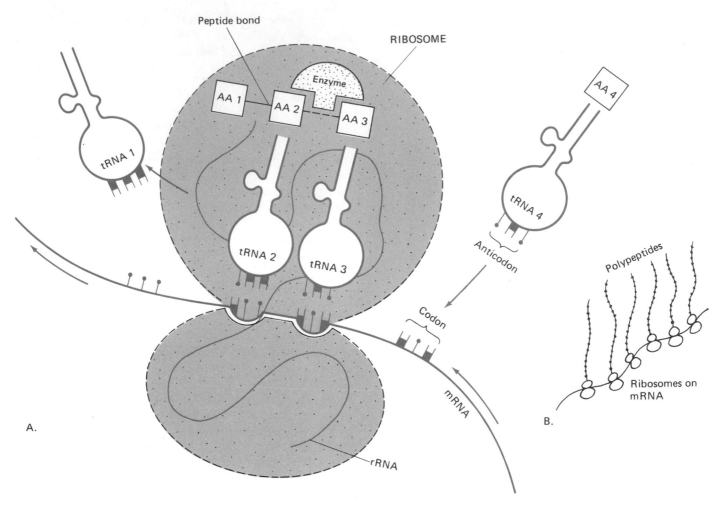

FIGURE 5-8
Relationship between ribosomal structure, tRNA attachment sites, synthesis of peptide bonds, and production of polypeptides. **A.** Relationship between a ribosome and tRNA. **B.** Polyribosome.

Sickle-cell anemia is a disease caused by a structural defect of red blood cells; the cells tend to shrivel up and become sickle-shaped and have a reduced capacity to transport oxygen (Fig. 5-9). This condition had been traced to a single nucleotide triplet in the gene that carries the code for the protein hemoglobin. The result of this change (mutation) in the gene is that the hemoglobin in sickle cells (**hemoglobin-S**) contains at one position the amino acid valine instead of glutamic acid. Here, the change of one nucletide triplet in a gene produces a serious structural and metabolic disorder.

On the basis of considerable evidence, it is thought that mutant genes are commonly the consequence of minor alterations in certain nucleotides. However, such changes do not always cause protein dysfunctions. Enzymes, for example, have active sites composed of a sequence of amino acids, any change in which will result

in a nonfunctioning or malfunctioning enzyme. However, in other regions of the enzyme, it does not seem to matter as much if the sequence of amino acids differs slightly. The result of such minor differences in a gene coding for an enzyme is that a series of enzymes called **isozymes,** each having essentially the same function, can exist in an organism. Sometimes isozymes are used to trace evolutionary relationships between species.

REPRESSION AND DEPRESSION OF GENES

Biologists studying the development and functioning of organisms long have been interested in the way genes are expressed at critical times. For instance, what determines the timing of the origin and outgrowths of limbs in a tadpole—what controls the timing for the synthesis

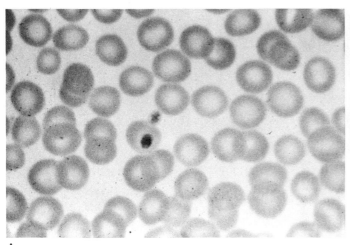
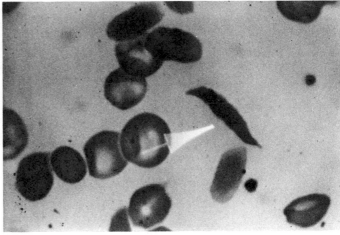

A.   B.

FIGURE 5-9
Sickle-cell anemia. **A.** Normal red blood cells. **B.** Sickle cell (to right of pointer). (Photos courtesy Center for Disease Control)

of enzymes relating to the breakdown and absorption of cells as the tail disappears when a tadpole becomes a frog? At a simpler level, what causes a bacterial cell to stop making a certain enzyme when the substance upon which the enzyme acts is depleted? Answers to questions such as these began to be worked out when the nature of the genetic code and its translation into enzymes became clarified.

Returning to the factory analogy in which computers were used to control production in relation to supply and demand, a parallel may be seen in the control of gene activity in cells. The colon bacterium *Escherichia coli* (abbreviated *E. coli*) is a favorite organism in modern genetics. It is easily cultured, lends itself to experimental genetics because many mutants are known, is usually nonpathogenic, and reproduces many times more rapidly than the fruitfly (in less than an hour, compared to about two weeks). A pair of French geneticists, F. Jacob and J. Monod, received the Nobel Prize in 1965 for their studies of the control of production of specific enzymes in cells of *E. coli*. In particular, the control of the synthesis of the enzyme **β-galactosidase** was studied. First, it was found that the cell does not produce much of this enzyme unless the sugar **lactose** is present. Only when lactose is present does the cell begin making β-galactosidase, the enzyme required for breaking down lactose into **galactose** and **glucose** (lactose is a disaccharide, composed of two hexoses). Enzymes that are made in response to the presence of a substrate molecule are called **inducible enzymes.**

Jacob and Monod found that a genetic mechanism alternately starts up or shuts down the transcription of the genes coding for β-galactosidase and two other associated enzymes (the latter catalyze the transport of lactose and its products into and within the cell). The mechanism involves two loci on the long circle of DNA that in *E. coli* sometimes is called a "chromosome." At one locus is a **regulator gene,** having the function of making an mRNA which is translated upon ribosomes into a small protein molecule called the **repressor.** The repressor has structural specificity that enables it to bind a molecule of lactose to itself. In the absence of lactose, however, the repressor instead binds to a second, **operator,** locus and inactivates transcription. But, when lactose is present, the operator is not inactivated by the repressor and functions as the starting point of the transcription of a "team" of genes that code for a long molecule of mRNA capable of producing several polypeptides. (The operator plus the "team" of genes constitutes the **operon.**) Under such circumstances, the operon is said to be **derepressed** and produces the mRNA which in turn codes for the "team" of enzymes, including β-galactosidase, required for the transport and hydrolysis of lactose (refer to Fig. 5-10 for a diagrammatic version of this process).

When lactose is absent from a medium in which *E. coli* is being cultured, it would not be economical in terms of energy for the cells to continue making β-galactosidase, and, as a matter of fact, the cells do not do so. Under this circumstance, the repressor will not become complexed with lactose but will bind instead to the operator. This inactivates the operon, making it incapable of transcription, in which case it is said to be **repressed.**

FIGURE 5-10
Repression or derepression of
the operon controlling synthesis
of β-galactosidase and associated enzymes. This operon is
called the **lac operon.**

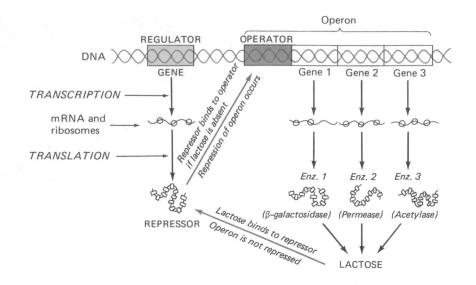

SPLIT GENES

One of the more remarkable recent discoveries in molecular genetics is that most genes of eukaryotic organisms are interrupted by noncoding base sequences. The implications of split genes are far-reaching. For one thing, eukaryotic organisms produce a large number of proteins (estimated to be between 30,000 and 150,000 for humans). Therefore, genes in pieces may contribute to the potential for variation in eukaryotes.

Analysis of eukaryotic genes shows that all except the histone genes are split. (Histones are basic proteins which stabilize the structure of nucleosomes in chromatin; Chapter 6.) The segments of split genes are called **exons** for expressed regions and **introns** for intervening sequences. As would be expected, split genes have one more exon than introns. Split genes vary in their complexity:

Globin gene	3 exons	2 introns
Ovalbumin gene	8 exons	7 introns
Conalbumin gene	17 exons	16 introns

In the mouse, the gene for one enzyme, DHFR (for dihydrofolate reductase), has six exons of about 2,000 base pairs each, and five introns that are more than 29,000 base pairs each. Thus, the DHFR gene has 12,000 base pairs of coding sequence and about 150,000 base pairs of noncoding sequence.

Two major ideas have been proposed for the functions of introns and exons:

1 Exons could combine in different ways through genetic rearrangement, thereby speeding up the rate of evolution (by a factor of 10^6 to 10^8 by some estimates). This is called exon shuffling. The exons specify functional regions or domains of proteins; joining different domains of proteins would be a much greater and faster change than the accumulation of many small or point mutations. Exon shuffling could be important in evolution of eukaryotes.

Control of Transcription in Eukaryotic Cells

Many processes occurring in eukaryotic cells require the "turning on" or "shutting off" of transcription. However, because the arrangement of DNA in eukaryotic chromosomes differs from that in prokaryotes, it is not known whether the Jacob-Monod model is applicable. At least some eukaryotic DNA is complexed with **histone** proteins and may be inaccessible to RNA polymerase. It is thought, therefore, that transcription of eukaryotic genes involves the alternate exposing and covering up of sections of DNA. Although no one knows how this occurs, it is suspected that histone-associated genes are activated or inactivated in larger gene combinations than is the case with bacterial genes and their repressors and derepressors. This could account for the greater complexity of cell processes in eukaryotes, where cells of the same origin differ markedly in their activities and development.

Lengthy mRNAs

Earlier in this chapter, in explaining the synthesis of a polypeptide by mRNA, tRNAs, and a ribosome, mRNA was considered to be equivalent to a single gene. As noted in the synthesis of β-galactosidase, however, some mRNAs contain the coding sequences of a series of genes, perhaps as many as the 10 enzymes required for the synthesis of histidine. It follows, therefore, that mRNA is more complex than just a series of nucleotide triplets coding for a chain of amino acids. In fact, a strand of mRNA may contain a start sequence composed of 100 or so nucleotides that function only as a starting

2   Splicing together of slightly different selections of exons would provide different gene products (proteins). Such a system is seen with the B-lymphocytes which are antibody-producing cells. As a B-lymphocyte differentiates, it produces specific antibody molecules (Chapter 15), which are anchored in its surface membrane: the tail is embedded in the membrane and the combining sites project into the extracellular fluid. When a foreign substance enters and is recognized by and bound by the combining site of a specific B-lymphocyte, the cell progresses to the next stage of differentiation: production of large numbers of antibody molecules that are released into the bloodstream. The tail of the anbibody must change so that it no longer will anchor in the cellular membrane. However, the rest of the antibody must remain the same so that the specificity for the foreign substances persists. This is accomplished by a switch between two exons encoding alternative tails for the antibody: anchored and not anchored. Here the significance of the genetic rearrangement is at the organismal level.

  Whatever the mechanism of aligning the exon and intron pieces into a gene, expression of a split gene involves copying the entire sequence from first exon to last exon into RNA. This produces a long messenger RNA precursor; then the RNA segments copied from exons are spliced together in the nucleus, and the RNA segments copied from introns are cut out and "thrown away." The messenger RNA, which is just exon sequences spliced together, crosses the nuclear membrane to the cytoplasm, where it is expressed in protein synthesis.

  Splicing processes that remove intervening sequences from precursors of messenger RNA must have predated or coevolved with introns. Recently, split genes have been found in archaebacteria, an ancient group of bacteria, and one case of a split gene has been reported in a true bacterium. Thus, interrupted or split genes seem to be quite widespread; additionally, viral genes, transfer RNA genes, and ribosomal RNA genes have been shown to be interrupted.

  For additional information on this subject, see the article by Chambon referenced at the end of this chapter.

VIRUSES

Viruses are particles composed of proteins and nucleic acids and are found within cells of all prokaryotes and eukaryotes (Fig. 1). Strictly speaking, viruses are not cells, and hence are not to be classified as prokaryotes or eukaryotes, although it has been suggested that they represent a very simple form of life. Viruses are submicroscopic particles; they are able to reproduce, but only within living host cells. This inability to reproduce independently is a major argument against considering viruses to be living organisms.

Virus particles have two major structural components: an outer shell composed of proteins and an inner fiber of either DNA or RNA. Their mode of life is to invade a cell and use the life processes of that cell to make more virus particles. It is a remarkable mechanism in which the DNA or RNA of the virus replaces some of the DNA or RNA of the host and directs the manufacture of viral protein and DNA or RNA. When sufficient viral components have accumulated, new virus particles assemble themselves, burst out of the cell, and infect other cells. Virtually every organism known, whether prokaryote or eukaryote, is subject to virus infection.

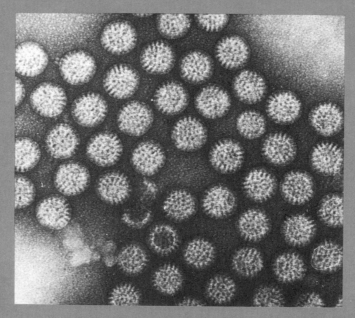

FIGURE 1
Electron micrograph of rotaviruses that cause a type of human gastroenteritis. The viruses are spherical, as are several other human viral pathogens, including those of influenza, the common cold, and polio. The viruses here are about 0.075 μm in diameter, about one-fiftieth the diameter of a streptococcus bacterium. (Photo courtesy S. Rodger and I. Holmes)

message to a ribosome. This portion of mRNA is called the **leader**, because it functions a bit like the leader of a motion picture film. The nucleotides coding for the first protein follow. This part may be 900 or more nucleotides long, depending on the size of the polypeptide. Then there may follow a nucleotide sequence telling the ribosome to stop assembling amino acids, to detach the polypeptide just made, and to get ready to make another. In this manner, a single molecule of mRNA may contain up to 10,000 nucleotides and code for several polypeptides before the end of the last message is translated.

RECOMBINANT DNA

The transfer of DNA between organisms is accomplished by several mechanisms. These include chromosomal exchanges during sexual reproduction, and, in prokaryotes, direct transfer of "naked" DNA during parasexual mating of cells (**bacterial conjugation**). DNA transfers also can occur between nonmating cells by the transport by viruses of nonviral DNA (**transduction**), and by the direct absorption by a cell of DNA from the substrate (transformation). Although transduction and transformation have been reported in eukaryotes as well as prokaryotes, they are more common in the latter.

Recombination of genes by sexual reproduction of eukaryotic organisms has been known for nearly a century and has been extremely helpful in understanding inheritance and evolution, in explaining and treating many medical disorders, and in producing new and often useful varieties of domesticated plants and animals. Knowledge of genetic recombination in prokaryotes, however, has a shorter history, dating as noted earlier in this chapter to transformation studies in bacteria by Avery, MacLeod, McCarty, and others in the early 1940s. Since then, bacterial genetics has become a cornerstone of modern biology and now is producing a new revolution in genetics under the general heading of **recombinant DNA,** one part of an interesting new branch of biological science called **genetic engineering.**

Viruses and Transduction

Transduction is the transfer of nonviral DNA by **viruses**. It is common among certain groups of bacteria and is known to occur also in eukaryotes.

Viruses are small infectious particles composed of protein and nucleic acids (usually DNA, but sometimes RNA). They are about the size of ribosomes and can reproduce themselves only within the cytoplasm of a host cell. (For this reason, it is arguable whether or not vi-

ruses should be considered living organisms, which by some definitions are always self-reproducing.) Their reproduction is dependent on the enzymes and ribosomes of the host cell. Viruses have the ability to inject their DNA (or RNA) into their host cells where the DNA (or RNA) acts as the code both for the synthesis of additional viral DNA (or RNA) and for the mRNA coding for viral protein. As a result, the virus is replicated many times over, the host cell becomes filled with viruses and dies, and the viruses escape. Cells of both prokaryotes and eukaryotes are subject to infection by viruses, of which there are many kinds.

Sometimes in the process of the replication of viral DNA in a host cell, a bit of host-cell DNA becomes incorporated into the viral DNA. When this happens, the new virus particles can introduce this attached host-cell DNA into still other hosts. Thus a transfer of host-cell genes may be accomplished, with the virus acting as the agent. Of course, if the virus then kills the host, such a transfer would have little effect insofar as genetic recombination in host cells is concerned, for there would be no survivors. However, sometimes when viruses infect a cell the viral DNA remains dormant (i.e., is not immediately transcribed into mRNA). Then, when the host cell divides, both the viral DNA and its attached host genes are replicated and passed on to the host daughter cells. It is in this way that chromosomal and plasmid genes of prokaryotes, and possibly of eukaryotes, are thought to be carried from cell to cell by viruses (Fig. 5-11).

Plasmids

A key factor in recombinant DNA research and technology is the existence in bacterial cells of small autonomously reproducing circles of DNA known as **plasmids** (Figs. 5-12, 5-13). Plasmids resemble the circular bacterial "**chromosome**" but are considerably smaller. They range in number from several to a hundred or so per cell, and a single plasmid type may be represented by multiple copies. When a bacterium divides, its plasmids are distributed to the newly formed cells; thus plasmids are perpetuated from generation to generation (Fig. 5-13). Plasmids are composed of genes that are transcribed and translated in polypeptides just as are chromosomal genes. Plasmid genes are important in the lives of bacteria, although bacteria may survive without plasmids.

Plasmids differ in size. The smallest may carry only two or three genes; larger ones may be composed of dozens of genes, among them **fertility (F) genes** which code for factors that can induce bacteria carrying them to mate (to **conjugate**). Other plasmid genes confer resistance to antibiotic drugs such as penicillin and strep-

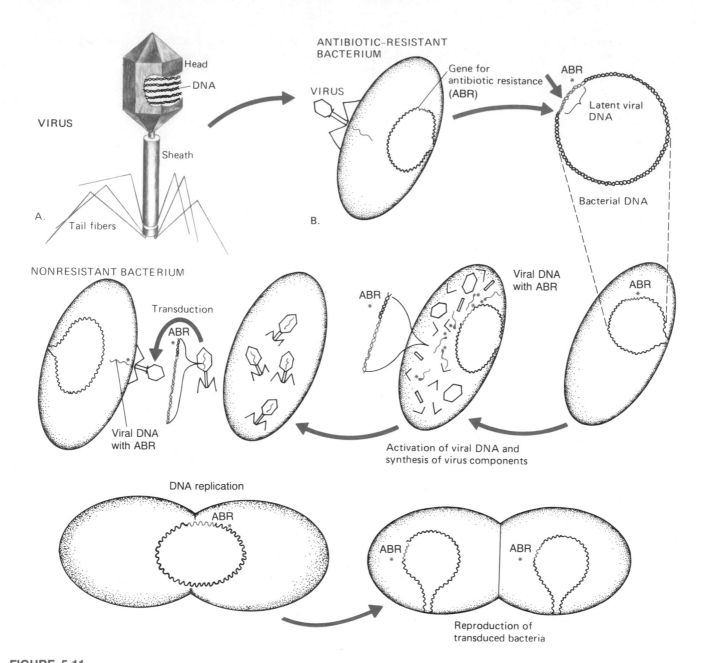

FIGURE 5-11
Transduction by a bacterial virus **(bacteriophage). A.** Bacteriophage. The virus becomes attached to a
bacterial cell by means of its tail fibers and injects its DNA through the core of its tail into the host cell.
B. Steps in the transduction of DNA. In these diagrams, the incorporation of bacterial DNA (black)—having
genes for antibiotic resistance (small arrows)—into viral DNA (color) is shown. Subsequent DNA transfers
into bacteria lacking antibiotic resistance, and the incorporation of virally transferred bacterial DNA, are
illustrated.

tomycin (Fig. 5-14); still others enable bacteria to metabolize a variety of substrates, including even such
substances as mercury compounds, long in use as anti-
septics.

It was once thought that all antibiotic-resistant
bacteria were recent mutants resulting from the use, and
misuse, of antibiotics to treat disease, but it now ap-

pears that most, perhaps all, contain **antibiotic-resistant
plasmids (R plasmids).** It is further thought that R plasmids may have evolved long ago in response to natural
antibiotics and other substances in the environment. In
many cases, development of antibiotic-resistant bacterial strains is the result of the survival of bacteria carrying R plasmids after exposure to medicinal antibiotics.

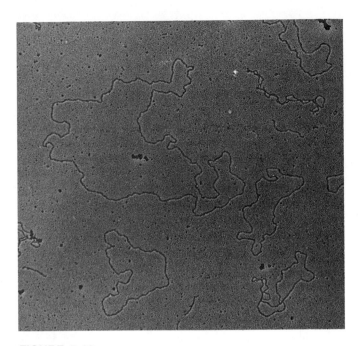

FIGURE 5-12
Electron micrograph of bacterial plasmids. Plasmids are small circles composed of a double strand (double helix) of DNA. (Photo by Rod Welch, Stanford University School of Medicine/Biological Photo Service)

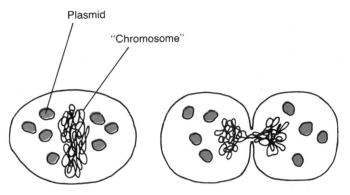

FIGURE 5-13
The distribution of plasmids in a bacterial cell. **A.** Plasmids are small circular double helices of DNA. There may be several kinds per cell and they may be in multiple copies. **B.** Plasmids are replicated in a cell independently of the bacterial "chromosome" and are distributed to the daughter cells during cell division.

Thus a bacterium does not have to be a mutant to acquire genes for resistance to a particular antibiotic or even to a combination of antibiotics. In fact, some bacteria have been discovered to have acquired resistance to as many as ten antibiotics!

Transposons

The fact that bacteria carry extra chromosomal genes in their cytoplasm in the form of plasmids is of more than passing interest. That these plasmid genes can confer resistance to antibiotics and the ability to break down toxic compounds has great significance. Even more remarkable, however, was the discovery in the 1960s that cytoplasmic genes of bacteria could somehow move into the bacterial chromosome and then be transferred to other bacteria by conjugation (Fig. 5-14). Such genes, called **episomes,** now are known to be equivalent to single genes or even to blocks of genes encoded in plasmid DNA. Among such plasmid genes are the F factor gene that induces bacterial conjugation, genes for antibiotic resistance, and, perhaps most remarkable of all, genes for breaking open plasmid and chromosomal DNA at certain specific sites. In the hands of modern-day geneticists, the ability to precisely cleave DNA of plasmids and chromosomes has proven to be a powerful tool for the manipulation of genes.

The ability of specific enzymes called **restriction endonucleases** to cleave DNA at highly specific points depends on their ability to "recognize" short base-pair sequences of DNA. A number of endonucleases now have been discovered and each exhibits specificity for a different base-pair sequence. Some, but not all, endonucleases cleave a site by a staggered cut as shown below. Here the base-pair sequence is:

```
- - - -GGATCC - - - -GGATCC - - - -
- - - -CCTAGG - - - -CCTAGG - - - -
```

Note that the two strands of DNA are complementary—the base-pair sequences are in reverse order. When restriction endonuclease is present, the DNA is cleaved in the following way:

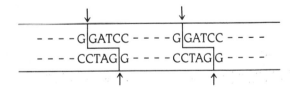

Note that this leaves a segment, or **transposon,** with two complementary ends:

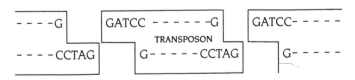

(Remember that in DNA, adenine [A] pairs with thymine [T], and guanine [G] pairs with cytosine [C].)

FIGURE 5-14

Plasmids and bacterial conjugation. The bacterial chromosome is relatively much longer than shown here. **A.** A plasmid bearing a gene, RA, for resistance to antibiotic A and a fertility (conjugation) gene F, has been inserted as a transposon in the chromosome of a bacterium ("RA"), inducing conjugation with a bacterium ("RB") bearing a plasmid with a gene, RB, for resistance to antibiotic B. **B.** After conjugation, the "RB" bacterium, now "RAB," contains plasmid genes for resistance to two different antibiotics.

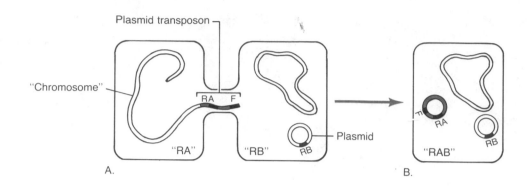

The bases in the two ends of the segment are unpaired and are said to be "sticky" because, if brought in contact with another diagonally cleaved identical base-pair sequence they will, in the presence of a repair enzyme called **ligase**, join up in a new complementary sequence. If, for example, the restriction endonuclease "recognized" and cleaved an identical base-pair se-quence in the bacterial chromosome, the plasmid segment could graft itself between the opened, "sticky" ends. This is the principle of recombinant DNA research (Fig. 5-15).

Referring to the diagrams in Fig. 5-14, imagine that a line of bacteria-bearing plasmids of the RA type are mingled with bacteria having RB plasmids (RA and RB

FIGURE 5-15

Transposons. **A.** In this sequence, a plasmid is opened up at a cleavage site by a restriction endonuclease enzyme (symbolized by scissors). The plasmid itself carries the gene (E) for production of the endonuclease, as well as genes for bacterial conjugation (F) and antibiotic resistance (R). **B.** When opened up, the plasmid has staggered coding sequences at each end. **C.** Plasmid endonuclease also cleaves the bacterial chromosome at the same coding sequence as in the plasmid. **D.** The plasmid section is inserted between the cleaved "sticky" ends of the bacterial chromosome, matching the staggered coding sequences, base pair for base pair. The cleaved ends of the plasmid and chromosome are "healed" by a ligase enzyme (symbolized by a needle and thread), for which the bacterial chromosome carries a gene (L). The plasmid insertion now exists in the bacterial chromosome as a transposon. The same principle applies to formation of recombinant plasmids. Two plasmids can be opened up and rejoined as a larger plasmid, perhaps carrying genes for resistance to two or more antibiotics as well as genes for other factors.

G—guanine
C—cytosine
T—thymine
A—adenine
F—fertility factor gene
R—antibiotic-resistance gene
E—endonuclease gene
L—ligase gene

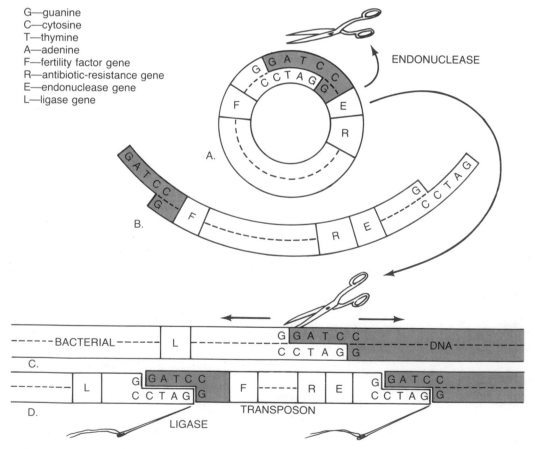

being antibiotic-resistance genes). Because RB plasmids carry the F factor, the RA bacteria (those carrying RA plasmids) can conjugate with other bacteria, including RB bacteria. Note in Fig. 5-14 that prior to conjugation the RA plasmid has opened up (in the manner shown in Fig. 5-15) and become a transposon inserted in the bacterial chromosome. This action induces conjugation, in which the part of the bacterium carrying the F gene is inserted into another bacterium (in this case one carrying the RB plasmid). In its new position, the RA plasmid can resume its original circular shape and be replicated many times over. The original RB bacterium now has acquired plasmid genes for resistance to both A and B antibiotics (say penicillin and streptomycin). Another possibility exists, namely, that the RA and RB plasmids might simultaneously open up and mutually recombine as a single larger plasmid bearing genes for resistance to two antibiotics. Then, when the bacterium divides, all the newly formed bacteria will be resistant to both antibiotics.

FIGURE 5-16
Dr. Barbara McClintock, discoverer of transposable elements (controlling elements). (Herb Parsons, Cold Spring Harbor Laboratory Research Library Archives)

Because plasmids readily can produce transposons, either from entire plasmids or from portions of plasmids, they have a marked tendency to produce new gene combinations.

Transposons were first described under a different name (**controlling elements**) by Barbara McClintock (Fig. 5-16), who was studying the genetics of a maize variety in which patches of color appeared in the developing kernels. (The variety was one of the "calico" types often used in Thanksgiving-time decorations.)

McClintock explained that the appearance and disappearance of pigmentation in kernels of calico corn was controlled by certain genes (controlling elements) which could move about on the chromosomes, suppressing the expression of adjacent genes, such as those for pigmentation. At the time of this discovery (1940s) little was known about the molecular nature of genes, which then were thought to be stable components of chromosomes. Some years later, when transposons were discovered in bacteria, the importance of McClintock's contributions to genetics was finally recognized and she was awarded the Nobel Prize (in 1983).

The controlling elements of maize and transposons of bacteria now appear to be manifestations of the same basic genetic phenomenon. In addition to bacteria and maize, transposons (also called **transposable elements**) have been found in DNA of fruitflies and in yeast, and it is believed they may be present in the DNAs of many other kinds of organisms.

GENETIC ENGINEERING

An aspect of genetics now receiving much attention is the transfer of genes into bacteria by transduction or transformation, sometimes by a combination of the two, and the propagation of those genes by bacterial conjugation and reproduction. The multiplication of such genes, called **gene cloning,** has become a methodology with dramatic potential for medicine and industry, as well as cause for concern lest it get out of hand.

The most common laboratory bacterium in recombinant DNA work is the relatively harmless E. coli. Because E. coli is a common human intestinal organism, some biologists and laypersons fear that DNA-transfer experiments might produce dangerous new forms of this microorganism (or other microorganisms). If these were to escape and infect human populations, or plants and other animals, it could be disastrous. This concern continues, although genetic recombination experiments currently are performed under carefully controlled guidelines and so far have proved to be safe.

Among the more dramatic products of recombinant DNA research and production are various pharmaceutical products, including vaccines, **interferons,** and insulin. Presently, viruses used in the manufacture of vaccines are grown in cultured animal cells or in animal serums. Recombinant-DNA technology will make it possible in the future to manufacture vaccines by using genetically engineered bacteria. Interferons are natural animal products thought to have value in the treatment of viral infections and possibly cancer. There are a number of types, of which at present only some appear useful. Interferons are produced naturally in very tiny amounts. Now, however, by use of gene-cloning techniques, *E. coli* is being used to produce large amounts of potentially useful human interferon.

Even more dramatic than the manufacture of synthetic interferon is the current production of human-type insulin. Until recently, insulin for use in treating diabetes was extracted from pancreases of cows, sheep, or pigs. These insulins differ from human insulin by one or more amino acids, and many patients are somewhat allergic to them. Recombinant-DNA technology and gene cloning has made it possible to produce human-type insulin in cultures of *E. coli.* To do this, a full-length DNA-sequence coding for human insulin is synthesized, then put in a plasmid carrying the gene for the enzyme β-galactosidase and introduced into *E. coli.* In *E. coli* cells, the plasmid produces a protein composed of combined insulin and enzyme. Pure insulin is removed from β-galactosidase by use of a cleaving enzyme. The basic principles of gene cloning are shown in Fig. 5-17.

Gene cloning is accomplished in bacterial cultures where a majority of the cells have acquired one or more foreign genes and the ability to replicate them. The requirements for gene cloning are:

1 Foreign DNA capable of incorporation into either a bacterial plasmid or a bacterial chromosome

2 A means of propagating the DNA once it is in the bacterial cell

3 Methods of harvesting either the cloned gene or its products (Fig. 5-17)

The isolation and identification of genes now is accomplished almost routinely in genetic laboratories. Several techniques are available:

1 Genes may be synthesized if the amino acid sequence of the desired protein is known. The nucleotide sequence of a gene can be deduced from the amino acid sequence, and a synthetic gene produced by adding the nucleotides in their proper order by presently known techniques.

2 Messenger RNA can be isolated and used as a template for the synthesis of **complementary DNA**

(**cDNA**) using the enzyme reverse transcriptase. The cDNA then can be copied using DNA polymerase, producing a double-stranded DNA which is identical to the original gene.

3 A gene may be dissected from a large fragment of DNA extracted from a cell. A number of restriction endonucleases having the ability to cleave DNA at specific points between genes have been identified, thus permitting isolation and selection of specific genes.

When a gene is available for cloning, it can be introduced into a suspension of bacterial plasmids which have been treated with restriction endonucleases corresponding to those used to extract the gene to be cloned. A ligase then is used to incorporate the induced gene into a plasmid in much the same manner as illustrated in Fig. 5-15. Commonly, such plasmids also have genes for resistance to an antibiotic and genes for a fertility factor capable of inducing conjugation once the plasmid is taken up by a bacterial cell as shown in Fig. 5-14. The recombinant plasmids are mixed with *E. coli* cells or some other bacterium, along with cell permeability factors (**permease enzymes**). At least some recombinant plasmids will be incorporated into bacterial cells and further spread through the population by conjugation. Resistant bacteria then can be selected by exposing the bacterial culture to antibiotics. Only bacteria that have taken up plasmids will survive, and these can be further screened for ability to produce the product of the cloned gene.

Gene-cloning techniques can be used to produce almost any substance for which genes can be identified and isolated. Thus, as noted earlier, transformed bacteria can produce insulin and other hormones, enzymes such as lysozymes, or other proteins. In addition to the direct production of proteins, the cloned genes themselves can be extracted and introduced into eukaryotic cells. Viral genes have been cloned and injected directly into the nuclei of amphibian egg cells. There they have been found capable of directing protein synthesis. Other cloned genes have been introduced into eukaryotic cells by transduction, using viruses as gene carriers. Experiments of this type hold forth the expectation that genetic defects someday can be cured by cloning nondefective genes and introducing them into cells of individuals suffering from genetic disease such as hemophilia, characterized by excessive bleeding, and sickle-cell anemia.

Genetic engineering is by no means restricted to production of pharmaceuticals or to correction of genetic mistakes. Considerable interest has been gener-

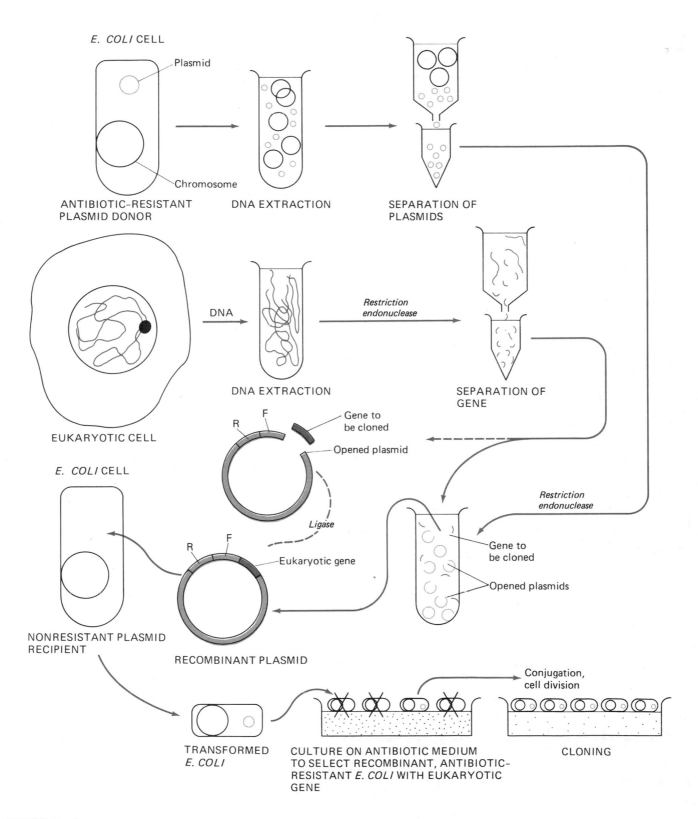

E. COLI CELL

Plasmid

Chromosome

ANTIBIOTIC-RESISTANT
PLASMID DONOR

DNA EXTRACTION

SEPARATION OF
PLASMIDS

EUKARYOTIC CELL

DNA

DNA EXTRACTION

Restriction
endonuclease

SEPARATION OF
GENE

Gene to
be cloned

Opened plasmid

R   F

Ligase

Restriction
endonuclease

Gene to
be cloned

Opened plasmids

E. COLI CELL

R   F

Eukaryotic gene

NONRESISTANT PLASMID
RECIPIENT

RECOMBINANT PLASMID

TRANSFORMED
E. COLI

Conjugation,
cell division

CULTURE ON ANTIBIOTIC MEDIUM
TO SELECT RECOMBINANT, ANTIBIOTIC-
RESISTANT E. COLI WITH EUKARYOTIC
GENE

CLONING

FIGURE 5-17
Gene-cloning experiment. F and R are plasmid genes. F is a fertility factor that can induce conjugation; R
is a gene confering antibiotic resistance.

ated by attempts to introduce N_2-fixation genes into plants or into fungi growing in association with plants (as discussed in Chapter 3). Required is a vector capable of introducing desirable transposons into plants and incorporating them into plant DNA. One such vector now used in experimental genetic engineering of plants is the **Ti plasmid** of an infectious agent, the **crown gall bacterium,** of certain plants. Using Ti plasmids of crown gall bacteria, a yeast gene for the enzyme alcohol dehydrogenase has been successfully introduced into tobacco plants. It is an unlikely combination, but a first step in what may eventually turn out to be a second "green revolution."

SUMMARY

A cell contains many kinds of molecules, and these for the most part are synthesized by the cell from simpler materials. Most of the molecules of the cell are made in reactions that are directed by particular kinds of proteins called enzymes. However, enzymes and other proteins are themselves produced only by nucleic acids (DNA and RNA). In other words, proteins are capable of catalyzing the formation of many of the cell's molecules, but are incapable of coding for other proteins. The key to nearly all the activities occurring in cells lies in the genetic code of DNA, which contains the instructions for making proteins.

A protein is composed of many amino acids linked together by peptide bonds into a chain. The DNA molecule consists of two parallel strands held together by hydrogen bonds between the opposed purines and pyrimidines that make up the nucleotide units of the double helix. The purines and pyrimidines of DNA form base pairs in which adenine and thymine, and guanine and cytosine, are always opposite each other. As a conse-

quence, when DNA is to be replicated in the cell, new strands can be formed by separation of the strands of the double helix and insertion of new nucleotides opposite the nucleotides of the original strand. When DNA is copied by mRNA, base pairing again ensures that the genetic message will be an exact copy of the original genetic code.

Messenger RNA employs ribosomes for the production of proteins. The genetic message is a code in which triplets of nucleotides correspond to individual amino acids. Another kind of RNA, transfer RNA (tRNA), conveys amino acids to the ribosome, where they are inserted into a growing protein chain at precisely the points specified by the nucleotide triplets of mRNA. There are 20 kinds of amino acids in proteins (not every protein contains all 20, however), and there is a different specific tRNA for each amino acid.

An important aspect of protein synthesis in cells is the control of genes. In most cells, genes do not direct the continuous production of proteins, but are "shut down" or "started up" according to genetic feedback systems. These feedback mechanisms are of several kinds. When the particular substrate is absent, the repressor protein "turns off" the genes that normally produce enzymes that would act upon the substrate. When the substrate is present, the repressor is tied up by the substrate and thus cannot shut down the production of enzymes by other genes; therefore metabolism of the substrate proceeds.

Advances in understanding the molecular nature of genes has led to genetic engineering—the manipulation of genes for useful purposes. Now it is possible to clone genes by isolating them and introducing them into bacteria. Often, small loops of DNA, known as plasmids, are used to transport the desired genes into bacterial cells where they can function to produce useful gene products.

KEY WORDS			
	information storage	nucleotide triplet	histone
	information retrieval	transcription	recombinant DNA
	DNA	messenger RNA	plasmid
	gene	transfer RNA	transduction
	transformation	codon	genetic engineering
	biochemical mutant	anticodon	genetic cloning
	polypeptide	translation	transposon
	double helix	ribosome	endonuclease
	genetic code	repression	ligase
	base pairing	derepression	

QUESTIONS FOR REVIEW AND DISCUSSION

1 If a cell were compared to a computerized factory in which a central file of computer programs is used to make computer tapes, which upon being fed to a machine result in the manufacture of specific products, what cellular structures would correspond to the programs, the tapes, the machine, and the products?

2 How do cells "turn on" and "turn off" genes so that the genes are operational only when their action is required?

3 Briefly describe the relationship between amino acids, tRNA molecules, mRNA, and the ribosome. A well-known geneticist has been quoted as saying about nucleic acids: "It takes one to know one." Explain.

4 What problems might be encountered in attempting to solve a genetic code if the master code consisted of proteins rather than nucleotides?

5 How does regulation of transcription differ in eukaryotes and prokaryotes?

6 What is the explanation of the "stickiness" of the ends of a plasmid opened up by an endonuclease?

7 What is the significance of the phenomenon described in question 6?

8 Briefly describe the genetic engineering procedure in preparation of synthetic human insulin.

SUGGESTED READING

ABELSON, J. 1980. A revolution in biology. *Science* 209:1319. (Overview of the discoveries made possible by using techniques for sequencing DNA.)

CHAMBON, P. 1981. Split genes. *Sci. Amer.* 244(5): 60–71. (Split genes enable a cell to synthesize innumerable unique proteins.)

CHILTON, M.D. 1983. A vector for introducing new genes into plants. *Sci. Amer.* 248(6): 50–59. (Use of Ti plasmid to introduce foreign genes into plants—fundamental.)

COHEN, S. N., and J. A. SHAPIRO. 1980. Transposable genetic elements. *Sci. Amer.* 242(3): 40–49. (Background material.)

GILBERT, W., and L. VILLA-KOMAROFF. 1980. Useful proteins from recombinant bacteria. *Sci. Amer.* 242(4):74–94. (Useful proteins such as insulin and interferons are made by bacteria after introducing mammalian DNA via plasmids or viruses. Anxiety over recombinant DNA is said to be lessening.)

GROBSTEIN, C. 1977. The recombinant-DNA debate. *Sci. Ameri.* 237(1):22–33. (Public fear of science is discussed, as well as elements of bacterial transformation.)

LAKE, J. A. 1981. The ribosome. *Sci. Amer.* 245(2):84–97. (Structure of ribosomes and binding of mRNA are discussed. Functions of ribosomal subunits are clarified.)

MILLER, O. L. 1973. The visualization of genes in action. *Sci. Amer.* 228(3):34–40. (A remarkable selection of electron micrographs showing mRNA translation, both in prokaryotes and in eukaryotes.)

MIRSKY, A. E. 1968. The discovery of DNA. *Sci. Amer.* 218(6):78–88. (A nice review of the chromosome theory of inheritance. Pictures are historical.)

NOVICK, R. P. 1980. Plasmids. *Sci. Amer.* 243(6):103–22. (Are plasmids organisms? An account of the nature and role of plasmids in antibiotic resistance and recombinant DNA technology.)

PESKA, S. 1983. The purification and manufacture of human interferons. *Sci. Amer.* 249(2): 36–43. (Describes recombinant-DNA techniques applied to manufacture of a pharmaceutical.)

SMITH, M., et al. 1977. DNA sequence at the C terminal of the overlapping genes A and B in the bacteriophage φX174. *Nature* 265.705. (Illustrates how nucleotide sequences of genes may overlap.)

WATSON, J. D. 1968. *The double helix*. New York: Atheneum Press. (A fascinating behind-the-scenes account of the discovery of the structure of DNA.)

———, J. TOOZE, and D. T. KURTZ. 1983. *Recombinant DNA. A short course*. New York: Scientific American Books, distributed by W. H. Freeman and Co. (A useful "catch-up" book for the interested reader, understandable by nonbiologists.)

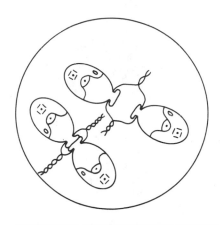

6

Cell Reproduction and Development

According to some theories of the origin of life, macromolecules accumulating in ancient seas eventually gave rise to simple living cells. Because life is dynamic, not static, cellular life from the start was characterized by growth. How large these first cells became is unknown, but we do know that growth in size has limitations and cannot go on indefinitely. Aside from problems of sheer maneuverability and support, a cell's surface increases by the square as its volume increases by the cube (Fig. 6-1). This brings about a relative reduction in surface area available for the transport of substances into and out of the cell and reduces efficiency. Division of a larger cell into two or more smaller ones alleviates such traffic problems. Probably quite early in the history of life on earth, mechanisms evolved for precise and regular cell divisions, thus providing not only for the maintenance of optimal cell size, but also for reproduction in numbers of cells and of individuals.

The simplest organisms known are **prokaryotes** ("before nucleus"), single cells that reproduce by a simple division process known as **fission**. Even at this level of life, however, a basic principle of reproduction is seen: reproduction involves both division of cytoplasm and also duplication and equal distribution of DNA. First the

double helix of DNA in the prokaryotic cell is duplicated **(replicated)** (Fig. 6-2). The cell then divides into two

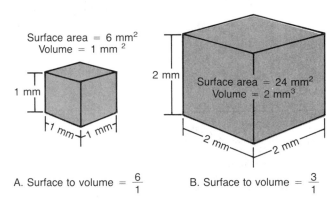

A. Surface to volume = $\frac{6}{1}$ B. Surface to volume = $\frac{3}{1}$

FIGURE 6-1
Surface-to-volume relationships of an expanding cube. In **A,** the cube measures 1 mm in each dimension and has a volume of 1 mm³ (e.g., 1 × 1 × 1 = 1). Each of its six sides is 1 mm² (1 × 1 = 1) and its surface area is 6 mm² (e.g., six sides, each 1 mm²). The surface-area-to-volume ratio of cube A is 6 (6/1). In **B,** the dimensions of the cube have doubled to 2 mm in each direction, and the volume has become 8 mm³ (2 × 2 × 2 = 8). Each of its six sides is 4 mm² (2 × 2 = 4) and its surface area is 24 mm² (six sides, each 4 mm²). The surface-to-volume ratio of cube B is 3 (24/8), relatively, half the surface of cube A. If the dimensions are tripled, the surface-to-volume ratio then becomes 2 (54/27), and so on.

FIGURE 6-2
Reproduction of a bacterial cell by
division.

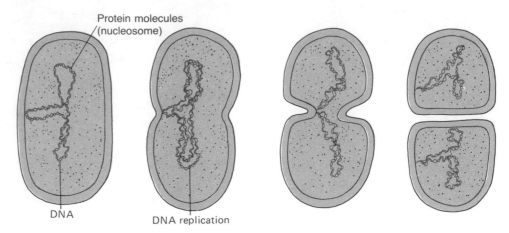

cells, each receiving one of the duplicated helices of DNA; both new cells thus possess the same genetic information. Although cell reproduction in **eukaryotes** ("true nucleus") is more complex, the basic principle is the same as in prokaryotes: each new cell has the same genetic information as the original cell. The presence of several to many **chromosomes** in eukaryotic cells is a complicating factor. (Note: the circular strand of DNA in a prokaryotic cell also is called a "chromosome," but chromosomes of eukaryotic cells are more complex and more numerous; i.e., more than one per cell.)

The eukaryotic chromosome during the time when the cell is not dividing **(interphase)** is believed to exist as a single long DNA molecule (Fig. 6-3). During interphase, the function of the chromosome is the transcription of portions of its DNA into mRNA. It is uncertain just how DNA is arranged in the interphase chromosome, but it is thought that the long double helix is complexed with histone protein in the form of loops of DNA wrapped about disclike protein molecules. This structure cannot be detected by ordinary microscopy, but electron microscope views of chromosomes show them to have a beaded appearance. The beads, which are called **nucleosomes,** correspond to the loops of DNA and histone (Fig. 6-3).

CELL REPRODUCTION

Two processes of nuclear and cell division take place in eukaryotic cells. In one process, chromosomes and their DNA are replicated and distributed equally between the **daughter cells** so that each possesses the same genetic information as the **mother cell** (same number of chromosomes, same amount of DNA). This nuclear process is mitosis (*mitos* = thread), which is involved in **asexual reproduction** of cells. Mitosis is the common mode of nuclear division in the multiplication of all eukaryotic cells and organisms, whether those of one-celled organisms such as amoebas or of body cells (also called **somatic** cells) of multicellular organisms. The second nuclear division process, **meiosis** (*meioun* = to make less), is associated with **sexual reproduction** and differs from

FIGURE 6-3
Structure of the chromosome in a nondividing (interphase) cell. The broken lines indicate that the chromosome is proportionately much longer than shown. It is thought that the double helix of DNA composing the chromosome is itself minutely convoluted or coiled, and that during mitosis it becomes very much coiled and shortened. Each chromosome has a centromere located somewhere along its length. The centromere is the body to which microtubules become attached during mitosis and meiosis. At frequent intervals, portions of DNA are looped around protein molecules, producing beadlike nucleosomes, as shown here.

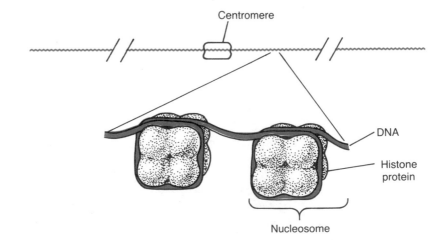

mitosis in that, after meiosis ends, the total number of chromosomes and the amount of DNA per nucleus in each daughter cell are half those in the original cell.

Meiosis occurs only in sexual reproduction of eukaryotes, where it is of fundamental importance. In all cases of sexual reproduction—whether occurring in one-celled organisms (unicells) or in multicellular plants and animals—sex cells (gametes) are produced. In animals and higher plants, the gametes are designated female and male—eggs and sperm, respectively (Fig. 11-3); in unicells and in some algae and fungi, the gametes are structurally similar but functionally different and usually are designated as plus and minus. Gametes do not remain independent entities very long but combine in pairs (sperm with egg or, in unicells, plus with minus), forming a larger cell, the zygote. If gametes and zygotes are examined microscopically after they have been stained to make their chromosomes visible, the gametes are seen to contain exactly half the number of chromosomes found in zygotes. Moreover, measurement of total DNA per cell shows that DNA content parallels chromosome number: gametes contain half the DNA of zygotes. It is customary to refer to the gametic or half-number of chromosomes as the $1n$ or haploid number, and to call the chromosome number found in zygotes the $2n$ or diploid number.

In addition to numerical differences, the chromosomes of gametes tend to differ from each other in relative size and in structural details, whereas those of zygotes consist of two sets of similar pairs. In human gametes, for example, there are 23 dissimilar chromosomes, and in human zygotes and cells derived from zygotes by mitosis, there are 23 pairs of chromosomes, or a diploid number of 46 (Fig. 6-4).

Although both mitosis and meiosis commonly involve division of entire cells, they are principally processes of nuclear division, even though division of the nucleus generally is accompanied by division of the cytoplasm (cytokinesis), so that complete cell division takes place. (In some organisms, only the nucleus divides, and the cytoplasm simply expands.) For the present discussion, however, only examples in which complete cell division occurs during mitosis will be considered.

In studying life processes, it sometimes helps to simplify the approach so that the essentials may be grasped more readily. Mitosis and meiosis are complicated processes that may be understood better if it is remembered that, in mitosis, the chromosomes in a nucleus become duplicated and then are separated into two new nuclei in such a way that each new nucleus has exactly the same number and kinds of chromosomes as the original nucleus.

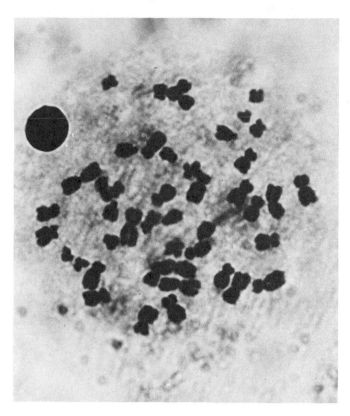

FIGURE 6-4
The 46 chromosomes of a human female cell. The cell, which is faintly visible in the preparation, was undergoing mitosis, hence the chromosomes are quite short and, after staining, very visible. (Photo courtesy B. L. Wismar)

In meiosis, on the other hand, the chromosomes, which occur in pairs, all replicate themselves and then separate in such a way that each of the newly formed nuclei contains one member of each original chromosome pair; the total chromosome number of that nucleus is one-half that of the original nucleus.

Mitosis—The Cell Cycle

Although the cells of each eukaryote contain a specific number of chromosomes, chromosome numbers differ widely, ranging from 4 in some worms and a species of desert plant to about 250 in crayfish and more than 1000 in a species of fern. Nevertheless, in all eukaryotes the structure of chromosomes is essentially the same. Typically, each chromosome is an elongated structure consisting of a single very long molecule of DNA.

Mitosis is sometimes called "duplication division" because the cells resulting from this type of nuclear division are exact duplicates of one another so far as genetic content is concerned. It is the usual method of cell reproduction in one-celled life and in the body cells of multicellular organisms. It should be noted, however, that the great majority of cells are not engaged in active mitosis at any one time. One-celled organisms at times may reproduce rapidly by mitosis and at other times rarely. In multicellular organisms, a majority of the cells are in mitotic division only during early stages of development of the individual. When the multicellular individual is fully formed, mitosis slows and in some parts of the organism may halt altogether. Nevertheless, nearly all cells of fully grown individuals retain the ability to divide; in some cases they do so almost continuously. In plants, for example, groups of cells at the tips of stems and roots divide continuously and provide for new growth. In animals, cells in blood-forming regions (bone marrow, spleen) also divide continuously; in other parts of the body, repair and replacement of cells requires frequent mitotic division. Only nerve cells of higher animals appear to have lost the ability to undergo mitosis.

Most studies of mitosis are done with cells which undergo rapid division, often dividing **synchronously** (all cells in the same stage of division, Fig. 6-9A). As one can imagine, studying cell division in a tissue in which cells divide at unpredictable intervals would be exceedingly difficult. For that reason, rapidly growing embryonic tissues often are used in studies of mitosis, but tissue cultures also are a source of information.

The chain of events in which cells divide by mitosis and replenish their DNA to the parental cell level is called the **cell cycle** (Fig. 6-5). The duration of the cell cycle in constantly dividing cells varies from around 10 hours to more than 20 hours. Analyses of cell cycles show that the longest portion of that time is occupied by DNA replication, which occurs in the interval between successive mitotic divisions of the chromosomes.

Immediately following the division of a cell, there occurs a brief period known as G_1, an abbreviation for "gap." This is a period of great variability depending on the cell type and the degree to which it has become a functioning unit of the adult organism (as contrasted to a rapidly reproducing unit of an embryo, a root tip, or a tissue culture; see Fig. 16-13). The G_1 may encompass days, possibly weeks or months or even years, or its duration may be measured in minutes. Indeed, in some rapidly dividing cells, the G_1 phase is bypassed. The G_1 phase is followed by the **S phase** (synthesis phase), a period lasting several hours (see Fig. 6-5), during which the DNA of each chromosome becomes duplicated. This occurs by separation of the two strands of the double helix and insertion of matching nucleotides, creating two new strands of DNA (Fig. 6-6). Following the S phase, a second gap in time, the G_2, occurs. In mammalian cells, the whole cell cycle takes about 18 hours, G_1 lasting about 8 hours, the S phase about 6 hours, and G_2 about 4.5 hours; mitosis itself occupies only about an hour.

Chromosomal Replication When the cell has completed the S phase of the cell cycle, the chromosome is

FIGURE 6-5
The cell cycle. Cell cycles differ in duration from species to species; the times given here are based on studies of mammalian cells in tissue culture.

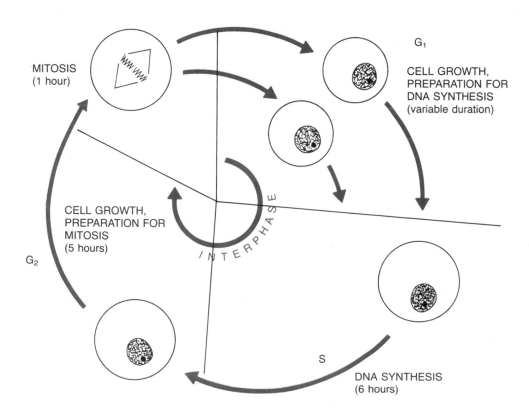

MITOSIS
(1 hour)

G_1

CELL GROWTH,
PREPARATION FOR
DNA SYNTHESIS
(variable duration)

CELL GROWTH,
PREPARATION FOR
MITOSIS
(5 hours)

G_2

INTERPHASE

S

DNA SYNTHESIS
(6 hours)

no longer a single threadlike body, but is longitudinally double, consisting of two parallel strands of DNA, each of which is a **chromatid** (Fig. 6-6). Thereafter the cell enters mitosis, its chromosomes each consisting of two chromatids, which then become visibly shorter and thicker. This shortening occurs by the coiling of the chromatids during **prophase,** as depicted in Fig. 6-8, A–D (p. 128). When coiling is completed, the chromosomes are many times shorter and considerably thicker than they were in interphase. Although the image most persons have of chromosomes is that of their shortened configuration, it should be remembered that this is not the form of chromosomes during the greater part of the life of the cell. Rather, the typical state of chromosomes resembles that shown in Fig. 6-3.

The Spindle Mechanism As mitosis progresses, a structure called a **spindle** develops in the nuclear region of the cell (Fig. 6-7). In most eukaryotes, the spindle is a bipolar basketlike structure composed of microtubules.

Two sets of microtubules comprise the spindle; one set extends between the two poles of the spindle, and the other set extends only from one or the other pole to the **centromeres** of chromosomes (Fig. 6-3). The pole-to-pole microtubules appear to be responsible for the shape of the spindle and for its orientation in the cell. They also may act as pathways for the movement of chromosomes in mitosis and meiosis. The chromosomal (pole-to-centromere) microtubules appear to be responsible for the orientation of chromosomes within the spindle and for their actual movement to the spindle poles.

In cells of animals and of many but not all algae and fungi, a small structure, the **centriole,** is found where the spindle microtubules converge at the poles (Fig. 6-7). Except for a few specialized cells (those forming motile male gametes, called **spermatozoids**), plant cells do not contain centrioles. Centrioles are thought to act as organizing centers for spindle microtubules. Although most plant cells lack centrioles, they appear to contain small polar aggregates of an undefined substance that may have the same spindle-organizing function as centrioles.

Chromosomal Movements For convenience, mitosis is divided into several stages: prophase, metaphase, anaphase, and telophase.

Prophase is characterized by progressive shortening of chromatids. In animal cells during prophase the centrioles are replicated and later on the daughter centrioles migrate to opposite poles of the cells (Fig. 6-8, A–D). In plant cells (Fig. 6-8, a–d) centrioles are absent and the poles are not clearly defined prior to spindle formation. Nevertheless, poles exist and aggregates of

FIGURE 6-6
The replication of a chromosome in the cell cycle. **A.** The chromosome at interphase contains a single long double helix of DNA which extends through a single centromere. **B.** In the S phase, the double helix of DNA opens up and functions as a template for two new strands of DNA. **C.** At the end of the S phase of the cell cycle, chromosomal replication is complete. The chromosome now consists of two copies of DNA, the chromatids, and two centromeres, which will adhere to each other until separated in anaphase of mitosis.

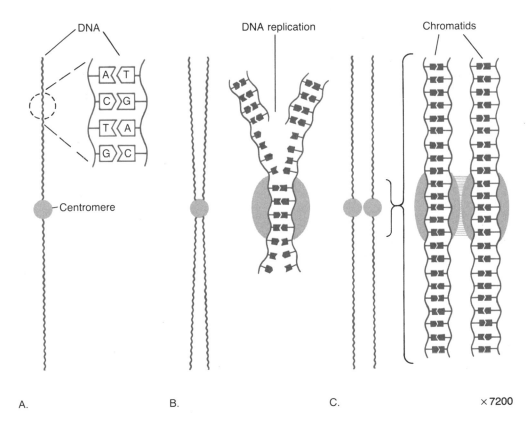

A.   B.   C.   ×7200

FIGURE 6-7
Spindle in a dividing animal cell.
The centrioles at each spindle
pole consist of paired structures
closely resembling basal bodies
of flagella and cilia. **A.** Electron
micrograph of an animal cell in
mitosis. (Photo courtesy David
Mason) **B.** Interpretive diagram
of a dividing animal cell.

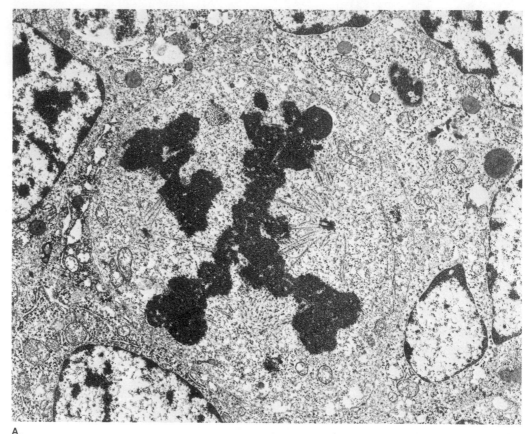

A.

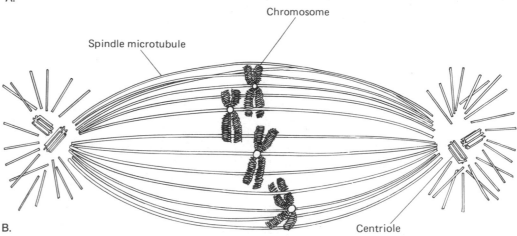

B.

granular material appear to function as focal points for
the deployment of spindle fibers to the chromosomes. In
animal cells the centrioles act as organizing centers for
spindle fibers.

The prophase stages of mitosis may be subdivided
into early, middle, and late prophase (Fig. 6-8B, C). Such
subdivision does not mean that mitosis proceeds by a
series of fits and starts; it is a continuous process. Indeed, cells often are found to be intermediate between
typical mitotic stages.

In early and middle prophase, chromosomes become visible as threadlike structures; by late prophase
the nucleolus has disappeared, the nuclear membrane
has broken down, and the chromatids of the chromosomes have reached their maximum degree of coiling
and shortening. Microtubules are attached to the centromere of each chromosome and extend toward one of
the spindle poles. Each chromatid has its own centromere held firmly to the centromere of its sister chromatid (Fig. 6-6). Now each chromosome, consisting of a

greatly shortened pair of chromatids, is pushed or pulled toward the middle of the spindle by elongation of spindle fibers; the exact mechanism is not understood, but in all cases the orientation of the spindle determines the plane of cell division; cytoplasmic division will be at a right angle to the orientation of the spindle poles.

At the end of prophase, when the chromosomes are positioned in the center of the spindle, a brief stage known as **metaphase** occurs (Fig. 6-8D). Sometimes it is said that in metaphase the chromosomes "split" down the middle, but that is only a figure of speech describing the separation of the chromatids.

Once the centromere has been duplicated, each new centromere has only one set of microtubules, directed toward only one of the spindle poles. This is of utmost importance because it means that the two new chromosomes, derived from a single original chromosome, will be attached to opposite spindle poles.

As soon as centromeres have separated, daughter chromosomes begin to migrate toward the poles, and the cell is said to be in **anaphase** (Fig. 6-8E, F). Microtubules appear to draw the chromosomes to the poles of the spindle. This implies that the microtubules have the ability to contract, but other studies show that not to be the case. More than likely, the centromere travels along the microtubules of the spindle toward the pole. The forces causing this movement have not been identified, but it is currently thought that microfilaments may be involved. Because they are composed of **actin**, a contractile protein, microfilaments may produce chromosomal movements along the spindle microtubules.

Anaphase persists until the chromosomes reach the poles, at which time the cell is said to begin **telophase** of mitosis (Fig. 6-8G and H). Telophase involves more than just the termination of the poleward migration of the chromosomes. It includes the dissolution of the spindle, the uncoiling of the chromosomes, the reformation of the nuclear envelope, and the reappearance of the nucleolus. The whole process of mitosis is completed in most cells within about two hours.

Summary of Mitosis Consider a cell in which the normal $2n$ number of chromosomes is six (three pairs). During prophase the chromosomes shorten and become centrally oriented in the spindle. In this phase each chromosome consists of two parallel strands, the chromatids, held together at one point by closely appressed centromeres.

At metaphase the duplicated chromosomes line up individually in the center of the spindle; each chromatid of a chromosome now is in fact a new daughter chromosome. There are therefore 12 chromosomes in the cell; the original chromosome number of 6 is restored by separation and segregation of the recently duplicated chromosomes during anaphase and telophase. The overall result of this is that each pole of the cell receives one of the two chromatids derived from each of the 6 original chromosomes (i.e., each new cell has three pairs of chromosomes, the original $2n$ number).

In addition to the chromosomes and their DNA, which are subdivided during mitosis, other cell components are shared between the new daughter cells. The division of cytoplasmic components is not as precise as that of chromosomes but is approximately equal in most instances. Therefore, in interphase there is a general rebuilding of the cell in which the cytoplasm and cellular organelles such as mitochondria, chloroplasts (if present), and Golgi bodies are replenished. In continuously dividing cells, the rate of growth during interphase regulates the onset of the next division.

Cytoplasmic Division In most cases, mitosis is accompanied by cytoplasmic division, but the manner in which new cells are produced varies. In plants, a cell wall encloses each cell, and cytoplasmic division results from the formation of a crosswall in the center of the dividing cell. This crosswall, called the **cell plate,** forms in the equatorial zone of the spindle and grows toward the sides of the cell. In telophase, the cell plate fuses with the old cell wall and partitioning is completed (Fig. 6-8g, h).

Animal cells divide by a furrowing process (cytokinesis) that constricts the middle of the dividing cell, eventually pinching it into two daughter cells (Fig. 6-8H). This process requires the synthesis of new plasma membrane material. Microfilaments are found in the region of furrowing, and probably act like the drawstring of a purse.

Other kinds of organisms, such as protozoans, algae, and fungi, divide by processes similar to one or the other of those just described.

Stimulation of Cells to Divide Scientists the world over are keenly interested in the factors that cause cells to begin or cease dividing by mitosis. A major reason for this interest is that cancer occurs when some cells of an organism (many organisms other than humans contract cancers) are stimulated into continuous and destructive cell division. In the normal body of an organism, cell division and cell differentiation are orderly and controlled processes. Animal cells in tissue culture freely proliferate for a time, but when they become numerous and begin to pile up, they cease to multiply. This phenomenon, sometimes called **contact inhibition,** also is known as **density dependent inhibition** (some biologists prefer to restrict the former term to inhibition of

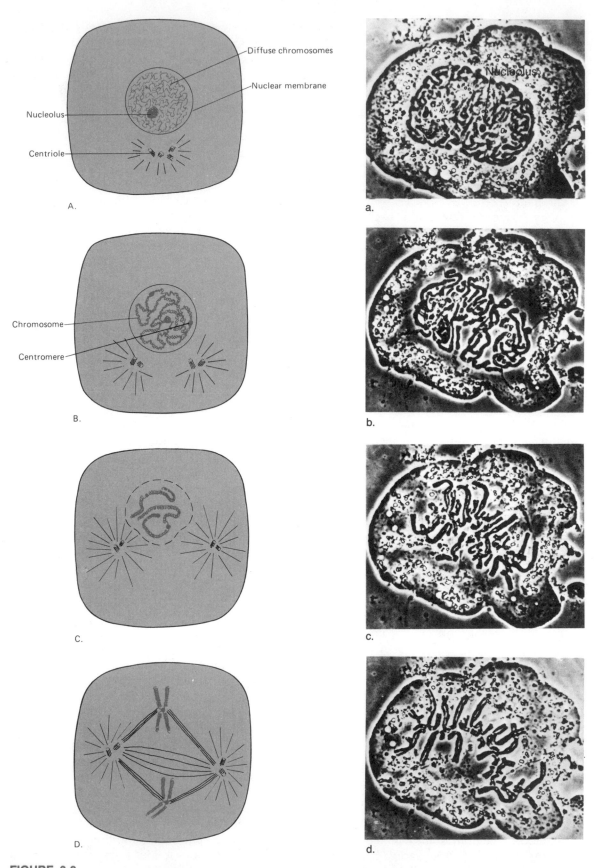

FIGURE 6-8

Mitosis: animal cells **(A–H)** are diagrammatic; plant cells **(a–h)** are phase microscope photographs. Drawings at left are representations of stages of mitosis in an animal cell; only one pair of chromosomes is included. Note the presence of centrioles. Photographs at right are of equivalent stages of mitosis. Note absence of centrioles, numerous chromosomes. The cells are of the African blood lily *(Haemanthus katherinae)* magnified about 300X. (Courtesy W. T. Jackson). **A. (a)** Interphase. **B. (b)** Early prophase. **C. (c)** Middle prophase. Note progressive shortening of chromosomes, which are visibly double, that is, composed of two chromatids. The nuclear membrane and nucleolus disappear.

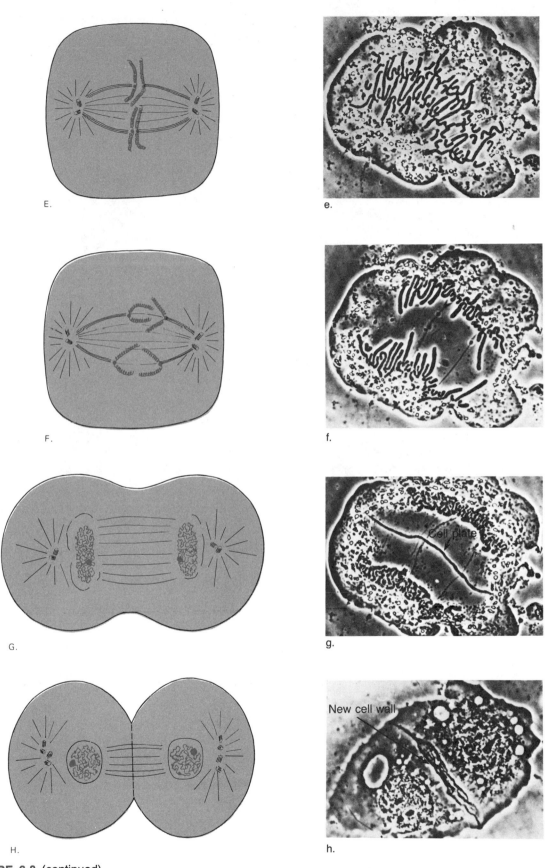

FIGURE 6-8 (continued)
D. (d) Metaphase. Spindle has formed but in plant cell is only faintly visible. **E. (e)** Early anaphase. The chromatids have separated and now are bona fide chromosomes. **F. (f)** Anaphase continued. In the plant cell, a cell plate, which will become the new cell wall, is forming. **G. (g)** Telophase. The chromosomes have reached the poles; a nuclear membrane is reforming. In the plant cell, the cell plate is quite apparent, but does not completely divide the cell. **H (h).** End of telophase, entrance into interphase. Division of the cytoplasm has been accomplished (cytokinesis). In the animal cell the centrioles have been duplicated. In the plant cell, the new cell wall is established.

ARTIFICIAL CHROMOSOMES

Recombinant-DNA techniques (Chapter 5) have been applied to determination of the structure and function of eukaryotic chromosomes (mainly of yeast). Four classes of functional DNA have been identified: **genes, origins for replication, centromeres,** and **telomeres** (end sequences of chromosomal DNA). Eukaryotic chromosomes consist of a long double helix of DNA associated with a variety of proteins. Researchers have been able to cleave chromosomal DNA with restriction endonuclease enzymes and put the various pieces together as "artificial chromosomes." Some of the results of such research include these findings:

1 Centromeric DNA is rich in A–T base pairs and has a structure that is "recognized" by and bound by microtubular proteins.

2 Telomeric DNA is essential for the regular movements of chromosomes in mitosis and meiosis; it contains simple repeating sequences of base pairs.

3 When genes, centromeric DNA, telomeric DNA, and replication origin pieces are linked together and introduced into yeast cells, the artificial chromosomes go through mitosis and meiosis.

4 Artificial chromosomes behave in yeast much like natural yeast chromosomes do, but they apparently are not perfect because they are lost more frequently than are natural chromosomes. That is, when the cells divide, the artificial chromosomes do not always segregate properly to the poles of the cells; at times they get left behind.

cell movements upon contact with each other). It is likely that density-dependent inhibition is one of the factors regulating cell division and growth in multicellular organisms. In cancer, cells are stimulated to divide in an uncontrolled fashion. Apparently unable to stop dividing, they form masses of cells which disrupt and strangle normal tissues and organs.

Another common condition associated with the reproduction of cells is aging. Unlike cancer, aging occurs when the rate of cell division is insufficient to replace worn-out cells. Scientists are becoming increasingly interested in factors related to aging processes, and it appears likely that some control of the rate of aging may be attained.

Normally, the division of cells is controlled. One factor controlling the rate of cell division is DNA synthesis. As noted previously, in metaphase a chromosome is divided into two new daughter chromosomes by division of the centromere and longitudinal separation of the two chromatids, but DNA is not replicated immediately.

Were the cell to divide again without increasing the DNA of its chromosomes, the new chromosomes would, in the next division, have only a fourth of the original quantity of DNA, and each division thereafter would further halve chromosomal DNA. Under actual conditions of normal cell division, a cell does not divide again until its chromosomal DNA has been replicated during interphase.

Cancer cells replenish their DNA so rapidly that there actually may be an excess of DNA. As a result, the chromosomes and nuclei may become abnormally enlarged and distorted. Such abnormalities may be seen in cancerous tissues when small slices of tissue removed from suspect organs are examined under a microscope. In this procedure, called a **biopsy**, skilled technicians detect nuclear and cytoplasmic abnormalities typical of cancer (Fig. 15-15).

Eukaryotic cells, growing continuously, may undergo mitosis once or twice a day. Such cells commonly divide synchronously, so most or all of the cells in a

5 Larger artificial chromosomes behave more properly than do the smaller ones; the largest artificial chromosome produced so far (55,000 base pairs) is about one-third the size of the smallest natural yeast chromosome (150,000 base pairs).

6 In meiosis, chromosome pairs come together (i.e., they synapse) prior to the reductional phase of meiosis. How the partners "recognize" each other is a question of great interest to cytologists and geneticists. Studies with artificial chromosomes indicate that the centromere does not have a role in recognition of homologous chromosomes in meiotic synapsis. Centromeric DNA from different yeast chromosomes may be substituted without decreasing the mitotic/meiotic performance of the artificial chromosome. Also, a centromeric section may be turned around without affecting the mitotic efficiency.

What is the significance of these studies involving the splicing of DNA sequences to produce artificial chromosomes? The ability to synthesize artificial chromosomes will allow researchers to study the details of mitosis and meiosis more closely. Various artificial chromosomes with differing amounts of the four kinds of chromosomal DNA will contribute to finding out the precise and special roles of each kind of DNA. These studies rely on the fact that a manageable procedure for introducing artificial chromosomes into yeast has been developed.

Another application will be in constructing artificial chromosomes that contain specific gene sequences and their control systems. If these artificial chromosomes are introduced into cells of individuals with genetic defects, the possibility of alleviating the symptoms of the defect exists. Phenylketonuria, sickle-cell anemia, and alcaptonuria are examples of genetic diseases which might be treated with artificial chromosomes bearing normal gene sequences.

group will be found to be in about the same stage of mitosis at a particular time of the day (Fig. 6-9). Synchronism seems to be due to a relationship between an internal rhythm (sometimes termed a **biorhythm,** or **biological clock**) and some environmental factor such as temperature or light. Nonsynchronous cells may be synchronized by exposing them to a sudden temperature change, a bright light, or some other rapid alteration in an environmental factor. If these newly synchronized cells are maintained in a uniform environment, they will continue to divide synchronously for a while.

It should be pointed out that once an organism attains its adult state, divisions of cells in many of its tissues become much less frequent, and in certain cases (e.g., nerve cells) cease altogether.

In plants and animals, organic molecules called **hormones** also regulate cell division in addition to their other functions. In animals, **growth hormone** secreted by the pituitary gland regulates cell division during childhood. If there is an excess of growth hormone, an individual will become abnormally large; if there is too little, subnormal stature results. In plants, several hormones act together to control cell division, principally in growth regions called **meristems** at the tips of stems and roots. Within animal cells, and perhaps plant cells, a molecule named **cyclic adenosine monophosphate (cyclic AMP)** also appears to regulate cell division (see Fig. 10-3).

Wounding of most kinds of tissue stimulates cell division. This is easily observed when a person is cut or scratched. It has been suggested that damaged cells may produce a substance that induces undamaged cells to undergo mitosis, but such a substance has not yet been isolated and identified. On the other hand, some kinds of cells have been found to contain mitotic inhibitors called **chalones,** which are thought to suppress mitosis in undamaged cells. When cells become damaged, their chalones are believed to "leak out"; this reduction in the concentration of chalones in the vicinity of a wound then may permit cells to divide. When a wound is healed, the normal levels of chalones would be restored, and the

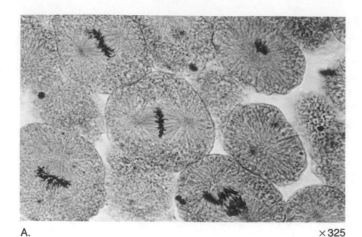

A. ×325

B. ×325

FIGURE 6-9
Mitosis in animal and plant cells. **A.** Synchronous divisions in cells of a whitefish early embryo. **B.** Nonsynchronous divisions in cells of an onion.

cells in effect would be "instructed" to stop dividing. This action is another example of feedback control of a life process. Probably all normal cell and tissue repair actions are regulated by similar controlling systems. Cancer is a result of the loss of control of cell division, but it is not yet understood how the control system works or what substances are involved.

Meiosis—Life Cycles

Often in discussions of mitosis, examples are used in which dividing cells have the $2n$ chromosome number. For instance, human somatic cells contain 46 chromosomes (the $2n$ number), and this number is maintained by mitosis in all succeeding mitotic divisions. In contrast, meiosis in human reproductive cells produces gametes (eggs and sperm) containing 23 chromosomes, the $1n$ number. This difference in chromosome number could lead to the erroneous assumption that mitosis al-

ways produces $2n$ cells. This indeed is the case in nearly all animal cells. However, in many kinds of plants, the $1n$ cells produced by meiosis are not gametes but asexual reproductive cells called spores. Spores divide many times by mitosis before some of them are transformed finally into gametes. Thus mitosis is a process that can occur in both $1n$ and $2n$ cells. The sometimes complicated relationships between generations of $1n$ cells resulting from meiosis, where chromosome numbers are reduced, and $2n$ cells resulting from fertilization (union of two gametes), where chromosome numbers are doubled, are called **life cycles** (Fig. 6-10).

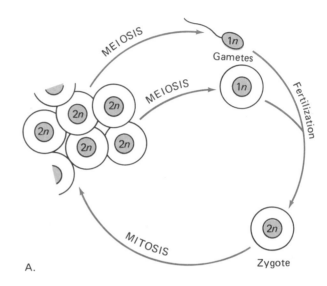

A.

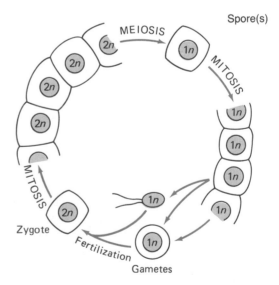

B.

FIGURE 6-10
Life cycles. **A.** Life cycle as found in humans, where meiosis occurs just before gametes are formed. **B.** Life cycle in which mitosis occurs in both $1n$ and $2n$ cells. Such a cycle, in which the first cells produced by meiosis function as asexual reproductive cells called spores, is found in many plants.

Sex The topic of sex never fails to generate interest, and poems, operas, tragic dramas, and even comedies have been written in tribute to the mysteries and powers of sexuality. Yet with all its mystery, romance, and intrigue, the subject of sex has a simple basis and can in fact be reduced to the level of DNA and what happens to it in reproduction.

The exchange of genetic information (the DNA code) between organisms is important because it results in the recombination of various genetic traits (genes) among many more individuals than would be possible if all reproduction occurred by mitosis alone. Sex cannot be dissociated from heredity and evolution; this interrelationship will be emphasized in this and following chapters.

The first cells in the world probably reproduced asexually by simple processes of cellular growth and cell division. Slowly these processes were replaced by more efficient mechanisms for the duplication and accurate distribution of DNA among new cells. As cells became more numerous in the ancient seas, competition increased for food supplies and space in which to live and grow, making efficient reproduction more and more important.

The growth and development of organisms is controlled by the DNA of genes. If a gene mutates (changes even a single nucleotide in its code), the product of the gene also will be altered in most cases. Most mutations are disadvantageous, although in some instances mutations confer an advantage.

Suppose, for example, that over a long period of time in a population of primitive cells there arose a series of mutations that resulted in the conversion of solar energy into the chemical energy of food. Obviously such cells would be highly effective competitors. Suppose further that in other primitive cells there occurred other mutations that eventually produced swimming appendages (flagella). Such cells also would have a competitive advantage, in that they could migrate to where the food was plentiful. These genes would be passed on to succeeding generations of new cells by DNA replication and cellular division. Then, if the cells could evolve a method of exchanging their genes, individuals might arise that would be both motile and photosynthetic. Presumably this would be even more advantageous. The example is hypothetical and simplified, but even prokaryotes have evolved a simple kind of sexuality by which genes are exchanged and thus combined in new arrangements **(genetic recombination)**. Genetic recombination is an essential result of sexual reproduction.

Sexual reproduction by gametes is unknown among prokaryotes. However, in the past several decades it has been discovered that genes are transferred between bacterial cells by **conjugation**. Because such exchange involves simply the transfer of DNA without the formation of sexual organs or specialized sex cells, it is known as **parasexuality**. As noted in Chapter 5, the parasexual process of bacterial conjugation is induced by a plasmid fertility factor (F). If a bacterium lacks the fertility factor, it will not conjugate. If a bacterium contains an F plasmid, it can conjugate by forming a cytoplasmic tube joining it to another bacterial cell. The DNA of the F-containing bacterium becomes replicated, and the extra copy of DNA then is inserted into the other bacterium through the cytoplasmic "bridge." It is a one-way process: only one cell receives DNA from the other. After the transfer, the cells separate and the DNA recipient then combines the new DNA with its own DNA (Fig. 5-14).

In one-celled forms of life, genetic exchange occurs between individual unspecialized cells rather than between specialized sex cells as happens in more complex groups. In prokaryotes only pieces of DNA are transferred between cells. In eukaryotes the nuclei of two cells are fused together, and often the cells become united as well.

It is not known what factors led to the first instances of nuclear fusion in evolving eukaryotes. In almost all present-day multicellular animals, gametes are $1n$ in chromosome number and hence have less DNA than do somatic cells. When gametes fuse together in fertilization processes, the result is a larger $2n$ cell, the zygote. Supposedly the zygote, and the $2n$ cells descended from it, would be physiologically stronger, having two sets of chromosomes, twice as much DNA, and consequently a greater capacity for biosynthesis. Whether this apparent advantage would have been sufficient to allow natural selection to operate in a direction favoring gamete formation and fertilization cannot be said. However, it is possible to write a scenario in which the evolution of sex would have been implemented by selection of cells equipped both for genetic exchange by $1n$ gametes and for survival as a consequence of being $2n$. Correlated with these evolutionary steps would be the origin of a division process capable of restoring the gametic chromosome number to $1n$ (meiosis).

Meiosis I and II Unlike mitosis, meiosis is a double nuclear division process. The first nuclear division, called **meiosis I** and often referred to as a **reductional division**, results in daughter nuclei having half the pairs of chromosomes of the mother cell nucleus. The second nuclear division, called **meiosis II** and often said to be the **equational division**, resembles mitosis. Earlier it was stated that meiosis is important for two reasons: (1) it reduces chromosomes from $2n$ to $1n$, which is essential if chromosome numbers are to remain constant in succeeding generations, and (2) it facilitates genetic recombination. It is the second of these that will now be considered.

Suppose that the 2n cells of a hypothetical organism contain two pairs of chromosomes, one pair designated A and A', the second pair B and B'. One chromosome of each pair originated in a gamete derived from a female parent; the others came from the male parent (Fig. 6-11). It might be suspected that when meiosis occurs, all the chromosomes originating from the female parent will go into the formation of female gametes, and those from the male parent will go into the formation of male gametes. This, however, is not the case. The specific distribution of chromosome pair members depends upon their orientation in the spindle during metaphase of the first meiotic division, called meiosis I (see Fig. 6-12). (Note that this differs from mitosis, in which individual, unpaired chromosomes line up at metaphase.)

The chromosomes in premeiotic cells, like those in premitotic cells, each consist of one double helix of DNA. We can call the chromosome in this state a **monad.** In meiosis as in mitosis, the single chromosome becomes duplicated prior to nuclear division, so that each chromosome is now composed of two chromatids held together at their centromeric region. We will refer to this configuration as a **dyad.** Then, during prophase of meiosis I, chromosome pairs known as **homologous chromosomes,** or simply as homologs, come together in a close association known as a **tetrad** (tetrad = four, meaning that four chromatids [two dyads] are arranged in parallel):

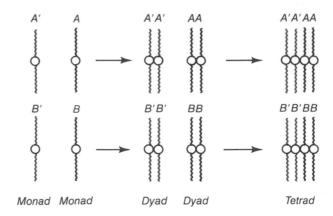

For the sake of simplicity, the shortening of the chromosomes occurring as meiosis progresses is not shown in the diagram above. A more accurate representation of chromosomal changes during meiosis is shown diagrammatically in Fig. 6-13, and is evident in photomicrographs of meiosis in grasshopper testes (Fig. 6-14).

Following metaphase of meiosis I, chromosome pair dyads are separated from each other: A'A' from AA, B'B' from BB, etc., and the dyads are drawn to opposite poles by action of spindle fibers. This ensures the separation of dyads (i.e., **paternal** from **maternal** chromosomes), but does not guarantee to which pole each dyad will go. That depends on the orientation of the tetrads in the spindle, as shown in Fig. 6-12.

So far as is known, tetrads are oriented in the spindle at metaphase by pure chance. Therefore, if we examine the products of meiosis, we find that about one-

FIGURE 6-11
A 2n cell of a hypothetical organism, showing origins of the chromosomes. This concept is continued in Fig. 6-12, where the assortment of chromosomes of the two parental types above is shown.

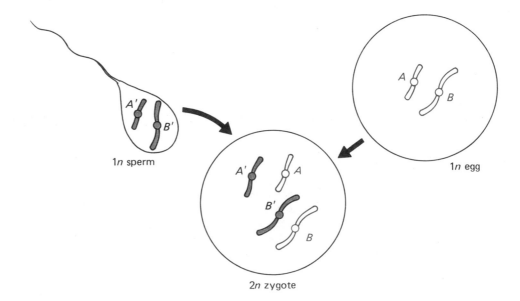

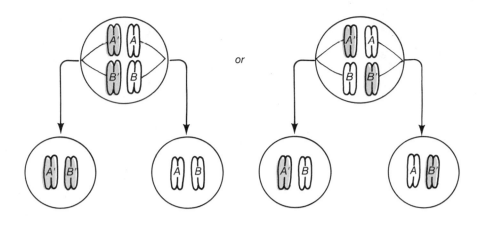

FIGURE 6-12
Possible assortment of chromosomes of parental origin occurring in cells of a son or daughter. For simplicity's sake, we show only two pairs of chromosomes. If more chromosomes are considered, the numbers of possible combinations of parental-type chromosomes are increased accordingly. In this diagram, four kinds of gametes are identifiable, based on chromosomal combinations. Note that this diagram is a continuation of information presented in Fig. 6-11, and that the color scheme is continued in Fig. 6-13.

fourth of the daughter cells will contain the combination of A and B, one-fourth will contain the combination A′ and B′, one-fourth will contain A′ and B, and one-fourth will have the A and B′ combination (this is illustrated also in Fig. 6-12). We shall see in the next chapter how important this is for the inheritance of genes.

One of the remarkable occurrences in prophase of meiosis I is a regularly occurring event called **crossing over**. Crossing over is just what its name implies, the actual crossing over of chromatids at certain, apparently preestablished, points. The result is a crossover or **chiasma** (pl. **chiasmata**), which is quite visible as the chromosomes continue to shorten (see Fig. 6-14D). Looking at crossing over diagrammatically, one can imagine it producing an actual exchange of chromosome material:

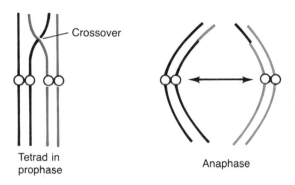

Tetrad in prophase

Crossover

Anaphase

Following metaphase in meiosis I, anaphase occurs; chromosome pair members become separated from each other and are conveyed along microtubules to the spindle poles. Telophase follows, and two daughter cells are formed by cytoplasmic division. In a few cases, division of the cytoplasm does not occur or is incomplete, but in all cases of meiosis I, two nuclei each are produced.

In summary, Meiosis I (Figs. 6-13, A–F, and 6-14) is a process of separation of chromosomes in a premeiotic cell into two sets, each containing half the homologous chromosomes present in the original nucleus. Thus at the end of meiosis I, each daughter nucleus will contain one dyad from each dyad pair (tetrad).

Meiosis II (Fig. 6-13G and H) usually follows meiosis I without pause. It resembles mitosis and need not be discussed in detail. Chromatid separation occurs in metaphase and anaphase as in mitosis, with the result that four $1n$ cells are produced. These can function as spores or gametes, depending on the organism, and can carry all possible combinations of the chromatids present in the original two pairs of chromosomes. If, on the other hand, a larger number of pairs are present in meiotic cells, a greater number of $1n$ cells must be examined to find all possible combinations of chromosomes. The significance of this will become apparent later when the subject of inheritance is considered in Chapter 7.

Gametes   The immediate products of meiosis in animals are gametes. Typically, the male gametes are tiny flagellated **sperm**; the female gametes are the much larger **eggs**. Each of the four cells produced by meiosis of a $2n$ **spermatocyte** (a male sex cell) is a **spermatid**, and becomes differentiated into a sperm by the development of a flagellum. Eggs originate by meiosis from an **oocyte** (the female equivalent of a spermatocyte), which in most animals produces one large functional gamete, the egg, and three tiny nonfunctional gametes called **polar bodies**.

The products of meiosis in protistans, fungi, and plants are various, depending on the life cycle. Sometimes they are nuclei rather than cells, but more often they are gametes, as is the case in animals. In fungi and

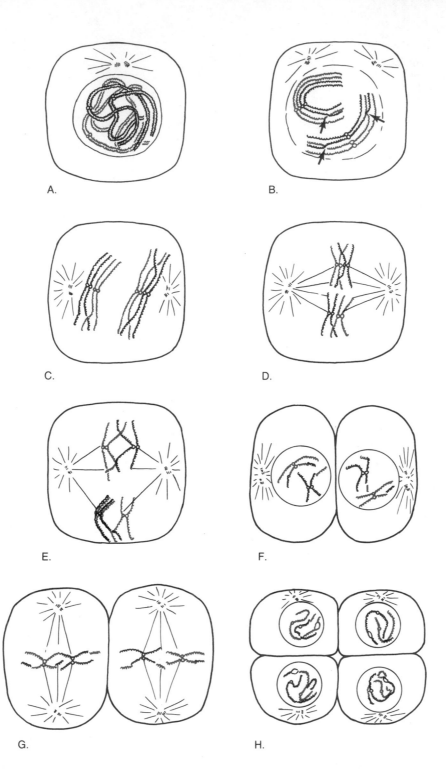

FIGURE 6-13
Summary of events in meiosis. **A–F.** Meiosis I. **A.** Early prophase. Chromosomes are very elongated. Chromatids are present. **B.** Mid-prophase. Chromosomes are shortening by coiling. Synapsis has occurred. Arrows indicate crossovers. **C.** Late prophase. Chromosomes are greatly shortened. Crossovers (chiasmata) are very obvious. **D.** Metaphase. Exchange of pieces of chromatids is shown. **E.** Anaphase. **F.** Telophase. (Two shades of color indicate the different parental origins of the chromosomes.) **G** and **H.** Meiosis II. **G.** Mitotic nature of meiosis II is shown. **H.** Telophase. Here the results of chromosomal crossing over are seen. These diagrams depict meiosis as it might occur in an animal cell. In plants, centrioles ordinarily would not be present and the cells would be enclosed within cell walls. Compare stages above with equivalent stages in photographs in Fig. 6-14.

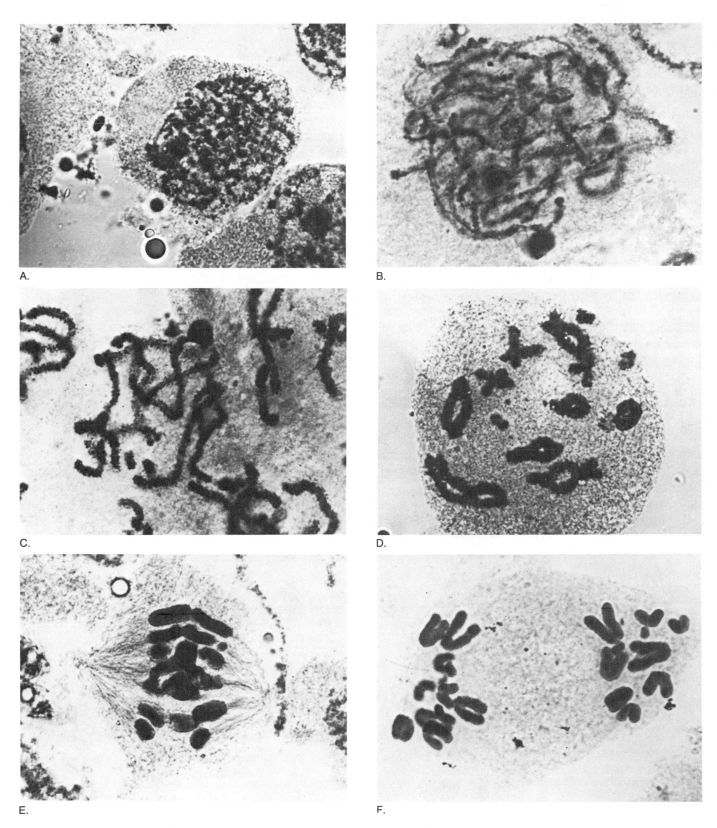

FIGURE 6-14

Selected stages of meiosis I in cells of the testes of the lubber grasshopper *(Romalea microptera).* **A.** Early prophase. Compare with Fig. 6-13A. **B.** Later prophase. Chromosomes are shortening. **C.** Mid-prophase. Synapsis. Compare with Fig. 6-13B. **D.** Late prophase. Crossovers are visible. Compare with Fig. 6-13C. **E.** Early anaphase. Compare with Fig. 6-13E. **F.** Telophase. A bit earlier than Fig. 6-13F.

plants, they commonly are **spores (meiospores)**, and gamete formation is delayed until later in the life cycle (Fig. 16-2).

Although it is not known how gametes originated in the first place, some of the factors that operate at present to cause cells to form gametes are well known.

Among many unicellular organisms, including one-celled algae, the normal state is the $1n$ condition, and gametes are merely transformed from nonsexual cells. The stimuli that induce this change usually are environmental factors, including decreases (or increases) in available nutrients, reduced (or elevated) temperatures, shorter (or longer) days, or a combination of these.

When $1n$ cells fuse, a $2n$ zygote results, and in animals the new individual produced by mitosis of the zygote also is $2n$. If such $2n$ individuals were to produce gametes by mitosis, there would be a redoubling of chromosomes in each generation. Not only is this impossible in terms of nuclear space, but the functional efficiency of most organisms does not appear to be increased when multiples of chromosome numbers much greater than $2n$ are present. Instead, $2n$ cells are stimulated, usually by hormones, to divide by meiosis and produce smaller, $1n$ cells (gametes), which, upon fertilization, restore the $2n$ number. In many plants and animals, the same environmental factors that induce gamete production in unicells are similarly active in inducing sexuality. Among the more important factors are changes in day length (**photoperiod**) and daily temperatures (**thermoperiod**).

The necessity for reduction of chromosome number from $2n$ to $1n$ is an important function of meiosis. Also important in meiosis is the sorting out of chromosomes that insures that gametes will contain various combinations of chromosomes and their genes. Thus genetic recombination depends upon meiosis as much as it does upon fertilization.

Sexuality is commonly associated with differences between gametes produced by organisms of the same species. These differences may be subtle or they may be quite obvious, depending on the nature of the organism. It is necessary for genetic recombination that gametes differ from each other in their genetic makeup, and it is also necessary that they be able to discriminate among one another so that exchange of DNA from different cells will occur. If gametes from the same organism simply recombined with each other, many of the advantages afforded by genetic recombination of DNA from different sources would be lost.

Most organisms have evolved mechanisms that replace self-fertilization. In many organisms, eggs are formed by some individuals and sperm are formed by others, and only male and female gametes from different individuals combine with each other. In others (usually simple organisms), the gametes look alike but are chemically dissimilar. Such gametes commonly are referred to as plus and minus, rather than egg and sperm. Plus gametes ordinarily do not fuse with other plus gametes, nor do minus gametes fuse with minus gametes. **Gametic fusion** occurs only between plus and minus gametes (Fig. 6-15). These differences in gametes insure that union of gametes will occur between different individuals rather than between cells of the same individual.

Probably the first step in the evolution of sex cells was the development of plus and minus strains of cells. Later on, the more distinctive male and female gametes

FIGURE 6-15
Gametic fusion in *Chlamydomonas*. Plus and minus cells mixed together on a slide are observed to associate in pairs.

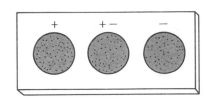

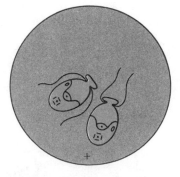
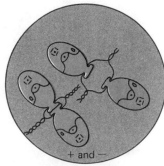
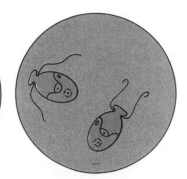

evolved, and with them individuals of each sex. (It is conventional to symbolize male as ♂ and female as ♀; ♂ is the sign of Mars [the shield and spear], and ♀ is the sign of Venus [the mirror].)

How do gametes recognize other gametes? When gametes of the plus mating type of an alga such as *Chlamydomonas* are mixed with minus gametes, mating occurs almost instantly. The explanation is that chemical differences occur between the two kinds of cells. These differences consist of recognition chemicals which operate on the key-in-lock principle, mentioned in connection with enzymes. Recognition proteins present in the outer membranes of the flagella of plus and minus cells of *Chlamydomonas* interlock upon contact and coupling occurs (Fig. 6-15). Then the cells move closer together, and finally their cytoplasm and nuclei fuse together. These recognition proteins are widespread among gametes at all levels of complexity.

BIOLOGICAL DEVELOPMENT

Biological development includes many processes. A **bacteriophage virus** infects a bacterial cell and is replicated many times over. An amoeba divides and develops into two new amoebas, each a duplicate of the other. A fertilized egg becomes a many-celled plant or animal. A grasshopper loses a leg and proceeds to grow a new one. A tree sheds its leaves in autumn and grows a new set the following spring. All of these, and many more, are examples of biological development. What, if anything, do they all have in common?

In most examples of development there is a change in the form of the cells involved, and in all there occurs a replication of structure. In addition, developmental processes are regulated by DNA and involve internal, DNA-controlled responses to external environmental factors. Some of these developmental processes will now be considered in greater detail.

Assembly of a Virus

A virus is a small particle composed of DNA (sometimes RNA) and proteins. Although viruses are not cells, some biologists consider them to be the simplest examples of life; others are reluctant to say that viruses are in any way alive. At any rate, viruses have the ability to reproduce, provided they can use the biosynthetic machinery of a host cell, and something can be learned about molecular aspects of development in living cells by studying viruses and their reproduction.

Each virus particle is composed of a number of protein "building blocks." These are synthesized in a host cell whose own DNA has been broken down by the virus, which then substitutes its own DNA for that of its host (Fig. 5-11). When a bacteriophage virus infects a bacterial cell, the protein units that constitute the virus body come together in a kind of assembly-line sequence. Apparently, as the pieces of the virus are produced and accumulate in the cytoplasmic "soup" of the host cell, they jostle each other and move about until they match up. Obviously the time element here is very important; the components must arrive in proper sequence so that they have other parts to conform to.

In terms of biological development, an important point to be garnered from this brief analysis of the replication of a virus is that the components of a complex structure can self-assemble.

Self-assembly and Cell Shape

Although organization of cells is vastly more complicated than that of viruses, it is probable that **self-assembly** of structural proteins occurs along the lines described for assembly of viruses and that such self-assembly is of considerable importance in determining the ultimate form of specific types of cells.

Microtubules, structural components of cells, were discussed in earlier chapters in relation to cell "skeletons" (**cytoskeletons**), cilia and flagella, and spindle fibers of dividing cells. Although such structures have different functions, the microtubules that compose them are basically alike; in fact, so are microtubules of all eukaryotic cells. The microtubules are in all cases about 25 nm in diameter, although of varying length (up to 25 μm within cells and 150 μm in flagella), and are composed of 13 rows of beadlike **protofilaments** (Fig. 6-16). The subunits of protofilaments are globular proteins called **tubulins**. Each tubulin molecule has two dissimilar binding sites, one at each end; these are functionally analogous to the balls and sockets of "snap beads."

Evidence from many experiments indicates that microtubules can assemble from subunits readily and can dissociate again into subunits. The appearance of a spindle in metaphase and its disappearance at telophase is a good example of this. Although the factors responsible for assembly and disassembly of microtubules are largely unknown, it is thought that organization sites called **microtubule organizing centers (MTOCs)** appear first. An MTOC then induces the assembly of tubulin subunits into protofilaments, which in turn form the microtubule itself. Centrioles, for example, function as MTOCs for spindle fibers, and basal bodies serve as assembly centers for cilia and flagella.

About 40 years ago it was observed that **colchicine**, a rather toxic extract of the autumn crocus (*Colchicum autumnale*) and long used to treat gout, could stop cells

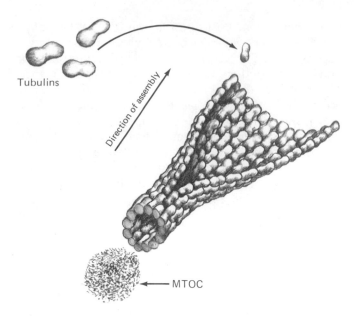

Tubulins

Direction of assembly

MTOC

FIGURE 6-16
Microtubules, subunits, and assembly.

from dividing. Examination of colchicine-treated cells showed that this occurred at metaphase by disruption of the spindle. Subsequent studies revealed that a colchicine molecule becomes attached to one of the two dissimilar binding sites of tubulin and thus prevents the self-assembly of protofilaments. Another substance, **vinblastine,** extracted from the periwinkle plant (*Vinca minor*), becomes attached to the second binding site of tubulin and likewise disrupts microtubule assembly. Vinblastine is less toxic than colchicine and is used to treat some cancers because it can prevent division of cancer cells (along with that of normal cells).

Because the chemicals colchicine and vinblastine can prevent cells from forming microtubules, they are extremely useful in experiments devised to study roles of microtubules in cells. Of particular interest in developmental studies is evidence regarding microtubules and cell shape. Many cells, in particular those of unicellular microorganisms, have distinctive and characteristic shapes. When such cells are treated with colchicine, they typically lose their unique form and become more or less shapeless blobs. Electron-microscopic examinations show that these blobs lack microtubules; however, if they are rinsed free of colchicine, they recover their unique shape and again may be found to contain microtubules.

Tubulin subunits can be obtained in test tube solutions by washing and centrifuging disrupted cells. Under some conditions (specific pH and calcium-ion concentration), dissociated tubulin subunits in such solutions can be induced to assemble into protofila-

ments and microtubules. An early step in microtubule assembly appears to be the formation of flat ribbons containing 13 protofilaments, followed by the curling of the ribbons into typical microtubules indistinguishable from those of living cells. Although it is by no means certain that microtubules develop in the cell by self-assembly, it is a fact that they appear and disappear rapidly, can be manipulated chemically, and are structurally consistent from cell type to cell type. Thus microtubules present evidence of the role large molecules play in the structure of cells. The lesson to be learned from this is that cell development has a molecular basis, and that at least some kinds of development can be discussed in fairly simple terms.

Development in Single-celled Organisms

Several patterns of reproduction and development are found among unicellular organisms. The *Amoeba* and similar unicells of no definite form simply continue to divide and grow in an irregular fashion. On the other hand, many other unicells, as well as multicellular organisms, have more complex patterns of growth. To a certain extent these can be described according to their structural form, or **symmetry** (see Fig. 8-2).

A small one-celled organism called *Stentor* (the trumpet) has been the subject for a number of studies of development at the unicellular level. *Stentor* belongs to a class of microorganisms (microscopic life) commonly called **protozoans.** Because it is one of the largest protozoans, approaching a millimeter in length, *Stentor* serves admirably for microsurgical experiments in which delicate scalpels and needles are used, with the aid of a microscope, to perform surgery on cells.

Stentor is a complex cell. Its form resembles a cone, tapered to a blunt point at one end and widened at the other end into a funnellike **mouth** (Fig. 6-17). A spiral row of cilia lines the funnel, and by adding dye particles to a culture of *Stentor* cells, the beating of cilia in the funnel can be seen to produce a minature whirlpool, or **vortex,** which carries dye particles down into the mouth. A thin flexible layer, the **pellicle,** protects the cell (in particular the outer region, or **cortex**) and presents a surface pattern of parallel strips of dense cytoplasm containing pigment granules alternating with colorless strips bearing rows of cilia. One zone of cilia, called the **fine line,** appears to play a dominating role in development. In the cytoplasm to the inside of the cortex, *Stentor* contains the usual assortment of cellular organelles, such as mitochondria, endoplasmic reticulum, Golgi bodies, nuclei, and contractile vacuoles.

The reproduction of *Stentor* involves a partial reorganization of many of the cellular components. Because

FIGURE 6-17
Grafting of the fine-line zone in
Stentor, with resulting **regeneration** of the mouth region. (After
Tartar, 1962)

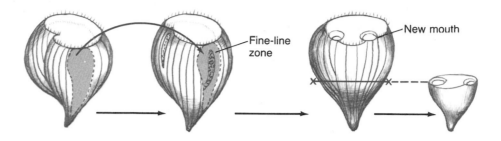

Stentor has a spiral form, it cannot divide itself into similar halves. Rather, it begins to constrict about the middle, assuming an hourglass shape, and a new spiral band of cilia begins to form about halfway down, just posterior to the constriction. Finally, the constriction is completed, the cells separate, and the spiral band of the posterior cell moves to a terminal position to create a new mouth and gullet. The anterior cell forms a new base.

The reproduction of *Stentor* differs from the reproduction of a virus in important ways. First, viral reproduction occurs within a host cell, whose biochemical machinery is "borrowed" for the occasion. Second, the replication of a virus is a completely new construction: not one part of the original virus is used to make a new virus. In *Stentor*, on the other hand, each daughter cell incorporates something of the parent cell.

If a *Stentor* is minced with a microscalpel into as many as 40 tiny interconnected pieces, and if the pieces are kept together, the organism will reassemble itself into a new *Stentor* within 24 hours, with the fine-line zone acting as the **organizer** for this reconstruction. As shall be shown, the concept of an organizer zone or region is important in many developmental patterns, both in unicellular and multicellular organisms.

If the mouth is cut out of a *Stentor*, a new mouth forms in the region of the fine-line zone. Moreover, if a patch of the cortex of one *Stentor*, including a fine-line zone, is grafted into another *Stentor*, the recipient cell will develop an extra mouth. If such a double-mouthed *Stentor* is replicated, either by cutting it in half or as a result of cell division, the mouthless half-cell, or daughter cell, will develop not one mouth, as would a normal *Stentor*, but two mouths (Fig. 6-17). Thus the cortex of *Stentor* somehow seems to lead a developmental life of its own, irrespective of its nuclear DNA coding for the normal cellular developmental patterns. However, if a double-mouthed *Stentor* is permitted to divide a number of times, the normal, single-mouthed pattern eventually is restored. The conclusion that may be drawn from grafting experiments with *Stentor* (as well as from experiments with other organisms, described later in the chapter) is

that the cytoplasm and nucleus act in concert to control development.

Development in some unicellular organisms is not quite so directly comprehensible as it is in *Stentor*, where a partial framework acts as an organizer in the restoration of the integrity of the whole organism. For example, in many unicells (and in many multicellular organisms as well), the organized structure of the individual seems to "dissolve" away during part of the reproductive process. For instance, the highly complex human body develops from a simple spherical cell, the zygote, in which there is no reminder of the complex body that produced it or of the individual to come (in contrast, see Fig. 11-1 for an early theory regarding the presence of a preformed human in the sperm or egg).

Nuclear–Cytoplasmic Interactions

It is noteworthy that both a nucleus and cytoplasm are necessary for cell replication, and many experiments have been performed to test this relationship. For instance, amoebas have a rather formless cortex capable of extension and retraction of **pseudopodia** ("false feet"), which are organelles of locomotion and ingestion. Using **micromanipulators,** tiny surgical instruments operated mechanically under the high magnification of a microscope, biologists have been able to remove the nucleus of the *Amoeba*. The *Amoeba* continues to survive and may actually undergo cytoplasmic division into two smaller amoebas, retaining some ability to form pseudopodia. However, these **enucleated** amoebas (amoebas from which nuclei have been removed) do not grow, but gradually shrink and die. This results because the nucleus contains the chromosomes, which in turn are composed of DNA in the form of genes. Without genes, protein synthesis and concomitant growth are impossible.

Acetabularia is a green marine alga of beautiful form. Its common name, the mermaid's wineglass, is quite descriptive, although in colonies its members look even more like delicate soft green parasols (Fig. 6-18). Most remarkably, *Acetabularia* is a single giant cell, often exceeding 5 cm in length; its parasol-like cap is about 1.5 cm across. The mature organism consists of a basal root-

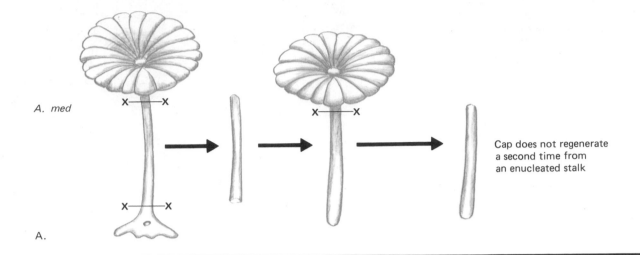

A. med

Cap does not regenerate
a second time from
an enucleated stalk

A.

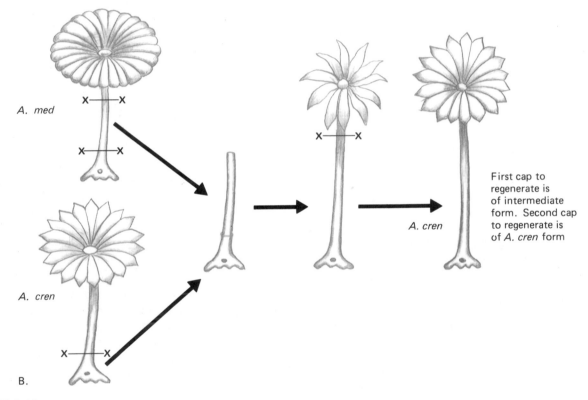

A. med

A. cren

A. cren

First cap to
regenerate is
of intermediate
form. Second cap
to regenerate is
of *A. cren* form

B.

FIGURE 6-18
Grafting experiments illustrating nuclear–cytoplasmic interrelationships in *Acetabularia*. **A.** *Acetabularia mediterranea.* Regeneration of stalk portion without nucleus. **B.** *A. crenulata* base with nucleus grafted with an anucleate stalk of *A. mediterranea.*

like anchorage organ (the **rhizoid**), a stalk, and the cap. Through most of the life of *Acetabularia*, there is a single large nucleus in the basal region. Only when reproduction occurs does this nucleus divide into many smaller nuclei, each of which is incorporated into a flagellated gamete.

More than two decades ago a remarkable series of experiments were done by Joachim Hämmerling, a German biologist. In one of his experiments, Hämmerling cut off the cap of an *Acetabularia* cell and found that the remaining stump of the stalk could regenerate a new cap. Furthermore, *Acetabularia* regeneration occurred even after the nucleus had been removed. In the latter circumstance, however, the powers of regeneration proved to be rather limited, and after a second decapping there was usually no further regeneration (Fig. 6-18A). The role of the nucleus in this case is much the same as that demonstrated for the *Amoeba*: it is required for the synthesis of new cell components.

Hämmerling did other experiments in which he grafted the basal portion (with a nucleus) of one species of *Acetabularia* to the capless, bottomless stalk (without a nucleus) of a second species (Fig. 6-18B). The two species differed in the structural pattern of the cap: the cap of *Acetabularia mediterranea*, often abbreviated A. *med*, is more regular in appearance than that of A. *crenulata*, abbreviated A. *cren*. Surprisingly, when this grafted composite individual regenerated, the cap was intermediate in form between A. *cren* and A. *med*, but after a second decapping the newly regenerated cap was much more typical of A. *med*, from which the nucleated base was obtained. The reverse experiment, in which a base of A. *cren* with its nucleus was grafted into an A. *med* stalk, produced essentially the same result: the first cap regenerated was of intermediate form, and later ones resembled those of A. *cren*. (Fig. 6-18A, B).

Although Hämmerling's experiments with *Acetabularia* very quickly were seen as a significant contribution to the study of biological development, it was some time before a convincing explanation of the action was found. Now it is recognized that the residual ability of the cytoplasm of an enucleated and decapped *Acetabularia* cell to regenerate a new cap is due to mRNA's remaining in the cytoplasm after the removal of the nucleus. For a time, this remaining RNA functions to synthesize sufficient structural proteins and enzymes to form a cap. However, mRNA usually is not a long-lived molecule, and replenishment by fresh transcription from DNA is required for continued development. Thus the exhaustion of the mRNA in the twice-decapped cell precludes further cap regeneration.

The regeneration of intermediate types of caps in grafted hybrids of A. *cren* nuclei and A. *med* cytoplasm, and vice versa, also can be explained in terms of mRNA. The first regeneration of a new cap used the A. *med* type of mRNA remaining in the cytoplasm of the stalk, plus new mRNA transcribed in the A. *cren* nucleus. Thus mRNAs of both species are translated upon the ribosomes of the host cytoplasm. The "message" of this experiment is clear: (1) both nucleus and cytoplasm are required for development, (2) the shape of a cell, and indeed of an organism, is dependent upon the synthesis of specific subunits, the cytoplasmic proteins, and (3) the cytoplasm of the cell has great flexibility in producing the components needed to build a specifically formed structure. The nucleus of a cell thus is analogous to an architect's blueprints, and cytoplasm may be compared to the workers who assemble the structure.

DNA and Development In Chapter 5 the Jacob-Monod model of control of protein synthesis by repression and derepression of an operator gene was presented. Similar mechanisms have been proposed for development in eukaryotic cells, but here the situation is more complicated and there are many unanswered questions. One difficulty is that the DNA of eukaryotic cells is complexed with protein in chromosomes instead of consisting of a single circle of "naked" DNA as in the prokaryotic cell. The diversity of cell types in a multicellular organism presents an additional complication that is particularly puzzling from a developmental point of view: even though all the cells of the organism originate from the same zygote (or spore) and consequently have identical DNA, this uniformity of DNA does not result in cellular uniformity. How, then, does the same kind of DNA produce so many different kinds of cells? The answer to this puzzle appears to be that only some regions of chromosomal DNA are active in each cell type.

Just how chromosomal DNA in multicellular eukaryotes is repressed and derepressed remains a mystery, as noted in Chapter 5, but it may involve a group of five different chromosomal proteins, the **histones**. Histones apparently operate by binding to regions of DNA, possibly blocks of genes, and rendering them inaccessible for transcription. Other, nonhistone proteins then might act as derepressors by causing histone-bound coils of DNA (nucleosomes) to be opened up for transcription. Also involved in gene activation in eukaryotes are certain hormones and at least two nucleotides: cyclic adenosine monophosphate (cyclic AMP) and **cyclic guanosine monophosphate (cyclic GMP)**. These nucleotides are made from ATP and GTP, respectively, by the removal of the two terminal phosphates and the bending back and attachment of the inner phosphate so that a small circle is formed; hence the cyclic designation (Fig. 10-3). The enzymes **adenylcyclase** and **guanylcyclase** catalyze these modifications.

Another kind of developmental stimulation by cAMP is seen in a much studied organism, *Dictyostelium discoideum*, a **cellular slime mold**. Cellular slime molds commonly occur as small, naked cells (**myxamoebas**) capable of amoeboid locomotion and phagocytosis. Eventually, myxamoebas undergo a behavioral change and secrete cAMP into their environment. This occurs when food supplies diminish, and cAMP serves as an attractant to other myxamoebas, which then congregate and join together, forming a larger protoplasmic mass called a **slug**. The slug moves slowly for a time, then forms an erect stalk topped by a spherical capsule. Small nonmotile cells, or spores, are formed within the capsule. The spores have thick walls and can survive many environmental stresses, but when environmental conditions improve, the spores open up and new myxamoebas creep out to begin the cycle of feeding, movement, and growth all over again. In this example, cAMP is involved as a messenger leading to the aggregation of protoplasm and its differentiation into a new form.

Cell Specialization

In single-celled organisms parts of cells become **specialized** during development, whereas in many-celled organisms development involves the specialization of entire cells to perform specific functions. Moreover, there are a multitude of cell types among the several kingdoms of organisms, reflecting not only the ecological circumstances in which organisms live but also the degree of organization and development of the organisms. In complex multicellular organisms such as plants and animals, there may be dozens to hundreds of unique and distinctive cell types. These cell types, however, can be divided into a few major categories called **tissues** (a group of associated cells of similar structure and function). In animals these include bone, muscle, nervous tissue, blood, epithelium, and others. In plants, major tissues are water-conducting xylem, food-transporting phloem, epidermis, and a few others.

SUMMARY

Cell division is a process that occurs among organisms, whether they are prokaryotes or eukaryotes. In prokaryotes cell division consists simply of replication of the "naked" DNA in a cell and its equal distribution between the two new cells. Mitosis is a process of nuclear division occurring in eukaryotes in which the chromosomes and their DNA are replicated and distributed in equal kinds and numbers to the two new nuclei.

A distinction must be made between mitosis, involved in asexual reproduction, and meiosis, which is a process of sexual reproduction. In mitosis, each new cell has the same number of chromosomes as its mother cell; in meiosis, the new cell has half the number of chromosomes present in the mother cell. In the human cell, the $2n$ number is 46; the $1n$ number is 23.

In mitosis changes in chromosome structure, number, and distribution occur. A spindle composed of microtubules is formed; chromosomes become attached to certain of the microtubules and are maneuvered to the center of the spindle, where each chromosome becomes divided. The new chromosomes then are conveyed to opposite ends of the spindle. Then the spindle disappears and each set of chromosomes becomes part of a new nucleus.

In eukaryotes sexual reproduction involves two nuclear and cellular processes: meiosis and fertilization. Meiosis may be described as a kind of nuclear division in which the chromosomes of a $2n$ cell are first replicated to form dyads, then separated into two sets, each set containing half as many dyads as the original set. Each new nucleus, therefore, contains half as many duplicated chromosomes (dyads) as the original cell. Then a second nuclear division which resembles mitosis occurs. This second division separates the dyads into monads. Usually, cytoplasmic division occurs along with nuclear division, so meiosis ends with the production of four cells from each original cell. However, in some organisms only the nuclei in cells undergo meiosis and there is no cytokinesis.

Development at the molecular level involves the control of gene transcription and the synthesis of proteins and other cell constituents, as well as their assembly into larger units. Studies of viruses and the way they reproduce in cells have revealed some aspects of the self-assembly of protein molecules into more complex virus particles.

The control of development may be studied by various types of surgery upon cells and organisms. For example, a requirement for cooperation between the nucleus and cytoplasm in the development of the individual form of an organism has been deduced by cell microsurgery. A revealing series of experiments done with the large uninucleate alga *Acetabularia* underscores the requirements for cooperation of nuclear DNA and cytoplasmic mRNA in development. Thus, although there are many unsolved riddles of development, studies of viruses and single cells show that DNA, through the actions of mRNA, ribosomes, and enzymes, can direct the production of complex organismal structures.

In multicellular plants and animals, developmental processes occurring within cells lead to the formation of tissues of various types.

KEY WORDS

asexual reproduction	anaphase	cyclic AMP
mitosis	telophase	hormone
somatic cell	cell cycle	cell specialization
chromosome	chiasma	cytokinesis
sexual reproduction	crossing-over	mother cell
gamete	parasexuality	daughter cell
zygote	meiosis I	tetrad
haploid	meiosis II	synapsis
diploid	self-assembly	dyad
life cycle	cytoskeleton	monad
interphase	microtubule organizing	homologous chromosomes
chromatid	center	maternal chromosome
nucleosome	*Stentor*	paternal chromosome
prophase	regeneration	
spindle	nuclear–cytoplasmic	
centriole	interaction	
metaphase	*Acetabularia*	

QUESTIONS FOR REVIEW AND DISCUSSION

1 Consider a 2*n* cell containing 24 chromosomes. Are one or two sets of chromosomes present? If the cell replicates by mitosis, what will be the nature (number and kind) of the chromosomes in the two new cells? If the cell undergoes meiosis, what will be the chromosomal number of the new cells? What is the origin of the 24 chromosomes in the 2*n* cell?

2 It once was thought that chromosomes became split during mitosis. Why do you think this misconception originated? What happens to chromosomes in the mitotic cell other than subdivision at metaphase?

3 When is DNA replicated in relation to chromosomal replication and division? Explain.

4 What are some causes of cell division? Does cell division ever become a problem in the lives of organisms? Explain.

5 What is the function of meiosis and fertilization in the long-range survival of a species?

6 Compare the behavior of chromosomes during mitosis and meiosis. Why does meiosis not occur in a 1*n* cell?

7 Describe the relation of protein and of tubulin to microtubules, and of microtubules to the shape of a cell.

8 How might the relationship between microtubules and cell shape be tested experimentally? Could such experiments also be used to evaluate the roles of microtubules in cell divisions?

9 What is the explanation for the ability of an enucleated, decapped *Acetabularia* to regenerate a new cap?

10 What is the significance of the regeneration of a cap that is intermediate between the A. *cren* and A. *med* types in nuclear transplants in *Acetabularia*? Why does the grafted individual not continue to produce intermediate caps after repeated decappings?

SUGGESTED READING

DUPRAW, E. J. 1970. DNA *and chromosomes*. New York: Holt, Rinehart and Winston. (Unusually fine diagrams and micrographs. Emphasis is on the molecular organization of chromosomes, and on mitotic mechanisms.)

GIBOR, A. 1966. *Acetabularia*: a useful giant cell. *Sci. Amer.* 215(5):118–24. (Probably the best illustrated popular account of the work of Hämmerling and J. Brachet on this interesting green alga.)

GRANT, P. 1978. *Biology of developing systems*. New York: Holt, Rinehart and Winston. (Fine discussion on aging and differentiation.)

KORNBERG, R. D., and A. KLUG. 1981. The nucleosome. *Sci. Amer.* 244(2):52–64. (Structural and functional aspects of chromosomal histones and DNA are discussed.)

MARX, J. L. 1983. A step toward artificial chromosomes. *Science* 222:41. (A short progress report.)

MAZIA, D. 1974. The cell cycle. *Sci. Amer.* 230(1):55–64. (The time frame of cell division is discussed, with emphasis on "readiness" of cells to divide.)

SPOONER, B. S. 1975. Microfilaments, microtubules, and extracellular materials in morphogenesis. *BioScience* 25:440–51. (Describes microtubules and microfilaments and their actions in cell movements and cell shapes.)

WOLLMAN, E. L., and F. JACOB. 1956. Sexuality in bacteria. *Sci. Amer.* 195(1):109–18. (Presents a short discussion of some features of bacterial conjugation, transformation, and transduction.)

ZAHL, P. A. 1949. The evolution of sex. *Sci. Amer.* 180(4):52–55. (An old but interesting speculative analysis of the origins of sex, together with some social commentary.)

The Biology of
Organisms

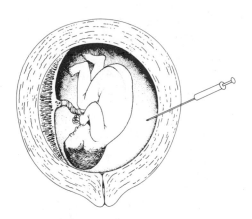

7

Inheritance

A little over 120 years ago, an event occurred which has come to have immense significance for the scientific world and for human society. On a cold February evening in the little town of Brunn, Austria, a rather portly Augustinian monk named **Gregor Johann Mendel** (1822–1884) (Fig. 7-1) spoke to a small audience of about forty friends and students on the subject of some breeding experiments he had done using the common garden pea. Over a period of eight years Mendel had cultivated, selected, and interbred some 10,000 individual pea plants. Earlier, he had obtained 34 varieties, differing from each other in such characteristics as height, seed coat color and seed shape, flower color and position, pod color and shape, and so on. From these he had selected 22 types for his experimentation and eventually narrowed his choice to just 7 varieties.

In his lecture Mendel talked about the results of experiments with two pure-breeding strains of peas. He told his listeners that he had crossed one strain, a tall form growing up to 2 m (6.5 ft) in height, with a dwarf strain less than 1 m (3 ft) high. Surprisingly, the offspring of this cross were not intermediate in height between the two parents but were all tall like the tall parent. Mendel then related that when the offspring of the tall–dwarf cross were permitted to self-pollinate, one-fourth of the next generation of progeny were dwarf, like the dwarf grandparent, and the remaining three-fourths were tall like the parents and the tall grandparent. (The actual count was 787 tall to 277 dwarf plants.) Mendel spent the remainder of the evening explaining these interesting results.

MENDELIAN GENETICS

In the example of the peas, each offspring of tall and dwarf parents inherits **factors** for both tallness and dwarfness but the factor for tallness is **dominant,** completely masking the factor for dwarfness, which therefore is **recessive.** This tall generation with its hidden recessive factors is the **hybrid** or F_1 **(first filial) generation.** (The generation of the parents is the P (parental) generation.) In addition to describing the relationship of inherited characteristics, Mendel invented a method, still in use today, of symbolizing the dominant and recessive factors in the F_1 generation. He represented the dominant factor for tallness as T, and the recessive factor for dwarfness as t. Dominant and recessive factors of this kind, which control a single inherited trait, now are

FIGURE 7-1
Gregor Mendel. (Drawing by James H. Hubbard)

called **alleles** ("counterparts"). An allele can be defined as one of two or more genetic factors occupying, or capable of occupying, the same position (**locus**, pl. **loci**) on each of a pair of homologous chromosomes. One of Mendel's greatest scientific contributions was his discovery of units of inheritance—as if his dominant and recessive factors were actual particles, as we now know them to be (i.e., units of DNA).

Mendel designated the hybrid generation of tall peas with the symbol Tt. However, he knew of no reproductive mechanism that would distribute hereditary factors so as to produce the precise ratios he had observed in his second-generation pea plants. He assumed, nevertheless, that in some way the genetic factors T and t, present in the hybrids, became separated from each other during reproduction so that each offspring received either a T or a t from each parent. Furthermore, Mendel showed mathematically that the separation of factors and their later recombination occurred by chance alone. The F_1 hybrids produced in equal numbers gametes containing either T or t. When fertilization by these gametes occurred, approximately one-fourth of the time a T gamete from the first parent combined with another T gamete, one-fourth of the time a t gamete from the first parent combined with a T gamete from the other parent, one-fourth of the time a T gamete from the first parent combined with a t gamete from the second parent, and one-fourth of the time t gametes from each parent were combined. Thus the F_2 (second filial) **generation** appeared in the ratio 1TT : 2Tt : 1tt. A ratio of this kind, in which the genetic factors of the offspring are shown, is called a **genotypic ratio**. The symbols TT and tt designate the **genotypes** of plants having factors only for tallness (TT) or dwarfness (tt); such individuals are referred to as **purebred** or **homozygous** ("same fusing"). The class of individuals represented by Tt have a **hybrid** or **heterozygous** (other fusing") genotype, but in their appearance, or **phenotype**, they are indistinguishable from the homozygous tall plants.

For the tall–dwarf cross described, the F_2 generation may be written as a 3 : 1 ratio of phenotypes, or as 1 : 2 : 1 ratio of genotypes. The 3 : 1 ratio is practical for sorting out the kinds of plants observed, but the 1 : 2 : 1 is more accurate in terms of a precise genetic analysis.

Probability and Genetics

The actual count of F_2 offspring in Mendel's crosses of tall and dwarf peas (787 tall plants and 277 dwarf plants) reduces to a ratio of 2.84 : 1. Why did Mendel's crosses not give an exact 3 : 1 ratio? Most people have at one time or another played games of chance and have come to realize that one seldom obtains exactly the expected combinations of numbers in throwing dice, or of heads or tails when tossing coins. For example, on one roll of a die, the chances are 1 in 6 of rolling a 6. When rolling a pair of dice, the chances of achieving two 6s are 1 in 36, or $\frac{1}{6} \times \frac{1}{6}$. Yet, in playing a game such as craps, 12s sometimes are rolled in succession, or the dice might be rolled more than 36 times without getting a 12. To come close to the expected ratio of 1 in 36, it probably would be necessary to roll the dice many hundreds of times, keeping an accurate account of all the combinations that turn up. In the same way, when tossing a coin one expects to obtain a 1 : 1 ratio of heads to tails; however, two successive tosses may produce two heads or two tails rather often, and only a large number of tosses will produce the expected 1 : 1 ratio.

Tossing two coins simultaneously simulates Mendel's experiments with tall and dwarf peas and his F_2 ratio of 1TT : 2Tt : 1 tt. Upon tossing the coins many times, it will be found that about one-fourth of the time both coins will come up heads (TT), one-fourth of the time both coins will be tails (tt), and two-fourths (one-half) of the time one coin will be heads and the other tails (Tt). This demonstrates that the **Mendelian ratios** may be considered to represent random or chance recombinations of gametes and their factors.

It is customary to sum up the results of such genetic experiments by using a diagrammatic device called a **Punnett square**. The Punnett square in Fig. 7-2 is a bit more elaborate than usual because sketches are included of the phenotypes of parents and offspring. In a Punnett square, the gametes produced by one F₁ parent are placed along the top of the square, and those produced by the other F₁ parent are located along the left side. The recombinations of genetic factors then are entered within the blocks of the square and given the expected ratios in which they will appear. The cross shown in Fig. 7-2 is called a **monohybrid cross** because the genetic behavior of only one pair of factors is analyzed.

Mendel's Laws

Thus far only examples of Mendelian inheritance involving a single trait have been considered. Mendel, however, worked with more than twenty traits in peas and selected seven for further experimentation, among them height, seed color, and texture. When Mendel crossed purebred (homozygous) peas bearing yellow, wrinkled seeds with homozygous peas characterized by green, round seeds, he found that all the F₁ generation bore yellow, round seeds. Thus yellow coat and round shape were governed by dominant factors (alleles); green and wrinkled characteristics were governed by recessive factors (alleles). When Mendel crossed F₁ plants, their offspring appeared in a phenotypic 9 : 3 : 3 : 1 ratio (9 yellow, round : 3 yellow, wrinkled : 3 green, round : 1 green, wrinkled). The Punnett square for this cross, which is called a **dihybrid cross** because it deals with two pairs of Mendelian factors (alleles), shows the reasons for Mendel's results (Fig. 7-3).

Mendel accounted for his results by stating that hereditary factors behave randomly during reproduction. That is, every allele behaves as an independent entity with respect to every other allele. The term given to this is **independent assortment** of alleles in gametes. Then, upon fertilization, the hereditary factors or alleles are recombined randomly to produce the possible combinations of the alleles present in the original parents. It is customary to summarize Mendel's contributions in the form of the following laws of heredity:

1 **Law of uniformity**, or **law of dominance**. All individuals of an F₁ generation will be equal and uniform. In most cases only one of a pair of alleles will be expressed (dominance); the other will not be expressed (recessiveness).

2 **Law of segregation**. Alleles become separated from each other during sexual reproduction. As a result, each gamete will receive only one of a pair of alleles

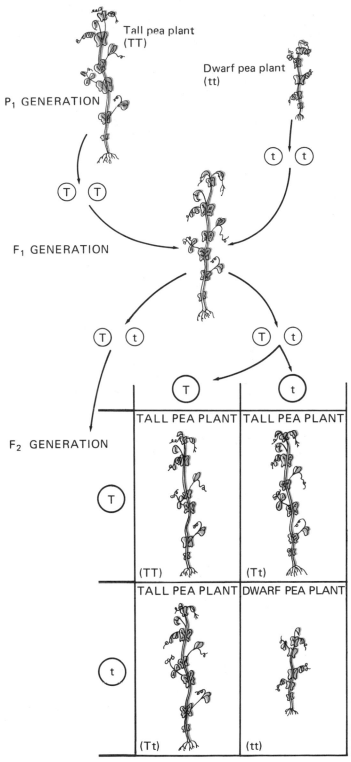

FIGURE 7-2
Monohybrid cross in garden peas involving factors determining the height of the plants. The Punnett square method is used to illustrate the recombination of genetic factors in the F₂ generation.

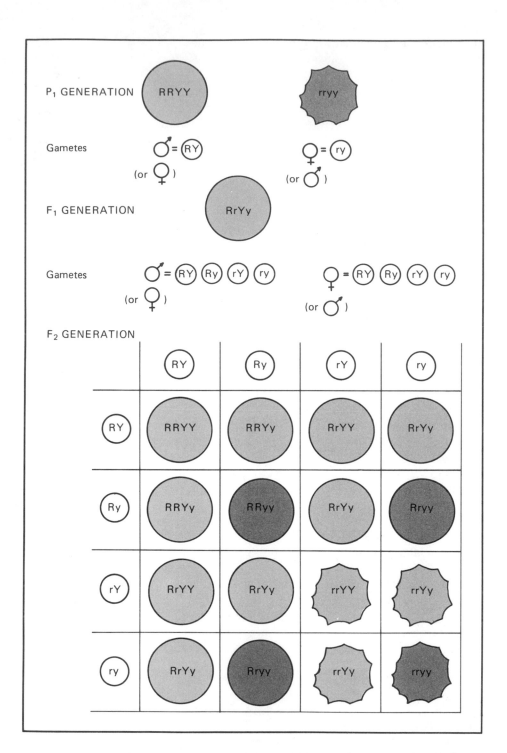

FIGURE 7-3
Dihybrid cross in garden peas. Factors for seed color and seed form are involved.

controlling a specific trait, and the two kinds of gametes will appear in a 1 : 1 ratio.

3 **Law of independent assortment.** When more than one pair of alleles are present, they will sort out independently of one another in gamete formation.

Mendel's theories, unfortunately, were not accepted by scientists of the nineteenth century. He tried, for example, to interest the great German botanist Karl Nageli in his work and in fact gave him a selection of his prized inbred lines of peas, but it is said that Nageli later threw these out on his trash heap; certainly he did no experiments either to verify or disprove Mendel's work. Mendel, disappointed, turned to astronomy and to service in the Catholic church and died not knowing he had been responsible for one of the greatest advances in the history of science. Then, in 1900, three biologists, working separately, came to exactly the same conclusions about inheritance as had Mendel, although at the time they were unaware of his work. Later on, upon learning of Mendel's experiments, these three—a Dutchman, Hugo De Vries; a German, Carl Correns; and an Austrian, Erich von Tschermak-Seysenegg—generously acknowledged Mendel as the discoverer of the laws of heredity.

Meiosis and Mendelian Genetics

As noted before, Mendel knew of no actual process by which his genetic factors could be sorted out and recombined during reproduction. The true nature of meiosis had not yet been discovered, although some microscopists had observed cells undergoing meiosis and had described some of its features. In particular, neither Mendel nor anyone else knew that reductional divison of chromosomes occurred. How simple it now seems to equate Mendel's principles with meiosis. All that needs to be assumed is that the Mendelian factors T and t are located on a particular pair of chromosomes of a hybrid pea plant, one factor on each of the chromosomes, and that the factors W and w similarly are located on another pair of chromosomes. Then when the members of these pairs of chromosomes are separated in the anaphase of meiosis I, each daughter cell will receive either a T or a t, together with either a W or a w. The second meiotic division will simply be a duplicate of the combinations of chromosomal types and their alleles present at the end of meiosis I, but will not alter the ratios further (see Fig. 7-4).

Genes (Mendelian Factors) and Gene Action Up to this point, biochemical definitions of Mendel's factors of inheritance have been avoided, but we have earlier noted that a piece of DNA, corresponding to the code for the synthesis of a specific polypeptide molecule in an enzyme, is a **gene**. Recall also that enzymes are protein molecules of one or more polypeptides and catalyze most of the chemical reactions in cells, including those in which complex organic molecules other than polypeptides are made.

To determine the steps involved in the development and expression of the many inherited traits that combine to make an individual, whether a pea plant or a human being, presently is impossible. Nevertheless, based on current understanding of the action of genes, it can be said that an inherited characteristic such as tallness in peas probably is dependent on many gene-controlled biochemical reactions occurring in the cells of

FIGURE 7-4
Meiosis and the inheritance of genetic factors. Alternate pathways of chromosomal and genetic factor distribution are shown. These depend on the orientation of the synapsed chromosome pairs (tetrads) in metaphase of meiosis I occurring in a mother cell.

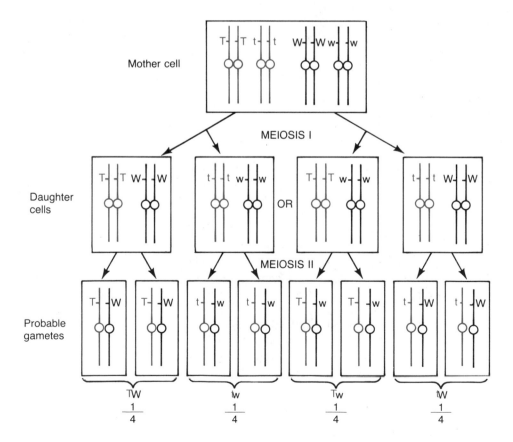

the organism, in this case the pea plant. Some insight into this may be gained by considering that dwarf peas may be made to grow to the same height as tall peas by spraying them with a chemical named **gibberellin.**

Gibberellin is a plant hormone known to occur in many kinds of plants. One of its important roles is to stimulate the elongation of cells, and this is the effect it has when applied to Mendelian dwarf plants. It is suspected that a Mendelian dwarf plant lacks the capacity to produce its own internal supply of gibberellin. If that should be the case, Mendel's factor for dwarfness in peas might correspond to that portion of the pea plant's DNA that determines a step in synthesis of gibberellin. Gibberellin is synthesized in cells by a number of chemical reactions, each of which is catalyzed by a specific enzyme. Earlier it was stated that enzymes are coded for by genes; therefore the Mendelian factor for dwarfness could be equivalent to a defective or mutant gene in the team of genes responsible for the synthesis of gibberellin.

Origin of Alleles As noted previously, Mendel thought that pairs of factors combined to bring about the hereditary traits observed in each generation of peas, and that these factors might occur as either dominant or recessive factors, now called alleles. One of the important questions that interested early geneticists was just how alleles originate. Biologists now think that alleles arise as **mutations** of existing genes. As discussed in Chapter 5, a mutation is a permanent change in a gene, but a broader definition includes changes not only in genes themselves but also in the locations of genes on chromosomes.

Hugo De Vries was the first to describe mutations (in the evening primrose), but Thomas Hunt Morgan of Columbia University was the first geneticist to explore the nature of mutations in the laboratory. He did this in the course of his research on the genetics of the fruitfly *Drosophila melanogaster.* Morgan used fruitflies in his work because they are small and easily handled; and because they reproduce rapidly (a generation may be raised to maturity in 10–11 days), he was able to observe a large number of individuals. Although **spontaneous** (natural) **mutations** are comparatively rare, occurring on the order of 1 per 10,000 to 1 per 1,000,000 per chromosomal locus, Morgan and his associates discovered a considerable number, which they used in their experiments. It was not until 1927 that the means of artificially causing mutations was discovered at the University of Texas by Hermann J. Muller. Muller announced in his paper "Artificial transmutation of the gene" that mutations could be caused by exposing fruitflies to X rays and that this rate of mutation was many times greater than the spontaneous rate. Of perhaps even greater importance was the discovery by Charlotte Auerbach in 1942 that mustard gas, the chemical used as a war gas in World War I, caused mutations. Since then, many other chemicals have been found to cause mutations. This work has been fundamental in recognition of the dangers of air, water, and food pollution. In addition to X rays and chemicals, ultraviolet light and even temperature extremes are known to act as agents of mutation, or **mutagens.**

Other Examples of Simple Mendelian Inheritance

Many examples of Mendelian inheritance operate on the same simple basis as dwarfness in peas. Among these are deafness in cocker spaniels, rough coat in guinea pigs, "silver-blu" mink, glossy broccoli, hairy tomatoes, notched ears in cattle, and albinism in humans and other animals.

Pigmentation versus albinism (lack of pigmentation) is determined by a pair of alleles, the recessive of which blocks the synthesis of the dark-brown pigment **melanin.** Varying concentrations of melanin, combined with other components of cells, produce many but not all of the varied coloring seen in vertebrate animals. Complete absence of melanin results in the albino condition. The genotypes are: normal pigmentation, AA and Aa; albino, aa (Fig. 7-5).

Because of the absence of melanin, albinos are very light-sensitive and sunburn easily. Why has this bad trait not been weeded out by natural selection? Probably for several reasons. Because the gene for albinism is recessive, it can be carried by individuals who appear normal but are heterozygous (Aa). As long as they do not mate with others like themselves (Aa × Aa), they will not produce albino progeny but will perpetuate the gene for albinism. In addition, in a population of one million humans, about 14 new recessive alleles for albinism will arise per generation by mutation of the gene for normal pigmentation, and it is likely that similar mutation rates prevail in populations of other vertebrates.

No doubt many albinos are protected from the weeding-out effects of natural selection by human social customs and by domestication. In certain primitive tribes, albinos are regarded with awe as children of the gods and are maintained in comfort as long as they live. In advanced societies, albino individuals can compete as effectively as pigmented individuals because survival no longer is dependent on skin coloration. Although albino animals are rarely seen in the wild, they occasionally occur. Usually, wild albinos do not live long because their distinctive coloration marks them for predation. In domestication, however, albino animals may be prized and bred for that characteristic alone.

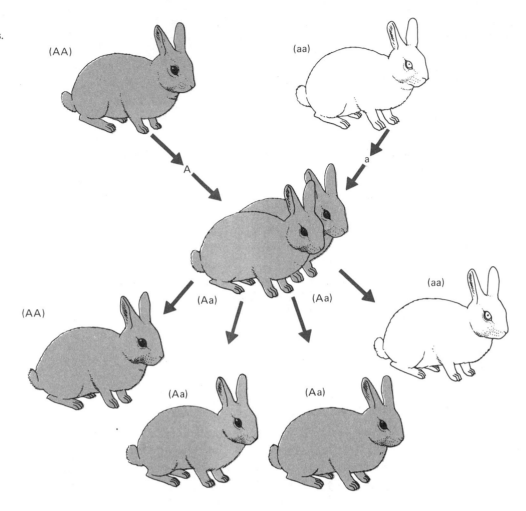

FIGURE 7-5
Inheritance of albinism in rabbits. The wild type, known as agouti, is represented by the dominant A; the albino factor is recessive and represented by a.

(AA)

(aa)

A

a

(AA)

(Aa)

(Aa)

(Aa)

(Aa)

(aa)

Test Cross or Back Cross Suppose you are given several rabbits of uncertain parentage. Is there any way you can tell if some or all carry the recessive allele for albinism? To test this, it will be necessary to make **back crosses,** mating each of your rabbits with an albino (homozygous recessive) of the opposite sex. If any one of your animals carries the recessive allele, then half of its offspring will have the albino phenotype (refer to Fig. 7-5). Animals not carrying the recessive allele will produce offspring of normal coat color, but all will carry and be able to transmit the allele for albinism. This crossing procedure is referred to as a back cross or **test cross.**

Lethal Genes Some genes, such as those for albinism, are deleterious but not necessarily deadly. Others, however, are very dangerous and may cause the death of individuals who inherit them. The recessive gene causing sickle-cell anemia, when in the homozygous condition, is such a gene (see Fig. 5-9). Other genes may be even

more deadly, killing offspring in infancy or even earlier in fetal stages of development. All such genes are called **lethal genes.** Lethal genes also occur in plants. An example is an allele of a gene (G) governing chlorophyll production. The recessive lethal allele (g) blocks chlorophyll production when in the homozygous state (gg). If two hybrids (Gg) are crossed, one-fourth of their progeny receive a double dose of the recessive allele (gg), and, being unable to photosynthesize, they die as soon as the food supply of the seed is exhausted.

Incomplete Dominance (Co-dominance) Not all alleles are either dominant or recessive. Some are neither, a condition called **incomplete dominance (co-dominance).** Flower colors often show incomplete dominance (Fig. 7-6). A cross between white (WW) and red (ww) snapdragons produces an F₁ generation that is pink (Ww). The F₂ genotypic and phenotypic ratios are the same: 1 white (WW) : 2 pink (Ww) : 1 red (ww).

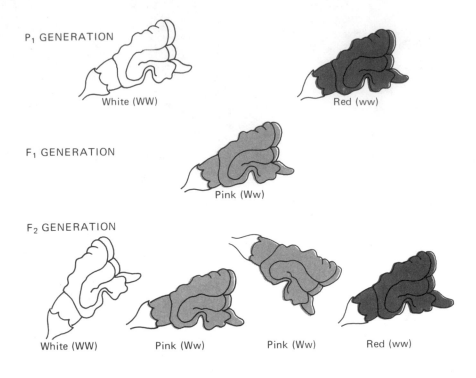

FIGURE 7-6
Incomplete dominance in snapdragons. Neither the allele for white (W) nor the allele for red (w) is dominant. (Redrawn with permission of Macmillan Publishing Co., Inc., from G. W. Burns, 1980, *The science of genetics,* 4th ed., New York: Macmillan Co. Copyright © 1972 by George W. Burns)

P₁ GENERATION

White (WW)   Red (ww)

F₁ GENERATION

Pink (Ww)

F₂ GENERATION

White (WW)   Pink (Ww)   Pink (Ww)   Red (ww)

In a somewhat similar fashion, coat colors in some animals show a blending of parental colors in the F₁ offspring.

Multiple Alleles   Certain characteristics in plants and animals are governed by a series of several alleles, any pair of which may occupy one chromosomal locus. Such genes are called **multiple alleles.**

In humans, the inheritance of A, AB, B, and O blood types is determined by three alleles, any two of which may be present in the genotype of an individual. In primroses, color patterns ranging from yellow to white are produced by three alleles. The plumage color of screech owls is controlled by three alleles. Gray is the more common color and is produced by a pair of recessive alleles (gg). Red is less common and is determined by a dominant allele (G). Intermediate color is produced by a third allele (G′), which is recessive to red but dominant over gray. The full range of possible phenotypes and genotypes are: red (GG, GG′, Gg), intermediate (G′G′, G′g), and gray (gg). Typical Mendelian ratios result when crosses are made between individuals of any of these genotypes.

Multiple Gene Inheritance   Our examples thus far have been of crosses in which one pair of alleles, T and t, for instance, occurring at a single locus on each of a pair of homologous chromosomes, govern a clear-cut trait (tall vs. dwarf, etc.). Even when a series of alleles govern a trait, as in human blood types, only one pair of

alleles occupy the loci on a pair of homologous chromosomes. Not all genetic traits are so sharply delimited, however. Early in the study of genetics, experimenters encountered conditions in which a large number of intermediate forms existed between the extreme expressions of a characteristic. Such traits are governed by many genes at different chromosomal loci and are examples of **multiple gene (polygene) inheritance.** Tallness in humans is a case of multiple gene inheritance. There is no single gene in humans for tall individuals or short individuals; rather, each individual's height reflects the total number of genes favoring tallness or shortness. Because a number of genes can collectively contribute to the phenotype, this example emphasizes the point that the totality of an individual's characteristics result from an interaction, or blending, of gene products.

Skin color in humans also is based on the inheritance of a group of alleles for pigmentation. Eye color as well is suspected to be determined by a complex interaction of polygenes, some of which are relatively recessive, others of which are relatively dominant (Table 7-1). This is deducible from the many shades of eye color possible in humans: light blue to medium blue, dark blue, gray, green, hazel, light brown, medium brown, and dark brown.

Chromosome and Genes

A review of terminology may be helpful at this point. Recall that a pair of chromosomes in a 2n cell are ho-

TABLE 7-1
Possible functional scheme relating to the inheritance of eye color in humans on the basis of multiple genes and incomplete dominance

Eye Color	Contributing Alleles
Dark brown	AA BB CC DD
Medium brown	Aa BB CC DD
Light brown	Aa Bb CC DD
Hazel	Aa Bb Cc DD
Green	Aa Bb Cc Dd
Gray	aa Bb Cc Dd
Dark blue	aa bb Cc Dd
Medium blue	aa bb cc Dd
Light blue	aa bb cc dd

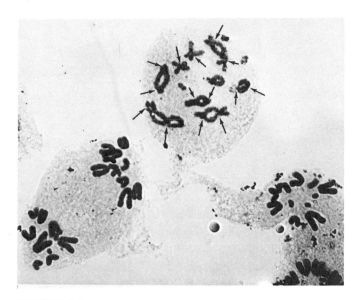

FIGURE 7-7
Preparation of meiotic chromosomes of a grasshopper, showing chiasmata (at arrows).

mologous chromosomes, or homologs. This implies that the two chromosomes carry genes for the same characteristics but are derived from different parents. However, they do not necessarily carry identical genes for those characteristics. In a pair of chromosomes of the fruitfly, for example, at a certain position or locus, there is a gene controlling eye color. Each chromosome of that pair will carry an eye-color gene at the same locus. One gene may be located on a chromosome inherited from a white-eyed parent; the other gene may be located on a chromosome inherited from a red-eyed parent. Thus one homolog would carry the gene for white eyes, and the other homolog would carry the gene for red eyes. Remember from our earlier discussion that such genes, at homologous loci and which are involved with the control of a single characteristic, are called alleles; many alleles, including those for characteristics other than eye color, would be expected along the length of a chromosome. Furthermore, all the alleles of a chromosome are said to be **linked**; that is, they tend to be inherited as a unit.

Recall also that when homologous chromosomes pair up during meiosis I, they undergo synapsis and form a tetrad consisting of four closely associated chromatids (see Figs. 6-13C, D and 7-8). During prophase of meiosis I, such chromatids often appear to be intertwined and to overlap at one or more points to produce **chiasmata**, as seen in Fig. 7-7. Chiasmata currently are considered to represent places where portions of adjacent chromatids have broken and exchanged places so that they have the appearance of having become crossed. The exchange might occur as shown in Fig. 7-8. The significance of chiasmata in genetics is, as we shall see, quite profound.

Linkage   Mendel demonstrated that certain pairs of genetic factors assorted themselves independently of others. However, soon after the rediscovery of Mendel's laws

in 1900, exceptions were found to the law of independent assortment of genes. One of the great pioneering workers in genetics, Thomas Hunt Morgan, mentioned earlier in connection with the discovery of mutations, noted that Mendel's law of independent assortment was not obeyed when crossing experiments involved more than one pair of alleles. Rather than pea plants, Morgan worked with the fruitfly Drosophilia, which has only four pairs of chromosomes, He soon found that many genes occurred together on each pair of these chromosomes. This occurrence of two or more pairs of genes in a single pair of chromosomes is called **linkage,** and the linked genes comprise a **linkage group.** In cases where two pairs of alleles occur on different pairs of chromosomes, the independent assortment of genes and of chromosomes during meiosis clearly may be correlated. However, if both sets of genes occur at different loci of one chromosome pair, the independent assortment of those two pairs of linked genes would seem highly unlikely.

However, shortly after linkage was discovered it was found that even linked genes tended to assort independently of each other (Fig. 7-8). As Morgan explained, during late prophase of meiosis I, chromosome pairs associate to form tetrads (synapsis), and their chromatids become joined at chiasmata (see Fig. 7-7). It was reasoned that the chiasmata were points at which an actual exchange of portions of chromatids and the genes present in those portions had occurred (**crossing over**). Crossing over occurs more frequently between loci widely separated on a chromosome pair than between those that are close together. The reason for this is that

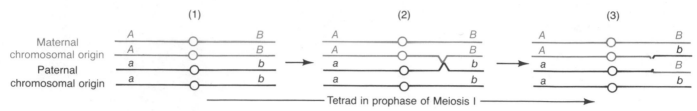

FIGURE 7-8
Linked genes and their recombination as a result of crossing over.

a chiasma is more likely to occur between two distant points than between two that are close to each other. Calculation of crossover frequencies among linked genes has enabled geneticists to prepare linkage maps of chromosomes and thus to identify many of the loci on chromosomes.

Sex and Sex-linked Genes   A number of factors are known to be associated with the sex chromosomes of animals. In most animals and in certain plants (e.g., mosses), a pair of chromosomes called the **sex chromosomes** determine sexual characteristics. Although the nature and behavior of sex chromosomes vary somewhat among species, many organisms possess a pair of dissimilar, sex-determining chromosomes referred to as X and Y. In humans, for example, there are 22 pairs of nonsex chromosomes, or **autosomal chromosomes,** and one pair of sex-determining chromosomes. An individual with a pair of X chromosomes (XX) is a female; an X and Y chromosome pattern (XY) produces a male. Many other organisms besides humans possess XX and XY sex-determining chromosomes, including such diverse examples as fruitflies (Fig. 7-9) and moss plants, in addition to all mammals. In birds, however, females are XY and males are designated XX.

Genes other than those controlling sex characteristics also are present on the sex chromosomes. In most

organisms the X chromosome is much longer than the Y chromosome and carries more genes. All genes occurring on sex chromosomes, whether controlling sex traits or not, are called **sex-linked genes**. A female carries a pair of sex-linked genes on her two X chromosomes, whereas in most cases a male has only one gene of this pair on his X chromosome and none on the Y chromosome.

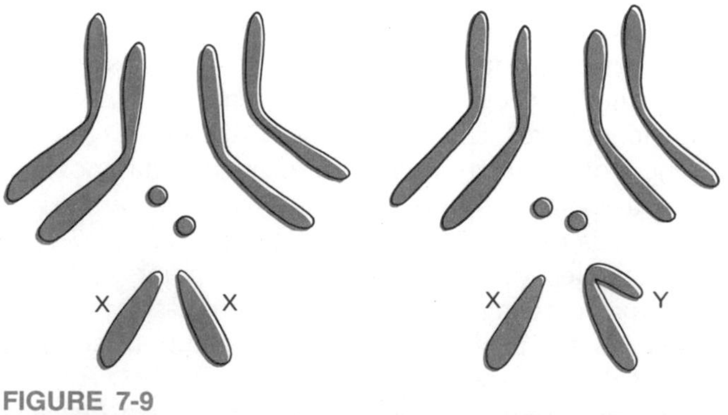

Sex-linked Inheritance   In *Drosophila*, the wild-type individuals have red eyes. A recessive allele causing white eyes is known. The alleles are carried on the X chromosomes (X^W = red-eyed wild type; X^w = white-eyed variant). In a typical cross, a red-eyed male (X^WY) (Y represents the Y chromosome lacking the locus for eye color) is mated with a white-eyed female (X^wX^w) (Fig. 7-10). The male offspring all are white eyed (X^wY); the daughters are red eyed (X^WX^w). In the F_2 generation half the sons will be red eyed (X^WY) and half will be white eyed (X^wY); half the daughters will be red eyed (X^WX^w) and half will be white eyed (X^wX^w). There will be no homozygous red-eyed daughters (X^WX^W) produced in this experimental cross. In addition to many known genes in fruitflies, as well as in other animals, a number of sex-linked genes are known to occur in humans.

GENETICS AND AGRICULTURE

Many variations on the theme of Mendelian genetics now are known to occur, and a goodly number of these have important agricultural consequences. Without knowledge of Mendelian genetics, the vast agricultural industry would be severely handicapped. For example, plant geneticists continually strive to select and interbreed plants having resistance to disease and insect pests. Even the form of plants and animals is genetically tailored to meet certain situations. Long-bodied lean hogs are selected and bred to produce more bacon. Cows with immense udders and milk-secreting capability are chosen for breeding with bulls known to transmit similarly desirable genes, although they do not themselves have such sex-related characteristics. On the basis

FIGURE 7-9
Metaphase chromosomes of male and female fruitflies *(Drosophila melanogaster).*

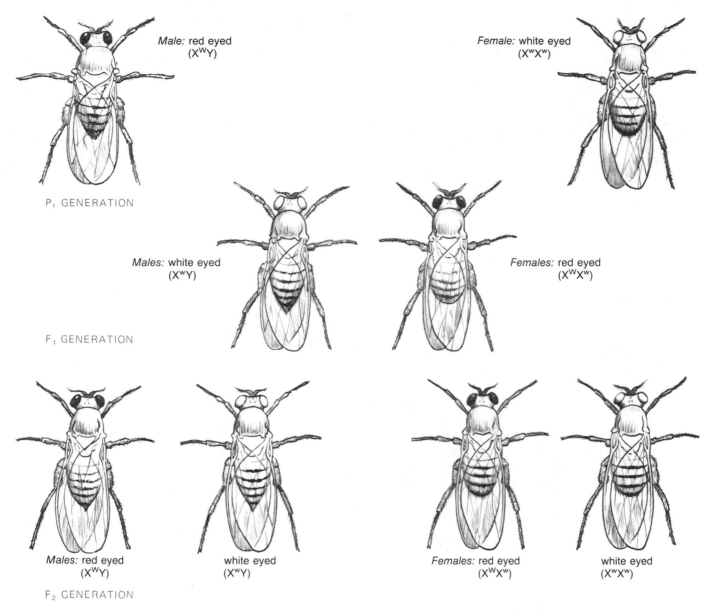

FIGURE 7-10
Inheritance of eye color in *Drosophila;* an example of sex-linked genes.

of their genetic charts, mink farmers produce furs of many shades from "platinum" to "blue frost."

The greatest success story of genetics, however, is breeding to produce **hybrid vigor.**

Certain crosses of unlike parents have long been known to produce offspring of unusual vigor. A classic example is the crossing of donkeys and horses to produce mules (Fig. 7-11). Generally speaking, horses have greater strength and speed than donkeys, but are not as hardy; the mule has much greater stamina than does the horse and is much stronger than the donkey. Mules usually are produced by breeding a male donkey with a fe-

male horse, because in the reciprocal cross the female donkey is too small to bear the large baby mule. However, both these crosses are known. The offspring of a female donkey and a male horse is a hinny.

Mules usually are not capable of interbreeding. The reason is that donkeys and horses have chromosomal differences such that in meiosis the "donkey" chromosomes occurring in the mule's gonads fail to synapse with the "horse" chromosomes. The end result of meiotic failure is the production of gametes that are almost exclusively nonviable. This does not mean that mules are incapable of sexual intercourse. Mules,

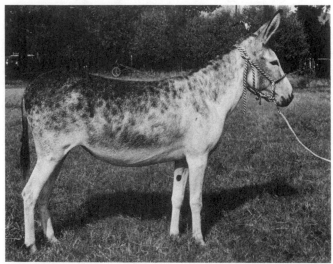

FIGURE 7-11
Horse, donkey, and their hybrid offspring, the mule.

donkeys, and horses can and do copulate with each other.

What is the nature of hybrid vigor? For one thing, hybrids have genotypic and phenotypic uniformity, which can be advantageous if a particular combination of alleles results in a vigorous constitution. Then all members of the F_1 generation will be similarly endowed and survival will be enhanced. One explanation of hybrid vigor is that the slightly different biochemical products (enzymes, hormones, etc.) of alleles derived from different parents act together in the F_1 individual more effectively than when alone, as in the parents or in the F_2 generations. It has been suggested also that the hybrid is better protected, or genetically "buffered," against environmental stresses because of its greater genetic diversity in comparison with individuals of greater homozygosity. Finally, hybrids seem to be less subject to the effects of deleterious recessive alleles, because many of these genes are masked by dominant alleles.

Hybrid maize is an example of the application of genetics to agriculture (Fig. 7-12). In the early years of this century, geneticists found it possible to self-pollinate maize for six or seven generations and produce rather pure-breeding strains called **inbreds**. Such inbred lines were themselves not especially productive or vigorous, but when two inbreds were crossed the result was a superior hybrid plant.

It should be understood that the ears and kernels of parental types of maize used in production of hybrids are not exceptional in quality or yield. The reason for this is that the ear and part of each kernel are maternal structures and not the products of pollination and fertilization. Only the embryos and the starchy parts of kernels are produced by pollination and fertilization, and it is the plants that develop from embryos in such kernels that are the actual hybrids.

In early attempts at hybridization it was difficult to produce much "seed" (the seed of maize is actually a type of fruit called a grain) because the parental strains were such poor producers, but later it was found that two productive hybrids from four different inbred lines could be crossed to produce large amounts of hybrid seed corn. Now, high-yielding inbred lines have eliminated the necessity of a two-generation double-cross procedure, and much of the seed corn is produced by a single-cross method.

The beauty of hybrid maize, from the seed company's point of view, is that farmers cannot use the kernels produced on their hybrid plants to insure high-yielding fields the following year. The kernels purchased from the seed company are true F_1 hybrids and grow into large, healthy, productive plants. However, the ears produced on these plants contain kernels that are the F_2 genera-

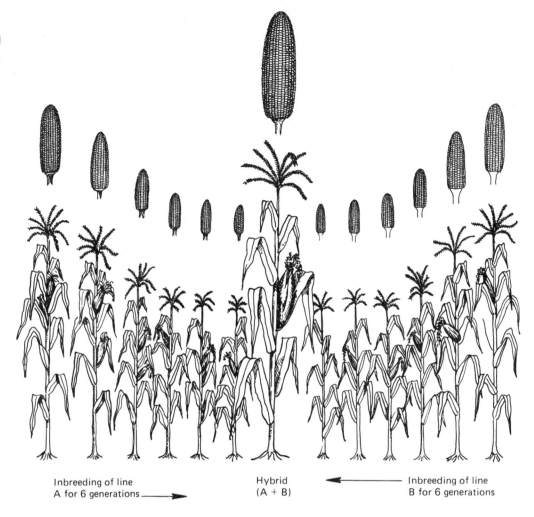

FIGURE 7-12
Inbred and hybrid maize, *Zea mays*. The production of a hybrid from two inbred lines is illustrated. (Adapted from *Plants and civilization*, 3rd ed., by H. G. Baker. © 1978 by Wadsworth Publishing Company, Inc. Reprinted by permission of Wadsworth Publishing Company, Belmont, California 94002)

Inbreeding of line
A for 6 generations ⟶

Hybrid
(A + B)

⟵ Inbreeding of line
B for 6 generations

tion. When planted, the F_2 kernels develop into a heterogeneous mixture of large plants and small ones, high yielders and low yielders, and disease-resistant and disease-susceptible individuals. Thus the farmers must return to the seed company each year for their seed. They have neither the resources nor the time to grow their own hybrid varieties.

It is more important to remember, however, that not all new kinds of crops are hybrids. A number of the higher yielding types, such as the "miracle" rices and wheats recently produced for use by Third World farmers, are inbred strains of high yield and high disease resistance. They are true-breeding types, and the farmers can grow them year after year if care is taken to avoid cross-pollination from neighboring fields of poorer scrubby types. In general, even the best inbred types do not yield as much as good hybrids. In the United States the best inbred lines of maize yield about 60 bushels of grain per acre, whereas good hybrids yield 100 and more bushels per acre.

GENETICS AND HUMANS

Thus far in our study of the principles of inheritance, we have primarily considered domesticated plants and animals such as peas, snapdragons, cattle, and rabbits. We therefore may have obtained a rather narrow view of genetics, restricted as our study has been to characteristics, such as albinism in rabbits or dwarfness in peas, which have been selected for over countless generations of domestication. This selection for uniformity of character expression tends to narrow the number of alleles in the population relative to the number present in a wild population. In the latter, while natural selection produces a certain characteristic range of phenotypes known as the "wild type," the number of alleles present in the population (the **gene pool**) tends to be great. The gene pool of a domestic variety by comparison tends to be relatively small (as in a certain strain of laboratory mice, all of which are thought to be descended from a single female ancestress).

In contrast to laboratory mice and almost any variety of cereal plant, human populations are genetically diverse; in terms of the gene pool they resemble wild animal populations. Yet in some ways human populations also can resemble those of domestic plants and animals. This is especially so when deleterious mutations accumulate as a result of inbreeding or in the absence of natural selection. Indeed, it has been estimated that each human being is the often-unsuspecting carrier of a half-dozen or so potentially lethal recessive genes. Usually, the occurrence of each of these in human populations is low, and the probability of homozygous recombinations is even lower. Such possibilities do exist, however, and despite knowledge of the often tragic consequences of genetic "mistakes," human society is late in recognizing the importance of genetic understanding and counseling.

Sex Chromosomes

Human Sex-linked Inheritance   There are a number of sex-linked recessive genes known in humans as well as in other animals (Table 7-2). A female will carry a pair of sex-linked genes in her two X chromosomes; a male, however, will carry only one of these genes because they are present only in the X chromosome. For this reason, if a sex-linked recessive allele is present in the male genotype, it is always expressed, but in the female geno-

TABLE 7-2
Some alleles occuring in human populations (see also Fig. 7-14)

	Allele Condition
Simple Mendelian Traits	
PTC (phenylthiocarbamide) tasting	Dominant
Huntington's chorea (degeneration of nervous system in adults)	Dominant
Polydactyly (extra fingers and toes)	Dominant
Brachydactyly (short fingers)	Dominant
Woolly hair (in Caucasians)	Dominant
Syndactyly (conjoined fingers)	Dominant
Alkaptonuria (dark urine, arthritis)	Recessive
PKU (phenylketonuria) (mental impairment)	Recessive
Albinism	Recessive
Sickle-cell anemia	Recessive
Infantile amaurotic idiocy (Tay-Sachs disease)	Recessive
Juvenile amaurotic idiocy (fatty degeneration of brain)	Recessive
Hereditary deafness	Recessive
Genetic goitrous cretinism (dwarfness, mental retardation)	Recessive
Sex-linked and Sex-influenced Genes	
Ichthyosis (skin dryness)	Sex-linked, recessive
Red-green color blindness	Sex-linked, recessive
Hemophilia	Sex-linked, recessive
Pseudohypertrophic muscular dystrophy	Sex-linked, recessive
Nystagmus (jerky eye movements)	Sex-linked, recessive
Pattern baldness	Sex-influenced; dominant in male, recessive in female
White forelock	Sex-influenced; dominant in male, recessive in female
Multiple Alleles and Polygenes	
Blood types (M, N, A, B, AB, O, Rh)	Multiple alleles
Tallness	Multiple genes
Skin pigmentation	Multiple genes
Intelligence	Multiple genes
Genetic Basis Not Clear	
Attached earlobe	Possibly simple recessive
Eye color	Probably multiple genes
Left-handedness	Possibly simple dominant

type, unless a pair of recessive genes are present, a normal phenotype occurs.

Hemophilia is caused by a sex-linked recessive allele. One of the most interesting case histories of hemophilia is presented by the descendants of Queen Victoria of England. Examination of the royal pedigrees amply illustrates the tendency of a sex-linked recessive gene to be carried by females and to be expressed predominantly in males (Fig. 7-13). Victoria was a recessive carrier; apparently she was a mutant, for there is no record of hemophilia in her ancestry. Because she did not display symptoms of hemophilia, it is assumed that she was heterozygous ($X^H X^h$) for the allele; males are either $X^H Y$ (normal) or $X^h Y$ (hemophiliac). In $X^h Y$ males the gene manifests itself in uncontrolled bleeding, resulting from the absence of blood-clotting factors; victims require constant care both to prevent cuts and bruises and to treat the symptoms, usually by repeated blood transfusions. Female hemophiliacs are unknown; presumably the gene is lethal early in fetal life.

Unwittingly, Victoria contaminated the inheritance of about half the royalty of Europe. It is said she bore some responsibility for the Russian Revolution because of the excesses of the Russian Empress, Victoria's granddaughter Alix, who in frantically seeking aid for her hemophiliac son may have prevented a sane solution by the Emperor to problems that precipitated the civil war of 1917.

Human Karyotype Until 1956 it was thought that the human chromosome number was 48, 2 more than are actually present. The reason for this mistake was that until recently the determination of chromosome numbers in cells was a laborious process requiring the slicing of cells into thin sections, the staining of these sections, and the attempt to reconstruct the chromosome pattern and number by microsopic study of the cell sections. Eventually, however, methods were found for staining whole cells and flattening them on microscope slides so that the chromosomes were spread out. When these new methods were used to study rapidly dividing human cells in bone marrow, embryonic tissues, and cultures of white blood cells, the true number of chromosomes was learned (see Fig. 7-20 for the methodology). Moreover, the chromosomes were so clearly displayed that it became possible to identify individual chromosome pairs, including the sex chromosomes as well as the autosomal chromosomes.

Photographic enlargements of human chromosomes can be snipped out with scissors and arranged by

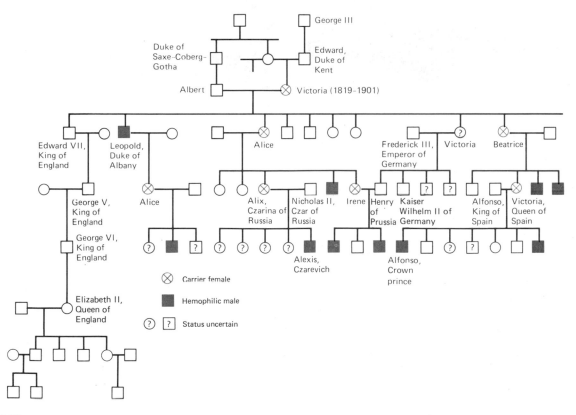

FIGURE 7-13
Case history of hemophilia in the descendants of Queen Victoria of England.

A

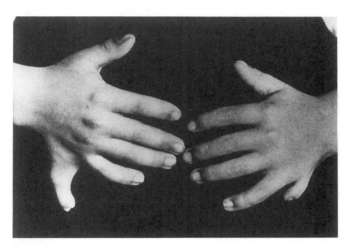

B

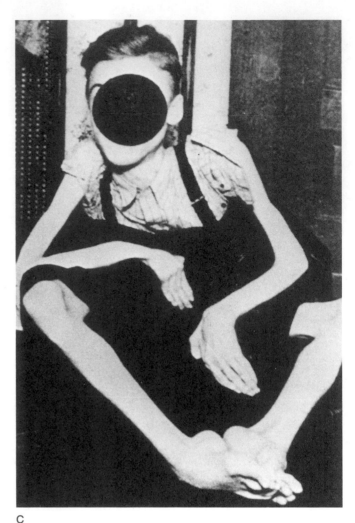

C

FIGURE 7-14
Simple Mendelian traits in humans. **A.** Albinism in a brother and sister whose parents were first cousins. **B.** Polydactyly. **C.** Advanced pseudohypertrophic muscular dystrophy. Refer to Table 7-2 for further information. (From A. M. Winchester and T. R. Mertens, 1983, *Human Genetics,* 4th ed., Columbus: Charles E. Merrill Publishing Co.)

pairs in a diagram called a **karyotype** (Fig. 7-15). Most analyses of human chromosomes are done in this manner, especially in cases of genetic diseases related to chromosomal defects.

Barr Bodies In 1949 Murray L. Barr, a Canadian, discovered a small, densely staining body in the nucleus of human female cells. The small nuclear body, the **Barr body,** has been determined to be one of the two X chromosomes and remains in the cell in a condensed condition. Later, in England, Mary Lyon found that the Barr body chromosome is genetically inactive during the life of the cell. Inactivation occurs during early development of the embryo, and chance alone dictates which X will

be inactivated. Thus some embryonic cells will have one X chromosome active; the others will have the second X chromosome in the active state. All the cells derived from each of these original cells will have the same X chromosome operative, so the individual becomes a genetic **mosaic.** A well-known mosaic is the calico color in cats. In this case, alleles producing black pigment (Y) and yellow pigment (y) can be present in the X chromosomes, and inactivation of one or the other of these in different regions of the body produces the mottled, calico effect of yellow and black patches.

The human female who has a sex-linked recessive allele for hemophilia ($X^H X^h$) also is a genetic mosaic. In about half her cells the recessive gene is active, and in

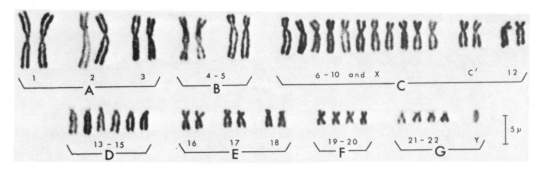

FIGURE 7-15
Human male karyotype. The 46 chromosomes of a human male cell are arranged serially in order of size and morphology. Note the single small Y chromosome; the larger X chromosome is one in group C. (Photo by Carolina Biological Supply Company)

half the dominant gene prevails. In this case the cells with the dominant gene produce enough clotting factor to overcome the effects of the recessive allele.

The presence of the Barr body in human females is the basis for a commonly applied test of femaleness. In the past it has been alleged that some competitors in international sports were really men masquerading as women. A simple test of skin cells from the mucous lining of the cheek will show whether or not Barr bodies are present in the nuclei and will indicate the sex of the person tested (Fig. 7-16).

Sex-influenced Genes   **Sex-influenced** genes are autosomal genes that are expressed differently in males and females. Such genes are related to sex but are not themselves present in the sex chromosomes. An example is human pattern baldness. The gene governing this characteristic is dominant in males and recessive in females. Men who are BB or Bb will develop a regular pattern of baldness as they mature; only men who are bb will be unaffected. However, women who are Bb and bb will have a normal hair pattern, and only those who are BB will become bald as they grow older. This explains why so many more men than women are bald.

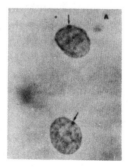

FIGURE 7-16
Epithelium cells from the lining of the mouth of a human female, showing Barr bodies against the nuclear membrane. (Photo courtesy Murray Barr)

A few cases of sex-influenced genes are known to occur in livestock. These include horns in sheep (dominant in males) and certain coat colors in cattle.

Blood Types

As stated previously, some human blood characteristics, or **blood groups**, correspond to a multiple allele pattern of inheritance. Those most commonly known are A, B, AB, and O, with corresponding genotypes A = I^AI^A, I^Ai; B = I^BI^B, I^Bi; AB = I^AI^B; and O = ii (the two dominant alleles are symbolized by superscripts).

Under ordinary circumstances, human blood genotypes create no known problems. However, in the early days of blood transfusion, seventy-five or so years ago, transfusions sometimes resulted in clumping, or **agglutination**, of red blood cells. Agglutination continues to be a serious problem, because clumps of red cells can block the flow of blood in the capillaries of the blood recipient, sometimes with fatal consequences. Further study of the problem revealed that four classes of people could be designated on the basis of cross-reactions between their red blood cells and foreign blood plasma. It should be noted that whenever a foreign protein (in this case, red blood cell protein) is injected into the bloodstream, the result is the production of a specific substance, present in the blood plasma, known as an **antibody**. This antibody attaches itself to the foreign protein, called the **antigen**, and forms an antibody–antigen complex, manifested in the present example as a clumping of the foreign red blood cells. The basis of cross-reactions in human blood typing is a set of antigen–antibody interactions.

Study of the blood transfusion problem also revealed that there are two antigens in the plasma membranes of human red blood cells and two corresponding antibodies in the blood. A person having **antigen A** is said to be blood type A, a person with **antigen B** is said to be blood type B, and a person having both antigens is type AB. An individual whose blood contains neither antigen A nor antigen B is type O. The inheritance of

CHROMOSOMES OF HUMANS AND CHIMPANZEES

In Chapter 21, relationships between humans and the great apes are depicted, and certain similarities between hominids and apes are discussed. In this connection, some scientists have suggested that chimpanzees may be more closely allied with humans than are any of the other primates. For example, base sequences of DNAs of chimpanzees and humans are more alike than those of humans and other primates. Comparative studies of amino acid sequences of many proteins also suggest there is a relatively close relationship between humans and chimpanzees, and now new evidence for this kinship has been brought to light.

Recently, a study of chromosome bands of humans and chimpanzees has shown startling similarities. In fact, cytologists J. J. Yunis, J. R. Sawyer, and K. Dunham report that the banding patterns are so alike that it is difficult to account for the many phenotypic differences of the two species.

Study of chromosome bands utilizes cell-culture techniques of the kind shown in Fig. 7-20 to obtain **prometaphase** ("near metaphase") and metaphase chromosomes. These chromosomes are pretreated with acids, alkali, or both and then stained in a complex dye mixture, **Giemsa's stain.** When Giemsa's stain is correctly applied, darkly stained bands are visible along the chromosomes; these bands are quite constant and specific and may be used for interchromosomal comparisons.

The karyotype of *Homo sapiens* (human) is composed of 46 chromosomes (23 pairs), including an X and a Y; that of *Pan troglodytes* (chimpanzee) consists of 48 chromosomes (24 pairs), including an X and a Y. The study of Yunis and his associates shows that many chromosomes of *Homo* and *Pan* have identical or nearly identical banding patterns and that two pairs of shorter *Pan* chromosomes are the equivalent of one pair of longer chromosomes of *Homo*. Apparently, somewhere in the ancestry of humans two shorter *Pan*-type chromosomes fused end to end to make the longer human chromosomes (see Fig. 1).

Also shown in Fig. 1 are X and Y chromosomes of both human and chimpanzee; here it is seen that banding patterns of X chromosomes in each

these blood types has a simple Mendelian basis, as illustrated in Fig. 7-17. Included in this figure are agglutination reactions between plasma and red blood cells of the several blood types. (The determination of blood types in human individuals is done by mixing their blood with reference blood serums on microscope slides and then examining the mixtures for agglutination of red blood cells. Serum is the liquid part of plasma remaining after clot formation.)

Similar to A, B, AB, and O blood groups, but differing in the nature of the antibody reactions, is the **Rh** blood group (discovered in the *Rhesus* monkey). A complex series of multiple alleles is known, but for conve-nience only a pair governing the Rh positive (Rh^+) and Rh negative (Rh^-) types will be considered here. About 85 percent of Americans are Rh^+, the remainder Rh^-. The Rh^- person has the potential for the Rh^+ antibody in his or her blood plasma, but does not actually produce the antibody unless exposed to Rh^+ blood by transfusion or some other means. Before Rh typing was done, it was possible for an Rh^- person to have received successive blood transfusions from Rh^+ blood donors. The first transfusion would initiate the production of Rh^+ antibodies, a process called **sensitization**. The second transfusion would produce agglutination of the introduced Rh^+ blood.

species are virtually identical; however, the Y chromosome of *Homo* is longer and has some bands not seen in the Y chromosome of *Pan*. It can be concluded that the mutual ancestors of *Homo* and *Pan* must also have had 48 chromosomes; possibly an early evolutionary step in the direction of modern humans was the fusion of a pair of chromosomes.

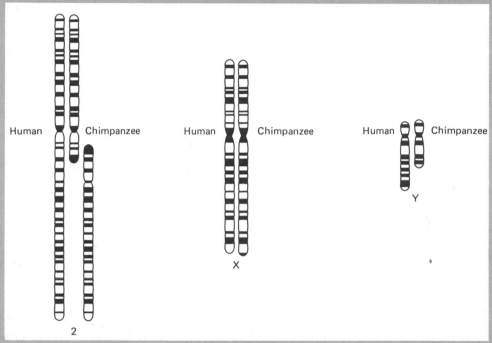

FIGURE 1

Giemsa-stained chromosomes of human and chimpanzee. Note the generally high degree of correspondence of banding patterns of the chromosomes shown: chromosome 2 of human, two shorter chromosomes of chimpanzee, and X and Y chromosomes of both species (for a complete comparison, see the original article, listed in the readings at the end of the chapter). The equivalent genetic information in human chromosome 2 apparently is present in two shorter chimpanzee chromosomes. The X chromosomes are nearly identical, but the human Y is longer than the chimpanzee Y and has some additional bands. The remaining chromosomes of human and chimpanzee show similar correspondences of banding patterns. (Redrawn with permission from J. J. Yunis, J. R. Sawyer, and K. Dunham, 1980, *Science* 208:1145. Copyright 1980 by the American Association for the Advancement of Science)

A serious difficulty may arise in matings between Rh⁺ males and Rh⁻ females in that they may have an Rh⁺ child (which will happen invariably if the father is homozygous Rh⁺). If the mother has previously become sensitized to Rh⁺, she will have developed Rh⁺ antibodies, and these may move across the placenta into the bloodstream of the Rh⁺ fetus and destroy its red blood cells. The result is a severe and often fatal disease called **erythroblastosis fetalis**. Another danger may occur during the trauma of birth. Some blood cells of the Rh⁺ fetus may find their way into the bloodstream and sensitize a previously unsensitized mother (one drop of Rh⁺ blood is sufficient). Then in a later pregnancy, Rh incompatibility will endanger the fetus. Fortunately, modern techniques can alleviate the symptoms of Rh incompatibilities.

Human blood types sometimes are used in settling complicated problems of human relationships. For example, in one case two mothers left a hospital on the same day, each with a newborn child. Upon reaching home, one mother discovered her baby to be tagged with the other mother's name. However, the second mother refused to exchange babies, and the case went to court. Upon examination of the blood types of the couple who claimed they had the wrong baby, it was found that both were type O; the tagged baby was type

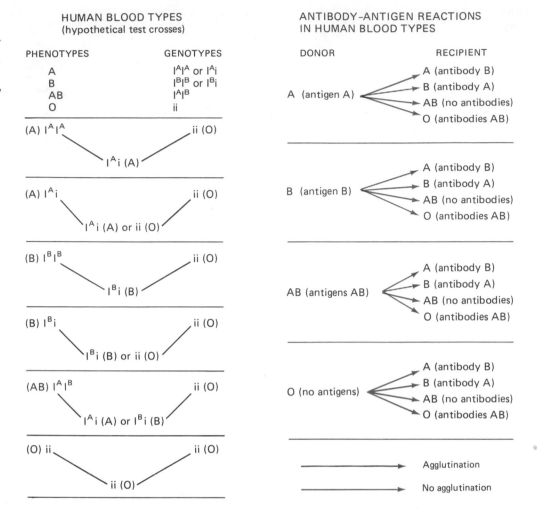

FIGURE 7-17
Human blood types. *Left,* hypothetical test crosses involving genotypes of the A, B, AB, and O series. *Right,* cross-reactions between red blood cells (antigens) and serums (antibodies) in the A, B, AB, and O series.

HUMAN BLOOD TYPES
(hypothetical test crosses)

PHENOTYPES	GENOTYPES
A	$I^A I^A$ or $I^A i$
B	$I^B I^B$ or $I^B i$
AB	$I^A I^B$
O	ii

(A) $I^A I^A$ ii (O)
$I^A i$ (A)

(A) $I^A i$ ii (O)
$I^A i$ (A) or ii (O)

(B) $I^B I^B$ ii (O)
$I^B i$ (B)

(B) $I^B i$ ii (O)
$I^B i$ (B) or ii (O)

(AB) $I^A I^B$ ii (O)
$I^A i$ (A) or $I^B i$ (B)

(O) ii ii (O)
ii (O)

ANTIBODY–ANTIGEN REACTIONS
IN HUMAN BLOOD TYPES

DONOR   RECIPIENT

A (antigen A)
- A (antibody B)
- B (antibody A)
- AB (no antibodies)
- O (antibodies AB)

B (antigen B)
- A (antibody B)
- B (antibody A)
- AB (no antibodies)
- O (antibodies AB)

AB (antigens AB)
- A (antibody B)
- B (antibody A)
- AB (no antibodies)
- O (antibodies AB)

O (no antigens)
- A (antibody B)
- B (antibody A)
- AB (no antibodies)
- O (antibodies AB)

→ Agglutination

→ No agglutination

A. Because there appeared to be no possible way that type-O individuals could produce a type-A baby, the judge ordered that the two babies be exchanged.

Cases of parentage also may be settled by means of blood typing. Although not a sure test, the system sometimes is useful: a type-O male, for example, could not be the father of an AB child. Another use of blood typing is in criminal situations where it sometimes is possible to determine the type designation of blood found on murder weapons, articles of clothing, automobile carpets, and so on. In addition to the A, B, AB, and O blood groupings, the medico-legal profession has about fifty other hemoglobin alleles to use in cases of doubtful identity.

Chromosomal Abnormalities

Specific chromosomal patterns and numbers are associated with every eukaryotic species, but for many species occasional irregularities have been described. Such abnormalities are called **chromosomal aberrations,** meaning that they are irregularities in the number of chromosomes per cell or are unusual structural alterations in one or more of the chromosomes.

Aneuploidy in Humans As stated earlier, chromosomes exist naturally in one of two alternate numerical states: the $1n$ (haploid) and $2n$ (diploid). In garden peas the $1n$ number is 7, the $2n$ number 14; in humans the $1n$ number is 23, the $2n$ number 46. However, in some cases the chromosome number is different than the natural $1n$ or $2n$ number. A condition in which there are one to several chromosomes more or fewer than $1n$ or $2n$ is called **aneuploidy.** Aneuploidy often produces abnormal phenotypes; that is, aneuploid organisms often display physical and chemical abnormalities, sometimes serious ones.

A number of abnormal conditions in humans are known to be the result of aneuploidy. In fact, a chromosomal abnormality shows up in about one of every 200 live births. Among these is **Klinefelter's syndrome** (Fig. 7-18A), a disorder of male humans characterized by the

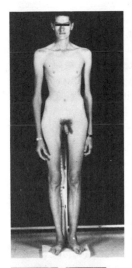
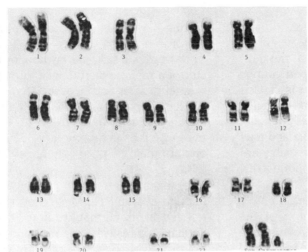
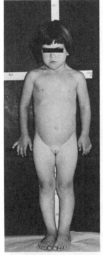
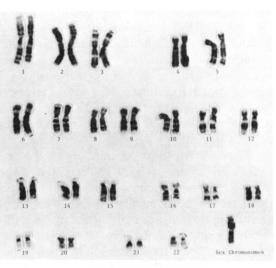
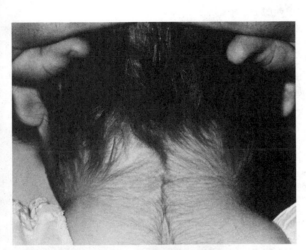
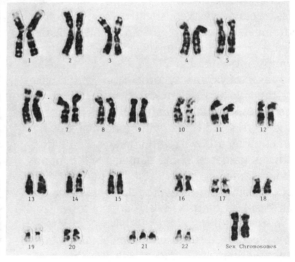

FIGURE 7-18

Disorders in humans caused by chromosomal aberrations. **A.** Klinefelter's syndrome, characterized by a slender, somewhat feminized male. Note the horizontal limit of pubic hair. **B.** Turner's syndrome. Webbing of the neck is characteristic. **C.** Down's syndrome. (A, photo courtesy Dr. Richard C. Juberg, The Children's Medical Center, Dayton, Ohio; karyotype courtesy Dr. Catherine G. Palmer, Indiana University School of Medicine; B, photos courtesy Dr. Stella B. Kontras, Children's Hospital, Columbus, Ohio; karyotype courtesy Dr. Catherine G. Palmer; C, photo courtesy Ted Shenenberger; karyotype courtesy Dr. Catherine G. Palmer)

presence of an extra X chromosome. These XXY individuals have 47 chromosomes and are sterile feminized males. Another is **Turner's syndrome,** in which the individual appears female but has underdeveloped ovaries and is sterile (Fig. 7-18B). Turner's syndrome is caused by the absence of an X chromosome ($2n = 45$). Females that are XXX also are known. They are clearly female, although some have poorly developed genitalia and may be somewhat retarded mentally. Most XXX females are fertile, however, and are cable of producing normal XX and XY children.

Particular attention has been focused on XYY males, sometimes called **supermales.** Humans of the XYY type are characteristically tall and may have severe facial acne during adolescence (acne, however, is not of itself evidence of the XYY condition). It has been claimed that XYY individuals tend to have a history of violence and to be predisposed toward criminal careers. A study in England in the 1960s found a higher percentage of white XYY males in English prisons (about 1 in 25) than in the populace as a whole (about 1 in 400). However, many XYY males lead normal lives, and a direct link between XYY and any tendency toward criminality is unclear.

One well-known genetic defect in humans is **Down's syndrome** (Fig. 7-18C). In the United States Down's syndrome occurs in about 1 in 600 births and is characterized by a smallish head; thick, fissured, and protruding tongue; stumpy fingers; and a fold of skin in the inner angle of the eye (the so-called **epicanthal fold**). Most Down's syndrome individuals suffer extreme mental retardation (IQ levels of 0–50), but in a few cases intelligence is average or even slightly above average. Down's syndrome is a result of **trisomy,** which means there is an extra chromosome, in this case an extra autosomal chromosome—number 21.

In some cases of Down's syndrome, one of the parents (usually the mother) will have 45 rather than 46 chromosomes, chromosomes 15 and 21 being conjoined. The phenotype of this individual will be normal, but he or she will function as a Down's syndrome carrier. When a carrier produces gametes, some gametes will contain an extra chromosome 21, some will be normal, and some will lack chromosome 21 altogether. If a gamete containing an extra chromosome 21 participates in zygote formation, the resulting child will have 47 chromosomes and be affected. Absence of chromosome 21 in a gamete acts as a lethal factor, and no offspring are produced. The remaining gametes with the normal $1n$ complement of 23 chromosomes have the potential for production of normal offspring. If Down's syndrome is known to have occurred in a family, the family members should have their chromosomes typed before contemplating having children. Karyotyping is a relatively simple procedure.

Down's syndrome is thought to arise also as a result of a meiotic disturbance during egg production in a normal mother and is associated with advancing age of the female parent. The incidence of Down's syndrome in children born of mothers over the age of 40 ranges between 2 times (for women of 40) and 25 times (45 years and up) that of a woman in her twenties (about 1 in 500).

Intersexes Intersexed individuals are found among humans as well as among other organisms and often are referred to as **hermaphrodites** (Fig. 7-19). True human hermaphrodites are very rare and are thought to result from the early presence of both XY and XX cells in the

FIGURE 7-19
Human hermaphrodite in an eighteenth-century painting. The near-normal breast development and intermediate male and female genitalia are accurately depicted. (Print from The Bettmann Archive, Inc.)

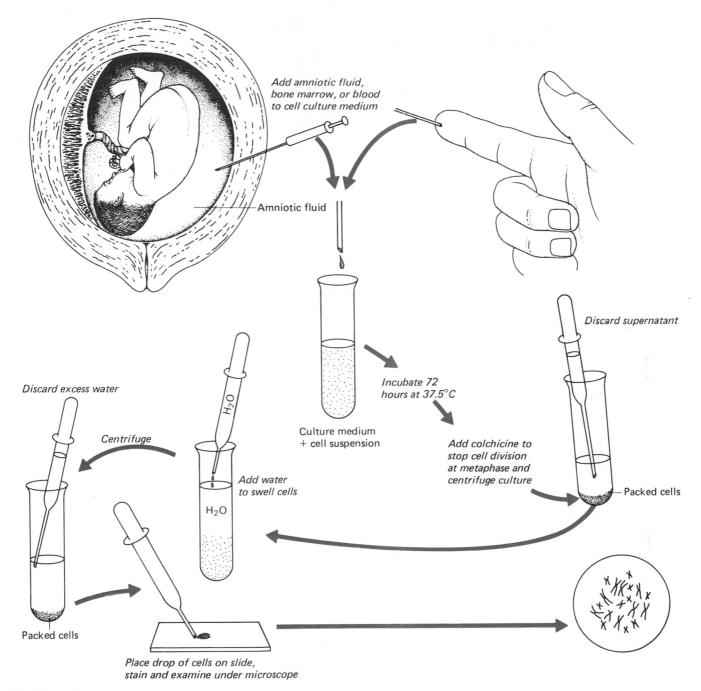

Add amniotic fluid,
bone marrow, or blood
to cell culture medium

Amniotic fluid

Discard supernatant

Discard excess water

Centrifuge

H₂O

Incubate 72
hours at 37.5°C

Culture medium
+ cell suspension

Add water
to swell cells

H₂O

Add colchicine to
stop cell division
at metaphase and
centrifuge culture

Packed cells

Packed cells

Place drop of cells on slide,
stain and examine under microscope

FIGURE 7-20
Human chromosomal analysis, or karyotype analysis. The initial source of cells may be bone marrow,
blood, or amniotic fluid obtained by amniocentesis. The cells generally are cultured in a special tissue-
culture medium, which induces them to divide, and then treated with colchicine to arrest mitosis.

developing embryo. As a result, both testicular and ovarian tissue may be present. The external genitalia tend to
be somewhat intermediate between male and female,
and although all hermaphrodites are sterile, most are
raised as females. Surgery and hormonal treatment can
accentuate feminine characteristics, but cannot restore

fertility. Masculinization treatment of hermaphrodites
has been less successful.

Genetic Counseling

The difficulties arising from the inheritance of a number
of deleterious genes and chromosomal abnormalities

are to a considerable extent avoidable through **genetic counseling** and medical treatment. Attention has been drawn to some of the more obvious genetic defects in humans, and most persons are aware of the importance of avoiding the consequences of transmission of such genes to future generations. No one is free of dangerous mutants; as pointed out earlier, it has been estimated that every human carries at least six deleterious recessive mutant genes.

Measures can be taken to avoid or to alleviate problems of inheritance of deleterious genes. When the subject is aware of a family history of real or apparently real hereditary disease, genetic counseling should be sought. The family physician is perhaps the most direct contact for such counseling and should be able to provide the necessary referrals. Often such counseling will dispel fears of transmitting bad genes to offspring. However, individuals should be prepared to avoid fathering or mothering children when there is danger of passing on certain very dangerous genes, such as those responsible for muscular dystrophy, cretinism, and hemophilia.

Milder genetic disorders can be successfully treated. For example, two fairly well-known metabolic defects, **alkaptonuria** and **phenylketonuria**, are due to mutant genes that block the metabolic utilization of amino acids. The symptoms of these disorders are successfully treated by the prescription of diets free of the offending amino acids. Some other defects, such as **syndactyly** and **polydactyly** (Fig. 7-14), are of a relatively minor nature and can be treated surgically when the child is an infant. Of course, parents must realize that the successful treatment of the effects of mutant genes does nothing to remove the genes from the cells of the child nor does it in any way prevent the further transmission of the genes when the child becomes an adult.

A relatively new technique, **amniocentesis**, permits an analysis for certain genetic defects in unborn children (Fig. 7-20). The procedure is fairly simple. During the early stages of pregnancy, a few drops of fluid in the amniotic sac surrounding the embryo are withdrawn through a sterile hypodermic needle. The fluid may be analyzed directly and used to reveal certain genetic problems such as Rh incompatibility or abnormal chromosome numbers. To test for chromosomal abnormalities, preparations are made of dividing lymph cells (**lymphocytes**), thus permitting an analysis of the chromosomes by means of a karyotype. Karyotype study will identify the sex of the embryo and also will reveal chromosomal irregularities such as those in Down's and Turner's syndromes, if they are present. In the event of chromosomal aberration, the pregnancy can be terminated upon advice of consulting geneticists and physicians, if it is the parents' wish to do so.

The risk of amniocentesis to the mother and fetus is low (although greater for the fetus than for the mother), and the procedure must not be confused with abortion.

Although amniocentesis can be a powerful medical tool, there is concern that it might be used frivolously. It would be most unfortunate if the sexual diagnosis of the unborn child were used as the basis of disposing of an unwanted male or female baby.

SUMMARY

The principles of inheritance were described by Gregor Mendel in 1865. He, for the first time, pointed out that inherited traits are passed on from generation to generation in the form of the hereditary units now called genes.

The simpler examples of inheritance are termed Mendelian, after their discoverer. One is the monohybrid cross, in which purebred parents AA and aa produce offspring that are Aa. The symbols A and a refer, respectively, to a dominant and a recessive allele. The offspring of AA and aa parents receive one gene for the particular trait from each parent; hence they are Aa, or hybrid. This generation, the F_1 (first filial) generation, produces gametes that are either A or a. When these combine to form the next generation, called the F_2 (second filial) generation, all possible combinations of A and a are formed in equal numbers, giving a genetic or genotypic ratio of 1AA : 2Aa : 1aa. This genotype is expressed in a phenotypic ratio of three dominant types to one recessive type. Many variations of these relationships between genes are known.

Combinations of dominant and recessive genes affecting the expression of some characteristic are called alleles. Early students of genetics wondered where alleles originated. Now it is recognized that they originate as changes in genes, called mutations. Mutations occur spontaneously with rare frequency, but can be induced at higher rates through the use of agents called mutagens. Among known mutagens are X rays, ultraviolet light, and mustard gas.

Modern genetics has had a profound influence on human existence. It has especially revolutionized agriculture. Livestock now are managed according to sound genetic principles, and crosses between different strains of pigs, chickens, cows, and so on, produce new, high-yielding types. Crop plants also have been vastly changed by the use of genetics in selective breeding.

In many organisms, including humans, the inheritance of the sex phenotype is governed by a pair of sex chromosomes. In humans, other mammals, and certain other organisms, these are designated XX (female) and XY (male). The X chromosome is the larger of the two sex chromosomes and can carry more genes. This means that the inheritance of dominant and recessive genes differs in males and females. Genes carried on sex chromosomes are sex-linked genes. A sex-linked gene carried on the single X chromosome of a male will be expressed whether it is dominant or recessive, whereas a recessive gene carried on the X chromosome of a female will not be expressed if the other X chromosome carries the dominant allele. Males generally exhibit the effects of deleterious sex-linked genes more frequently than do females. An example of this is the occurrence of hemo-philia in males of the descendants of Queen Victoria.

In humans, as well as other organisms, variations in the number and structure of chromosomes can occur. These are called chromosomal aberrations. Aneuploidy, the occurrence of one or more extra or fewer chromosomes than the normal complement, has important consequences in human heredity. Sterility, Down's syndrome, and other kinds of abnormalities result from an extra X chromsome, the absence of an X chromosome, or other abnormal chromosomal numbers. Human genetic counseling should be employed when it is suspected that individuals carry genes that may cause trouble.

KEY WORDS

Gregor Mendel	Mendelian factor	Barr body
dominant gene	mutation	sex-influenced genes
recessive gene	alleles	blood group
hybrid	lethal gene	antibody
F_1 generation	incomplete dominance	antigen
F_2 generation	multiple alleles	Rh factor
phenotype	polygenes	chromosomal aberration
genotype	linked genes	aneuploidy
homozygous	chiasma	Klinefelter's syndrome
heterozygous	crossing over	Turner's syndrome
probability	sex-linked genes	supermale
monohybrid	hybrid vigor	Down's syndrome
dihybrid	hemophilia	genetic counseling
Mendel's laws	sex chromosome	amniocentesis
Mendelian ratio	karyotype	

QUESTIONS FOR REVIEW AND DISCUSSION

1 What is the genotype of a tall pea plant whose offspring from a mating with a dwarf plant number 123 tall plants and 119 dwarf plants? Why is this kind of genetic cross useful?

2 What is the relationship between genes and Mendelian factors? Could a Mendelian recessive be caused by a mutation in more than one gene? Explain.

3 Explain linkage in genes. How would a dihybrid cross differ if in one case the genes were completely linked and in the other they were not linked? How can recombinations of linked genes occur?

4 Define "sex linkage." How is sex determined genetically in animals and many plants?

5 Could a red-eyed male fruitfly, mated with a white-eyed female, produce red-eyed male offspring? Explain using a simple diagram.

6 What is hybrid vigor? What is its genetic basis? How is hybrid maize produced?

7 In the case of the sex-linked gene responsible for hemophilia, a hemophiliac father never transmits hemophilia to his sons. Explain.

8 A bald man, whose father was not bald, marries a woman who is not bald but whose mother was bald. They have four sons and four daughters. Calculate the genotypes of the parents and grandparents and the probable genotypes and phenotypes of the sons and daughters.

9 Mary Roe, the mother of a child whose blood type is O, claims that the father of her child is John Doe. John Doe denies paternity of the child, claiming that because his blood type is A, he could not possibly be the father of this child. Does John Doe have a good case?

10 What is Down's syndrome? Can anything be done to prevent its occurrence or to relieve the problem when it does occur? Explain fully.

SUGGESTED READING

AYALA, F. J., and J. A. KIGER, JR. 1984. *Modern genetics*, 2nd ed. Menlo Park, Calif.: Benjamin-Cummings Publishing Co. (A textbook covering the basics, with emphasis on population genetics.)

BAER, A. S. 1977. *The genetic perspective*. Philadelphia: W. B. Saunders Co. (An up-to-date readable textbook emphasizing human aspects of Mendelian and molecular genetics.)

BEARN, A. G., and J. L. GERMAN, III. 1961. Chromsomes and disease. *Sci. Amer.* 205(5):66–76. (Presents basic information on the preparation of human karyotypes.)

CAVALLI-SFORZA, L. L. 1974. The genetics of human populations. *Sci. Amer.* 231(3):89–99. (Gene frequencies and gene selection are discussed, with emphasis on sickle cells and racial history.)

FUCHS, F. 1980. Genetic amniocentesis. *Sci. Amer.* 242(6):47–53. (Discusses the detection of human hereditary defects and its medical and legal implications.)

HEISER, C. B., JR. 1973. *Seed to civilization: the story of man's food*. San Francisco: W. H. Freeman and Co. (Discusses food growing in both ancient and modern contexts.)

HENDIN, D., and J. MARKS. 1979. *The genetic connection*. New York: New American Library. (A comprehensive reference tool, with information on a wide variety of topics.)

ILTIS, H. 1932. *Life of Mendel*. New York: W. W. Norton and Co. (An interesting account of the life of the great pioneer of the science of heredity.)

JUDSON, H. F. 1979. *The eighth day of creation: makers of the revolution in biology*. New York: Simon and Schuster. (The history of molecular biology from the 1930s through the mid-1970s is presented in a fascinating format—through the eyes of the history makers.)

KLUG, W. S., and M. R. CUMMINGS. 1983. *Concepts of genetics*. Columbus: Charles E. Merrill Publishing Co. (A well-illustrated basic genetics text with emphasis on molecular genetics and development.)

McKUSICK, V. A. 1971. The mapping of human chromosomes. *Sci. Amer.* 224(4):104–12. (Gives an interesting account of the relationship of genes and chromosomes, linkage groups, inheritance, and so on.)

MENDEL, G. 1865. *Experiments in plant hybridization*. Reprinted in J. A. Peters, ed. *Classic papers in genetics*. Englewood Cliffs, N.J.: Prentice-Hall, 1959, pp. 1–20. (This translation of Mendel's original paper is readily comprehensible by the student of introductory biology, and is well worth reading.)

MOODY, P. A. 1975. *Genetics of man*. New York: W. W. Norton and Co. (A readable presentation of human genetics for the student of introductory biology.)

WINCHESTER, A. M., and T. R. MERTENS. 1983. *Human Genetics*, 4th ed. Columbus: Charles E. Merrill Publishing Co. (A paperback containing useful and interesting information on reproduction and genetics of humans. Human genetic abnormalities and diseases, including cancer, are discussed.)

YUNIS, J. J., J. R. SAWYER, and K. DUNHAM. 1980. The striking resemblance of high-resolution G-banded chromosomes of man and chimpanzee. *Science* 208:1145–48. (An interesting discussion of similarities between human and chimpanzee chromosomes.)

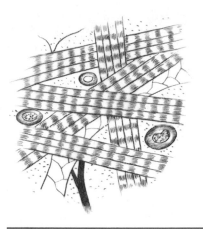

8 Multicellularity

Early in the history of life on earth, all organisms were unicellular. The unicell plan, however, has one major drawback: size. Presumably, an increase in size would confer some advantages in that the larger unicell could eat more smaller unicells with less risk of being eaten itself. The favored increase in size could be brought about by a simple growth to, let us say, double or triple the original dimensions. However, there seems to be a key problem here: as the unicell increases in size, there is relatively a smaller and smaller surface area to serve the needs of the cytoplasm. As discussed in the opening paragraph of Chapter 6, when a cell's surface increases by the square, its volume increases by the cube (see Fig. 6-1). The result is a relative loss of membrane capacity for the transport of food, oxygen, and other substances into the cell and for the elimination of wastes. One way out of the surface-to-volume dilemma is to become multicellular.

Multicellularity may be achieved by divisions and subsequent aggregation of simple, autonomous cell types, such as those described in Chapter 2 (see, for example, Fig. 2-9). In this arrangement, the colony, each cell remains pretty much an independently functioning unit, fully capable of all the processes required for sustaining life, but attains some benefits. For one thing, a collection of closely associated individuals is sheltered to an extent from environmental hazards. Among these are rapid and severe changes in such factors as cold, heat, drying, and possibly light. Then too, the colony is less susceptible to being gobbled up by small predatory unicells (for a vivid example of this, see Fig. 19-6, which shows a unicell, *Didinium*, eating another unicell larger than itself). There are many examples of the colonial form of multicellularity. One of the more remarkable is the green alga *Volvox*, an emerald-green sphere composed of more than a thousand tiny flagellated cells. Among animals, perhaps the best known examples are sponges. Sponges are composed of similar unicells, all of which function independently in food getting and food processing. However, the sponge cells are combined in a regular and characteristic form, so that one can describe such an aggregation as an individual as well as a colony. Actually, the lines between a colony of individuals, such as a *Volvox*, and a truly multicellular individual tend to be blurred when one considers multicellularity at its simpler levels. True multicellularity requires at least some division of labor and interdependence between cells.

In plants, animals, and many fungi, an evolutionary trend leading from colonies of like cells to genuine individuals constructed of several to many distinct cell types can be inferred. Cellular division of labor is the hallmark of true multicellularity.

EVOLUTIONARY TREND:
INCREASING SPECIALIZATION
OF CELLS, TISSUES, AND ORGANS

Primitive or ancestral   Advanced or derived
multicellular organism ⟶ multicellular organism

In classification schemes multicellular animals are referred to as **Metazoa**, multicellular plants as **Metaphyta**. Both groups are composed of many specialized cells. With increasing complexity, cells are arranged in **tissues** having similar structure and functions: **skin, muscle, bone, cork, fibers,** and so forth. Several different tissues go into the makeup of even more complex **organs** such as **brain, liver, root,** and **leaf**. The entire body of an animal or plant can be viewed as a functioning association of cells, tissues, and organs.

METAZOAN CHARACTERISTICS

As food seekers and consumers, animals are completely dependent for their energy needs on outside food sources, because none has the ability to photosynthesize. The fact that animals must seek out, pursue, and capture food has been largely responsible for the natural selection of their major structural and integrative characteristics (Fig. 8-1). These include such features as body shape, how the animal moves about, feeds, and breathes, whether or not it has a skeleton, the arrangement of its internal parts, and a number of other characteristics.

Shape

The shape of an animal is related directly to its life style, and vice versa. Animals that actively pursue food in an open sea are streamlined (sharks and porpoises); animals that are fixed to the bottom and wait for food to come to them are more often than not tubular in form with extensible tentacles (corals and sea anemones). In each case the shape is recognizable and distinctive.

The distinctive shapes of animals may be described in terms of **symmetry** (Fig. 8-2). A measure of symmetry is whether or not an object can either literally or figuratively be cut so as to produce identical halves. Some objects such as a sphere may be halved along innumerable planes with the same result: each half will be like the other. A sphere, therefore, has no front or back, no top or bottom, no right or left. This kind of symmetry is **spherical** or **universal symmetry**. Few organisms are completely spherical; among animals, spherical symmetry is restricted to eggs, zygotes, and very young developmental stages **(embryos)**.

More complex objects have fewer planes of symmetry. A vase, for example, has **radial symmetry**. It can be cut symmetrically along many longitudinal planes, but if it is cut across the middle, the top and bottom halves will not be identical. Radial symmetry is seen principally in some groups of aquatic **filter feeders**, animals that are sedentary and that filter their food from passing water currents.

Humans also are symmetrical objects, but have only one plane that cuts them into mirror-image halves; they have **bilateral symmetry**. Most animals have bilateral symmetry; their right and left halves are alike, but not the front, rear, top or bottom halves. The section

FIGURE 8-1
Composite drawing of major structural and integrative features of metazoans.

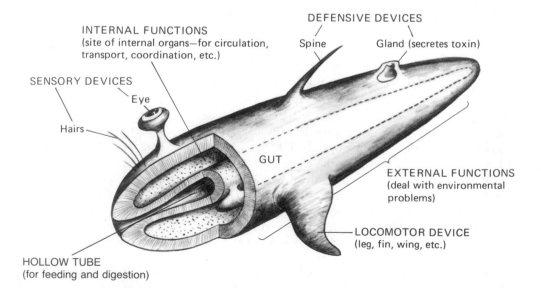

INTERNAL FUNCTIONS
(site of internal organs—for circulation, transport, coordination, etc.)

DEFENSIVE DEVICES
Spine   Gland (secretes toxin)

SENSORY DEVICES
Eye
Hairs

GUT

EXTERNAL FUNCTIONS
(deal with environmental problems)

LOCOMOTOR DEVICE
(leg, fin, wing, etc.)

HOLLOW TUBE
(for feeding and digestion)

FIGURE 8-2
Symmetry and asymmetry in animals.

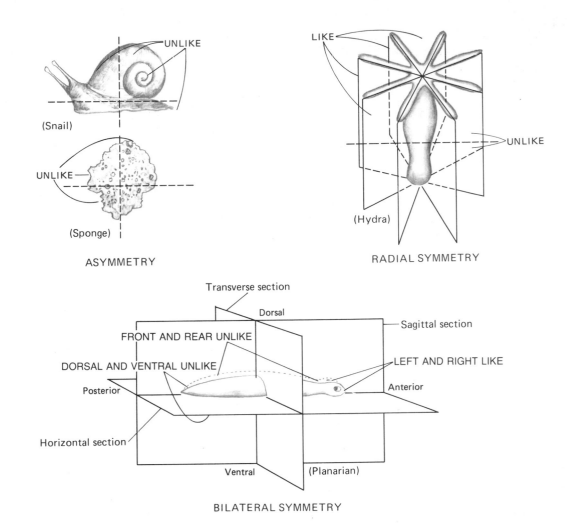

(Snail)

UNLIKE

(Sponge)

UNLIKE

ASYMMETRY

LIKE

UNLIKE

(Hydra)

RADIAL SYMMETRY

Transverse section

Dorsal

Sagittal section

FRONT AND REAR UNLIKE

DORSAL AND VENTRAL UNLIKE

LEFT AND RIGHT LIKE

Posterior

Anterior

Horizontal section

Ventral

(Planarian)

BILATERAL SYMMETRY

producing symmetrical halves is called a **sagittal section**. A horizontal section at right angles to the sagittal section produces dissimilar upper or back (**dorsal**) and lower or front (**ventral**) sides. The head (**anterior**) and the rear (**posterior**) ends also are unlike (Fig. 8-2).

A few kinds of animals have no symmetry. Sponges, particularly large ones, are of rather irregular form and hence **asymmetrical**.

Skeletons

Closely related to the distinctive shapes of many organisms is the presence or absence of a supporting framework or **skeleton** in their bodies. The presence or absence of a skeleton distinguishes major groups of animals. For example, the **invertebrates,** one large group of almost innumerable species and numbers, either have no skeleton at all (jellyfish, certain kinds of worms, etc.) or else have a variety of internal or external skeletal elements (clams, snails, starfish, lobsters, insects, etc.). In no invertebrate, however, is there a "**backbone.**" In contrast with invertebrates are the animals with backbones, the **chordates**. Of these the more numerous are the **vertebrates,** chordates with jointed backbones composed of **vertebrae.**

Locomotion

Movement and pursuit of food are the animal way of life. Various structural adaptations to a life of locomotion and predation can be observed among different animal groups. Swimming animals may move by undulations of the entire body (sea snakes), by movements of fins (fish), or by jets of water (squid). Locomotion over sea bottoms or land results from actions of multitudes of cilia, tentacles, accordionlike expansion and contraction of the body, water-filled tube feet, or jointed legs. A diversity of animals have mastered the air; they travel by gliding or by beating wings. Locomotion is an important feature in classifying animals.

Not all animals are motile, however. Many invertebrates, including sponges, corals, oysters, and barnacles, remain fixed in place, trapping food rather than pursuing it. Sponges circulate water by the vigorous beating of flagellated cells lining internal passages; corals use tentacles to encircle and entangle whatever prey comes near; oysters move seawater past ciliated gills, straining out and feeding upon plankton; barnacles employ leglike appendages to gather plankton. Although not motile as adults, all these animals have motile stages in their life cycles. Free-swimming larvae of sponges, corals, oysters, and barnacles account for their ability to colonize habitats rapidly.

A feature common to many sedentary animals that wait for food to come to them from any direction is radial symmetry. Motile, predatory animals, on the other hand, characteristically are bilaterally symmetrical, with a head and a tail end. An additional trend in bilaterally symmetrical animals is **cephalization,** an evolutionary process that results in the concentration of sensory devices and nervous tissue at the head end for effective monitoring of environmental stimuli.

Digestive Tracts

Metazoans have internal cavities within which digestion takes place. Animals may be classified according to the structural complexities of digestive systems (Fig. 8-3). Sponges lack digestive cavities. Jellyfish and simple worms have a **gastrovascular cavity** which serves as an organ of digestion and distribution of digestive products (**circulation**). The cavity's single opening functions both for food intake (**mouth**) and waste elimination (**anus**). Other animals have, as an organ of food processing, a tubular **digestive tract,** or **gut,** open at both ends. In simple animals a gut may function in circulation as well as digestion, but in most cases food distribution is carried on by a separate **circulatory system.**

Body Layers

The simplest and most primitive body organization is a mass or aggregate of cells. Sponges are living examples, and it is thought that the ancestors of metazoans were no more complex. As animals evolved, cells became specialized and organized into groups and layers having specific functions. The simplest animals in which this is evident have two layers: an outer skin and the lining of a gastrovascular cavity. These layers are broadly designated as **ectoderm** ("outer skin") and **endodoerm** ("inner skin"), respectively. Obviously, this simple organization has limitations if animals are to grow in size or further complexity. Nearly all animals have evolved a third, middle layer, the **mesoderm** ("middle skin"), from which muscles, circulatory constituents, and skeletal elements are derived (Fig. 8-4).

FIGURE 8-3
Digestive systems of animals.

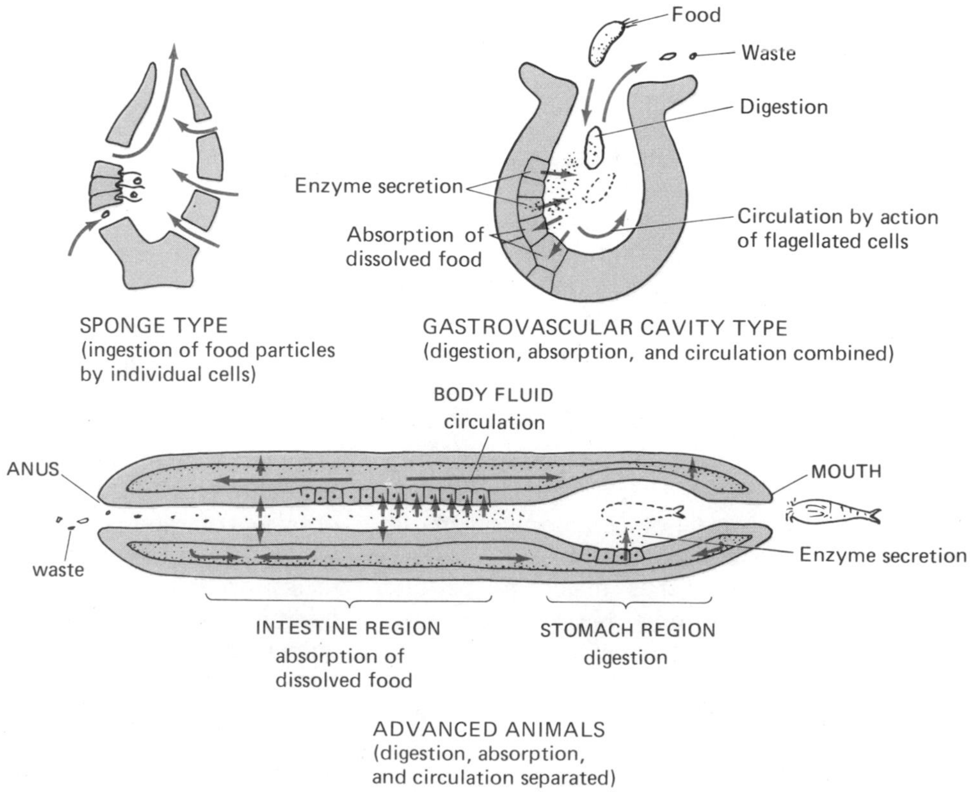

Food
Waste
Digestion
Enzyme secretion
Absorption of dissolved food
Circulation by action of flagellated cells

SPONGE TYPE
(ingestion of food particles by individual cells)

GASTROVASCULAR CAVITY TYPE
(digestion, absorption, and circulation combined)

BODY FLUID circulation

ANUS
waste
MOUTH
Enzyme secretion

INTESTINE REGION
absorption of dissolved food

STOMACH REGION
digestion

ADVANCED ANIMALS
(digestion, absorption, and circulation separated)

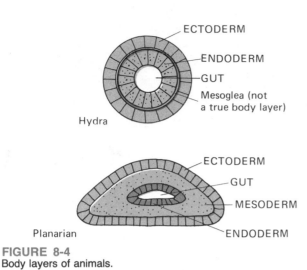

FIGURE 8-4
Body layers of animals.

Body Cavities—The Coelom

A major advance in the evolution of animal life was the development of a fluid-filled cavity, the **coelom,** between the digestive tract and outer body wall. In simple animals coelomic fluid may serve as an elementary circulatory system for transport of nutrients, gases, and wastes. More importantly, a coelom allows for the evolution of a true circulatory system with pumping structures (**hearts**) and separate transport vessels (**arteries** and **veins**). Furthermore, the coelom provides for separation of digestive tract and body wall musculature as well as space for other internal organs.

Jellyfish exemplify the solid body type called **acoelomate** ("without coelom"). In two other types the coelom occurs between the digestive tract and the outside. If this space is lined with a type of skin called **peritoneum,** the animal is said to be **coelomate** ("with coelom"). If the body cavity lacks a peritoneum, the term **pseudocoelomate** ("false coelom") is applied (see Fig. 8-5).

Circulatory Systems

Once food has been digested, products of that digestion are available for distribution to all the cells of the animal body. In the simple, soft-bodied animals such as jellyfish and certain worms, no cell is very far removed from the digestive tract, and thus food is readily accessible. However, in more complex animals characterized by many layers of cells outside the digestive canal, circulatory systems are necessary for food distribution and waste collection. Circulatory systems also maintain a flow of oxygen from the atmosphere to the cells of the body, as well as a reverse flow of carbon dioxide, a major waste product of metabolism.

Circulatory systems are of two general types: **open systems,** in which the circulating fluid (**blood**) bathes cells and tissues surrounding spaces known as **sinuses;** and **closed systems,** in which blood circulates through vessels (Fig. 8-6). In both types, one or more hearts propel the blood along its course.

Excretory Systems

In primitive animals the requirements for waste elimination are met by direct **excretion** from cells into the environment (water) and by expulsion from a gastrovascular cavity, if one is present, of debris remaining after digestion. Even in the simplest systems, two distinct types of waste matter and two methods of waste disposal are recognizable. One is the solid waste matter (**feces**) remaining as a residue of digestion of raw food; the second is waste resulting from wear and tear on cells and chemical activities occuring in cells (**metabolic wastes**).

In more complex animals digestive wastes are collected in the lower gut and evacuated through the anus. Metabolic wastes accumulate in body fluids and are excreted in several ways (Fig. 8-7). Ciliated tubules collect wastes from the coelomic fluid and empty either to the outside or into a gut. In vertebrates, metabolic wastes are collected from cells by blood, filtered out of the blood by kidneys, and excreted in the form of **urine.**

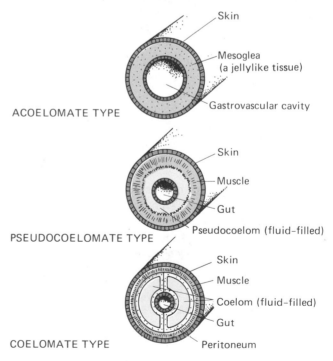

FIGURE 8-5
Acoelomate, pseudocoelomate, and coelomate structure.

FIGURE 8-6
Metazoan circulatory systems.
A. Open. **B.** Closed.

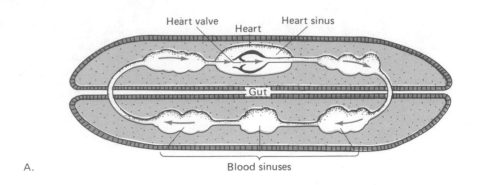

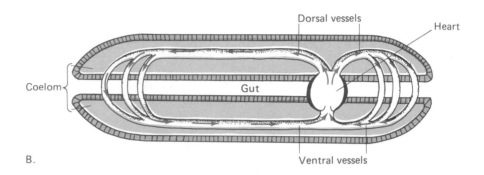

Respiration

Respiration includes two vital processes. One is liberation of food energy in cells by oxidation (burning). The second is the exchange of oxygen and carbon dioxide between an organism and its environment. (Oxidation in cells was covered in Chapter 4. In this and the next several chapters, references to respiration will be in the context of gas exchange [Fig. 8-7].)

At its simplest, gas exchange occurs at the cellular level, and all cells of an organism may exchange gases directly with the environment. Sponges are an example, because in them no cell is remote from water. In more complex animals, body fluids transfer gases from the environment to interior cells. Gas exchange occurs exclusively through the skin in small aquatic animals, but in larger aquatic animals **gills** having comparatively great surface areas are major sites of gas exchange. On land, **lungs** or other air passages serve the same function as gills.

Reproduction

An early step in the evolution of multicellularity was the specialization of certain cells for reproduction. These cells, known as **germ cells,** became localized in reproductive organs called **gonads.** As animals became more highly organized, accessory sex organs evolved; reproduction in many animals has become rather complicated, as will be seen in following chapters. Nevertheless sexual reproduction in all animals involves the essential step of production of male and female sex cells, or **gametes** (Fig. 8-8), which unite to produce a single cell, the **zygote.** The zygote then undergoes a series of cell divisions and **cell specializations** to form the adult individual. Thus in each generation the organism returns to the unicellular state, then becomes a colony of cells, and finally develops into a highly organized adult.

CELL SPECIALIZATION AND ORGANIZATION

In single-celled organisms (**unicells**) parts of cells become specialized during development, whereas in multicellular organisms development involves the specialization of entire cells to perform specific functions. Moreover, there is a multitude of cell types among the several kingdoms of organisms, reflecting not only the ecological circumstances in which organisms live but also their degree of organization and development. In complex multicellular organisms such as plants and animals, there may be dozens to hundreds of unique and distinctive cell types. These cell types, however, can be

FIGURE 8-7
Excretion and respiration in animals.

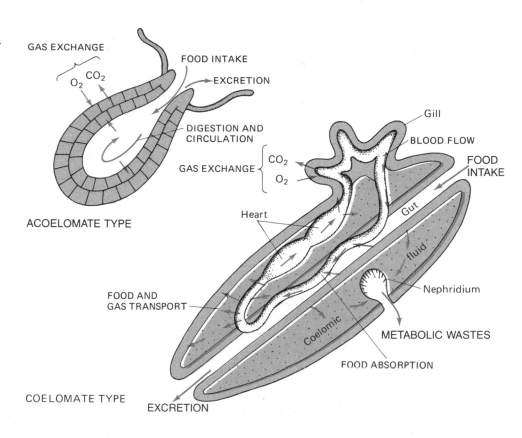

GAS EXCHANGE

O_2 CO_2

FOOD INTAKE

EXCRETION

DIGESTION AND CIRCULATION

ACOELOMATE TYPE

Gill

BLOOD FLOW

GAS EXCHANGE { CO_2 / O_2 }

FOOD INTAKE

Heart

Gut

fluid

Nephridium

FOOD AND GAS TRANSPORT

Coelomic

METABOLIC WASTES

FOOD ABSORPTION

COELOMATE TYPE

EXCRETION

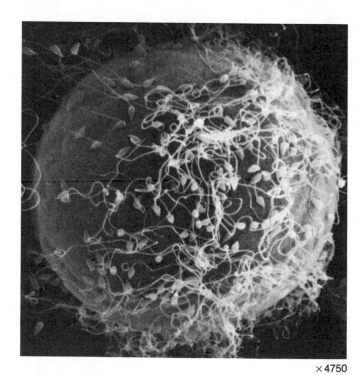

×4750

FIGURE 8-8
Sperm and egg of the sea urchin at the time of fertilization. Note the immense size differential between sperm and egg. (Photo courtesy David Epel and Mia Tegner)

divided into fewer, major categories referred to as tissues and defined as aggregates of cells having similar structures and functions.

Plant and animal tissues are of two basic types: **somatic tissues** (*soma*, "body") that function in the day-to-day maintenance of the individual, and **reproductive tissues**, whose function is the production of the next generation of individuals. Discussion of reproductive tissues will be deferred to Chapter 11.

ANIMAL TISSUES

The principal somatic tissues of animals are **epithelium, connective tissue, nerve tissue,** and **muscle.**

Epithelium

Epithelium forms the outer covering and inner lining of all animals more complex than sponges (Fig. 8-9). One type of epithelium, called **squamous** or **pavement epithelium,** is, as the name implies, a single layer of flattened cells (See Fig. 8-9A). The skins of simple animals are composed of squamous epithelium, as are the linings of body cavities, arteries, and veins; capillaries are simple tubes of squamous epithelium.

FIGURE 8-9

Types of epithelium.
A. Simple squamous epithelium.
B. Simple cuboidal epithelium.
C. Simple columnar epithelium.
D. Stratified squamous epithelium. **E.** Stratified columnar epithelium.

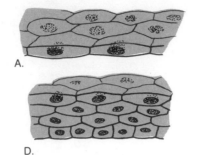
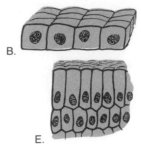

Similar to squamous epithelium is **simple cuboidal epithelium,** which is composed of a single layer of cubical cells (Figs. 8-9B and 8-12B). Ducts and canals commonly are composed of cuboidal epithelium. A specialized form of cuboidal epithelium forms glandular tissue, in which epithelial cells are specialized for the production and secretion of hormones, enzymes, and other products (Fig. 8-12C). **Simple columnar epithelium** (Fig. 8-9C) resembles the cuboidal type and sometimes intergrades with it, as in kidney tubules. The digestive cavities of many lower animals are lined with simple columnar epithelium.

Even more complex types are the **stratified epithelia,** composed of layers of cells, the newest formed at the bottom or inside, the older at the top or outside. As the outer cells wear out, they are replaced from underneath. Several kinds of stratified epithelium are distinguished by the form of their surface cells. If these cells are flat, the type is **stratified squamous epithelium** (Fig. 8-9D); if surface cells are columnar, the type is **stratified columnar epithelium** (Fig. 8-9E). Subtypes of these exist in which the outer cells are ciliated, as in respiratory passages and reproductive ducts, or equipped with microscopic fingerlike projections called **microvilli,** characteristic of the absorbing epithelium of the small intestine.

Connective Tissue

Connective tissues include **cartilage, bone, fibrous tissues,** and blood, which form the supportive elements of bodies of higher animals (vertebrates), including fish, amphibians, reptiles, birds, and mammals. All connective tissue consists of cells embodied in a **matrix.** In cartilage the matrix is a gelatinous solid; in bone the matrix is calcified and stonelike; in fibrous tissue the matrix is gelatinous; in blood the matrix is liquid. In all cases living cells are constantly present.

Cartilage is a connective tissue in which the cells are widely separated by a matrix of fine fibers composed of the simple protein **collagen,** and a gel composed· of **chondrin,** a protein-polysaccharide (Fig. 8-10A). Sharks and some other primitive fish have skeletons made solely

of cartilage. Other vertebrate animals, including humans, have both cartilage and bony elements in their skeletons.

In the embryos of humans and other bony vertebrates, the elements of the cartilaginous skeleton are gradually replaced by bone, a process called **ossification.** Ossification normally is not completed until the individual attains its adult size. In the human infant, for example, the bones are not ossified at their ends, where cartilaginous regions permit continuing growth in length. In addition, in both infants and adults the bone itself continually undergoes dissolution and restructuring.

Bone, like cartilage, is cellular, but the intracellular regions are filled with calcium carbonate and phosphate, which are in effect limestone, an almost universal building material found in skeletons of many animals. Microscopic examination of bone shows its structure to consist of more or less parallel bony cylinders, each constructed radially around a canal, the **Haversian canal,** which encloses an artery, a vein, and sometimes a nerve fiber (Fig. 8-10B).

Another type of connective tissue is **loose connective tissue,** a random array of collagen fibers and gelatinous protein (Fig. 8-10C). Whereas bone and cartilage generally are restricted to vertebrates, loose connective tissues are found in all animals. The jelly layer of the jellyfish, for example, is composed of **mesoglea,** a type of loose connective tissue. Another variation of loose connective tissue is fatty tissue (**adipose tissue**), which contains many fat storage cells. In vertebrates loose connective tissue also constitutes the fibrous material that binds together the cells of many organs, including the liver, kidneys, gonads, and glands.

Blood also is considered to be a type of loose connective tissue, but one in which the fibrous elements are in solution and the cells, which include red blood cells and several kinds of white cells, float freely in a liquid matrix (Figs. 8-10D and 15-10).

A somewhat more dense connective tissue, called **compact connective tissue,** composes the deep layer, or **dermis,** of the skins of vertebrates. Glands, blood vessels, and nerves are found in the dermis, which is covered by the **epidermis** (Fig. 8-9E).

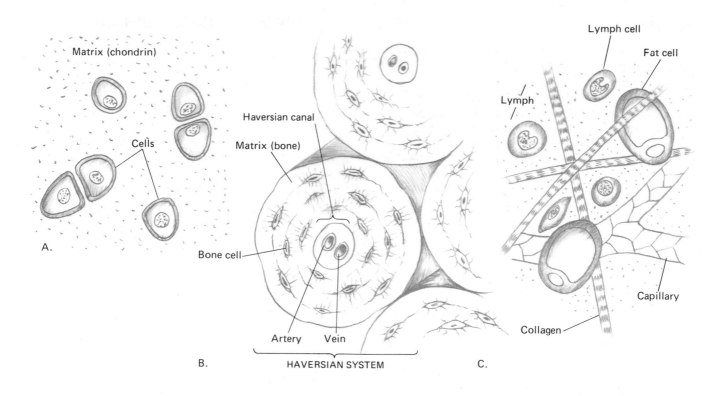

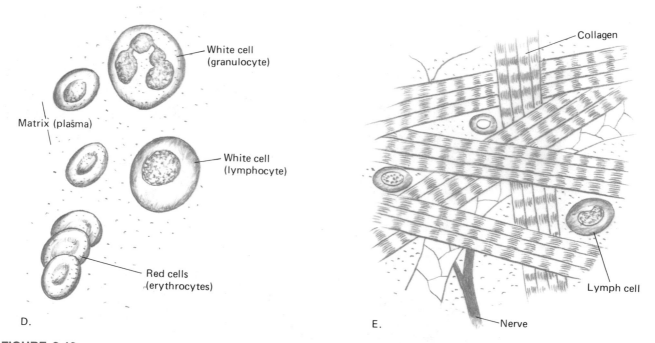

FIGURE 8-10
Connective tissue. **A.** Cartilage. **B.** Bone. **C.** Loose connective tissue. **D.** Blood.
E. Compact connective tissue.

Nerve Tissue

The nerve tissues of the multicellular body are those of the brain, spinal cord, peripheral nerves, and sensory receptors. Although a number of cell types are present, the fundamental cell in nerve tissue is the **neuron** (Fig. 13-2). The neuron consists of a cell body and one or more fibers (**axons** and **dendrites**) extending from the cell body. Structural and functional aspects of nerve tissues are discussed in several chapters and are a major topic in Chapter 13.

Muscle

The structure and function of muscle was discussed in Chapter 4, where the energy relationships of muscle contraction were presented. All muscle cells, of which there are three types—**smooth (involuntary), cardiac,** and **skeletal (striated)**—are contractile (see Figs. 8-11 and 8-12A). Cardiac and skeletal muscles have longitudinally disposed microfilaments of actin and myosin arranged so that they present a banded pattern. Skeletal muscle cells are very long and have multinucleate cells. Cardiac muscle cells are relatively short and have single nuclei. Both are comparatively rapid-acting in contrast to smooth muscle. The actin and myosin filaments of smooth muscle cells are less regularly oriented, and banding is absent.

METAPHYTE CHARACTERISTICS

Shape

With the exception of certain motile algae, plants are stationary photosynthetic organisms; their shapes reflect this mode of existence, just as the shapes of animals provide clues to their predatory life style. Because the "plant strategy" is to capture sunlight maximally, plants tend to exhibit radial symmetry (consider a barrel cactus, or pines and palms).

All plants have requirements for nutrient absorption, gas exchange, transport between cells, support and deployment of photosynthetic cells, and dissemination of reproductive units (spores, gametes, seeds, and fruit). In simple unicells and colonies of algae, these requirements result in minimal conflict—every cell performs all functions, but multicellular plants, being larger, have cells and tissues specialized for quite specific functions. This is especially true of land plants. Root cells buried in soil obviously can't photosynthesize, leaf cells can't absorb soil nutrients; both require an interconnecting transport system. Thus land plants have tissues and organs specialized for anchorage and absorption (roots), support and transport (stems), and photosynthesis and gas exchange (leaves).

Because plants are nonmotile organisms, their "strategy" for obtaining nourishment is one of continual

FIGURE 8-11
Muscle tissue types. **A.** Smooth muscle. **B.** Cardiac muscle. **C.** Skeletal muscle.

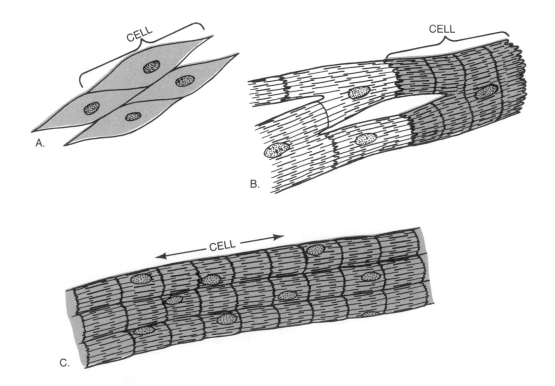

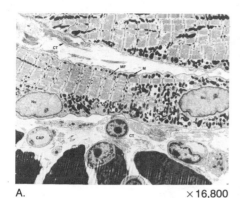

A.   ×16,800

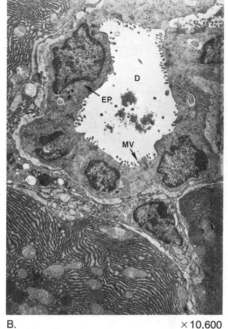

B.   ×10,600

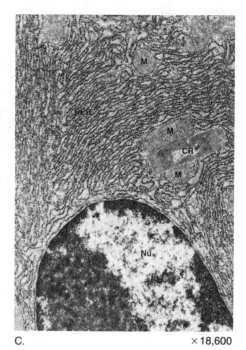

C.   ×18,600

FIGURE 8-12

Electron micrographs of some mammalian cell and tissue types. **A.** Striated or skeletal muscle cells and related tissues. Note the regular bands of contractile units (sarcomeres). **B.** Epithelial cells of a pancreatic secretory duct. **C.** A portion of a pancreatic glandular cell. Note that the cytoplasm is composed in part of numerous parallel membranes of rough endoplasmic reticulum. The numerous very small dark bodies are ribosomes. Cap, a capillary cross section; CR, a crystalline body; CT, connective tissue fibers (collagen); D, canal of secretory duct; EP, an epithelial cell; M, mitochondrion; MF, muscle filaments (refer to Fig. 14-8); MV, microvilli; Nu, nucleus; RER, rough endoplasmic reticulum. (A, photo courtesy David Mason; B and C, photos courtesy Dian Molsen)

growth—probing the soil for nutrients and water, competing for sunlight with neighboring plants. As a result of competition, plants have evolved a different kind of development than that exhibited by the majority of animals. Plants are said to have **indeterminate development,** meaning that as long as they continue to live, they also continue to grow. Contrast this with the development of an animal such as a cat, or for that matter, an elephant, in which the individual normally attains a specific size and shape at maturity. This is known as **determinate development** and is achieved by control of cell divisions occurring in tissues throughout the body. The indeterminate growth of a plant, on the other hand, results from divisions of cells in localized regions or **meristems** (*meristos*, "divided") at the tips of stems and roots or just beneath their surfaces. Therefore, plant growth is additive (as one notes in the growth rings of a tree stump), and plants are theoretically immortal (never in actuality, of course, but certain individuals among redwoods and other conifers are the oldest organisms known).

Organization of a Plant

Reduced to its simplest configuration, the body of a land plant consists of an elongated cylinder tapering to meristems at each end. Meristems provide for growth in length; the region in between, which we may call the **root-shoot axis,** is composed of organs having functions of absorption, transport, support, photosynthesis, and gas exchange. Specifically, the **root** is the organ mainly responsible for absorption and transport of substances from the soil, the **stem** has major functions of food and water transport and support of leaves, and **leaves** are the major organs of photosynthesis and gas exchange. However, some of these functions can be shared, as in the case of plants having green, photosynthetic stems.

Body Layers

In mature regions of the root-shoot axis, three main tissues can be discerned in cross-sectional views (Fig. 8-13): an **epidermis** (one cell thick), the **cortex** (several to many cells thick), and the **vascular cylinder** (many

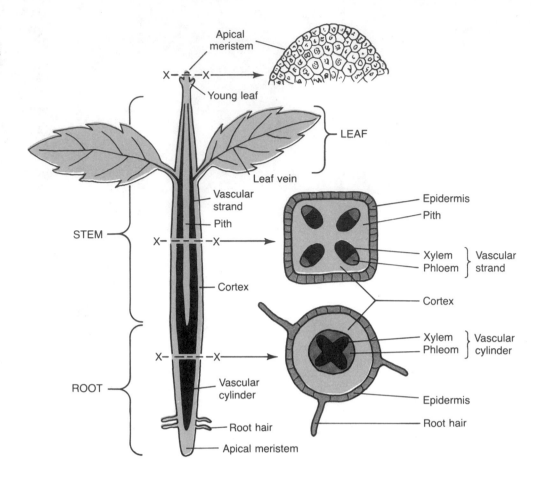

FIGURE 8-13
The organs and body layers of a mint plant (stems square in cross section are characteristic of the mint family).

Apical meristem

Young leaf

LEAF

Leaf vein

Vascular strand

Pith

STEM

Cortex

Epidermis

Pith

Xylem
Phloem } Vascular strand

Cortex

Xylem
Phleom } Vascular cylinder

Epidermis

Root hair

ROOT

Vascular cylinder

Root hair

Apical meristem

cells thick). In older parts of the stem and root, the epidermis may be replaced with several layers of cork cells (Fig. 8-13B), and the vascular cylinder may be augmented by additional layers of conductive tissues, as in the growth rings seen in a tree trunk. The vascular cylinder is continuous from the root tip to the stem tip, giving off branches (veins) into the leaves. In the stem, the cylinder consists of **vascular strands,** separated by **parenchyma** cells (Figs. 8-13; 8-14; 8-15) and enclosing a central region, the **pith,** composed of parenchyma cells. In the root, the vascular cylinder is a solid core without a pith region. Usually the cortex is thicker in roots than in stems, especially so in roots having storage functions (carrots, beets, and turnips).

Leaves also are basically three-layered, having an outer layer of epidermis, within which is a packing of photosynthetic cell tissue and branches of the vascular cylinder (refer to Fig. 3-7).

Reproduction

Plant reproduction differs from animal reproduction in that plants can reproduce asexually by **spores** and sex-

ually by **gametes.** Spores (Fig. 8-16B) are small, often thick-walled reproductive cells produced, frequently in large numbers, within organs called **sporangia** (sing. **sporangium**). Plant gametes differ from plant group to plant group. In many algae, the gametes look very similar and cannot be said to be either male or female; they are often simply designated **plus** and **minus.** In land plants, the gametes are the male **sperm** and the female **egg,** variously produced in organs of several kinds, the most familiar of which are **cones** and **flowers.** In some land plants, the male gametes are equipped with flagella and are motile (Fig. 8-15B); in others (most seed plants) the male gametes are not flagellated (Fig. 16-4).

PLANT TISSUES

As in animals, the functions of the plant are carried out by the integrated activities of tissues and organs. Plant functions are not as diverse as those of animals because plants are nonmotile food-producers, not motile food-pursuers. Digestive and excretory systems are not found in plants because plants do not ingest food particles nor

FIGURE 8-14
Plant tissues **A.** Epidermis.
B. Cork. **C.** Supportive tissues.
D. Vascular tissues. **E.** Parenchyma tissue. Epidermal cells
are flat and lack chloroplasts, except for guard cells surrounding
the stomata. Guard cells have
chloroplasts. The xylem shown is
a vessel, a cell type having open
ends. The phloem cells are the
sieve tube member, part of an
elongated conduit system, the
sieve tube, and an accessory cell
type called the companion cell.
Companion cells have nuclei,
sieve tube members do not. Xylem cell, sclereids, and cork cells
are dead, empty cells. Parenchyma cells are principal storage
cells of plants.

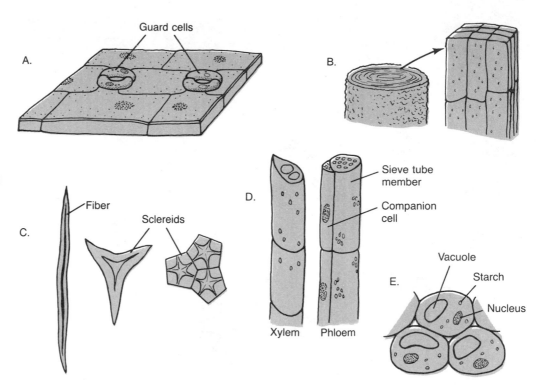

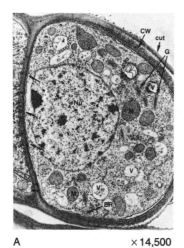 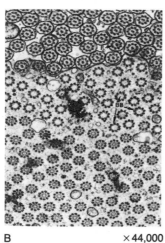 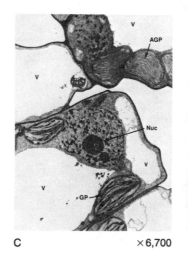 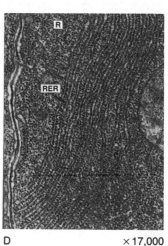

A ×14,500   B ×44,000   C ×6,700   D ×17,000

FIGURE 8-15
Electron micrographs of some plant tissue and cell types. **A.** An epidermal gland cell of a salt marsh grass
(*Cynodon*). **B.** A portion of a flagellated male gamete of a primitive seed plant, a cycad (refer to Fig.
20-31). The cell has thousands of flagella, each supported by a cytoplasmic basal body. **C.** Portions of
photosynthetic cells (chlorenchyma) of a maize leaf. The two kinds of chloroplasts of this C-4 plant are
easily distinguishable (i.e., granal mesophyll chloroplasts and agranal bundle-sheath chloroplasts). **D.** A
portion of the cytoplasm of a microsporangium of jack pine. The cell is very actively biosynthesizing, as
indicated by the numerous membranes of rough endoplasmic reticulum (compare with Fig. 8-12C). Note
also the large numbers of ribosomes (very small dark bodies). AGP, agranal chloroplast; BB, flagellar
basal bodies; cut, cuticle; CW, cell wall; ER, endoplasmic reticulum; F, flagella; G, Golgi bodies; GP,
granal plastid; M, mitochondrion; Nu, nucleus; Nuc, nucleolus; R, ribosomes; RER, rough endoplasmic
reticulum; V, vacuole. (A, photo from J. W. Oross and W. W. Thompson, 1982, *Amer. J. Bot.* 69:
939–949; C. photo courtesy of Dian Molsen; D, photo from R. A. Cecich, 1984, *Amer. J. Bot.* 71: 851–864)

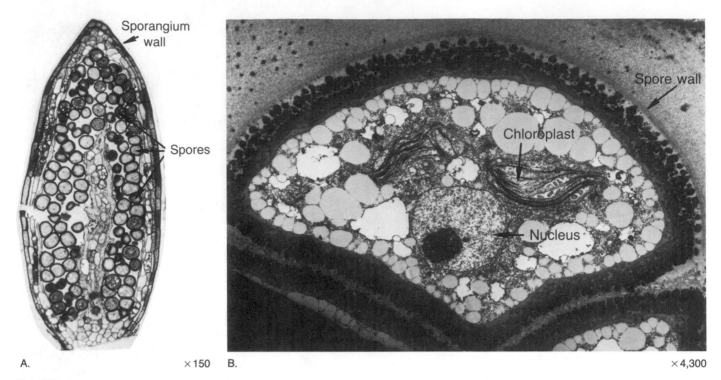

Sporangium wall

Spores

Spore wall

Chloroplast

Nucleus

A. ×150 B. ×4,300

FIGURE 8-16
The sporangium and spores of a moss *(Andreaea rothii)*. **A.** Longitudinal section of the sporangium with its spores. **B.** A sectioned spore. The spores have three angular faces where they have developed in contact with three other spores. The four spores are produced from one spore mother cell by meiosis. (Photos reprinted with permission from R. E. Brown and B. E. Lemmon, 1984, *Amer. J. Bot.* 71:412)

excrete particulate wastes. (Note: Insectivorous plants do indeed prey on animals but do not ingest them. See box, page 560). All uptake, transport, and elimination of substances occurs at the molecular level in plants. Functions of support and conduction tend to be combined. Water is conducted by thick-walled hollow tubes of **xylem** tissue. Xylem, augmented by **fiber** cells, also acts as the major skeletal tissue of the plant body. Food is transported by thin-walled living **phloem** cells. Together, xylem and phloem constitute the vascular tissue of the plant. Transport of substances between the plant and the environment is the function of the epidermis. In roots, water and soil nutrients enter through epidermal cells by active and passive transport (refer to Chapter 2). **Root hairs,** extensions of root epidermal cells, add to the uptake efficiency of this tissue system. In stems and leaves, the epidermal cells are coated with water-resistant cuticle, so movement of water vapor and oxygen and carbon dioxide occurs through stomata (*stoma,* "mouth") composed of liplike pairs of **guard cells** (Fig. 17-8). The function of photosynthesis, as well as food and water storage, is carried on by parenchyma—photosynthesis by a chloroplast-containing parenchyma tissue called chlorenchyma.

Epidermis

The epidermis of plants consists of a layer of flattened cells present as a continuous covering of the outside of roots, stems, and leaves (Figs. 8-14A; 8-15A). Root epidermal cells are permeable to water and solutes and in some cases bear elongated root hairs which add to the absorptive surface. Epidermal cells of stems and leaves are covered by a cuticle, which renders them impervious to water. Guard cells of stomata contain chloroplasts but other epidermal cells do not and therefore transmit light readily to photosynthetic tissue beneath. Other specialized epidermal cells of stems and leaves are glands and trichomes (Fig. 8-15A), thought to have protective functions.

Parenchyma

Parenchyma (*para chein,* "fill in") tissues consist of roughly spherical or cuboidal cells (Figs. 8-14E) that function in food and water storage and localized cell-to-cell transport. Parenchyma cells are found in the cortex of stems and roots, between vascular strands in the pith of stems (Fig. 8-13), and in many seeds and fruits. They are characterized by relatively thin walls, large vacuoles, abundant starch granules, and often oil droplets. A type of parenchyma called **chlorenchyma** is composed of chloroplast-containing cells and is the photosynthetic tissue of leaves, green stems, and the green parts of flowers and fruits (Fig. 8-15C).

Cork

The dead cells of cork that Robert Hooke viewed with his primitive microscope constitute an important tissue in mature plants (Fig. 8-14B). Cork tissue replaces the epidermis of older stems and roots. It is waterproofed by a lipid material called **suberin**. Here and there in the surface of cork are air passages, the **lenticels**. When bottle corks are manufactured, the cork is cut so that the lenticels are crosswise; otherwise the cork would leak.

Supporting Tissues

Sclerenchyma and collenchyma are strengthening tissues in plant organs (refer to Fig. 8-14C). The veins of leaves are rather rigid because they have a girderlike construction in which sclerenchyma cells are the strengthening elements. Schlerenchyma also composes the **fibers** of stems of such plants as flax, bamboo, and palms. The cells of sclerenchyma commonly are very elongate, thick-walled, and hardened by deposits of **lignin,** a complex organic molecule. At maturity these cells have lost their protoplasm and are dead. Some types of sclerenchyma are not elongate but instead are quite irregular in shape. Among these are **sclereids,** which are found in the pulp of pears and are responsible for their gritty texture. Very irregular, star-shaped sclereids are present in the leaves of some plants. Collenchyma cells are long and slender, but unlike sclerenchyma cells they remain alive and are soft and unlignified. Celery stalks contain a good bit of collenchyma.

Vascular Tissue

The vascular tissues of plants are a composite of several cell types, and include parenchyma as well as sclerenchyma, but the basic cell types are those of the water-conducting tissue (xylem) and the food-conducting tissue (phloem) (Fig 8-14D). Their structure and functions are discussed in Chapter 17 (see especially Figs. 17-6, 17-9, and 17-10).

Meristematic Tissue

As we have noted, the growth of plants differs from animal growth in that it is largely confined to areas called meristems, whereas in most cases growth of animals occurs by cell division throughout the body. Meristems located at the tips of stems and roots and in certain other localized areas of cell division are responsible for the production of all the basic tissues of the plant. Many higher plants also possess lateral meristems, or cambiums, which by division of their cells produce thickenings of stems and roots. A specialized lateral cambium, the cork cambium, is present in the outer bark of woody plants.

Meristematic cells generally are thin-walled cells with dense cytosplasm and tiny vacuoles. Apical meristems, those of stem and root tips, tend to be cuboidal (Fig. 8-13); cambial cells are flattened and elongated.

SUMMARY

Multicellular plants and animals, metaphytes and metazoans, are characterized by the presence of tissues and organ systems. Organs are structures that compartmentalize the several functions of an organism. Among these are support (skeleton in animals, roots and stem in plants), transport (circulatory system in animals, vascular cylinder in plants), and nutrition (the gut in animals, leaves in plants). These are functions which in one-celled organisms are the responsibility of such cell organelles as vacuoles, microtubules, and the endoplasmic reticulum.

Plants and animals have characteristic shapes that reflect their mode of existence. Animals that are sedentary feeders fixed to the sea bottom demonstrate radial symmetry; those that actively pursue food show fore-and-aft specialization and are said to have bilateral symmetry. Plants that live on land typically have radial symmetry. Both multicellular plants (Metaphyta) and animals (Metazoa) are characterized by the presence of tissues and organs. Organs are composed of tissues. The stomach, for example, is composed of epithelial tissue, muscle tissue, connective tissue, and glands. A leaf is composed of epidermal tissue, chlorenchyma tissue, and the vascular tissues, xylem and phloem. The arrangement of tissues and organs in the body of a plant or animal is such that distinct layers or regions can be discerned. In the animal body, a common arrangement is tubular and three-layered: outer, ectoderm; middle, mesoderm; inner, endoderm. Plants also can be described as three-layered: outer, epidermis; middle, cortex; inner, vascular cylinder.

multicellularity	respiration	stems
Metazoa	germ cells	leaves
Metaphyta	gametes	vascular cylinder
symmetry	zygote	cortex
skeleton	cell specialization	pith
invertebrate	unicell	xylem
chordate	somatic tissue	phloem
vertebrate	reproductive tissue	epidermis
gastrovascular cavity	animal tissue	cork
circulatory system	epithelium	parenchyma
ectoderm	connective tissue	supporting tissue
mesoderm	nerve tissue	meristematic tissue
endoderm	glandular tissue	spores
coelom	muscle tissue	sperm
open circulation	indeterminate development	egg
closed circulation	determinate development	
excretion	roots	

QUESTIONS FOR REVIEW AND DISCUSSION

1 Illustrate the major types of symmetry with diagrams of common inanimate objects. How does symmetry of animals relate to life strategies?

2 List the major animal functions and describe how each is accomplished by tissues and organ systems.

3 Distinguish between digestive wastes and metabolic wastes. Which organelles, cells, tissues, and organs may be involved?

4 What is the evolutionary and developmental significance of body layers? Of a coelom?

5 What is the distinction between somatic tissue and reproductive tissue?

6 Contrast the development of a plant with that of an animal. Which is determinate? Which is indeterminate?

7 What major body regions or layers are present in a plant?

8 How does the stem differ structurally from the root of a plant?

9 Name and give functions for the major plant and animal tissues.

SUGGESTED READING

ESAU, K. 1977. *Anatomy of seed plants*, 2nd ed. New York: John Wiley and Sons. (This is a first-class book on plant structure!)

FINGERMAN, M. 1976. *Animal diversity*, 2nd ed. New York: Holt, Rinehart and Winston. (Principles of classification are outlined and major groups defined.)

HANSON, E. D. 1972. *Animal diversity*, 3rd ed. Englewood Cliffs, N.J.: Prentice-Hall. (Emphasis is on origin and evolution of invertebrates.)

ROMER, A. S. 1968. *The procession of life.* Cleveland: World Publishing Co. (Readable introduction to animal diversity. In particular, chapters entitled "Worms" and "Whence Came the Vertebrates" are thought-provoking.)

VALENTINE, J. W. 1978. The evolution of multicellular plants and animals. *Sci. Amer.* 239(3): 140–58. (The entire issue is given over to evolutionary biology.)

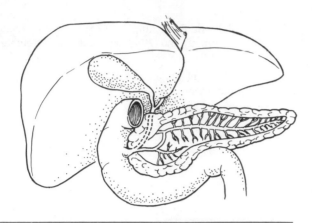

Food and Waste Processing in the Animal Body

9

It is a not uncommon experience to observe that overnight a pleasant clearwater lake has been transformed into peasoup consistency by billions upon billions of unicellular algae. Such population explosions come about when nutrients precipitously increase and temperature and light conditions are optimum. In such circumstances, algae can be an inconvenience and can sometimes cause environmental problems. Regardless, these rapid proliferations of enormous numbers of algae, bacteria, or other microorganisms show that unicells can be remarkably effective metabolizers and reproducers indeed. Given that kind of efficiency, one might wonder what forces in the ancient world drove microorganisms to evolve into larger and more complex plants and animals in which organ systems composed of multitudes of specialized cells are required to accomplish basically the same processes as are carried on in free-living cells. Perhaps microorganisms actually are less efficient than multicellular organisms. But if that is so, how does one explain the long-term survival and present abundance of numerous species of unicellular life? In actuality, evolution does not eliminate simple life forms and replace them with more complex ones; rather, it is a matter of opening up new plateaus of existence, so that the history of life on earth has been and continues to be one of increasing coexistence of organisms of all levels of complexity. True, in many cases larger organisms prey upon smaller ones, but the reverse also occurs. The advent of newer and larger organisms provides new opportunities for smaller organisms to exploit by death and decay or by symbiotic relationships.

Look analytically at the unicell and how it works. It swims or floats in a sea of suspended food particles and of dissolved nutrients, minerals, and gases. Its life consists of absorbing these substances, using them, and eliminating the leftover by-products. Thus it survives from day to day and, if it has been sufficiently competitive and efficient, enough energy has been stored up in structural and food molecules to enable it to grow and reproduce. Consider next another kind of cell, one residing deep in the tissues of an animal body—a liver cell, perhaps. In what way does its existence differ from that of the free-living cell? Simply this: the liver cell must have its surrounding sea of nutrients transported to it and its wastes carried away and eliminated. However, what the liver cell does internally with its absorbed nutrients, minerals, and gases, and with their products and by-products, is not basically different from what occurs in the cytoplasm of the free cell. This chapter will explore processes by which nutrients and other substances are made available and transported to animal cells and by which waste products of metabolism are eliminated from the cellular environment.

AUTOTROPHY AND HETEROTROPHY

A word about food—in the simplest of terms, there are two kinds of organisms: those that make their own food, usually by photosynthesis (**autotrophs,** "self-feeders") and those that depend upon an outside-the-cell food source (**heterotrophs,** "other-feeders"). The autotrophs include a few kinds of bacteria, some one-celled eukaryotes (protistans), and all green plants. The heterotrophs encompass most bacteria, many protistans, all fungi, and all animals. Because this chapter is about animal nutrition, attention first will be given to examining the nature of food, then to how food is made available to cells.

There are, as you know, three classes of foods: **carbohydrates, proteins,** and **lipids** (fats, oils, etc.); with the exception of monosaccharides such as glucose, all are large molecules that must be broken down by enzymatic hydrolysis into their soluble subunits before they can be absorbed by the cell. This involves either passive or active transport through the cell's membranes, followed either by incorporation into cytoplasmic structures, conversion into storage products, or oxidation and energy utilization. Because no heterotrophic cell is capable of manufacturing carbohydrates from inorganic raw materials, all are dependent on autotrophs for their carbohydrates and in many cases also for proteins and lipids. Animals cannot make amino groups, but can transfer them from an amino acid to an organic acid molecule and in this way produce other amino acids. Nevertheless, of the twenty amino acids found in proteins, nine must be obtained from food and ultimately from plants. These nine, called **essential amino acids,** include leucine, lycine, tryptophan, valine, histidine, threonine, methionine, isoleucine, and phenylalanine. Because animals can make their own lipids from carbohydrates, there is no absolute lipid requirement, but for many species lipids (principally fats and oils) are important high-energy foods.

Refined Foods

In most cases a heterotroph is confronted with the prospect of extracting nutrients from the bulk of its food. In these instances the crude food matter is torn, crushed, and ground into fine particles, then subjected to digestion. Arthropod mandibles, vertebrate jaws and teeth, and the talons and gizzard stones of birds are all adaptations for the refining of food prior to its digestion. Some organisms however, have evolved feeding strategies that avoid the necessity of either grinding or digesting their food. These are the parasitic species such as tapeworms, hookworms, liver and blood flukes, and others that feed solely on predigested food supplied by their hosts.

Digestion

Digestion is the process of breaking down food particles into soluble fractions: carbohydrates into **simple sugars,** proteins into **amino acids,** and lipids into **alcohols** and **fatty acids.** Note, however, that not all foods require digestion prior to uptake by cells. A number of sugars readily pass through cell membranes and are important foods. These include the simple sugars (**monosaccharides**) **glucose,** or grape sugar, and **fructose** (fruit sugar) and the **double sugars (disaccharides) sucrose** (table sugar), **lactose** (milk sugar), and **maltose** (malt sugar). Although disaccharides are readily assimilated by cells they are digested within cells to produce simple sugars: sucrose $\rightarrow$ glucose + fructose; lactose $\rightarrow$ glucose + galactose; maltose $\rightarrow$ glucose + glucose.

For some animals, such as nectar-feeding insects and humming birds, as well as for fruit-eating species, sugars are all-important foods. Many other animals, however, obtain most or all of their carbohydrates in the form of long-chain polysaccharides, such as **starch, glycogen,** and (in some instances) **cellulose.** These require digestion, as do proteins and lipids. The large food molecules—polysaccharides, proteins, and lipids—differ markedly in their chemistry, but they share a common principle of digestion: they all are broken down by enzymatic **hydrolysis** (splitting by addition of water). (Note: You may wish at this point to refresh your memory on this subject by referring to Fig. 1-22.)

Because the bonds that hold subunits together in each of the three classes of food are similarly formed by removal of molecules of water from between adjacent subunits, it follows that such bonds can be broken by adding back the water that was deleted during bond formation. And that is exactly what does occur; digestion proceeds by enzymatic addition of water to **oxygen bridges** in polysaccharides, to **peptide bonds** in proteins, and to **ester linkages** in lipids. Because there are a great many varieties of carbohydrates, lipids, and proteins, and because enzymes are very specific in their action, there are as many enzymes as there are substrates and reactions (for example, see Fig. 1-29).

How Foods Are Used

A century ago a few simple equations served to describe all that then was known about the way animals used their foods. Processes of digestion were just beginning to be known and enzymes had yet to be discovered. Now the metabolic pathways by which carbohydrates, proteins, lipids, and other organic molecules are processed in cells fill complicated charts from edge to edge, and the products, by-products, reactions, and enzyme systems involved are almost countless.

When food molecules are taken into a cell, they can be used in a variety of ways. Amino acids can be assembled in many distinctive kinds of protein, such as the structural elements of microtubules and microfilaments, or can be added to the battery of enzymes that carries on the work of the cell. Fatty acids may be reconstituted into lipids and employed in making new cell membranes or stored as fats and oils of food reserves. Sugars can be reassembled into storage polysaccharides such as glycogen or combined with proteins to make components of membranes and secretory products. Finally, but by no means of least importance, the chemical energy stored in lipid, protein, and carbohydrate molecules can be liberated in the respiratory pathways of the cell.

It will be remembered that in the glycolysis stage of biological oxidation, one glucose molecule upon oxidation yields a certain amount of energy in the form of ATP plus two 2-carbon residues of acetate complexed with coenzyme-A (acetyl Co-A) and two molecules of carbon dioxide. Subsequently, each acetyl Co-A is further oxidized in the Krebs cycle to yield a significantly greater amount of ATP and additional carbon dioxide. Thus glucose, and indeed other carbohydrates composed of glucose or other sugars, are capable of producing a considerable amount of useful energy. Proteins yield about as much energy as carbohydrates. First they are broken down into amino acids; then the amino groups are removed and the organic acids that remain enter the Krebs cycle and are oxidized. The fats in food contain long-chain fatty acids which are broken down during respiration into many 2-carbon fragments of acetic acid. These form acetyl Co-A and are oxidized in the Krebs cycle (see Fig. 9-1). Because of their ability to yield comparatively large numbers of acetyl fragments, fats are high-energy foods.

Intracellular Digestion

Certain kinds of cells, such as amoebas, paramecia, and white blood cells, eat food particles, including small organisms, by first enwrapping the particle in a small pocket of plasma membrane and drawing it into the cytoplasm. This process is **phagocytosis** (cell feeding, see Fig. 2-28); it is followed by formation of a **food vacuole** in which digestion can occur. Up to this point, the contents of the vacuole still are outside the cytoplasm, separated from it by a vacuolar membrane. Digestion now takes place within the food

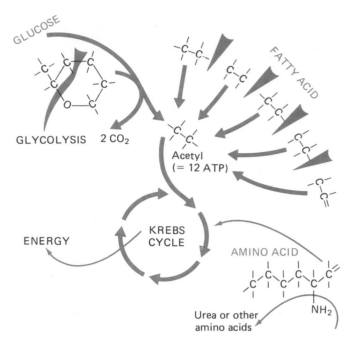

FIGURE 9-1
Metabolism of glucose, fatty acids, and amino acids.

vacuole, through the activities of a variety of enzymes, and the soluble components of the food move through the vacuolar membrane and into the cytoplasm. This process, called **intracellular digestion,** occurs in many heterotrophic microorganisms and also in the white blood cells of humans and many other animals.

Extracellular Digestion

If food is digested outside the cell, the process is known as **extracellular digestion.** It is the only means bacteria and fungi have of digesting and absorbing food, and it is the common means of digestion and nutrient uptake of almost all animals (the exceptions being those earlier mentioned parasites that feed on predigested food within the body of a host animal). In bacteria and fungi, enzymes are secreted into the immediate environment, where food particles are present. Then the food is digested and the solutes are absorbed into the cells. Note, however, that digestion and absorption cannot occur unless a thin film of water is present around the cells. Soluble nutrients accumulate in this film, forming a solution from which nutrients pass into the cell by active and passive uptake processes described in Chapter 2. This same principle applies also to nutrient uptake by cells of multicellular animals. In this case, the cells, although in close contact, are to a greater or lesser degree surrounded by films of **interstitial fluid.** The interstitial fluid is the equivalent of the fluid environment of the free-living single cell.

In all multicellular animals, except sponges and certain parasites, digestion is extracellular, occurring by secretions of enzymes into a digestive cavity or canal within the animal's body. Such a cavity or canal also is referred to simply but quite properly as a **gut**. Sponges lack a gut and their digestion is intracellular; certain parasites also lack a gut for previously mentioned reasons.

Gastrovascular Cavities   A tiny, less than 1-cm-long water animal named *Hydra*, after the ferocious many-headed serpent of Greek mythology, represents an early step in the evolution of food-processing systems in animals (see Color Plate 10, Fig. b). *Hydra* has a beautifully simple multicellular organization, doing many things with its tentacled, vaselike body. It gets its name from the snakelike or whiplike tentacles that serve as its organs of defense and feeding. With its tentacles armed by batteries of stinging cells, it is able to subdue and capture other small animals, then pass them through its mouth and into its **gastrovascular cavity** (Fig. 9-2). Although small, *Hydra* is an efficient predator, capable of catching even recently hatched fish. Population explosions of hydras have been known to devastate trout hatcheries.

Once food is ingested by a *Hydra*, secretory cells discharge their enzymes into the gastrovascular cavity, reducing the digestible parts of the prey to soluble products. These are circulated to all the inner cells of the body by contractions or pulsations of the body wall, so that no cell is more than one cell removed from a food supply. Respiration is likewise simple: direct exchange of oxygen and carbon dioxide occurs between cells and the environment. Metabolic wastes are secreted into the environment or into the gastrovascular cavity; all wastes eventually are evacuated through the mouth. Here then we see four processes, **digestion** of food, **circulation** of products of digestion, **respiration** (gas exchange and transport), and waste **excretion**, that have a very simple basis: the gastrovascular system.

Elaboration of a Simple Plan—Flatworms

One of the solutions to problems of absorption and transport one finds repeatedly in organisms at all levels of sophistication is coiling or branching of structures, a process that enlarges the surface area without greatly increasing overall bulk. Thus plant roots greatly increase their absorptive surface by the development of innumerable root hairs; gills of fishes and other aquatic animals are branching and featherlike, and digestive systems are branched, looped, and coiled. The digestive systems of a group of simple ribbonlike worms, known generally as flatworms, demonstrate one solution to the problem of efficiently processing and circulating nutrients in a many-celled body. A flatworm, known by its common name, **planaria**, or its generic name, *Dugesia*, illustrates a

FIGURE 9-2
Body plan of *Hydra.* The different cell types are shown. Flagellated cells of the inner body wall aid in circulation of food and in some cases also can take in food by phagocytosis. Gland secretory cells produce digestive enzymes. In the outer body wall, stinging cells **(cnidocytes)** contain coiled-up poison darts or adhesive hairs that on discharge can immobilize prey (refer to Fig. 21-3). Outer epithelio-muscular cells produce body movements. A gluelike **mesoglea** separates outer and inner cells but permits diffusion of food, gases, and wastes.

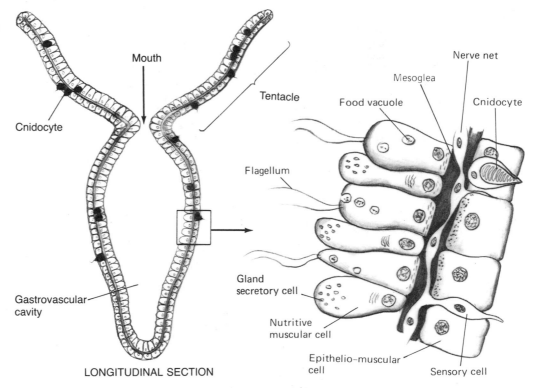

LONGITUDINAL SECTION

rather complex gastrovascular cavity (Fig. 9-3) relative to that of *Hydra.* In planaria we see that the problem of food distribution is solved by a digestive tract which is branched in such a way that no body cell is far removed from a supply of nutrients. Nevertheless the basic plan is the same as that of *Hydra*—a body cavity with one opening, which serves as both mouth and anus.

In planarians as in hydras, solid wastes are discharged through the **mouth–anus,** but in the former there is in addition a simple water-excretory system consisting of bulblike cells (Fig. 9-2) attached to canals that branch throughout the body and exit through pores to the outside. Collection of excess water, and possibly soluble wastes, from interior cells is accomplished by **flame bulbs** containing beating flagella which propel water through the excretory channels to the outside. We will encounter this principle applied to excretion in other, more advanced animals.

Diversification—Separation of Digestion and Circulation

The multipurpose gastrovascular tract of hydras and flatworms is quite functional at that level of complexity, but is not seemingly applicable in larger animals. Imagine applying the principle of the branching gut to a large animal; the convolutions, bends, and blind endings would be truly monumental. In most animal phyla we find, instead of a ramified gut, a second distribution system interposed between the digestive tract and the cells of the body. As a result, the digestive tract can be more compact, the chore of distribution being taken over by the fluid of the **circulatory system,** which then distributes soluble substances to the cells. In addition, the circulatory system also can collect and transport oxygen from the external environment and internally produced carbon dioxide. Finally, wastes can be collected from cells and transported to excretory organs. The principle of the circulatory system is illustrated diagrammatically in Fig. 9-4.

Further Specializations

Once circulatory systems had evolved, the path became clear for additional evolutionary advances and for further elaboration of animal forms. Animals grew both in size and complexity. Moreover, specialized organs of gas exchange, such as gills, tracheae, and lungs, eliminated the requirement for overall permeability of the body surface (refer to Fig. 8-7 for a diagram showing this feature). Thus aquatic animals could develop protective body armor (exoskeletons of crabs and lobsters, for example), and life in the relatively dry atmosphere of the land's surface became feasible. Land animals were in this way

FIGURE 9-3
Longitudinal and cross-sectional views of *Dugesia,* a planarian flatworm. The mouth terminates the pharynx, a muscular sucking organ. There is a simple nervous system, described in Chapter 13. Two flame cells (flame bulbs) are shown. Planarians move both by muscle contractions and by the beating of a ventral surface layer of epidermal cilia.

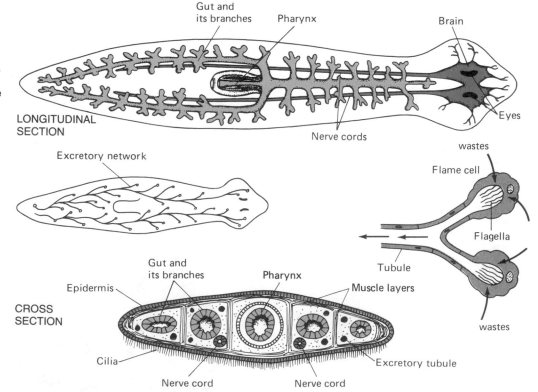

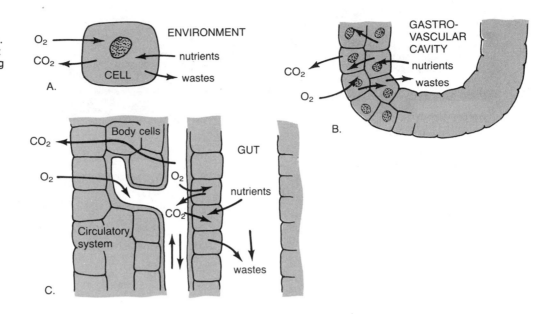

FIGURE 9-4
Principles of nutrient circulation.
A. Nutrient and waste transport in the single cell. **B.** Functioning of a gastrovascular cavity. **C.** How the circulatory system relates to the gut, the environment, and the cell.

able to take in oxygen and emit carbon dioxide despite **integuments** that guarded against water loss. (The integument is a term describing all the outer layers of an animal's body: the skin with its scales, feathers, hair, etc.) In a like manner, development of excretory systems permitted collection and condensation of soluble wastes from body fluids and localized elimination of the wastes through pores rather than by diffusion through the body wall. In invertebrates, the origins of excretory systems are seen in the flame bulbs of planarians (Fig. 9-3). Annelid worms employ somewhat analogous ciliated **nephridia** to collect wastes from the coelomic fluid (Fig. 8-7), and the **kidneys** of vertebrates are intricate assemblages of capillaries and collecting tubules which efficiently collect wastes while conserving water and useful solutes. For life on land, water-conserving excretory systems are an obvious necessity.

The basic features of digestive, respiratory, circulatory, and excretory systems have been described here and in Chapter 8 (refer especially to Figs. 8-3 to 8-7), and we turn now to an examination of vertebrate systems.

ORGAN SYSTEMS OF VERTEBRATES

Digestive Systems

The structure of the vertebrate gut differs according to evolutionary status and ecological niche (primitive versus advanced, aquatic versus terrestrial, herbivore versus carnivore). At the anterior of the gut is the mouth, behind which lie the **mouth cavity** and **pharynx** (Fig. 9-5). The latter is a comparatively large chamber equipped with **gill slits;** it serves as a water passage in external

respiration (gas exchange) as well as a food passage. (In air-breathing vertebrates the pharnyx is of reduced size and has no slits.)

Food passes from the mouth into the **esophagus,** a short muscular tube; then into the **stomach,** where digestion begins. Digestion continues in the **intestine,** and food absorption commences. Between the stomach and intestine is the **pyloric sphincter,** which opens and shuts to allow or stop passage of food. Intestinal organization differs among vertebrates, but there are many common features.

Two glands, the **pancreas** and the **liver,** aid digestion by secreting fluids containing enzymes and other substances into the intestine. The liver also functions as a food-storage, chemical-processing, and excretory organ. It supplies products of **food processing** to the circulatory system and excretes waste products into the intestine via the **bile duct.** Undigested matter passes into an enlarged posterior part of the intestine, the **large intestine.** Here waste products are compacted in the form of **feces** and passed outside through the anus.

Circulatory and Respiratory Systems

Some vertebrates discussed in this chapter are called **cold-blooded** animals. This is a misleading term, because their blood is approximately the same temperature as their surroundings—warm if the surroundings are warm, cold in a cold environment. A more accurate term is **ectothermic** ("outside temperature"). Fishes, amphibians, and reptiles are all ectothermic animals. Birds and mammals, however, are said to be **warm blooded:** their body temperatures are maintained at a constant level irrespective of the outside temperature. They are more

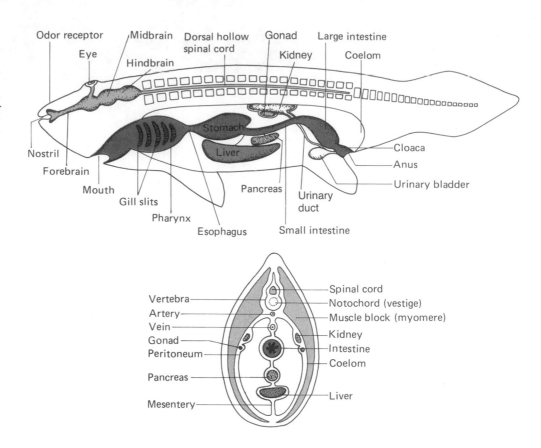

FIGURE 9-5
Nervous and digestive systems of a generalized vertebrate. The vestigial notochord, shown in the cross-sectional view, is considered to be a remnant of the primitive rodlike backbone of earliest vertebrates. In man it persists as the intervertebral disks.

Odor receptor — Midbrain — Dorsal hollow spinal cord — Gonad — Large intestine — Coelom — Kidney
Eye — Hindbrain
Stomach
Liver
Nostril — Cloaca — Anus — Urinary bladder
Forebrain
Mouth — Pancreas — Urinary duct
Gill slits
Pharynx
Esophagus — Small intestine

Vertebra — Spinal cord
Artery — Notochord (vestige)
Vein — Muscle block (myomere)
Gonad — Kidney
Peritoneum — Intestine
Coelom
Pancreas — Liver
Mesentery

correctly termed **endothermic** ("inside temperature") animals.

    Circulatory systems serve as a medium of exchange between the organism and its environment and between the cells of the organism. Lower, water-dwelling vertebrates have relatively simple circulatory and respiratory systems in which a two-chambered **heart** pumps a fluid (**blood**) through **gills**, where oxygen and carbon dioxide are exchanged, then to other parts of the body.

    Oxygen-transporting red blood cells (**erythrocytes**) are a unique feature of vertebrate blood. The first heart chamber (**ventricle**) is a muscular organ which pumps blood into a major artery, the **ventral aorta.** Blood flows into gill capillaries and then into the **dorsal aorta** to var-

ious regions. Wastes are removed as blood passes through kidneys; food is collected during the return flow of blood from intestines through the liver to the second heart chamber (**atrium**; pl. **atria**) (Fig. 9-6).

Urogenital System

In vertebrates excretory organs (kidneys and excretory ducts) are closely associated with reproductive organs (gonads and reproductive ducts) (Figs. 9-2, 9-7). Together they comprise the **urogenital system.** Kidneys are pairs of oval to cylindrical structures embedded in the body wall beneath the **peritoneum,** a thin lining of epidermal cells. They are composed of multitudes of **nephrons** (see

FIGURE 9-6
Circulatory system of an aquatic vertebrate.

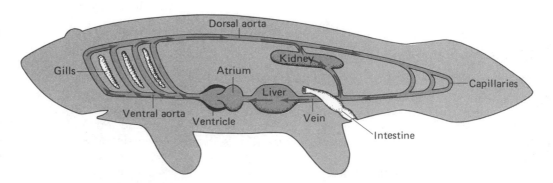

Dorsal aorta
Kidney
Gills — Atrium — Capillaries
Ventral aorta — Liver — Vein
Ventricle — Intestine

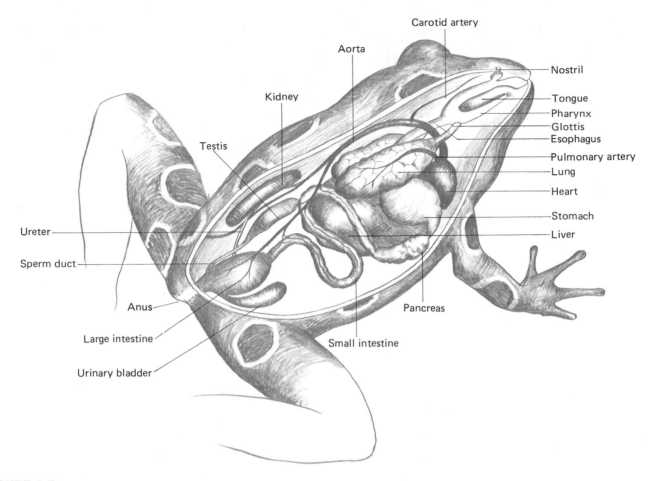

Carotid artery
Aorta
Nostril
Kidney
Tongue
Pharynx
Glottis
Esophagus
Testis
Pulmonary artery
Lung
Heart
Stomach
Ureter
Liver
Sperm duct
Anus
Pancreas
Large intestine
Small intestine
Urinary bladder

FIGURE 9-7
The frog, a transitional vertebrate. Frogs are amphibians that have made an almost complete break with their aquatic ancestry. However, as discussed in previous chapters, their reproductive cycles are water based and their larval stages are equipped with gills rather than lungs. Frogs must be regarded as specialized rather than representative amphibians.

Fig. 9-21), the functional units of kidney structure. Nephrons regulate the composition of body fluids by excreting substances that are in excess concentrations and retaining substances that are needed.

Land Vertebrates

When vertebrates began to exploit the land environment some 275 million years ago, major developments in circulatory and respiratory systems occurred. Gills were replaced by lungs, with accompanying modifications in head and pharyngeal regions to provide air passages (Figs. 9-7, 9-10). Digestive and excretory systems in land vertebrates, however, are not greatly different from those of the generalized vertebrate described earlier (Fig. 9-5) except that provisions for storage of urine and intestinal residues are better developed. Also, because land vertebrates are not continually bathed in water, problems of

water conservation occur. Amphibians for the most part get around this difficulty by living in humid environments; reptiles, birds, and mammals have developed impervious and in many cases insulative modifications of their integuments, and their kidneys generally are more efficient in water retention than those of amphibians and other vertebrates that dwell exclusively in water.

Air-breathing vertebrates evolved **lungs** as organs of gas exchange. To accommodate this feature, a second circulatory route evolved in which **systemic** (general) circulation is partially or completely separated from **pulmonary** (lung) circulation. In amphibians this is accomplished somewhat inefficiently by two atria and a single ventricle, in which some mixing of oxygenated and unoxygenated blood occurs. Birds, mammals, and a few reptiles have four-chambered hearts, and separation of pulmonary and systemic circulation is complete or nearly so (Fig. 9-8).

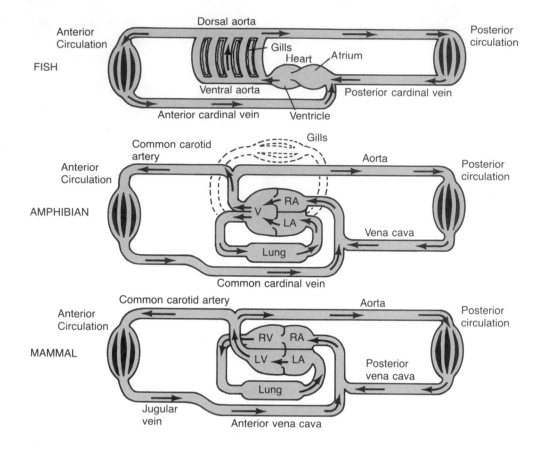

FIGURE 9-8
Circulatory systems of fishes, amphibians, and mammals compared. Some amphibians possess both gills and lungs at the same time (mudpuppies, Fig. 21-29); the majority do not, but all have gills in the larval stage of their development. The reptilian system is intermediate between that of amphibians and mammals; circulatory systems of birds are similar to those of mammals.

Amphibians Lungs of frogs and other amphibians are connected with the pharnyx by a short windpipe, or **trachea**. The opening of the trachea into the pharynx is called the **glottis**; it can be closed when the frog is eating, to prevent choking the lungs with food particles. Frogs lack a chest cavity; air is taken into the lungs by gulping and swallowing. **Vocal cords** are located in the trachea, and the amplification of the croaking, twittering, or peeping sounds of frogs is accomplished with the aid of a bellowslike pair of **vocal sacs** on either side of the pharnyx.

Lungs necessitate an extra heart chamber, the **left atrium**, which receives oxygenated blood from the lung via the **pulmonary vein** (Fig. 9-8). Blood deficient in oxygen is received by the **right atrium** from the **hepatic vein** and **vena cava**. The ventricle receives blood from both atria. Partial separation of blood flow in the ventricle is achieved by means of a **spiral flap** of tissue that shunts oxygenated blood into the general circulation of the body, by way of the **aorta** and its branches. Oxygen-deficient blood is shunted into the pulmonary arteries to the lungs. Mixing of oxygenated and oxygen-deficient blood in the ventricle tends not to be a serious problem, however, because gills (not present in adult frogs and

toads) and the skin also serve as organs of gas exchange and supplement the oxygen supplied by the lungs.

Reptiles The living reptiles are ectothermic tetrapods ("four legs") as are amphibians, but they have achieved greater independence of water in their lives and reproduction. A majority are carnivorous and have digestive systems similar to the generalized amphibian system. Although some modern reptiles are semiaquatic, most are terrestrial and have developed a tough waterproof integument and efficient kidneys that effectively conserve water.

The circulatory system of reptiles is further modified in the direction of complete separation of oxygenated and nonoxygenated blood. A wall partially divides the ventricle into two chambers, one of which (the **right ventricle**) circulates unoxygenated blood to the lungs; the other (the **left ventricle**) delivers oxygenated blood to the rest of the body.

Birds Birds have toothless jaws and do not chew their food, although the sharp beaks of carnivorous species (hawks, owls, etc.) are used to tear prey into pieces. The digestive systems of birds show adaptations to sustained flight, including a storage sac **(crop)** capable of containing more than a day's supply of food. The stomach consists of two chambers: an upper secretory portion, the **proventriculus**; and a lower muscular part, the **gizzard**, or **ventriculus**, which in seed-eating birds contains small pebbles, "grit," eaten by the bird and used as grinding stones. Food, after grinding and partial digestion, passes into the **small intestine**, which is convoluted and may be more than eight times the bird's length (Fig. 9-9). (Interestingly, some dinosaur skeletons contain fairly large stones, called **gastroliths** ["stomach stones"], which probably were used in the same way as gizzard stones of chickens and other birds; sometimes gastroliths are highly polished, indicating long-term use as grindstones.)

Waste matter collects in a short, terminal part of the intestine called the **colon**, where water is extracted. Solidified wastes then pass into the **cloaca**, a common excretory and genital chamber, and out the cloacal opening. Kidney wastes also pass into the cloaca; with the exception of ostriches, birds lack urinary bladders and excrete a pastelike urine.

Circulatory systems of birds essentially follow the reptilian plan but have a complete separation of pulmonary and systemic circulation, which prevents mixing of oxygenated and oxygen-depleted blood.

Metabolic Rates Activities of animals often are described in terms of the rate at which they consume oxygen. This rate is a measure of their use of food to support life processes and is called the **metabolic rate**. Metabolic rates of birds are significantly higher than that of any cold-blooded vertebrate and also higher than those of most mammals. Body temperature is a good indicator of metabolic rate. In birds, body temperatures are maintained at levels as high as 46°C (116°F); in comparison, human body temperature is 37°C (98°F). If this difference seems inconsequential, consider that, in the human body, a bird temperature would amount to a fatally high fever. Heartbeat rates in birds are correspondingly high (up to 600 beats per minute in hummingbirds) in contrast to the human heart, which beats about 70 times a minute. Hummingbirds use about as much fuel (food) on a per-weight basis as does a helicopter, conserving energy at night by lowering their body temperature to about 24°C (75°F); at this temperature they are so torpid that they can be picked off their perches without a struggle.

Mammalian Features

Mammals are that branch of the vertebrate family tree (subphylum Vertebrata, Class Mammalia) that is characterized by giving "live birth" to young, which then are nourished by a unique glandular secretion called **milk**, the product of **mammary glands** (*mamma*, "breast"). It is not a particularly large assemblage of species, only about 4,000, compared with some 10,000 species of birds, 128,000 of mollusks, and over 900,000 of arthropods, yet in its ability to exploit the resources of the environment it stands second to none.

Digestive System Characteristically, mouths of mammals are equipped with cutting, tearing, and grinding teeth, but numerical and structural differences are com-

FIGURE 9-9
Features of the digestive tract of the chicken. Also shown are the locations of urogenital structures in the male bird.

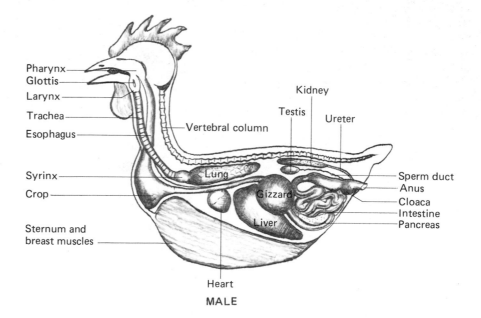

Pharynx
Glottis
Larynx
Trachea
Esophagus
Syrinx
Crop
Sternum and breast muscles
Vertebral column
Lung
Gizzard
Liver
Heart
MALE
Kidney
Testis
Ureter
Sperm duct
Anus
Cloaca
Intestine
Pancreas

mon. The gut generally resembles those of birds and reptiles, consisting of an esophagus, a stomach, a small intestine, and a large intestine. Perhaps the greatest difference between mammalian systems and reptilian and avian systems is the absence of the common genital and excretory passage, the cloaca (except in the egg-laying monotremes, where a cloaca is present, and in marsupials, where there is a rudimentary one). The mammalian large intestine terminates posteriorly in a short nonabsorptive chamber, the **rectum,** which ends at the anus.

Respiration and Circulation A principal difference between mammals and reptiles and birds is in the enlargement of the braincase of mammals. In addition to having more space for larger brains, mammalian skulls have well-developed **nasal passages** separated from the mouth cavity by the **palate** (Fig. 9-10). Nasal passages provide a warming and moisturizing chamber for air prior to being taken into the lungs and are thought to have evolved as an adaptation to life in cold climates.

As noted for amphibians, air-breathing vertebrates require a food–air shunt to keep from getting food into their lungs. In mammals, the opening into the trachea ("windpipe"), the glottis, is guarded by a valvelike flap called the **epiglottis** (Fig. 9-10), which guards the glottis against entry of liquids and food particles. The epiglottis closes the glottis, the so-called glottal reflex, when one

begins to swallow. Sometimes if one is simultaneously eating and talking or laughing, the glottal reflex may not act quickly enough and one may inhale a bit of food. Choking can result. If the bit of food is large, it can block the glottis or larynyx and asphyxiation may occur. The **Heimlich maneuver,** a first aid action which can save lives in such situations, should be learned by everyone old enough to be helpful in such emergencies. Essentially, the Heimlich maneuver consists of approaching the choking person from the rear, encircling his or her body with one's arms, and applying one or more strong hugs to the upper abdomen, just below the diaphragm. Instruction in use of the maneuver is important so that the correct application of pressure is understood.

Reference to the diagrams in Fig. 9-8 will show the basic modifications in circulation required in the transition from gill respiration in fishes, through intermediate stages in amphibians, in which both gills and lungs may be present, to exclusive lung respiration in reptiles, birds, and mammals. A more complete version of mammalian circulation, as exemplified by the human circulatory system, is presented in Fig. 9-17.

Blood Up to this point little has been said about the blood of vertebrates except that it contains hemoglobin and transports oxygen, food, carbon dioxide, and wastes. Blood is a tissue in which the principal cell types are

FIGURE 9-10
The mammalian pharynx. Separation of the nasal passages from the mouth is shown in **A.** Note this depicts the inhalation of air. The epiglottis blocks entry to the esophagus. In **B,** the glottis is closed by the epiglottis and passage of food into the esophagus can occur.

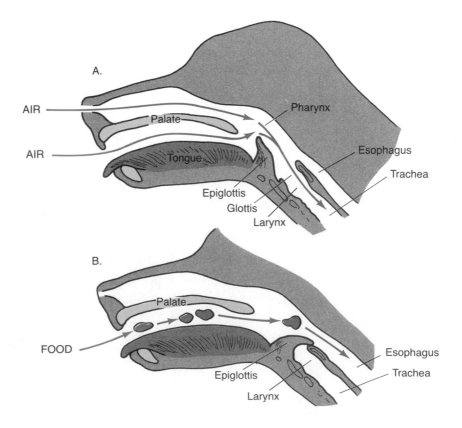

erythrocytes (red cells) and leukocytes (white cells). Erythrocytes are by far the most abundant, there being about five million in every cubic millimeter of human blood volume; about 5000 leukocytes are present in the same volume. Approximately 250,000 platelets, spherical or ovoid bodies about 3 μm in diameter, also are present in each cubic millimeter of blood. Platelets are produced by fragmentation of large cells called megakaryocytes (see Fig. 15-10). Together, blood cells and platelets comprise about 45 percent of blood volume; the remaining 55 percent is composed of plasma, the fluid component of blood. Plasma is a complex mixture of water (92 percent), proteins (7 percent), and minerals (1 percent).

Erythrocytes of most vertebrates are nucleated; those of mammals are not. Mammalian erythrocytes lose their nuclei during development and mature into flattened disks containing mostly hemoglobin. There are several types of leukocytes, and all are nucleated. Blood cells are not produced in the blood itself but are formed by mitosis of cells in the spleen, bone marrow, and lymph glands.

Transport of oxygen by physical binding to hemoglobin is a major function of erythrocytes. Erythrocytes also contain an enzyme instrumental in the transport of carbon dioxide by blood. The five types of leukocytes combat infections, dispose of waste matter, and produce antibodies which inactivate foreign substances such as bacterial toxins. Antibodies also cause the rejection phenomenon in human organ transplants.

All body cells are bathed in a thin film of liquid, the interstitial fluid, which constitutes the external environment of the cells. The function of blood is to maintain a supply of oxygen and nutrients in the interstitial fluid and to remove excessive waste substances. This is

BLOOD CLOTTING

When the tissues of the body are wounded, several events occur that resist the loss of blood from cut vessels and begin the healing processes. At the wound site, blood platelets swell and burst, releasing chemicals that cause the arteries to constrict and thus slow the loss of blood. The platelets also initiate a process that results in activation of prothrombin activator, an enzyme that converts a blood-protein constituent called prothrombin to thrombin. In the presence of calcium ions, thrombin catalyzes the conversion of another blood-protein component, fibrinogen, into fibrin threads. The fibrin threads form a fine meshwork in the blood and together with trapped cells and other blood particles make a clot and block the further escape of blood from the wound (see figure). In the hereditary disease hemophilia, discussed in Chapter 7, there is a deficiency in the factors that activate the formation of prothrombin activator; the hemophiliac lacks the ability to form clots at the point of wounding and may bleed to death from an apparently minor cut or abrasion.

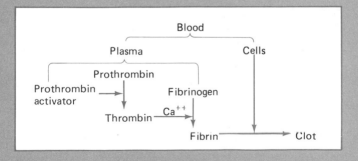

accomplished by diffusion of substances into and out of blood contained in **capillaries**, the smallest and most numerous blood vessels, which are found close to all body cells (see Figs. 8-12A and 15-9). The combined surface of all capillaries of the human body is about 1.5 acres, and their combined length is on the order of 100,000 km (60,000 mi.). Capillary walls are composed of a single layer of flattened epithelial cells through which nutrients diffuse readily. Leukocytes also can penetrate capillary walls, but erythrocytes cannot. As blood circulates through the capillaries, liquids tend to leak out into the interstitial fluid. Surplus interstitial fluid, which otherwise would waterlog the body, is drained away into highly permeable **lymph capillaries**, then into larger **lymph ducts**, and finally into a large vein near the heart.

Temperature Regulation In mammals as in birds, body temperatures tend to be maintained in a steady state, but characteristically in the neighborhood of 100°F, somewhat lower than for birds. Most of the mammalian body heat comes from internal combustion of food (**biological oxidation**), and is radiated from the skin and lungs; some also is lost in urine and feces. The rate of heat loss is controlled to a large degree by adjustments in blood circulation at the body surface, under the control of a "thermostat" located in the **hypothalamus** of the brain. Some animals are able to "turn down" their hypothalamus during the winter and go into a period of greatly slowed metabolism called **hibernation**.

Urogenital System A pair of kidneys function as the chief excretory organs in mammals as they do in other vertebrates. Each is typically ovoid in shape and contains thousands of nephrons which drain into an internal cavity, the **renal pelvis** (a mammalian feature) (Fig. 9-11). From here, urine passes into a urinary duct, the **ureter**. The ureters transport urine from each kidney to the **urinary bladder**, where it is stored temporarily before draining to the outside of the body through the **urethra**. The urethra ends at the tip of the **penis** in males and near the ventral margin of the opening to the genital tract (**vagina**) in females (see Fig. 9–11).

FIGURE 9-11
Urogenital systems of human males and females.

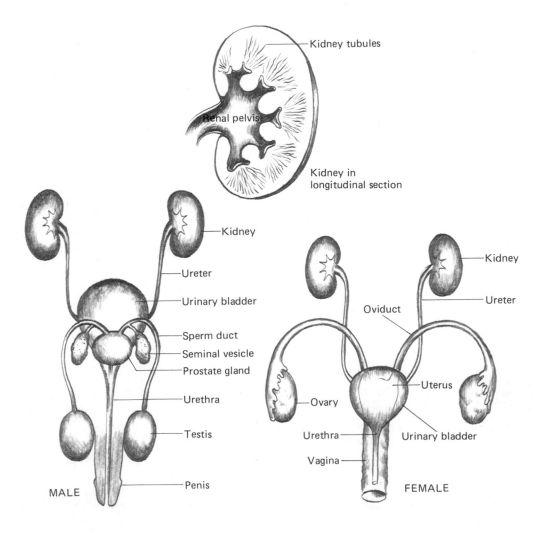

FOOD AND WASTE PROCESSING IN THE HUMAN BODY

The Digestive System

Digestion   The processes of digestion begin the moment food is ingested. As solid food is chewed, it is mixed with **saliva**, a **salivary gland** secretion containing water and mucus, which lubricate food and make it slippery enough to swallow. In addition, saliva contains the enzyme **salivary amylase**, also called **ptyalin**, which starts the digestion of starch. When food is swallowed it passes through the **esophagus**, a muscular tube about 30 cm (1 ft) long, and then into the **stomach**, where it is further processed by stomach secretions. (Refer to Fig. 9-12 for a diagram of the digestive tract.)

The Stomach   Stomach action was first observed by an American doctor, William Beaumont, who in 1822 was a surgeon at a fort on Mackinac Island, Michigan. A drunken young French trapper was shot in the stomach one night in a tavern and brought to Beaumont, who treated him but did not expect him to live. To Beaumont's surprise the patient recovered almost completely, except for an opening from his stomach to the outside (a gastric fistula). Beaumont was able to observe the digestion of food in his patient's stomach by prying open the fistula and looking inside. He also could withdraw food through the fistula to follow the progress of digestion and analyze stomach fluids, and he observed the effects of irritants and stress on stomach activity. (He was the first physician to observe the effects of alcohol on the stomach lining.) Beaumont's work is still cited, although more recent observations and experiments have added greatly to the understanding of digestive processes.

The stomach is delimited at its upper and lower ends by rings of muscles called **sphincters**, which act like drawstrings, opening and closing in response to swallowing, belching, vomiting, and movement of partly digested foods into the small intestine. The upper sphincter is the **cardiac sphincter**; the lower one is the **pyloric sphincter**.

The human stomach is a large muscular sac capable of considerable distention and has a fluid capacity of about 2 liters (2 quarts) in a person of normal development and eating habits. When food is present, successive contractions of the muscles in the stomach wall mix food with gastric secretions. The stomach lining contains several kinds of secretory cells, including some that secrete hydrochloric acid and others that produce enzymes; an important one is protein-digesting **pepsin**. In addition, mucus-secreting cells coat the stomach lining with mucus and prevent it from digesting itself.

Protein digestion is the main enzymatic action occurring in the stomach. As digestion progresses, the food becomes a soupy liquid known as **chyme**. Chyme measures about 2 on the pH scale, so acidic that if one vomits it irritates the throat and is quite discomforting. Normally, the chyme is injected in spurts through the pyloric sphincter into the **duodenum**, the upper part of the small intestine. Here digestion continues, facilitated by secretions from the pancreas, liver, and intestinal glands.

The Pancreas   The pancreas is a large mixed gland located in the curvature of the duodenum and connected to it by a secretory duct (Fig. 9-13). When chyme passes from the stomach into the duodenum, its acidic nature induces certain endocrine cells of the duodenum to secrete hormones into the bloodstream.

The endocrine functions are carried on in clusters of cells called **islets of Langerhans**, which produce and secrete **insulin** and **glucagon**. These two hormones act antagonistically in a feedback system controlling blood-sugar levels as will be discussed in Chapter 10.

The Liver   The liver also secretes into the duodenum; its secretion, **bile**, is temporarily stored in the gallbladder (Fig. 9-13) and is secreted into the duodenum in response to the hormone **CCK-PZ**, as discussed in the next chapter. Bile is a complex mixture of bile salts, cholesterol, lecithin, and bile pigments; the last give bile a greenish yellow color and are breakdown products of red blood cells and hemoglobin.

Bile is not a hormone nor does it contain enzymes, yet it facilitates digestion and absorption of food in the

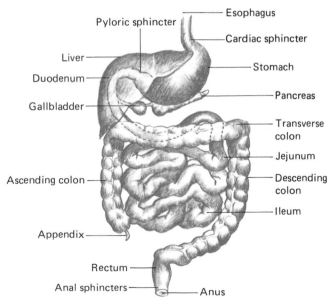

FIGURE 9-12
Human digestive tract and associated organs and glands.

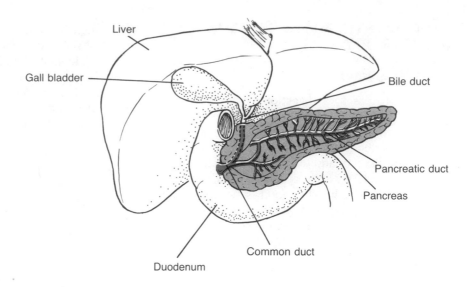

FIGURE 9-13
The pancreas in relation to the liver and duodenum. The pancreatic duct and the bile duct form the common duct which empties into the duodenum. (Reprinted with permission from Byron A. Schottelius and Dorothy D. Schottelius)

Liver

Gall bladder

Bile duct

Pancreatic duct

Pancreas

Common duct

Duodenum

intestine. Some of its components combine with fat droplets in the chyme to produce a kind of mixture called an **emulsion**. Emulsions are mixtures in which normally immiscible substances are kept suspended. A household detergent, for example, can produce an emulsion when added to oil and water and well shaken.

Sometimes the bile duct becomes plugged, usually by a gallstone composed of aggregated cholesterol. When this happens, bile backs up in the liver and enters the bloodstream, giving an individual a yellowish complexion known as **jaundice**. In such cases removal of a gallstone often is necessary, and damage to the gallbladder may require its removal as well. Without a gallbladder, bile is secreted continuously and strict regulation of the diet is required so that enough bile will be present to emulsify dietary fats. Usually a low-fat diet must be followed, and meals must be small and frequent.

The Small Intestine The human small intestine is about 4 cm (1.5 in) in diameter and about 7 m (23 ft) long. The duodenum is the initial 20 cm (8 in) or so, and the remainder consists of a middle section, the **jejunum**, which is about 1 m (3 ft) long, and a terminal portion, the **ileum**. Circularly arranged smooth muscle in the wall of the small intestine produces rhythmic waves of contractions **(peristalsis)** which move the chyme along. Glands in the intestinal wall secrete mucus and sodium bicarbonate, which facilitate the movement of chyme and complete its neutralization.

The small intestine is lined with projecting, finger-like structures called **villi** (sing. **villus**, "tuft") (Figs. 9-14 and 9-15). Each villus encloses a network of capillaries and a lymph duct called a **lacteal**. The outside of each villus is covered with columnar epithelial cells whose outer borders have small projections known as **microvilli**

(Figs. 9-14 and 9-16). The membranes of microvilli contain enzymes that break down food molecules further and facilitate their transport into intestinal cells.

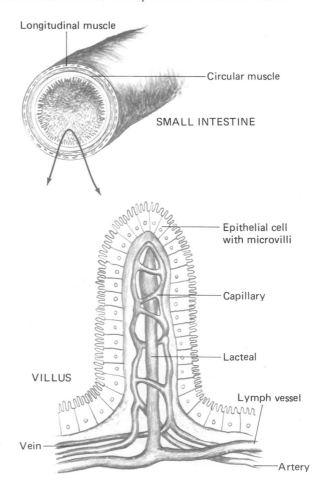

Longitudinal muscle

Circular muscle

SMALL INTESTINE

Epithelial cell with microvilli

Capillary

Lacteal

Lymph vessel

VILLUS

Vein

Artery

FIGURE 9-14
Cross section of the human small intestine and of a villus.

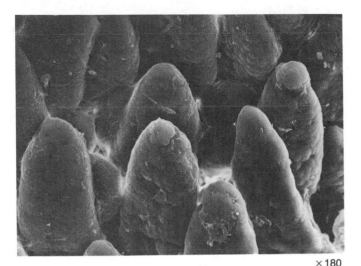

FIGURE 9-15
Scanning electron micrograph of villi. (Photo by G. T. Cole, University of Texas/Biological Photo Service)

× 180

When food enters the duodenum, its digestion is far from complete. Most of the components of the chyme are digested in the small intestine with the aid of pancreatic and intestinal secretions.

Absorption and Assimilation As chyme is moved down the small intestine, ions and soluble molecules (sugars, amino acids, fatty acids, vitamins, emulsified fats, etc.) are absorbed by microvilli of the surface epithelial cells and then passed on to other cells in the interior of the villus. From here, all smaller molecules are absorbed into the capillaries. The emulsified fats and some vitamins, termed **fat-soluble vitamins,** are absorbed by the lacteal and moved into the lymphatic system. One vitamin—vitamin B_{12}—is a special case in that it has numerous ionic charges that inhibit its movement through cell membranes by diffusion. A protein named **intrinsic factor** attaches itself to vitamin B_{12}, and together they move through cell membranes in the ileum. The specific function of intrinsic factor is not known, but one possibility is that it assists in the uptake of vitamin B_{12} by pinocytosis (refer to Fig. 2-29).

A great deal of water enters the alimentary canal every day in food and drink and up to 10 liters (10 quarts) per day enters the small intestine; only about a half liter (half-quart) per day passes on into the large intestine. Consequently, absorption of water is an important function.

The Colon Whatever of the original raw food remains in the small intestine passes into the **colon,** or large intestine, a tube about 1.3 m (4.25 ft) long and 6.5 cm (2.5 in) in diameter. About 0.5 quart of chyme enters the colon each

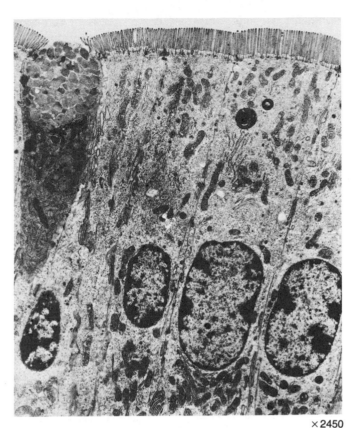

FIGURE 9-16
Columnar epithelial cells from a portion of a small intestinal villus. A mucous cell (goblet cell) is at the left. Other cells have an outer border of microvilli (at top). (Photo courtesy David Mason)

× 2450

day, and its principal actions are absorption of sodium and water, secretion of mucus, and formation of **stools,** or feces. The large intestine also is the site of action of groups of bacteria called **intestinal flora.** These bacteria further break down the chyme, derive energy from any remaining food, and secrete small amounts of vitamins which are absorbed by the large intestine. Some bacteria form a gaseous mixture of nitrogen, hydrogen sulfide, and flammable hydrogen and methane. The accumulated gas, called **flatus,** is eliminated via the anus.

Waste Disposal About 150 g (5.25 oz) of feces are eliminated from the colon per day. The feces normally are about two-thirds water and one-third solid matter composed of undigested food particles, cellulose, bacteria, bile pigments, cell debris, and mucus. Feces accumulate in the lower end of the colon, a region called the rectum. When the rectum becomes distended with feces, a nervous reflex produces a rectal contraction and a relaxation of the **inner anal sphincter,** a ring of smooth muscle encircling the anus. This action is involuntary, but an outer ring of striated muscle, the **outer anal**

sphincter, is under voluntary control. Loss of voluntary control of defecation may happen in extreme fright or because of great pressure in the rectum resulting from **diarrhea** (watery stools). The opposite case, **constipation,** in which regular defecation does not occur, seems to be a major preoccupation of television advertisements but is usually not a serious health problem.

Actually, accumulation of toxic products in the colon and rectum is rare, and natural laxatives such as vegetable fiber usually are adequate to control constipation. Fiber increases the motility of the intestines by stretching the smooth muscles and giving them bulk to work against. Harsh laxatives, such as castor oil, act by irritating the smooth muscles of the intestine, thus increasing motility. Others, including milk of magnesia, prevent the reabsorption of water by the intestine and therefore increase the fluidity of the feces.

The Cardiovascular System

Transport of Nutrients As a result of digestion and absorption of food, the blood becomes charged with soluble food molecules which are distributed by the blood to all the cells of the body. A major food-distribution and -processing organ is the liver. The liver also serves to remove toxic materials absorbed into the blood from the small intestine. In addition, it is a major storage organ. Glucose molecules are assembled there in long chains of glycogen, the animal equivalent of starch, and then stored. The liver also processes and stores fats, although most fats are transported in the lymph and deposited in adipose tissue in various parts of the body.

Blood and Lymph The principles of blood circulation in humans are the same as in other vertebrates. Blood is pumped by the heart to organs and tissues of the body, where substances are either removed or added.

FIGURE 9-17

Human circulatory system. (Redrawn from *General zoology,* 5th ed., by Claude A. Villee, Warren F. Walker, Jr., and Robert D. Barnes. Copyright © 1978 by W. B. Saunders Company. Reprinted by permission of W. B. Saunders Company, CBS College Publishing)

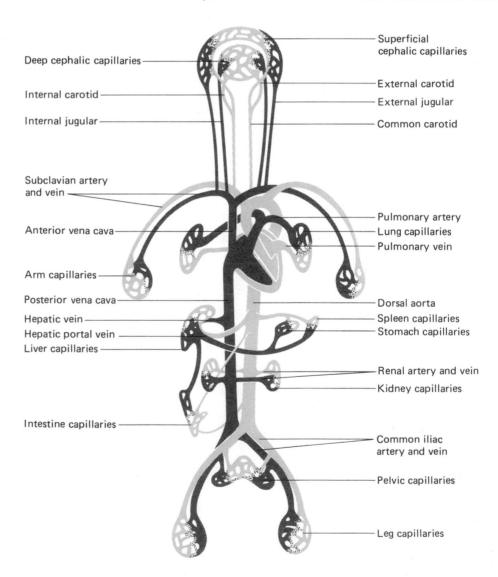

Figure 9–17 is a diagrammatic version of the human circulatory system in which **arteries** distribute food-rich oxygenated blood from the left ventricle through the aorta to all regions of the body, including the kidneys, where certain wastes are removed. Through **veins,** blood returns to the heart from cells and tissues of the body carrying carbon dioxide and other metabolic wastes, as well as food absorbed in the small intestine and processed in the liver. This venous blood is collected by the right atrium of the heart and passes into the right ventricle. From here the blood is pumped into the lungs, where it is relieved of carbon dioxide and picks up a load of oxygen. The newly oxygenated blood is transported into the left atrium via the pulmonary vein, then moves into the left ventricle and starts a new circulatory cycle.

Most of the fluid in the spaces between the cells of the body has filtered out of the capillaries and is the direct transporting agent between the blood and the cells. The fluid, called interstitial fluid, does not accumulate for long in the body spaces (except in degenerative diseases such as dropsy, caused by congestive heart failure), but passes into lymphatic capillaries and joins a slow stream of lymph in the lymph vessels. Lymph is a transparent yellowish liquid which is nearly identical to blood plasma. Lymph vessels are equipped with valves that cause a one-way flow of lymph away from the tissues and back to the bloodstream. Here and there, lymph vessels enter **lymph nodes** which act as filtering stations, removing foreign particles, bacteria, and other contaminants. Finally, the lymph passes into the large thoracic lymph duct which passes upward along the backbone and empties into the large vein from the left arm.

The Heart   The heart is a muscular organ roughly the size of two closed fists. During every minute it contracts 70–90 times, so in the average human lifetime it will have contracted two billion times and will have pumped 200 million liters of blood. It can do this because unlike any mechanical invention, it is able to maintain and repair itself without losing function. It is not uncommon for a physician, at the bedside of an elderly patient wasted by cancer or another debilitating disease, to find the heart still beating sturdily until other systems break down and the heart finally must stop.

Heartbeat   The heartbeat is a complex action. It starts in cells in the wall of the right atrium, in the **sinoatrial (S-A) node,** commonly called the **pacemaker.** A wave of impulses spreads rapidly from muscle cell to muscle cell in both atria, causing them to contract almost simultaneously. About 0.15 second later the atrial impulses reach a second control center, the **atrioventricular (A-V) node,** which relays the impulses to specialized muscle cells (**Purkinje fibers**) in the ventricles responsible for simultaneous contraction of both ventricles. This double set of contractions, first of the atria and then of the ventricles, is accompanied by opening and closing of heart valves, which give the heartbeat its characteristic "kathump" sound. The beating of the heart is an autonomous action, but rate of heartbeat is regulated by the nervous system.

Artificial pacemakers are almost routinely installed in cases in which the normal initiation of heartbeat impulses is impaired. Unlike open-heart surgery to repair heart damage, implantation of a pacemaker is a relatively simple operation. A small, battery-driven device is placed beneath the skin of the chest and delivers a small, regular electrical stimulus to the heart muscle.

The one-way movement of blood through the heart chambers is controlled by valves. Valves in the left ventricle sustain the highest pressures; if these have been damaged by disease, the consequences are incapacitating. A fairly common practice is open-heart surgery to repair damaged valves or, in some cases, to replace them with artificial valves.

Blood Pressure   Each heartbeat consists of a contraction phase, or **systole,** followed by a relaxation phase, or **diastole.** The blood pressure in an adult human heart will rise during systole to about 120 mm of mercury. During diastole, the pressure recedes to that of the blood vessels, normally about 70 mm of mercury. Blood-pressure readings much higher or lower than these may be a cause for concern.

Transport of Gases   It may seem that the lungs supply the power for the drawing of a breath, but this is not the case. The muscular **diaphragm, intercostal** (rib) **muscles,** and muscles in the abdomen are responsible for the breathing effort. Lining the inside of the **rib cage (thoracic cavity)** are two sacs called the **pleura.** The lungs lie within the pleura and are bathed in **pleural fluid.** During an exhalation the muscles are relaxed, while abdominal muscles and internal intercostal muscles are contracted. This applies pressure to the pleural fluid, which in turn squeezes the lungs and expels air. Inhalation is the result of relaxation of abdominal and internal intercostal muscles, while the external intercostal muscles and muscles of the diaphragm contract, relieving pressure on the pleural fluid. This increases the volume of the chest, and atmospheric pressure causes the flow of air into the lungs (Fig. 9-18). Sometimes by accident or disease, air gets into the pleura. When this occurs, the lungs collapse. If this should happen to both pleura, breathing becomes impossible and death results unless artificial respiration is begun.

Oxidation As has been learned, food is burned in the cells of the body by processes that are generally termed biological oxidation, and more specifically glycolysis, fermentation, the Krebs cycle, and the electron transport system. The food molecules in the blood are transported to all cells, but a supply of oxygen also is required if the Krebs cycle and the electron transport system are to operate and produce ATP. Oxygen, as has been stated, is absorbed from air in the lungs and diffuses into the capillaries from the air sacs of the lungs. The lungs are not simply air bags but contain many interior air passages ending in small air sacs called **alveoli** (sing. **alveolus**), each of which is richly supplied with capillaries and serves as a site of gas exchange between the air and blood (Fig. 9-19).

Gas Exchange in the Blood Recall that diffusion occurs in the direction of lower molecular concentrations. If, for example, the concentration of molecules of gas A is high on the outside of a membrane and low on the inside, while the concentration of molecules of gas B is high on the inside and low on the outside, then a two-way traffic of diffusing molecules of A and B will occur. In essence this is the explanation for the movement of oxygen from the air in the alveoli across the alveolar and capillary membranes into the blood and of carbon dioxide from the blood into the air of the alveoli. There is, however, a problem in transport of carbon dioxide and oxygen that relates specifically to their solubilities. Neither gas is very soluble, and they do not diffuse rapidly into liquids; however, in blood of both invertebrates and vertebrates, molecules capable of absorbing and transporting oxygen and carbon dioxide have evolved. Hemoglobin, the best known of these, occurs universally in vertebrates and in many invertebrates.

About 98 percent of the oxygen carried by blood is transported by hemoglobin present in red blood cells. Why must hemoglobin be contained in cells? Actually, it

FIGURE 9-18
Act of breathing.

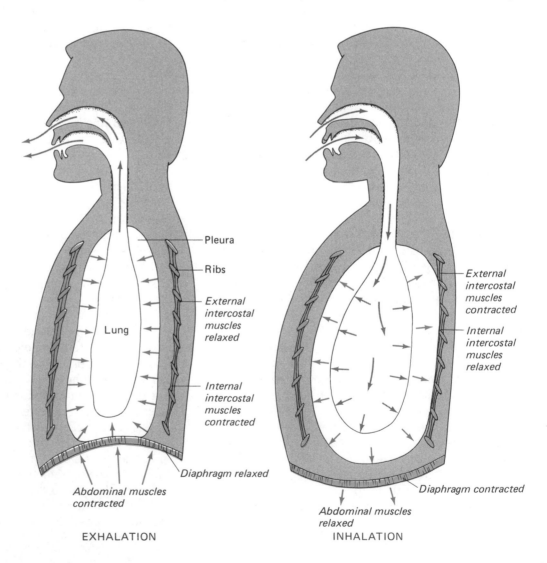

Pleura

Ribs

External intercostal muscles relaxed

Lung

Internal intercostal muscles contracted

Abdominal muscles contracted

Diaphragm relaxed

EXHALATION

External intercostal muscles contracted

Internal intercostal muscles relaxed

Abdominal muscles relaxed

Diaphragm contracted

INHALATION

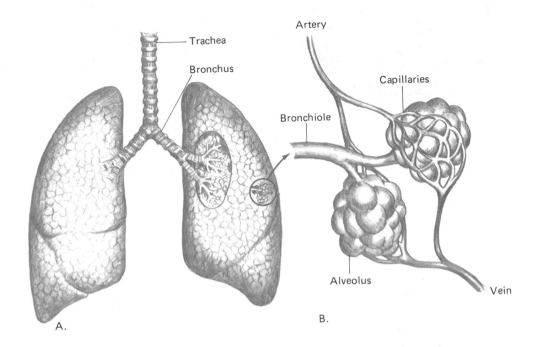

Trachea

Bronchus

Artery

Capillaries

Bronchiole

Alveolus

Vein

A.

B.

has been found that the hemoglobin does not have to be in blood cells to do its work. Experiments done years ago, in which red blood cells were gradually filtered from the blood of cats and replaced with naked hemoglobin molecules, showed that such animals had no difficulty transporting oxygen in their blood. However, after a day or so, they became anemic because their hemoglobin was excreted by their liver and kidneys. A role of the red blood cell is to prevent this excretion.

When blood circulates in capillaries surrounding alveoli of the lungs, oxygen first diffuses into the blood plasma. It then diffuses through the membrane of the red blood cell and is immediately tied up by an iron atom of a hemoglobin molecule, forming **oxyhemoglobin** (oxygenated hemoglobin). In terms of diffusion, such bound oxygen is inactive, so additional diffusion of oxygen continues until all of the hemoglobin is oxygenated (Fig. 9-20). Each red blood cell can hold about one billion oxygen molecules. Oxygenated red blood cells then are transported to other parts of the body where oxygen is required and is in short supply. The low concentration of oxygen in the interstitial fluid results in diffusion of oxygen out of red blood cells. Oxygen is not transferred directly from the capillaries to oxygen-depleted cells. First, blood plasma diffuses out into the interstitial fluid, carrying along oxygen. The oxygen diffuses out of the interstitial fluid and into actively respiring cells. Additional oxygen is released from the hemoglobin and diffuses into the blood plasma and then into the interstitial fluid to make up the deficit. The net result is that the body cells get much of the oxygen carried by the blood, and

the hemoglobin molecules are once again able to bind more oxygen on the next passage through the alveoli (Fig. 9-20). Deoxygenated hemoglobin is called **reduced hemoglobin** and is a darker red (but not blue) than bright-red, oxygenated hemoglobin.

The blood plasma does not carry much dissolved carbon dioxide, as such. Most carbon dioxide is transported in the form of the bicarbonate ion (HCO_3^-), which is quite soluble in blood plasma. In addition to that dissolved in the plasma, about 18 percent of the bicarbonate ions are carried by hemoglobin (Fig. 9-20). Carbon dioxide in water (or in blood) combines with water molecules to form carbonic acid ($CO_2 + H_2O \rightarrow H_2CO_3$). Carbonic acid is a weak acid and normally does not dissociate much to form hydrogen and bicarbonate ions. However, both the formation of carbonic acid and its dissociation are speeded up in the red blood cells by the enzyme **carbonic anhydrase (CA)**, and this enables more carbon dioxide to diffuse into the blood, replacing that converted to bicarbonate ions ($CO_2 + H_2O \underset{}{\overset{CA}{\rightleftharpoons}} H_2CO_3 \rightleftharpoons H^+ + HCO_3^-$). Upon return of blood to the alveoli, bicarbonate ions are reconverted to water and carbon dioxide, and carbon dioxide then diffuses out of the red blood cells, into the blood plasma, and into the air of the alveoli (Fig. 9-20). All the reactions described are reversible; they can proceed in either direction, but in each case the factor governing the direction in which the reaction will go is the relative concentrations of carbon dioxide and oxygen in the blood and outside the blood.

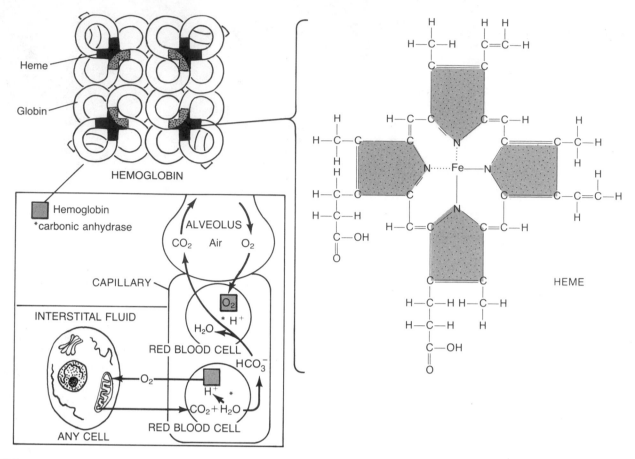

FIGURE 9-20
Transport of oxygen and carbon dioxide in the blood. The hemoglobin molecule (upper left) is composed of four polypeptides (globins), each bearing one heme molecule (upper right). (Hemoglobin model based on R. E. Dickerson and I. Geis, 1969, *The structure and action of proteins*, Menlo Park, Calif.: The Benjamin/Cummings Publishing Company)

The Kidneys

Waste Transport In addition to carbon dioxide, there are many waste products of cell metabolism removed from the body's cells and disposed of by excretion from the lungs. The skin also functions as an excretory organ; some wastes diffuse into the sweat glands and are eliminated, and some gas exchange occurs through the skin itself. However, most of the wastes from cells are excreted by the liver and kidneys. Kidneys serve as major homeostatic organs, controlling concentrations of sodium, potassium, hydrogen, calcium, and carbonate ions in blood and urine. Their principal actions deal with the conservation of water, mineral salts, and foods such as glucose and amino acids; they also have endocrine functions. One important role is the collection and disposal of **urea,** a breakdown product of proteins and amino acids. Urea is formed in the liver by the removal from amino acids of amino groups (—NH$_2$), which are con-

verted to ammonia (NH$_3$). The ammonia then is reacted with carbon dioxide to make urea:

$$NH_2-\underset{\underset{O}{\|}}{C}-NH_2$$

Kidneys have many detoxifying functions, eliminating poisons such as pesticides and other man-made molecules.

Kidney Structure The human kidneys are a pair of fist-sized organs lying to either side of the backbone, about 20 cm (8 in) above the hip line and bulging slightly into the coelom. Each has a tough outer covering of connective tissue, the **capsule,** within which is a "pulp," or matrix, containing about one million tubular filtration units called **nephrons,** together with a rich supply of arteries, veins, and their connecting network of capillaries. Nephrons are aggregated in conical structures, the **renal pyramids,** which project into the collect-

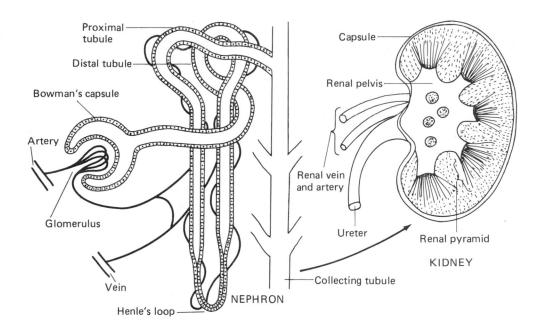

FIGURE 9-21
Human kidney and nephron.

Proximal tubule

Distal tubule

Bowman's capsule

Artery

Glomerulus

Vein

Henle's loop

NEPHRON

Collecting tubule

Capsule

Renal pelvis

Renal vein and artery

Ureter

Renal pyramid

KIDNEY

ing cavity, the renal pelvis, of the kidney (Fig. 9-21). The function of the nephrons is the filtration of blood plasma, from which wastes are collected, and the conservation of water, ions, and other blood constituents, which are returned to the bloodstream. The wastes are concentrated by the nephrons in their excretory product, urine.

Figure 9-21 shows the five regions of the nephron: **Bowman's capsule, proximal tubule, Henle's loop, distal tubule,** and **collecting tubule.** Urine formation starts with the filtration of plasma (by dialysis, accelerated by a pressure differential between the capillary blood and the fluid in Bowman's capsule) through the capillaries into Bowman's capsule. In humans, about 180 liters of fluid pass into the nephrons per day, and all but 1–2 liters of this are reabsorbed. Since the total volume of blood fluids is only about 3 liters, it follows that the blood must be filtered and cleansed about 60 times every day.

In the fluid that filters into the nephrons (i.e., **glomerular filtrate**) are metabolic wastes such as urea as well as conserved molecules including glucose, amino acids, and anions and cations of mineral salts. (Blood proteins and other macromolecules do not, however, diffuse into Bowman's capsule.) In addition to diffusion, other molecules are secreted into the urine by active transport. These can include ions as well as toxic substances. Many of the latter are not normally found in the body. Penicillin and dyes are examples. As the urine passes down the tubules, ions, glucose, amino acids, and other solutes are removed by diffusion and active transport (energy-requiring processes) principally in the proximal tubule and Henle's loop. Water is removed

from the nephron by osmosis, primarily in the collecting tubule.

The flow of blood through the kidneys, as well as the concentrations of solutes in the urine, is controlled by sensitive feedback mechanisms. Particularly important is the regulation of sodium levels in the extracellular fluid and blood, because sodium concentration controls the reabsorption of water from the nephrons. If sodium levels in blood and interstitial fluid are low, blood pressure falls; if sodium is too high, blood pressure rises, possibly to the point at which the tissues become waterlogged and circulatory difficulties (strokes, heart attacks) may arise.

The nephrons drain into the renal pelvis, which in turn drains urine into the urinary duct, the ureter (Fig. 9-21). The ureters transport the urine to the urinary bladder, where it is stored temporarily. The bladder is drained by the urethra, which opens to the outside at the tip of the penis in males and just in front of the vaginal opening in females.

SUMMARY

Animal nutrition is a complex subject, because the animal, unlike the plant, cannot make food from inorganic matter. Acquisition of nutrients by cells requires digestion and transport of food, of oxygen, and of waste products of metabolism.

The evolution of food-processing and food-transport systems in animals has its origin in extracellular digestion, in which enzymes, secreted by cells, dissolve carbohydrates, proteins, and lipids into soluble molecules which then can be absorbed by cells. Primitive animals such as hydras and flatworms have gastrovascular cavities in which functions of digestion and also circulation are carried on. More complex organisms have evolved circulatory systems containing fluids (usually blood) capable of transporting nutrients and oxygen to all cells of the body and of transporting carbon dioxide and soluble, metabolic wastes to excretory organs from which these waste products are excreted into the external environment.

Digestion occurs in the digestive tract, first in the mouth, where saliva starts the breakdown of starch, and next in the stomach, where hydrochloric acid and pepsin begin protein digestion. The partially digested stomach contents, called chyme, pass into the small intestine, where digestion is completed and where nutrient assimilation occurs. Nutrients then are transported by the blood to all the cells of the body. Waste residues in the small intestine pass into the large intestine, where water and salts are absorbed, leaving the feces to collect in the upper colon before passing to the rectum and to the outside through the anus.

The lungs are the principal organs of gas exchange between the animal body and the atmosphere. Oxygen is collected by capillary blood in the lung alveoli, and carbon dioxide at the same time diffuses from the blood into air in the alveoli. The reverse action occurs in the tissues: oxygen leaves the blood and carbon dioxide enters it. Neither gas is very soluble in the blood plasma. Both oxygen and carbon dioxide are transported by the hemoglobin of the red blood cells, but most carbon dioxide is transported in the form of the carbonate ion in the plasma.

Kidneys are major organs of waste collection and disposal, filtering out metabolic wastes such as urea and also collecting and excreting toxic substances of foreign origin. About 180 quarts of blood fluids are filtered and cleansed of impurities each day by the kidneys.

KEY WORDS

autotroph	pyloric sphincter	peristalsis
heterotroph	pancreas	villus
essential amino acid	liver	lacteal
double sugar	urogenital system	intrinsic factor
sucrose	glottis	colon
maltose	vocal sacs	pacemaker
lactose	gastrolith	systole
intracellular digestion	mammary glands	diastole
extracellular digestion	palate	oxyhemoglobin
interstitial fluid	epiglottis	reduced hemoglobin
gastrovascular cavity	Heimlich maneuver	carbonic anhydrase
flame bulb	hypothalamus	Bowman's capsule
nephridium	islets of Langerhans	glomerulus
integument	insulin	proximal tubule
kidneys	glucagon	Henle's loop
pharynx	CCK-PZ	collecting tubule
gill slits	duodenum	ureter
esophagus	jejunum	urinary bladder
stomach	ileum	urethra

QUESTIONS FOR REVIEW AND DISCUSSION

1   Contrast nutrition, digestion, transport, and waste disposal in unicells and metazoans (multicellular animals).

2   Explain relationships between autotrophy and heterotrophy.

3   Briefly discuss enzymatic hydrolysis of polysaccharides, proteins, and lipids.

4   How are fats assimilated? Why do fats yield more energy than carbohydrates? What are some relationships of fats to heart action?

5   What is the action of hormone secretion in the digestive process? Cite an example.

6   Describe how carbon dioxide is transported from the cells to the lungs.

7   Explain the mechanical action of the human lung.

8   What is the role of interestitial fluid in nutrient and gas transport in metazoans and, particularly, in land vertebrates?

9   Briefly describe the structure and function of the nephron.

10 About how many liters of fluid are filtered through the human kidneys per day? How much of this fluid is conserved? About how much is urine?

SUGGESTED READING

COMROE, J. H., JR. 1966. The lung. *Sci. Amer.* 214(2):57–68. (Summarizes relationships of structure and function.)

ECKERT, R., and D. RANDALL. 1978. *Animal physiology*. San Francisco: W. H. Freeman and Co. (An especially well-written and beautifully illustrated basic text.)

GRIFFIN, D. R. 1962. *Animal structure and function*. New York: Holt, Rinehart and Winston. (A simple exposition of basic information on digestion, metabolism, and transport.)

MERRILL, J. P. 1961. The artificial kidney. *Sci. Amer.* 205:(1):56–64. (Gives insight into kidney function and dialysis.)

PERUTZ, M. F. 1978. Hemoglobin structure and respiratory transport. *Sci. Amer.* 239(6):92–125. (Discusses molecular structure and functions of hemoglobin.)

SCHOTTELIUS, B. W., and D. D. SCHOTTELIUS. 1978. *Textbook of physiology*, 18th ed. St. Louis: The C. V. Mosby Co. (A well-written elementary human physiology text.)

SMITH, H. W. 1953. The kidney. *Sci. Amer.* 188(1):40–48. (Excellent diagrams.)

VANDER, A.; J. SHERMAN; and D. LUCIANO. 1980. *Human physiology: the mechanisms of body function*, 3rd ed. New York: McGraw-Hill Book Co. (Human organ systems are the subject of this lucid introductory textbook.)

WIGGERS, C. J. 1957. The heart. *Sci. Amer.* 196(5):74–87. (A basic introduction to elements of heart functions.)

ZUCKER, M. B. 1980. The functioning of blood platelets. *Sci. Amer.* 242(6):86–103. (Functions and disorders of platelets are discussed, including the roles of platelets in arterial disease.)

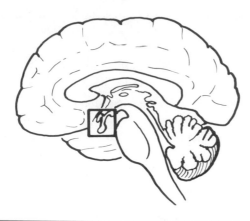

Hormones 10

An organism, regardless of kind, lives in a constantly changing environment to which it must adjust if it is to compete successfully with other organisms. This requires continual sensing of environmental factors and repeated adjustments in spatial orientation and physiological activities—responses which may be classified under the general heading of **homeostasis.**

Sensing of the environmental factors and regulation of body functions require many kinds of integrated actions by cells, tissues, and organs. In animals, this integration is accomplished by two kinds of intercellular communication: electrochemical **nerve impulses** and chemical messages in the form of **hormones.** Plants also must adjust to environmental conditions, although their activities are not of the same kind as those of animals; plants are nonmotile and autotrophic, whereas animals are in most cases motile heterotrophs. These differences are reflected in the comparatively greater complexity of regulatory systems of animals, in which interactions between integrated systems of nerve cells **(neurons)** and hormone-secreting **glands** regulate body functions. In plants the regulation of functions is accomplished principally by hormones.

HORMONAL CONTROL

Hormones usually are defined as "messenger" molecules produced in one part of the body and transported by the bloodstream to certain cells, where they produce quite specific effects. Hormones are better defined by function than by chemical structure, for unlike enzymes they fall into a number of biochemical categories. Adrenaline (epinephrine), noradrenaline (norepinephrine), and histamine, for example, are modified amino acids and relatively small molecules; sex hormones and a number of other similar hormones referred to as **steroids** are somewhat larger molecules derived from cholesterol. Other hormones such as insulin are peptides or polypeptides composed of amino acids in chains of short to intermediate length; still others, including several pituitary hormones, are proteins.

Proteins, as noted in Chapter 1, are composed of one or more polypeptides folded three-dimensionally in such a way as to confer a high degree of specificity (see Fig. 1-28 for an example). Whether a hormone is a protein, peptide, amino acid, or some other organic molecule, its action depends on structural specificity for "recognition" by cells of the tissues and organs in which the hormones induce specific reactions.

Kinds of Hormones

Two broad categories of animal hormones are recognized: **regulatory** and **developmental** hormones. Regulatory hormones control energy transactions and responses to environmental situations (such as stresses related to fear or aggression or responses to cyclical events, including day-length changes and seasonal temperatures). Developmental hormones control nearly all reproductive and developmental processes in the animal body. (For a partial listing of both kinds of hormone, refer to Table 10-1.)

TABLE 10-1
Selected mammalian hormones

Source	Hormone	Chemical Nature	Effect(s)
Pituitary (anterior lobe)	Growth hormone (GH)	Protein	Stimulates bone growth
	Thyroid-stimulating hormone (TSH)	Protein	Stimulates thyroid to produce thyroxin
	Adrenocorticotrophic hormone (ACTH)	Protein	Stimulates adrenal glands to produce cortisol
	Luteinizing hormone (LH)	Protein	Stimulates gonads
	Follicle-stimulating hormone (FSH)	Protein	Stimulates gonads
	Prolactin	Protein	Stimulates milk production
Pineal body	Melatonin	Peptide	Regulates sexual maturation
Pituitary (posterior lobe)	Oxytocin	Polypeptide	Stimulates uterine contractions and milk release
	Vasopressin	Polypeptide	Stimulates water retention in kidneys
Adrenal cortex	Cortisol	Steroid	Controls sex characteristics
	Estrogens*	Steroid	See Ovary
	Testosterone*	Steroid	See Testes
	Aldosterone	Steroid	Regulates K^+ and Na^+ concentrations
	Progesterone	Steroid	See Corpus luteum
Adrenal medulla	Epinephrine (adrenaline)	Amino acid	Stimulates conversion of glycogen to glucose
	Norepinephrine (noradrenaline)	Amino acid	Constricts blood vessels
Thyroid	Thyroxin	Amino acid	Increases metabolism
Parathyroid	Parathormone	Protein	Regulates blood calcium
Pancreas (islets of Langerhans)	Glucagon	Peptide	Increases blood-glucose levels
	Insulin	Peptide	Decreases blood-glucose levels
Duodenum	Cholecystokinin-pancreozymin (CCK-PZ)	Protein	Stimulates release of bile
	Secretin	Polypeptide	Stimulates secretion of pancreatic enzymes
Testes	Testosterone	Steroid	Stimulates development of male sex characteristics
Ovary	Estrogens	Steroid	Stimulates and maintains female sex characteristics and development of uterus
Corpus luteum	Progesterone	Steroid	Inhibits LH; prepares uterus for embryo
Placenta	Relaxin (releasin)	Polypeptide	Relaxes ligaments and smooth muscles during childbirth
	Estrogens, progesterone	Steroid	Suppresses secretion of LH and FSH during pregnancy; maintains placenta
Damaged cells	Histamine	Amino acid	Relaxes walls of blood vessels and makes them more permeable; contracts bronchioles

*Primary sources are gonads.

Hormone Synthesis and Secretion

Amino acid, small peptide, steroid, and fatty acid hormones are synthesized by cytoplasmic enzymes in gland cells; larger peptide and protein hormones are synthesized on ribosomes. Until secreted by the cell in which they originate, most, if not all, hormones are enclosed by cytoplasmic membrane vesicles (Golgi vesicles) and are discharged through the cell membrane by a process that is essentially the reverse of phagocytosis (refer to Fig. 2-28). The membrane-enclosed hormone is conveyed to the periphery of the cell, where its bounding membrane fuses with the plasma membrane. The hormone then is emptied to the outside of the cell, leaving behind its bit of membrane, which is recycled by the cell. After leaving the cell, the hormone diffuses into the interstitial fluid (fluid around cells), then into the blood stream.

A number of hormones are produced by **endocrine glands,** commonly called **ductless glands,** which secrete their product directly into the bloodstream (Fig. 10-1). In addition to endocrine glands (pituitary, thyroid, adrenals, and others), some organs (stomach, intestine, kidney) have endocrine functions, producing and secreting hormones as well as other products (Fig. 10-2). Other glands such as sweat, tear, and salivary glands exemplify

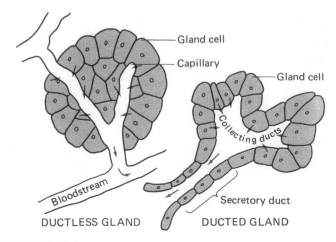

FIGURE 10-1
Structure of endocrine (ductless) and exocrine (ducted) glands.

exocrine glands, or ducted glands, which do not secrete hormones but instead secrete enzymes, mucus, and other products through ducts leading into body passages or to the exterior (Fig. 10-1). The pancreas, ovaries, and testes have mixed endocrine and exocrine functions and are therefore called **mixed glands** (Fig. 10-2).

FIGURE 10-2
Endocrine and mixed glands of the human female and male.

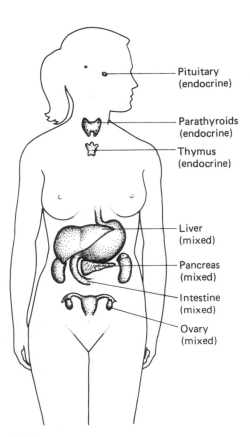

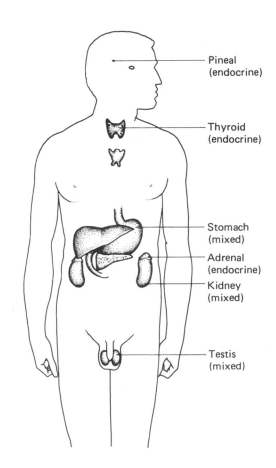

Target Cells

Definitions of hormone action usually include transport of hormones by the bloodstream to certain specific destinations, known as **target cells,** or **target organs.** This definition is appropriate but should not be interpreted to mean that hormones are directed specifically to some cells and organs and not to others. In actuality, hormones reach cells, tissues, and organs throughout the body but are capable of acting selectively on cells of the target alone. This kind of specificity implies an ability on the part of target cells to "recognize" a specific hormone, and vice versa. Larger hormone molecules, such as protein and polypeptide hormones, do not enter target cells but are bound to the outside of the target-cell membrane. Such binding implies the existence of specific acceptor molecules, to which hormones fit by key-in-lock action. The combined hormone and acceptor apparently stimulates the production within the target cell of internal "messengers," possibly cyclic AMP, the so-called "second messenger" of the cell's interior (see Fig. 10-3). Cyclic AMP (cAMP), together with cyclic GMP (cGMP), can act as a chemical messenger within cells and sometimes as a messenger outside certain cells. One way in which cAMP acts is to bind to certain activator proteins. A combined cAMP and activator then may bind to a very

specific site on DNA and increase the transcription of genes responsible for the synthesis of important molecules. When cAMP levels are high, the cellular functions are primarily **metabolic** (energy conversions). However, when cAMP levels are relatively low and cGMP levels are correspondingly high, the cell becomes channeled in other directions, in some cases including division and development. The levels of cAMP, and perhaps cGMP, seem to be controlled in animal cells by interactions with hormones. Hormones carry "messages" between cells and in at least some cases bring about changes in cellular levels of cAMP.

Smaller hormone molecules, such as steroid, fatty acid, and amino acid hormones, apparently diffuse equally well into target and nontarget cells, but in target cells appear to bind to specific cytoplasmic acceptor molecules. The complex of hormone and acceptor then is thought to move to the nucleus, where it becomes attached to specific repressor regions of DNA, which prior to the binding suppress the transcription of certain genes. When the hormone–receptor complex binds to a gene repressor, the gene is derepressed; mRNA is produced and then directs enzyme synthesis. (For examples and a diagram of the principle of gene repression and derepression, refer to Chapter 5).

Hormones and Cell Differentiation

Hormones are molecules capable of eliciting many specific responses in cells. For example, the hormone **insulin,** produced in the pancreas and carried in the blood to various parts of the body, controls changes in the metabolism of sugar. Specifically, insulin binds to a receptor protein in the cell membrane. There it blocks the action of adenylcyclase, and the level of cAMP is lowered. This in turn leads to an increased uptake and utilization of glucose in the cell and contributes in a general way to cell growth and development. In this case, the hormone does not act within the cytoplasm of the cell. Other hormones, however, including those controlling sexuality and developmental pathways, may act within cells and may have quite specific developmental effects.

An interesting example of hormonal action is seen in control of the development of fruitflies through a series of larval and pupal stages. In this case, the hormone **ecdysone** controls molting of the outer protective coat of the larvae as they change from stage to stage. At a specific time in this development, determined by external factors such as temperature, food supply, and internal timing (**biological clocks,** discussed in Chapter 24), the nerve cells of the larvae secrete ecdysone into the bloodstream. Its action may be seen by examining certain chromosomal changes that occur simultaneously.

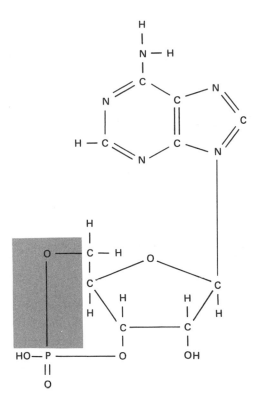

FIGURE 10-3
Structure of cyclic AMP. The colored area designates the cyclic portion of the molecule.

In the salivary glands of fruitfly larvae are groups of very large cells. These cells contain **giant chromosomes** in which transverse patterns of banding are clearly visible (Fig. 10-4). As the fruitfly larva passes from one developmental stage to the next, different bands in the giant chromosomes expand to form **chromosomal puffs.** Careful microscopic studies of the puffs show that they are regions in which loops of transcribed mRNA extend outward into the nucleoplasm.

When ecdysone is present in a salivary gland cell of a fruitfly larva, a specific band region of a giant chromosome is activated. The DNA in this band transcribes mRNA, which extends outward from the band as a puff (Fig. 10-5). This mRNA is translated into protein, which in turn initiates a sequence of additional puffs, each of which sequentially produces its distinctive protein. The result is a precisely timed sequence of gene activations. Although little is yet known about the way in which such activations control the differentiation of cells and influence the development of an entire organism, there is strong evidence that an important role of hormones is the activation of genes controlling developmental processes.

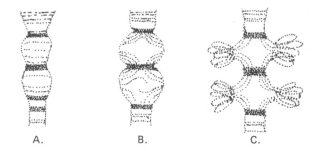

FIGURE 10-5
Chromosomal puffs of the giant salivary gland chromsomes of *Drosophila.* **A.** Before administration of ecdysone. **B.** Thirty minutes after administration of ecdysone. **C.** Three hours after administration of ecdysone. (After M. W. Strickberger, 1976. *Genetics,* 2nd ed. New York: Macmillan Co.)

Regulatory Hormones

Examples abound of hormonal actions and of interactions between hormones and the nervous systems of vertebrates and invertebrates. An often-cited one is the production, transport, and action of regulatory hormones in the control of blood-sugar concentrations. The overall sensing and regulatory center for blood glucose lies in the **hypothalamus,** a region of the vertebrate brain (see Fig. 10-9) which relays its information on blood-glucose concentrations over nerve pathways to the liver and pancreas. There, depending on the input of information, actions occur that either increase or decrease concentrations of glucose in the blood. These actions are a classic example of hormonal regulation, as well as of feedback responses in the production (in the pancreas) of two hormones, insulin and **glucagon** (Fig. 10-6).

When blood-glucose levels are low, glucagon is secreted by alpha cells of the pancreas, diffuses into the bloodstream, and upon reaching liver cells stimulates the conversion of glycogen to glucose. This increases the concentration of blood glucose, making it available for cellular respiration. As blood-glucose levels increase, insulin is secreted by beta cells of the pancreas and transported by the bloodstream to cells throughout the body, making their membranes more permeable to glucose and thus increasing its utilization. Thus insulin tends to cause depletion of blood glucose, and glucagon tends to increase it.

If too much glucose is present in the blood, a condition resulting from insulin deficiency (**diabetes mellitus**) occurs, and there is underutilization of glucose, excess sugar in the urine, and general malnutrition. If too little blood sugar prevails (a condition called **hypoglycemia,** which may be due to malnutrition or to an imbalance of glucagon and insulin), cell starvation, coma, and death may result. Diabetic comas sometimes occur

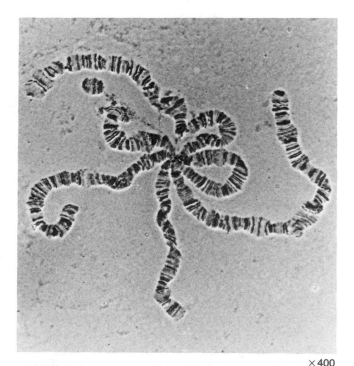

FIGURE 10-4
Giant chromosomes of the fruitfly *Drosophila melanogaster,* stained to show the banding pattern. The chromosomes adhere to certain regions; each "arm" corresponds to a chromosome or part of a chromosome. (Photo courtesy J. S. Yoon, University of Texas, Austin)

×400

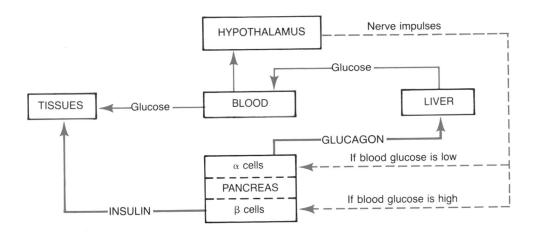

FIGURE 10-6
Regulation of blood glucose concentrations.

in persons who require insulin injections to maintain blood-glucose levels. If they overdose themselves, blood-glucose levels may become dangerously low. Diabetics usually carry sugar lumps to remedy an attack of insulin-induced hypoglycemia; they also may carry cards giving emergency instructions should they collapse in public. Tragedies sometimes occur when diabetics are mistaken for public drunkards and hauled off to jail without getting needed medical attention.

Developmental Hormones and Metamorphosis

In Chapter 11, two well-known reproductive cycles are described. One, that of the silk moth (Fig. 11-5), illustrates in an invertebrate animal the interesting phenomenon of metamorphosis. The second cycle exemplifies **metamorphosis** also, but in a vertebrate animal: the development of a frog from its larval stage, the tadpole (Fig. 1, p. 239). Both examples illustrate processes in which developmental hormones interact with nervous systems in regulating growth and development.

Insect Metamorphosis Many insects pass through larval, pupal, and breeding adult stages during their lifetimes. The silk moth, whose cycle is shown in Fig. 11-5, spends the greater part of its short lifetime as a **larva** or "caterpillar" feeding voraciously on leaves. During this time it undergoes five molts, each time growing rapidly before its newly secreted exoskeleton becomes hardened and resistant to further expansion. After the fifth molting, a very different and especially hard-shelled exoskeleton, the **pupa**, is formed. Then, during some months of **pupation,** a complete reorganization of tissues and organs occurs, after which the adult moth emerges, breeds, and shortly thereafter dies.

In a series of elegant experiments, notably by V. B. Wigglesworth in England beginning in the 1930s and

more recently by C. Williams of Harvard University, it has been learned that two hormones, ecdysone (*ekdyein*, "to shed") and **juvenile hormone,** operate antagonistically to control molting and pupation in insects. A third, as yet undefined hormone, secreted by cells of the larval brain, thus named **brain hormone,** controls the production of ecdysone by the **prothoracic gland** in the anterior thorax (Fig. 10-7). Ecdysone has several functions. It stimulates growth, it induces molting, and it controls pupation and metamorphosis. Which of these occurs depends in part on the presence or absence of juvenile hormone, the antagonist of ecdysone. A pair of glands, the **corpora allata,** located near the brain of the larva, produce and secrete juvenile hormone during larval development in concentrations sufficient to restrict the action of ecdysone to stimulation of growth and molting; metamorphosis will not be induced. However, as the larva grows, juvenile hormone levels decrease to the point where, at the fifth molt, ecdysone overcomes the inhibiting effect of juvenile hormone and is able to induce pupation. (Ecdysone, itself, is thought to act at the cellular and molecular level by activating genes in specific chromosomal regions in a manner similar to that described earlier; Fig. 10-5.) After pupation, the production of juvenile hormone ceases altogether and under the continuing influence of ecdysone, metamorphosis goes to completion.

Delicate surgery performed on caterpillars at several stages of development have added to our understanding of the way in which developmental hormones interact. Removal of the prothoracic gland maintains the larval state indefinitely; removal of the corpora allata results in immediate pupation and metamorphosis, producing an abnormally small adult. Conversely, grafting of additional active corpora allata results in a delayed pupation, continued larval development, an additional molt, and an abnormally large adult.

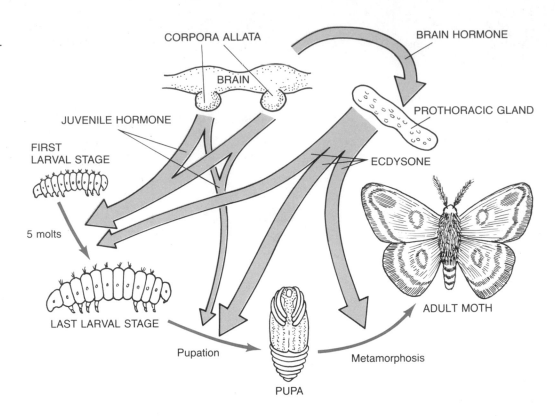

FIGURE 10-7
Hormonal regulation of development in silk moths.

CORPORA ALLATA

BRAIN HORMONE

BRAIN

PROTHORACIC GLAND

JUVENILE HORMONE

FIRST LARVAL STAGE

ECDYSONE

5 molts

ADULT MOTH

LAST LARVAL STAGE

Pupation

Metamorphosis

PUPA

An interesting sidelight to studies of insect meta-morphosis is seen in recent discoveries that some plants (balsam fir is one) produce juvenile hormonelike com-pounds that suppress metamorphosis in insects feeding upon them. The larval insect continues its feeding, caus-ing a further loss of leaf tissue to the host plant but eliminating the next and presumably more numerous and destructive generation.

It has been suggested that juvenile hormone, which has been isolated and possibly could be synthe-sized in appreciable amounts, be sprayed wholesale over growing crop plants. Attractive as this might seem at first glance, it would threaten the existence of beneficial insects as well as pestiferous ones. Juvenile hormone lacks the specificity of a really safe and successful bio-logical control weapon.

Amphibian Metamorphosis Metamorphosis is quite common among invertebrates, but noticeably less so among vertebrates. A notable exception, however, occurs among members of the Amphibia. Amphibians begin life in the water as free-swimming larvae, undergo metamor-phosis, and emerge as air-breathing reproductive adults. Probably the best-known example is the transition from a tadpole to an adult frog (Fig. 1, p. 239).

The changes that occur in the body of a tadpole during its metamorphosis are quite profound—the tail is absorbed, gills disappear, lungs develop, the intestine

becomes greatly shortened, major blood vessels are re-routed, the shape of the mouth changes, a tongue and teeth are formed, a middle ear develops, eyes move to the top of the head and eyelids develop, limbs are formed, and a new pattern of skin pigmentation appears. These and other changes can be related quite specifi-cally to production, secretion, and interaction of devel-opmental hormones, three of which are of particular importance: **thyrotropic hormone, thyroxin,** and **prolac-tin.**[*] Thyrotropic hormone and prolactin are secreted by the **hypophysis (pituitary),** a small endocrine gland at the base of the brain; thyroxin is a product of the **thy-roid,** also an endocrine gland. The thyroid is located an-teriorly, in the throat region of the tadpole.

Some appreciation of the role of the hypophysis and thyroid in metamorphosis of tadpoles can be gained, as in the example of insect metamorphosis just discussed, by removing either or both glands from living subjects. If the hypophysis is removed (**hypophysec-tomy**), the tadpole remains a tadpole indefinitely, never undergoing metamorphosis but continuing to grow, sometimes to double the size of the normal tadpole. If the thyroid is removed (**thyroidectomy**), essentially the same result will be noted. If, however, thyroxin is added to the water in which hypophysectomized and thyroidec-

---

[*]Prolactin is identified with the secretion of milk in mammals; here it has a different function.

tomized tadpoles are swimming, the tadpoles go through the complete metamorphosis. The role of thyrotropic hormone thus is seen to be one of inducing the thyroid to secrete thyroxin. During larval development prolactin counteracts thyrotropic hormone and thereby reduces the amount of thyroxin secreted by the thyroid. That prolactin lowers thyroid concentrations can be tested by adding extra thyroxin to the water in which normal tadpoles are kept. In this case premature metamorphosis takes place, producing perfectly formed but abnormally small adult frogs.

When tadpoles secrete thyroxin, it circulates freely in their blood, so that all the tissues are exposed to it. Nevertheless, only those tissues and organs that are altered in metamorphosis react to it. This has been shown quite dramatically by a relatively simple operation. If an eye from one tadpole is excised and grafted into the tail of a second tadpole, the eye remains healthy though nonfunctional. Then, under the influence of thyroxin the tail (the target) shrinks, but the eye (nontarget) remains unaffected and eventually migrates to the rear end of the young frog (Fig. 10-8). This illustrates the concept of target vs. nontarget cells and tissues relative to hormonal regulation.

HORMONES IN THE HUMAN BODY

It is appropriate that an account of hormonal regulation in the human body begins with the **pituitary gland** and the **hypothalamus.** Together these glands integrate the actions of many hormones with processes controlled principally by the nervous system. In its role of a control center, the hypothalamus not only acts as a coordinating agent of the nervous system, but itself has genuine hormonal functions. Its neurons secrete a number of small-

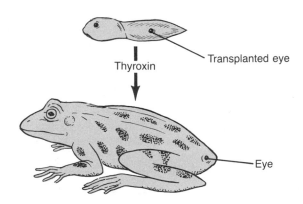

FIGURE 10-8
Differential responses of target and nontarget tissues as illustrated by eye-transplant experiments. (Based on the work of J. Schwind, 1933, as reported in *J. Exp. Zool.* 66:1–14)

peptide hormones, called **releasing factors,** which are transported in the bloodstream to the pituitary via a direct vascular connective. The pituitary produces—and in response to releasing factors, secretes—nine recognized hormones, which control secretions of regulatory hormones by the thyroid gland, mediate the action of insulin, stimulate the **adrenals** to secrete a variety of "stress" hormones, induce the production and secretion of **sex hormones,** and regulate a number of growth processes. Because of this multiplicity of hormonal functioning, the pituitary often is called the **master gland** of the body. It interacts with many other glands, such as the thyroid, parathyroids, adrenals, and gonads, to control both regulatory and developmental functions of the body.

Growth Hormones

Pituitary and Hypothalamus The pituitary gland, also known as the hypophysis, is located beneath the cerebrum at the base of the hypothalamus. It is a double gland composed of anterior and posterior lobes (Fig. 10-9). The posterior lobe develops from brain cells and remains connected to the hypothalamus by a short stalk, an arrangement that permits integration of pituitary functions with actions of the nervous system. The anterior lobe has a separate origin from that of the posterior lobe, developing in the embryo from tissues in the roof of the mouth and subsequently forming an intimate structural and functional partnership with the posterior lobe. The anterior lobe produces and secretes six hormones; the posterior lobe is known to produce and secrete three hormones, two of which have nearly identical functions.

In response to specific kinds of nervous and chemical stimulation, certain neurons of the hypothalamus secrete a variety of releasing factors which reach the anterior pituitary through the bloodstream. For example, in stress situations, a releasing factor is secreted and induces the secretion of **ACTH (adrenocorticotrophic hormone)** by the pituitary. ACTH in turn induces the secretion of cortisol by the adrenal gland and readies the body for stress situations. Also, under the influence of the hypothalamus and its releasing factors, the anterior pituitary secretes gonad-stimulating hormones and **growth hormone (GH),** which regulates overall growth (refer to Table 10-1 for a partial listing of pituitary hormones). Growth hormone affects development in a couple of ways. It stimulates metabolism by promoting incorporation of amino acids into proteins and inhibiting storage of fatty acids as fat, and it promotes the growth of bones. Severe undersecretion of growth hormone during childhood results in the production of a true **midget,**

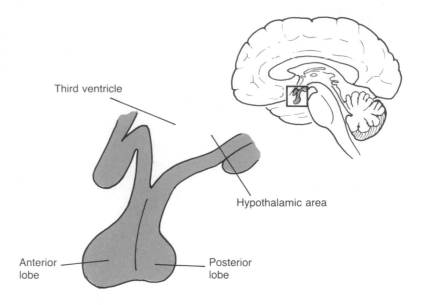

FIGURE 10-9
Pituitary and hypothalamus. (See Fig. 13-18 for additional details of brain structure.)

Third ventricle

Hypothalamic area

Anterior lobe

Posterior lobe

perfectly proportioned but tiny. Oversecretion of growth hormone in childhood can produce a very tall individual of the **giant** type (Fig. 10-10). Sometimes, after an appar-

FIGURE 10-10
Some effects of pituitary growth hormone. Captain Gulliver, 8 ft, 1 in tall, with Major Mite, 3 ft tall, and Tiny, 2 ft tall. (Photo from Bettmann Archive, Inc.)

ently normal childhood and adolescence, the output of growth hormone increases abnormally and produces a gross overdevelopment of the lower jaw, nose, forehead, hands, and feet—a condition called **acromegaly**.

Thyroid Among the many effects of pituitary hormones is that of stimulating development and maintaining activity of the thyroid, an H-shaped endocrine gland located in front of the trachea just below the larynx. The thyroid secretes two hormones: thyroxin and **calcitonin**. Thyroxin regulates some aspects of cellular metabolism and influences development. One well-known action of thyroxin is its ability to induce metamorphosis in tadpoles. In humans a severe thyroid deficiency in childhood may create a mentally deficient **cretin dwarf**.

Thyroxin production is controlled by **thyrotrophic releasing factor** secreted by the hypothalamus. This releasing factor induces the pituitary to secrete **thyroid-stimulating hormone (TSH)**. When an increase in metabolism is called for, for instance, when the body is putting forth maximum mechanical effort, the pituitary increases secretion of TSH; this stimulates secretion of thyroxin by the thyroid. As thyroxin levels in the blood increase, the hypothalamus decreases output of thyrotrophic releasing factor, and the pituitary then cuts back production of thyroid-stimulating hormone. Consequently, the production of thyroxin decreases. However, when thyroxin concentrations in the blood again fall to a point at which they are insufficient to sustain adequate levels of cellular metabolism, TSH secretion resumes, and thyroxin levels go up again. Such oscillations are characteristic of many feedback control systems.

The thyroxin molecule contains iodine; when iodine is lacking in the diet, thyroxin production decreases. As compensation for decreased thyroxin production, the thyroid gland greatly increases in size, resulting in a condition known as **simple goiter**. An individual suffering from simple goiter is listless and may be obese. The addition of iodine to dietary salt is all that is necessary to prevent simple goiter. It is ironic that unrefined salt contains plenty of iodine; only when the use of highly refined salt became prevalent did simple goiter become a problem.

Hyperthyroidism is another condition of the thyroid that can be a health problem. It results from overdevelopment and overactivity of the thyroid gland; thyroxin output increases and the victim becomes hyperactive. Hyperthyroid individuals often lose weight, become tense, and may have protruding eyeballs (**exophthalmia**), a condition known as **toxic goiter** (Fig. 10-11). A remedy for hyperthyroidism is surgical removal of part of the thyroid gland or the prescription of drugs that block thyroxin secretion. (The radioactive isotope ^{131}I is no longer used to treat hyperthyroidism because of high incidences of thyroid cancer and genetic damage.)

The second thyroid hormone, calcitonin, also called **thyrocalcitonin**, is involved in the control of calcium levels in the blood, and its action is to decrease the absorption of calcium by growing bone. It is secreted in response to reduced levels of calcium in blood serum.

Thymus   The **thymus** is an irregularly shaped gland lying below the thyroid (Fig. 10-2). It usually is well developed in children but tends to diminish in size after puberty. For years the thymus was something of a mystery; its removal seemed to have no specific negative effects. It is now known, however, that the thymus produces at least one hormone—**thymosin**—which is thought to stimulate the development of factors involved in immune responses. Removal of the thymus in early postnatal development impairs production of antibodies during the individual's entire lifetime.

Regulatory Hormones

Parathyroids   Embedded in the thyroid gland are four tiny glands, the **parathyroids** (Fig. 10-2). In the early days of thyroid surgery to correct hyperthyroidism, the parathyroids sometimes were unknowingly removed. This resulted in a marked decrease in serum-calcium levels and subsequent violent muscle contraction. This condition, known as **tetany**, often resulted in the death of the patient. The parathyroids, through their secretion of **parathormone**, help maintain blood-calcium levels. Parathormone and calcitonin (from the thyroid) act together in regulating excretion, absorption, and reabsorption of calcium ions by bone, kidneys, and intestines, thereby

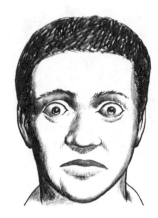

FIGURE 10-11
Toxic goiter results from an overactive thyroid gland (hyperthyroidism).

maintaining serum-calcium levels within a narrow physiological range.

Adrenal Glands   The **adrenal glands** lie next to the kidneys and are actually two glands in one (Fig. 10-2). The outer part, or adrenal cortex, produces more than 40 **steroid hormones** of quite similar chemical structure; however, with the exception of **cortisol** and a few others, their functions are only beginning to be understood. Cortisol, and probably other similar cortical steroids, are important in helping the body adjust to stress conditions, particularly effects of long-term stress such as malnutrition, overcrowding, and other fatiguing situations. Other cortical hormones regulate the activity of cells in the convoluted tubules of the kidneys, which in turn regulate the balance of sodium and potassium in the blood and urine. In addition, the cortex produces small amounts of steroid sex hormones.

The inner part of the adrenal gland, the **adrenal medulla**, secretes two hormones, **adrenaline (epinephrine)** and **noradrenaline (norepinephrine)**, both involved in short-term "fight-or-flight" responses. Under their influence the heartbeat increases, blood pressure rises, production of glucose from glycogen increases, and the smooth muscles of hair follicles contract, causing the hair to become erect and producing "goose flesh."

Pineal Body   The tuatara, a primitive, rare lizard found only in a few islands near New Zealand, has a remarkable middle eye in its forehead. Although vestigial and covered by skin, this eye has structural characteristics of functional eyes, and it is thought, on this and other evidence, that at least some ancient amphibians and reptiles may have had functional middle eyes.

GREETINGS, SIR . . . ER . . . ER . . . MADAM?

The family Hyaenidae comprises four species: the aardwolf and three species of hyenas, although some biologists classify the aardwolf in a family of its own. Of the hyenas (spotted, striped, and brown), the best known is the spotted, due mainly to the sustained field researches of one scientist, Hans Kruuk.

The much-maligned spotted hyena has long been considered a sneaky, skulking scavenger, snatching bits of carrion from prey brought down by other carnivores. Kruuk's evidence is contrary to this widely held notion. He shows that hyenas do scavenge opportunistically but kill the bulk of their own food via a communal clan system. Spotted hyena clans, some numbering up to 80, maintain and vigorously defend their own territories. Prey brought down by the social spotted hyena may even be scavenged by African lions.

Studies of clan social organization led Kruuk to dispel another long-standing hyena myth—that they are hermaphrodites! It is almost impossible to tell the sexes of spotted hyenas apart in the field because the genitalia of the adult females so closely resemble those of males. The clitoris of the female resembles the male's penis, even being capable of elongation and full erection. Also, the labia form a pseudoscrotum that even contains fatty lumps that closely mimic testicles. Additionally, females are larger than males and are dominant in most social interactions.

A look at the spotted hyena's "meeting ceremony" may give a clue to a possible function of male-mimicking genitalia of female spotted hyenas. When two hyenas meet they normally follow these "rules of the meeting game" regardless of their sex:

1 Each sniffs the other's head region.

2 They stand parallel with heads in opposite directions.

3 Both lift their inside hind leg, the subordinate first.

4 They mutually sniff and lick each other's now-erect penis or pseudopenis.

5 They both resume the four-footed stance and walk away.

What developmental pathways might lead to female spotted hyena genitalia mimicking those of males? Two British biologists, Racey and Skinner, decided to investigate the hormones of hyenas in order to determine their role, if any, in sexual mimicry. One could expect higher levels of male-producing androgens in adult testes than in adult ovaries. Not surprisingly, this is what the researchers

In other living vertebrates, it is theorized that the middle eye is reduced to a small dorsal projection of the midbrain known as the **pineal body** (Fig. 13-18). In lower vertebrates, the pineal body is photoreceptive; its cells have some characteristics in common with rods and cones and are capable of sensing changes in light intensity. In response to increased light, the pineal bdy produces the hormone **melatonin,** which induces contraction of dermal pigment cells, causing the animal's skin to become lighter in color. In mammals, the pineal body

FIGURE 1
Spotted hyena. (Photo courtesy South African Tourism Board)

found. However, androgen levels in blood plasma were the same in adult males and females. Even female fetuses of spotted hyenas contained about the same testosterone levels in their blood as did their mothers.

Commonly, the early mammalian embryo is primed to develop either male or female genitalia, but after gonadal differentiation, the testes, if present, secrete androgens. Thus at high androgen levels, male genitalia develop; at low levels, or none, female genitalia develop. But spotted hyena embryos, regardless of sex, are exposed to high androgen levels, hence they become masculinized. Subsequently, at mating time, the masculinized genitalia of adult female spotted hyenas undergo rapid feminization and function reproductively as in other female mammals.

What about brown and striped hyenas? Neither species shows female mimicry of male genitalia. In these two species, Racey and Skinner found androgen levels in the blood plasma of adult females to be lower than in males.

is not photoreceptive but still functions as an endocrine gland, producing melatonin. It is thought that the pineal gland responds to cycles of light and dark as perceived by the eyes and secretes melatonin inversely to light intensity. Lessening melatonin concentrations may influ- ence the pituitary to increase production of gonad-stimulating hormones (gonadotrophic hormones).

The Pancreas, Stomach, and Duodenum The pancreas is a large mixed gland located in the curvature of the upper small intestine (**duodenum**) and connected to it by a secretory duct (Figs. 9-13 and 10-2). When partially digested food, known as **chyme**, passes from the stomach into the duodenum, its acidic nature induces certain endocrine glands of the duodenum to secrete hormones into the bloodstream. Among these are **secretin** and **cholecystokinin-pancreozymin (CCK-PZ)**, which act on their target organ, the pancreas, and **motilin** and **villikinin**, which stimulate muscle movements in the stomach and intestine. The pancreas is induced by CCK-PZ to secrete a mixture of **proenzymes** (unactivated enzymes) and enzymes into the duodenum, which completes the digestive processes begun in the stomach and to some extent by action of salivary enzymes in the mouth.

The endocrine functions of the pancreas are carried on in clusters of cells called **islets of Langerhans**, which produce and secrete insulin and glucagon. These two hormones act antagonistically in a feedback system controlling blood sugar levels, as discussed earlier in the chapter.

It should be noted here that the stomach also has endocrine functions. Glands in the stomach wall produce and secrete the hormone **gastrin** into the bloodstream, from which it is absorbed by cells in the stomach lining, inducing them to secrete **hydrochloric acid**. The stomach thus functions not only as an endocrine gland but also as its own target organ. Subsequently, secretin induces the pancreas to secrete **sodium bicarbonate** into the duodenum. This addition of base begins the neutralization of the acidic chyme received from the stomach.

Sex Hormones

Human sexual characteristics, like those of other animals, are under the control of hormones. In vertebrates, these are secreted by the pituitary, the **gonads (ovaries and testes)**, the adrenals, and (in mammals) the **placenta** of the pregnant female. The hypothalamus is the main integrating center for nerve impulses relating to hormonal control of sex. Hypothalamic hormone-releasing factors stimulate the pituitary, which in response secretes gonad-stimulating hormones.

Gonads The gonads have two major functions: production of gametes and synthesis and secretion of hormones (refer to Fig. 10-2 for a diagram of glands related to sex). Testes produce sperm and the male sex hormone, **testosterone**. Testosterone and testosterone-like hormones are collectively known as **androgens**. The ovaries produce eggs and synthesize and secrete very similar hormones known as **estrogen** and **progesterone**. These male and female hormones are responsible for expression of secondary sexual characteristics and in the female also control maturation of eggs, the menstrual cycle, pregnancy, and childbirth. The placenta also functions during pregnancy as a source of estrogen and progesterone.

Prostaglandins A group of substances with hormone-like properties, the **prostaglandins**, exhibit a variety of stimulatory and inhibitory actions and are receiving increasing attention in medical research and practice. They are produced by many kinds of cells and can be transported in the bloodstream, but also are present in fluids other than blood. For example, the secretions of accessory male sex glands are rich sources of prostaglandins. (Prostaglandins, in fact, got their name from the **prostate gland**, in which they originally were found. However, it turns out to be a misnomer; they actually originate as secretions of the **seminal vesicles**.) At present, 14 different prostaglandins have been described. All are 20-carbon compounds derived from polyunsaturated fatty acids, and they produce a variety of responses in the human body: increasing or decreasing blood pressure, inhibiting gastric secretions, increasing urine flow, and either intensifying or relieving pain. The efficacy of aspirin is thought to be due to its ability to block one or more prostaglandins.

Two prostaglandins, referred to as PGE_2 and PGF_2, are powerful stimulators of smooth-muscle contractions and are produced by the uterine wall during childbirth. They have been used as abortifacients (abortion inducers) and have proved to be safer than other methods, but have raised legal and ethical questions because they sometimes produce living aborted fetuses. This results in problems of keeping the fetuses alive and introduces possibilities of homicide charges if they die.

SUMMARY

Hormones are molecules of several kinds (steroids, amino acids, peptides, fatty acids, or proteins) synthesized in gland cells and transported in the blood to target organs in which they induce specific responses. The overall controlling center for hormonal actions in vertebrates is the hypothalamus region of the brain. In invertebrates, the brain also can regulate hormone production and secretion, as we see in the hormones controlling metamorphosis in insects.

Often pairs of hormones act oppositely to each other. Glucagon and insulin act antagonistically to regulate sugar levels in the blood. Metamorphosis in insects and in amphibians is similarly regulated by oppositely acting hormones (juvenile hormone and ecdysone in insects control molting and metamorphosis).

Hormones may be designated as regulatory or developmental on the basis of the activities they regulate. Among regulatory hormones are insulin, glucagon, cortisol, adrenaline, gastrin, secretin, and many others. Among developmental hormones are thyroxin, growth hormone, juvenile hormone, ecdysone, and sex hormones such as androgens and estrogens, as well as many others.

Hormones are produced by endocrine tissues in endocrine glands such as the thyroid and thymus, in organs such as the stomach and duodenum, which have glandular and nonglandular functions, or in mixed glands, such as the pancreas, testes, and ovaries. In all cases, they are secreted into the blood stream, unlike the secretions of exocrine glands, such as sweat glands, which are conveyed by ducts to the outside of the body or into interior cavities.

KEY WORDS

regulatory hormones	pupa	thyroid
developmental hormones	metamorphosis	thymus
steroid	ecdysone	parathyroids
endocrine gland	juvenile hormone	adrenals
exocrine gland	thyrotrophic hormone	pineal body
mixed gland	thyroxin	pancreas
target cell	prolactin	duodenum
hypothalamus	releasing factor	gonad
insulin	stress hormone	androgen
glucagon	sex hormones	estrogen
larva	pituitary	prostaglandin

QUESTIONS FOR REVIEW AND DISCUSSION

1 Describe several important hormones on the bases of broad functions, their chemical nature, and their specific regulatory actions.

2 How does an exocrine gland differ structurally and in terms of secretion product from an endocrine gland? Are there glands having both endocrine and exocrine functions? If so, identify one.

3 In what chemical and/or structural aspect might target cells differ from nontarget cells?

4 The hypothalamus is a major hormonal regulatory center in vertebrates. Explain how it functions in regulation of blood glucose concentrations.

5 Is it possible, by manipulation of glands and hormones, to control the adult size of insects? Of adult frogs? Explain.

6 In metamorphosis of the tadpole into the adult frog, the changes observed are superficial rather than fundamental. True or false? Explain your conclusion.

7 If thyroxin is added to water in which tadpoles are swimming, what, if any, changes may occur in the tadpoles as a consequence?

8 What is meant by the term "hypothalamus releasing factor"? Are there comparable factors secreted by insect brain cells? If so, what might they be?

9 Why is the pituitary sometimes referred to as a master gland?

10 Explain why the stomach is said to be its own target organ.

SUGGESTED READING

BINKLEY, S. 1979. A timekeeping enzyme in the pineal gland. *Sci. Amer.* 240(4):66–71. (The pineal hormone melatonin correlates gonadal activities, changes in skin pigmentation, and internal rhythms with external changes in day length.)

ETKIN, W. 1966. How a tadpole becomes a frog. *Sci. Amer.* 214(5):76–88. (Roles of hypothalamus, pituitary, and thyroid glands are deduced by experimental gland transplants.)

GOULD, S. J. 1981. Hyena myths and realities. *Nat. Hist.* 90(2):16–24. (A popular account of the intriguing male mimicry of the female spotted hyena.)

GUILLEMIN, R., and R. BURGUS. 1972. The hormones of the hypothalamus. *Sci. Amer.* 227(5): 24–31. (Hormone secretions of the hypothalamus originate in secretory neurons which release their product directly into the blood or into intermediary storage cells.)

KRUUK, H. 1972. *The spotted hyena.* Chicago: The University of Chicago Press. (Technical but readable study of sociobiology of the spotted hyena.)

O'MALLEY, B. W., and W. T. SCHRADER. 1976. The receptors of steroid hormones. *Sci. Amer.* 234(2):32–43. (Discusses the nature of hormones and their transport and action within target cells.)

PASTAN, I. 1972. Cyclic AMP. *Sci. Amer.* 227(2):97–104. (Describes how cAMP may function as a second messenger in target cells.)

SAMSON, W. K., and G. P. KOZLOWSKI. 1981. Nerve cells that double as endocrine cells. *BioScience* 31(6):445–48. (Releasing factors are hormones produced in the hypothalamus and secreted into the bloodstream of the pituitary portal system.)

SCHALLY, A. V., A. J. KASTIN, and A. ARIMURA. 1977. Hypothalamic hormones: the link between brain and body. *Amer. Scientist* 65(6):712–19. (A clear description of relationships between releasing factors of the hypothalamus, pituitary hormones, and peripheral endocrine glands.)

WIGGLESWORTH, V. B. 1959. Insect metamorphosis. *Sci. Amer.* 200(2):100–10. (Describes experiments in which reversals of insect metamorphosis are attempted—with partial success.)

WILLIAMS, C. W. 1950. The metamorphosis of insects. *Sci. Amer.* 182(4):24–28. (Fascinating experiments in determining actions of hormones in metamorphosis of silk moths.)

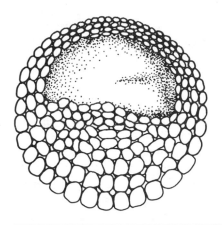

11 Animal Reproduction and Developtment

The zygote of a plant or animal, when viewed with a light microscope, is really not a very complex-looking cell, nor does it give any strong hint of the complex organism it can become. Curiously, however, in the early days of microscopy some scientists thought they saw miniature human beings in the heads of sperm, and others "found" similar tiny individuals in egg cells (Fig. 11-1). Each faction believed that the adult human owed its unique qualities either to the egg or to the sperm; they called themselves **ovists** and **spermists**, respectively.

Obviously, if the sperm or the egg actually contained the minature body of the adult, development would require only the continued growth of the already existing fully formed individual. Of course, it is now known that the zygote does not contain any such preformed body but instead is an undeveloped cell of great potential. Its development into the adult individual occurs in phases, each adding to the final form.

One great and obvious difference between the development of one-celled and many-celled organisms is in reproduction. In Chapter 6 we learned that a simple unicellular organism such as *Stentor*, *Paramecium*, or *Chlamydomonas* may reproduce by a simple division of its total structure and then reconstruct what has been lost as a result of that division. However, reproduction in multicellular organisms is not quite that simple, especially when the development involves sexual reproduction, as it does in most cases. In the sexual cycle the organism literally is reduced to a single cell, the **zygote** (fertilized egg), and then must completely rebuild its multicellular body. It does this by producing a stage called the **embryo**, in which progressively more complex specializations of groups of cells can

FIGURE 11-1
Highly imaginative drawing of a tiny human figure, or *homunculus*, once thought to be present in the head of a sperm.

be seen. This process usually is called differentiation, and the development of the zygote into a multicellular individual is called **embryogenesis**.

REPRODUCTION

Many organisms have the ability to reproduce both sexually and asexually, the latter being common among unicells and simple colonial forms of life. Among vertebrate animals, mainly sexual reproduction is known, but in plants and among many animal groups other than vertebrates, both sexual and asexual reproduction occur. Some simple animals, including sponges, jellyfish, and tapeworms, regularly produce reproductive buds which grow into new individuals. Asexual reproduction by fragmentation also may occur; if an animal is broken or cut into pieces, each piece may be capable of developing into a new individual. This is known as **regeneration**. Asexual reproduction, whether by budding or regeneration, is, however, rare among animals of complex structure and diversified organ systems.

Many variations in sexual reproduction of animals exist, but all involve **gametes** produced in specific sex organs called **gonads**. Animal gametes are the **sperm** and the **egg;** these are produced, respectively, in gonads called the **testis** and the **ovary** (Fig. 11-2). Usually testes and ovaries occur in different individuals (i.e., male and female), the **dioecious** ("two houses") condition. However, some animals such as earthworms are **monoecious** ("one house"), and both testes and ovaries are present in one individual, although cross-fertilization occurs even in these.

Animal gametes are produced by meiotic divisions occuring in diploid ($2n$) germ cells of the gonads: oocytes (oh-oh-cytes) in ovaries and spermatocytes in testes (Fig. 11-3). Oocytes produce four haploid ($1n$) cells: a comparatively large egg (refer to Fig. 8-8 for size comparison with sperm) and three diminutive polar bodies. The latter degenerate and do not participate in fertilization. Spermatocytes also produce haploid cells, called spermatids. These develop flagella and become functional sperm.

FIGURE 11-2
Generalized reproductive processes in animals. **A.** External fertilization. **B.** Copulation and internal fertilization.

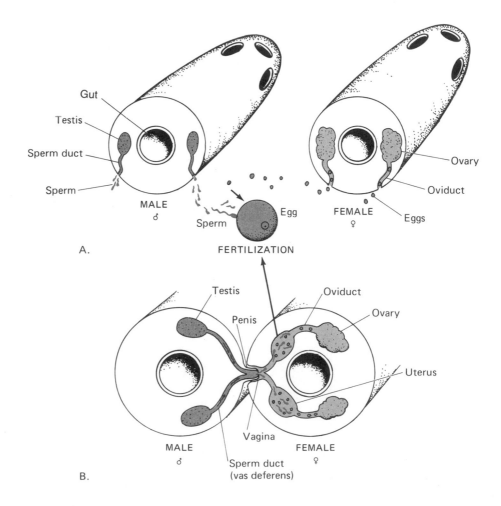

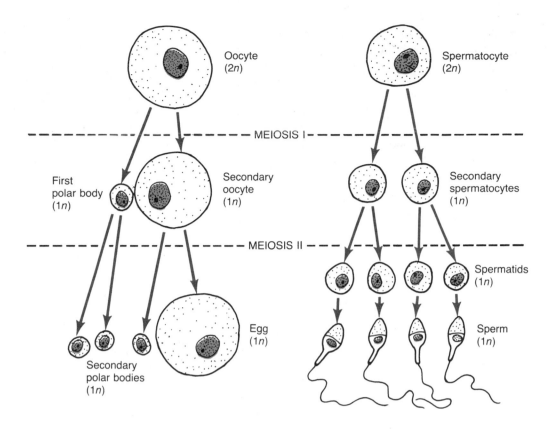

FIGURE 11-3
Production of animal sperm and eggs by meiosis.

Oocyte (2n)

Spermatocyte (2n)

------ MEIOSIS I ------

First polar body (1n)

Secondary oocyte (1n)

Secondary spermatocytes (1n)

------ MEIOSIS II ------

Spermatids (1n)

Egg (1n)

Sperm (1n)

Secondary polar bodies (1n)

In many aquatic animals fertilization is external; the sperm and egg are both shed into water, where fertilization takes place and a zygote is formed. The zygote soon divides to form a small multicellular embryo. Because the embryo, in cases where the fertilization is external, is not in direct association with its parents, it requires its own food supply. Sometimes the egg from which an embryo develops contains a rich supply of food; in other cases the embryo becomes self-feeding at an early age and may be quite unlike its mature state. Often it is a ciliated **larva** superficially resembling some of the more complex ciliated protozoans. Life is precarious for a young larva required to "go it alone" from the time of fertilization.

Animals higher up the trunk and branches of the evolutionary "tree" show progressively more complex forms of reproduction. Life on land, for example, often requires special accessory sex organs for the transport and storage of gametes and for the sheltering and nourishing of embryos. This is accomplished

1 by **copulation** (coupling), during which direct introduction of sperm by males into females takes place;

2 by development of embryos within shelled eggs; or

3 by retention of eggs and embryos within the body of the female until the young animals become self-sufficient.

Complex courtship behavior may precede copulation in advanced animals.

In a typical case of internal fertilization, the sexes are separate (refer to Fig. 11-2). Sperm, formed in a testis, pass down a **sperm duct (vas deferens)** and are injected into the female sex organ **(vagina)** by the male sex organ **(penis)**. Eggs in a female are formed in an ovary, pass down an egg duct, the **oviduct,** and into an egg sac, the **uterus.** Fertilization may occur either in the oviduct or in the uterus. Embryos may develop directly into small individuals of adult form or into larval feeding stages which later become adults.

EMBRYOGENESIS

Although there is considerable variation in the details of embryogenesis among animals, in most cases the early stages of development are quite consistent. The first mitotic divisions of the zygote are called **cleavages.** In them the zygote becomes divided into more, ever-smaller cells; an actual increase in the overall size of the embryo may be delayed until the cells begin to assimilate food either from the environment or from yolk material originally present in the egg.

Cleavage divisions produce a mulberry- or blackberry-shaped solid ball of a few cells, the **morula** ("mulberry"); later, as cell division continues, the ball becomes hollow, a stage known as the **blastula** ("embryo stage") (the cavity within the blastula is the **blastocoele**; see Fig. 11-4). Shortly thereafter, by one means or another (depending on the organism), the hollow ball of cells becomes cup-shaped, sometimes by a simple invagination process, in other cases by more complicated overgrowing folds of cells. The embryo at this stage (which is called the **gastrula**, "stomach stage") consists of an outer **ectoderm** layer of cells and an **endoderm** (Fig. 11-4).

Invertebrate Development

Simple Organisms   For some simple organisms like *Hydra* the cup-shaped gastrula stage is near the end of the developmental process. *Hydra* consists of an elongated gastrulalike body equipped with tentacles about its open end, which functions as a mouth. The cavity (in embryological terminology, the **archenteron**) within a *Hydra* functions as a simple digestive tract, or gut. Food is taken in through the mouth in typical animal fashion, digested in the gut, and the waste residues are expelled through the mouth.

In animals more advanced than *Hydra*, there follow additional stages leading to the development of a more complex body form. Echinoderms, which include animals with radial symmetry such as starfish, sea urchins, sand dollars, and sea cucumbers, form a regular gastrula which then develops into a ciliated larval stage called a **pluteus** (Fig. 11-4). The pluteus is a feeding organism and seems to be a necessary interruption in development, because the eggs of echinoderms do not contain much stored food. The pluteus feeds, grows, and finally develops into an adult echinoderm.

Although the pluteus is relatively simple, it is more advanced than the gastrula stage represented by *Hydra*. In addition to endoderm and ectoderm, **mesoderm** ("middle skin") is present. The three cell layers (ectoderm, mesoderm, and endoderm) are called **germ layers**. All subsequent layers and parts of an adult body including the epidermis, nervous system, muscle, bone (if present), vascular system (blood vessels and blood), and digestive and excretory organs can be traced back to one or another of the germ layers.

The Arthropods   Invertebrates having jointed legs, sometimes wings, and always a tough external skeleton, the exoskeleton, are placed in the Phylum **Arthropoda** ("jointed legs"). The exoskeletons of this most numerous and diversified group of animals provide both protection

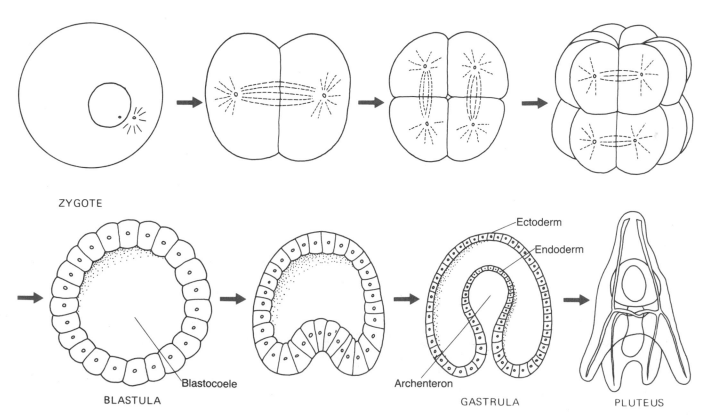

ZYGOTE

Blastocoele

BLASTULA

Ectoderm

Endoderm

Archenteron

GASTRULA

PLUTEUS

FIGURE 11-4
Development of a starfish embryo. The pluteus develops into a young starfish.

and mobility, but because of their rigidity, complicate growth and development.

As would be expected in a large phylum whose members occupy diverse habitats, reproduction varies considerably in arthropods. Among aquatic species, fertilization may be external or internal (Fig. 11-2); in land arthropods fertilization is internal. However, all arthropods, whether centipede, lobster, scorpion, insect, or other, go through a series of moltings as they develop. At each molting, the old exoskeleton is shed and replaced by a new and larger one. In many arthropods, developmental changes also accompany the moltings (Fig. 11-5). In some species the changes are quite remarkable, as seen in the **complete metamorphosis** of the maggot into the mature housefly or of the caterpillar into the colorful moth or butterfly. Other arthropods, including grasshoppers and locusts, change less drastically from molt to molt (**incomplete** or **gradual metamorphosis**).

Vertebrate Reproduction and Development

Fertilization in vertebrates may be either external (in **fishes, amphibians**) or internal (**reptiles, birds, and mammals**), and the sexes characteristically are separate.

FIGURE 11-5
Molting and metamorphosis of grasshopper and moth compared. (Redrawn from *General Zoology*, 5th ed., by Claude A. Villee, Warren F. Walker, Jr., and Robert D. Barnes. Copyright © 1978 by W. B. Saunders Company. Reprinted by permission of W. B. Saunders Company, CBS College Publishing)

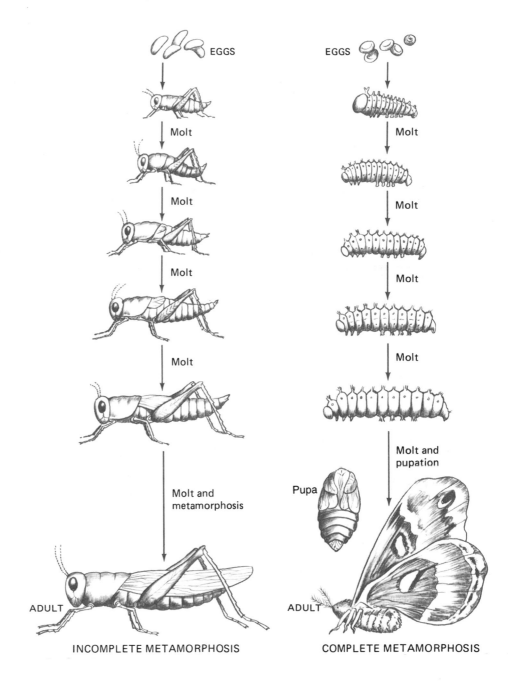

EGGS   EGGS

Molt   Molt

Molt   Molt

Molt   Molt

Molt   Molt

Molt   Molt and pupation

Molt and metamorphosis   Pupa

ADULT   ADULT

INCOMPLETE METAMORPHOSIS   COMPLETE METAMORPHOSIS

Vertebrates may be divided into two subgroups: **egg layers** and **womb bearers**. The former include fish, amphibians, reptiles, and birds; the latter are mammals. Although eggs, zygotes, and embryos of all these groups have been studied extensively, those of amphibians (frogs, especially) often are singled out for observation and for experimentation.

In vertebrates, reproductive organs (gonads and reproductive ducts) are closely associated with excretory organs (kidneys, urinary bladder, excretory ducts). Together they comprise the urogenital system (Fig. 11-6). Kidneys are pairs of oval to cylindrical structures embedded in the body wall beneath the peritoneum, a thin epithelial lining of the coelom. The kidneys empty into the urinary bladder via the ureters. From the bladder urine passes through the urethra either directly to the outside or into a common excretory and genital chamber, the cloaca, at the posterior of the gut.

At its simplest, the vertebrate reproductive system consists, in the male, of paired testes (sing. **testis**) in which sperm are produced and from which sperm pass to the outside through sperm ducts (Fig. 11-6) and, in the female, of a pair of ovaries in which eggs are formed. The eggs are shed into the coelomic fluid and are collected by the oviducts (which have ciliated, funnel-like openings adjacent to the ovaries), then passed to the outside.

Fishes and Amphibians   Fishes, as is well known, reproduce by spawning, in some cases after long and hazardous journeys from the sea into freshwater streams (examples: eels and salmon). The male gametes, com-

monly called "milt," are produced in enormous quantities by the paired testes and are conveyed by ducts to an opening near the anus. Eggs also are produced in great numbers, collectively the "roe," by the paired ovaries and released into the coelomic fluid. They are collected immediately from the coelom by the ciliated, funnel-like openings of the paired oviducts, in which they may be stored temporarily. The oviducts pass the eggs into the cloaca, then to the outside (directly to the outside if a cloaca is not present).

The majority of fishes deposit eggs and sperm into the water, usually in a spawning bed or nest. A smaller number of species are "live-bearers," incubating their young within the oviducts or in brood pouches. The seahorse is a particularly interesting example of a live-bearing species because it is the male who becomes "pregnant." He collects the eggs from the female, fertilizes them, and carries them about in an abdominal brood pouch until they hatch. Note that this kind of "live birth" is not equivalent to the gestation period of mammals. The young of live-bearing fish are nourished by egg yolk; the young of mammals are nourished by substances extracted from the mother's blood through a placenta.

The reproductive structures and processes in amphibians such as frogs, toads, and salamanders are quite similar to those of fish. Even those species that are principally land dwellers require deposition of eggs and sperm in water for successful fertilization and embryo development. Most species deposit eggs and sperm in ponds and streams, but a few find unique habitats for their reproduction (see "Parental care in anurans," p. 238).

Eggs, zygotes, and embryos of amphibians are relatively large and, because they develop in water, quite accessible. The typical amphibian egg is 2–4 mm in diameter and rich in stored food in the form of yolk granules; fertilization is external and embryos develop in open water rather than inside the female parent as in mammals.

Typically, an amphibian zygote has an upper, dark-colored hemisphere—the **animal pole**—and a lower, lighter colored hemisphere—the **vegetal pole**—in which yolk material tends to be concentrated. Jellylike capsules cover eggs, zygotes, and embryos, protecting them from infection and damage. At the time of fertilization, a narrow grayish band called the **gray crescent** develops on one side of the zygote in the animal pole just above its equator (Fig. 11-7A). The gray crescent appears to function as an organizer in the sense that it in some way determines the plane of the first cleavage division, which precisely bisects the zygote through the center of the crescent (Fig. 11-7B).

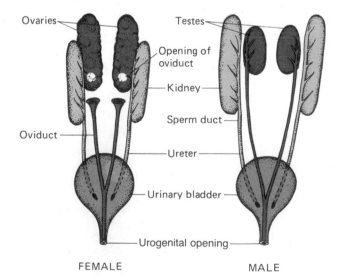

FEMALE   MALE

FIGURE 11-6
Vertebrate urogenital system.

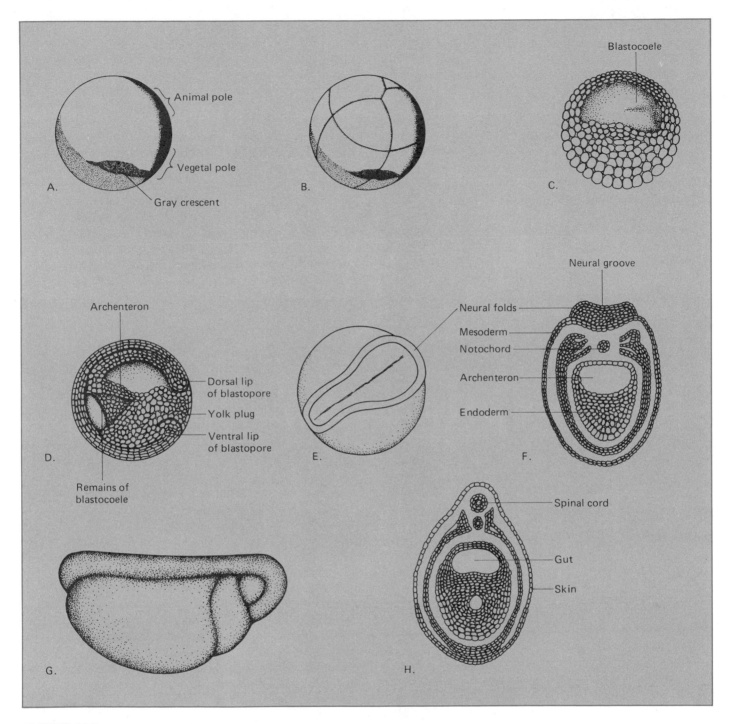

FIGURE 11-7
Early embryology of the frog. **A.** Zygote. **B.** Eight-celled embryo. **C.** Blastula. **D.** Gastrula. **E.** Neural fold stage, surface view. **F.** Neural fold stage, cross section. **G.** Surface view of later embryo. **H.** Cross section of later embryo.

Examination of a developing frog embryo shows that the cleavage of the zygote produces a somewhat "bottom-heavy" blastula due to the larger yolk cells of the vegetal hemisphere. Because of this disproportion of cells, the formation of morula and blastula stages involves a greater participation in the invagination process by the smaller cells of the animal hemisphere (Fig. 11-7B and C). Subsequently, a gastrula is formed when a fold of dorsal cells, called the **dorsal lip,** develops in the gray crescent region and acts as an organizer for later development of the embryo (Fig. 11-7D).

As a result of invagination and subsequent expansion of the dorsal lip, most of the yolk cells of the embryo become enclosed within the gastrula, and the germ layers are established. In the next step of development, the ectoderm in the vicinity of the dorsal lip forms a dorsal elliptical fold, the **neural fold** (Fig. 11-7E and F), and a valley between the folds called the **neural groove.** Eventually, the neural fold arches over the neural groove, meeting in a midline and producing a **neural tube** which later develops into the brain and spinal cord. Meanwhile, mesoderm develops between the ectoderm and endoderm and then separates into a central rod of cells that becomes the embryonic backbone, or **notochord,** and lateral blocks of cells that later become muscle and bone. The endoderm develops first into a primitive gut (archenteron) and later into the digestive and excretory organs of the embryo. At this stage the form and symmetry of the young frog become apparent. Included are a head, a tail region, and back (dorsal) and belly (ventral) areas (Fig. 11-7G and H). The archenteron develops a new mouth at one end, and the opening of the gastrula, the **blastopore,** contributes to the formation of the excretory opening. Shortly after this stage in development has been attained, the embryo develops external gills and becomes a self-sustaining larva, commonly known as a tadpole. The tadpole, depending on the species, spends a few weeks to nearly a year as an aquatic animal before absorbing its gills and tail, developing legs and lungs, and completing its metamorphosis.

PARENTAL CARE IN ANURANS

When frog and toad reproduction is mentioned, Fig. 1 clearly illustrates what usually comes to mind. A male frog, for example *Rana catesbiana,* the common bullfrog, takes a position in a shallow pond and gives its familiar deep-throated call. This serves in part to attract a female of the same species. The male grasps the female, and in mating by **amplexus** ("entwine") (Fig. 1), both sexes shed their numerous gametes directly into the water. There is no copulation and no parental care. The total parental investment in the next generation consists of sperm and egg contributions. The zygotes develop into an aquatic larval stage, the so-called tadpole. Larvae are tailed and have external gills. After a period of growth, the larvae metamorphose into small editions of adults, now capable of a terrestrial existence. Note that the life history is almost entirely an aquatic one and that reproduction is clearly tied to a watery medium. A survey of anuran reproduction indicates that about 90 percent of the world's species show this "classic" pattern of reproduction. The remaining 10 percent of the species, however, show a tremendous diversity of reproductive modes involving some kind of parental care.

A first step in the evolution of parental care is shown by *Nectophryne afra,* a species in which the male attends the eggs and larvae. While in attendance, the male swims in place and thereby directs water currents toward the developing eggs. Poor water circulation and lessened aeration would lead to abnormalities during development.

Building some sort of protective device, such as a foam or mud nest, or laying the eggs in a crevice or burrow, is another evolutionary step toward more complex parental care. The territorial male of *Hyla rosenbergi* actively defends his depression in the mud. The female lays her eggs in the mud nest, and larval development takes place there as well. In this way the larvae are given a measure of protection from their numerous enemies in the aquatic environment.

Hand in hand with parental care goes reduced clutch size. Female *Leptodactylus ocellatus* take larval protection one step further. When the tadpoles

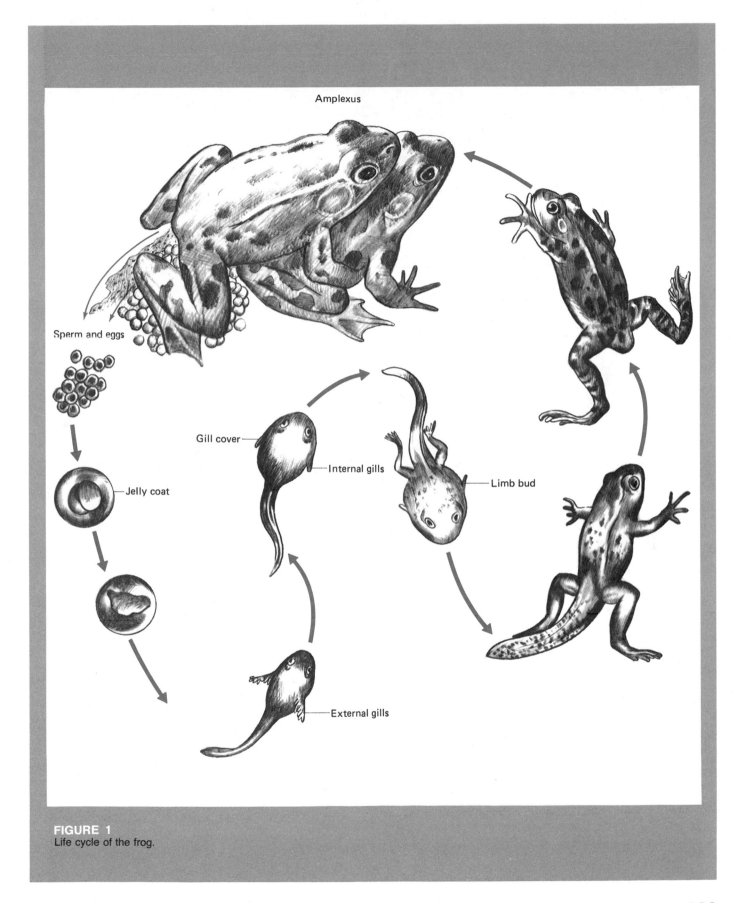

Amplexus

Sperm and eggs

Jelly coat

Gill cover

Internal gills

Limb bud

External gills

FIGURE 1
Life cycle of the frog.

leave their foam nest they form a school which feeds by ranging around the natal pond. Wherever the school goes, so goes the female in her attendant role as protector of the larvae, attacking would-be predators as they approach the tadpole school. This seems to be the only known case of parental protection of a free-swimming school of larvae. Although outmatched in egg production by the common bullfrog, the female *L. ocellatus* insures greater survival of her lesser number of young by vigilant antipredator behavior.

Parental attendance is not restricted to the aquatic environment. In the glass frogs *Centronella colymbiphyllum* and *C. valerioi* of Costa Rica, males of both species are nocturnal, and during the breeding season they defend territories consisting of leaves overhanging a stream. Attracted females lay their eggs on the underside of the leaves. At dawn, *C. colymbiphyllum* males leave their eggs and sleep during the day in nearby vegetation. *C. valerioi* males, by contrast, keep vigil near their eggs during daylight hours. The greater parental care shown by *C. valerioi* males pays off; more of their eggs reach the larval stage than do those of their close relatives.

Further refinement of parental care is shown by those species with terrestrial eggs and larval transport by either the male or the female. The female *Gastrotheca marsupiata,* the marsupial frog, has a pouch of loose skin on her back, and she places her eggs (assisted by the male) in the pouch, where development takes place. When emergence of the young is at hand, the female assists by further opening a slit in her pouch, thus allowing the young access to the water.

Charles Darwin, during the voyage of the *Beagle,* discovered a frog now named in his honor—*Rhinoderma darwini,* an inhabitant of Chile and Argentina. The male of this remarkable species guards the developing eggs until they are about to hatch. With his tongue the male then inserts several eggs into his grossly enlarged vocal pouch, where development continues. Metamorphosis takes place inside this pouch, and soon the offspring emerge as tiny replicas of their parents.

Only a small number of anuran species show parental care, but this brief review serves to indicate how diverse and increasingly complex is such added investment by the adults.

Reptiles and Birds Despite many differences, reptiles and birds share a number of common anatomical features. Both groups have achieved independence of water for reproduction. Their fertilization is internal and their embryos develop on land within shelled eggs, which generally are few in number in contrast to the thousands produced by female fish and amphibians.

A characteristic of the embryology of reptiles and birds is that the embryo develops within a small sac of fluid, the **amnion**, formed by the embryo itself as it grows (Fig. 11-8). The amnion is a watery cushion that protects the delicate embryo during its development, and it represents an important milestone in the transition of animal life from water to land. Instead of requiring a stream or pond in which to grow, the reptilian embryo in effect provides itself with a tiny aquarium in which to live during its development.

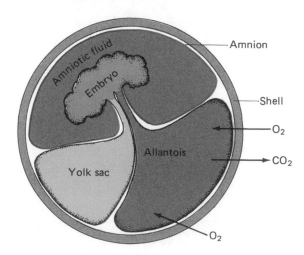

FIGURE 11-8
Reptilian egg and embryo.

Mammalian Reproduction Because of the requirements for bearing of young, reproductive organs in mammals, especially of the female, are considerably modified in comparison with those of lower vertebrates. With the exception of egg-laying monotremes (duckbilled platypus and echidna), mammalian reproductive systems are basically similar.

Reproduction in monotremes resembles that in reptiles and birds: a cloaca is present in which sperm is deposited by the male, and shelled eggs are laid and incubated. Female organs among all mammals include a pair of ovaries (one functional ovary in the duckbilled platypus), and paired oviducts, also called **fallopian tubes,** which lead to a uterus in which eggs produced by the ovaries are deposited and in which (with the exception of monotremes) they develop into embryos.

The mammalian uterus in most groups is partially divided into two chambers, but in primates is a single chamber. (Fig. 11-9). In all cases the uterus lies at the inner end of the vagina, a tube that functions both as a receptacle for the penis during copulation and as a birth canal. The outer opening of the vagina is surrounded by the external genitalia of the female, which include the **clitoris** and folds of skin termed **labia** (sing. **labium**). Collectively, the structures of the female external genitalia are called the **vulva.**

Male reproductive organs of mammals include a pair of testes suspended outside the coelom in a sac (**scrotum**); paired sperm ducts that conduct sperm and seminal fluids produced in part by the testes; **seminal, Cowper's,** and **prostate glands;** and the penis, through which the sperm duct passes (see Fig. 11-9). The penis is the male organ of copulation and serves to deposit sperm in the uterus via the vagina. Sperm of mammals, like those of most animals, are uniflagellate cells produced in great numbers (nearly a half-billion are released by the human male at each copulation). Following copulation, sperm deposited at the inner end of the vagina swim through the uterus and enter the oviducts, where eggs, if present, are fertilized.

Mammalian Embryology Ovaries of most mammals (monotremes are exceptions) produce eggs with little yolk. The eggs are shed periodically into the coelomic fluid, from which they immedately are collected by beating cilia in the funnel-like opening of the oviduct. As they pass down the oviduct, they are fertilized, provided sperm are present. Early embryological development occurs in the oviducts, and young embryos pass into the uterus, where they become **implanted** in uterine walls.

Recent human embryo transplants bypass fertilization and early embryo development in oviducts. Eggs are collected from the ovaries by minor surgery, placed in test tubes containing fluids like that in the coelom and oviducts, fertilized with sperm, and permitted to undergo several mitotic divisions before artificial implantation in the uterus.

Following implantation, the embryo forms an envelope, called the **chorion,** which develops fingerlike projections that invade the uterine wall and participate in forming a placenta. Within the chorion are three membranous sacs associated with the developing embryo: an **amnion,** which forms a fluid-filled cushion around the embryo; a **yolk sac;** and the **allantois** (Fig. 11-10). The yolk sac and allantois function as sources of food and oxygen for embryos of birds and reptiles, but in mammals, except for monotremes and marsupials, the yolk sac is nonfunctional and the allantois takes on the pro-

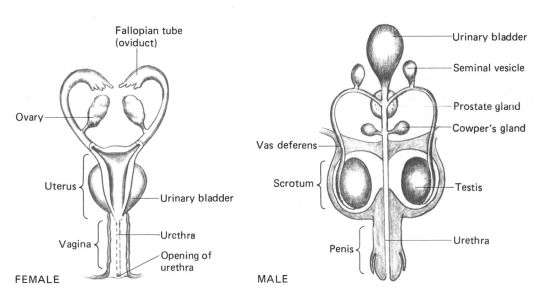

FIGURE 11-9
Mammalian reproductive systems. (From R. F. Oram, P. J. Hummer, Jr., and R. C. Smoot, 1979, *Biology: Living systems,* Columbus, Ohio: Charles E. Merrill Publishing Co.)

Fallopian tube (oviduct)
Ovary
Uterus
Vagina
Urinary bladder
Urethra
Opening of urethra
FEMALE

Urinary bladder
Seminal vesicle
Prostate gland
Cowper's gland
Vas deferens
Scrotum
Testis
Penis
Urethra
MALE

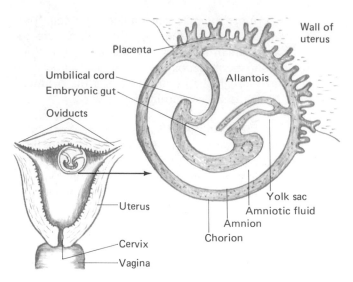

FIGURE 11-10
Early stage of the human embryo.

Labels: Placenta, Umbilical cord, Embryonic gut, Oviducts, Uterus, Cervix, Vagina, Wall of uterus, Allantois, Yolk sac, Amniotic fluid, Amnion, Chorion

cesses of transport of food and oxygen between the placenta and the body of the embryo.

The period of development following implantation is called **gestation**. Length of gestation is quite variable among the different mammalian groups, ranging from a few weeks in mice and other rodents to nearly two years in the elephant. In humans it is approximately nine months.

At birth the offspring is expelled from the uterus by powerful contractions of smooth muscle fibers in the uterine wall. The amnion ruptures, and **amniotic fluid,** the so-called birth water, escapes. The placenta is loosened and detaches from the uterus. The offspring, still connected by its **umbilical cord** to the placenta, is expelled through the vagina to the outside of the mother's body. The mother then severs the umbilical cord and (with the exception of human mothers) proceeds to eat the placenta and cord, thus regaining some of the nutrients expended during pregnancy.

In general, the state of development of the offspring at birth reflects the length of gestation; in those species with short gestation periods, the baby mammal is little more than a larva and requires constant care. Marsupial babies such as those of kangaroos and opossums are just a fraction of an inch long when born but crawl into the mother's pouch, aided by their precociously developed forelegs and possibly their sense of smell. When they reach the relative security of their mother's pouch, they locate and attach themselves to a nipple (actual fusion of the baby's lips and the nipple occurs) and continue their development for several more months. On the other hand, large animals such as horses, cows, deer, and other herbivores produce babies that within a few hours after birth are ambulatory and able to keep up with their parents and other adults. In this way they are better able to escape predation than if they were not so precocious and had to be left unguarded while their parents foraged. Conversely, the young of predators are commonly quite helpless at birth and require a great deal of parental protection.

Embryonic Homologies

The concept of **homology** is one of the guiding principles of biology. Structures having a common evolutionary origin, wings of birds and arms of humans for example, are said to be **homologous**. Both are modified forelimbs.

An interesting aspect of vertebrate embryology is that in their early development, vertebrate embryos look very much alike (Fig. 11-11). The lower vertebrates (fish and amphibians) and higher vertebrates (reptiles, birds, and mammals) all have similar embryological states. Later, the embryos of each group become more specialized and diversified, the more so in the higher forms. Biologists interpret these basic embryological similarities as evidence of a common evolutionary origin. In fact, at one time it was thought that all vertebrate embryos rather inflexibly followed the same embryological routes of development. This concept was dignified with the name **biogenetic law,** which in effect stated that in its embryological development an organism retraces its evolutionary history. That is, the early embryo of a human resembles a young fish because humans are descended from fish, and so forth. In reality, the human embryo never looks like an adult fish, or an adult amphibian, or any adult of any presumed ancestral type. Rather, all vertebrates diversify from homologous early embryonic states. What the biogenetic law does suggest is that evolutionary processes tend to be conservative, and it appears that developmental stages are not lost in evolution if they continue to be functional in development.

EXPERIMENTAL EMBRYOLOGY

As noted earlier, a burning issue in the pioneering days of microscopy was whether or not the egg and zygote were preformed. Even though it was soon realized that the zygote does not actually contain a minature sea urchin, starfish, or frog, many biologists of the day thought that the zygote was preformed in the sense that it had regions, each of which would eventually develop into a particular region or structure of the adult body.

FIGURE 11-11
Comparison of stages in the de-
velopment of five representative
vertebrate animals. (From R. F.
Oram, P. J. Hummer, Jr., and R.
C. Smoot, 1979, *Biology: Living
systems,* Columbus, Ohio:
Charles E. Merrill Publishing Co.)

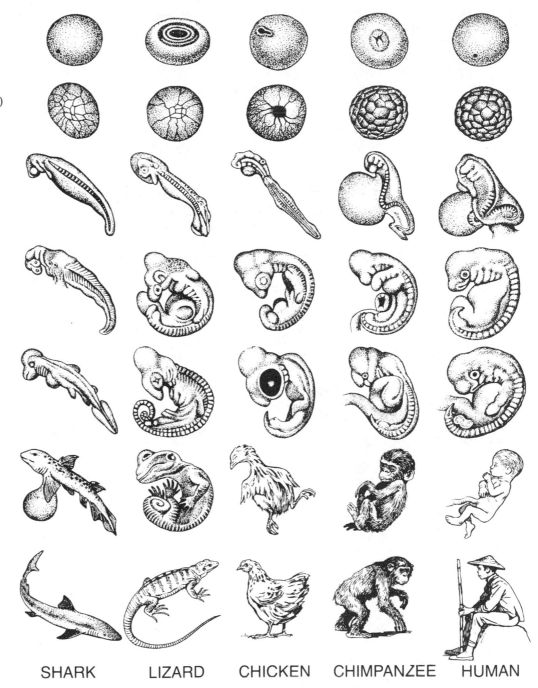

SHARK   LIZARD   CHICKEN   CHIMPANZEE   HUMAN

Theories of **preformation,** as well as alternative hypotheses, have been carefully examined and tested over
the years, resulting in considerable insight into developmental processes. A number of animal embryos have
served as model systems for such developmental studies, including *Hydra,* a relatively simple organism; echinoderms; arthropods; and of course vertebrates such as
frogs, chickens, and certain mammals. Echinoderms (sea

urchins in particular) were used in some important pioneering experiments in developmental biology. They
are readily obtainable, produce large numbers of eggs
which can be fertilized *in vitro,* and are easily studied under the microscope. They were used in some simple but
important experiments done in the early 1890s by a German named Hans Driesch (1867–1941) and continue to
be used in research.

DIVERSITY IN ANIMAL DEVELOPMENT

Based on early cleavage divisions, animals can be classified into two major groups. A majority of animal phyla exhibit **spiral cleavage** (Fig. 1) in which the plane of each cell division is offset slightly to the planes of previous divisions. Others exhibit **radial cleavage,** in which the planes of successive divisions are in a parallel line with or at right angles to those of earlier divisions.

Embryonic cleavage patterns provide some information about evolutionary relationships. All protostomes (flatworms, mollusks, annelids, etc.) exhibit determinate spiral cleavage, whereas all deuterostomes (echinoderms, protochordates, and vertebrates) exhibit indeterminate cleavage. This division is supported by comparative embryology.

The arthropods display a rather distinctive kind of cleavage in which the divisions of the zygote are superficial and the yolk mass at the zygote's center remains uncleaved. This type of development is thought to represent a special form of spiral or determinate cleavage and to relate the arthropods to the mollusk–annelid line of evolution.

Following development of a ball-shaped morula, the embryo becomes hollow (blastula), then indented, and forms a cuplike gastrula (Fig. 2). The identation **(invagination)** of the blastula that produces a gastrula results in formation of ectoderm and endoderm and produces the digestive cavity. Its open end, the **blastopore,** in jellyfish and other simple forms, may function as both mouth and

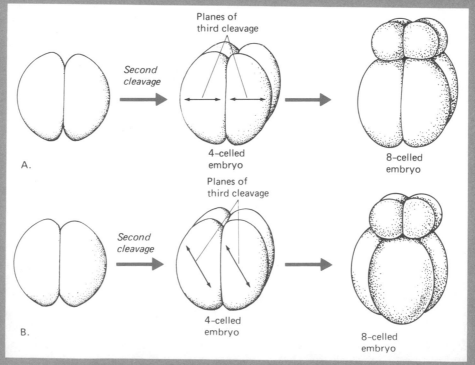

FIGURE 1
Cleavage of embryos. **A.** Radial. **B.** Spiral.

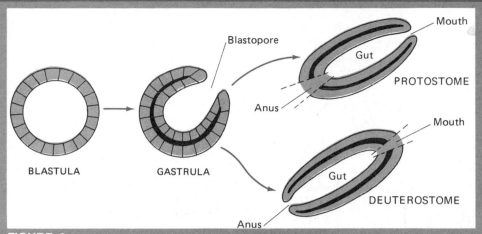

FIGURE 2
Gastrula formation and development of protostomes and deuterostomes.

anus. More complex animals go through additional developmental stages in which a second opening is formed. In one great group of animals, the blastopore becomes the mouth and the new opening is an anus. These animals are called **protostomes** ("first mouth") and include most of the invertebrates. The second group, called **deuterostomes** ("second mouth"), includes the chordates and a few members of the invertebrates. In deuterostomes the embryo develops a new mouth and the blastopore becomes the anus.

Up to and usually including the gastrula stage, animal embryos bear a marked resemblance to one another despite differences in cleavage patterns. Subsequently, deuterostome and protostome embryos become more and more diversified. Protostome embryos form mesoderm from a solid mass of cells produced by a single initial cell near the blastopore. Formation of a coelom (if present) occurs by splitting within the mesoderm mass. In deuterostome embryos, on the other hand, mesoderm arises in the form of pouches of the archenteron (primitive gut), and coelom forms from the saclike cavities of the mesoderm pouches (Fig. 3).

In later stages of animal development, many variations can be seen. Certain of these have been mentioned in earlier chapters, for instance, the ciliated larvae of mollusks, echinoderms, and protochordates; the metamorphosis of arthropods and in particular the dramatic changes in form exhibited by certain insects; and the adaptations in embryo development exhibited by placental mammals.

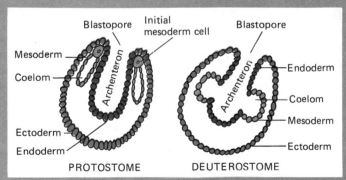

FIGURE 3
Origin of mesoderm and coelom in protostomes and deuterostomes.

Preformation and Epigenesis

Driesch demonstrated that in the sea urchin, at least, preformation did not occur, but rather that development was a building process, or **epigenesis** ("building upon"). The first cleavage divisions of a sea urchin zygote (and of many other animal zygotes) divide the zygote into halves, quarters, eighths, and sixteenths, and usually these divisions are very regular. Driesch and others wondered if such cleavages divided the zygote into cells of more and more limited developmental potential. For example, did the first cleavage divide the zygote into halves, each of which could form only the right or the left half of the animal? To answer this question, Driesch gently shook suspensions of sea urchin embryos (from the first cleavage on, the cells are considered to be part of the embryo) in successive cleavage division stages. The immediate result was that the cells separated from each other, apparently undamaged. Driesch then found that each cell of a two-cell embryo could develop into an entire sea urchin; moreover, this ability to reproduce an entire individual persisted through the 32-cell stage

of the embryo. Thereafter, the cells formed only badly distorted embryo fragments.

It should be noted here that not all species of animal eggs, zygotes, or early embryos are undifferentiated at the onset of development. Some eggs and zygotes, notably those of arthropods, are regionally preformed in the sense that each region gives rise to a specific part of the embryo. The term **mosaic** is applied to such eggs. Destruction or removal of a portion of the egg will result in a corresponding loss of part of the embryo. If after the first cleavage of a mosaic zygote the two cells are separated, each cell produces an incomplete embryo. Eggs of echinoderms and chordates are nonmosaic and hence lend themselves to the kind of experimentation pioneered by Driesch.

Hans Spemann (1869–1941), a brilliant student of Driesch, later extended the work of experimental embryology to amphibians, principally newts. Some earlier experiments with amphibian eggs and zygotes had suggested that they were developmentally preformed, and Spemann decided to test this further. He carefully tied loops of baby hair around newt zygotes, constricting

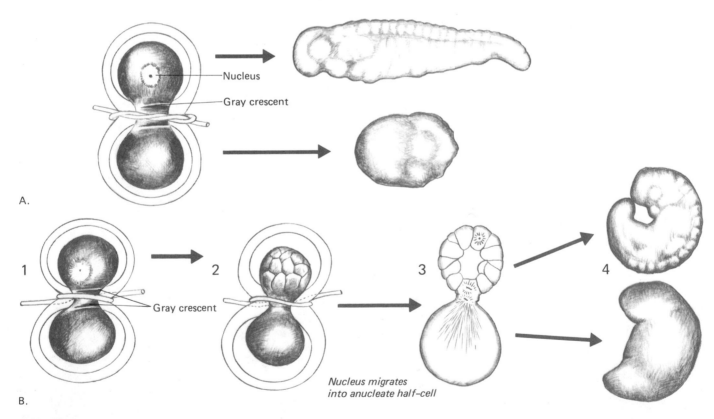

FIGURE 11-12
Spemann's grafting experiments in embryos of newts. **A.** Both nucleus and gray crescent are isolated in the same half-cell. **B.** Gray crescent is divided by a loop so that both half-cells have some gray crescent material.

"SCARCE AS HENS' TEETH"

Toothlessness is characteristic of all birds, although *Archaeopteryx,* a presumed ancestral bird, had teeth of the reptilian pattern. Recently, Edward Kollar and Christopher Fisher, of the University of Connecticut Dental College, have shown that chicken embryo cells have retained genes for tooth-enamel production even though chickens never produce teeth.

In toothed animals, two kinds of embryonic tissue participate in tooth development: **mesenchyme,** which produces the inner **dentin** of the tooth, and **epithelium,** which lays down the covering of **enamel.** Kollar and Fisher dissected epithelium from the head area of five-day-old chicken embryos and laid it over mesenchymal tooth-bud tissue from mouse embryos. The "sandwich" was implanted inside the eye of a "nude athymic mouse" (a strain used in grafting experiments because it does not reject implanted tissues). Here the implant grew and formed a complete tooth, demonstrating that chicken cells have genes for tooth production even though chickens lack teeth. Kollar and Fisher suggest that bird tooth genes normally are not expressed because of developmental changes that halt tooth formation. Earlier work has indicated that a dental mandible starts to form naturally in early stages of chick embryo development, then disappears.

Remarkably, teeth formed by combined chick–mouse tissue lacked the cusps typical of mouse molars and instead were of the peglike reptilian type.

them in two halves, and subsequently found that one half formed a complete embryo whereas the other did not. Significantly, the half-zygote retaining the capacity for normal development included the gray crescent, which was known to develop opposite the point at which the sperm penetrated the egg in fertilization. It seemed, therefore, that the sperm itself functioned as the preliminary **organizer** for later development; it since has been found that even an unfertilized amphibian egg can be made to develop by pricking it with a needle. Thus a rather simple external event may in some circumstances serve to initiate the development of an organism. In the language of embryology, such developmental stimuli are called **inductions.**

Shortly after fertilization, the amphibian zygote divides vertically, bisecting the animal and vegetal hemispheres and splitting the gray crescent zone into right and left halves. This is the first of many divisions of the zygote that eventually produce the multicellular adult. Spemann did experiments to test the significance of the orientation of this first division. He did this by tightening a noose of baby hair around the new zygote so as to bisect the gray crescent and produce two half-cells. Each

half-cell contained some of the gray crescent. but only one half-cell contained the single nucleus of the zygote (Fig. 11-12). The result, in this case, was mitotic division of the nucleated half-cell and subsequent normal embryological development. The half-cell lacking a nucleus did not divide immediately, but later a nucleus from one of the cells derived from the nucleated half-cell migrated into the nonnucleated half-cell. This half-cell then also divided many times to form an embryo. Spemann therefore reasoned that the potential for forming a complete animal was not solely a property of the zygote. This experiment also demonstrated the importance of the gray crescent as an organizer for normal embryo development.

In subsequent experiments Spemann exchanged pieces of embryos by **reciprocal grafts** (see Fig. 11-13). He found that if a bit of the tissue from the vegetal hemisphere of a newt embryo were exchanged with a bit from the animal hemisphere, each transplant would form a portion of the embryo that was in perfect agreement with its new position. Thus a piece from the vegetal region could participate in the formation of the dorsal part of the embryo instead of forming belly tissue as it might have done had it remained in its original location. How-

FIGURE 11-13
Reciprocal grafts in embryos of
amphibians. **A.** Reciprocal grafts
of younger, nondetermined cells.
B. Graft of cells from a presumptive gill region of a larva into a
non-gill-forming area of a second
larva.

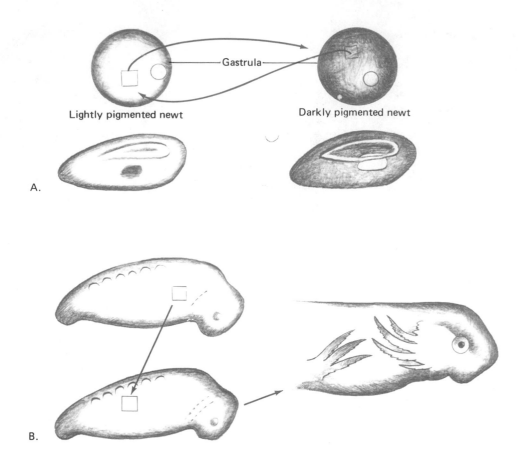

ever, when Spemann grafted cells containing gray crescent material from one embryo into another, so that now
the embryo had two gray crescent zones, "Siamese
twins" eventually developed. This showed that the gray
crescent zone has the ability to organize not only the
cells in its own position in an embryo but also in any
other position in an embryo into which it might be
grafted. The influence of the gray crescent is a classic
example of an organizer and is very reminiscent of the
action of the fine-line zone in a grafted *Stentor* (Fig. 6-2).

The interchangeability of bits of zygote persists
even in layers of cells in older embryos. Spemann found
that when patches of **presumptive** (i.e., destined normally to be a certain type of cell or tissue) belly skin of
a dark-skinned newt were grafted into a presumptive
head region of a light-skinned newt, the result was the
formation of a dark patch in the head region of the light-
skinned newt (Fig. 11-13A). The presumptive belly cells
obviously had not yet been programmed, that is, **determined,** to produce only belly skin. However, as the embryos grew older this flexibility of development of the
cells in various regions of the embryo tended to vanish;
cells seemed to become channeled into very specific
destinies. This is demonstrated by experiments in which
cells from a presumptive gill region are grafted into the
belly of the embryo. Under these circumstances a gill
will form in the region of the graft (Fig. 11-13B). Presumptive eye regions also can be grafted into unlikely
regions and will form mislocated eyes, and so on.

Nuclear Transplants

What happens to cells in older tissues that lessens their
adaptability when grafted into embryonic zones other
than their own? It is apparent that when cells have
reached the developmental stage in which they form the
kind of organ they were destined to produce, they have
become specialized (**differentiated**) and have lost much
of their ability to diversify.

At one time it was thought that irreversible
changes in the nuclei or in the chromosomes had occurred in determined cells. This appeared to be a logical
conclusion because the role of the nucleus in development seemed to be clear-cut. The nucleus is, after all,
the site of all or at least a majority of the genetic information of the cell; hence a change in the genetic nature
of the nucleus should control the destiny of a cell. However, a series of experiments, initiated by R. W. Briggs
and T. J. King in 1952 at the University of Indiana and
extended by them and many other workers, have sug-

gested that the nucleus may have less to do with the determined state of development than formerly thought. Briggs and King by delicate dissections were able to remove the nucleus from an unfertilized frog's egg and substitute the nucleus of a blastula cell. In many such cases, eggs that received nuclear transplants went through all the normal stages of embryo development and formed normal adult frogs. However, in other experiments in which nuclei from cells of older, well-differentiated embryos were used, Briggs and King found that normal embryological development did not occur. Somehow, such older nuclei had lost their ability to support the whole program of development of the frog embryo. There is general agreement among animal embryologists that the nuclei of differentiated cells can lose some capacity to control all the steps of early embryological development. Indeed, there are recent reports that certain genes and blocks of genes (transposable elements) can shift positions on chromosomes. Such shifts can alter patterns of gene expression during development (see Chapter 5). On the other hand, some experiments have indicated that the nuclei of certain differentiated amphibian cells may retain full capability to support the growth and development of amphibian eggs.

J. B. Gurdon, of Oxford University, has found that nuclei of intestinal cells of tadpoles of the South African clawed frog, *Xenopus laevis*, sometimes permit normal embryological development when transplanted into unfertilized enucleated eggs (see Fig. 11-14). Because it is not always possible to be sure that enucleation of an egg is complete, a strain of donor nuclei having a **genetic marker** was used. In this case the genetic marker was the presence of only one nucleolus per nucleus in a mutant strain of *Xenopus* (the normal nucleus in *Xenopus* has two nucleoli). After nuclei of the uninucleolate mutant type were transplanted into eggs of the normal strain, researchers had only to examine cells of tadpoles developed from such eggs to tell whether their development actually had been directed by transplanted nuclei. If only nuclei having single nucleoli were present, then the transplant had been successful.

In Gurdon's studies, only about 1.5 percent of eggs with transplanted intestinal nuclei developed into adult frogs, and some critics of the work have suggested that these nuclei may have come from undifferentiated germ cells (i.e., cells that had retained some of the properties of undetermined cells) present among the differentiated intestinal cells. In any event, not all types of differentiated cells induced normal development of embryos in Gurdon's experiments. Yet even brain cells, which never divide or synthesize DNA, resumed some of the nuclear functions of zygote nuclei when injected into eggs. The assumption in such experiments is that the regulation of

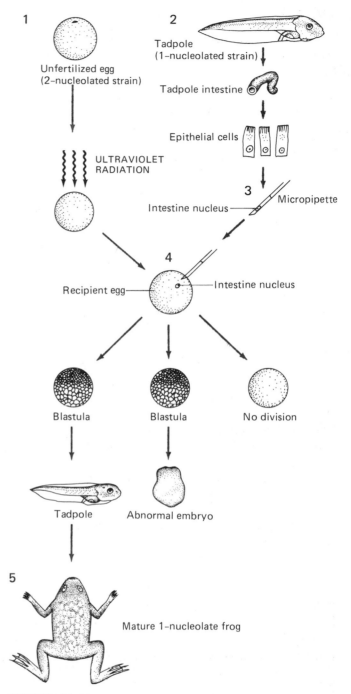

FIGURE 11-14
Nuclear transplant experiments with the African clawed frog, *Xenopus laevis*. (From J. B. Gurdon. Transplanted nuclei and cell differentiation. Copyright © 1968 by Scientific American, Inc. All rights reserved.)

DNA synthesis is controlled to some degree by the state of the cytoplasm. Cytoplasm in eggs, zygotes, or early embryos may be able to rejuvenate older nuclei from well-differentiated cells.

Cloning

Turning out identical copies of an organism is called **cloning.** If, for example, cells of an individual could be separated and cultured so as to develop as normal embryos, the result would be a **clone** of identical individuals. Such cloning has not been achieved in animals. However, nuclei from cells of a mouse embryo have been exchanged for nuclei of several mouse zygotes. These have been transplanted into a mouse uterus and have developed into identical young mice. This step toward cloning has many important, and possibly controversial, implications for human development. Successful test-tube fertilizations and uterine implants of human eggs have been done; the next step—the transplantation of nuclei from a human donor cell—is technically feasible.

Regeneration and Polarity in Animals

Regeneration occurs in multicellular organisms as well as in one-celled ones and seems more understandable than embryogenesis because there is an organized starting point in development.

If the "head" of a *Hydra*, with its mouth and circle of tentacles, is cut off and both the head region and the headless body are kept under observation, the body stump will be seen to grow a new head region, and the head piece will grow a new body. Thus the production of two hydras from one has been accomplished, in effect a simple clone. If hydras with rather long bodies are selected and cut in half at different levels, something else about development will be noted: no matter where the cut is made, cells at the cut surface of the body stump always form a new head, and those on the posterior end of the head fragment form a new base. This demonstrates that the same cells have the capacity to form either a head or a base, depending on their location with respect to the cut (Fig. 11-15). The conclusion, therefore, is that there are no cells in *Hydra* destined invariably to form heads or to form bodies. Instead, their course in regeneration depends solely on their position with respect to the rest of the regenerating *Hydra*. A term often used to describe this positional relationship is **polarity.**

It is believed that a **concentration gradient** of hypothetical "head-forming" substance exists in the body of *Hydra*. That is, this substance, whatever it may turn out to be, is thought to be more concentrated in the head region and gradually less concentrated in the direction of the foot region. The specific term used in developmental biology for a longitudinally distributed concentration gradient of some form-inducing substance, whether real or hypothetical, is **axial gradient.** If both the head and foot of a *Hydra* were to be cut off, which end would regenerate the new head? According to the con-

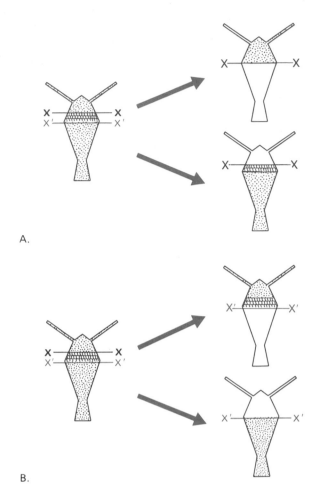

FIGURE 11-15
Regeneration in *Hydra.* In these diagrams, the effect of position upon the developmental pattern is illustrated. Cells in the regions between X——X and X'——X' exhibit potential to form either heads or bodies, depending upon their position in relation to the cut. If, as in **A,** the cut is at X——X, those cells form a new head, and the detached body forms a new head. If, as in **B,** the cut is at X'——X', those cells will instead form a new body, and the detached head will form a new head.

cepts of polarity and axial gradient, a new head would develop in the region where the head-forming substance is the higher, nearest the position of the former head end. And this, of course, is what actually happens; the headless, footless stump never "forgets" which end is which. Note, however, that a relative amount rather than any absolute or critical amount of head-forming substance is required for head formation. This explains why the same region of cells may in one circumstance form a base and in another case form a new head.

Animals even more advanced than *Hydra* retain impressive powers of regeneration. For example, starfish may be cut into several pieces, and each piece that retains a portion of the central body can grow into a whole new animal. Many fishermen kill starfish because they prey upon valuable shellfish and because they become entangled in nets. If fishermen chop up the starfish and throw the pieces back in the sea, as they have been reported to do, what have they accomplished in the way of eliminating their enemy?

Tadpoles are immature frogs; as they grow they form tiny legs which sometimes may be cut off accidentally. When this happens, the tadpole simply grows new ones. Adult frogs, however, cannot regenerate new limbs, and this loss of developmental potential seems to hold true for all higher animals, including humans. The young of advanced vertebrates such as birds and mammals also lack the power of regeneration exhibited by amphibians. However, an English physician, Cynthia Illingworth, recently has reported that the accidentally crushed fingers of young children can regenerate fingertips, nails included, provided the crushed fingertip is not surgically removed but is simply disinfected, bandaged, and left alone.

Adult humans cannot regenerate missing parts of limbs, but some regenerative capacity is retained by certain organs. For example, three-fourths of the liver may be cut away in a surgical operation, and as long as it has not been too seriously damaged by old age or disease, it will regenerate to normal proportions. The human kidney will not regenerate if removed totally or in part, but if one kidney is removed the remaining one may enlarge and successfully cope with the functions of both. A dream of physicians and surgeons is that someday the means will be discovered for regenerating all damaged or lost organs.

SENESCENCE

It has been said that death begins at birth; this of course is an exaggeration, even though some evidence of cell death can be observed in a newborn infant. However, it is no exaggeration that gradual death, or **senescence,** begins when an individual's period of growth and development is largely over (in humans, usually in the mid-twenties but sometimes earlier).

Senescence can be observed in every multicellular organism of determinate development, and even the open-ended or indeterminate development of such organisms as trees shows evidence of the wear and tear that eventually brings life to an end. The study of human senescence is of great public interest because it is a societal as well as an individual phenomenon. Why do some individuals age less rapidly than others? What factors contribute to senescence? Can it be delayed or even prevented?

Today it is well known that radiation from X rays, atomic explosions, and solar energy can accelerate the mutation rate of cells and lead to accelerated aging. Random gene mutations probably also lead to senescence, as cells accumulate more and more mutant genes and bodies carry an increasing burden of less efficient cells. It has been suggested that every individual organism possesses a biological time clock in the form of random genetic mutations, which accumulate over time and eventually produce death of the organism. Not all such mutations are spontaneous; many are caused by radiation, and others are the result of exposure to mutagenic chemicals in the environment.

Cells are known to have cellular repair mechanisms capable of excising damaged DNA and splicing in new DNA, and evidence now is accumulating that some organisms live longer because of more efficient or more active DNA repair. It is possible that such differences in genetic repair may account for variations in human longevity.

Aside from avoiding exposure to mutagens, or otherwise living temperately, what can be done about aging? Very little, it seems; hormones, gland extracts, and other treatments appear to have little delaying effect on long-term degenerative processes. Some hope has been expressed that use of antioxidant substances including vitamins C and E may suppress intracellular oxidations associated with aging, but evidence of this is lacking. It is ironic that science in the most advanced cultures is giving increased attention to delaying the aging process at a time when human populations in most of the world's nations are already in excess of resources needed to support them.

SUMMARY

As a result of nearly a century of observation and experimentation, a number of developmental concepts are beginning to emerge. Obviously, the study of the development of organisms is a complicated one, involving genetics, biochemistry, physiology, anatomy, and classification.

Study of the development of zygotes of many organisms has shown that the development of a multicellular organism is gradual, and that the first cells each may have the potential to form a complete organism, provided they are separated from one another.

At what point in development do cells become programmed in such a way that they will produce only certain parts of the whole organism? Simple organisms seem to have cells that are developmentally plastic, but in advanced animals, especially, a point in development is reached at which cells have only a limited developmental potential. This is shown particularly well in humans, whose amputated arms, legs, and so on, are not regenerated. Nevertheless, certain organs such as the liver have the ability to regenerate as much as three-fourths of lost cells. In amphibians such as frogs, salamanders, and newts, young individuals retain a limited power of regenerating lost limbs and other parts; these capacities are not present in adults.

Nuclear transplant studies have shown that the nuclei of certain cells in relatively later developmental stages have some capacity to direct the development of the whole organism. In general, it is thought that as cells mature, they lose much of their potential for unlimited development. However, in some organisms isolated cells from mature tissues still may divide and form new individuals. Plant cells, in particular, have been shown to retain almost unlimited capacity for renewed development. Although the culture of human cells has not yet led to similar developments of highly organized multicellular states, it is hoped that eventually it may be possible by such processes to at least replace lost or damaged body parts.

KEY WORDS

penis	embryogenesis	biogenetic law
sperm duct	zygote	epigenesis
ovary	cleavage division	induction
testis	morula	nuclear transplant
vagina	blastula	cloning
arthropod	gastrula	regeneration
reptile	germ layer	polarity
bird	egg layer	axial gradient
mammal	womb bearer	protostome
amphibian	gray crescent	deuterostome
preformation	dorsal lip	senescence
differentiation	organizer	

QUESTIONS FOR REVIEW AND DISCUSSION

1 What is meant by "preformation" of an organism? Give a fictitious example.

2 How do the inner and outer layers of a multicellular animal originate?

3 What is meant by the term "germ layer"? What are the principal germ layers of the animal and into what do they develop?

4 What developmental function is demonstrated by the dorsal lip of a developing frog embryo?

5 Can human cells be cultured and used to grow new humans in the same way as plant cells may be cultured to give new plants? Discuss this both in biological and sociological terms.

6 What is the significance of the nuclear transplant work done with *Xenopus* by Gurdon and his associates?

7 Can humans regenerate missing structures? Explain.

8 What are some significant aspects of spiral versus radial cleavage in embryos?

SUGGESTED READING

BALINSKY, B. I. 1981. An *introduction to embryology*, 5th ed. Philadelphia: Saunders College Publishing. (A comprehensive, up-to-date textbook on animal embryology.)

BORGENS, R. B.; J. W. VANABLE, JR.; and L. F. JAFFE. 1979. Bioelectricity and regeneration. *BioScience* 29(8):468–74. (Electrical currents and potentials play an important role in the development of multicellular organisms and also in the regeneration of amputated salamander limbs.)

GIERER, A. 1974. *Hydra* as a model for the development of biological form. *Sci. Amer.* 231(6):44–54. (Interesting experimental approach. Strong advocacy of biochemical gradients in establishing polarity.)

GORDON, R., and A. G. JACOBSON. 1978. The shaping of tissues in embryos. *Sci. Amer.* 238(6):106–13. (Computer simulations of development aid in the analysis of forces at work in shaping the patterns of embryo growth.)

GRAY, G. W. 1957. The organizer. *Sci. Amer.* 197(5):79–88. (Excellent simple diagrams of amphibian development. Good reference to early work of Driesch, Spemann, and others.)

GURDON, J. B. 1968. Transplanted nuclei and cell differentiation. *Sci. Amer.* 219(6):24–35. (An account of nuclear transplant experiments in *Xenopus* using genetic markers and labeled DNA.)

HOPPER, A. F., and N. H. Hart. 1980. *Foundations of animal development*. New York: Oxford University Press. (A well-developed exposition of experimental and descriptive animal embryology.)

POLICANSKY, D. 1982. The asymmetry of flounders. *Sci. Amer.* 246(5):1116–22. (Describes an interesting case of secondary imposition of asymmetry upon a basic bilaterally symmetrical organism.)

WOLPERT, L. 1978. Pattern formation in biological development. *Sci. Amer.* 239(4):154–64. (Describes the role of cell interactions in determination of developmental pathways.)

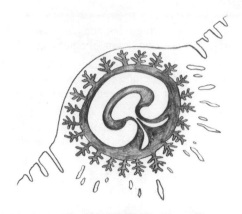

Human Reproduction 12

S everal years ago *Pioneer 10* was sent into deep space with the intention that it would wander eternally in the outermost reaches of the universe. On the chance that it might be intercepted by some alien civilization, perhaps greatly unlike earth's, a message was engraved on a sheet of titanium attached to its side. This message is reproduced in Fig. 12-1. Included was a map of our solar system, with the earth indicated as the rocket's origin. Then, to tell the strangers what we looked like, naked male and female figures were shown. These figures illustrate not only the physical features of the human race but also the distinctive physical characteristics of the human male and female. These characteristics and underlying sexual processes are the subject of great interest, curiosity, speculation, scientific investigation, commercial exploitation, controversy, and legal action.

HUMAN SEXUALITY

Human males and females differ in a number of sexual characteristics, which are divided into **primary** and **secondary** traits. Primary characteristics are those characteristics specifically related to structure and functioning of ovaries and testes. Secondary sex characteristics are

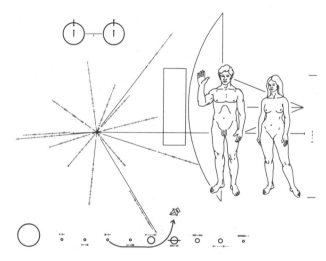

FIGURE 12-1
Message of *Pioneer 10* space probe.

those traits common to both sexes but expressed differently in each sex. Among them are the amount and distribution of body hair, differences in the distribution of body fat deposits, general development of bones and muscle, the pitch of the voice, and more specific differences in breast development and pelvic width.

Primary and secondary sexual characteristics are under the control of hormones secreted by the pituitary, the gonads (ovaries and testes), the adrenals, and the placenta of the pregnant female. The hypothalamus is the main integrating center for nerve impulses relating to hormonal control of sex. Hypothalamic hormone-releasing factors stimulate the pituitary, which in response secretes gonadal-stimulating hormones.

Sex Chromosomes

The initial development of sexual characteristics in humans, as in most animals, is determined by genes present in female- and male-determining sex chromosomes designated X (female) and Y (male) (see Chapter 7). The X chromosome is common to both sexes, and each sex normally has only one active X per cell. The male has a single X (is XY), whereas the female has two (is XX).

Pituitary Secretions and Sex

Two important pituitary hormones, secreted in both males and females, are follicle-stimulating hormone (FSH) and luteinizing hormone (LH). Both terms describe responses in which the ovary is the target organ; however, follicle-stimulating hormone also stimulates sperm production in the testes, and luteinizing hormone stimulates the development of androgen-secreting interstitial cells of the testes. Because follicle-stimulating and luteinizing hormones act upon gonads of males as well as those of females, they also are known as human gonadotrophic hormones (HGHs). In females, there is a third pituitary hormone, prolactin, which induces the synthesis of milk following childbirth. Prolactin also is secreted in males, but the function is unknown.

Gonads

As noted in Chapter 10, the gonads are mixed glands which under the control of the pituitary produce and secrete sex hormones. The testes, in addition to sperm production, produce and secrete androgens, the most important of which is testosterone. The ovaries, in addition to egg production, produce and secrete estrogen and progesterone. The roles of these hormones in reproduction are discussed in the following sections.

Sex Hormones and the Adrenals

The adrenal cortex produces and secretes small quantities of estrogens, androgens, and progestogens, but usually not in sufficient amounts to influence sexual development by themselves. However, in some recorded cases in which overgrowth of the adrenal cortex has occurred, overmasculinization of males and females has resulted.

The presence of adrenal estrogens is one reason that castration of young boys results in some degree of feminization. At one time castration was done with the object of producing just such results—for maintaining soprano voices of boys trained to sing in cathedral choirs. Such castrations have not been practiced for many decades anywhere in the world, but damage by accident or disease sometimes necessitates removal of the testes. In such cases the male hormone, testosterone, can be administered to compensate for loss of natural androgens.

FEMALE REPRODUCTIVE SYSTEM

Structure and Function

The reproductive system of the human female includes the clitoris, vulva, vagina, uterus, oviducts, and ovaries (Fig. 12-2). Externally, the female genital organs are the vulva and clitoris. The vulva is composed of two parallel folds of flesh, the labia majora, which are united at the front in a fleshy mound, called the mons pubis or mons veneris, and are confluent at the sides and back with the perineum, an area of skin, connective tissue, and smooth muscle.

The labia majora partially enclose two smaller folds, the labia minora, which extend forward and form a hood around the clitoris, which is developmentally homologous to the male penis. Just to the rear of the clitoris is the urethral orifice, which is the opening of the urinary canal (urethra) from the bladder. The urinary bladder itself is placed lower in the abdominal cavity and is smaller in women than in men. The vaginal opening is located to the rear of the urethral orifice, in the cleft between the labia minora. In women who have not had sexual intercourse, the opening may be partially closed by a thin membrane, the hymen, which differs in size and durability from individual to individual. The hymen is ruptured at the first sexual intercourse, but can be broken by accident or removed surgically.

Internally, the vagina, an extensible, muscular tube about 10–13 cm (4–5 in) long, extends upward and somewhat rearward from the vulva to the uterus. The uterus is about the size and shape of a pear and projects slightly into the vagina. This projecting end, the cervix, contains the opening into the uterus. Two oviducts (also called Fallopian tubes), each about 10 cm (4 in) long, have fringed, funnel-shaped openings in the body cavity and lead to the upper end of the uterus (Figs. 12-2B and 12-7). Their openings in the body cavity are lined with ciliated epithelium and collect eggs released by the ovaries. The two ovaries, to the right and left of the uterus, are each roughly grape-sized and are attached to a thin

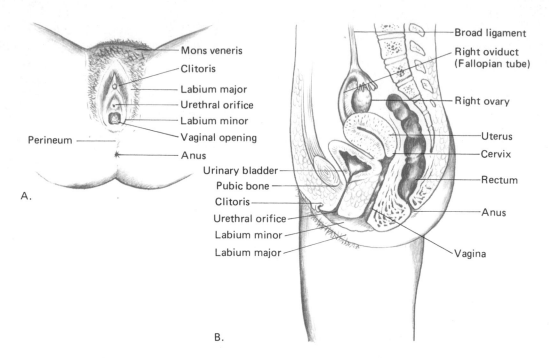

FIGURE 12-2
Reproductive system of the human female. **A.** Superficial view. **B.** Sagittal section.

A.

Mons veneris
Clitoris
Labium major
Urethral orifice
Labium minor
Vaginal opening
Perineum
Anus

Broad ligament
Right oviduct (Fallopian tube)
Right ovary
Uterus
Cervix
Rectum
Anus
Urinary bladder
Pubic bone
Clitoris
Urethral orifice
Labium minor
Labium major
Vagina

B.

fold of epithelium called the **broad ligament,** which also supports the oviducts.

Both male and female reproductive systems are similar during the early development of the human embryo (Fig. 12-3A) and consist of a **genital tubercle** and a pair of **genital (labio-scrotal) folds** extending to the rear of the genital swelling on either side. The genital swelling is cleft by the urethral orifice. In males this cleavage closes up, forming a tube (the **penis**), with an opening at the tip (**glans**), and the genital folds fuse together to form the **scrotum** (sac in which the testes are enclosed) (Fig. 12-3B). The scrotum, however, does not at this time contain the testes; these form in the abdominal cavity

and pass through the **inguinal canal** (a duct between the body cavity and scrotal sac) into the scrotum just before birth. In the developing female, the genital cleft lengthens and becomes the labia minora, and the labio-scrotal folds become the labia majora. The tip, or glans, of the genital tubercle persists and becomes the clitoris (Fig. 12-3C).

Formation of Eggs

The ovaries of a female baby at birth already contain a lifetime supply of egg-forming cells. These cells, the **primary oocytes,** number about a quarter of a million in

FIGURE 12-3
Development of human genitalia. **A.** Unspecialized stage in a 29 mm (1.15 in) human embryo (beginning of third month). **B.** Male fetus at about four months. **C.** Female fetus at about four months. (Redrawn from *An introduction to embryology,* 4th ed., by B. I. Balinsky. Copyright © 1975 by W. B. Saunders Company. Reprinted by permission of W. B. Saunders Company, CBS College Publishing)

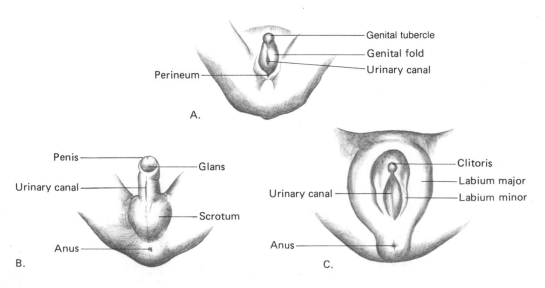

Genital tubercle
Genital fold
Urinary canal
Perineum
A.

Penis
Urinary canal
Glans
Scrotum
Anus
B.

Clitoris
Labium major
Labium minor
Urinary canal
Anus
C.

each ovary, obviously many more than will be utilized during the lifetime of an individual. Normally, one oocyte gives rise to one egg each month over a period of about 35 years, except during pregnancy when none at all are formed.

During the monthly course of the development of an egg (the **ovarian cycle,** also called the **estrous cycle;** see Fig. 12-4), a primary oocyte undergoes the first of two meiotic divisions. At the onset of this development, the primary oocyte is enclosed within a layer of cells, an assemblage known as an **ovarian follicle.** During the first 14 days, the primary oocyte divides once meiotically, forming a **secondary oocyte** and a small **polar body.** Meanwhile, the ovarian follicle rapidly expands and becomes a fluid-filled sac. At this point, on or about day 14, the follicle breaks and releases the secondary oocyte and its attached polar body. This is called **ovulation.** The oocyte is swept into the oviduct, and there—if sexual intercourse has occurred within the past few days—encounters sperm. If a sperm penetrates it, the oocyte undergoes the second meiotic division, forming an **ovum** (egg) and a second polar body. The polar bodies are nonfunctional and soon disappear. Shortly after the second meiosis, the egg nucleus fuses with the sperm nu-

cleus and a **zygote** is produced. The zygote begins to develop as an **embryo** while traveling down the oviduct, and by the time it reaches the uterus and is implanted in the uterine wall, it has reached the **blastula** stage (see Figs. 11-4 and 11-7).

Meanwhile, the ruptured follicle begins to differentiate into another endocrine tissue. After healing over, it becomes a hormonal secretory structure called the **corpus luteum,** which continues to grow for 10–12 days. As it develops, it secretes hormones that stimulate the uterine wall to become receptive to implantation of the embryo. If, however, fertilization has not occurred and there is no embryo to become implanted, the corpus luteum ceases to produce hormones and quickly degenerates. This in turn causes the lining of the uterus to break down and be discharged through the cervical canal and vagina, a process called **menstruation.**

Hormonal Control of the Ovarian Cycle

The ovarian cycle is regulated by hormones produced in the pituitary gland, the ovaries, and the uterus. Adrenal hormones may modify the cycle but do not play a major role in its normal regulation. At the beginning of **pu-**

FIGURE 12-4
Human ovarian cycle.

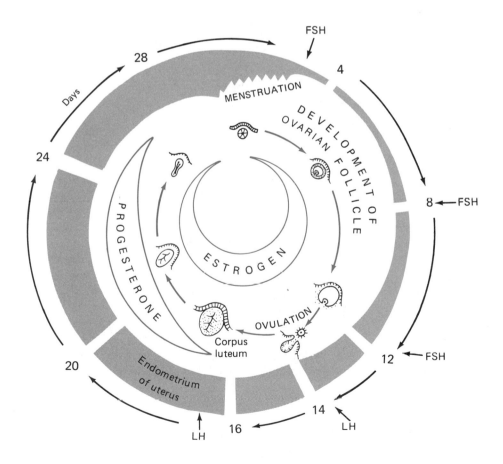

berty (literally, the development of hair, i.e., pubic hair), usually between the ages of 9 and 13, three sex hormones are secreted by the anterior pituitary gland, under the control of the hypothalamus. They are follicle-stimulating hormone, luteinizing hormone, and prolactin (Table 10-1).

Follicle-stimulating hormone (FSH) promotes the growth of ovarian follicles. The follicle nourishes an oocyte and also produces and secretes estrogen, which causes the inner wall of the uterus, the endometrium, to begin to thicken and become spongy. As the level of estrogen continues to rise, it reaches a concentration in the blood sufficiently high to cause the pituitary to shut down its FSH production. At the same time, the high estrogen content of the blood triggers the release of luteinizing hormone (LH) from the pituitary. Due to the acute release (or surge) of luteinizing hormone, the ovarian follicle ruptures and an oocyte is released. Under the influence of this hormone, the cells of the ruptured follicle differentiate into lutein cells to form a corpus luteum.

A corpus luteum secretes estrogen and progesterone, which cause the endometrium to thicken further and secrete uterine "milk." As the level of progesterone increases, the production of luteinizing hormone and follicle-stimulating hormone is halted. If fertilization does not occur, the corpus luteum degenerates and progesterone and estrogen secretion decrease. In the absence of high levels of estrogen and progesterone, the endometrium breaks down and the resulting blood and debris is discharged as menstrual fluid. The menstrual fluid contains cellular debris, interstitial tissue fluid, and other secretions as well as blood. Normally, the total fluid is about 250 milliliters (1 cup), of which about one-third actually is blood.

Menopause

At age 45–55, a gradual shift in the production of sex hormones occurs in a woman's body, and she enters the end of her reproductive years. This hormonal change usually is signaled by irregularities in the menstrual cycle, leading to the cessation of menstruation, or menopause. The ovaries no longer respond to follicle-stimulating hormone and luteinizing hormone, levels of estrogen and progesterone decline, and FSH and LH concentrations rise. These changes in hormonal levels may cause physiological and psychological symptoms until the body becomes adjusted to the new set of conditions. In only about 10 percent of the female population are these symptoms severe enough to require medical treatment.

MALE REPRODUCTIVE SYSTEM

Structure and Function

The external male genitalia are the penis and scrotum. The latter contains a pair of testes (sing. testis), also called testicles. Each testis consists of pie-shaped lobes which contain highly coiled seminiferous tubules, within which sperm are produced. Between the seminiferous tubules are interstitial cells that produce and secrete testosterone. When sperm leave the testes, they enter and collect in a convoluted duct called the epididymis, located on the borders of the testes. The sperm are stored there in a nonmotile state until ejected during sexual excitement. The epididymis of each testis connects with a sperm duct, or vas deferens, which extends from the interior of the scrotum through the inguinal canal into the body cavity (Fig. 12-5A).

Within the body cavity, the paired vasa deferentia join with ducts from a pair of accessory sex glands, the seminal vesicles, to form a common sperm duct that enters another accessory sex gland, the prostate, where it joins with the duct of the urinary bladder. The urinary duct, or urethra, conveys sperm and urine, although not at the same time. During periods of sexual excitement, waves of smooth-muscle contractions convey sperm along the vasa deferentia to the urethra, where they are mixed with secretions of the seminal vesicles, the prostate, and a pair of Cowper's glands at the base of the penis. The result is semen, a viscous mixture containing motile sperm, lubricants, buffers that regulate pH, and sugars that supply energy for sperm motility.

In the penis the urethra passes through the center of a cylinder of spongy tissue, the corpus spongeosum (Fig. 12-5B). Two other cylinders of similar spongy nature, the corpora cavernosa, also are present in the penis, and all three become engorged with blood during sexual excitement. As a result, the penis elongates and stiffens, producing an erection. The nonerect penis varies greatly in size from individual to individual and even in the same individual, depending on emotional conditions and other factors. When erect, the average penis measures about 15 cm (6 in); the range is 12–30 cm (4.5–12 in).

The penis consists of the shaft and a hood (the prepuce, or foreskin), which covers the bulbous glans at the end. The glans is thin-skinned and richly supplied with nerve endings. A common surgical operation, called circumcision, removes the prepuce and leaves the glans exposed. The major reason for doing this, aside from certain religious practices, is that secretions from cells in the glans, accumulating under the prepuce, may provide

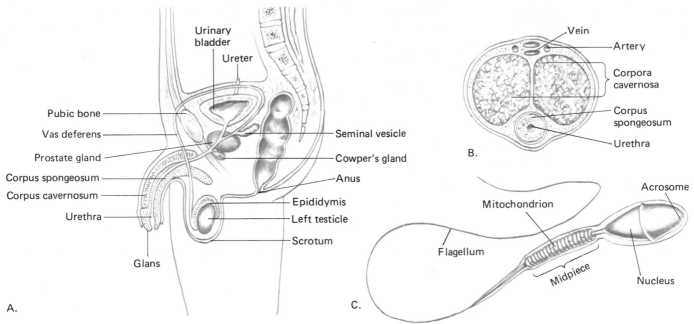

FIGURE 12-5
Reproductive system of the human male. **A.** Sagittal section. **B.** Cross section of the penis.
C. Human sperm.

a medium for bacterial growth, causing inflammation and adhesions. The importance of circumcision from a health standpoint probably has been overemphasized, as the prepuce can easily be kept clean.

Formation of Sperm

The seminiferous tubules are lined with cells in various stages of development. In cross section, from the periphery inward, the **spermatogonia** can be seen. These are proliferating cells that give rise to additional spermatogonia and also to **primary spermatocytes**. These divisions are mitotic. Primary spermatocytes divide meiotically (meiosis I) to form a layer of **secondary spermatocytes**, each $2n$ primary spermatocyte producing two $1n$ secondary spermatocytes (refer to the diagram of meiosis presented in Fig. 6-13). Each secondary spermatocyte undergoes meiosis II to form two $1n$ **spermatids**. Spermatids are spherical $1n$ cells that become differentiated progressively into sperm, a layer of which lines the seminiferous tubule. During this differentiation, a long tail is formed that consists of a basal part encircled with mitochondria (the **midpiece**) and a long flagellum. The head of the sperm is flattened and somewhat elongate. It is almost entirely composed of a nucleus enclosed within a cap (**acrosome**) of enzymes. The acrosome is a specialized Golgi body in which a concentration of enzymes is localized.

SEXUAL INTERCOURSE

Some form of erotic behavior usually occurs as an introduction to the sex act itself. This initially may be subtle, but eventually is characterized by overt actions. A state of readiness for intercourse is attained in males when the penis is erect. At this time, the scrotal muscles contract and pull the testes upward at the base of the penis; the testes become enlarged by as much as 50 percent.

In the female, secretion of a mucuslike fluid by the vagina and labia minora occurs, the labia majora and minora become engorged with blood and enlarge, and the clitoris may enlarge slightly. The vagina elongates, and the uterus becomes slightly elevated. **Copulation** then may begin, ending in a nervous and muscular reaction, or **orgasm** (also called a **climax**).

Male orgasm is necessary for the biological completion of sexual intercourse, because it is accompanied by the discharge of sperm (**ejaculation**). During the male orgasm, muscular contractions at the base of the penis and in the scrotum and prostate occur in a series of rapid spasms, with the result that spurts of semen are ejected. The female orgasm consists of a series of muscle contractions in the uterine wall and the lower vagina. The upper vagina expands in a process known as the **tenting effect** that is believed to aid in the transport of sperm into the uterus. The female orgasm, unaccompanied by ejaculation of fluid, is not required for **fertilization**.

ARTIFICIAL INSEMINATION AND TEST-TUBE BABIES

Artificial insemination has long been practiced in the livestock industry, making possible the mating of a prize bull or boar with hundreds of cows or sows. At first, the process was simply one of collecting the semen of the donor, diluting it with a physiological solution and injecting it into receptive females. This practice also was and still continues to be employed with human females who wish to conceive and bear a child. Some 20 years ago, it was discovered that if sperm were mixed with glycerol, they could be quick-frozen at $-79°C$ (by placing a drop of the sperm suspension on a block of solid CO_2 ["dry ice"]) and stored in their frozen state. Upon thawing, the frozen sperm become active and capable of fertilizing an egg, as though nothing had intervened. A refinement of this method is to store the quick-frozen sperm in liquid nitrogen at $-196°C$; they can be kept for many years.

As one can imagine, quick-frozen sperm have been invaluable to the livestock industry. A prize bull or stallion is a perishable asset, but his sperm can be collected and preserved long after the animal is gone. This procedure also is becoming important in preservation of endangered wild animals, where males and females may be separated by great distances in zoos and preserves. Most interesting from the standpoint of human reproduction is that frozen sperm can be collected from a potential father for future use—after a vasectomy, for example, if he should so desire. The Nobel laureate H. J. Muller, codiscoverer of radiation-induced gene mutation, proposed that sperm banks from great men might be set up to be available to women who might want a child by a Nobel Prize winner, an Olympic athlete, or a President. One such "Nobel Prize" bank was recently set up in California and the first child of it, a girl, was born in 1982. A son born to another sperm recipient was recently reported to be showing signs of musical precocity. Such schemes are sure to attract attention and may be genetically quite sound, although sounding a bit like the plot of the book "The Boys from Brazil," with its clone of youthful Hitlers. Some problems come to mind: (1) No one has yet defined genius on a genetic basis, nor can anyone predict which of the countless numbers of sperm involved will bear the "right combination" of genes. (2) If such plans are widely practiced, the human gene pool could be considerably reduced; quite possibly frequencies of deleterious genes could be amplified and advantageous ones could be reduced or lost.

Test-tube fertilization has in the past several years grown from the laboratory-experiment stage to a practical reality. The procedure involves several manipulations, the first of which is inducing supraovulation in the female by administering gonadotrophic hormones. (This procedure by itself has been used to "cure" some cases of female sterility; it also has produced a number of multiple births.) Test-tube fertilization then is employed to bypass conditions that prevent normal transport and fertilization of the egg (blockage of oviducts, chemical imbalances, etc.) This procedure is outlined in Fig. 1.

As the figure shows, eggs (actually secondary oocytes—the stage in which release occurs) are collected, either from the oviducts or, if these are nonfunctional, directly from the ovarian follicles. The eggs are then placed in a vessel of tissue-culture medium where they mature further. Meanwhile sperm are collected from the father-to-be and activated by placing them briefly in the mother-to-be's (or some other) uterus, then removing them and adding them to the culture

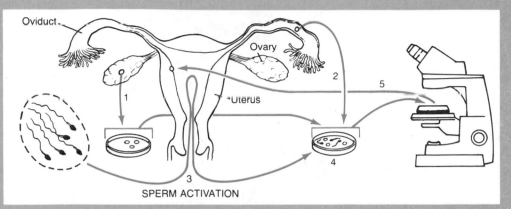

FIGURE 1

Test-tube fertilization. Multiple ovulation is induced by injection of gonadotrophins (1 and 2). The eggs may be collected from the ovaries (1) or, after ovulation, from oviducts (2); if (1), they require a maturation period in a tissue-culture medium. Sperm is collected from male donor, placed in uterus briefly for activation (3), then added to eggs from either (1) or (2) in culture medium where fertilization can occur (4). Culture of eggs and sperm is then examined for presence of zygotes, and if any are present a "healthy" looking zygote is introduced into the uterus of the mother where implantation may occur. From this point on, a typical pregnancy can follow.

of eggs. The eggs then are examined for evidence of fertilization and if fertilized are placed in the uterus of the mother. Since the embryo will be derived from the eggs and sperm of the desired parents, it is theoretically immaterial whether the "real" mother or a substitute (surrogate) mother receives the implanted embryo and nurtures it in her womb. The latter has given rise to the possibility that the practice of "wombs for rent" may become common without any stigma attached to either the real or the surrogate mother. It is, however, paradoxical that in an era in which widespread overpopulation, infant mortality, and childhood deprivation are common in many parts of the world, millions are spent on research and medical care to produce a numerically insignificant number of babies.

An apparently unforeseen complication involving test-tube fertilization and human embryo transplantation recently came to light. A wealthy American couple had deposited an embryo with the custodians of a deep-freeze embryo "bank" in Melbourne, Australia, apparently with the thought that it might be incubated later either in the womb of its real mother or in that of a surrogate. In the meantime the parents were both killed in an accident and the disposition of their embryonic offspring became the center of a legal question of far-reaching significance. Does this "orphaned" embryo have a "right to life"? Would disposing of it constitute a homicide? If incubated in a surrogate mother, would it inherit the family fortune?

PREGNANCY AND CHILDBIRTH

When the sperm are ejected into the upper vaginal tract, they move through the cervix into the uterus and from there into the distal (upper) ends of the Fallopian tubes. This journey covers a distance of about 15 cm (6 in) and may take as little as an hour. It is aided by **peristaltic action** of the oviduct, which pushes the sperm distally in that structure; as many as several thousand sperm may complete this journey. Peristaltic action, or peristalsis, consists of waves of muscular contractions passing along a tubular structure.

If a receptive egg is present, fertilization will occur. If an egg is not there, the sperm will remain viable for as long as 24–48 hours, so if an egg is produced during that time, fertilization may still occur. Even in the presence of both egg and sperm, successful fertilization resulting in the formation of a viable zygote does not always take place.

The egg (strictly speaking, still a secondary oocyte) at the time of fertilization has a coating composed of adhering follicle cells, the **corona** (Fig. 12-6), and a clear

FIGURE 12-6

A living human egg surrounded by its corona of adhering follicle material. The egg is being penetrated by a sperm in the process of fertilization. Note surrounding sperm (small arrows) and a polar body (large arrow). At this stage the "egg" is actually a secondary oocyte; meiosis II, which produces the true egg, or ovum, occurs at the time of fertilization. (Courtesy Landrum Shettles)

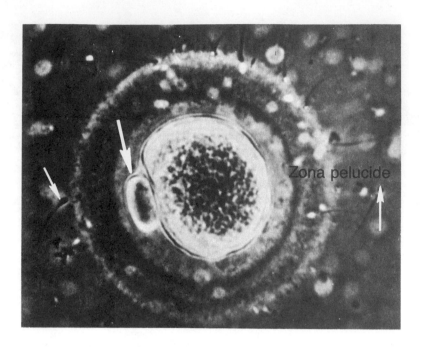

Zona pelucide

TABLE 12-1

Growth of the human embryo and fetus

Time After Fertilization	Length of Embryo (fetus)	Characteristics of Embryo (fetus)
0 days	0.15 mm (zygote)	Fertilization of ovum* occurs in oviduct
5–7 days	0.5 mm	Embryo is implanted in wall of uterus. Embryo is in blastula stage (blastocyst)
14 days	1.5 mm	Germ layers develop, embryo elongates
4 weeks	4.5 mm	Spinal cord, brain, eyes, and digestive tract are differentiated. Heartbeat begins. Limb buds appear (Figs. 12-7 and 12-8)
6 weeks	12 mm	Major organs are formed. Head is about ⅓ the length of body. Fingers are apparent. Period between 4th and 8th week is that of greatest susceptibility to drug- and disease-induced malformation
8 weeks	25 mm	Embryo now called a fetus; human in appearance. Eyelids are present (closed). Sexual differences are apparent. Bone formation begins
12 weeks	75 mm	Fetus begins to move. Breathing (of amniotic fluid) begins. Fetus is less susceptible to drugs, disease (End of first trimester)
16 weeks	125 mm	"Fetal position" is assumed. Hair, fingernails, toenails form
20 weeks	25 cm (250 mm)	Weight is about 1 kg, can sometimes survive premature birth
38 weeks (40 weeks following last menstruation)	60 cm	3–4-kg fetus assumes head-down birth position (Fig. 12-9). Birth occurs

*It is customary to call this cell the ovum or egg, but it actually is a secondary oocyte—meiosis II occurs shortly after sperm entry.

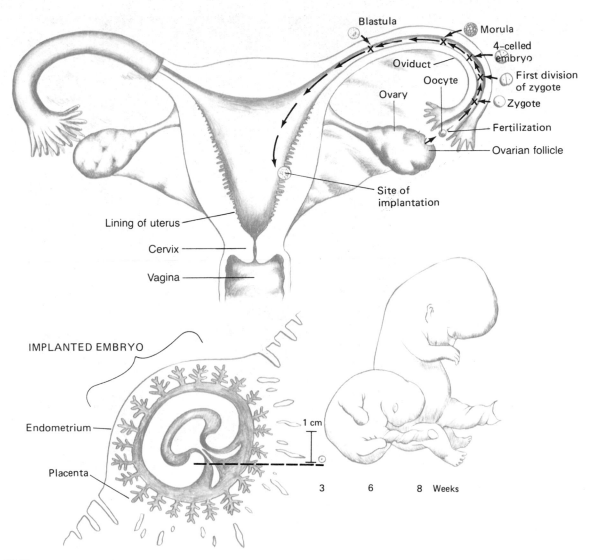

FIGURE 12-7
Implantation and embryo development in the human female.

layer called the **zona pellucida.** Aided by enzymes secreted by the acrosome, the sperm pierces the zona pellucida and enters the egg cytoplasm, losing its tail in the process. Through an incompletely understood process, the entry of a single sperm causes the zona pellucida to become impenetrable to other sperm. After fertilization, the oocyte completes meiosis and the 1n nuclei of the egg and the sperm fuse. The resulting zygote continues its passage down the oviduct, a trip taking three or four days. During this time, the zygote has divided and become an embryo. When it reaches the uterus, the embryo releases enzymes that digest a cavity in the endometrium, and it **implants.** A **placenta** then begins to develop (Fig. 12-7). For the first eight weeks the developing child is considered an embryo; after that, and until

birth, it is called a **fetus** (Table 12-1, Fig. 12-8).

When implantation occurs, the corpus luteum continues its secretions of progesterone and estrogen, but the placenta increasingly takes over as the major source of these hormones. Under their influence the mother experiences an increased appetite and her metabolic rate increases. The placenta also secretes a hormone called **human chorionic gonadotrophin (HCG),** which helps maintain the corpus luteum. Human chorionic gonadotrophin is the basis of certain pregnancy tests because it is excreted by the kidneys in an active form and can be detected in very small amounts by a laboratory test called an **immunoassay.** Formerly, the presence of this hormone was detected by injecting a small amount of urine into test animals. If human chorionic gonadotro-

FIGURE 12-8
Human embryo thirty-two days after fertilization. The large swelling under the head contains the heart, which must develop early so it can pump blood through the umbilical cord to the placenta. (Courtesy James Ebert)

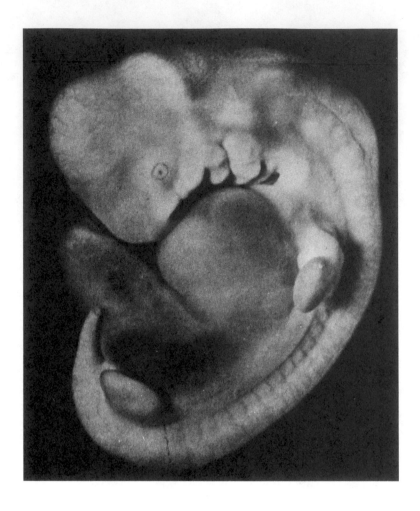

phin was present, it induced ovulation in female rabbits and release of sperm by male frogs.

Childbirth

A normal pregnancy lasts 266 ± 10 days. During the last weeks of pregnancy, the fetus shifts downward in the uterus, so that in a normal pregnancy its head is in contact with the cervix (Fig. 12-9). At birth, uterine contractions begin at intervals of 10–15 minutes, and steadily increase in intensity and frequency. As a result, the head of the fetus acts as a wedge forcing the cervix open, and further contractions move the baby through the cervix and vagina. The **umbilical cord** functions until delivery of the baby. Further uterine contractions deliver the placenta.

The onset of birth is stimulated by multiple factors, including **prostaglandins** produced by cells in the uterine wall, hormones produced by the fetus, and lowered progesterone levels in the blood, probably due to the aging of the placenta. Under nerve stimulation from the cervix, a hormone called **oxytocin** is released from the posterior pituitary and further stimulates uterine contraction. Oxytocin also may be secreted by the pituitary of the fetus, with the same effect. A hormone called **relaxin** is produced by the ovary and the placenta; its function is to soften the muscles and ligaments of the birth canal and allow them to stretch.

Multiple Births

Multiple births—twins, triplets, and higher multiples—can have two origins: the early embryo can split and form two identical fetuses **(one-egg twins)**, or multiple ova can be simultaneously released, fertilized, and implanted to develop as fraternal **(two-egg)** twins. Triplets and higher multiples can occur by either of these processes or a combination of them. The Dionne quintuplets are believed to have been derived from a single zygote, hence are **monozygotic**. Monozygotic multiples are identical, sharing the same genetic inheritance. Fraternal multiples are **dizygotic, trizygotic,** and so on, and

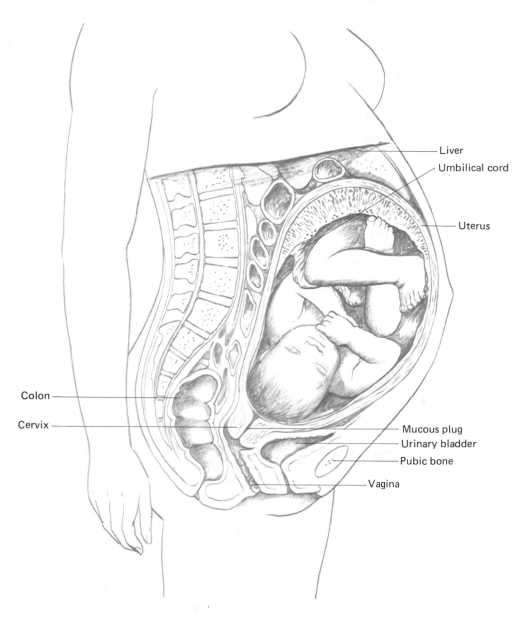

FIGURE 12-9
Position of the human fetus at eight months.

Liver
Umbilical cord
Uterus
Colon
Cervix
Mucous plug
Urinary bladder
Pubic bone
Vagina

therefore are no more closely related genetically than separately conceived siblings. Multiple births of the latter kind have increased dramatically, due to the use of gonadotrophic (gonad-stimulating) hormones to stimulate fertility. The result sometimes is **supraovulation,** the simultaneous release of two or more eggs, which when fertilized result in multiple fetuses.

Fraternal multiples each have their own placenta, amnion, and amniotic sac, but monozygotic multiples commonly share the same placenta and amnion unless their separation from a common early embryo occurs at the 2- to 16-cell stage, before the placental and amniotic membranes are formed. In that case each will have its own set of membranes. Sometimes separation of monozygotic twins is incomplete and the two fetuses are par-

tially fused, giving rise to **Siamese twins.** This, however, is a rare event.

Lactation

In 1910 one of a pair of Siamese twins became pregnant and delivered a normal child. Both the mother and her nonpregnant twin sister began lactating after birth of the baby. Because Siamese twins have a common blood supply, they also share hormones. In this example, hormones of the pregnant sister circulated in her twin, and the hormonal control of milk production was demonstrated.

During puberty, both estrogens and progesterone induce the development of the breasts, and during pregnancy increased hormonal concentrations further stimulate breast development.

Prolactin, a pituitary hormone, stimulates milk production, but its action is blocked prior to and during pregnancy by low levels of estrogen and high levels of progesterone. After childbirth, the placenta no longer is present to maintain high levels of progesterone, and its inhibition of milk secretions ceases. Further stimulation of prolactin production and the secretion of oxytocin, which triggers the release of milk, is maintained by stimulation of the nipples by the nursing infant. When nursing and nipple stimulation cease, lactation stops within a few days.

Infertility

Infertility, or failure to conceive, can have two main causes. The first of these is nonproduction of viable eggs or sperm. Many cases of infertility in women are due to failure of the ovarian follicles to mature or rupture. Similarly, in men, failure of sperm to develop or mature is a cause of infertility. Often this is due to damage of sperm-forming tissues of the testes by injury or disease, but elevated temperatures in the scrotal region (hot weather, heavy binding clothing) can also lower sperm production. Ordinarily, reduced sperm production rather than complete absence of sperm is a factor in male infertility. Although millions of sperm may be ejaculated during intercourse, comparatively few reach the vicinity of the egg in the oviduct. Only one sperm fertilizes the egg, but a number must be present for sperm penetration. A concentration of acrosomal (sperm-head) enzymes is required to soften the egg membrane sufficiently for the penetration of a single sperm, and one sperm alone cannot fulfill the enzyme requirement. Low sperm counts and ovulation failures can often be counteracted by fertility drugs, collectively called gonadotrophins.

DRUGS AND OTHER BIRTH HAZARDS

It is now nearly 25 years since the **thalidomide** scandal shook governments, drug companies, and societies in Europe and America. Thalidomide, a tranquilizer prescribed to relieve tensions in pregnant women, produced severe birth defects in thousands of babies in Germany, England, and Sweden, and to a lesser degree in Canada and Australia. In the United States, fewer than a dozen were affected. The drug had been tested in European laboratories before being released, and the tests had indicated that the drug was safe for human use at the prescribed dosages. In the United States the Food and Drug Administration (FDA) had not followed the Europeans in releasing thalidomide. The few women who had used it had obtained it abroad. Often criticized for being slow to release new drugs and for requiring "unreasonably" high safety standards, the FDA in this case had prevented thousands of tragedies.

Later on, after thalidomide had been taken off the market everywhere, new tests showed that laboratory mice and other animals were far less susceptible to thalidomide than were humans—laboratory mice were affected only by dosages 60 times higher than those producing birth defects in humans. Such differences in dosage responses are the main reasons that large safety factors must be allowed in testing drugs for human use.

A number of substances other than thalidomide have been identified as capable of damaging the human fetus, especially during the first trimester of pregnancy, when major parts of the body are being formed (see Table 12-1). Among these are antibiotics, especially tetracycline, antihistamines, weed killers (2,4-D, 2,4,5-T—"agent orange"), and synthetic sex and growth hormones (anabolic steroids). Tobacco and alcohol also can damage the fetus, the latter sometimes inducing a condition known as fetal alcohol syndrome (Fig. 1).

A second cause of infertility is mechanical. Damage to either the Fallopian tubes (oviducts) or sperm ducts can prevent either the sperm or the egg from coming in contact with each other and the zygote from passing into the uterus. Sometimes these conditions are correctable by surgical procedures, and more recently "test-tube baby" techniques have been developed to bypass the oviducts (see Box, p. 260).

BIRTH CONTROL

Preventive Methods

Birth control, or **contraception**, is practiced in one form or another by most cultures in the world. It has a long history, going back to the ancient Egyptians, who used various substances including lemon juice, butter, honey, and gum arabic as vaginal lotions to inhibit sperm survival and motility. The ancient Greeks practiced both contraception and abortion. Many contraceptive methods used today were practiced in the Middle Ages: **coitus interruptus** (withdrawal of the penis prior to ejaculation), condoms (then made of fish skin), and douching. These latter methods are still used, along with others that allow control over pregnancy never before possible.

Coitus Interruptus   Coitus interruptus is perhaps the most ancient birth-control method. It is, however, an ineffective method of contraception; the pregnancy rate using this method alone is 20–25 per 100 women per year (compared to 50 pregnancies per 100 women per year when no birth-control methods are used).

Douching   **Douching** consists of injecting spermicidal solutions into the vagina following intercourse. Because the active sperm reach the uterus within 90 seconds af-

Diseases such as measles (rubella), syphilis, venereal herpes, mumps, influenza, and others also are known to produce birth defects, either by directly infecting the fetus or as a result of the mother's high fever. Many other environmental hazards can cause damage to the unborn child. Pregnant mothers should avoid X rays (once used to diagnose twins) of the abdominal region. Exposure to chemical fumes such as paint thinners, gasoline, and cleaning fluids should be avoided also, as should exposure to pesticide vapors. In many instances exposure risks may seem insignificant, but the soundest policy is for the pregnant mother not to take chances with exposure to any avoidable disease, drug, or pollutant.

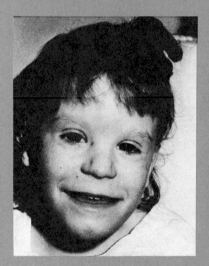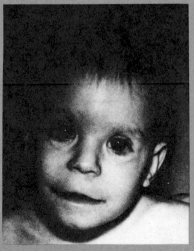

FIGURE 1
The fetal alcohol syndrome. The mothers of these two children consumed alcoholic beverages excessively during early pregnancy. Both children are greatly retarded mentally and have facial features characteristic of this syndrome. (Courtesy James W. Hanson)

ter intercourse, this method is not very effective. Furthermore, douching with harsh chemicals can do permanent damage to the vaginal lining and cervix.

Rhythm Method

The **rhythm method,** which requires calculation of the time of ovulation, is a very ineffective birth-control measure (failure rate about 35 percent), because to be effective it requires precise timing of the individual female's ovarian cycle. As noted earlier, the egg is shed about day 14 of the menstrual cycle. For about 24–48 hours thereafter, it is receptive to fertilization. Until about 50 years ago it was believed that ovulation occurred immediately after menstruation and that the "safe period" began about a week later and lasted until the next menstruation. The charts had to be revised when it was realized that ovulation actually occurs during the presumed "safe period." Ovulation is often advanced or delayed by days, so the "safe period" may vary considerably. Because a slight rise in body temperature occurs at ovulation, careful reading and recording of the daily sequence of temperatures can assist in determining the "safe" and "unsafe" times of the month. However, even with these refinements, the rhythm method is said to be unreliable.

Spermicidal Agents

Spermicidal jellies, creams, and **foams** are available under a variety of trade names; the foams are the most effective of these (failure rate about 25 percent). All are meant to be used just prior to intercourse, allowing a few minutes for them to spread and diffuse. They are more effective than the rhythm method or douching, but less effective than condoms, diaphragms, IUDs, oral contraceptives, or sterilization.

Condoms and Diaphragms

Condoms and **diaphragms** are rubber articles designed to block mechanically the entrance of sperm into the uterus. Condoms are thin rubber sheaths worn over the penis during intercourse (see Fig. 12-10). They are effective (failure rate about 10 percent), but sometimes may break during use. Diaphragms are rubber caps placed over the cervix prior to intercourse and are quite effective when properly fitted and used with a spermicidal cream (failure rate about 18 percent).

Intrauterine Devices

In the nineteenth century, it was known that **intrauterine devices,** or **IUDs,** prevented pregnancy, but the idea was condemned by the medical profession as harmful; it is only in the last couple of decades that this contraceptive method has been used on any scale.

It is not known why an IUD prevents implantation of an early embryo; possibly the endometrium is changed in such a way as to reject both sperm and fertilized eggs. There are several types of IUDs; all are inserted into the uterus through the cervical canal by use of an applicator (Fig. 12-10). A trained technician such as a physician or health nurse must insert the IUD to avoid accidental damage or infection to the uterus. IUDs may be painful, particularly to women who have never experienced pregnancy, and they may be expelled spontaneously. They also can cause damage to the tissues of

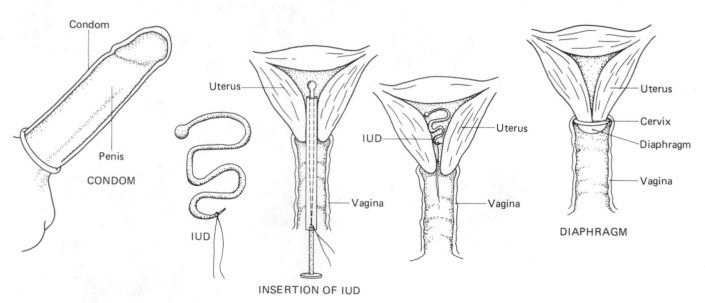

FIGURE 12-10
Condom, IUD, and diaphragm.

the uterus. However, they are cheap as well as effective (failure rate about 4 percent). This makes their use attractive in countries with serious overpopulation problems; unfortunately, routine medical supervision of their use often is not readily available.

Oral Contraceptives Oral contraceptives were developed in the United States during the 1950s and are used widely despite real and alleged dangers. They are synthetic estrogens and progestogens and, with the exception of sterilization, are the most effective contraceptive method known (failure rate less than 2 percent). They inhibit FSH and LH production by the anterior pituitary and thus suppress the development of the ovarian follicle and the release of eggs. The pills are taken every day for 20–22 days, depending on the specific pill, then are stopped. During the next 6–8 days menstruation occurs; then pill taking is resumed.

There are several versions of "the pill," and a woman may be advised to use one or another, depending on her reaction to them. All require care in use, and they must be taken faithfully in order to work. A lapse of even one day may result in pregnancy.

There seems to be a slightly increased risk for formation of blood clots and heart attacks in women who are "on the pill." In addition, it has been suggested that prolonged use of oral contraceptives may produce cervical abnormalities, but there is no proof of this. (Cervical abnormalities can be detected by the **Papanicolaou [Pap] smear test,** in which a tiny bit of cervical tissue is removed and examined for the presence of abnormal cells. All women, whether they use oral contraceptives or not, should have routine Pap tests.)

A male contraceptive pill is in the process of development. Male sterility pills combine estrogenlike compounds to suppress sperm production with synthetic androgens to maintain sexual potency. Early experimental pills were effective in suppressing sperm production but had feminizing effects.

Sterilization Currently, voluntary male sterilization is accomplished by a simple operation called a **vasectomy,** in which the vasa deferentia are tied and cut. The operation is painless, can be done in a physician's office, and recovery is rapid. The procedure is increasingly popular, and it is a safe and completely effective birth-control method. However, it is not immediately effective, because living sperm remain in the vas deferens for several weeks after the vasectomy.

Female sterilization normally involves ligating the oviducts, a procedure called **tubal ligation,** or **laparoscopy.** At one time this operation required abdominal surgery, but now it is done using an instrument called a laparoscope. This is a thin tube equipped with a light, a lens system, and a means of introducing delicate surgical instruments into the abdominal cavity through a very small incision. The ligation can be done under a local anesthetic and does not require postoperative hospitalization.

To date, restoration of fertility by rejoining the cut vasa deferentia or ligated oviducts has not been very successful.

Abortion

Methods Under natural circumstances, the embryo sometimes dies and is rejected and discharged by the uterus. This is called a **spontaneous abortion,** or **miscarriage.** The deliberate termination of a pregnancy is an **induced abortion.** The Chinese practiced induced abortion 4000 years ago. In ancient Rome women regularly relied on induced abortion to relieve themselves of unwanted children. In recent years the Japanese have checked their rate of population increase largely by abortion, and abortion is widely practiced in the Soviet Union. Only in the last decade has induced abortion become an open practice in the United States because of repeal of abortion laws in many states and a Supreme Court decision in 1973 that upheld the right of a woman to seek an abortion.

There are several comparatively safe methods of abortion in current use. **Uterine aspiration,** in which the embryo or fetus is removed by suction through a tube, is the most widely used method in abortion clinics during the first three months of pregnancy (the **first trimester).** Another method is **dilation and curettage (D and C),** in which the cervix is dilated and the uterine wall is scraped to remove the embryo or fetus. Both methods require skill and absolute asepsis to prevent damage and infections. Many abortions are similarly performed by unauthorized persons, but often with very crude instruments, including sharpened teaspoons, knitting needles, and wire coat hangers. Under such circumstances there is great danger of perforation of the uterus and infection.

In later stages of pregnancy, after the third month, a common method of inducing abortion is injection with a long hypodermic needle of salt water **(saline solution)** into the amniotic sac of a fetus. Usually this induces labor within a few hours and the fetus is expelled. Drugs also may be used to induce abortion and have been employed throughout the ages.

Abortions may be induced up to the twenty-sixth week of pregnancy, but become increasingly complicated. Moreover, during the later stages of pregnancy the fetus is fully formed and may be capable of survival outside the uterus. If that is the case, its death can leave those involved in the act open to criminal charges.

Abortions may be requested for many reasons, including incest, rape, or the possibility that the fetus may be defective. The use of amniocentesis, described in Fig. 7-20, makes it possible to detect certain kinds of fetal defects, particularly those involving chromosomal abnormalities. The possibility that a fetus is defective is today considered a valid reason for an induced abortion. Other generally accepted reasons for legal abortions are serious risk of damage to the health of the mother and pregnancies resulting from incest or rape.

The Question of Abortion Unwanted babies have always been a problem for society. In China, even in the present century, girl babies were cast into the Yellow River because they were considered a liability to poverty-stricken families. They could not earn money by their manual labor, and if they were reared to marriageable age, a dowry would be expected of the parents. In eighteenth- and nineteenth-century Europe, for similar reasons, unwanted babies were strangled by their parents or abandoned to die. Even today, babies are found in trash cans, supermarket carts, and back alleys, and welfare agencies are overloaded with abandoned children.

Abortion raises a number of ethical questions, particularly with respect to the legal rights of the fetus. This introduces arguments about whether the fetus is a person, when sentient life begins, and whether the zygote also must be protected. The other side of the coin is that overpopulation in the less-well-developed nations often leads to the starvation and death of children. In the famines that periodically sweep across Asia and Africa, the first victims are the children. Therefore, it is argued that it is better to abort a fetus than to have the child experience later misery and death.

An argument often advanced for legalized abortion is that pregnant women will have abortions whether or not they are legal. If a legal abortion is not available, the mother's health may be risked by clumsy and dangerous practitioners of illegal abortion. It is said also that only poor women will suffer if legal abortions are not available, and that wealthy women can afford contraception or have ways of getting abortions whether or not they are legal.

It is difficult and even presumptuous for anyone not personally involved in the question of abortion to make such an important decision for another individual. Careful guidance and counseling is required, preferably involving others who have been faced with similar decisions. Far more important, however, is being able to prevent pregnancy when it is unwanted.

SUMMARY

Sexual characteristics of human males and females are controlled by gonadal hormones. Male and female gonads are determined by the sex chromosomes (XX = female, XY = male), but the development of sex-related structures and features and the control of processes relating to the functioning of sex organs, of pregnancy, and of childbirth are hormonally controlled.

About every 28 days the human female goes through an ovarian cycle. During the first 14 days, an ovarian follicle matures and releases an egg. The ruptured follicle heals over and becomes the progesterone-secreting corpus luteum, which prepares the uterus to receive the embryo. Meanwhile, during a period of about 4 or 5 days, the egg travels down the oviduct toward the uterus. If it is fertilized by sperm following sexual intercourse, it begins to develop into an embryo. When it reaches the uterus, the embryo becomes implanted in the wall of the uterus and develops into a fetus. If fertilization does not occur, the egg is lost, the corpus luteum degenerates, and the uterine wall returns to its state at the beginning of the cycle. This requires the breakdown of blood vessels and cells in its lining, and these together with blood are discharged in a process called menstruation.

Many methods of preventing pregnancy have been tried over the ages. The most effective of the procedures commonly used today are condoms, diaphragms, IUDs, oral contraceptives, and sterilization. Other methods such as male withdrawal, douching, the rhythm method, and spermicidal agents are less than successful. Several of the methods have undesirable side effects.

Once pregnancy has occurred, an abortion may be performed. Abortion, a safe operation done in early pregnancy by a skilled practitioner, has many social, ethical, and legal implications, and remains an extremely controversial subject.

KEY WORDS

sex chromosome	ovum	orgasm
gonad	polar body	fertilization
follicle-stimulating hormone (FSH)	corpus luteum	supraovulation
luteinizing hormone (LH)	endometrium	implantation
human gonadotrophic hormone (HGH)	progesterone	human chorionic gonadotrophin (HCG)
prolactin	menstruation	fetus
prostaglandin	testis	childbirth
estrogen	testosterone	monozygotic twins
ovary	seminiferous tubule	dizygotic twins
uterus	spermatocyte	Siamese twins
Fallopian tube	sperm	abortion
oocyte	penis	contraception
ovarian follicle	copulation	

QUESTIONS FOR REVIEW AND DISCUSSION

1 Differentiate between primary and secondary sex characteristics of human males and females. Describe the common origins of male and female genitalia.

2 Describe the roles of luteinizing hormone (LH) and follicle-stimulating hormone (FSH). Are these hormones present just in females?

3 What is the corpus luteum? What does it do?

4 What is the relationship of the ovarian cycle to the period of maximum fertility?

5 Where in the male body are sperm formed? Be quite specific. Trace the development of sperm and their transport.

6 Describe the methods of safe induced abortion.

7 List some arguments for and against abortion.

8 Discuss amniocentesis and its medical, social, and ethical implications.

SUGGESTED READING

BEACONSFIELD, P.; G. BIRDWOOD; and R. BEACONSFIELD. 1980. The placenta. Sci. Amer. 243(2):94–102. (Covers structure and function of this embryonic feature.)

GRABOWSKI, C. T. 1983. Human reproduction and development. Philadelphia: Saunders College Publishing. (Up-to-date paperback covering important aspects of human reproduction and genetics. Nice diagrams!)

PENGELLEY, E. T. 1974. Sex and human life. Reading, Mass.: Addison-Wesley Publishing Co. (A simple, well-written, and readable account covering many physiological and social aspects of human sex.)

SCHALLY, A. V.; A. J. KASTIN; and A. ARIMURA. 1977. Hypothalamic hormones: the link between the brain and the body. Am. Scientist 65:712–19. (Discusses the relationships of hypothalamic releasing factors and the secretion of pituitary hormones.)

SCHOTTELIUS, B. A., and D. D. SCHOTTELIUS. 1978. *Physiology*. St. Louis: C. V. Mosby Co. (A readable general textbook with a well-illustrated chapter on human reproduction.)

SEGAL, S. J. 1975. The physiology of human reproduction. *Sci. Amer.* 231(3):52–62. (Emphasizes nervous and hormonal regulation of sex.)

TIETZE, C., and S. LEWITT. 1977. Legal abortion. *Sci. Amer.* 236(1):21–27. (A review of the state of legalized abortion in the world.)

VILLEE, D. B. 1975. *Human endocrinology*. Philadelphia: W. B. Saunders Co. (Provides much information on fertilization, early embryology, and sexual maturation.)

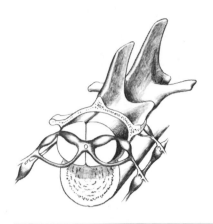

13

The Nervous System

Organisms, regardless of kind, live in constantly changing environments to which they must adjust if they are to compete successfully with other organisms and survive. Imagine, for a moment, the chain of events set into motion when a predator spots a potential prey. Receptors of sight, sound, and smell are activated and messages travel over sensory nerves to the brain, where they are interpreted and checked against the memory bank. Rabbit or cat? Food or nonfood? Safe or dangerous? Another predator in sight? Best approach? Instantly, the course of action is decided; nerve impulses are generated and speed down motor nerves to a multitude of effector structures. Adrenals are stimulated and adrenaline flows into the bloodstream. Salivary glands are activated and the mouth drools in anticipation of food. The intricately balanced system of muscles and bones is set in motion—one muscle set contracts on cue, simultaneously a second set relaxes (otherwise the body will lock up rigidly in midstride). A flow of motion ensues, so complicated that its understanding has challenged the intellects of generations of scientists.

Sensing of environmental factors and regulation of body functions require many kinds of integrated actions by cells, tissues, and organs. In animals this integration is accomplished by two kinds of intercellular communication: electrochemical nerve impulses and chemical messages in the form of hormones. In many cases the messages themselves are also of two kinds: **inhibitors** and **inducers,** acting oppositely to maintain the so-called internal "steady state" or homeostasis. Glucagon and insulin act in opposition to regulate blood sugar (as noted in Chapter 10); likewise, different nerve pathways induce opposite actions on the part of effector structures.

Plants also must adjust to environmental conditions, although their activities are not of the same kind as animals. Plants are nonmotile food-producers, whereas animals are in most cases motile food-consumers. These differences are reflected in the comparatively greater complexity of organ systems of animals, in which interactions between integrated systems of nerve cells **(neurons)** and hormone-secreting glands regulate body functions. In turn, muscle sets are stimulated, organ functions regulated, enzymes secreted, and complex behavioral responses occur. In plants, the regulation of functions is accomplished principally by hormones, and behavioral responses usually are slow movements resulting from differential growth of cells.

273

COMPLEXITY OF NERVOUS SYSTEMS

In their simplest forms, nervous systems of animals consist of networks of neurons over which electrochemical messages known as **nerve impulses** travel. In *Hydra*, a small (< 1 cm long), tentacled, aquatic animal of simple tubular form (Fig. 9-2), for example, the nervous sytem is a network of multibranched neurons over which impulses travel in several directions (Fig. 13-1A). (In more advanced organisms impulses are unidirectional.) Any neuron in *Hydra* can act as a sensory receptor and is capable of initiating impulses which then are transmitted to other neurons or directly to **effector cells,** such as the contractile elements in the body wall, that are induced to carry on a specific activity. The intensity of the response depends on how many neurons are initially stimulated and how many effectors are activated. The result is movement of part or all of the body toward or away from the source of the stimulus. Although quite simple in organization, the nervous sytem of *Hydra* coordinates feeding response, avoidance reactions, and fairly elaborate locomotion.

In flatworms, at a slightly higher level of animal complexity than *Hydra*, trends toward development of a **central nervous system (CNS)** are discernable (Fig. 13-1B). In free-living planarian flatworms (see Fig. 9-3), the nervous sytem has a longitudinal orientation, with two major **nerve cords** (a nerve cord is a large bundle of neurons) that give off branches called **nerve trunks,** or simply **nerves** (a nerve is a bundle of nerve-cell extensions). At the anterior end of the paired nerve cords of the planarian *Dugesia*, two conjoined masses of nerve-cell bodies, or **ganglia** (sing. **ganglion,** "swelling"), act as a coordinating center for sensory reception and nervous stimulation of muscles and other effectors—in effect a simple brain. Thus the brain ganglia and nerve cords constitute the axis of a simple CNS, and the branches— nerve trunks that extend peripherally to all parts of the body—compose the **peripheral nervous system (PNS).** Two kinds of neurons occur in nerve trunks: **sensory neurons,** which transmit impulses toward the CNS, and **motor neurons,** which transmit impulses from the CNS to muscles in the body wall.

Dugesia has a distinct head bearing a pair of dorsal eyes of simple construction (Fig. 13-1B); chemical and tactile sensory receptors also tend to be concentrated in the head region. These anterior specializations are consistent with the overall trend toward cephalization, evident in the evolution of both invertebrates and vertebrates. Therefore, in comparison with the nerve net of *Hydra*, the nervous system of *Dugesia* demonstrates several advanced features, including: (1) central and peripheral nervous systems, (2) concentrations of sensory and coordinating neurons in the head region, and (3) differentiation of sensory and motor pathways for nerve impulses. Further elaboration of these features is seen in higher invertebrates and in all vertebrates.

The Vertebrate Central Nervous System

Nervous systems of animals tend to show a progressive linearity and anterior specialization (cephalization— development of the head and brain region, as in Fig. 13-1B). In invertebrates, such as earthworms and arthropods (centipedes, lobsters, insects, etc.), there is a double, longitudinal, solid, ventral nerve cord, ending anteriorly in a brain. This same linearity and cephalization is apparent in vertebrates, but now the structure and orientation of the brain and nerve cord is different; both are hollow rather than solid, and they are dorsal rather than ventrally located structures.

The origin of the vertebrate nerve cord and brain (the central nervous system, or CNS) was described in Chapter 11 (Fig. 11-7). This origin from a longitudinal furrow in the ectoderm of the early embryo explains the hollow nature of the vertebrate CNS. As the vertebrate embryo continues to grow, three major brain regions become differentiated. They are the **forebrain** (associated with sensory reception, behavior, memory, and learning), the **midbrain** (associated with certain visual functions and nerve tracts between the forebrain and hindbrain), and the **hindbrain** (associated with the coordination of motor functions). Nerve impulses are transmitted to and from the CNS by paired **cranial** and **spinal nerves** (refer to Fig. 9-5 for a generalized diagram, and to Figs. 13-5 to 13-18 for the human CNS).

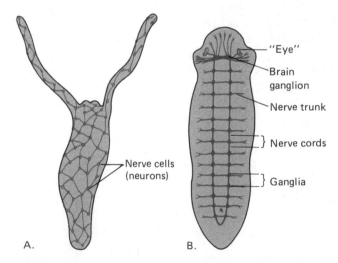

FIGURE 13-1
Simple nervous systems. **A.** Nerve net of *Hydra*. **B.** Nerve cords of *Dugesia*, a planarian.

Neurons

Electrochemical impulses can be initiated and conducted by many kinds of cells, not just by neurons. In most cases, however, such actions are neither highly coordinated nor very rapid. Neurons, on the other hand, are specialized for directional conduction of impulses, and because they are elongated cells, such impulses usually are transmitted for some distance. In extreme cases, such as the neurons that extend from the brain of the giraffe to its midsection, a neuron may be between 2 and 3 m (6 and 10 ft) long. Even longer neurons—perhaps 10 m or more—must have been present in the enormous bodies of some extinct dinosaurs.

All neurons have a cell body, containing a nucleus and other organelles, and one or more elongated cytoplasmic extensions generally called **nerve fibers**. Mitochondria and other organelles are present in the fibers just as in the cell body, and the cell body together with its extending fibers constitutes a complete cell.

The uniqueness of neurons is that they do not divide and therefore are not replenished as other cells are. This has implications in many kinds of disease (polio, for example) as well as in accidental damage and the gradual loss of some senses with aging.

Nerve Fibers

In the nerve net of *Hydra*, neurons have several fibers and can conduct impulses in more than one direction, depending on the point of initial stimulation. In more advanced nervous systems, however, neurons have only one or two major nerve fibers and transmit nerve impulses in only one direction. Sensory neurons transmit impulses toward the CNS from the PNS, and hence are called **afferent** ("carrying toward") **neurons**; motor neurons carry impulses from the CNS toward the PNS and are known as **efferent** ("carrying away from") **neurons**. A third kind of neuron, found in ganglia and nerve cords, is the **connector** or **association neuron,** and as its name implies, it acts as a link between sensory and motor neurons.

Two kinds of nerve fibers are recognized in neurons: unbranched or sparsely branched **axons** ("axis") and branching **dendrites** ("tree"). Motor and sensory neurons have long axons and short or no dendrites; connector neurons have axons and dendrites of approximately equal length (Fig. 13-2).

Motor and sensory neurons of vertebrates are associated with flattened **Schwann cells** that form a wrapping around adjacent sections of axons (Fig. 13-3B). Each Schwann cell can be visualized as a thin, sheetlike cell wrapped around an axon one to many times, depending on the function of the neuron—many wrappings if the function is rapid conduction of nerve impulses, only one wrapping if the function is slow conduction. In cases of many wrappings, a thick, fatty covering called the **myelin sheath** is produced; the axon and its sheath comprise a **myelinated fiber**.

The nature of the myelin sheath was not recognized until the advent of electron microscopy; formerly, the waxy portion was thought to be a secretion of the Schwann cell, but electron microscopy revealed that its fatty nature is due to many wrappings of the lipoprotein Schwann cell plasma membrane. The manner in which this is accomplished is intriguing. The Schwann cell first envelops a section of an axon, then forms a lip which grows around the axon many times (Fig. 13-3B). Slight gaps are present in the myelin sheath at junctions of adjacent Schwann cells. These gaps, called **nodes of Ranvier,** are important in rapid conduction; nerve impulses jump from one node of Ranvier to the next, greatly speeding the impulse conduction.

FIGURE 13-2
Neurons.

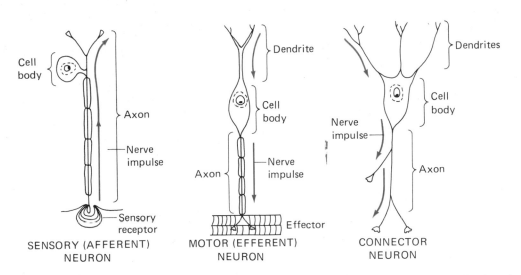

SENSORY (AFFERENT) NEURON

MOTOR (EFFERENT) NEURON

CONNECTOR NEURON

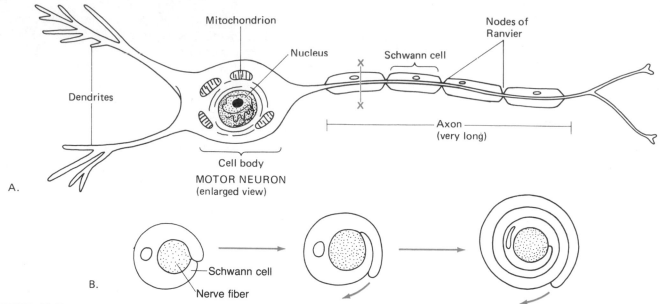

FIGURE 13-3
Myelinated fiber and its Schwann cells.

The Nerve Impulse

Years ago, when the telephone was invented, it was considered possible that neurons conducted electricity in much the same way as telephone wires. In the telephone system a message travels as an electric current, actually as a stream of electrons moving from one end of a wire to the other. However, when it became possible to measure the speed of a nerve impulse, that speed was found to be much slower than that of an electric current—at most about 120 m (400 ft) per second, as compared to 300 million m (1 billion ft) per second for electricity. Sensitive instruments capable of measuring the transmission of nerve impulses have revealed that they consist of a relatively slow-moving wave of electrochemical changes in plasma membranes of nerve cells. The speed of the impulse wave is on the order of 120 m (400 ft) per second in myelinated neurons of the CNS and about 10–15 m (30–50 ft) per second or less in slower acting, unmyelinated neurons of the autonomic nervous system (**ANS**; see Fig. 13–17).

Although the flow of electricity through wires and the flow of electrochemical nerve impulses through neurons are different phenomena, both involve electrically charged ions. In electrical wires electrons flow from atoms of one kind to those of another kind, thus creating ions (i.e., oxidation–reduction reactions), whereas in neurons ions flow from one place in the cell to another. In both cases, however, the flow—either of electrons or of ions—can be measured by electronic devices.

Positive and negative ions cannot be extracted and separated from a compound; that is, it is impossible to have a reagent bottle full of sodium ions alone. However, anions and cations can be segregated within a solution by placing a barrier in the system so as to create compartments separating them. The result of such separation of oppositely charged particles is the production of a form of potential energy known as an **electrical potential**. In a storage battery, this electrical potential can be released by connecting positive and negative compartments (the positive and negative poles) with a wire so that a flow of electricity occurs. In a living cell, an analogous relationship exists in which there is an unequal—in effect compartmentalized—distribution of anions and cations (K^+, Na^+, and Cl^-) on either side of the plasma membrane; the outside of the membrane is more positively charged than the inside. In this case the electrical potential is known as the **resting potential**. The resting potential can be measured with a sensitive voltmeter and in most cells is about -70 to -90 millivolts (mV).

Much of the research relating to electrical potentials and nerve impulses of neurons has been done using giant motor neurons of squids. These cells are nearly 1 mm (0.04 in) in diameter, large enough that electrodes can be implanted in them. As a result of such experimentation, it has been possible to correlate nerve impulses with localized flows of sodium and potassium ions across the plasma membrane of the neuron.

Action Potential Normally, there is an accumulation of positively charged sodium ions on the outside of the plasma membrane of a neuron, making the outside relatively more positively charged than the inside:

$$\text{(outside)} \quad Na^+ \ Na^+ \ Na^+ \ Na^+ \ Na^+ \ Na^+ \ Na^+$$

$$\text{(inside)} \quad Cl^- \ Cl^- \ Cl^- \ Cl^- \ Cl^- \ Cl^- \ Cl^-$$
$$K^+ \ Cl^- \ K^+ \ Cl^- \ K^+ \ Cl^- \ K^+ \ Cl^- \ K^+ \ Cl^- \ K^+ \ Cl^-$$

When a neuron is stimulated, its membrane temporarily becomes permeable to sodium ions which then rapidly pass to the inside:

$$\text{(outside)} \qquad Na^+ \qquad Na^+$$

$$\text{(inside)} Cl^- \ K^+ \ Cl^- \ Na^+ \ Cl^- \ Na^+ \ Cl^- \ Na^+ \ Cl^- \ K^+ \ Cl^-$$
$$Cl^- \ K^+ \ Cl^- \ K^+ \ Cl^- \ K^+ \ Cl^- \ K^+ \ Cl^-$$

The result is a reversal of polarity, called an **action potential**. This reversal is immediately followed by a counterflow of potassium ions through the plasma membrane so that the outside of the membrane once again is positively charged:

$$\text{(outside)} \qquad K^+ \ K^+ \ K^+ \quad K^+ \ K^+ \ K^+$$

$$\text{(inside)} \quad Cl^- \ Cl^- \ Cl^- \ Cl^- \ K^+ \ K^+ \ Cl^- \ Cl^- \ Cl^- \ Cl^-$$
$$Cl^- \ Na^+ \ Cl^- \ Na^+ \ Cl^- \ Na^+ \ Cl^- \ Na^+ \ Cl^- \ Na^+$$

As a result of the counterflow of potassium ions, the outside of the plasma membrane is again electropositive. However, in this state it is incapable of generating an action potential, which requires an accumulation of sodium ions on the outside of the cell membrane and potassium ions within it. Restoration of a resting potential quickly follows an action potential because ion transport molecules in the membrane, known as **sodium/potassium-ion pumps**, exchange sodium and potassium ions. The membrane then is again capable of generating an action potential. Work must be done by the neuron to restore and maintain a resting potential. This work, which is accomplished by the ion pumps, uses energy derived from ATP.

An action potential is capable of inducing further polarity reversals in adjacent regions of the plasma membrane of a neuron, so that a wave of self-propagated action potentials can pass from one end of the neuron to the other. This wave is the nerve impulse (Fig. 13-4). Because of the rapidity with which diffusion and active transport of ions occurs in the nerve-cell membrane, a rapid sequence of nerve impulses can traverse the neuron. This, however, requires a constant supply of ATP energy, an observation that correlates well with the high oxygen requirements of the nervous system.

Directionality An action potential can be generated at any point on a cell membrane, so that if the axon of a neuron is sufficiently stimulated in some intermediate region, impulses will be transmitted toward both ends of the neuron. What, then, accounts for the directionality of impulse conduction in sensory and motor neurons? The answer lies in the nature of neuron endings and in the way impulses are generated in sensory receptor endings and at synapses. Sensory neurons have specialized receptor endings (Fig. 13-7), which upon stimulation induce an action potential. The resulting wave of action potentials (nerve impulse) moves toward the end of the axon, where there are one or more bulblike endings, known as **synaptic knobs** (Fig. 13-5). When an action potential is induced at a synaptic knob, it causes tiny sacs **(synaptic vesicles)** of **neurotransmitter** chemical (there are several kinds; acetylcholine is a common one) to move through the membrane of the knob and discharge into the **synaptic cleft**. (The membrane on the secretory side of the synapse is called the **presynaptic membrane**; the membrane on the other side of the synapse, upon which the neurotransmitter chemical initiates an action potential, is the **postsynaptic membrane**.) The synapse acts as a one-way valve for the transmittance of nerve impulses from neuron to neuron; only the presynaptic membrane secretes neurotransmitter chemicals, and only the postsynaptic membrane is stimulated to induce an action potential. The generation of impulses at a synapse is not reversible.

At the point where a motor axon meets an effector cell, such as a muscle, there also is a synapse. Here, the postsynaptic membrane of course is not that of another

FIGURE 13-4
Nerve impulse.

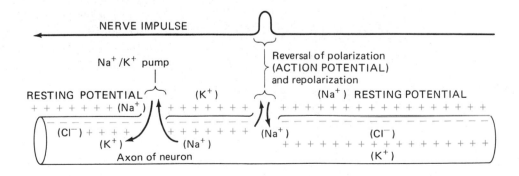

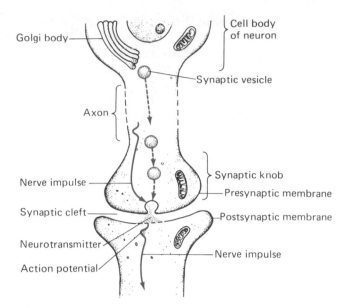

Golgi body

Cell body of neuron

Synaptic vesicle

Axon

Nerve impulse

Synaptic knob

Presynaptic membrane

Synaptic cleft

Postsynaptic membrane

Neurotransmitter

Nerve impulse

Action potential

FIGURE 13-5
Synapse. Synaptic vesicles are formed in the nerve cell body and migrate to the synaptic knob, where their contents are discharged as shown.

neuron but that of the somatic cell. In an example commonly cited—the neuromuscular junction in vertebrates—the axon ending consists of an aggregation of presynaptic knobs and postsynaptic membrane called a **motor end plate.** When an impulse reaches the motor end plate, synaptic vesicles containing acetylcholine are discharged into the synaptic cleft and induce an action potential on the muscle-cell membrane. This, in turn, results in the sliding of actomyosin filaments (see Figs. 14-7 and 14-11)

Acetylcholine is the neurotransmitter in many other synapses, but, as noted previously, is not the only one. About 40 different neurotransmitters have been identified in the CNS and ANS, including dopamine, norepinephrine, serotonin, histamine, enkephalins, and several amino acids.

When a neurotransmitter is discharged into a synaptic cleft, it apparently does not then pass through the membrane of the postsynaptic cell, but is thought to be bound temporarily to a specific acceptor molecule in the postsynaptic membrane. Neurotransmitter and acceptor molecules are believed to fit together on the key-in-lock principle, often alluded to in enzyme actions.

Opiate Receptors

One of the most remarkable recent discoveries of neurophysiology is that the human body apparently produces its own **opiates** (pain-relieving substances) in the form of molecules capable of inhibiting transmission of

impulses from pain-receptor neurons. The action probably is not directly on the pain receptor but rather seems to be on certain connector neurons in the CNS. In the brain, several substances called **enkephalins** have been isolated and are thought to act as natural opiates. Similar substances, called **endorphins,** are produced by the pituitary gland and also appear to have opiate properties. Enkephalins and endorphins seem to be produced in stress situations and probably constitute part of the "fight-or-flight" syndrome discussed later in this chapter and in Chapter 10. This, perhaps, is the reason that some persons involved in stressful situations do not notice injury until some time after the stress is past. It is thought also that **acupuncture,** a form of localized stress, may owe its pain-relieving effectiveness to production of endorphins or enkephalins.

Long-distance runners often experience a sensation of well-being called a "runner's high." It is thought that this feeling is not simply a result of fresh air and exercise, but may be due at least in part to endorphins. When the body is taxed by exertion, a situation somewhat like the "fight-or-flight" response occurs and the body becomes temporarily insensitive to pain. Because the stress is not incapacitating, the feeling of well-being induced by endorphins is all the more noticeable. In contact sports, where injuries sometimes can be severe, cases are known of athletes continuing to play with broken bones, apparently not feeling the pain until the game is ended.

The discovery of enkephalins came about as a result of investigations of the principles of morphine and other pain-relieving drugs; apparently, such drugs can substitute for enkephalins at receptor sites and thus relieve pain.

A phenomenon known as **competitive inhibition** is recognized in enzyme actions. Molecules having a configuration similar to those of the substrate can fit the substrate acceptor site of an enzyme and thus "compete" with substrate molecules (you may wish to refer to Fig. 1-29). In many cases the fit is partial and the nonsubstrate molecule is neither modified by enzymatic action nor released from the enzyme. If many nonsubstrates become bound to enzymes, enzyme–substrate molecules will be inhibited or blocked by competitive inhibition. In an analogous way, certain drugs are thought to compete with neurotransmitter molecules for receptor sites in postsynaptic membranes and either substitute their action for that of the neurotransmitter or inactivate the synapse completely. If the drug is quite similar to the neurotransmitter, it may induce nerve impulses in the postsynaptic membrane, thus increasing the response of an effector structure. If the drug is less similar, it may simply inhibit the action of the effector. Lysergic acid

(LSD) is a powerful competitor for serotonin binding sites in certain connector neurons of the brain, and its effect is to disrupt synaptic functions. Amphetamines interact with norepinephrine (noradrenalin) at synapses and overstimulate effectors, as does nicotine in reacting with acetylcholine binding sites. Morphine, heroin, and methadone act as substitute enkephalins, depressing transmission of dull pain sensations. Alcohol, however, seems to act more generally as a depressant of neurons in the frontal lobes. Tetrahydrocannabinol (THC), the active ingredient of marijuana, also acts as a general depressant, but the mechanism is not well understood. (See Chapter 15 for a further discussion of drug interactions.)

Reflex Arcs

A **reflex arc** is a complete impulse-conducting pathway between a sensory neuron and an effector structure (e.g., a muscle) via a motor neuron. The result of stimulating a sensory receptor of a reflex arc is a rapid and automatic response called a **reflex**. Reflexes are protective responses that enable the body to react without requiring time for a voluntary command from the brain. Only a sensory neuron and a motor neuron are required for the simplest kind of reflex arc (the arc shown in Fig. 13-6 has a connector neuron between a motor and a sensory neuron). An example of a reflex is the knee-jerk response to a tap just below the kneecap. In this specific reflex, pressure-receptor neurons in the kneecap ligament are stimulated, inducing nerve impulses that travel in the afferent fibers of sensory neurons to their endings in the spinal cord. Here the impulses are induced in dendrites of motor neurons, and these travel through the efferent fibers of motor neurons to endings in cells of an extensor muscle in the thigh. The neurons involved in a reflex arc are not in actual contact with each other;

rather, the end of the axon of one neuron is separated by a narrow gap, the **synapse**, from the dendrite of the associated neuron.

Many reflexes also involve one or more connector neurons. The connector neurons are located in reflex centers in the CNS and relay impulses to higher awareness and coordination centers in the brain (Fig. 13-6). For instance, some connector neurons may stimulate neurons in the spinal cord that lead to pain centers in the brain. Other impulses may travel over other neurons associated with the reflex arc and produce compensatory actions. For example, a person steps on a tack and an almost instantaneous reflex response is to pull up the injured foot. If this were all that happened, he would almost instantly topple over; however, connector neurons send messages to the other leg from balance centers in the inner ear, and to many other muscles of the body, so that a one-legged stance can be maintained.

SENSORY RECEPTION

Sensory receptors are neurons having one of several kinds of endings capable of specific stimulation by pressure (touch, deep pressure, sound waves), light, irritants (pain), temperature (hot or cold), or chemicals (taste and smell) (Fig. 13-7). Further specializations within each category provide a great range of senses.

At this point it should be clear that there is a great variety of sensory neurons, and that detailing their individual actions would be an almost endless task. Therefore, only general principles of sensory reception, transmission, and recognition will be considered here. First, regardless of the stimulus or the kind of receptor involved, the message conveyed to the CNS is the same in all cases, namely, a nerve impulse. Second, receptors

FIGURE 13-6
Reflex arc.

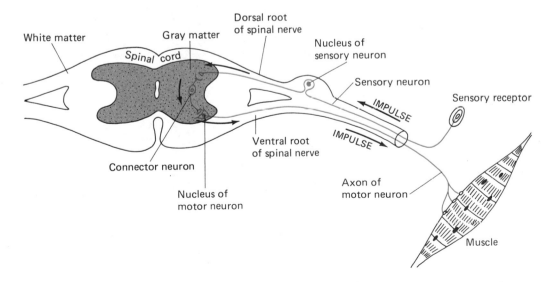

produce an **all-or-none impulse**. That is, receptors do not send a weak impulse to a weak stimulus and a strong impulse to a strong stimulus. Each receptor type has a threshold below which it is not stimulated and above which it generates an impulse in a nerve fiber. If, however, a stimulus is strong, it is likely that a rapid succession of impulses will be transmitted to the CNS. Also, the intensity of a stimulus is discriminated on the basis of different threshold values of receptors.

Because the nerve impulse is the same regardless of the stimulation, interpretation of the message must occur in the brain rather than at the site of the receptors. Specific sensory areas in the brain reconstruct the information presented by the impulses routed to specific points in those areas via sensory and connector neurons. That this is the case for vision is demonstrated by certain kinds of brain damage that cause partial or complete blindness even though the eye and its photoreceptors are intact.

Four kinds of sensory nerve endings are present in the skin, functioning as receptors of heat, cold, touch and pressure, and pain (refer to Figs. 13-7 and 13-8). They are more numerous in some regions than in others. The patterns of their distribution can be ascertained easily by touching various body surfaces with a bristle (for touch and pressure) or with hot or cold rods (for temperature). Pain receptors are the most numerous and are found in tissues and organs throughout the body, with the exception of the central nervous system. Tumors have been removed from the brains of conscious patients without causing pain.

The perception of orientation in space is produced by a type of deep-pressure nerve ending called a *proprio-*

ceptor. Proprioceptors are spindle-shaped structures lying among muscle fibers and tendons; these structures give an individual a sense of what muscles and limbs are doing even though unseen.

Light Perception

The stimulation of any sensory neuron results in a change in the electrical properties of the plasma membrane of the cell. The sensory receptors of the vertebrate eye are pigment-containing cells of the **retina** (Figs. 13-9 and 13-10). Here there are two kinds of photoreceptor cells: **rods** and **cones**. Rods respond to all wavelengths of visible light (in effect, give vision in shades of gray), whereas cones, of which there are three subtypes, respond selectively to red, blue, or green wavelengths, thus adding the element of color vision to the eye.

In the eyes of primates (monkeys, apes, and humans), cones are particularly numerous in a circular re-

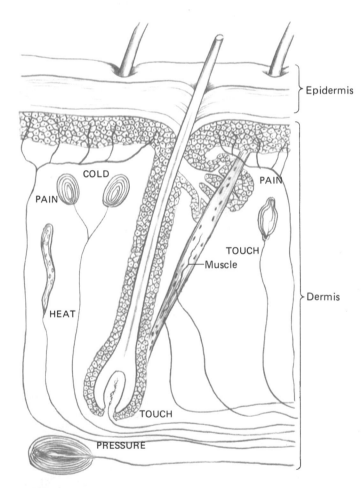

FIGURE 13-8
Sensory receptors in the human skin. (Redrawn from W. R. Amberson and D. C. Smith, 1939, *Outline of physiology*, New York: F. S. Crofts and Co. Copyright © 1939 by Williams & Wilkins Co., Baltimore)

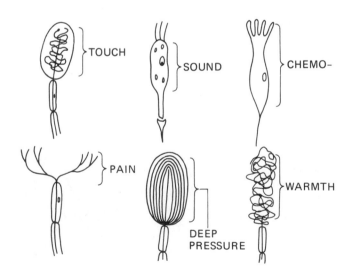

FIGURE 13-7
Sensory receptors.

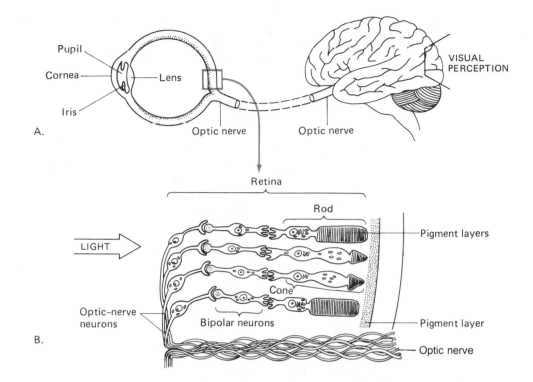

FIGURE 13-9
Light reception of the vertebrate eye.

gion called the **fovea** located at the focal point of the lens. Rods and cones are highly modified ciliated cells in which the shaft of the cilium is enlarged and filled with transverse membranes.

Light impinging on the visual pigments in the interior of the rods or cones produces changes in the pigment molecules (the pigment in rods is **rhodopsin**, also called **visual purple**; the cone pigment is a slightly different molecule, **iodopsin**). In turn, changes in pigment molecules produce a change in the plasma membrane of the photoreceptor cell. This results in generation of an impulse which travels to a synapse with a **bipolar neuron** (a type of connector neuron), where a second impulse is generated. That impulse travels to a synapse with an optic-nerve neuron, generating a third impulse, which is transmitted to the visual area of the brain, where a perception of a visual image is reconstructed (see Fig. 13-9).

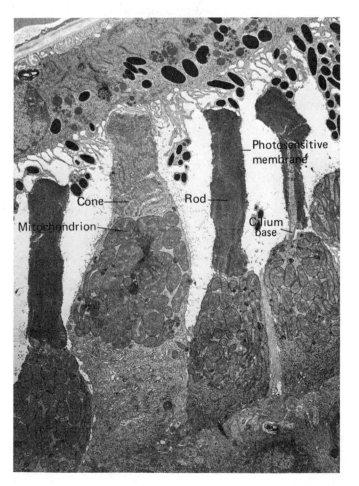

FIGURE 13-10
Electron micrograph showing rods and cones of a squirrel retina. (Squirrels have color vision, unlike many mammals.) Rods function in dim light; cones function in bright light and color vision. (Photo from D. H. Anderson and S. K. Fisher, 1976, The photoreceptors of diurnal squirrels: outer segment structure, disc shedding, and protein renewal, *J. Ultrastr. Res.* 55:119–41. Reprinted by permission of Academic Press, Inc.)

Sound Perception

The sensory activities of retinal rod and cone cells of the eye illustrate a principle of sensory reception that also applies to actions of other sensory receptors. Although the nature of the stimulus and the mechanism of reception differ in the various kinds of sensory receptor neurons, in each case a stimulus induces a membrane change in a neuronal sensory ending; this in turn induces an impulse which is transmitted in the neuron. Regardless of the initial stimulus, the nature of the impulse is exactly the same in all sensory neurons. Take for example sound perception, which involves sensing of sound waves and therefore is a special case of touch perception. Vibrating air (sound waves) impinging on the eardrum generates equivalent vibrations in a linked series of middle-ear bones that transmits the vibrations to liquid in the **cochlea**, a snail-like chamber of the inner ear. Here, hairs on the surfaces of **hair cells** move in response to vibrations in the cochlear fluid, with the result that an impulse is transmitted to auditory centers of the brain where sound perception is registered (Fig. 13-11; refer also to Fig. 13-12). The hairs, from which hair cells get their name, are modified, nonmotile cilia. When these hairs are disturbed by sound waves or other contact, the hair cells are capable of inducing nerve impulses in surrounding sensory neurons.

Sensory organs of mammals generally are highly developed, those of sound perception more so than in most other vertebrates. External ears are well developed in nearly all mammals, in contrast to the simple ear

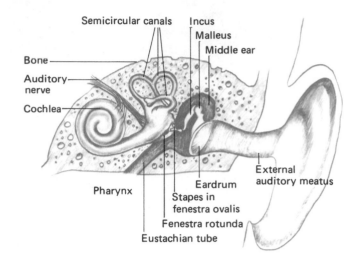

FIGURE 13-12
Human ear.

holes of birds and reptiles. An ear canal, the **external auditory meatus**, terminates inward at the **eardrum**. The **middle ear** is a chamber in which three bones known as the **hammer (malleus), anvil (incus),** and **stirrup (stapes)** conduct sound vibrations to the fluid of the inner-ear canals and sacs through an oval window, the **fenestra ovalis**. The middle-ear chamber is air-filled and is connected with the pharynx by the **eustachian** tube. Changes in outside air pressure normally are adjusted by air movements through the eustachian tube and the middle-ear chamber. If the tube is plugged as a result of

FIGURE 13-11
Sound perception in higher vertebrates.

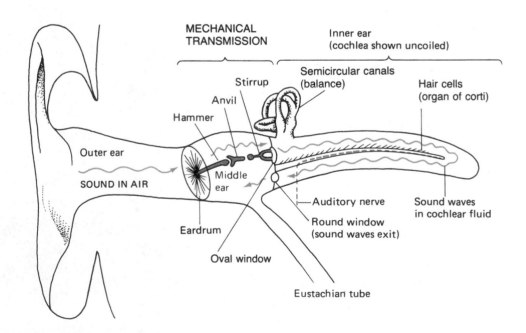

infection or excess mucus, uncomfortable pressure can result.

The inner ear consists of three **semicircular canals** and the snail-shaped cochlea, which contains the ciliated sensory patches of a sound receptor organ called the **organ of Corti** and is associated with a pressure-regulating opening between the middle and inner ear, the **fenestra rotunda ("round window")** (Figs. 13-11 and 13-12).

In addition to sound perception, the vertebrate ear has a second function: sensing the body's position with respect to gravity. In the upper part of the inner ear are three fluid-filled semicircular canals, each connected to a bulblike chamber containing hair cells and crystals of calcium carbonate. Inclination of the head in any direction causes fluid to exert pressure on the crystals, which then impinge on the hairs of a set of hair cells. This results in the generation of nerve impulses in associated sensory neurons. The impulses are registered in the brain and, if the impression is one of imbalance, impulses are relayed to skeletal muscles by way of motor neurons, and corrective action results.

Odor and Taste Perception

The sense of smell is not readily dissociated from taste, and in fact many sensations described as taste also combine perceptions of odors. Chemoreceptors in the back of the nasal passages react to gases dissolved in the air and generate nerve impulses which travel over sensory neurons to the forebrain. Similarly, taste receptors, principally in the tongue, react to chemicals dissolved in food and water, generating nerve impulses that are perceived in the brain principally as sweetness or sourness (refer to Fig. 13-13). The fact that many substances may combine chemicals capable of stimulating several different chemoreceptors makes the discriminatory capability of the nervous system seem almost infinite.

Smell reception occurs when molecules of odoriferous substances come in contact with chemoreceptor neurons in membranes of nasal passages. There the odor molecules become bound to acceptor molecules in the plasma membrane of the chemoreceptor by some action resembling the key-in-lock principle.

The sense of smell is thought to involve seven different kinds of receptors present in mammalian nasal passages, including those capable of differentiating between the following odors: musk, flowerlike, minty, pungent, camphorlike, etherlike, and putrid. It has been suggested that these receptors differentiate between odors by "recognition" of molecular shapes. Molecules of quite

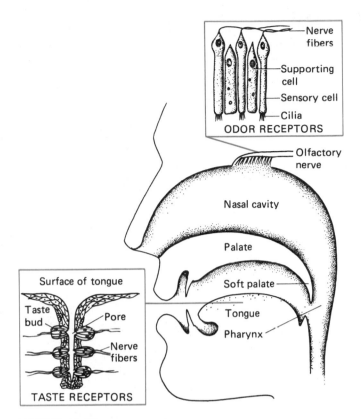

FIGURE 13-13
Human smell and taste receptors.

dissimilar chemistry may be perceived as having identical odor if the overall shapes of the molecules are such that they "fit" the conformation of a particular kind of acceptor molecule in a binding site in the plasma membrane of a chemoreceptor. The sense of taste is less discriminating, being based on four kinds of chemoreceptors—sweet, bitter, salt, and acid—located in the tip, base, and sides of the tongue. A gourmet palate necessarily depends on both taste receptors and odor receptors, and if nasal passages are blocked as a result of a cold or some allergy, food tends to seem rather tasteless. The aging process also diminishes taste sensitivity, and repeated damage to taste receptors (hot drinks, etc.) reduces taste discrimination.

To sum up sensory perceptions, it appears that regardless of the nature of the initial stimulus or kind of receptor, a stimulus induces an electrochemical change in the plasma membrane of a receptor neuron, producing a nerve impulse which moves through afferent fibers to the CNS. Some stimuli induce reflex actions as well as awareness, and others produce an awareness of an outside set of circumstances but do not necessarily produce reflex actions. In all cases, perception and awareness depend on integrative functions of circuits of neurons in the CNS.

CENTRAL, SOMATIC, AND AUTONOMIC NERVOUS SYSTEMS

The human nervous system, and indeed the nervous system of vertebrates in general, can be divided into the CNS and the PNS. The latter system is further subdivided into **somatic** and **autonomic** systems. These divisions and subdivisions reflect both structural and functional distinctions between the sets of neurons involved, but it should be remembered that they are not exclusive structural and functional entities. Rather, they are integrated with one another by interconnecting neurons.

The CNS, in conjunction with the somatic division of the PNS, exerts both voluntary and involuntary control over actions of the skeletal muscles. Sensory and connecting motor neurons of the CNS and somatic system also participate with the autonomic nervous system (ANS) in regulating involuntary actions, which include glandular secretion and actions of smooth muscles of visceral organs, arteries, hair follicles, the eye, and other structures. However, the motor neurons that directly stimulate these structures are components of the ANS. The cell bodies of sensory neurons are located in the brain and in the dorsal-root ganglia of spinal nerve trunks (Figs. 13-14 and 13-15). Somatic motor neurons have their cell bodies in the brain and spinal cord, as do motor neurons that connect the CNS with ganglia of the ANS. The ANS is composed of motor neurons only, the cell bodies of which are located in ganglia in various parts of the body.

Human Central Nervous System

Structure   Structurally, the human CNS consists of a brain from which 12 pairs of cranial nerve trunks are given off, and a spinal cord to which 31 pairs of spinal nerve trunks are appended (Fig. 13-14). Both brain and spinal cord are hollow organs, originating by infolding of the embryonic neural groove (see Fig. 11-7E and F). Both are basically bilateral in organization, being composed of right and left halves; together they contain an uncountable number of neurons, perhaps as many as a trillion. Both brain and spinal cord are enclosed within fluid-filled membranes, the **meninges**. The **cerebrospinal fluid** within the meninges protects the brain and spinal cord from damage by external impact and also has a nutritive function.

The human brain, like the brains of other vertebrates, consists of three regions: forebrain, midbrain, and hindbrain. The mammalian forebrain is relatively

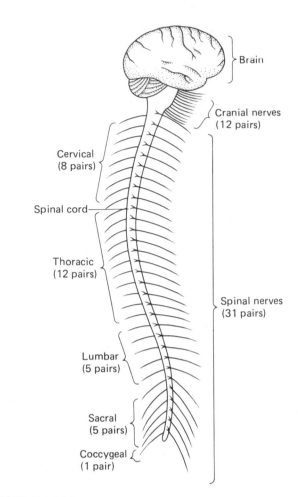

FIGURE 13-14
Human central and peripheral nervous systems in outline.

larger and more complex than those of other vertebrates. (*Brontosaurus*, the gigantic "thunder lizard," had a brain no larger than that of a cat!) This development of the mammalian brain is due principally to expansion of the **cerebral hemispheres** of the forebrain, in which a number of centers of sensory perception, motor skills, learning, and memory are located.

Expansion of the mammalian cerebrum has occurred both by an increase in the total mass and by development of surface folds. In humans, the average weight of the cerebral hemispheres is about 1250 g (44 oz) (out of a total brain weight of approximately 1650 g [58 oz]), and the surface convolutions provide an area of about 0.14 m^2 (1.5 ft^2). The increased area is significant because it provides room for a great number of cerebral neurons and **glial cells**. (The function of glial cells is unknown, but it is thought that they support or possibly protect neurons.) The presence in the **cortex** (outer part of the brain) of cell bodies of sensory and motor neurons, as well as unmyelinated connector neurons, gives

IS YOUR BRAIN REALLY NECESSARY?

Professor John Lober, a neurologist at Sheffield University, in England, asked a conference of pediatricians, "Is your brain really necessary?" and went on to say, "There is a young student at this university who has an IQ of 126, has gained first-class honors in mathematics, and is socially completely normal. Yet the boy has virtually no brain. His cranium is filled mainly with cerebrospinal fluid." How can this strange paradox be explained? The student exemplifies a type of hydrocephaly in which increased pressure of cerebrospinal fluid has expanded the brain ventricles, filling 70–90 percent of the cranium. Many such individuals are mentally retarded, but about half have IQs greater than 100, and a few have apparently normal brain function, as in the instance cited. Professor Lober concludes that "there must be a tremendous amount of redundancy or spare capacity in the brain, just as there is in the liver and kidney."

it a gray color; hence the brain's exterior is known as **gray matter**. Beneath the gray matter is a region composed principally of myelinated neurons that interconnect the brain neurons.

Spinal Cord The spinal cord, like the brain, has a bilateral organization, but here the position of gray and **white matter** is reversed. (White matter is composed of myelinated extensions of brain and spinal cord neurons.) A cross section of the spinal cord (Figs. 13-6 and 13-15) shows a butterfly-shaped gray region enclosed within white matter. On each side of the cord, the proximal parts, or dorsal and ventral roots of nerve trunks, can be distinguished. Dorsal roots contain sensory neurons with cell bodies in ganglia near the junction with the spinal cord. Ventral roots contain axons of motor nerves (whose cell bodies lie in the gray region of the spinal cord) within a short distance of the spinal cord. On either side of the spinal cord, the dorsal and ventral roots merge to form pairs of spinal nerve trunks, which extend outward through gaps between adjoining vertebrae (Fig. 13-15). Both spinal and cranial nerve trunks contain motor and sensory axons and therefore are known as **mixed nerves**.

Functionally, the CNS is a center of sensory perception, awareness, learning, and many motor functions. As noted earlier, many of the motor functions are located in the spinal cord, existing as reflex arcs. Although reflex responses are of an involuntary nature, awareness

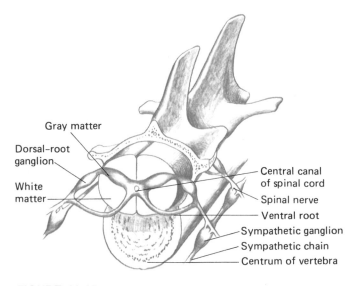

FIGURE 13-15
Human vertebral column showing arrangements of spinal nerves and bilateral symmetry of the spinal cord and its branches. (Redrawn from A. M. Elliot and D. E. Outka, *Zoology*, 5th ed., © 1976, p. 343. Reprinted by permission of Prentice-Hall, Inc., Englewood Cliffs, N.J.)

of those involving sensory receptors in the somatic system of the PNS resides in the brain. In the knee-jerk reflex, for example, a person is usually aware both of the sensory stimulus and of the motor response. Many other kinds of reflexes in the body, however, do not produce awareness. Among them are secretions of glands, opening and closing of the pupil of the eye, movements of

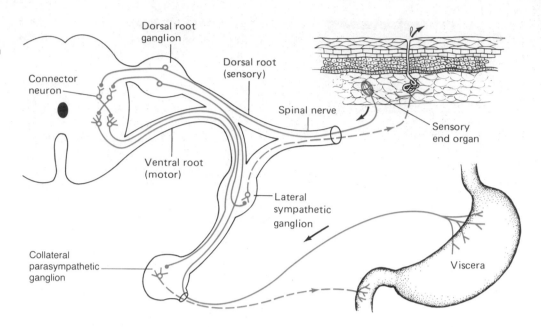

FIGURE 13-16
Nerve pathways of the autonomic nervous system. (Redrawn from A. M. Elliott and D. E. Outka, *Zoology*, 5th ed., © 1976, p. 344. Reprinted by permission of Prentice-Hall, Inc., Englewood Cliffs, N.J.)

stomach and intestinal muscles, and minor changes in the rate of heartbeat. The nerve pathways regulating these kinds of action are part of the autonomic nervous system (Figs. 13-7 and 13-16).

Autonomic Nervous System

Functionally, the ANS consists of two sets of motor neurons that either accelerate or depress actions of glands, organs, and other structures they regulate. For example, the **sympathetic system** of neurons may accelerate actions of an effector structure; the other set, the **parasympathetic system** of neurons, may repress actions of that effector. In each specific activity of an effector structure, the two systems therefore can act antagonistically (Fig. 13-17). As a result of this kind of feedback regulation, organs, glands, and other structures can respond to situations requiring either extra effort or relative inaction. Secretions of sweat glands, for instance, are stepped up when the body is exposed to higher air temperatures and inhibited at lower temperatures when cooling is not required. In much the same manner, the rate of breathing is increased during periods of stress and decreased when the body is resting. This latter example is exceptional, however, because the CNS exercises a measure of control over the action, although it cannot override the ANS completely. If a person consciously stops breathing to the point of suffocation, the ANS takes control and restores breathing action.

Structure   Structurally, the ANS consists of two sets of ganglia. One set is composed of two rows of **lateral sympathetic ganglia** on each side of the spinal cord and large sympathetic abdominal ganglia (the **solar plexus** is one). The other set consists of **collateral parasympathetic ganglia** located near the structures they serve (Figs. 13-16 and 13-17). Both groups of ganglia are connected with the CNS by motor neurons in spinal or cranial nerve trunks, and both receive impulses originating in sensory neurons of the PNS and relayed via synapses in the brain and spinal cord to motor neurons in cranial or spinal nerve trunks. These motor neurons, in turn, form synapses with sympathetic and parasympathetic neurons in ganglia of the ANS.

Fight-or-Flight Response   The fight-or-flight response was alluded to in Chapter 10 in connection with the adrenal hormones (adrenaline and norepinephrine). This response, although hormone related, is initially generated by sensory receptors (sight, sound, odor, taste—any or all). The interpretation of the message by the brain and the response to the message can be varied: Danger? Flee? Pursue? Hide? Surrender? We humans experience fight-or-flight sensation in many emergencies—a near miss or a hit in a traffic incident, a loud noise behind our backs, and so on. The heartbeat increases, "goose bumps" appear, the eye pupils dilate and things appear brighter, the body tenses, and pain, if present, disappears. This action of readying the body to respond in an emergency involves a complex interaction between the nervous system, hormones, and effectors (muscles, glands, etc.) Assessing the situation requires action by the central nervous system; readying the body for action involves the autonomic nervous system, in particular the sympathetic pathways (refer to Fig. 13-16), and hormones. In an emergency, the sympathetic system inhib-

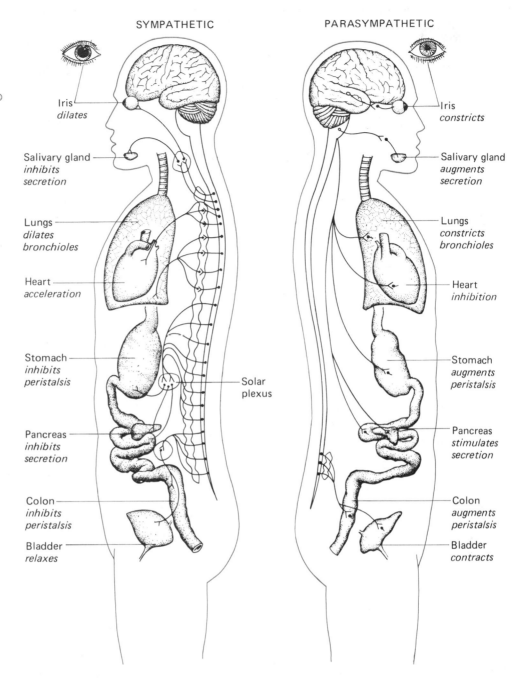

FIGURE 13-17
Sympathetic and parasympathetic nervous systems. The actions of each system upon the structures served are specified. (Redrawn from A. M. Elliott and D. E. Outka, *Zoology,* 5th ed., © 1976, p. 346. Reprinted by permission of Prentice-Hall, Inc., Englewood Cliffs, N.J.)

SYMPATHETIC

Iris
dilates

Salivary gland
*inhibits
secretion*

Lungs
*dilates
bronchioles*

Heart
acceleration

Stomach
*inhibits
peristalsis*

Pancreas
*inhibits
secretion*

Colon
*inhibits
peristalsis*

Bladder
relaxes

Solar
plexus

PARASYMPATHETIC

Iris
constricts

Salivary gland
*augments
secretion*

Lungs
*constricts
bronchioles*

Heart
inhibition

Stomach
*augments
peristalsis*

Pancreas
*stimulates
secretion*

Colon
*augments
peristalsis*

Bladder
contracts

its some functions and accelerates others, causing smooth muscles in hair follicles (see Fig. 13-8) to contract ("goose bumps"), arteries to constrict (blood pressure rises), stomach to slow down (that "sinking feeling"), eye pupils to dilate (vision appears sharper), and the heart to beat faster. These actions all are part of the body's readiness to face an emergency.

Next, based on the brain's assembly and interpretation of all the available information, the appropriate course of action is decided upon. Then, when the danger has passed, the neurons of the parasympathetic path-

ways take action to restore a more or less relaxed state: heartbeat slows, stomach and intestinal peristalsis resumes, eye pupils constrict, muscles of hair follicles relax, and so on.

THE BRAIN

The invertebrate brain is scarcely more than the anteriormost and largest of a sequence of ganglia (concentrations of neurons) distributed the length of a nerve cord

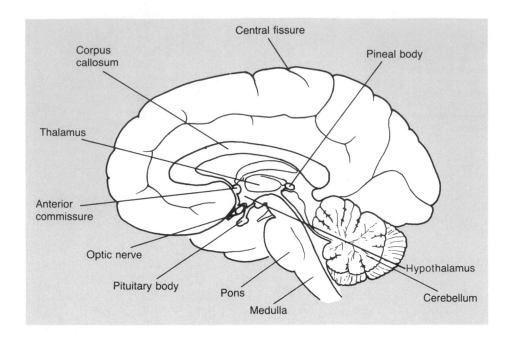

FIGURE 13-18
Human brain, sectional view.

Central fissure

Corpus callosum

Pineal body

Thalamus

Anterior commissure

Optic nerve

Pituitary body

Pons

Medulla

Hypothalamus

Cerebellum

(Fig. 13-1B). Vertebrates, on the other hand, possess relatively well-developed brains composed of the forebrain, midbrain, and hindbrain. The forebrain consists of an anterior **telencephalon** and posterior **diencephalon**. The former is associated with senses of smell and sight and in birds and mammals is greatly enlarged into two cerebral hemispheres which cover the diencephalon (Fig. 13-18).

The Forebrain

The greatest expansion of the mammalian forebrain has occurred in the cerebral hemispheres. Further study of the human cerebrum shows that in addition to right and left halves, it is partially divided into anterior and posterior regions by a deep fissure. It is the anterior of these, the **frontal lobes,** that controls many motor functions and thought processes. Three posterior lobes also are present in each hemisphere: **parietal lobes,** in which sensory and speech centers are located; **occipital lobes,** associated with visual perception; and **temporal lobes,** where sound-perception centers are located.

Scientific understanding of the functioning of different brain regions is far from complete. Much of the present knowledge of human brain function has come from destruction of portions of the brain by accident or disease, with corresponding disruption of body functions. Additional information has resulted from the use of electrical probes to stimulate various brain centers during surgery. Finally, experiments with laboratory mammals, monkeys and apes in particular, have given insight into probable human brain functions.

As a consequence of accidental damage or brain surgery, or by inference from experiments with laboratory animals, it has been learned that each cerebral hemisphere is capable, to a degree, of independent functioning. Both hemispheres, however, normally are coordinated by nerve impulses passing between them through a dense basal band of neurons, the **corpus callosum** (Fig. 13-18).

Some years ago, before the availability of effective medication, one of the treatments for severe epilepsy was surgical destruction of the corpus callosum. In many such cases there appeared little superficial change in behavior of the patient, except for virtual disappearance of the symptoms of epilepsy. However, when these patients were positioned so that the visual field (i.e., what the eye sees) of the right eye was screened from the left eye, it was found that they could not describe what was seen by the right eye. Conversely, they had difficulty in performing manual tasks in which only the left eye could see objects. (It should be pointed out that the optic nerves of the right and left eyes cross before they enter the brain; what the right eye sees is relayed to visual centers in the left hemisphere, and vice versa.) From such experiments, as well as from other evidence, it is known that the cerebral hemispheres are not functional duplicates. For example, the main language center is in the left hemisphere, and spatial abilities reside in the right hemisphere.

The diencephalon contains the **thalamus** and **hypothalamus,** which regulate temperature, sleep, hunger, and certain sex hormones. Two small bodies, the **pituitary gland** and the **pineal body,** project from the diencephalon and are active in the hormonal regulation of numerous physiological activities.

The thalamus is a primary relay point for incoming sensory impulses; the hypothalamus is a control center, regulating a number of body functions, including temperature responses, water balance, appetite, blood pressure, and sleep. The hypothalamus also exerts control over secretions of the pituitary gland, sometimes referred to as the master gland of the body.

Experiments have been done in which electrical probes were inserted in or near the hypothalamus or the greater **limbic system,** a part of the forebrain which modulates such behavioral patterns as eating, drinking, and sexual activities. Application of very small electric currents to such probes sometimes has produced responses interpreted to indicate extreme pleasure; hence the term **pleasure center** has been applied to points producing these responses. The pleasure centers of rats have been wired with probes which were attached to switches that the animal could trip. The result has been a tireless tripping of the switch, to the point of exhaustion.

The Midbrain

The human midbrain consists largely of neurons connecting the forebrain with other parts of the brain and spinal cord. In lower vertebrates the midbrain is the chief visual center of the brain, but in mammals, that function has been shifted to the cerebrum, and the midbrain functions as a pathway to the **cerebrum** for nerve impulses transmitted from the eye by a pair of **optic nerves.**

The Hindbrain

The hindbrain consists of two regions: the **cerebellum** and the **medulla oblongata.** The cerebellum is the seat of coordinated control of muscle functions. It has a gray convoluted dorsal surface and a ventral enlargement, the **pons,** which consists of nerve fibers connecting the cerebellum with sensory areas of the cerebrum.

Damage to the cerebellum causes loss of equilibrium, jerky movements, and tremors; its total removal from laboratory animals results in completely unsynchronized movement but does not immediately result in death. The medulla oblongata is an expansion of the forward end of the spinal cord and is an important reflex center, coordinating and controlling heartbeat, breathing, gastric movements, and other functions.

Learning and Memory

Storage and retrieval of information in the brain constitutes **memory,** which can be defined as the ability to recall a past experience. At a very primitive level of brain function, behavior of organisms apparently is independent of memory and may be interpreted as the result of a number of integrated reflexes. This commonly is called **innate behavior,** or **instinct.** However, other behavioral actions which seem to be based on experience can be observed even in such simple animals as planarians (see Fig. 13-1), and this kind of behavior, which seems at least partially based on experience, is called **learning,** or **learned behavior.** Both innate and learned behavior will be discussed more extensively in Chapter 22; for the present, attention will be given only to a discussion, some of it quite speculative, of the nature of memory and learning.

Types of Memory How is information stored in the brain? It is tempting to compare the brain to a computer, which has the ability to store information for later recall and application. Even a simple hand-held calculator can store a limited amount of information so that the "memory" button can be pressed, resulting in retrieval of certain mathematical functions.

Although computers seem very complicated to many people, and indeed are marvels of electronic circuitry, they all have a simple basis, for they consist of a multitude of bits, or on–off switches, connected by circuit wires. Such bits and circuits may be compared functionally with neurons and synapses, although the circuitry of the human brain is infinitely more complex, both structurally and functionally, than the largest of computers. Nevertheless, one appropriate question is whether the storage of memory in the brain has a physical basis analogous to that of a computer. Does memory consist of opening and closing the equivalents of on–off switches so that impulses follow highly specific pathways, or does memory reside in some other form, possibly encoded in complex molecules such as nucleic acids and proteins?

Two kinds of human memory are recognized: **short-term** and **long-term.** Short-term memory fades rapidly, as for instance the length of time one remembers a telephone number after looking it up and dialing it. Long-term memory, on the other hand, may be retained maximally for a lifetime, minimally at least a few days.

Both short-term and long-term memory are thought to involve establishment of specific nerve impulse pathways known as **memory loops,** or **traces.** One explanation of memory is that an experience is registered in the brain by passage of an impulse through a circuit of neurons by way of a specific series of synapses. Because the brain's connector neurons may have as many as 1000 synapses each, the potential for establishing such unique impulse pathways is great.

It is not known if short-term memory is totally lost. In human brain surgery, electrical probes sometimes are used to analyze brain function, and on occasion they produce recollections of seemingly trivial incidents from the past.

Whether or not short-term memory is stored, it is known that transfer from short-term to long-term memory can be accomplished by reinforcement of an experience. For example, many repetitions of words, music, or manipulations effect this transition. So too does trauma associated with experience—a child touching a hot stove, for example. This suggests that an impulse pathway in the brain (a memory loop) is strengthened by repeated passages of impulses through it or by the intensity of an experience. It is not known, however, if these induce permanent changes in synapses so that, for example, they more readily produce neurotransmitters, or whether alternate impulse pathways become blocked.

Memory Molecules It has been suggested that memory may be encoded in macromolecules stored either in neurons or in accessory glial cells of the brain. There is some evidence favoring this concept in that chemicals that block the transcription of DNA into mRNA also block acquisition of long-term memory. This suggests that long-term memory may exist either in the form of mRNA, which has the ability to encode considerable information, or in protein molecules synthesized by translation of mRNA. Some rather puzzling experiments have been done in which brain extracts containing DNA or RNA were taken from trained laboratory animals. When such extracts were fed to or injected into untrained animals, the untrained animals appeared to duplicate some of the simple responses learned by the donor animals.

Research workers, however, have had difficulty obtaining confirmation of these results in duplicate experiments, and there is considerable scepticism about the actual existence of so-called memory molecules.

In spite of doubts as to whether there are such things as memory molecules, it is possible that long-term memory may involve synthesis of specific kinds of mRNA and proteins. For instance, establishment of a certain set of synapses might involve production of enzymes responsible for the synthesis of certain neurotransmitters or the inactivation of others. Because enzymes are highly individual proteins produced by the translation of specific mRNAs by ribosomes, a link between memory and synthesis of macromolecules cannot be dismissed completely.

SUMMARY

Coordination of the functions of cells, tissues, and organs in the animal body is accomplished by the nervous system and hormones.

Messages travel as nerve impulses. These impulses are produced by localized changes in the electrical charges of the membrane of the neuron. These charges are propagated in waves the length of the nerve fiber. Impulses typically travel from dendrite to axon and stimulate the release of chemicals that diffuse across a synapse, which associates the axon ending with a dendrite of the next neuron.

The center of nervous coordination is the brain. Here sensory and motor neurons are associated in such a way as to control complex functions such as coordinated movements, memory, and learning.

KEY WORDS

neuron	action potential	somatic nervous system
nerve impulse	sodium/potassium-ion pump	autonomic nervous system
central nervous system	synaptic vesicle	crainial nerve
nerve cord	synaptic cleft	spinal nerve
nerve trunk	neurotransmitter	sympathetic nervous system
peripheral nervous system	opiate receptor	parasympathetic nervous
sensory neuron	synapse	system
motor neuron	sensory receptor	brain lobe
connector neuron	retina	thalamus
axon	rod	hypothalamus
dendrite	cone	learning
Schwann cell	cochlea	memory
reflex arc	hair cell	

QUESTIONS FOR REVIEW AND DISCUSSION

1 What determines the direction of the nerve impulse in a neuron? Explain.

2 What appears to be the major function of the Schwann cell?

3 Describe the "fight-or-flight" response, and tell what systems of the body are involved.

4 What is the nature of the nerve impulse? How rapidly does it move? What is its effect when it reaches the end of a nerve fiber?

5 What part of the brain is the center of conscious thought?

6 What is the possible nature of learning? Of memory?

7 Contrast the actions of the sympathetic and parasympathetic systems.

8 What are opiate receptors? How do narcotics act on neurons?

SUGGESTED READING

HELLER, H. C.; L. I. CRAWSHAW; and H. T. HAMMEL. 1978. The thermostat of vertebrate animals. *Sci. Amer.* 239(2):102–13. (Compares ectothermic and endothermic vertebrates and describes their thermoregulation.)

HUDSPETH, A. J. 1983. The hair cells of the inner ear. *Sci. Amer.* 248(1): 54–64. (Great micrographs and diagrams! Tells how the cilia of cells translate mechanical action into nerve impulses.)

KEYNES, R. D. 1979. Ion channels in the nerve-cell membrane. *Sci. Amer.* 240(3):126–35. (Discusses nerve impulse propagation relative to neuronal membranes.)

LESTER, H. A. 1977. The response to acetylcholine. *Sci. Amer.* 236(2):106–18. (Describes an important neurotransmitter and its action in neuromuscular synapses.)

O'MALLEY, B. W., and W. T. SCHRADER. 1976. The receptors of steroid hormones. *Sci. Amer.* 234(2):32–43. (Discusses the nature of hormones and their transport and action within target cells.)

PARKER, D. E. 1980. The vestibular apparatus. *Sci. Amer.* 243(4):118–35. (Beautiful diagrams and photographs. Discusses many aspects of balance and orientation, motion sickness, and weightlessness in space flight.)

ROUTTENBERG, A. 1978. The reward system of the brain. *Sci. Amer.* 239(5):154–64. (Discusses pleasure centers of the brain.)

SCIENTIFIC AMERICAN. September 1979. Vol. 241, no. 3. (The entire issue is devoted to articles on the brain. See especially "The Neuron," by C. F. Stevens, and "The Chemistry of the Brain," by L. L. Iversen.)

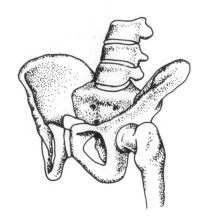

Movement in Animals—Skeletons and Muscles

14

How do animals move? The contraction of muscle cells offers only a partial explanation, although alternate contraction and relaxation of muscles is a common denominator in all animal movements. A jellyfish swims languidly by rhythmically contracting a ring of muscles in the rim of its bell-like body. Squids use jet propulsion, squirting water out of a nozzle protruding from a mantle cavity. Here also muscle contraction is the motive force even though water is the propulsive agent. It is noteworthy that neither jellyfish nor squids possess skeletons, although the latter have vestigial internal shells. As you may know, one of the principles of animal locomotion holds that muscles must have a support of some kind against which to work (imagine rowing a boat without oarlocks!). Jellyfish and squids are of course completely aquatic animals and, in both, the support or "skeleton," is nothing more or less than water. This is not as preposterous as it might sound because, as you know, water is incompressible—it has no "give" as long as it is within a container of some kind. One group of animals, the annelids (to which earthworms and leaches belong), have even applied this principle to life on land. Their compartmented bodies contain a watery fluid, and pressure exerted

against this fluid by sets of body muscles produces the accordianlike locomotion for which these organisms are well-known. In this example, the internal body fluid functions as a skeleton and is in fact referred to as a hydroskeleton.

Although many aquatic organisms have soft, unskeletonized bodies, many others have skeletons composed of one or another kind of structural material. So also do almost all land animals.

SKELETAL SYSTEMS

Ages ago, when organisms lived exclusively in the sea, many are thought to have lacked skeletons, relying instead on the buoying effect of water, as do present-day jellyfish and even giant squids many meters in length. That was all well and good as long as life was confined to water, but if stranded on a beach such soft-bodied creatures undoubtedly found life hazardous. Eventually however, during the long course of evolution, some animals developed primitive skeletal features such as hard spines and shells. Perhaps at first these were merely protective devices, but subsequently some of these structures were further modified and adapted as aids to locomotion.

Functions of Skeletons

Skeletons serve three basic functions. These are: (1) protection, (2) support, (3) aids in locomotion. In some simple organisms, such as nonmotile sponges, skeletons are supportive. Crystalline particles and/or a network of leathery spongin fibers function mainly to maintain the integrity of a complicated system of water passages. Farther up the scale of animal life, we see that mollusks (clams, snails, etc.), though generally considered soft-bodied animals, have durable external skeletons (**exoskeletons**). Here again the functions of the skeleton are protection and support, but in some cases there may be the third function of locomotion (as in scallops; see Fig. 14-1).

It should not escape your notice that a basic principle of muscle action is illustrated by the **jointed skeletons** of clams and scallops. True, the shell of a clam has only one joint and thus is far less complicated than the skeletons of walking, running, and flying animals. Nevertheless the basic principle applies, that of a muscle attached to skeletal elements on either side of a **moveable joint**. In clams and scallops, the two shells or valves are hinged at the side by an elastic ligament that tends to spring the shells apart and are equipped with strong **adductor** muscles (adductor, "pull toward"). These muscles in clams are smooth muscle and, although they act relatively slowly, they can exert a very strong force and maintain it for a considerable time. In scallops the adductor muscle is divided into a second set of striated muscle, which can contract rapidly and repeatedly to produce the clapping action familiar to scallop gatherers. At each clap, a jetlike spurt of water jumps the scallop a halfmeter or so along an erratic path.

Animals with more sophisticated jointed skeletons than clams and scallops have two or even more sets of muscles at each joint. These sets occur in oppositely acting sets; one set produces a movement in one direction, the other in the opposite direction.

The Arthropod Skeleton

Arthropods are animals with jointed **exoskeletons**. They are the most numerous of all animal groups, comprising more than 900,000 species (compared, for example, with about 38,000 vertebrate species). Included are aquatic forms such as crabs, lobsters, barnacles, and the extinct trilobites. Land arthropods include a great diversity of forms, among them spiders and scorpions, centipedes, millipedes, and most numerous of all, insects. The arthropod exoskeleton is composed of a complex protein–carbohydrate substance, chitin, which may be further toughened by impregnation with calcium salts. Because the exoskeleton is not expandable, growing arthropods must undergo periodic moltings, shedding the too-small exoskeleton and secreting a new one.

We are interested principally in the relationship of the arthropod exoskeleton and its musculature. Arthropods are the only invertebrate group to possess multi-jointed locomotory appendages; they are accomplished swimmers, runners and jumpers, and even flyers (insects only).

The arthropod skeleton consists basically of a series of segments (Fig. 14-2B), each composed of a dorsal plate, two lateral plates, and a ventral plate. Each segment, except the first and the last, has a pair of jointed appendages. In more primitive arthropods, the segments are numerous and identical, or nearly identical, as in centipedes, for example. In more advanced arthropods evolutionary reductions in the number of segments and appendages and specializations of the remainder have produced such derivative body architecture as that seen in insects and spiders. All, however, preserve elements of the basic appendage structure in which internally attached muscles account for mobility (Fig. 14-2C).

In an arthropod leg, such as the femur (proximal or upper element) of a grasshopper, a pair of muscles are attached to the inside of the exoskeleton in such a way as to pull the lower leg element in two directions, **extension** and **flexion**, hence the muscles are called **extensors** and **flexors** (Fig. 14-2C). Note that flexors and extensors must extend across a joint and be anchored at both ends. If one could imagine an arthropod turned inside

FIGURE 14-1
Locomotion and muscle action in the scallop.

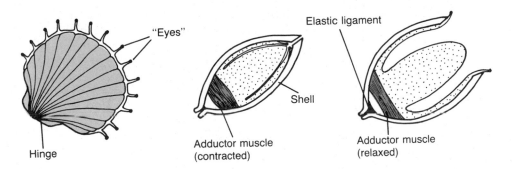

"Eyes"

Hinge

Elastic ligament

Shell

Adductor muscle
(contracted)

Adductor muscle
(relaxed)

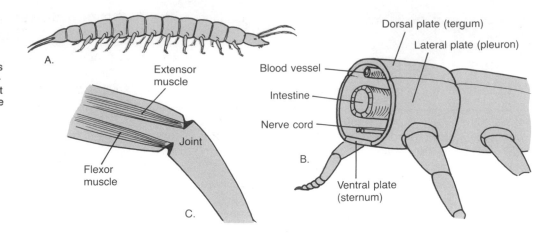

FIGURE 14-2
The arthropod exoskeleton.
A. Primitive arthropod body plan—many similar segments, many appendages. **B.** Elements of the exoskeleton. **C.** Musculature of an appendage. Note that the joint itself is simply a flexible place in the exoskeleton.

out, then the arrangement of muscles and skeletal parts would resemble those of vertebrates in which the skeletal elements are internal and the musculature is external.

The Vertebrate Skeleton

The vertebrate skeletal system is an internal one (endoskeleton) and is the supporting framework of the body. Much is known about its evolutionary history because it often is the only part of an animal to be preserved as a fossil. Skeletons are composed of cartilaginous or bony elements, or combinations of cartilage and bone. Both cartilage and bone are varieties of connective tissue, one of the basic tissue types of animal bodies (see Fig. 8-10). Cartilage is a tough, fibrous, rather elastic material; bone is composed of connective tissue impregnated with mineral salts.

All vertebrates possess an **axial skeleton** composed of a **braincase** and **vertebral column** (Fig. 14-3). In the most primitive vertebrates (**cyclostomes** and **sharks**), the braincase is composed of cartilage; jaws, if present, are separate structures. In more advanced vertebrates (**bony fish, amphibians, reptiles, birds,** and **mammals**), a skull (braincase plus upper jaw) composed

of bony plates is present; a lower jaw is attached to the skull by bands of connective tissue (**ligaments**).

In addition to an axial skeleton, present-day vertebrates except cyclostomes have an **appendicular skeleton** consisting of **pectoral** and **pelvic girdles,** and attached pairs of appendages (fins, legs, or wings) (Fig. 14-3).

Vertebrates evolved in the sea and the skeletons of marine forms show their adaptations to swimming locomotion: long bodies, many vertebrae, fins as lateral appendages. When vertebrates became adapted to life on land, major changes in skeletal structures occurred. Among these were transition of lateral fins into legs, appearance of the rib cage, elaboration of the neck vertebrae, enlargement of the skull and reduction in numbers of skull bones, development of strong pelvic and pectoral girdles to which the fore and hind limbs were attached, and development of the rib cage. These developments were accompanied by changes in the organ systems, including development of lungs and loss of gills, with corresponding changes in the circulatory systems. Except among amphibians, in which certain fishlike characteristics still prevail, modifications in reproductive organs commensurate with land life also occurred.

FIGURE 14-3
Skeleton and muscles of a generalized vertebrate animal.

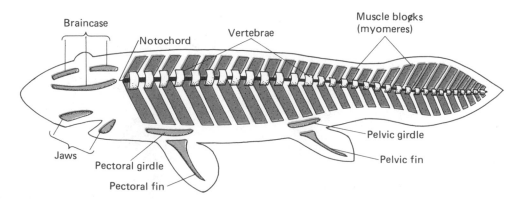

While the **endoskeletons** of vertebrates vary, their composition (**cartilage** and **bone**) is uniform. Bone is composed of 20–25 percent organic material (principally the protein collagen, comprising the connective tissue component), 45–50 percent inorganic compounds (mainly **hydroxylapatite,** a mixture of calcium phosphate and calcium hydroxide), and about 25 percent water. Bones can remain relatively intact long after the decay of soft parts of a vertebrate body, but it would be unfortunate if one retained the impression that bone therefore is always nonliving. In truth, the bones of a living vertebrate are in a state of continual renewal and repair. Populations of living bone cells dissolve old and damaged bone while laying down replacement material. A system of living canals, the **Haversian canals,** each containing an artery, a vein, and usually a nerve fiber, extend throughout the bones of the body (refer to Fig. 8-10B). Bones are classified in two main categories: **dermal bone,** which develops in the skin, and **endoskeletal bone,** which develops internally and is preformed in cartilage. Dermal bone composed the body armor of primitive fishes and persists in the bony plates of skulls, fish scales, and the body armor of turtles. Primitive vertebrates such as sharks and rays have endoskeletons composed solely of cartilage, a connective tissue composed of fibers embedded in **chondrin** (essentially gelatin). In advanced vertebrates, the skeleton is cartilaginous in embryo stages of development; later most of the cartilage is replaced by bone, the cartilage persisting only in joints between long bones and in the form of **intervertebral disks.** Cartilage furnishes a smooth gliding surface much as does Teflon in artifical joints. The long bones of the vertebrate body are not, of course, solid rods; that would make them far too heavy. Instead, many of them are hollow, analogous to steel tubing, the interiors occupied by **marrow,** a blood-forming tissue. In addition, parts of larger bones are composed of spongy bone covered with compact bone, thus preserving strength without undue weight (Fig. 14-5D).

Vertebrate Integument

Skin and associated structures such as scales, feathers, and hair compose the **integument** of vertebrates. Skin (**epidermis**) is not a single layer of cells but a deep and tough coat composed of outer layers of flat to cuboidal cells (**epithelium**) covering a zone (**dermis**) of interwoven connective tissue fibers containing nerves and their sensory endings, lymph, fat cells, and other components, including in some instances (fishes) plates of dermal bone (Fig. 14-4). Protection is a major function of the integument; it also has secretory and respiratory roles.

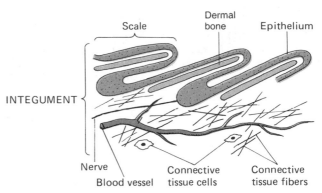

FIGURE 14-4
Integument of a representative vertebrate (fish).

Evolution of the Vertebrate Skeleton

Chordates In architectural history, the transition from medieval castles and cathedrals, with supporting stone or brick walls many feet thick, to modern skyscrapers, with relatively thin walls of glass and aluminum, resulted from the invention a century ago of inner skeletons of steel girders. The same principle, the endoskeleton, made it possible for land vertebrates to attain in some cases truly awesome proportions, whereas the largest land invertebrates of any era are measured in inches.

Animals with backbones constitute the Phylum **Chordata,** a group whose bodies are characterized by the presence of a dorsal rod of tough, elastic connective tissue in at least some stage of development. This rod, called a **notochord,** is considered a first step in the evolution of a **backbone,** and some primitive chordates have not progressed beyond the notochord stage. In **vertebrates,** which are advanced chordates, the notochord is present in embryos, but then is replaced wholly or in part by a **jointed backbone** composed of **vertebrae.** Vestiges of the notochord persist in adults as intervertebral disks (Fig. 14-6B).

Cyclostomes The most primitive vertebrates are the **jawless fish (Agnatha).** Abundant as fossils in early Paleozoic sediments, they are represented today only by a few species of eel-like fish called cyclostomes ("round mouths," hagfish and lampreys; see Fig. 21-24).

Sharks In the history of vertebrate evolution, jawless fish came first and later gave rise to fish having jaws (Fig. 21-26). The best known of the primitive jawed fish, a group called the **placoderms,** had a body sheathed by bony scales and plates. A large shield-shaped helmet of bone covered the head and gill region. The largest placoderm fish were nearly 9 m (30 ft) long. No placoderm fish are known to exist today, but some, upon loss of their bony plates, may have given rise to certain sharks.

A second group of **ancient jawed fish,** primitive members of the class **Chondrichthyes** ("cartilage fish"), were more direct ancestors of present-day sharks. Like modern sharks, they had jaws, cartilaginous skeletons, paired front **(pectoral)** and rear **(pelvic)** fins, and gills interposed between slits in the pharynx and the body wall. They first appeared around 400 million years ago, in the Devonian period of the Paleozoic era, and gave rise to modern sharks, skates, and rays.

Bony Fish   The majority of fish alive today are **bony fish,** or Osteichthyes (see Fig. 21-28). These terms derive from the fact that the fish have a skeleton composed of bone in addition to cartilage. Other features also differentiate them from sharks and sharklike fish: the brain is enclosed within a skull covered by bony plates (dermal bones).

Among the early bony fish were some which possessed lungs and were capable of living on land. These lungfish were not truly terrestrial, but had lobed, paddle-like pectoral and pelvic fins and probably were capable

of locomotion across mudflats and other wetlands. Some of these, it is conjectured, became more and more adapted to life on land. By transformation of their pectoral and pelvic fins into front and back legs and development of water-conserving features (efficient kidneys, resistant skin, etc.), they became the first amphibians.

Amphibians   The earliest amphibians were salamanderlike creatures that probably spent most if not all of their time in water and differed from present-day amphibians in having scales and bonier heads. Being **tetrapods** (four-legged animals), they were capable of crawling on land. From them have descended the two major groups of present-day amphibians: the tailed **urodeles** (mudpuppies [Fig. 21-29], salamanders, newts, etc.) and the tailless **anurans** (toads and frogs) (Fig. 21-29). (Even though frogs and toads are tailless as adults, most have an earlier tadpole stage characterized by both tails and gills.)

Bodies of urodeles retain some features of fish, including numerous vertebrae and muscle blocks in the

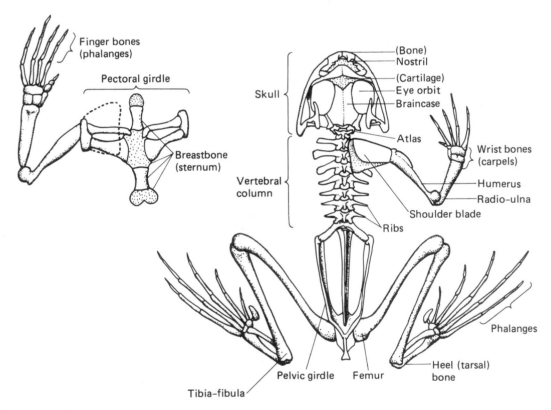

FIGURE 14-5
The skeleton of the frog, an amphibian. Frogs are rather specialized amphibians in comparison with urodeles (mudpuppies, newts, salamanders—the tailed amphibians), and their skeletons are indicative of this specialization—for land life and leaping. Compare with the human skeleton in Fig. 14-6, and note that the frog lacks a rib cage, that the frog has but one neck bone, the atlas, whereas mammals have several (seven), and that the skull of the frog has relatively less room for brain development.

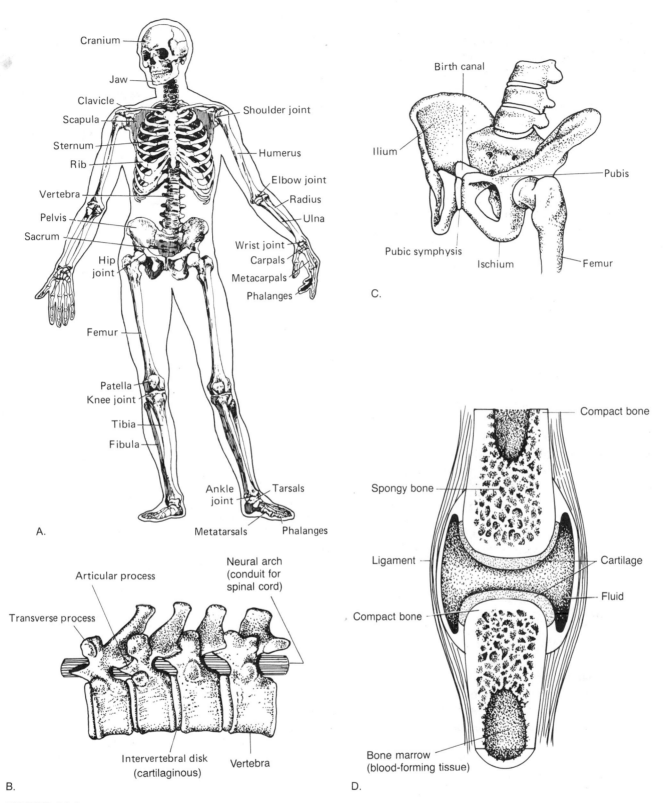

FIGURE 14-6

Elements of the vertebrate skeleton—the human skeleton. **A.** Overall view. **B.** A portion of the vertebral column. **C.** Female pelvic girdle. **D.** A joint. (A from R. F. Oram, P. J. Hummer, Jr., and R. C. Smoot, 1979, *Biology: Living systems,* Columbus, Ohio: Charles E. Merrill)

tail, but bodies of frogs are not fishlike: the vertebrae are very few, and muscle segments are not prominent. Unlike fish, amphibians have a neck bone (the **atlas**; see Fig. 14-5), and hence can turn their heads, although only to a limited extent. Nearly all amphibians have well-defined sets of arm and leg muscles which produce a variety of movements.

Reptiles and Birds   Reptiles and birds are thought to be closely related, as will be discussed in Chapter 21. Their skeletons have many common features, especially those of the strongly bipedal dinosaurs, which are probably the nearest ancestors of birds.

In comparison with amphibians, major advances in skeletons of reptiles are the further elaborations of neck vertebrae, ribs, and pectoral and pelvic girdles.

Scales of reptiles are unlike fish scales; they consist of folds of skin having an outer horny layer composed of a tough protein called **keratin**. They are not therefore homologous with fish scales, which contain dermal bone. Hair and feathers are thought to have evolved from reptilian scales, as all are composed of keratin. Some reptiles, such as turtles, have bony plates (dermal bones) embedded in the skin. This feature also was prevalent in certain extinct reptiles characterized by elaborate body armor.

The principal adaptations of the bird skeleton are a stiff and shortened trunk with fused vertebrae; a strong rib cage and a deep breastbone (**sternum**); a long, flexible neck; strong hind legs variously adapted for running, landing, grasping, offense, and defense; long, lightweight forelimbs adapted for flight; and a short, flexible tail. The flexibility of the neck is especially important, for birds use their heads in attack and defense, feeding, preening, and balancing in flight.

Mammals   A principal difference between mammalian skeletons and those of reptiles and birds is in the enlargement of the braincase in mammals (compare Figs. 14-4 and 14-5). In addition to having more space for larger brains, mammalian skulls have well-developed **nasal passages** separated from the mouth cavity by the **palate**. Nasal passages provide a warming and moisturizing chamber for air prior to its being taken into the lungs and are thought to have evolved as an adaptation to life in cold climates.

Mammalian teeth are more specialized and fewer in number than those of reptiles. Characteristically, mammals have cutting teeth (**incisors**) at the front of the jaws, tearing teeth (**canines**) next to the incisors, and grinding teeth (**molars**) at the rear of the jawbones.

A second major difference between the skeleton of mammals and reptiles is seen in the **pelvic girdles** (Fig.

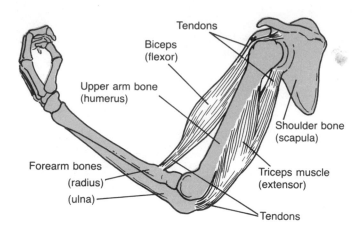

FIGURE 14-7
Muscles and bones of the arm. The biceps is a flexor muscle; the triceps is an extensor.

14-6C). Consider for a moment that a crocodile weighing nearly a ton lays an egg about the size of a hen's egg, whereas a cow, which weighs about half as much, gives birth to a calf the size of a large dog. The passage of such proportionately large offspring from the uterus through the pelvic girdle requires a large pelvic canal. Elastic ligaments between the pelvic bones permit further expansion of the canal to accommodate the passage of the fetus.

DEVELOPMENTAL ORIGIN OF BONE AND MUSCLE

Relatively early in the development of nearly all animal embryos, three cell layers or **germ layers** are formed (the exceptions are sponges, with none, and cnidarians with but two—ectoderm, endoderm). The three germ layers are: an outer, **ectoderm**; a middle, **mesoderm**; and an inner, **endoderm**. As the embryo continues its development, each germ layer produces a characteristic group of the animal's tissues and organs. As one might expect, the ectoderm gives rise to the covering cells of the body: in vertebrates to the integument and some elements of the skeletal system (dermal bone), and also to the central nervous system (see Fig. 11-7F, H). The mesoderm (Figs. 2 and 3, p. 245, and Fig. 11-7F, H) gives rise to muscles; to connective tissues, including blood, cartilage and bone; to the urogenital system; and to the membranes lining the coelom (peritoneum and mesenteries). The endoderm gives rise to the visceral organs.

Function of Bones and Muscles

In addition to providing support and protection for soft parts, bones of the vertebrate body, together with muscles, provide for movement and locomotion. As in arthropods, the locomotory principle is that of muscular forces applied to levers (skeletal elements, in this case bones) across fulcrums (joints). Here again, the principle of oppositely acting sets of muscles applies. Joints, as we have noted, have their bearing surfaces covered with a pad of cartilage and are held together with **ligaments,** as the bone-to-bone connectives are called. Sets of muscles are attached by **tendons** (muscle-to-bone connectives), as between the shoulder and the upper arm bone (**humerus**) and the upper ends of the two forearm bones (**radius** and **ulna**) (Fig. 14-7).

Consider the movement of the human arm in the act of bowling. Lifting the ball is accomplished by contraction of the **biceps.** This is a flexing action; therefore the biceps is a flexor muscle. In swinging the arm backward preparatory to rolling the ball down the alley, the biceps relaxes and the **triceps,** at the back of the arm, contracts; therefore the triceps is an extensor muscle. Of course, bowling involves many other muscles as well. Putting a spin on the ball involves a pair of upper muscles which rotate the radius bone on its axis; forearm muscles, acting through long tendons extending along the finger bones, produce many other complicated movements of wrist and arm. In other sports, and indeed in all one's activities, an absolutely astonishing interplay of sensory organs, brain regions, and motor nerves is involved, to say nothing of actions of the autonomic ner-

vous system, glands, hormones, and all the other systems of the body. What these actions accomplish in terms of food gathering, defense, aggression, reproduction, and social interactions is **behavior,** the topic explored in Chapter 22.

The Muscle Cell — Huxley's Sliding Filament Model

In Chapter 4 we learned that the bond energy of ATP could be translated into the mechanical action of muscle fibers. This use of chemical fuel by the body to produce movement (i.e., to do work) is rather analogous to burning gasoline in your automobile to push pistons, work gears, and turn wheels to carry you and your friends from one place to another. All this activity is accompanied by a good bit of mechanical commotion in comparison with the quiet and efficient way your body moves.

Biologists have been familiar with the general principles of muscle action and have known for more than a century that different types of muscle cells exist, but it is only in recent years that an understanding of the way in which such cells work has been reached. Our present knowledge of muscle cells comes from the studies of many scientists, but two Englishmen, A. F. and H. E. Huxley (not related, by the way) are credited with developing the now widely accepted **sliding filament model,** in which muscle contraction is produced by an interac-

FIGURE 14-8
The arrangement of actin and myosin filaments in striated (skeletal) muscle. **A.** A muscle, showing relationship of cell in B. **B.** A single striated muscle cell, which is much longer than shown in the diagram. A motor nerve is included. **C.** The entire cell is composed of repeating subunits (sarcomeres), one of which is shown in a relaxed state. Actin filaments are anchored to the Z-lines at the boundaries of the sarcomere; myosin filaments are interspersed among the actin filaments as shown in E. **D.** Contracted sarcomere. **E.** Cross section of sarcomere.

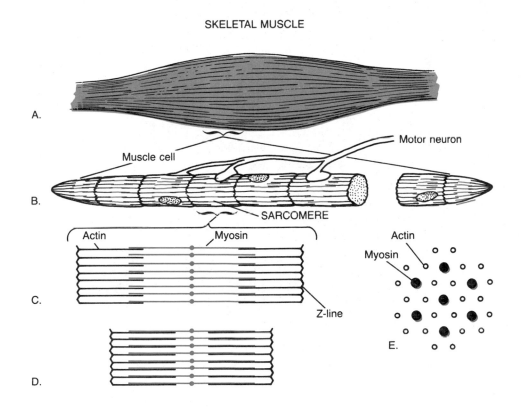

SKELETAL MUSCLE

tion between filaments of two kinds of protein, **actin** and **myosin**. These two men independently observed that when a kind of muscle known as **striated** (because of its prominent transverse banding) underwent contraction, some of its bands remained the same width while others became considerably narrower (see Fig. 14-8). It had been observed also that even though muscle cells became shorter when fully contracted, they did not become appreciably thicker. Putting this information together with observations made using the newly perfected electron microscope (this was in the early 1950s), the Huxleys explained that neither actin nor myosin filaments actually become shorter during muscle-cell contraction; instead the filaments slide past each other. The motive force was thought to be the result of movements of peg-like extensions attached to the myosin filaments and acting in such a way as to pull (or push) the filaments past each other.

Neuromuscular Actions

There are three kinds of muscle tissue in the bodies of vertebrate animals: relatively slow-acting **smooth muscle**; continuously active **cardiac muscle** (i.e., heart muscle); and fast-acting striated or **skeletal muscle**. In invertebrate bodies, either or both smooth and striated muscles are present, but cardiac muscle is absent.

Smooth Muscle Smooth muscle cells are uninucleate and spindle shaped, without striations (hence "smooth," Fig. 14-9). They are present in structures and organs such as the visceral organs, in which involuntary movements occur, in walls of blood vessels, in hair follicles, and in the iris of the eye. Smooth muscles in the viscera are not under nervous control, but others commonly are served by two sets of nerve endings: one set from the sympathetic system, inducing contraction, the second set from the parasympathetic system, inhibiting contraction. In any case, smooth muscle actions are involuntary; one cannot "think" them. An example follows.

Smooth Muscle and the Pupillary Reflex You no doubt have noticed that the pupils of one's eyes become wider in dim light and narrower in bright light. This re-

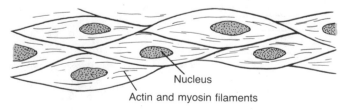

Nucleus

Actin and myosin filaments

FIGURE 14-9
Smooth muscle cells.

sponse, called the **pupillary reflex,** is regulated by nerves serving two sets of smooth muscles in the iris of the eye. One set of muscles, known as **constrictors,** circle the rim of the iris; when they contract the pupil narrows. A second set of **dilator** muscles run like spokes from the rim of the iris outward to its border; when they contract, the pupil widens. Naturally, any response to light intensity requires a light-sensing device, which in this case is the retina at the back of the eye. Nerve impulses from the retina are conveyed to the light reflex center in the midbrain and then appropriately relayed to the corresponding iris muscles via sympathetic or parasympathetic neurons (Figs. 13-17, 14-10).

Drugs, emotional disturbances, muscular exertion, and asphyxiation can also cause changes in pupillary diameter, whether or not light is present. Opium, morphine, and some other drugs induce great pupillary constriction. Atropine (belladonna), used during eye examinations, paralyzes the endings of the third cranial nerve, and because certain fibers in this nerve trunk connect with the iris constrictor muscles, the pupils become very much dilated. You may have noticed that a burst of strenuous exertion produces a momentary sensation of brightness. This is due to pupillary dilation, an effect produced also by fright or by asphyxiation. Before the advent of sophisticated monitoring devices, pupillary dilation was used to judge the depth of anesthesia—when a patient is not getting sufficient oxygen, excessive CO_2 in the blood causes the pupils to dilate.

Cardiac Muscle Cardiac muscle is somewhere between smooth and skeletal muscle in cytoplasmic organization and action. The cells are uninucleate and richly supplied with mitochondria. They are banded transversely, but the banding pattern is not as conspicuous as that of skeletal muscle (compare Figs. 14-8, 14-11, and 14-12). Unlike smooth or skeletal muscle, the cells of cardiac muscle are branched and form a network of interlocking cells that appear separated at intervals by densely staining disks or bars. With the use of electron microscopy, it is seen that these bars are the tightly interdigitated membranes of adjacent cells, further strengthened by numerous desmosomes (Fig. 14-11). This very intimate cell-to-cell contact is thought to explain the highly coordinated contractions of cardiac muscle cells.

By taking a tiny bit of chick embryo heart and placing it in a drop of tissue-culture solution, one can maintain a culture of living heart cells. As long as the nutrients are replenished and contamination does not occur, the cells will continue to grow, divide, and metabolize. If now we place a drop of nutrient containing a few cultured cells in a depression slide (slide with a hol-

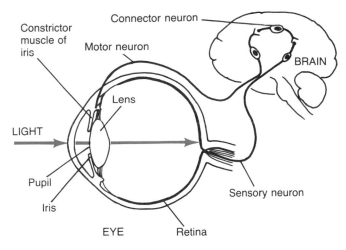

FIGURE 14-10
The pupillary reflex. Bright light impinging on the retina generates nerve impulses that are relayed through connector neurons in the brain to motor neurons that induce contraction of the iris constrictor muscles. Modern automatic cameras work in much the same manner, with light generating a flow of electrons (electricity) in a photocell which in turn controls the aperture of the iris diaphragm and thus the exposure of the film.

lowed-out place under the cover slip) and examine it with a microscope, we will see quite an amazing sight. The cells are rhythmically contracting and relaxing—just as if they were still part of a living heart. Even one isolated cell of cardiac muscle will continue beating, notably in the complete absence of any kind of nervous or hormonal stimulation. This autonomous action on the part of cardiac muscle is in fact the reason why heart transplants are possible—the heart is a relatively independent organ and will continue to pump even after its nervous connections have been cut.

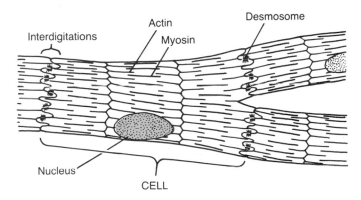

FIGURE 14-11
Cardiac muscle. Unlike skeletal muscle, cardiac muscle cells, though similarly banded, are uninucleate. They also are branching, as shown in this figure.

Skeletal (Striated) Muscle Skeletal muscle, as the name implies, is found in animals in which moveable skeletal elements are present. The cells are usually much elongated and exhibit a pronounced and regular banding pattern. Unlike those of either smooth or cardiac muscle, skeletal muscle cells are multinucleated; they are remarkably complex cells composed of many longitudinal subunits called **sarcomeres**. Each sarcomere may be considered a basic contractile unit, and each is composed of several to many bundles (**myofibrils**) containing, in turn, a great many parallel filaments of actin and myosin (see Fig. 14-8). Although skeletal muscles can participate in involuntary responses, such as the knee-jerk reflex, the individual is aware of them, and, in fact, of most skeletal muscle actions. For this reason the terms striated, skeletal, and voluntary are considered synonyms for this muscle type.

Earlier, the relationship between muscle filaments and ATP was mentioned. Also noted was that the arrangement of actin and myosin filaments was such that it accounted not only for the regular banding pattern of striated muscle, but also for shifts in the pattern during contraction or relaxation (Fig. 14-8). Now we turn to the intimate details of the skeletal muscle sarcomere in an attempt to understand its inner workings. Figure 14-12 shows a chain of reactions occurring within the time frame of initial stimulation of a muscle cell by a motor neuron and contraction of a sarcomere. In essence, an impulse in a motor neuron results in release of **acetylcholine**, a neurotransmitter molecule, whose action at a neuronal synapse is described in Chapter 13 and illustrated in Fig. 13-5. Acetylcholine has basically the same effect on the muscle-cell membrane as on a postsynaptic nerve membrane: it induces a new action potential. The new impulse spreads rapidly throughout the muscle cell, propagating along its membranes and inducing, in turn, release of calcium ions by membranes of a specialized kind of endoplasmic reticulum, the **sarcoplasmic reticulum**. Within milliseconds of the arrival of the impulse at the motor neuron-muscle cell junction, the sarcoplasmic reticulum releases calcium ions throughout the sarcomeres. In a relaxed sarcomere, two molecules, **troponin** and **tropomyosin**, are combined with each other and with actin in such a way as to inhibit the formation of links or "bridges" between actin and myosin filaments. Troponin and tropomyosin, on contact with calcium ions, are envisioned as shifting position along the actin filament, with the result that myosin bridges can then attach to binding sites on the actin filament.

Each myofibril of a skeletal muscle cell is a bundle of actin and myosin filaments, seen in cross section to have a hexagonal packing pattern (Fig. 14-8E). Longitudinally, the myofibrils are partitioned by **Z-lines**, to

FIGURE 14-12

Response of a muscle cell to a nerve impulse. In the upper part of the drawing, a nerve impulse causes release of acetylcholine, a neurotransmitter, into the muscle cell membrane. A new impulse propagates rapidly through the membrane of the muscle cell and causes the sarcoplasmic reticulum (a type of endoplasmic reticulum) to rapidly release calcium ions (Ca^{2+}) among the filaments of actin and myosin. Myosin bridges are shown in extended position in **A** and are depicted as blocked from contact with actin filaments by molecules of troponin-tropomyosin. **B.** Within milliseconds after nerve stimulation, the calcium ions combine with troponin-tropomyosin and result in the removal of this block to muscle contraction. Meanwhile, ATP becomes enzymatically bound to myosin bridges and its energy is transferred to the bridges by hydrolysis of ATP to ADP and P_i (inorganic phosphate). The result is a conformational change in the myosin bridge, depicted as an inward movement. This causes actin and myosin filaments to slide past each other, producing muscle cell contraction.

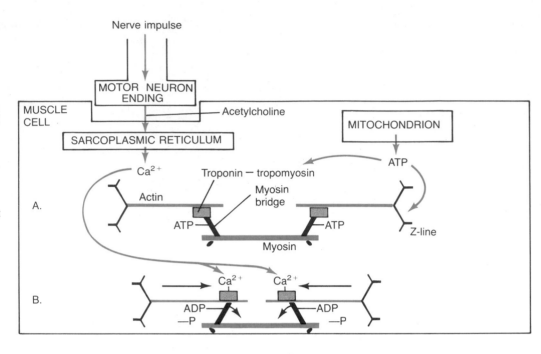

which actin filaments are anchored as shown in Figs. 14-8 and 14-12. Thicker myosin filaments lie midway between the Z-lines, intermeshing with actin filaments and linked to them by myosin bridges, which can exert a "rowing" action against actin filaments and produce the filament sliding postulated in the Huxley model.

In addition to inactivation of the troponin-tropomyosin block, attachment of the myosin bridge to actin also requires binding of ATP to myosin. This is catalyzed by the enzyme **ATPase,** which hydrolyzes ATP to ADP and inorganic phosphate (P_i), an action that transfers energy to myosin. As a result of this energy transfer, the myosin bridge undergoes a conformational change, which is envisioned as a bending or rowing action that causes the actin filaments to move toward the centerline of the sarcomere, as shown diagrammatically in Fig. 14-12B. This shortens the sarcomere a bit and it becomes further shortened when another ATP comes along and replaces the ADP still remaining bound to the myosin bridge after the earlier contraction. If no more ATP is available, the muscle may remain in a "locked-up" state, unable either to contract further or to relax. This is the explanation for *rigor mortis*, which occurs when death puts an end to ATP production in the cells of an animal. Normally, repeated

stimulations by a motor neuron are required for the maintenance of the contracted state; otherwise calcium-ion pumps in the membranes of the sarcoplasmic reticulum rapidly withdraw calcium ions from the myofibrils, the troponin-tropomyosin blocks are restored, and the muscle cell relaxes.

We have now examined the manner in which body movements are produced by interactions between sets of muscles. Sometimes, as in the movements of soft-bodied invertebrates or of visceral structures in both vertebrates and invertebrates, the actions do not involve skeletal elements. In vertebrates, however, many movements, particularly those resulting in changes of body orientation, result from interactions between muscles and skeletal structures. In Chapter 22, dealing with the behavior of organisms, we will relate what we have learned to interactions between organisms and of organisms with their environments.

SUMMARY

Neurons and muscle cells interact in producing movements. Usually these are in response to external and internal stimuli, although some, such as those of visceral organs, are automatic and not turned on or shut off by nerve impulses. Muscles in the vertebrate body are of three types: smooth, cardiac, and skeletal. Smooth muscle is involuntary in action and is present in visceral organs, blood vessel walls, hair follicles, the iris, and other structures in which voluntary control is absent. Cardiac muscle cells are striated and somewhat resemble skeletal muscle, which is composed of multinucleated, striated cells. Skeletal muscle is voluntary in action and under control of motor neurons. The skeletal muscle cell is a complex cell in which subunits called sarcomeres contract by an interaction between ATP, as the source of energy, and two kinds of filaments, actin and myosin. Contraction occurs by a "rowing" action of peglike myosin structures (bridges) against adjacent actin filaments. The same action applies also to cardiac and smooth muscle.

Movements involving muscle contractions require the employment of opposing forces, such as sets of muscles acting against a resisting object, in effect, the principle of the lever and fulcrum. In some aquatic animals, such as jellyfish, surrounding water acts to oppose muscle force, but most other animals employ a skeleton in that capacity. Some animals, such as earthworms, use body fluids as a hydroskeleton, but two important groups, the arthropods and the vertebrates, have jointed skeletons, in which muscle pairs, one capable of flexing (bending) and the other of extending skeletal structures, produce movement. The two groups differ in that arthropods have an external skeleton (exoskeleton), while vertebrates have an internal skeleton (endoskeleton).

KEY WORDS

skeleton	Haversian canal	cardiac muscle
adductor muscle	dermal bone	skeletal muscle
exoskeleton	cartilage	pupillary reflex
extensor muscle	chondrin	acetylcholine
flexor muscle	chordate	sarcoplasmic reticulum
axial skeleton	vertebrate	troponin
appendicular skeleton	germ layer	tropomyosin
pectoral girdle	sliding filament model	Z-line
pelvic girdle	actin and myosin	ATPase
hydroxyapatite	smooth muscle	rigor mortis

QUESTIONS FOR REVIEW AND DISCUSSION

1 Describe the basic principle of movement in relation to actions of sets of muscles. Compare movement of an arthropod leg with that of a vertebrate leg.

2 In what ways is the rib cage important to the life of land vertebrates?

3 Briefly state the germ layer origins of the following vertebrate structures: The spinal cord and brain. The backbone. The bony plates of the human skull. The muscles of the body wall.

4 What is the relationship between pectoral and pelvic fins and the limbs of tetrapods?

5 What are actin and myosin? Where are they found? What is their significance?

6 Compare the structural appearance of smooth, cardiac, and skeletal muscle. Which is (are) voluntary?

7 How does the pupillary reflex work?

8 Show by diagrams the relationship between the motor neuron and the sarcoplasmic reticulum and how they interact in producing a contraction in a muscle cell.

SUGGESTED READING

EVARTS, E. V. 1979. Brain mechanisms of movement. *Sci. Amer.* 241(3):164–79. (How the brain and spinal cord control movements in the body—relationship of motor units, units of muscle cells served by one motor neuron, and muscle tension are discussed.)

HUXLEY, H. E. 1958. The contraction of muscle. *Sci. Amer.* 199(5): 66–82. (Sliding filament hypothesis presented.)

_____. 1965. The mechanism of muscle contraction. *Sci. Amer.* 213(6): 18–27. (More on the sliding filament story, excellent electron micrographs and diagrams.)

LAZARDIES, E., and J. P. REVEL. 1979. The molecular basis of cell movement. *Sci. Amer.* 240(5):100–13. (Discusses the role of actin and myosin filaments in many kinds of cell movements. Includes description of muscle and gives one an impression of arrangement of contractile filaments in nonstriated cells.)

LESTER, H. A. 1977. The response to acetylcholine. *Sci. Amer.* 236(2):106–18. (All about the relationship between the motor neuron, neurotransmitters, and muscle contraction.)

MURRAY, J. M., and A. WEBER 1974. The cooperative action of muscle proteins. *Sci. Amer.* 230(2):59–71. (How ATP interacts with myosin "heads" in muscle contraction.)

RAILBERT, M. H., and I. E. SUTHERLAND. 1983. Machines that walk. *Sci. Amer.* 248(1):44–53. (How many legs are best? Principles of locomotion described.)

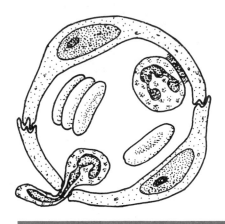

15 Health

Many factors are involved in the maintenance of good health. If hurt in an automobile accident, one depends on the combined abilities of medical professionals to diagnose the injuries, set broken bones, fix damaged teeth, replace lost blood, and repair torn tissues. If this help is prompt and efficient, the chances are excellent for recovery from wounds that 50 years ago would have been fatal. Much the same can be said for any of an assortment of life-threatening accidents and diseases to which one can become exposed. Nevertheless, many factors other than *post facto* medical care are involved in the maintenance of good health. Maintaining a state of good health requires adherence to a whole program of preventive care that includes good nutrition, controlling stressful situations, avoiding environmental poisons, wise use or complete avoidance of intoxicating substances, immunization against disease organisms, and getting sufficient rest and exercise.

Perhaps the greatest advance in adding years to the human life span has been proper sanitation. The treatment of disease by immunization, antibiotics, and hospital care has saved countless lives, but even more have been saved by preventive measures such as water and sewage purification, pest control, and general cleanliness. It is no idle claim that social groups who are educated in an understanding of the importance of proper foods, good sanitation, and preventive health care not only outlive those who are ignorant of the importance of these factors, but enjoy better health while doing so. It is a popular fallacy that the prosperous enjoy better health because they can afford to pay more for medical service. On the whole, those who are careful about what they eat and drink and who practice proper sanitation undergo fewer major surgical procedures and spend fewer days in hospitals than those who are ignorant of even rudimentary health practices.

NUTRITION AND HEALTH

Good health is necessary not only for performing one's daily activities but also for maintenance of a satisfactory emotional state. Nutrition is a vital factor. Poor nutrition is a problem in many parts of the world. In some regions the problem is simply lack of food, but even in prosperous societies, where the total intake may be high, eating the wrong foods can bring on malnourishment. In fact, while children and adults in some countries worry about getting enough to eat, in other countries books are written exhorting people to eat less for better health. We are becoming more and more aware of the factor of diet in human health. Foods rich in certain kinds of fats have been linked to heart disease, and diets high in carbohydrates and low in proteins do not permit normal growth and development. Lack of essential vitamins can produce nervous and other disorders, and foods poor in minerals may result in weak bones, poor teeth, and irritability. With so many problems of overeating and undernourishment, what are the yardsticks for maintaining good health?

In planning a diet, one needs to know human nutritional requirements and what foods should be eaten to supply those requirements. A prime consideration is the amount of energy a given food provides. Food must suffice to meet the ATP requirements of body cells, for although foods do not themselves contain much ATP, ATP is generated when the foods are digested and oxidized by the cells of the body (refer to Fig. 9-1). **Fats**, **proteins**, and **carbohydrates** are all good energy sources. Proteins and carbohydrates, on a per-weight basis, yield about equal amounts of energy in biological oxidations. Fats yield considerably more energy on oxidation than do proteins and carbohydrates but currently are not highly regarded as bulk foods because of health considerations. However, all three classes of foods are necessary for proper nutrtion.

A common dietary question today is whether a vegetarian can get a balanced diet of proteins, fats, and carbohydrates. The answer is yes, but with some qualification. To attain good nutrition, vegetarians and nonvegetarians alike must consume food having a correct balance of proteins, fats, and carbohydrates, as well as **vitamins** and **minerals**. This can be achieved provided some basic principles of nutrition are recognized (refer to Fig. 15-1).

Components of Diet

Protein   The diets of many humans, whether or not they are vegetarians, are deficient in certain kinds of proteins. The nutritional quality of a protein depends upon its amino acid composition. Of the twenty amino acids that make up proteins, eight cannot be made in the human body. These eight are the **essential amino acids: leucine, lysine, isoleucine, methionine, phenylalanine, threonine, tryptophane,** and **valine.** Plants, which are the original source of all amino acids, are a primary source of the essential eight. These amino acids can also be provided by eating certain animal proteins.

A few examples will demonstrate some important characteristics of dietary proteins and the essential amino acids they supply. Gelatin is an animal protein

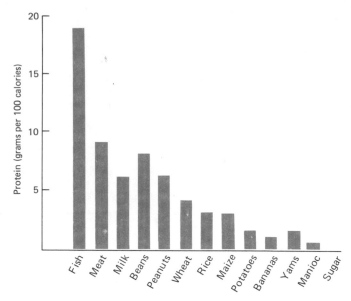

FIGURE 15-1
Protein content of foods. (Redrawn from C. B. Heiser Jr. 1973. *Seed to civilization: The story of man's food.* San Francisco: W. H. Freeman and Co.)

and a useful food supplement. However, it is composed almost entirely of repeating units of just six amino acids (alanine, arginine, aspartic acid, glutamic acid, glycine, and proline); it has no tryptophane and only small amounts of the seven other essential amino acids. Obviously, one's health would suffer on a diet of gelatin. Beef contains all the essential amino acids, but is low in tryptophane and methionine. Egg white contains a good range of amino acids, although it also is somewhat low in tryptophane and methionine. Fish protein is high in all essential amino acids and is perhaps the best source of protein for the human diet. Among plants, maize (corn) protein has no lysine, but is a relatively good source of other amino acids and is very high in leucine. Soybean protein has a good range of amino acids and is not deficient in any essential amino acid. Putting this information together, it is quite possible to subsist on a meat-free diet provided a good balance of vegetable proteins is eaten (see Fig. 15-1). The traditional Mexican Indian diet of *tortillas* (maize cakes) and *frijoles* (beans) includes two sources of plant proteins that complement each other. The practice in many overpopulated lands of supplementing a largely vegetable diet with fish meal also can provide an adequate diet.

Calories and What They Mean   Many Americans are calorie-conscious, often without knowing just what a calorie is. No doubt some believe it to be a fat unit. However, a calorie is not a substance that can be eaten or weighed but rather a unit of energy. Specifically, a calorie (cal) is that amount of energy required to raise the temperature of 1 g of water 1°C. The calories listed on food labels, however, are not the calories of this definition but rather **kilocalories** (kcal) (1000 calories).

Vitamins   Vitamins are not classified as food, but they are necessary elements of the diet because they function as coenzymes in many vital metabolic processes (Table 15-1). Generally speaking, vitamin deprivation is not a serious health problem in the United States, but carelessness in choosing foods, as well as other dietary inadequacies, can lead to difficulties. American diets usually are sufficiently varied that at least minimal vitamin requirements are met. In the past, restricted diets in some regions of the United States caused **pellagra**, a result of severe niacin deprivation, and currently in some lands vitamin deficiencies constitute serious threats to human health.

In China and other countries of the Orient, rice is the staple food in an almost completely vegetarian diet. At one time, meals of peasant farmers included quantities of crude rice directly from the threshing floor, each kernel covered by a brown seed coat. Then rice-polishing machines were invented to remove the brown coat and produce polished rice with its nice, fluffy, white appearance when cooked. Polished rice became the vogue. Af-

ter a short time, coolies who pulled rickshaws and did other manual labor in the big cities began collapsing at work. They developed **beriberi** ("extreme weakness"), a kind of paralysis, and sometimes did not recover. Beriberi was a mystery until a Dutch physician named Eijkman, who worked in the East Indies, observed that chickens fed on scraps of polished rice became paralyzed, but recovered when fed unpolished rice. Putting two and two together and realizing that eating polished rice was the common factor in paralysis both of chickens and humans, Eijkman concluded that polishing the rice kernel removed an important nutritional factor. Now it is recognized that the rice hull, and also the hulls or seed coats of wheat and other cereals, are good sources of **vitamin** B_1, also known as **thiamine**, the anti-beriberi vitamin. Vitamin B_1, as well as other vitamins, are routinely added to bread today, in part to make up for the loss of vitamins during production of white flour.

Vitamin C is one of the vitamins about which much is written and claimed. It is present in fruits and vegetables and in practically all animal tissues, but is readily destroyed by cooking. Most animals are able to synthesize vitamin C in their bodies, but humans cannot. Sailors in Columbus's day, during long sea voyages, usually developed **scurvy**, a debilitating and sometimes fatal disease. The typical diet was pickled beef, salt pork, and

TABLE 15-1
Some important vitamins and their requirements by humans

Vitamin	Minimum Daily Requirement	Source	Deficiency Symptoms
A—Retinol	5000 I.U.*	Egg yolk, green and yellow vegetables, fruits	Night blindness, skin lesions
B_1—Thiamine	1.5 mg	Milk, eggs, liver, whole cereals	Polyneuritis, beriberi
B_2—Riboflavin	1.5–2.0 mg	Liver, whole grains	Skin disorders, cataracts
B_6—Pyridoxine	1–2 mg	Whole grains	Dermatitis, nervous disorders
B_{12}—Cyanocobalamin	2–5 mg	Liver, intestinal bacteria	Pernicious anemia
Biotin	150–300 mg	Eggs, intestinal bacteria	Dermatitis, pain and weakness of muscles
C—Ascorbic acid	75 mg	Fresh fruits, vegetables	Scurvy
D—Calciferol	400 I.U.	Fish oils, liver	Rickets
E—Tocopherol	?	Green leafy vegetables	Blood disorders, possibly impaired sexual functions in males
Folic acid	500 µg or less	Meats	Anemia
K—Naphthoquinone	?	Intestinal bacteria	Failure of blood to coagulate
Niacin	17–20 mg	Whole grains	Pellagra
P—Rutin	?	?	Capillary fragility
Pantothenic acid	8.5–10 mg	Meats?	Nervousness

*International unit.

hardtack. Then it was discovered that scurvy could be prevented by eating fresh fruit. Limes were carried on British warships during the 1700s, and the crew members were required to suck them regularly to prevent scurvy. As a consequence, British sailors were called "Limeys."

Although nearly everyone now gets enough vitamin C to prevent scurvy, some individuals think that the diets of many individuals do not contain sufficient vitamin C to prevent other problems, such as the common cold. Thus far there is no solid evidence regarding benefits to be derived from large daily doses (100 mg to several grams) of vitamin C, but such amounts do not appear to be harmful.

Minerals Mineral compounds containing iron, potassium, calcium, phosphorus, and other elements are required in the diet (see Table 1-1 for a complete list). The minerals most often found deficient in the human diet are iron and calcium. Leafy green vegetables are good sources of both iron and calcium; dairy products also provide adequate amounts of calcium; iron requirements generally are satisfied by red meats. Nevertheless, in some cases it may be necessary to supplement the diet by taking calcium and iron pills, especially when fresh vegetables are scarce or unavailable.

Roughage People on bulkless diets often suffer intestinal upsets due to a lack of inert fiber in the food. This is a problem with overrefined foods, in which tough, stringy matter is removed in processing. The seed coat of cereal grains (the bran) is removed during the production of white flour, eliminating a good source not only of vitamins but also of indigestible cellulose fibers, which serve as an important source of bulk. In the last century, an American physician named Sylvester Graham invented the graham cracker to supply bran in the diet. However, whole wheat bread is an even better source of roughage, as are fresh and cooked fruits and vegetables.

Losing Weight

It should be obvious from even this brief discussion that diets designed to produce a weight loss must be carefully regulated to provide necessary vitamins, minerals, and amino acids. Recently, liquid protein food, often made from dissolved cowhides, has been the base of high-protein diets designed to produce rapid weight loss. The theory behind high-protein diets is that fasting humans first lose water, then begin breaking down cellular proteins, and only later begin to break down stored fat. If only proteins are eaten, the theory goes, then the body will burn only fats. The difficulty with this theory is that fats are required for normal metabolism, and some dietary fat is necessary for health. Prolonged fasting or unrestricted dieting in any case is risky, even with protein supplements, because the body begins to draw on muscle proteins. If heart muscle is broken down, as sometimes happens in unsupervised long-term diets, heart malfunction and perhaps death may result. The best diets are those in which a balance of foods is eaten and those in which one food does not predominate.

CARDIAC DISEASE

Although many quarts of blood pass through the heart each day, the heart cannot utilize the blood within itself, but rather is sustained by branches of arteries originating externally. These coronary arteries can become blocked by the deposition of fatty material (including cholesterol), connective tissue fibers, and calcium deposited in the vessel walls. If the blockage is severe, numerous heart muscle cells become oxygen starved and the heart stops. In less severe cases, only part of the heart muscle dies as a result of the blockage. The dead muscle is replaced by scar tissue and the heart continues to function, though often less efficiently. Diet, stress, smoking, and alcohol all contribute to the problem of heart attacks. Interestingly, although fats in the diet often are blamed for heart attacks, about 70 percent of the heart's energy requirement normally is derived from oxidation of fatty acids.

TABLE 15–2
Daily human caloric requirements

	Age	Weight (kilograms)	Height (centimeters)	Requirement (kilocalories)
Females and males combined	0–1	4–9	72	kg × 110
	1–2	12	81	1100
	2–3	14	91	1250
	3–4	16	100	1400
	4–6	19	110	1600
	6–8	23	121	2000
	8–10	28	131	2200
Females	10–12	35	142	2250
	12–14	44	154	2300
	14–16	52	157	2400
	16–18	54	160	2300
	18–22	58	163	2000
	22–35	58	163	2000
	35–55	58	160	1850
	55 and over	58	157	1700
	Pregnant			+200
	Lactating			+1000
Males	10–12	35	140	2500
	12–14	43	151	2700
	14–18	59	170	3000
	18–22	67	175	2800
	22–35	70	175	2800
	35–55	70	173	2600
	55 and over	70	171	2400
	Hard physical labor			+1000

Humans differ greatly in their energy needs. These needs depend on sex, age, weight, kind of work, and individual metabolic rate (see Table 15-2). The specific energy requirements of an individual at rest and at work can be determined by measuring the metabolic rate. This is done by determining heat lost from the body during rest and when active, in proportion to the amount and kind of food eaten. On the average, an adult human requires about 35–40 kcal per day for each kilogram of body weight (1 kg is about 2.2 lb). A middle-aged man weighing 70 kg would need to take in food equivalent to 2500 kcal to maintain himself in good condition. If individual caloric requirements were carefully measured and the diet contained just the right number of calories to sustain this requirement, there would be neither a weight gain nor a weight loss. However, if the diet provides more food than required to generate those calories, the excess food is stored as fat. If the diet is insufficient, then carbohydrates, fats, and proteins in the body are broken down and burned to make up the deficit. The latter process is the simple explanation behind all weight-control diets. However, current research indicates that the whole process of weight regulation may be more complex than previously thought.

Diet and Cardiovascular Disease

Diseases of the heart and circulatory system are the leading cause of nonaccidental deaths in the United States, which has the unenviable record of one of the highest rates of heart disease in the world. In 1983 more than a million Americans suffered heart attacks, from which more than half died. **Cardiovascular diseases** include **hypertension** (high blood pressure), **cerebral hemorrhage** (stroke), and **coronary artery disease** (heart attack).

Most cardiovascular disease is due to diminished capacity of arteries to carry blood because of internal narrowing as a result of deposition of **plaque,** which is composed of lumps of **cholesterol** and fats, calcium deposits, and proliferations of connective tissue and smooth muscle cells in the artery wall just beneath the

FIGURE 15-2
A normal artery, in cross section at the left, compared with an artery in which a deposit of plaque has narrowed the interior space. Plaque deposits contain cholesterol, other lipids, smooth muscle, and connective tissue.

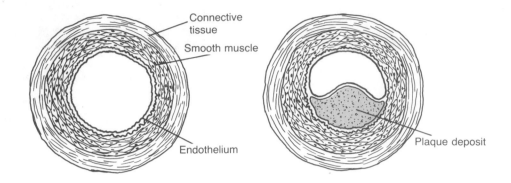

Connective tissue

Smooth muscle

Endothelium

Plaque deposit

endothelium (Figs. 15-2, 15-3, 15-4). Popular opinion usually assumes that this degenerative process, called **atherosclerosis,** is a problem for middle-aged and elderly persons, but actually it can begin rather early in life. Postmortems performed on American soldiers killed in the Korean war showed that 77 percent had the beginning symptoms of atherosclerosis; their average age was 22. Simultaneously, autopsies done on Korean dead of the same age group showed no evidence of the onset of atherosclerosis. Aside from ethnic differences, Korean soldiers consumed a diet low in red meat and fats and high in vegetables and cereals; the diet of young Americans was, and still is, almost the exact opposite. Evidence derived from studies of Europeans in enemy-occupied nations during World War II has been compared with evidence of their post-war health and has

FIGURE 15-3
Cross section of a normal human artery **(A)** and of a diseased artery partially plugged with plaque **(B).** (A, © Charles Bailey, 1984; B, © Camera M. D. Studios, 1973)

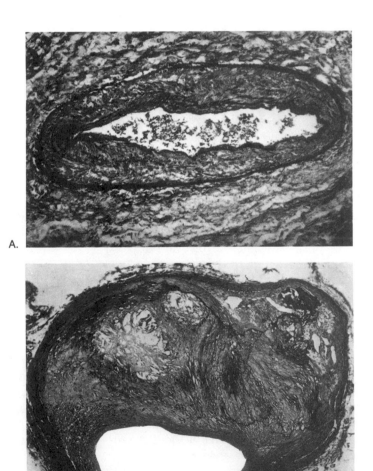

A.

B.

FIGURE 15-4
Chemical structures of cholesterol, saturated and unsaturated
fatty acids, and fats. The cholesterol (A) molecule belongs to a
class of substances called steroids, to which several sex hormones also belong. Tristearin (B)
is a representative saturated animal fat. Linolinic acid (C) is a
representative plant oil. It is polyunsaturated, meaning more than
one double (unsaturated) carbon-
to-carbon bond is present (arrows) in the molecule. Stearic
acid (D) is a saturated fatty acid,
having no double carbon-to-car-
bon bonds.

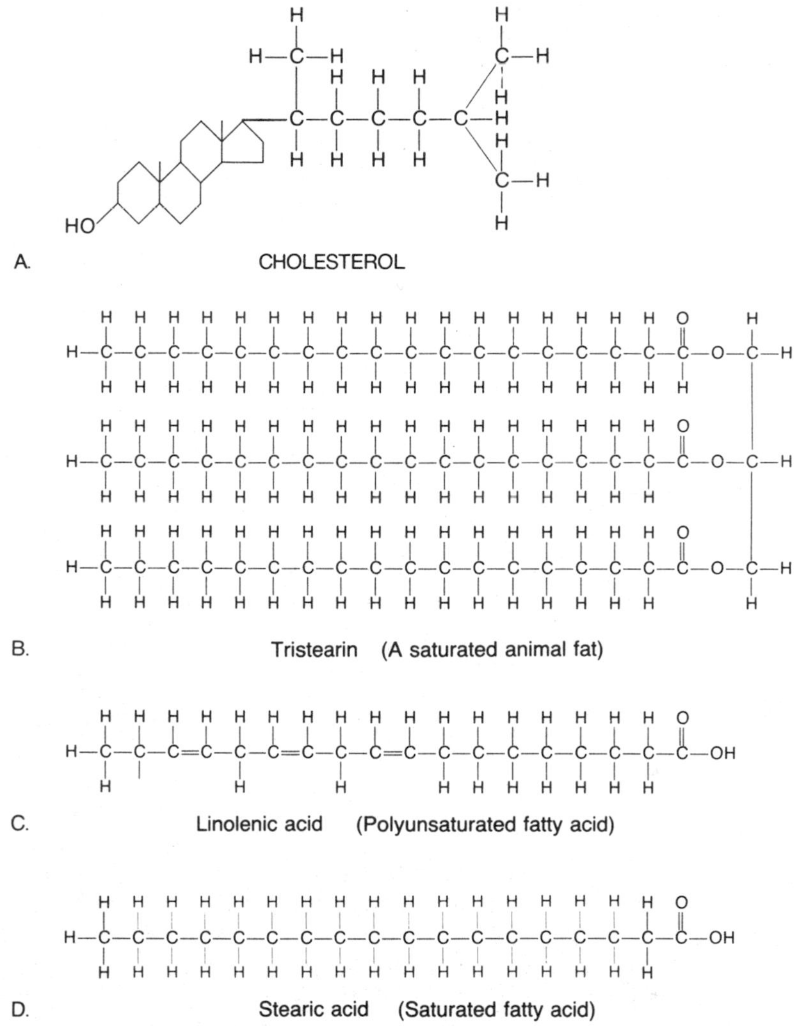

A. CHOLESTEROL

B. Tristearin (A saturated animal fat)

C. Linolenic acid (Polyunsaturated fatty acid)

D. Stearic acid (Saturated fatty acid)

also implicated rich diets in the growing prevalence of
atherosclerosis.

Some years ago, the National Heart, Lung, and
Blood Institute (NHLBI) began a study of 3806 American
men aged 35–59 in order to determine the relationship,
if any, between blood cholesterol levels and heart attacks. Half of the men were given a cholesterol-lowering
drug, half were given a placebo (plah-see-bo—an inert
and harmless substitute resembling the real drug). Now,
10 years later, the treated group has been found to average 8.5 percent lower blood cholesterol and to have
had a remarkable 24 percent fewer deaths from cardiovascular diseases compared with the placebo group. Although drugs were employed in this study, cholesterol
levels can also be lowered effectively by one's choice of
diet (refer to Table 15-3) and simply by reducing one's
total intake of food in excess of daily requirements.

Cholesterol circulates in the blood complexed with
one or another of two carrier substances: HDL, for high-

density lipoprotein (in high concentrations in the presence of unsaturated fats), and LDL, for low-density lipoprotein (high when saturated fats are present). HDL
appears to be helpful in removing cholesterol from circulation in the blood; LDL enhances deposition of cholesterol in arteries (plaque formation) and therefore increases the risk of heart attacks.

Stress and Disease

Defense mechanisms against environmental hazards
have evolved among all organisms, and certainly this is
so in human evolution. Among such mechanisms is the
"fight-or-flight" response (described in Chapters 10 and
13), in which the body responds to emergencies by generating hormone-associated stress responses. The metabolic changes occurring in a "fight-or-flight" situation, or
in any other stressful situation, are quite profound and
include elevated heartbeat rate, elevated blood pressure,

TABLE 15-3
Cholesterol and saturated-fat
content of some foods

Food	Cholesterol Content	Saturated-fat Content
MEATS	(mg/100 g)	(g/100 g)
Beef liver	440	3.0
Egg (one)	325	2.0
Veal	100	4.7
Pork	95	3.8
Chicken: dark (light)	97 (90)	3.2 (1.5)
Lean beef	66	2.8
Fish	70	0.35–1.4
Shrimp	57	0.2
DAIRY PRODUCTS	(mg/100 g)	(mg/10 g)
Butter	190	4965
Cheddar cheese	105	2100
Ice cream	25	370
Milk	14	200
Yogurt	4.6	70
OILS	(mg/10 ml)	(g/10 ml)
Coconut	0	8
Palm	0	4.5
Olive	0	1.2
Corn	0	1.1
Safflower	0	0.8

changes in blood-sugar and hormone concentrations in the blood, hypersensitivity to environmental substances, allergic reactions, and other effects.

We have no way of knowing how often primitive humans were severely stressed in their daily lives. Perhaps stress was a normal state for them, but in that era stress was probably more a life-saving than a life-threatening response. The primitive human life span was very brief, not much over 20 years, and long-term effects of stress on body functions probably did not have time to develop. In modern society, however, stresses are usually not induced by "fight-or-flight" situations, which, however deadly, are over rather quickly. Instead the stresses often are the much more insidious ones of the marketplace, the classroom, the traffic snarl, and overpopulation, to name a few. In our era, stress is the common denominator in cardiovascular disease, as it is also in many forms of mental maladjustment, drug abuse, and antisocial behavior. In a preceding section on cardiovascular disease, the consequences of malnutrition relative to the onset of atherosclerosis have been described. Intake of cholesterol and fatty acids has been directly implicated in the development of atherosclerosis and heart attacks; even so, stress may be the chief predisposing factor. Bad nutrition, overeating, and obesity, which contribute to cardiovascular diseases, often are re-sponses to feelings of stress and insecurity. Stress and cardiovascular disease are now also considered to be positively correlated with lack of exercise. Insufficient daily exercise can contribute to stress, and regular exercise, versus a sedentary life style, is known to result in stress reduction and lowering of the incidence of heart attacks.

DRUG USE AND DRUG ABUSE

When individuals lead a stressful existence, they may turn to drugs, which afford temporary relief from their problems or provide a means of coping with a given situation. Among such drugs are caffeine, alcohol, nicotine, and, increasingly, opiates, barbiturates, tranquilizers, and hallucinogens. Some of these may be prescribed by a physician and may afford relief from symptoms of stress; often, however, such drugs are taken without prescription and lead to serious consequences.

Psychoactive drugs generally are classified as **stimulants, depressants,** or **hallucinogens** (Table 15-4). Additionally, they often are defined as being either **addictive** or **nonaddictive,** depending on whether an individual suffers withdrawal symptoms on termination of their use. Actually, hard-and-fast definitions pertaining to drug use and addiction are rather unsatisfactory.

TABLE 15-4
A partial listing of substances classified as psychoactive drugs

Substance	Source	Potential for Dependence		Overall potential for tolerance	Effects	Potential for abuse	Effects of abuse
		Physiological	Psychological				
STIMULANTS							
Caffeine	Coffee	Low	Moderate	Moderate	Alertness, wakefulness	Low	Headache, insomnia, heart palpitations
Nicotine	Tobacco	High	High	High	Accelerated responses*	High	Tremors, irritability, cardiovascular disease, cancer
Cocaine	Coca plant	Moderate	High	High	Euphoria, mood enhancement	High	Intoxication, nasal damage, hallucination, convulsions
Amphetamines (benzedrine, dexedrine, methedrine)	Synthetic	Low	High	High	Accelerated responses, euphoria	High	Intoxication, mental disturbances, aggressiveness, delusions, hallucination, convulsions, brain damage
DEPRESSANTS							
Alcohol	Fermentation	High	High	High	Initial euphoria, then impaired motor functions, drowsiness, sleep	High	Intoxication, mental disturbances, aggressiveness, delirium, coma, brain damage, liver damage
Opiates (opium, codeine, morphine, heroin)	Poppy, synthetic derivatives	Very high	Very high	High	Initial euphoria, narcosis, drowsiness, pupillary constriction	High	Intoxication, antisocial behavior, mental disturbances, coma, severe withdrawal symptoms
Barbiturates (phenobarbital, seconal, tuinal, nembutal, amytal)	Synthetic	High	Moderate	High	Drowsiness, sleep	High	Irritability, impaired reactions, potential for overdose, coma; second highest mode of suicide (first is carbon monoxide)
Tranquilizers, mild (Valium, Librium)	Mostly synthetic (reserpine from Indian snakeroot plant)	Low	Moderate	Moderate	Anxiety relief, drowsiness	Moderate	Danger in combination with other drugs, coma, respiratory failure
Tranquilizers, heavy (Thorazine, Stelazine, reserpine)		High	High	Moderate	Sedation, drowsiness	High	Danger in combination with other drugs, coma, schizophrenia or schizophrenialike withdrawal symptoms
HALLUCINOGENS							
THC (tetrahydrocannabinol)	Marijuana (*Cannabis* plant)	Low	Low	Low	Mood enhancement, euphoria, loss of temporal judgment pupillary dilation	Moderate	Loss of short-term memory transfer to long-term memory
LSD and others (mescaline, psilocybin)	Synthetic ergot derivative (cactus, mushrooms)	Low	Moderate to high(?)	Low	Altered perceptions, anxiety, hallucination, pupil dilation	High	Loss of memory, irrational behavior, precipitation of psychoses, nausea, flashbacks†

*While an effect of nicotine on the CNS is one of acceleration of responses, carbon monoxide inhaled with smoke from burning tobacco can reduce the capacity of blood to carry oxygen and thereby cause a depression of the CNS.

†One of the experiences that can occur with use of hallucinogens is flashback; that is, a recurrence of a hallucination some time after using the drug. Flashbacks can recur several to many times even though use of the drug has been discontinued.

Technically, addiction is equated with the development of severe withdrawal symptoms such as pain, hallucination, and convulsions. Even in the absence of such physiological symptoms of withdrawal, an individual may have become so behaviorally habituated that stopping the use of a drug can be extremely difficult.

Drugs and Neurotransmitters

Although the specific mode of action of certain drugs remains unknown, for others it seems clear that the principle of their actions is based on their similarities to **neurotransmitter molecules (NTs),** which function in the nervous system at synapses and neuromuscular junctions (refer to Figs. 13-17 and 13-25). Recall that at axon endings, an action potential (nerve impulse) arriving at the presynaptic membrane causes the release of NT into the synaptic cleft. Each NT then attaches to a specific receptor site in the postsynaptic membrane of a neuron (or muscle cell membrane) and causes a change in ion permeability. The result is the generation of a new action potential. Because a neuron may synapse with several to many other neurons, and because it may be capable of producing several kinds of NTs and may be equipped with a corresponding variety of NT receptor sites, many different neurological functions can be regulated.

Many, perhaps all, drugs act upon NTs or NT-receptors in one or another of the following ways: (1) They may block or stimulate NT release. (2) They may block the synthesis of an NT by an enzyme, or enzymes. (3) They may block an NT-receptor site. (4) They may mimic NT molecules, that is, be so similar to the NT as to fit the same receptor site and produce the same response as the NT (Fig. 15-5). (5) They may block the action of enzymes that normally break down NTs, so that NT action may be prolonged. (6) They may block the reuptake of NTs at the presynaptic membrane so that NT action is prolonged.

Stimulants

We have encountered examples of the "key-in-lock" principle of biochemical "recognition" and action in connection with enzymes and with the sensing of odors by chemoreceptors in our nasal passages. Now we encounter it once again in connection with the action of certain drugs. As an example, we turn to the action of benzedrine, one of a group of similar stimulatory drugs known as **amphetamines** (Table 15-4). Benzedrine resembles norepinephrine (Fig. 15-5), an NT associated with the collection of responses, including acceleration of metabolism, alertness and wakefulness, and reduced pain sensitivity, known as "fight or flight." When benzedrine is taken, either in capsule form or by injection, and reaches the central nervous system (CNS), it triggers the release of the NT norepinephrine (also known as noradrenalin) from presynaptic neuronal membranes. Nicotine may act along similar lines. Cocaine apparently blocks the reuptake of norepinephrine by presynaptic membranes. The result is a continuing presence of norepinephrine which prolongs the stimulation. On the other hand, caffeine, a comparatively mild stimulant, is thought to act by causing increased levels of cyclic AMP in NT target cells.

Drugs classified as stimulants (energizers) accelerate metabolic processes and produce a feeling of well-being (mood enhancement). Their effects range from mild stimulation (caffeine and nicotine) to strong stimulation (cocaine, amphetamines). Generally, their potential for psychological dependence (habit formation) is high, as in the case of nicotine, but they are not considered to cause severe physiological withdrawal symptoms. They do, however, have a potential for the development of tolerance. (Drug tolerance is the development of resistance to the action of a drug so that increasingly larger doses are needed to achieve the same results.)

In addition to the direct effects of drugs and their potentials for addiction, one must also be aware of side effects associated with the way they are used. Nicotine is absorbed either by smoking or chewing tobacco, and substances produced by the curing of tobacco leaves or by burning them in pipes, cigars, and cigarettes are known to cause cancer. Lung cancer is significantly greater in smokers than in nonsmokers and is among the least responsive kinds of cancer to treatment (cure rate less than 10 percent). Sniffing of cocaine ("snorting")

FIGURE 15-5
Molecules of the neurotransmitter norepinephrine (also called noradrenalin) and an amphetamine. Points of difference are circled.

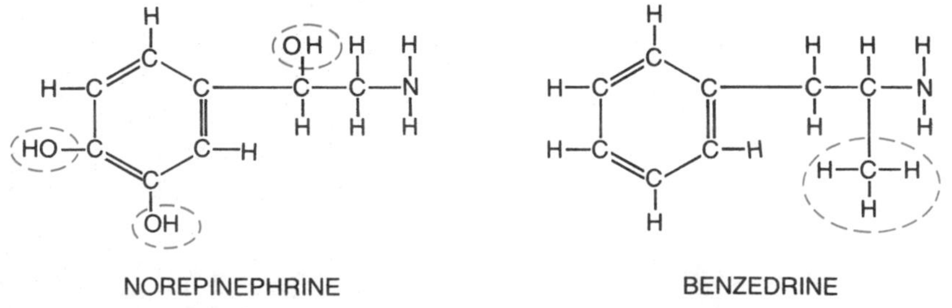

NOREPINEPHRINE

BENZEDRINE

can bring about degeneration of the nasal septum (flat nose), and if it is injected ("main-lining"), infections from dirty needles, as well as dangerous overdosing, can occur.

Depressants

It is not unusual to hear drugs referred to as either "uppers" (stimulants) or "downers" (depressants), terms aptly describing their effects on mood, personality, and behavior. Among downers, alcohol is the most widely used and also the most often abused. That alcohol is among the downers or depressants may seem at odds with the euphoria (pleasure, well-being) and conviviality experienced with moderate usage. The first effect of depressants, including **alcohol, opiates (opium, codeine, morphine, heroin), barbituates,** and **tranquilizers,** is anxiety reduction. Because insecurity and anxiety often are factors leading to use of these drugs, it is not strange that relief from these stress symptoms results in feelings of well-being and "mood enhancement." The general effect of depressants on the body, however, is that of depression of the CNS. It should, therefore, not surprise one to learn that one function of depressants is to block NTs. Alcohol, for example, has a general depressant effect on the CNS, interfering with and possibly blocking NTs. Although moderate amounts of depressants induce a sense of well-being, they interfere even at low levels with motor skills; greater amounts induce drowsiness and sleep, and excessive amounts produce paralysis, coma, and death by respiratory failure.

Alcohol is a leading cause for confinement in mental institutions; it is the leading contributing cause of highway accidents (more than half of those occurring in the United States each year are attributed to drunk drivers); it is a major contributing factor in divorce. Its consequences are far-reaching, ranging from suspected mental retardation in babies born of mothers who drink alcoholic beverages during pregnancy (see Fig. 1, p. 267) to cardiovascular disease, liver degeneration, and brain damage in chronic alcoholics.

Certain drugs are capable of producing **narcosis** (a condition of deep unconsciousness) and are referred to as **narcotics,** although that term now has restricted usage because of legal implications. The pain-relieving action of opiates (natural and synthetic opiumlike substances) is due to their ability to mimic a group of natural painkillers, the **enkephalins** and **endorphins,** which act by inhibition of certain NTs (see Chapter 13, page 277 for additional information). Opiates also appear to block NTs regulating thought and consciousness patterns in the brain.

It is not certain how depressants interact with NTs, but studies of schizophrenics and victims of **Parkinson's** disease offer some clues. There is evidence that levels of the NT **dopamine** become elevated prior to episodes of **schizophrenia,** and that tranquilizers, which are helpful in reducing the severity of schizophrenic episodes, may do so by inhibiting or blocking dopamine—by blocking either dopamine synthesis or dopamine receptors, or both. Schizophrenia (split personality) is not a single mental disease, but anxiety, altered perceptions, delusions, and feelings of persecution are common symptoms. Experimental evidence of the role of dopamine in schizophrenia has come from studies showing that **L-dopa,** a dopamine precursor, can induce early onset of schizophrenic symptoms in individuals predisposed to some forms of schizophrenia (dopamine itself cannot be administered because it will not pass through the blood–brain barrier; L-dopa does so; also see diagram, p. 316).

There is reason to think that some tranquilizers owe their effects in reducing anxiety to interfering with dopamine. **Heavy tranquilizers,** such as chlorpromazine (Thorazine) and trifupromazine (Stelazine), used to treat schizophrenics, can lower dopamine levels, which appear to be abnormally high in some types of schizophrenia. Conversely, individuals who have taken heavy tranquilizers to reduce anxiety may develop temporary schizophrenialike symptoms when they stop using them. Reserpine, the first of the heavy tranquilizers to be put into use in mental hospitals, apparently owes its effectiveness to its ability to reduce the output of norepinephrine in the central nervous system. The effectiveness of milder tranquilizers also is due to interactions with NTs or their receptors. Valium, the most widely prescribed tranquilizer, blocks GABA-receptor sites in postsynaptic membranes of brain neurons. Gamma aminobutyric acid (GABA) is the most abundant NT in the brain.

In some cases inhibitory and stimulatory NTs may act together to regulate certain kinds of behavior (Fig. 15-6). Dopamine and **acetylcholine (ACh)** together regulate certain motor functions coordinated in the brain. Dopamine accelerates neurotransmission in some CNS pathways and ACh inhibits it; together they can produce a balanced coordination.

Parkinson's disease (henceforth referred to simply as Parkinson's) is characterized by a shuffling walk, uncontrolled tremors of head and limbs, and generally uncoordinated movements. Some cases evidently have a genetic origin, but others are a delayed consequence of a severe viral infection. Parkinson's is a disease in which an enzyme catalyzing the conversion of the amino acid tyrosine into L-dopa, the immediate precursor of dopamine, is inactivated. As a result dopamine levels are reduced and the symptoms of Parkinson's appear (a similar but not so profound set of symptoms can appear with

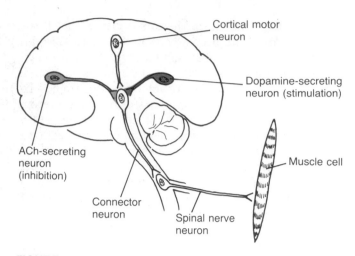

FIGURE 15-6
Relationship of dopamine and acetylecholine, ACh, in regulating motor function.

heavy tranquilizers, which act by blocking dopamine). Dopamine cannot be given directly, as noted previously, but L-dopa produces dramatic relief from symptoms of Parkinson's. The principle of this treatment is as follows:

TYROSINE $\xrightarrow[\text{enzyme}]{\text{Blocked}}$ L-DOPA $\xrightarrow[\text{enzyme}]{\text{Functional}}$ DOPAMINE
(amino acid)   (precursor)   (neurotransmitter)

                         ↑

Parkinson's   Administered in   Relief of symptoms
           treatment of Parkinson's   of Parkinson's

In addition to restoring normal or near-normal motor functions in victims of Parkinson's, L-dopa appears to act as a general stimulant. No doubt some of the resulting feeling of well-being is due simply to relief over restoration of a semblance of health. Nevertheless, L-dopa treatment has been known to produce hyperactivity in some patients and also to enhance their sexual drive.

Hallucinogens

The principal **hallucinogens** used as mood-altering drugs are **THC** (**tetrahydrocannabinol,** the principal psychoactive component of marijuana), **mescaline** (from the peyote cactus), **psilocybin** (from certain mushrooms), and **LSD** (**lysergic acid diethylamide,** from ergot fungus of cereals). PCP (also known as angel dust) is another drug with hallucinogenic effects being increasingly used today. There is some question about the inclusion of THC in this grouping; its mode of action differs from that of other hallucinogens, as do some of its effects. However, **marijuana** (the common source of THC) and LSD both interfere with transfer of information from short-term memory to long-term memory. Altered perceptions also occur with both but are much less pronounced with use of marijuana. The effects of even large dosages of mari-

juana are much milder than those of LSD or, for that matter, other hallucinogens.

LSD is colorless, tasteless, and odorless and is effective in extremely small dosages (microgram quantities). It is derived from ergot, a fungus growth on kernels of cereals, commonly rye. The first reported symptom of its use (it was ingested accidentally) was that everything appeared distorted, as if viewed in a faulty mirror. Because in early experiments it seemed to mimic schizophrenia, hope was expressed that it would serve as a powerful tool in exploring the biochemical aspects of mental disease. These hopes have not been realized and the current view is that it is dangerous, principally because its effects are unpredictable. The molecules of mescaline and psilocybin are similar to that of serotonin, an NT associated with sensory perceptions and sleep; that of LSD also is similar. All three drugs are thought to act by blocking serotonin-receptor sites in the CNS.

So far we have considered substances we willingly ingest—in maintaining health, in what we might call recreational aspects, and to a lesser degree in treating bodily disorders. Now we examine the impact on our health of viruses and disease organisms.

INFECTIOUS DISEASE, ANTIGENS, AND THE IMMUNE RESPONSE

Parasitism

All infections of a living organism by another organism or by a virus can be considered forms of symbiosis (living together) in which the host organism is more or less imposed on by the infecting organism or virus. This kind of symbiosis is **parasitism**—symbiosis in which one symbiont lives at the expense of another. In mild cases of parasitism, the host may suffer little harm. *Escherichia coli,* the well-known coliform bacterium, commonly occurs in the human colon and usually is not pathogenic. However, in infants, particularly in bottle-fed babies, it can cause infantile diarrhea, a usually nonfatal illness, and subtypes of E. *coli* to which an individual has not previously been exposed are responsible for some forms of "traveller's diarrhea." In explaining the observation that many parasitic infections are nonfatal, it is often said that it is "not in the best interest" of a parasite either to kill or to seriously debilitate its host and that the "most successful parasite" is the one that goes unnoticed. Every organism is host to many such parasites; in fact, parasitism can be avoided only by rearing the newborn in complete and sterile isolation. "Athlete's foot" fungus, "tooth-decay" bacteria, "crab" lice, and other parasites are an annoyance, to be sure, but human life goes on despite them.

Viruses Naturally we are very concerned about those parasites that cause seriously debilitating illnesses and premature deaths. Although **viruses** are not considered to be organisms in the sense of being cellular, they do exist as parasites on cells. Among them are many that parasitize human cells, causing influenza, poliomyelitis, rabies, smallpox, and yellow fever, as well as less fatal debilitating or sometimes merely annoying infections such as measles, mumps, hepatitis, herpes (venereal herpes and cold sores), and the common cold. Viruses invade our bodies through breaks in the skin or are inhaled or ingested and find their way through the more delicate lining membranes of internal passages. Unlike microorganisms such as bacteria and protozoans, viruses actually invade our cells and "live" within them. Within a host cell, an invading virus takes over the DNA transcription and protein-synthesizing processes of the cell in a manner described in Chapters 5 and 6, and shown diagrammatically in Fig. 5-11. The result is that the invaded cell becomes converted essentially into a "bag" full of newly formed virus particles (Fig. 15-7), from which the infection can spread to other cells. Depending on the infectiveness (virulence) of the virus and the strength of the host's defenses against the infection, the disease will be mild or severe, and recovery will be complete, partial, or nonexistent. In addition, certain viruses can remain in living cells as a latent infection, breaking out anew from time to time. The common cold sore, caused by a herpes virus, is an example (Fig. 15-7).

Bacteria and Disease Over most of the span of human existence, the causes of disease have been unknown, and its treatment has ranged from ludicrous to

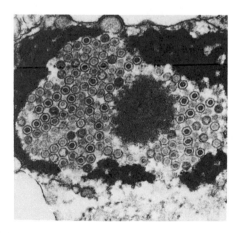

FIGURE 15-7
A human cell infected by herpes virus. (Electron micrograph by C. McLaren and F. Siegel, Courtesy of Burroughs Wellcome Company, Research Triangle Park, N. C., in F. C. Ross, 1983, *Introductory Microbiology*, Columbus, Ohio: Charles E. Merrill Publishing Co., Fig. 19.14, p. 463)

destructive. For example, one of the physicians in London during the great bubonic plague of 1665 wore a dried toad around his neck to ward off the disease, and physicians treating George Washington's last illness, apparently a streptococcus infection, contributed to his death by bleeding him excessively.

In the latter half of the nineteenth century the great French chemist Louis Pasteur finally explained the connection between **bacteria** and disease. Recognition of this relationship has led to the conquering of many infections and diseases, so that now most Americans are relatively safe from pathogenic microorganisms.

Bacteria cause disease principally by invading a host organism, multiplying rapidly, and producing poisons, or **toxins**. If reproduction of the bacterium is not checked by the defense mechanisms of the host, bacterial toxins can cause illness and sometimes death.

The *Salmonella* bacterium responsible for a form of food poisoning offers a rather commonplace example of the role of toxins in human disease. *Salmonella* lives and reproduces in sewage and polluted water and also can infect foods. If its growth in foods is not stopped by heating or refrigeration, the toxins it secretes under nearly all conditions accumulate and contaminate the food. Certain foods provide an especially suitable environment for *Salmonella* (e.g., chicken and turkey dressing or cream pie) and can poison those who consume them, causing considerable discomfort, although seldom death (however, a recent epidemic of *Salmonella* poisoning among elderly patients in a British hospital caused 40 deaths). *Salmonella* poisonings are fairly common at large banquets where quantities of food are prepared well ahead of time and often not properly stored.

A human disease more serious than *Salmonella* poisoning is produced by the diphtheria bacterium, *Corynebacterium diphtheriae*, which invades the human body, proliferates, and secretes a toxin capable of killing and then dissolving the cells of the host. Diphtheria once was a leading cause of death among children but has been largely eliminated by good sanitation and immunization.

Other Parasitic Organisms Viruses and bacteria account for a majority of infectious diseases of humans living in advanced cultures. In less advanced societies, however, eukaryotic parasites such as the protozoans causing malaria, amoebic dysentery, and African sleeping sickness often are the cause of seriously debilitating and even fatal illnesses. Many such infections are the consequence of poor sanitation practices in which water for domestic use is a source of waterborne disease, and vectors of infection such as flies, fleas, ticks, and mosquitos find ample opportunity to feed and breed (for an example, refer to Fig. 19-17, the life cycle of malaria).

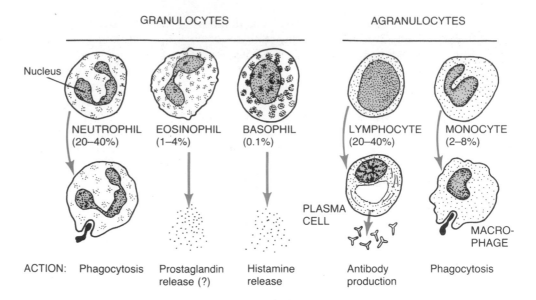

FIGURE 15-8
The major types of white cells in the human body. Percentages given below each cell type refer to the percentage representation of that cell type in the total leukocyte population. Note that each cell type has a characteristic nuclear configuration.

GRANULOCYTES

AGRANULOCYTES

Nucleus

NEUTROPHIL
(20–40%)

EOSINOPHIL
(1–4%)

BASOPHIL
(0.1%)

LYMPHOCYTE
(20–40%)

MONOCYTE
(2–8%)

PLASMA
CELL

MACRO-
PHAGE

ACTION: Phagocytosis   Prostaglandin
release (?)   Histamine
release   Antibody
production   Phagocytosis

Larger, multicellular parasites, such as tapeworms, liver and blood flukes, hookworms, and others also are associated with poor sanitation. While outbreaks of these can and do on occasion occur in advanced societies, they too are more prevalent in less advanced countries, most often in the subtropics and the tropics. For this reason, travellers to such regions should be aware of the dangers of infection, not only by the relatively mild pathogens of "traveller's diarrhea," but also by some severely debilitating parasites.

First Line of Defense—Nonspecific Mechanisms

The body's first line of defense against infection is nonspecific and is not directed against any particular kind of invading microorganism, virus, or toxic substance. On the outer, exposed, surfaces of the body, tough epidermal cell layers resist penetration by viruses, bacteria, and toxic substances. More delicate membranes are protected by secretions such as tears and mucus which, in addition to providing lubrication, contain the enzyme **lysozyme,** which can destroy many kinds of microorganisms. Although the skin and mucous membranes are impervious to infection when intact, they can, when breached by scratches or more serious wounds, allow viruses, bacteria, and other microorganisms to gain entry and become a focal point of infection.

Inflammation   Shortly after an infectious agent gains entrance to the tissues through a break in the skin, **inflammation**—a complex of reactions of a nonspecific nature—begins. Blood vessels in the vicinity of the infection become dilated (**vasodilation**) and more permeable

to liquids and large molecules. As a consequence the tissues become infiltrated with fluids, the temperature rises locally, and swelling occurs. In addition, the walls of capillaries in the zone of infection undergo a modification that makes them sticky to **neutrophils** (Fig. 15-8), a phagocytic type of white cell. As a result, numbers of neutrophils collect, then become mobile and migrate out of the capillaries into the surrounding tissues (Fig. 15-9).

Neutrophils belong to a class of **white cells (leukocytes)** containing conspicuously staining cytoplasmic granules; for that reason they are known also as **granulocytes.** In addition to neutrophils, which are pink staining, there are two other kinds of granulocytes: red-staining **eosinophils** and blue-staining **basophils.** The last two types also collect at an infection site but do not become phagocytic. All three types are produced in the

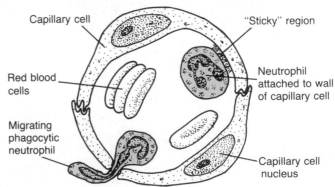

Capillary cell

"Sticky" region

Red blood cells

Neutrophil attached to wall of capillary cell

Migrating phagocytic neutrophil

Capillary cell nucleus

FIGURE 15-9
Factors relating to aggregation of neutrophils in capillaries at a site of infection. The precise mechanism for attraction of leukocytes and their adherence to capillary walls is unknown. They then become mobile and migrate out of the capillaries through gaps between capillary-wall cells.

bone marrow. There is uncertainty about the role of eosinophils, but they appear to be associated with release of prostaglandins at sites of inflammation. Prostaglandins stimulate vasodilation and also lower the threshold of pain sensitivity, causing an infected area to be "tender" and sensitive, and so tending to immobilize and protect the wounded part. Basophils release **histamine,** which has several functions, including vasodilation and increasing cell permeability.

Increased permeability of capillaries and tissues in a region of inflammation not only permits the initial influx of granulocytes and subsequently other kinds of leukocytes but also permits fibrinogen to infiltrate an infected area and then to form fibrin clots and seal off an infection. The hoped-for result of this is confinement and localization of an infection as well as its subsequent suppression.

Fever Fever is a symptom of infection and ranges from localized temperature increases in a zone of inflammation to general whole-body responses. Mild fevers may be beneficial in stimulating mobilization of first-line defenses, but high fevers (greater than 39°C [102°F]) are harmful, and very high fevers can cause convulsions and death.

Interferon Interferon, actually a group of similar substances, is a protein which inhibits the multiplication of viruses in host cells. It is produced by several types of leukocytes in response to a viral infection and binds to membranes of uninfected cells, inducing them to produce antiviral substances. Among these antiviral substances are enzymes that can prevent the translation of viral mRNA into viral proteins. At one time interferon was available only from natural sources in only very tiny amounts and at great cost, but now synthetic human interferon is manufactured by recombinant DNA technology and appears promising for treatment of some virus infections.

Complement Complement is a name given to a system of blood proteins triggered into action by infection. Complement factors enhance inflammation, mobilization of leukocytes, vasodilation, histamine release, and permeability of cells and tissues to proteins (including **antibodies,** described later). One complement factor attaches to bacterial surfaces and somehow forms tunnels

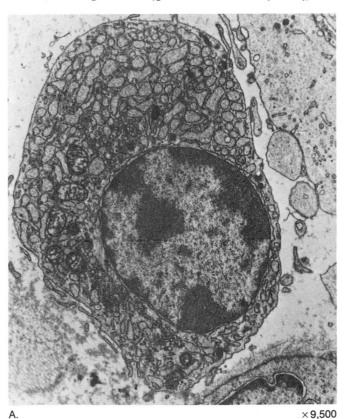

A.   ×9,500

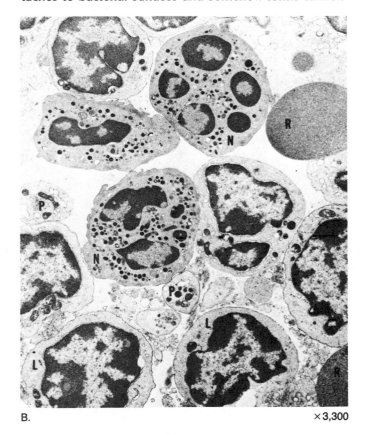

B.   ×3,300

FIGURE 15-10
Electron micrographs of some human white-cell types. **A.** A plasma cell; note its extensive internal membrane system. **B.** Neutrophils (N), lymphocytes (L), erythrocytes (R). Also shown are platelets (P). The latter are not cells but are cytoplasmic elements produced by megakaryocytes. (Courtesy of David Mason)

into the interior of the bacterial cell; this results in a fatal "leakiness" of the cytoplasm. Another complement factor adheres to foreign bodies, marking them as "targets" for phagocytosis by leukocytes.

Phagocytosis Phagocytosis is the most important nonspecific defense mechanism of the body. It is carried on by certain members of the white-cell population of the blood and lymph. About 60 percent of this population is composed of neutrophils, which are the first phagocytic cells to go into action against an agent of infection. When neutrophils collect at a site of an infection, they do so in response to concentration gradients of substances diffusing into the blood from a wound site—a response referred to as **chemotaxis**. Subsequently they become capable of amoeboid locomotion (Fig. 15-9), squeeze through spaces in the capillary walls, and move into the tissues, where they actively combat infection by engulfing invading microorganisms (refer to Fig. 2-28 for a diagram of phagocytosis). Neutrophils are able to ingest almost any kind of particle and are assisted by a second type of phagocyte, the **macrophage** (Figs. 15-8, 15-10). The latter are produced when agranular leukocytes called **monocytes** migrate into a zone of inflammation and become transformed into amoeboid macrophages. Both neutrophils and macrophages can distinguish between "self" and "nonself" because they carry **compatability factors** in their plasma membranes that enable them to "recognize" the body's own cells and also because invading cells or substances may become "tagged" for identification by complement factors.

Usually neutrophils are most numerous early in an infection; then their numbers begin to fall off and they are replaced by macrophages. The latter are very active, engulfing bacteria and toxins as well as damaged and dead body cells. Worn out phagocytes, damaged and dead body cells, and fluids constitute **pus**, which can accumulate at the focal point of an infection, later to drain away, as when a pimple or boil erupts, or to be reabsorbed by the healing tissues.

ANTIBIOTICS

The subject of **antibiotics** can be discussed in relation to several groups of organisms, because antibiotics are produced by bacteria, fungi, and probably some plants. A typical antibiotic is produced within an organism and secreted into the environment, where it has an inhibiting effect on the growth of other kinds of life (Fig. 1). It is becoming increasingly clear that antibiotics have an ecological strategy of reducing competition by other organisms for energy and space.

Penicillin, a product of the fungus *Penicillium notatum,* was the first antibiotic to be used in medicine. Discovered in 1929 by Alexander Fleming but not purified and used until 1941, penicillin has saved more lives than all other drugs in the history of medicine. Its action is to suppress cell-wall formation in dividing bacteria. As a result, bacteria are produced that lack a wall capable of restraining internal pressures resulting from normal intake of liquids; consequently, they swell up and burst. Other antibiotics used in treatment of bacterial disease have different modes of action, but all have direct effects on reproduction of the infective bacterium, rather than treating symptoms of disease as do more traditional medicines.

A serious problem in the use of antibiotics is the evolution of increasingly resistant mutant strains of bacteria. (This, of course, is a powerful example of the relationship between gene mutation, selection, and evolution.) For example, a new, dangerous strain of gonorrhea coccus has evolved among prostitutes in the Philippine islands and is spreading rapidly to other lands. The new bacterial strain apparently developed as a consequence of continual dosing with penicillin in attempts to remain free of venereal disease. Unfortunately, the dosages were not sufficiently great to kill all the bacteria—only the least resistant ones. New, more

If the body's first line of defense remains intact, the infection is confined to the original site of invasion, the agents of infection are destroyed, and healing processes begin. If, however, the first-line defenses are breached, the infection may spread in one or both of two ways. First, bacterial toxins may diffuse into the tissues and lymph even though the bacteria that released the toxins remain confined. If the toxins are extremely poisonous, as in the case of tetanus, death may result even though the invaders remain at the site of infection. Second, invading microorganisms or viruses may multiply so rapidly that they break out of the first-line defenses and the infection becomes generalized. Should either or both of these events occur, agents of first-line defenses continue to operate and phagocytosis continues, especially in the lymph nodes, where macrophages are produced in large numbers and function as "road blocks" to a spreading infection. However, when an infection becomes generalized, first-line defenses may not be effective and the body's second line of defense, the **immune system**, takes over.

Second Line of Defense—Specific Immune Responses

Specific immune responses are those defensive actions that are directed against a particular kind of invading organism, virus, or substance of foreign origin (e.g., cell component, toxin, enzyme, or breakdown product). The immune system is not, however, activated by ions or small molecules, but rather by polypeptides and proteins or substances containing them at least in part. These foreign proteinaceous substances are known as **antigens,** and the body's response is to produce antigen-inactivating **antibodies.**

There are literally millions of substances in our environment capable of acting as antigens, and the task of producing a specific antibody against each of them would seem insurmountable. How the body copes with this array of potentially destructive substances has only in recent years begun to be understood. The first step in the immune response involves a type of agranular leu-

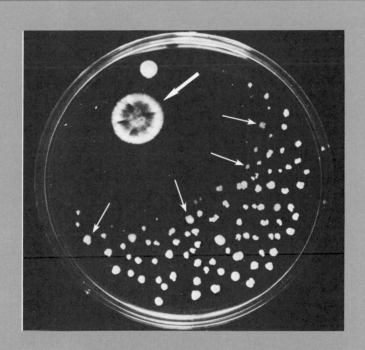

FIGURE 1

The inhibiting effect of an antibiotic released from a colony of *Penicillium* mold (larger arrow) upon growth of colonies of the bacterium *Staphylococcus* (smaller arrows). (Courtesy of Ronald Hare and reprinted from *Chemistry* (now *SciQuest*), vol. 51, no. 7, 1978)

highly penicillin-resistant mutant strains arose. This same relationship between antibiotics and microorganisms is an ever-present danger in many diseases.

kocyte, the **lymphocyte**. Lymphocytes are produced both in bone marrow and lymphatic tissues (lymph nodes, spleen, thymus). They constitute 30 to 40 percent of the body's population of white cells and are of two types: **B-lymphocytes** and **T-lymphocytes**. B-lymphocytes function chiefly in the production of antibodies active in blood and lymph; T-lymphocytes function chiefly in the tissues.

B-lymphocytes play an important role in **antibody-mediated immunity** (AMI). When B-lymphocytes come in contact with an antigen, some of them enlarge and begin to divide, producing a clone of cells capable of responding to that specific antigen. These cells then become transformed into **plasma cells** (Fig. 15-10), which are the chief antibody-producing cells of the body.

T-lymphocytes produce a range of chemicals known as **lymphokines**, with which they are able to kill foreign cells, such as those of invading microorganisms or of transplanted tissues and organs (Fig. 15-14). T-lymphocytes, like B-lymphocytes, require prior exposure to an antigen before becoming activated.

Antibody Production Because antibodies act against specific antigens and only against specific antigens, it follows that lymphocytes must be prepared to make any one of a multitude of antibody types (because there are so many substances in the environment capable of acting as antigens). One way to meet this challenge would be for the lymphocyte to have on hand the total arsenal of antibodies, ready to release upon infection. This obviously would mean manufacture and storage of a great

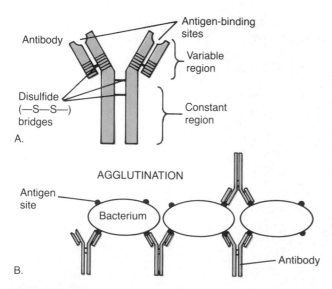

FIGURE 15-11
A. A diagram of an antibody, illustrating its composition of four polypeptides held together with disulfide bonds. **B.** A diagram of the manner in which antibodies can produce agglutination.

many antibodies, the majority of which would never be used. Instead, lymphocytes respond to the challenge of a specific antigen by "analyzing" that antigen and then manufacturing a "tailor-made" antibody. Unless the body has previously been exposed to an antigen, the lymphocyte must "design" and produce a totally new kind of antibody; if, however, there has been a previous challenge by that antigen, then the machinery for making the specific antibody is on hand and antibody production can begin forthwith. This ability to respond quickly to an antigenic challenge is referred to as "immunological memory," and is the basic principle behind immunization against disease.

It is believed that a variety of subtypes of lymphocytes exist, each subtype capable of reacting only to a "family" of antigens. When one of these subtype lymphocytes comes in contact with an antigen that it "recognizes," probably by some kind of "key-in-lock" principle, genes for antibody production are genetically derepressed and the lymphocyte proceeds to make antibodies.

An antibody is a Y-shaped molecule composed of four polypeptides (Fig. 15-11). Each arm of the Y is a duplicate of the other arm. The "handle" of the Y is the same in all antibodies; only the arms of the fork differ and constitute the antigen-specific part of the antibody. The arms themselves are composed of variable subunits, synthesized by different genes; therefore the possible combinations of arm components are essentially limitless and account for the great variety of antibody types as well as the specificity of antibodies.

When an arm of an antibody comes in contact with an antigen, arm and antigen adhere to each other to form a complex. These complexes differ according to the size and nature of the antigen and are described as either **agglutination, precipitation, lysis, opsonization,** or **neutralization**. Agglutination and precipitation reactions reveal the remarkable effectiveness of the Y-shaped configuration of an antibody. If the antigen is, for example, an incompatible red blood cell (refer to blood typing in Chapter 7), each arm of the Y attaches to a different blood cell; a cross-linked group of blood cells forms a clump and agglutination occurs (Figs. 15-11, 15-12). If the antigen is a large molecule, but not a cell, cross links form also between antibodies and antigens, but the reaction is slower and the clumps are smaller—precipitation is the result.

Lysis is the consequence of attachment of antibodies to cell walls and membranes of invading cells, making them leak and eventually killing them. Opsonization occurs when antibodies, in cooperation with complement factors, form a coating on the surface of an invading microorganism, marking it as a "target" for phago-

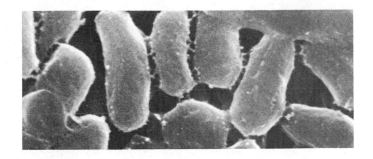

FIGURE 15-12
Cells of the bacillary dysentery organism, *Shigella flexneri,* clumped (agglutinated) as a result of exposure to antibodies of *flexneri* polyvalent antisera. ×13,000. (Electron micrograph by F. Siegel, Courtesy of Burroughs Wellcome Company, Research Triangle Park, N. C.; in F. C. Ross, 1983, *Introductory Microbiology,* Columbus, Ohio: Charles E. Merrill Publishing Co., p. 384)

cytosis by neutrophils and macrophages. The fifth kind of antigen–antibody reaction, neutralization, occurs between antibodies known as antitoxins and soluble toxin molecules or viruses present in body fluids. The **toxin–antitoxin** reaction blocks the active sites of the toxin or virus and renders it harmless (i.e., neutralizes it), but does not produce either agglutination or precipitation.

Immunization The account of how a British physician, Edward Jenner (1749–1823), discovered **immunization** against smallpox is a classic. Jenner had noticed that milkmaids, who often were exposed to a mild infection known as cowpox (in medical terms, *vaccinae*) did not come down with smallpox and its disfiguring pustules and scars (Fig. 15-13). It then occurred to him that

inoculating (or "vaccinating" with *vaccinae*) individuals with cowpox pus could also confer immunity against smallpox just as being a milkmaid did. It worked, and today **vaccination** has all but eliminated the disease. (The World Health Organization has announced smallpox to be extinct, but given severe overcrowding in some parts of the world plus dangerously inadequate sanitary regulations, it may be best to reserve judgment for a while yet.)

The principle behind the success of Jenner's procedure, as well as immunization against other diseases, is an interesting one. The viruses of smallpox and cowpox are apparently the same, but when the former infect a cow they become weakened (the medical term is **attenuated**) and so can safely be used to immunize humans.

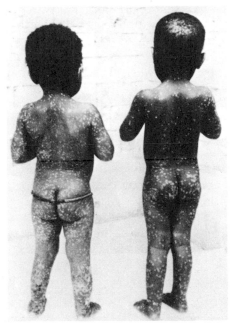

A.

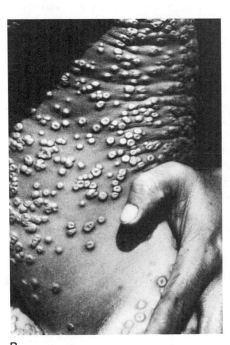

B.

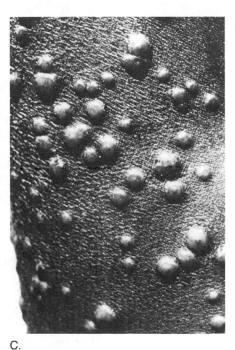

C.

FIGURE 15-13
Smallpox, a virus disease. **A.** The 21st day of the rash. **B.** Frontal view of the pustules (pox). **C.** Smallpox pits. (A and B courtesy of the Center for Disease Control, Atlanta, Ga; C, courtesy of the World Health Organization, Diagnosis of Smallpox Slide Series, 1968)

Similarly, Pasteur attenuated rabies (hydrophobia) virus by infecting rabbits with it, and Sabin perfected oral polio vaccine by first infecting monkey-cell cultures with live virus.

Active and Passive Immunity It has no doubt occurred to you that if one is exposed for the first time to a deadly disease, say bubonic plague, and is immediately immunized, but soon after becomes deathly ill, the body may not have had enough time to develop immunity. Therefore there will be no antibodies available to combat the infection and to develop **active immunity**. In such cases, a solution is to "borrow" someone else's antibodies—someone else sometimes being a horse or some other animal capable of developing antibodies against an agent of human infection. This is the principle of **passive immunity,** in which antibodies are carefully removed and concentrated from the blood serum of a donor, another human perhaps, but often a laboratory animal.

Autoimmune Disease It happens sometimes, fortunately rarely, that an individual develops antibodies against his or her own cells. It happens because some kind of cellular event, perhaps a gene mutation or a viral infection, causes the body's immune system to develop a clone of T-lymphocytes which are programmed to produce antibodies against "self" cells. The symptoms of this "self-rejection" are referred to as **autoimmune disease** and require inactivation of the immune system by drugs or radiation therapy, much as is done in the case of organ transplants.

During embryonic stages of development, the immune system of the fetus is inactive, thus permitting tissue and organ differentiation to take place without inducing rejection phenomena. Later, when differentiation is essentially completed, the immune system also is differentiated and operative. However, before the immune system develops in an embryo, it has been found (in mice only at this stage of research) that foreign cells can be transplanted into the embryo and will thenceforth be accepted as "self," so that later on a graft from the donor individual can safely be done without incurring graft rejection. Whether this principle will have any human application remains to be seen, but it opens up some interesting possibilities.

Allergy Allergies are annoying though seldom fatal instances of antibody action against relatively nontoxic antigens such as ragweed pollen, chicken-feather protein, and horsehair. When these substances adhere to the epidermal cells of the body, especially the more delicate mucous membranes of respiratory passages, they induce the formation of a type of antibody that causes the release of histamine by certain cells called **mast cells**. Histamines are responsible for the discomfort of allergies: mucus discharge, swelling of tissues, and inflammation. The treatment commonly is to take an **antihistamine**; in severe cases cortisol is administered. While allergic responses normally run the gamut from mild to serious discomfort, some individuals become unduly sensitive; even a single bee sting can send them into shock and even death. For severe allergic reactions, a desensitizing program of increasing exposure to an **allergen** may result in production of blocking antibodies against the allergen. These prevent the release of histamine by mast cells so that allergic symptoms do not occur.

Graft Rejection Organ and tissue transplants have become almost commonplace in recent years; kidneys, hearts, and even lungs and livers have been transplanted with varying degrees of success. Nearly everyone is familiar with the problems of **graft rejection** and is aware that great care must be taken to match the tissues of the donor and recipient—a procedure similar to blood typing. Nearly always some incompatibility exists, and because the grafts are of cells, a cell-mediated immune reaction involving T-lymphocytes takes place. T-lymphocytes, as noted earlier, can release killer substances that dissolve foreign cells, or they can target foreign cells for phagocytosis by macrophages (Fig. 15-14). T-lymphocytes also carry antibodies on their surface membranes and can combine with foreign cells directly, then kill them by secreting lytic substances. As a sidelight, it should be mentioned here that T-lymphocytes also kill the body's own cells when those cells contain virus particles and therefore are antigenically altered; some kinds of cancer cells are similarly attacked and killed by T-lymphocytes.

Graft rejection is a function of the body's normal immune-response system and can be overcome only by interfering with the immune system itself. Drugs capable of doing so have been developed, and those capable of differentiating between T-lymphocytes and B-lymphocytes have been particularly sought after. Even with the circulating antibodies formed by B-lymphocytes left intact, an individual remains very susceptible to infectious disease.

CANCER

On the basis of current statistics, one out of four Americans can expect to suffer from one or another form of **cancer**—over 1000 men, women, and children die from cancer each day, and the death rate is rising. Some kinds

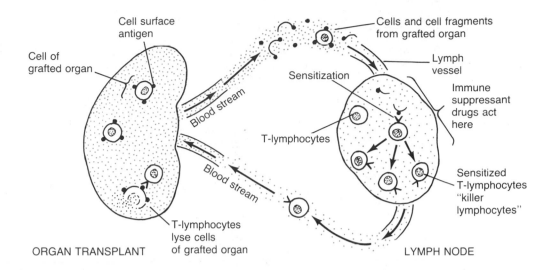

FIGURE 15-14
The immune response to an organ transplant such as a kidney transplant.

Cell surface antigen

Cell of grafted organ

Cells and cell fragments from grafted organ

Lymph vessel

Sensitization

T-lymphocytes

Immune suppressant drugs act here

Blood stream

Blood stream

Sensitized T-lymphocytes "killer lymphocytes"

T-lymphocytes lyse cells of grafted organ

ORGAN TRANSPLANT

LYMPH NODE

of cancer are yielding to treatment, but others remain profoundly life-threatening—fewer than 10 percent of those afflicted with lung cancer can expect to recover from it, a particularly agonizing statistic because we know the leading contributing factors (smoking, air pollution). It is estimated that about 80 percent of human cancer is induced by cancer-causing substances (**carcinogens**) in the air we breath, the water we drink, and the food we eat. While we cannot avoid these carcinogenic substances completely, we can minimize our exposure to them and so reduce the risk of cancer.

Kinds of Cancer

Cancer is a disease in which cells somehow become activated into uncontrolled multiplication and thus produce an overgrowth, or **tumor**, composed of malformed, malignant cells (Fig. 15-15). Cancerous tumors can occur in almost any tissue of the body, although some are more often affected than others. Three general kinds of cancer, named after the tissues most often affected, are recognized: **carcinomas** (from *karkinoma*, the Greek word for cancer), which commonly involve epithelial tissue (epidermis and lining membranes); **sarcomas** (Greek *sarkoma*, "flesh"), which affect mainly connective tissues, including bone; and **leukemias**, which start in the bone marrow and lymphatic tissues and spread in the blood and lymph.

Causes of Cancer—Oncogenes

There now is general agreement that the onset of cancer can be correlated with exposure to certain kinds of environmental pollutants, as noted in the relationship between smoking and lung cancer, but absolute correlations are difficult to prove. Laboratory animals exposed to suspected carcinogens may invariably develop cancer, but the inability, for obvious reasons, of cancer specialists to do similar cause-and-effect experiments with human subjects leaves a theoretical loophole that has caused various interest groups to reject the implications of laboratory studies.

Although cancer can be caused by a variety of agents, including carcinogens, viruses, X rays, and ultraviolet light, there is a growing consensus that the disease is basically a cellular and genetic phenomenon, possibly the consequence of activation of cancer genes (**oncogenes**) which may be part of every cell's genetic heritage and which may also be related to virus infections.

The existence of tumor viruses first was demonstrated in 1910 when Peyton Rous found that a cell-free filtrate of an extract obtained from a certain kind of cancer of chickens could induce similar tumors when injected into cancer-free chickens. The active agent in this infection was later identified as a virus (the **Rous sarcoma virus**). Subsequent discovery of other cancer-causing viruses has established the relationship between some cancers and certain specific viruses.

Even though many cancers appear to have no direct relationship to virus infection, a basic principle of cancer development is emerging from the study of virus-induced cancer. Apparently some viruses have a single gene that can rapidly transform a normal cell into a cancerous one; other viruses appear to induce cancers more slowly. In both situations, however, viral genes are thought to become integrated into the host cell's DNA. Some virus genes appear to act as oncogenes and directly transform a normal cell into a cancerous one; others appear to act indirectly by causing mutation of nearby host-cell genes into oncogenes. In some in-

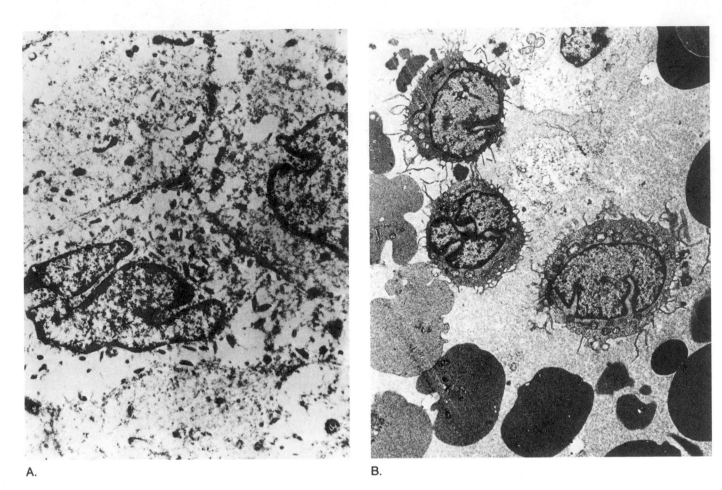

FIGURE 15-15
Human cancer cells. **A.** Melanoma, the small dark-pigment granules (melanosomes) are characteristic of this cancer. Lobed nuclei in cells are evidence of great metabolic activity and often are characteristic of cancer cells. Not all cells having lobed nuclei are cancerous, however; see Fig. 15-10. **B.** Hairy-cell leukemia, characterized by the fibrous "hairs" extending from the cell membranes of these leukocytes. (Courtesy of David Mason)

stances, oncogenes (whether of viral or host-cell origin) remain quiescent until conditions in the cell's environment changes. They then become activated and transform the cell's reproductive "machine," possibly by producing novel enzymes. It is believed that oncogenes can act as cellular "time bombs," triggered to go off only when conditions inside and/or outside the cell change in certain directions.

It has been suggested that oncogenes occur in a majority of human cells, perhaps all, a legacy of ancient as well as recent viral infections. Normally, they remain quiet in the cell. When stimulated into action by a carcinogen, however, they can convert the cell into a cancerous phase of growth and reproduction. The identification of oncogenes as fundamental causes of cancer, if borne out by future research, will represent a major step in detection, prevention, and treatment of the disease.

The Body's Defenses Against Cancer

Cancer cells do not divide any faster than normal body cells, but do so without stopping. As we have seen in the case of bacterial populations (Figs. 19-3 and 19-4), such nonstop reproduction can soon result in an enormous number of cells, which in the case of cancer can then migrate throughout the body, starting new tumors. This process is known as **metastasis**. Accompanying this proliferation of cancer cells are changes in cell-surface proteins, which then can function as antigens. These antigens induce formation of antibodies (apparently both B- and T-lymphocytes participate), which can interact with and destroy cancer cells (Fig. 15-16). It is thought that most newly formed cancer cells are soon destroyed by lymphocytes and their antibodies, and that tumor formation may in relative terms be fairly rare.

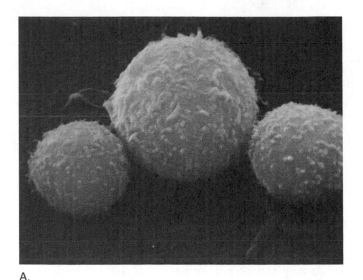

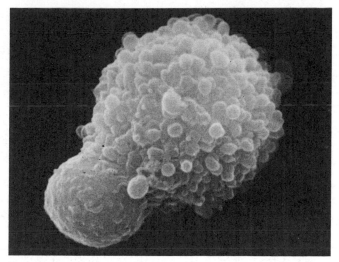

A.

B.

FIGURE 15-16
A. In this scanning electron micrograph, lymphocytes attack a cancer cell. The cancer cell (large spheroid, center) has molecular "labels" (antigens) on its surface that enable one of the sensitized lymphocytes (smaller cells) to selectively attack and kill it. **B.** Death of the cancer cell is indicated by the blebs, or deep folds, appearing on its surface membrane. The mechanism by which lymphocytes kill tumor cells is unclear but probably involves the release of a toxic factor that disrupts the cell membrane. (Photos by A. Liepins)

Cancer Treatment

Ideally the treatment of cancer should have its beginning in prevention. Once cancer has been detected, however, the cancerous cells must be removed surgically or destroyed in place. Surgery is the usual treatment if cancer is detected early, and considerable attention has been given to ensure that the so-called "cancer signals" are known and watched for. These include *any prolonged or unusual irritability of the skin, respiratory passages, or gastrointestinal tract, any change in bowel or bladder habits,* and *any unusual bleeding from any body opening. The occurrence of lumps beneath the skin,* especially in the female breasts, should be considered suspicious; *growth and irritability of warts or birthmarks* also should be reported to one's physician. In short, any suspicious symptom that does not disappear in a week's time should be carefully monitored. Above all, one should not allow fear of the discovery of cancer to keep one from the doctor's office. Certain kinds of cancers are readily operable if discovered in time.

When cancer cells have metastasized, a combination of surgery with radiation treatment and chemotherapy is usually employed. Thus far no chemicals have been discovered that are capable of seeking out and destroying only cancer cells, but combinations of chemicals and of chemicals with radiation are effective in treating some cancers. Meanwhile the search goes on.

SUMMARY

The key to good health is good nutrition and the maintenance of good sanitary habits. Important to good nutrition is the consumption of the three principal foods: fats, proteins, and carbohydrates. Foods should be chosen also to provide necessary vitamins and minerals, although these usually are provided by a diet containing a selection of meats, fruits, and vegetables. A well-chosen vegetarian diet will suffice also.

Improper diet not only can lead to malnourishment and vitamin and mineral deficiencies but also can contribute to cardiovascular disease. The chief culprits in the latter are cholesterol and saturated fats; unsaturated fats, conversely, can inhibit the deposition of cholesterol and fat deposits (plaque) which plug arteries in atherosclerosis. Also contributing to malnutrition and cardiovascular disease is stress, characterized by hypertension and anxiety. Stress symptoms can be treated by the administration of appropriate drugs, including certain stimulants and depressants. Stress may lead also to drug abuse: harmful amounts may be taken in an effort to alleviate the condition. Probably all drugs are addictive to a greater or lesser degree, but some, including opiates, tobacco, and alcohol, are especially so.

The basis for action of both stimulant and depressant drugs is their interaction with neurotransmitter (NT) molecules in the central nervous system. Some drugs mimic the effect of NTs by occupying NT-receptor sites and producing the same effect as the NT they mimic. Others, such as certain heavy tranquilizers used to treat schizophrenia, act by blocking NT production. Some hallucinogens also interact with NTs (LSD with serotonin, for example), but the actions of others (marijuana) are unknown.

Many health problems are associated with parasitism, commonly by viruses and bacteria. Viruses act by taking over the reproductive "machinery" of cells and using it for production of new virus particles. Bacteria more often damage cells, tissues, and organs by the production of toxins. Foreign substances, whether virus components, toxins, parts of bacterial cells, or allergens, act as antigens when they invade the human body. The body's defense is activation of the immune system and production of antibodies, which in one way or another inactivate or destroy antigens. Any foreign cell can act as an antigen and activate the immune system. This is the basis of graft rejection and organ-transplant problems. In order to prevent such rejection, immune-suppressant drugs are employed.

Cancer is a disease in which cells continuously reproduce, overgrowing normal tissues. The transformation of a normal cell into a cancerous one can be triggered by environmental substances known as carcinogens. Current thinking is that carcinogens activate cancer genes (oncogenes) which may be generally present in many, perhaps all, body cells. Because cancer cells differ antigenically from normal body cells, it is believed that in many cases the body's immune system disposes of them before they can get out of control. For those that survive and develop into tumors, the major recourses at present are surgery, radiation treatment, and chemotherapy.

KEY WORDS

vitamin	mescaline	antigen
mineral	psilocybin	antibody
essential amino acid	LSD (lysergic acid diethylamide)	B-lymphocyte
kilocalorie	parasitism	plasma cell
cardiovascular disease	virus	T-lymphocyte
hypertension	bacterium	agglutination
plaque	toxin	precipitation
cholesterol	immune response system	lysis
placebo	lysozyme	opsonization
HDL (high-density lipoprotein)	inflammation	neutralization
LDL (low-density lipoprotein)	vasodilation	immunization
neurotransmitter (NT)	neutrophil	vaccination
stimulant	phagocytosis	attenuation
depressant	leukocyte	active immunity
amphetamine	histamine	passive immunity
narcosis	interferon	autoimmune disease
enkephalin	fever	allergen
endorphin	complement	cancer
schizophrenia	antibiotic	oncogene
heavy tranquilizer	chemotaxis	
THC (tetrahydrocannabinol)	specific immune response	

QUESTIONS FOR REVIEW AND DISCUSSION

1 Design a vegetarian diet providing adequate amounts of fats, proteins, and carbohydrates.

2 Why is gelatin by itself not an adequate source of food even though it is pure protein?

3 Describe the relationship between cholesterol, saturated fat, and unsaturated fat in the diet and in atherosclerosis.

4 Explain the action of stimulant and/or depressant drugs in terms of their relationship with neurotransmitters.

5 Why is L-dopa administered rather than dopamine in cases of Parkinson's disease? What is the principle of its action?

6 Why does the common cold sore usually recur?

7 Explain the action of neutrophils, lymphocytes, and monocytes in a bacterial infection.

8 What is the principle of antibody specificity?

9 How do antibiotics work?

10 What are oncogenes? What is their possible involvement in cancer?

SUGGESTED READING

BISHOP, J. M. 1982. Oncogenes. *Sci. Amer.* 246(3):80–92. (Is there a common pathway to cancer?)

CUNNINGHAM, B. A. 1977. The structure and function of histocompatability antigens. *Sci. Amer.* 237(4):96–107. (Excellent discourse on antigen properties—similarities between antigen and antibody molecules are described.)

CUNNINGHAM, J. D. 1983. *Human biology.* New York: Harper and Row. (A nicely illustrated textbook covering many aspects of human anatomy and physiology.)

KOLATA, G. 1983. Drug transforms transplant medicine. *Science* 221:40–43. (Cyclosporin suppresses graft rejection, apparently by preventing T-lymphocytes from dividing.)

MARX, J. L. 1981. Antibodies: getting their genes together. *Science* 212:1015–17. (A brief discussion of antibody components, their assembly, and the genes which code for them.)

————. 1984. What do oncogenes do? *Science* 223:673–76. (Lucid summary of recent findings.)

NICHOLSON, G. 1979. Cancer metastasis. *Sci. Amer.* 240(3):66–76. (Characteristics of cancer cells and their spread are discussed—role of lymph system, nature of tumors, etc.)

SNYDER, S. H. 1984. Drug and neurotransmitter receptors in the brain. *Science* 224:22–31. (A rather detailed and technical article, but one that describes the role of receptor multiplicity in actions of neurotransmitters.)

VAN DYKE, C., and R. BYCK. 1982. Cocaine. *Sci. Amer.* 246(3):128–41. (Cocaine action in blocking norepinephrine uptake is discussed—includes surprising new findings on cocaine usage.)

Plant Reproduction and Development

16

Plants and animals alike have their origins in ancient oceans; as would be expected, their reproductive systems give evidence of that ancestry. Motile gametes probably hark back to eras when all cells lived a motile existence; later, when some cells became specialized for reproduction, cilia and flagella were retained for locomotion and fertilization. Thus, while plants and animals have become multicellular organisms and either have developed new modes of locomotion (in animals) or are sedentary organisms (plants), their gametes often have retained the kind of motility which we associate with unicellular life. In those plants and animals that have remained aquatic, gametes usually are free-swimming. But in those plants and animals that long ago migrated to land, free-swimming gametes are found only in those species that spend at least parts of their lives in wet areas. Dry-land plants and animals have evolved protected gametes which either swim in fluids secreted by sex organs or, in the case of flowering plants, are passively transported (refer to Table 20-1 for major characteristics of plant groups).

ALTERNATION OF GENERATIONS

Botanists are not sure about the sequence, or sequences, of events in plant evolution that led to the es-

tablishment of plant life on land. They are, however, agreed that the ancestors of green land plants were simple, probably filamentous (chains of cells) organisms known as **green algae** (Fig. 16-1). There are several reasons for this supposition. First, green algae and green land plants have the same kinds of photosynthetic pigments: chlorophylls a and b (green), and carotenoids (yellow-orange). Second, the cells have similar walls, in both cases composed of cellulose cemented together with the carbohydrate material pectin. There are other similarities, including enzyme systems and ultrastructural details both groups have in common. An important difference, however, between the majority of green algae and green land plants is in their reproduction, that of green algae (Fig. 16-1 and 16-2A) being relatively simple as compared with that of green land plants (Fig. 16-2B and 16-3).

The sticking point in explaining land-plant evolution seems to be the origin of the typical reproductive cycles of all green land plants, a process known as **alternation of generations**. Simply put, green land plants exist as alternating haploid and diploid generations of individuals (Fig. 16-2 and 16-3). Thus, for example, a large diploid fern plant does not reproduce identical large diploid plants at each generation. Instead, it produces **spores** which develop into tiny, usually heart-shaped,

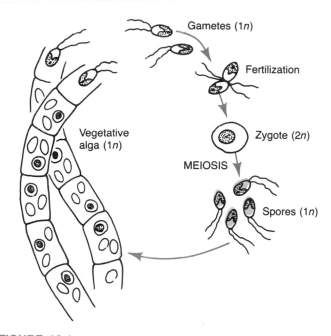

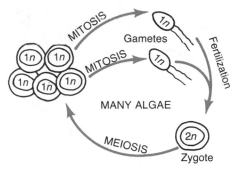

FIGURE 16-1
Typical reproductive cycle of a green alga. Meiosis follows fertilization and haploid spores are produced. These develop directly into new algal filaments.

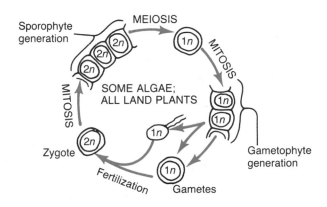

B. ALTERNATION OF GENERATIONS

FIGURE 16-2
Reproductive cycles of some types of plant. **A.** This reproductive cycle is typical of many green algae. The vegetative plant is haploid and produces haploid gametes. The zygote, which is the only diploid stage, directly undergoes meiosis, giving rise again to the vegetative stage. This diagram corresponds to Fig. 16-1. **B.** This reproductive cycle is typical of some algae and all green land plants. There is a diploid generation called the sporophyte which produces spores by meiosis. The haploid spores produce haploid gametophytes which reproduce by gametes. The zygote does not undergo meiosis but produces the sporophyte generation by mitosis.

green haploid plants. These heart-shaped fern plants, which often are no larger than a matchhead, reproduce by gametes, not spores. After fertilization, however, the zygote does not develop again into a heart-shaped plant, but into the large plant we think of as a "typical" fern. Of course, if we reflect on it, the tiny heart-shaped plant is as much a fern as is the much larger leafy plant.

We need add to this discussion of fern reproduction only that the large, spore-producing fern is called the **sporophyte** ("*spore-*[producing]*plant*"); the small heart-shaped plant is called the **gametophyte** ("*gamete-*[producing]*plant*"). The sporophyte produces spores by meiotic divisions within a spore-producing structure termed a **sporangium**. The spores, which are the products of meiosis, are therefore haploid **meiospores**; the gametophytes they produce are also haploid, as are the sperm and egg that are produced by the gametophytes.

You may wonder what the functional significance might be in having such totally different individuals occurring in a reproductive cycle. There are several explanations. One is that plants are nonmotile organisms: employment of wind-dispersed spores provides a means of colonizing new habitats. Another explanation is that land plants evolved from water-dwelling algae, and many of them have retained free-swimming male gametes as an important part of their reproduction. We see therefore that to interpret the significance of structures and func-

tions in plant reproduction, we must take into account both ecological and evolutionary information.

FLOWERING PLANTS

Plant life is quite diversified and various groups of plants, including algae and ferns, have flourished in the past and continue to do so in the present. Most of us, however, think first of **flowering plants** when "plants" are mentioned. Most of our economic plants are flowering plants: we obtain nearly all our vegetables, fruits, and cereals from them, to say nothing of fibers, drugs, spices, and so on. Moreover, the flower, which is the characteristic structure of this plant group, is itself often highly prized for its beauty and is much admired.

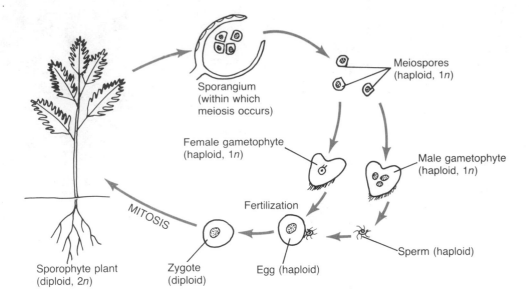

FIGURE 16-3
Alternation of generations in a green land plant. This reproductive cycle does not correspond to that of any specific plant, but it most resembles that of ferns.

Sporangium
(within which
meiosis occurs)

Meiospores
(haploid, 1*n*)

Female gametophyte
(haploid, 1*n*)

Male gametophyte
(haploid, 1*n*)

Fertilization

MITOSIS

Sperm (haploid)

Sporophyte plant
(diploid, 2*n*)

Zygote
(diploid)

Egg (haploid)

The Flower

Flowering plants belong to a major plant division generally referred to as **angiosperms**. The name signifies that the seeds produced by floral reproduction are enclosed within a fruit; angiosperm means "enclosed seeds."

The plant that bears flowers is the sporophyte generation, and the spores it produces are formed within organs of the flower. The spores, themselves, however, never see the light of day; if one is to see them one must dissect away the parts of the flower in which they are produced.

A flower is considered by many botanists to be a highly modified leafy branch; accordingly, each flower part corresponds to a modified leaf. At the base of the flower are modified sterile leaves called **sepals**. These may be either green or colored. Collectively, they are the **calyx** and compose the outer covering of the flower bud. **Petals**, located within the calyx, are commonly colorful and also represent sterile leaves; together they are the **corolla**. Within the petals are **stamens**. Stamens are considered a type of fertile leaf called a **microsporophyll** (*micro-* ["small"] because **microspores** are produced within **microsporangia**; sporophyll means "spore-bearing leaf"). (It should be explained here that seed-producing plants have two kinds of spores: microspores, which produce **male gametophytes**, and **megaspores** [*mega-*, "big"], which produce **female gametophytes**.) The microsporangia of a stamen are fused together to form the **anther** (Fig. 16-4). Stamens can be very colorful and serve as pollinator attractants.

Within a circle of stamens is a **pistil** (sometimes more than one). It is composed of one or more modified megasporophylls (also called **carpels**), each bearing one or more **ovules** (Fig. 16-4), within which megaspores are produced. The ovule is a vase-shaped structure having a small opening called a **micropyle** at its free end.

The stamens and pistils are the essential sexual parts of flowers; sepals and petals are accessory parts and differ from flower type to flower type depending on the manner in which the flowers are pollinated. Many kinds of flowers are wind pollinated; that is, the pollen is transported from stamens of one flower to pistils of other flowers by wind. Wind-pollinated flowers usually do not have large or highly colored petals and sepals. On the other hand, flowers that are pollinated by insects, or by birds and even bats, have expanded, "target-type" petals. They also are likely to be scented, brightly colored, and to produce nectar.

Gametophytes

The male gametophyte of a flowering plant is the usually two-celled **pollen grain** which develops within the anther from a microspore. The female gametophyte is the usually seven-celled **embryo sac** which develops within the ovule from a megaspore. Pollen grains are shed from the anthers at the time the flower bud opens. If the flower is wind pollinated, chance alone dictates if the pollen will be carried to the receptive part of a flower of the same species; if the flower is animal pollinated, the transport usually is more specific and direct. In any event, when a pollen grain lands on the tip or **stigma** of a pistil (Fig. 16-5), it produces a **pollen tube** having two sperm (Fig. 16-4). The pollen tube grows down within the neck **(style)** of the pistil and enters its basal chamber, the

FIGURE 16-4

Flower diagram. The flower is composed of male and female parts: stamens and pistils. Stamens are equivalents of microsporophylls; anthers are considered to be microsporangia. Pistils are equivalents to megasporophylls (one or more per pistil) and contain one or more ovules. The germinated pollen grain is the male gametophyte and contains two sperm cells; the embryo sac is the female gametophyte and contains one egg, a central cell with two polar nuclei, two synergid cells, and usually three antipodal cells.

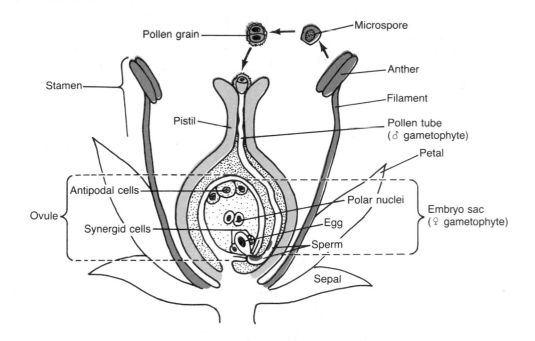

A.

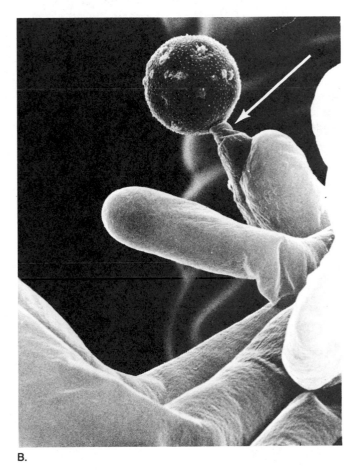

B.

FIGURE 16-5

Pollination of *Arenaria uniflora* (sandwort). The flower has three styles thickly set with stigmatic projections to which pollen grains adhere. **A.** The three styles and stigmas are loaded with pollen grains. ×300. **B.** A single germinating pollen grain, pollen tube indicated by arrow. (Photos courtesy Robert Wyatt)

ovary. Here the tube is attracted to the **micropyle** of an **ovule** and grows inside it (Fig. 16-4).

The female gametophyte consists of the following: one **egg cell**, two **synergid cells** next to the egg, a large **central cell** containing two **polar nuclei**, and three **antipodal cells**. The synergid cells are thought to induce the directional growth of the pollen tube, which usually grows through one of the synergids, destroying it. At that point, the pollen tube discharges its two **sperm** cells, one of which fertilizes the egg. The other sperm cell fertilizes the central cell. The zygote then becomes the **embryo** of the seed and the fertilized central cell develops into a multicellular food-storage tissue called **endosperm**. The antipodal cells have no known functions. In some plants, the food of the endosperm is transferred into the developing embryo, which produces fleshy food-storage leaves called **cotyledons** (Fig. 16-6). In other plants, the endosperm is not taken up immediately by the embryo but remains in the seed and sustains the growing seedling during germination.

Upon completion of fertilization, the development of the fruit begins. The petals and sometimes the sepals wither away and are shed (in a few cases portions of them remain, or become fleshy and incorporated into the fruit). In most flowers, only the pistil along with its seeds becomes the fruit (Fig. 16-6). Fruit formation is accomplished by extensive growth of the basal part of the pistil (the ovary) and maturation of the enclosed seeds. Depending on the fruit type, the wall of the pistil may become a fibrous pod, a fleshy fruit, a hardened nut,

or a combination of fleshy and stony layers, as in plums and peaches. Further information about fruits and seeds and their importance in plant ecology is found in Chapter 24.

The reduction of the gametophyte stage of land plants, culminating in the pollen tube and the embryo sac of the flowering plant, has enabled such plants to colonize comparatively dry habitats for which ferns and similar plants are not well adapted. As we have noted, those plants require the presence of free water for fertilization. Once plants evolved enclosed fertilization, they had attained the same degree of independence of water for reproduction that we find in reptiles, birds, and mammals (as compared with fish and amphibians). The further evolution of flowers and fruit by the flowering plants appears to many botanists to have occurred by **coevolution** of certain animals and plants. Coevolution is a subject that will be discussed in later chapters; here we will define it as the evolution of cooperative behavior and of compatible organs in two different species of organisms in such a way as to foster interdependence. Examples are nectar-producing flowers and nectar-gathering insect pollinators of flowers or the production of nutritious and attractive fruit by plants and the consumption of those fruits by fruit-eating and seed-dispersing animals.

PLANT GROWTH AND DEVELOPMENT

The growth of plants is an interesting but often misunderstood phenomenon. Adult plants are not simply large seedlings in the sense that adult animals are grown-up babies. Instead, finding traces of the seedling plant in the adult tree necessitates looking deep within its tissues, because plant growth is a matter of accretion; the addition of many layers of cells and tissues to the outside and ends of the seedling plant often conceals its original form. Even plants with short lifespans, such as common garden annuals, grow by addition to the stem and root tips and by lateral addition of layers of tissues by the vascular cambium. However, even though the growth of plants is continuous and the adult shape may be quite different from the young seedling, the overall pattern of plant development is apparent in the embryo. To understand plant development, it is necessary to begin with the embryo, just as animal development is comprehensible only in embryological terms.

Embryo Development

Multicellular plants, including mosses, liverworts, ferns, club mosses, gymnosperms, and flowering plants, have reproductive cycles that differ considerably; neverthe-

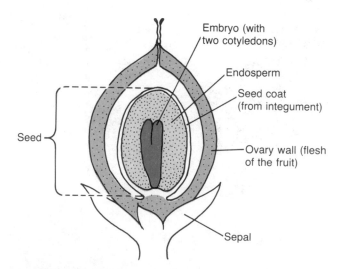

FIGURE 16-6
Fruit, seed, and embryo of a flowering plant. This figure follows Fig. 16-4. The petals and stamens wither and fall away. The ovary part of the pistil becomes the fruit. The ovule has become a seed-containing endosperm (developed from the fertilized polar nuclei and cytoplasm) and an embryo (developed from the zygote).

less, their zygotes and early embryonic stages are remarkably alike. In this way, at least, plant development resembles that of animals. Plant embryology, however, is not as complicated as animal embryology, because plant activities do not require as many highly specialized organ systems. In plants there are no central nervous systems, digestive tracts, glands, muscles, articulated skeletons, and so on, and later embryonic development consequently is relatively uncomplicated. Nevertheless, many of the basic questions regarding cell specialization and organization are major concerns of both plant and animal embryologists.

Plant development, like animal development, starts with the zygote. Early in the life of the plant embryo, the basic pattern of development becomes established by a "decision" as to which end will be up and which will be down (i.e., which will be the root end and which will be the stem end). In many plants, the first division of the zygote is transverse, and this division establishes the polarity of the embryo.

A simple example of the induction of polarity is seen in the development of the brown alga *Fucus* (Fig. 16-7). The *Fucus* zygote is a large cell which settles on the sea bottom and then develops into a mature seaweed, anchored to some solid object by a rootlike organ called the **holdfast**. It has been found that the polarity of the zygote and the later embryo is light induced. The side of the zygote away from the light forms the holdfast; the lighted side develops into the shoot of the seaweed. The forces or factors responsible for the plane of the first

division of a seed-plant embryo are not known, but it is suspected that the polar organization of the embryo sac and biochemical gradients inside the ovule are involved (see Fig. 16-8).

Establishment of polarity in all plant zygotes seems to require some kind of one-sided exposure to environmental factors (light, gravity, contact, nutrient gradients, hormones, etc.), but the actual mechanism of polarization is rather obscure. It has been suggested that the cell, as a result of induction by environmental factors, develops an internal gradient of electrical potential so that one side is more positive, the other side more negative, and the cell becomes electrically polarized (electrical polarization has been measured in *Fucus* zygotes prior to their division and differentiation). Once the zygote has become electrically polarized, enzymes, which like other proteins bear electrical charges, would migrate toward the positive or negative poles of the cell according to their own electrical charge. As a result of this differential migration, called **electrophoresis,** one

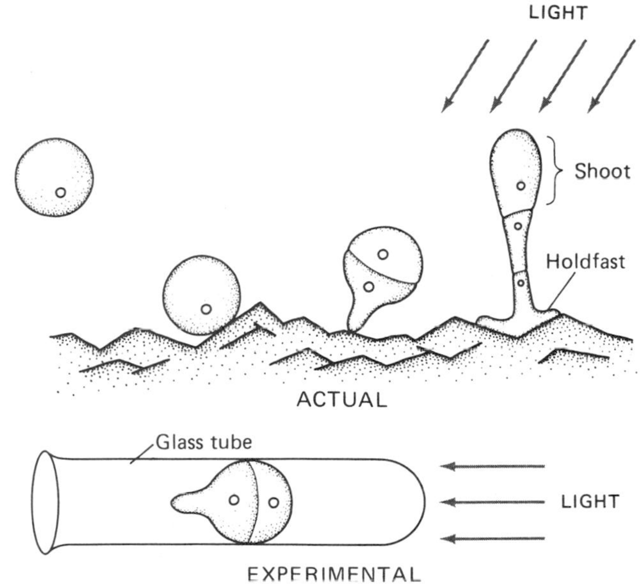

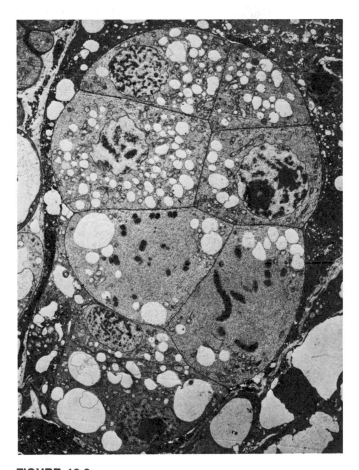

FIGURE 16-8
Young barley embryo. Note evidence of the establishment of polarity in the elongate shape and planes of cell division.

FIGURE 16-7
Induction of polarity in the zygote of the brown alga *Fucus.*

LIGHT

Shoot

Holdfast

ACTUAL

Glass tube

LIGHT

EXPERIMENTAL

side of the cell would have a capacity for different enzyme-catalyzed developmental functions than would the other side.

In some plants, embryo polarity becomes apparent only after several divisions, but in all cases the determination of polarity is an early event in plant development (Fig. 16-8), thus determining the axis of the embryo. Soon root and shoot meristems are established at the ends of the axis. Thereafter, development is largely a matter of division of cells in the meristems and differentiation of cells into tissues and organs.

Monocotyledons and Dicotyledons

Flowering plants are divided in classification into two broad groups known as **monocotyledons** and **dicotyledons,** or just as **monocots** and **dicots.** Many characteristics separate the monocots and dicots, as will be explained in Chapter 20. The distinction that gives monocots and dicots their names, however, is the number of first leaves (one or two) to be developed by the embryo.

The full-term embryo of a flowering plant consists of a short axis with either one or two primary leaves (cotyledons) (Fig. 16-9). The part of the embryo above the cotyledons is the shoot bud, or **epicotyl,** and that below the cotyledons is the stem of the embryo, a region called the **hypocotyl.** At the lower end, a root meristem forms and gives rise to the **radicle,** the first root of the embryo.

The functional role of the embryo is twofold: (1) it reestablishes the basic organization of the plant, and (2) together with the structures of the seed that envelops it, it serves as a survival and dispersal device. It is, in effect, a survival package, demonstrating such ecological strategies as dormancy, long-term survival, and transport by wind, water, and animal actions.

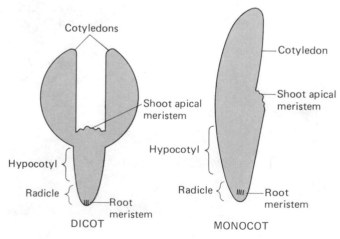

FIGURE 16-9
Monocot and dicot embryos.

Seed Germination

Seeds usually become very dry late in their development. The green plant may contain more than 90 percent water, but seeds often are bone dry. Because of this, seeds often survive for many years without sprouting. However, when exposed to moisture, the seed and embryo absorb quantities of water and swell; the seed coats split open and the embryonic organs begin to elongate (Fig. 16-10). Enzymatic activities are greatly increased, foods stored in the endosperm or in the cotyledons are digested and transported to the rapidly dividing cells in the shoot and root apices, and the embryo becomes a seedling.

Apical Meristem Development

A plant embryo can be described as consisting essentially of opposed stem and root meristems. Meristems have full capacity to produce all the tissue and organ

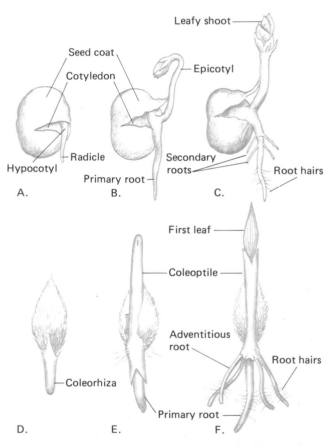

FIGURE 16-10
Seeds and seedlings of monocots and dicots. **A.** Pea seed (dicot) just beginning its germination. **B.** Later stage in pea germination. **C.** Young pea seedling. **D.** Oat seed (monocot) beginning its germination. The **coleorhiza** is a root sheath of the embryo. **E.** Elongation of the coleoptile, the shoot sheath of the embryo, and emergence of the primary root from the coleorhiza. **F.** Young oat seedling.

DR. BEAL'S 100-YEAR-OLD SEED-GERMINATION EXPERIMENT

In 1879 W. J. Beal, a professor of botany and forestry at Michigan Agricultural College (now Michigan State University) began a long-term experiment to test the survival of weed seeds in the soil. Dr. Beal described his work in the following way:

> In the autumn of 1879, I began the following experiments, with the view of learning something more in regard to the length of time the seeds of some of our most common plants would remain dormant in the soil and yet germinate when exposed to favorable conditions. I selected fifty freshly grown seeds from each of twenty-three different kinds of plants. Twenty such lots were prepared with the view of testing them at different times in the future. Each lot or set of seeds was well mixed in moderately moist sand, just as it was taken from three feet below the surface, where the land had never been plowed. The seeds of each set were well mixed with the sand and placed in a pint bottle, the bottle being filled and left uncorked, and placed with the mouth slanting downward so that water could not accumulate about the seeds. These bottles were buried on a sandy knoll in a row running east and west.

Dr. Beal's intention was that his seed bottles be dug up at 5-year intervals for germination tests. After 35 years, tests were performed every 10 years. Beal conducted the first seven, and since his death others have continued the exhumations. The 100th year for the experiment was 1980, at which time seeds of three species germinated, two at very low percentages. *Verbascum blattaria* (moth mullein), however, has retained a 42 percent viability. One of the remaining two, *Malva rotundifolia* (common mallow), last germinated in 1899. The other species, *Verbascum thapsis* (common mullein), last germinated in 1914. Obviously, it is possible that a few seeds of the other 20 species are still viable. The experiment can be continued another 60 years at the present sampling rate.

One interesting opportunity presented by the recovery of viable seed a hundred and more years old is that of examining the effects of long-term dormancy on the characteristics of seedlings and mature plants obtained from such seeds. A study now under way will explore mutational frequencies of plants obtained from Beal's 100-year-old seeds and their progeny.

systems of a mature plant; hence the nature of meristems and the manner in which they initiate and produce tissues and organs are of great interest and importance. (The subject of organization of shoot and root meristems was introduced in Chapter 8.)

Apical Growth Sporophytes of vascular plants grow in length by divisions and enlargement of cells in the tips of stems and roots, rather than throughout the organism as is the case with most animals and some algae and fungi. Terminal growth regions, called **apical meristems,** are the principal sites of longitudinal growth in all vascular plants; they not only add to the length of a plant but also are the source of new tissues and organs (Fig. 16-11).

If the growing stem tip of a vascular plant is examined under a microscope, a dome-shaped **apex** (also called an **apical dome**) may be seen (Fig. 16-12), surrounded by leaves in various stages of development. The younger leaves, called **leaf primordia,** are at the base of the dome; progressively older and larger leaves are farther down the developing stem. Although cell divisions

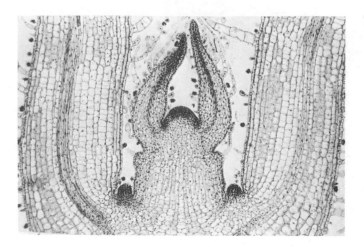

FIGURE 16-11
Apical meristem and leaf primordia of a *Coleus* plant.

occur in the stem and leaf primordia below the extreme tip of the stem, the apical dome is the initial source of all stem cells, and experiments show that if the dome is destroyed, the stem ceases its growth. Root tips are composed of a similar apical meristem but differ in the absence of leaf primordia and in the presence of a protective **root cap** over the apical dome (Fig. 16-13). In both stem tips and root tips, the apical meristem is confined to the terminal half-inch or inch; below that, addi-

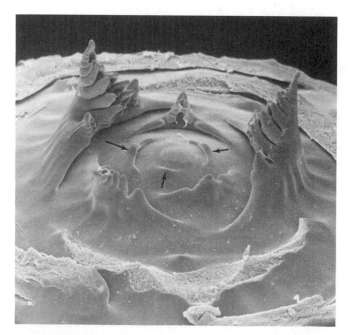

FIGURE 16-12
Scanning electron micrograph of shoot apex of celery. Celery has alternate, compound leaves, shown here originating as leaf primordia (arrows). Older leaves were cut away. (Photo courtesy R. D. Meicenheimer)

tional growth occurs principally by cell enlargement, usually cell elongation. In roots, the root hairs are concentrated in this region; in stems, leaf expansion and development of vascular tissues are found.

Once the seedling is established and its meristems are active, its transformation into a mature plant can be interpreted in terms of the maturation of tissues formed in the shoot and root apices (Figs. 16-13 and 16-14). In the growing shoot, there are three basic tissue types: the **protoderm**, the **ground meristem**, and **strands** of **procambium**. These, in turn, produce other tissues. The protoderm becomes the epidermis, the ground meristem gives rise to the parenchyma of the pith and cortex, and the procambial strands become bundles of vascular tissues.

Vascular Tissues Vascular tissues are restricted to strands or bundles running the length of roots and stems and branching out as **veins** into leaves. Sometimes just one bundle is present, in the form of a solid central cylinder known as a **protostele**. This is the condition in stems of primitive plants and in roots of nearly all vascular plants.

The stem of a dicotyledonous flowering plant (dicot), when young, exhibits a circle of vascular bundles interconnected laterally by a continuous layer of vascular cambium (monocot stems normally lack a vascular cambium). Outer and inner parenchymal regions, the **cortex** and **pith**, respectively, enclose vascular bundles. The outer covering of the young stem is the **epidermis**. Vascular cambium within the bundles adds **xylem** and **phloem** to that already present; cambium between vascular bundles fills in the region between bundles with new xylem and phloem.

Xylem and phloem, the two kinds of transport tissue in vascular bundles, both have cells that are cylindrical and arranged end to end. Xylem, the principal water-transporting tissue, is composed of elongate, hollow cells with thin areas **(pits)**. When mature, xylem cells are dead and consist solely of cell walls. In primitive plants the principal xylem cells are **tracheids**, characterized by closed, pointed ends and the presence of lateral pits. Advanced plants, principally flowering plants, have xylem **vessels** composed of cylindrical cells **(vessel elements)** with open ends and lateral pits (Fig. 16-15A, B). They are joined end to end and are more efficient water conduits than the tracheids of lower vascular plants, in which the ends are closed and water must follow a more circuitous route through the pits in the lateral walls. The presence of large vessels in the springwood of a dicot produces a porous appearance, as shown in Fig. 16-16.

The second vascular tissue, phloem, is composed of living, food-transporting **sieve cells** having clusters

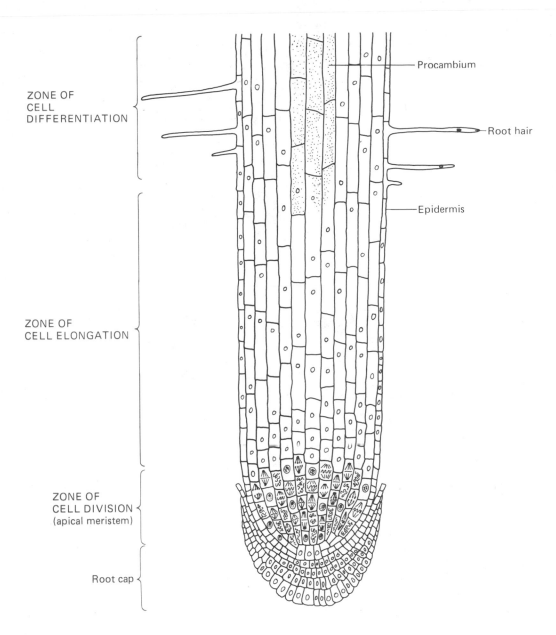

FIGURE 16-13
Growth regions of a young root.

ZONE OF
CELL
DIFFERENTIATION

ZONE OF
CELL ELONGATION

ZONE OF
CELL DIVISION
(apical meristem)

Root cap

Procambium

Root hair

Epidermis

of **pores (sieve areas)** in their end and side walls (Fig. 16-15C). Sieve cells usually are arranged end to end, forming **sieve tubes.** In that case, the individual sieve cell also may be called a **sieve tube member** (Fig. 16-15D). This differs from a sieve cell of other vascular plants in having in its end walls clusters of large pores called **sieve plates,** through which food in the form of soluble organic molecules passes from sieve member to sieve member. Sieve tube members are living cells, although lacking nuclei and thus genetically nonfunctional. A partnership exists between each sieve tube member and an adjacent nucleated **companion cell,** which is thought to coordinate with a sieve tube member in carrying on food transport.

Buds Branches originate as **axillary buds** at the bases of leaves in a region called the **axil.** The axillary buds are formed in the shoot apex at about the same time as leaf primordia are formed (see Fig. 16-11). In deciduous plants, axillary buds usually are inactive their first season, unless something happens to the apical bud, in which case they become active. Normally, the buds are activated the following spring and develop into leafy branches.

Leaves

Leaves are formed at the apices of stems and branches. Within the bud of a flowering plant are progressively

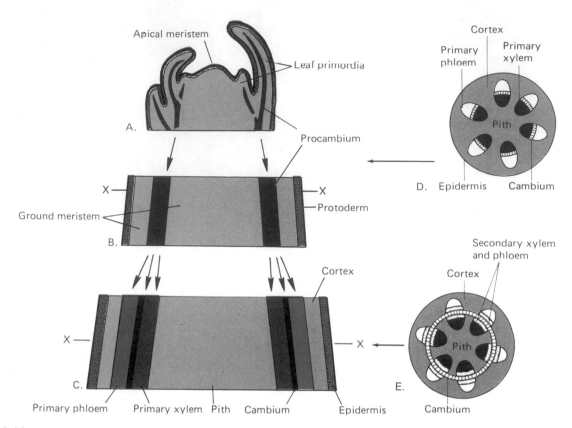

FIGURE 16-14

Organization of the young shoot apex of a dicot. **A.** Apical region, composed of the apical meristem, leaf primordia, and differentiating tissues. **B.** Region of continuing cell division, cell elongation, and cell differentiation. **C.** Region of tissue specialization. **D.** Cross section of part B at X—X. The vascular bundles are arranged in a ring. **E.** Cross section at part C. A continuous cambium has been formed interconnecting the vascular bundles and their cambium layers.

FIGURE 16-15

Cells and elements of vascular tissues. **A.** Tracheids, not the pointed ends, lateral pits. **B.** A part of a vessel; the individual cells are vessel elements. **C.** A sieve tube of the phloem of a nonflowering vascular plant. The cells, unlike those of xylem, are living and have nuclei. **D.** A part of a sieve tube of a flowering plant. The sieve tube members lack nuclei and are bordered by companion cells.

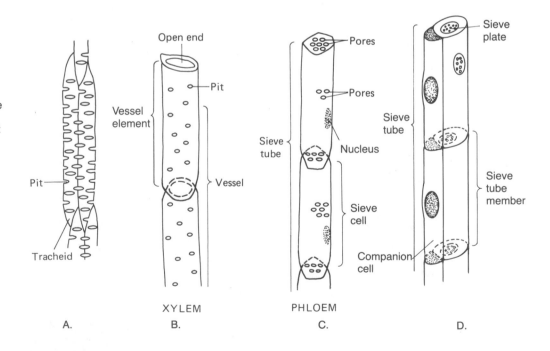

FIGURE 16-16
Scanning electron micrographs of the wood of elm *(Ulmus).* The large vessels of springwood are clearly visible. (Photos courtesy Institute of Paper Chemistry, Appleton, Wisconsin)

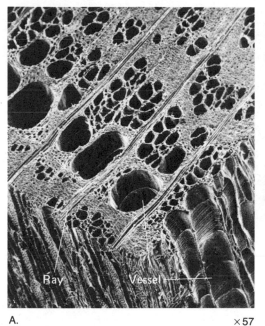

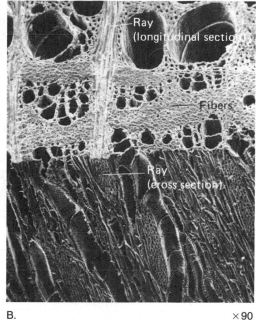

A. ×57   B. ×90

younger leaves, and finally the actual shoot meristem. It is here that all the cells of the stem and its leaves originate. Small mounds of cells regularly spaced about the apical dome are the first visible evidence of leaf formation (see Figs. 16-11 and 16-12). These tiny protuberances, the leaf primordia, eventually grow into leaves, and their spacing determines the pattern of leaf arrangement on the stem. These patterns vary from species to species. In all cases, however, the final arrangement depends on the pattern in which leaf primordia are initiated.

In perennial plants with seasonal growth, the shoot apex and its leaf primordia form a dormant bud. An outer layer of modified scale leaves protects the dormant bud until the next season's growing period. Herbaceous plants have continually growing shoot apices and do not have protected buds. In both dormant and nondormant buds, however, the arrangements of the apical meristem and the leaf primordia are the same.

Origin and Arrangement   Some leaf shapes and arrangements can be attributed to ecological influences, but others seem not to serve any specific function. There is no convincing explanation why some leaves have a toothed or lobed margin while others do not. However, the arrangement of leaves on a stem is significant from the standpoint of energy utilization; leaves are placed so that they intersect a maximum of light. Examination of the leaf arrangement of a plant shows that no leaf is located precisely above or below the next one. If a plant has **opposite** leaves, in which two leaves are oppositely attached at each node, one pair of leaves will be ori-

ented at right angles to those immediately above and below. If more than two leaves occur at each node, in the **whorled** arrangement, each whorl of leaves is positioned at a different angle than those above and below. A common leaf arrangement is the **alternate** leaf type, in which each leaf is attached to a node so as to form an ascending (or descending) spiral (Fig. 16-17; see also Fig. 16-12 for the origin of the alternate leaf arrangement).

The efficiency of photosynthesis may depend on leaf arrangement; therefore, ecologists as well as plant geneticists are very interested in analyzing the orientation of leaves in space. Plant breeders, for example, have produced crop plants with leaf arrangements adapted for close cultivation and high productivity.

Development   The specific pattern of leaf arrangement is genetically regulated and is important in plant classification. The common house plant *Coleus* has opposite leaves. This is a characteristic of the mint family, to which it belongs. Further study of the shoot apex shows that *Coleus* leaf primordia of the same age are evenly spaced from each other and form opposite sets. Leaf primordia of other species likewise are spaced in patterns characteristic of the families to which they belong.

DEVELOPMENTAL POTENTIAL

Manipulation of Plant Life Cycles

With the discovery of meiosis in the late nineteenth century, it seemed reasonable to think that 2*n* sporophytes

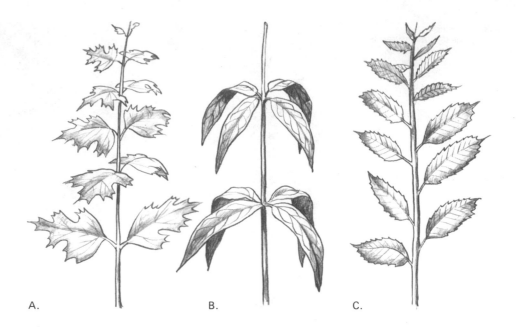

FIGURE 16-17
Patterns of leaf arrangement.
A. Opposite. **B.** Whorled. **C.** Alternate.

A.   B.   C.

and 1*n* gametophytes in plant life cycles were irreversibly set apart by their different chromosome numbers. However, in the early 1900s W. H. Lang observed that when fern gametophytes were cultured under dry conditions, they sometimes produced sporophytes by direct development from 1*n* gametophyte tissues. This phenomenon, which occurred in the absence of sexual reproduction (it will be recalled that fern gametes require water in fertilization), is called **apogamy** ("without fertilization"). Somewhat similar experiments by Elva Lawton show that cultured fern leaflets sometimes produce cellular outgrowths that develop into 2*n* prothalli (gametophytes), and these have the capacity to form antheridia and archegonia, undergo fertilization, and produce 4*n* sporophytes. The direct production of gametophytes from sporophytes, without intervening meiosis and meiospore production, is termed **apospory** ("without spores").

Shortly after Lang reported apogamy in ferns, E. Marchal performed a series of experiments with mosses, demonstrating conclusively that chromosome number is not the major determinant of gametophyte and sporophyte structure. These experiments, illustrated diagrammatically in Fig. 16-18, showed that slices of a young sporophyte (taken from near the base of the developing sporangium, a region called the **apophysis**), when cultured on a suitable nutrient medium, would form cellular filaments. These filaments were identical to **protonema,** the first stage in moss-spore germination; like normal protonema, the filaments formed buds that became gametophyte plants (this was apospory). However, there was one major difference between the experimentally produced gametophytes and normal gametophytes: the experimental gametophytes were 2*n*, and when they formed antheridia and archegonia and reproduced sexually, 4*n* (tetraploid) sporophytes were formed. Marchal found that the process could be repeated several times, up until 32*n* sporophytes were formed. Beyond this, irregular, malformed plants resulted.

The significance of apospory and apogamy seems to be that the determining factors of plant form and function are at least partly environmental. This has considerable significance for the experimental control of plant growth for, as will be seen, it enables botanists and agronomists to manipulate plant development toward certain useful ends.

Cellular Totipotency

It has been shown that cells in adult organs of plants often retain their ability to produce entire new plants if they are separated from each other and cultured in nutrient solutions. F. C. Steward, of Cornell University in Ithaca, New York, removed small blocks of cells from roots of a carrot plant and cultured the cells of these blocks in a liquid tissue-culture medium, using special rotating flasks that permitted the growth of isolated, free cells (Fig. 16-19). In many cases such free cells divided and formed small clumps, sometimes called **embryoids,** which later developed into entire new carrot plants. Steward thus demonstrated the concept of **cellular totipotency,** defined as the ability of a vegetative plant cell to produce an entire plant by cell division and cell differentiation.

FIGURE 16-18
Apogamy in mosses. The diagram depicts four repetitions of an experiment in which slices of sporophyte tissue placed in culture produce protonema buds, sexual gametophytes, and, after sexual reproduction, sporophytes of twice the prior chromosome number.

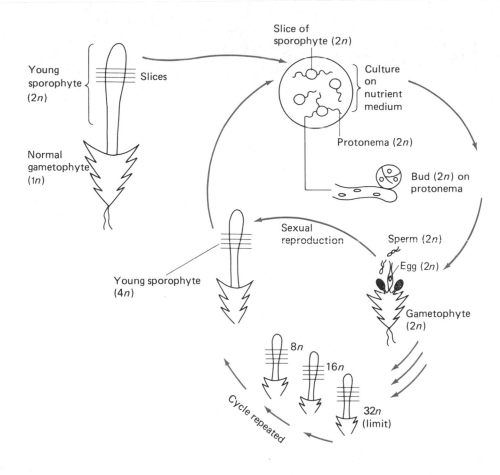

Cloning

Plant-tissue cultures of the type just described are now routinely done to grow **clones** of genetically identical plants. **Cloning** of orchids, gladiolas, geraniums, bromeliads, anthuriums, and other decorative house plants is a flourishing business in many countries. Production of disease-free plants is an important aspect of plant cloning. Potatoes, sugarcane, gladiolas, tulips, and other plants propagated by cuttings, tubers, or bulbs often become infected with viruses and bacteria. In recent years, plant-tissue culture has enabled horticulturalists to produce vigorous, disease-free varieties.

The animal equivalent of plant totipotency would be demonstrated by the removal of a few human cells, their culture in test tubes of nutrients, subsequent formation *in vitro* of human embryos, and finally, via incubators, the rearing of many duplicate individuals of the original cell donor. Contrary to science fiction or alarmist writers of social commentary, true cloning of vertebrates has never been done, although certain nuclear transplants approach true cloning, as noted in Chapter 11. However, even the lesser objective of growing human organs in tissue cultures would be of great value, as it would be an important aid in solving the rejection prob-

lem of organ transplants. At present, transplants, except those between identical twins, are supplied by donors of different genotypes.

Haploid Plants

Microspores are reproductive cells occurring in the floral organs of flowering plants or in sporangia of other plants (refer to Fig. 16-4). Because they are produced by meiosis, microspores are $1n$. Those of flowering plants and gymnosperms develop into pollen grains.

In one of the most significant advances in the experimental embryology of plants, S. Guha and S. C. Maheshwari of the University of Delhi, India, cultured the anthers of *Datura* (jimson weed) and found that the microspores, instead of forming pollen, developed into embryoids (nonsexual embryos). Because microspores are $1n$, the embryoids also were $1n$, and Guha and Maheshwari were able to grow mature $1n$ *Datura* plants from embryoids developed in tissue culture. This is a special case of apogamy—the formation of a sporophyte plant from a gametophyte stage, in this case the meiospore.

In recent years botanists have found it possible to grow $1n$ plants of several species by culture of microspores and, in some cases, pollen. The most successful

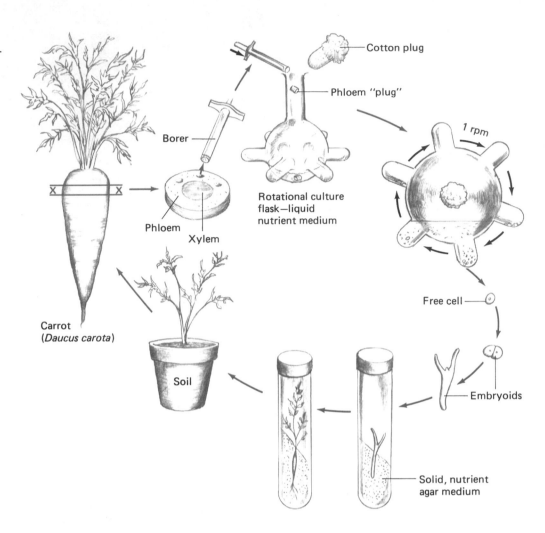

FIGURE 16-19
Tissue and cell cultures of a carrot. With suitable nutrient media and technique, the life cycle of the carrot plant can be completed asexually *in vitro*. (Based on the work of F. C. Steward)

Cotton plug

Phloem "plug"

Borer

1 rpm

Rotational culture flask—liquid nutrient medium

Phloem

Xylem

Free cell

Carrot (*Daucus carota*)

Soil

Embryoids

Solid, nutrient agar medium

of these experiments have used plants belonging to the family Solanaceae (potato, tobacco, tomato, and *Datura*), and one of the more productive methods has been developed by the French botanists Jean and Collette Nitsch (Fig. 16-20). By adjusting the components of their tissue-culture media, the Nitsches induced the direct formation of embryoids from tobacco microspores and grew 1*n* plants that eventually flowered. These were about a third smaller than the normal 2*n* plants, and were sterile. However, when 1*n* embryoids were grown on media containing a plant growth hormone (specifically kinetin), their chromosomes were doubled and they grew into 2*n* plants. The significance of this work for genetics and plant breeding is far-reaching.

It will be recalled that the "one gene, one enzyme" hypothesis was formulated by Beadle and Tatum, who worked with the 1*n* fungus *Neurospora*. The study of genetic mutations in *Neurospora* was possible because the genes in such 1*n* organisms are unpaired. Recessive mutants, therefore, are immediately expressed. In the same way, it is hoped that the use of 1*n* mutants of agricul-

tural plants will speed up the study of mutations. Another important aspect of 1*n* plants is that purebreds may be produced very simply. All that is needed is to produce 1*n* embryoids and double their chromosomes. The new, 2*n* plants will always be purebred because each chromosome, and hence each gene, is faithfully duplicated. It is estimated that the use of the microspore culture technique, when applied to many agricultural species, will greatly reduce the effort of producing inbred lines for use in the production of hybrid seed (see also Chapter 7).

Regeneration and Polarity in Plants

All multicellular organisms, including plants, possess at least limited powers of **regeneration**. If a small, cylindrical piece is cut from a poplar twig and kept in a moist chamber, roots eventually may form at one end of the cut twig and leafy buds at the other. Significantly, leafy buds always are produced at the "top" end, nearest the terminus of the branch, and roots always form at the end

344 PART TWO | THE BIOLOGY OF ORGANISMS

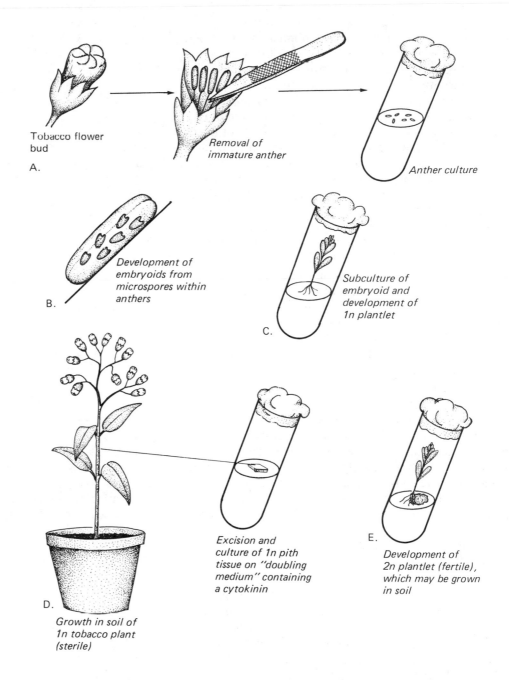

FIGURE 16-20
Steps in the formation of plants from cultured anthers of tobacco. (Based on J. P. Nitsch, 1969, Experimental androgenesis, in *Nicotiana*, *Phytomorphology* 19:389–404. Redrawn from *Plant biology* by Knut Norstog and Robert W. Long. Copyright © 1976 by W. B. Saunders Company. Reprinted by permission of W. B. Saunders Company, CBS College Publishing)

Tobacco flower bud

A.

Removal of immature anther

Anther culture

Development of embryoids from microspores within anthers

B.

Subculture of embryoid and development of 1n plantlet

C.

Excision and culture of 1n pith tissue on "doubling medium" containing a cytokinin

E.

Development of 2n plantlet (fertile), which may be grown in soil

D.

Growth in soil of 1n tobacco plant (sterile)

of the piece of twig nearest the root end of the plant from which it came.

Even if the position of the piece of twig in a moist chamber is reversed so that the "root end" is uppermost, roots will be formed by that end. It is evident, therefore, that gravity alone does not dictate the origin of roots, but rather an internal polarity controls the kind of development that will occur. Furthermore, as was learned in the regeneration experiments with *Hydra*, the relative position of cells is all-important in determining whether the cells will follow one developmental pathway or another. Only merest chance may dictate where the two

cuts are made that remove a piece of twig from a branch. The cells that form roots might just as well have formed leafy shoots, had the cuts been a fraction of an inch nearer the twig's apex. Again, the lesson is that the pattern of cell development is influenced by cell location with respect to neighboring cells.

Protoplasts

In recent years the culture of plant cells and tissues has gained importance in studies of plant development and in plant breeding, but some experiments have been impossible to do because plant cells normally are enclosed by a wall. For example, fusion of cells of two kinds, accomplished in such animal cells as those of mouse and human, or introduction of foreign DNA or other matter into living cells, has not been possible. For this reason, considerable attention has been given to the removal of walls from living plant cells, with the objective of maintaining such "naked" cells, or **protoplasts,** without walls long enough to do cell fusions or DNA transfers. After many failures, plant-tissue culturists in several laboratories have attained proficiency in protoplast production, and cell-fusion experiments are under way. The secret of these successes has been the use of cellulase enzymes, extracted from certain wood-dissolving fungi or from the cellulose-digesting organs of snails, to digest away cell walls while leaving the plasma membranes intact. Part of the problem has been that wall-less plant cells become very sensitive to osmotic pressures of culture media, swelling and bursting. Therefore, concentrations of solutes in culture media have to be carefully adjusted and maintained at constant levels. This now is accomplished with the aid of **propylene glycol,** an osmotically active but metabolically inert fluid. Cellulase is retained in protoplast cultures until experiments have been completed; transfer of protoplasts to cellulase-free media usually results in the synthesis of new walls by protoplasts. Such reconstituted cells have been observed to divide and, in some cases, to form new plants. Thus far, protoplast fusions have had only limited success, but the potential for production of useful somatic hybrids is great. In particular, the possibility of introducing genes from widely different genera and species is attractive, combining, for example, characteristics of maize and wheat or introducing the ability of legumes to form nitrogen-fixing associations with bacteria. Another possibility is the introduction of specific segments of DNA—for example, the coding for nitrogen-fixing symbiosis. Most recombinant DNA experiments thus far have been directed toward introducing eukaryotic genes into bacterial cells, as for instance those coding for insulin production or other useful products of eukaryotic cells. However, attention now is being given to transferring genes from one kind of plant into another using bacterial plasmids as gene carriers (see the section on recombinant DNA in Chapter 5).

SENESCENCE

Plant Senescence

Many plants are long-lived, some amazingly so. Redwoods survive for as long as 3000 years, the bristlecone pine (Color plate 8E) for 5000, and trailing club mosses, which grow at one end and die at the other, have even

greater longevity. On the other hand, some desert annuals complete their cycles of growth, flowering, and seed production in a couple of weeks, then wither and die.

Plants can be classified approximately into three categories: **annual** (one-year life cycle), **biennial** (two-year life cycle), and **perennial** (repeated life cycles in the lifetime of a plant). Reproductive processes in these categories are essentially alike, but growth rates and duration differ.

Common garden plants such as peas, radishes, and tomatoes are regarded as annuals by those who cultivate them, but only peas are true annuals. The pea plant grows, flowers, bears fruit, and dies in 2–3 months. Prolonging the growing season will not alter the pattern of this life cycle. In many areas the tomato also flowers and bears fruit in a single summer, the first frost ending its life, but in the tropics tomato plants grow as perennials. Thus in temperate regions tomatoes are only functional annuals. Radishes are grown for their roots and normally are harvested before flowering. However, if they are left in the ground until the following year, they will be seen to flower, produce fruit, and die in a two-year cycle (they are biennials, as are cabbages and a number of other garden vegetables and flowers).

A majority of vascular plants are long-lived. Ferns, club mosses, and gymnosperms, for instance, are perennial plants, reproducing annually or sometimes biennially but nearly always surviving each reproductive cycle. Their growth pattern is not finite and is said to be **indeterminate.** Among flowering plants, tomatoes, potatoes, lawn grasses, oaks, maples, and many others are indeterminate in growth; however, annuals such as garden peas, biennials such as radishes and cabbages, and many weeds, including crabgrass, burdock, and ragweed, have short-lived, finite growth patterns referred to as **determinate.**

In the tropics, where frosts do not occur to end the lives of plants, some determinate varieties become quite old before flowering and dying. Bamboo is an example. Many bamboos are on lengthy reproductive cycles; in some cases they flower after 50 years and then die. A peculiarity is that nearly all bamboos of a given species flower and die the same season, so that after 40 or 50 years mass death of entire forests of bamboo occurs. In China such a period of flowering with resulting senescence is at hand. Because the giant panda feeds mainly on bamboo shoots and foliage, there is fear that this animal, already threatened with extinction, may not survive in the wild when its major food source dwindles away.

The century plant, *Agave sisalana,* flowers after a dozen or so years, not a hundred as its name implies. However, after flowering and fruiting it undergoes senescence and dies, for like bamboo it is a plant of determinate growth (Fig. 16-21). An even more spectacular example of determinate growth is the immense talipot palm of India (*Corypha elata*), which attains a height of 30 m (100 ft) or more during its 50–60-year life, produces a flower cluster, or **inflorescence,** weighing a ton, and then dies. More modest examples of determinate growth are plants such as wheat and maize.

The senescence and death of plants have two basic causes, which do not necessarily occur together. In peas and other plants, growth tends to be indeterminate, flowers are produced laterally along stems and branches, and apical meristems continue to be active during flowering. Pinching off blossoms as they arise can prolong survival of some varieties of peas, but eventually the apical meristems become inactive and the plant undergoes senescence. One explanation for this kind of senescence is that the flowering "exhausts" the plant's nutrients and other growth factors, and death of nonreproductive cells and tissues occurs. It also has been observed that cytokinins, normally present in roots of growing plants, disappear in plants undergoing senescence; hence senescence may be hormonally controlled.

Senescence in most biennials as well as in the century plant, wheat, maize, bamboo, talipot palms, and similar examples is a consequence of the conversion of the entire stem apical meristem from leaf production to flower production. After flowering, death is inevitable because the apical meristem has been used up.

Many plants exhibit senescence intermediate between the extreme examples just cited. For example, above-ground parts of Irish potatoes die, but underground stems (tubers) survive and propagate the following year. Similar senescence of above-ground parts and survival of underground structures is seen in irises, lilies, perennial grasses, and many other species.

Leaf Senescence

Perennial woody plants do not undergo senescence in any manner previously described. They are potentially immortal, as demonstrated by Lombardy poplars, all of which are descended from one sterile mutant poplar that originated in the province of Lombardy, Italy, in the eighteenth century. All Lombardy poplars are descended in an unbroken line by vegetative propagation of the original; hence they are clones. As long as they continue to be propagated, the original individual in a real sense continues to live. Normally, however, death and decay end the lives of even the hardiest and most long-lived trees, and all undergo partial senescence either on an annual cycle or every few years in the form of leaf senescence and leaf-fall by **abscission.**

Abscission of leaves, and fruits as well, occurs as a consequence of development of a layer of cells, the **abscission layer,** at the bases of leaves or fruits. In many trees, especially those of temperate regions, leaf abscission is an annual affair brought on by shortening autumnal day length and cooling temperatures. Leaf abscission also occurs in tropical and temperate evergreen plants, but on a more random basis; an individual leaf may last several years before abscission results.

Abscission is accompanied by conversion of leaf chlorophyll to red and yellow carotenoids, accumulation of red and blue leaf anthocyanins, withdrawal of leaf nutrients, and profound changes in leaf hormones.

Senescence and abscission of leaves and fruits are correlative phenomena involving interactions of several hormones whose relative concentrations in leaves, fruits, and stems are induced and regulated by environmental as well as internal factors.

FIGURE 16-21
Sisal hemp plant, also called the century plant. Four plants are in flower; the fifth (leaning to the right) has completed flowering and is dead. This plant exemplifies a determinate kind of growth in which the plant becomes mature, reproduces, and dies.

SUMMARY

Reproduction of plants reflects the environment in which they live. Algae have simple reproductive cycles in which free-swimming gametes unite to form zygotes. The zygotes then undergo meiosis (zygotic meiosis) and produce a haploid vegetative stage. Land plants differ considerably. They have a two-stage reproductive cycle. The gametophyte stage is haploid and reproduces by gametes. The resulting zygote develops into a diploid sporophyte. It produces meiospores (by meiosis) that are air transported. This is an adaption to the dry-land environment. Meiospores develop into gametophytes. Those of ferns and similar plants require water for their fertilization and, therefore, are restricted to wetland environments.

Seed plants have evolved internal fertilization and thus escape the requirement for surface water for their reproduction. Their gametophytes are enclosed within ovules, which after fertilization become seeds. The most advanced seed plants have their ovules further enclosed within pistils, which after fertilization become the fruits of flowering plants.

Plant development differs from animal development in that most plants retain embryological functions (in meristematic regions) throughout their lives. The location of meristems is determined early in the development of the plant embryo, and subsequent growth occurs by continuing division of cells in the shoot and root meristems. An early step in the growth of a plant from a zygote is the determination of polarity to form the axis of the embryo and establish the meristems at opposite ends of the axis. Two basic kinds of embryos occur in flowering plants: monocotyledons and dicotyledons. In dicots the principal parts are the epicotyl (shoot tip), the cotyledons, the hypocotyl (embryonic stem), and the radicle (first root). Monocot embryos are basically similar to dicot embryos but are nutritionally more dependent on stored food in seeds.

Experiments involving tissue culture have shown that plant cells retain embryonic potential even though present in specialized tissue and organ systems. If plant cells are isolated in tissue culture, they demonstrate the capacity to produce new, complete plants. This capacity for nonsexual reproduction has considerable value in plant propagation related to agronomy and horticulture.

Senescence occurs in plants as well as in animals, and although some plants are potentially very long-lived (e.g., redwoods), others grow, reproduce, and die in a few weeks. The various forms of senescence fall into two categories: those types occurring in determinate plants, where programmed senescence is apparent (e.g., the century plant), and those in which growth is indeterminate (e.g., oak tree, redwood). Even in the latter, limited senescence is ongoing in the form of leaf senescence and abscission.

KEY WORDS

alternation of generations	endosperm	leaf primordium
gametophyte	cotyledon	totipotency
sporophyte	monocotyledon	apogamy
sporangium	dicotyledon	apospory
angiosperm	epicotyl	cloning
flower	hypocotyl	regeneration
sepal	radicle	protoplast
petal	seed	senescence
stamen	germination	annual
pistil	apical meristem	biennial
carpel	protoderm	perennial
ovule	ground meristem	determinate growth
embryo sac	procambium	indeterminate growth
embryo		

1   Compare the reproductive cycle of a green alga with that of a green land plant. How might the latter have evolved from the former?

2   What is the nature of the gamete-producing plant of a fern? After fertilization, what does the fern zygote develop into?

3   Describe the gametophyte generation of angiosperms.

4   Is the flower a gametophytic or sporophytic structure? Explain.

5   Does the flowering plant produce spores? Does it have sporangia? If so, what are they and where are they?

6   What, if any, functions do sepals and petals perform?

7   What is the double fertilization occurring in flowers?

8   What is the function of the endosperm? Of the seed? Of the fruit?

9   Describe the principal plant meristems and their functions.

10 What are the three basic tissue types of a vascular plant? Do they compare in any way with the principal germ layers of animals? How do the basic plant tissues originate?

11 Discuss totipotency in relation to plant growth and development. Give an example.

12 What is apogamy? What is its significance with respect to developmental potentials of plants?

13 Some plants have determinate growth, others indeterminate. Differentiate between these two modes of development, and give some specific examples of each.

SUGGESTED READING

ANDREWS, H. N. 1963. Early seed plants. *Science* 151:925–31. (Presents convincing evidence for the origin and evolution of seeds.)

BOLD, H. C.; C. J. ALEXOPOULOS; and T. DELEVORYAS. 1980. *Morphology of plants and fungi*, 4th ed. New York: Harper and Row. (An excellent comprehensive textbook of morphology, emphasizing algal evolution.)

EBERT, J. D., and I. M. SUSSEX. 1970. *Interacting systems in development*, 2nd ed. New York: Holt, Rinehart and Winston. (Presents a broad review of embryo development in plants and animals, emphasizing experimental analysis of development. Strong with respect to regulation of development.)

GRAHAM, C. F., and P. F. WAREING. 1976. *The developmental biology of plants and animals*. Philadelphia: W. B. Saunders Co. (Integrates plant and animal development. Good reference but a bit complex for students of introductory biology.)

LEOPOLD, A. C. 1975. *Plant growth and development*, 2nd ed. New York: McGraw-Hill Book Co. (An excellent survey of plant growth, with emphasis on hormonal control. The chapter on senescence is especially good.)

MEINS, F., JR., and A. N. BINNS. 1979. Cell determination in plant development. *Bio-Science* 29:221–25.
(In particular, refer to the discussion of limited developmental potential in the Norfolk Island pine.)

SHEPARD, J. F. 1982. The regeneration of potato plants from leaf-cell protoplasts. *Sci. Amer.* 246(5):154–66. (Of many interesting aspects of tissue culture and agriculture, the selection of new somatic mutant cell lines, using cell-culture methods, promises a new era for agriculture.)

WHITTIER, D. P. 1971. The value of ferns in an understanding of the alternation of generations. *BioScience* 21.225–27. (One of a series of fern articles in this journal, it presents the experimental approach to questions of development and evolution.)

Nutrition, Transport, and Regulation in Plants

17

Various aspects of metabolism, as well as actions of cells and tissues associated with movements of metabolic substances in plants, were covered in earlier chapters. In this chapter, some of these important processes are reviewed so that the correlated activities of cells, tissues, and organ systems of plants will be better understood and appreciated.

The plant kingdom includes simple plants such as algae and mosses, in which many or all cells and tissues of the organism are in intimate contact with its environment. In algae, for example, the uptake of nutrients from the environment (in this case, water), as well as the transport of those nutrients and the discharge of metabolic waste products, proceeds by diffusion and by **active transport** into, out of, and between cells. In such simple plants, complex problems of nutrient assimilation and transport are not often encountered. However, as plants invaded the land from the sea, uptake and transport of nutrients became more critical and special tissue systems and organs evolved.

PLANT NUTRIENTS

Plant nutrition is not comparable to animal nutrition except in a very general way. Plants, unlike animals, are self-feeding autotrophs. The process of **photosynthesis** provides plants with an internal supply of food. The nutrients derived from outside the plant therefore are not foods but rather the raw materials from which foods, as well as other constituents of plant cells, are made. These raw materials include water and carbon dioxide, which are the raw materials of photosynthesis and also the essential elements that comprise protoplasm.

Soil

The primary sources of mineral nutrients for the plant are found in the soil–water solution and in the atmosphere. Soil is a composite material of decayed plant matter (called **humus**), rock particles of various sizes, and anions and cations. The anions and cations compose the nutrient elements taken up by the plant; they originate by dissolution of small rock particles by water, as metabolic by-products of living organisms, and from the death and decay of organisms. Living microorganisms such as bacteria, fungi, algae, and protozoans sometimes are considered to be soil components because their activities add to soil fertility.

Rock particles commonly are classified as **gravel, sand, silt,** and **clay,** in descending order of particle size.

If one were to start with a boulder and a hammer and pound the boulder into smaller and smaller pieces, all the classes of soil particles would eventually be produced. Nature does the same thing through its agents of gravity, freezing and thawing, heating and chilling, wave and stream erosion, and the sand-blasting effect of wind.

Further breakdown of rock particles occurs by chemical action and provides the anions and cations that constitute the mineral nutrients of the plant. Phosphate ions, for example, are formed from phosphate rocks, and calcium ions are derived from limestone (calcium carbonate). The other anions and cations are similarly derived from various kinds of rocks.

Each component of the soil contributes to soil fertility. Gravel and sand provide air spaces, giving underground organisms and roots access to the oxygen required in their metabolism. However, if a soil were composed only of sand and gravel, it would not hold water very effectively. Humus, which has an affinity for water, provides this ability to hold water, as do finer rock particles such as silt and clay. Around these small particles of mineral and organic matter, water is present in the form of a thin film of **capillary water**. This thin film provides much of the water and minerals taken up by plant roots.

Clay particles, the finest of soil particles, hold water effectively but do not give it up easily. For this reason, a heavy clay soil cannot support the growth of even very hardy plants. Yet clay in lesser amounts is an important component of good soil, for clay particles bind mineral ions. Localized concentrations of electrons are present on the surfaces of clay particles, to which cations such as those of calcium, potassium, and sodium are attracted and held, but not so tightly that the plant cannot, in effect, pry them loose. The plant, in fact, trades one kind of ion for another, a process called **cation exchange**. The root, for instance, trades hydrogen ions for potassium, calcium, magnesium, iron, and some other essential cations. Anions such as sulfate (SO_4^{2-}), nitrate (NO_3^-), and phosphate (PO_4^{3-}) are present in the soil water. Because they are not bound to clay particles, they tend to be dissolved out of the soil more rapidly than are cations.

Long ago it was discovered that plant growth could be improved by adding certain substances to the soil. The farmers of ancient Rome knew that application of limestone, wood ashes, compost, and manure to soil improved plant growth. Today, although refined versions of these materials in the form of commercial fertilizers are used for the same reason, many farmers and gardeners still use natural fertilizers such as compost and manure.

Because soil becomes depleted when crops are grown continually on it, it might be reasoned that plants actually consume quantities of soil. Perhaps because of this notion, beween 1630 and 1750 a search began for substances in nature that cause plants to grow. More than 300 years ago a Dutch scientist, Jean van Helmont (1577–1644), set out to discover, as precisely as was then possible, what it was that plants took from their environment during their growth.

As accurately as he could, van Helmont weighed out 91 kg (200 lb) of dry soil, in which he planted a young willow shoot weighing exactly 2.3 kg (5 lb). During the next five years the willow was watered with clean rainwater, and at the end of this time the plant, which by then had become a small tree, was separated from the soil. Both soil and tree were weighed; it was found that the tree had gained 74.5 kg (164 lb) but that the soil had lost only 57.1 g (2 oz) (Fig. 17-1). Clearly, the plant had consumed little if any soil, and van Helmont concluded that the willow had gained most of its weight from water.

Van Helmont and other scholars of his day had no understanding of the role of water as a raw material in photosynthesis. The willow had the capability, as do all green plants, of converting water and carbon dioxide (from the atmosphere) into sugar; by processing sugar, a plant can manufacture many other components of living matter.

In addition to its role in photosynthesis, water makes up 60–90 percent of the weight of both plants and animals, and constitutes the fluid medium of living mat-

FIGURE 17-1
Van Helmont's seventeenth century plant-nutrition experiment. A willow sapling was grown in a measured quantity of soil for 5 years, watered only with rainwater, then soil and plant were weighed separately. (Redrawn with permission from J. W. Bonner and A. W. Galston, 1952, *Principles of plant physiology*, San Francisco: W. H. Freeman and Co., p. 47)

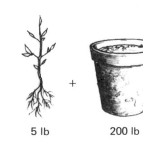 5 lb 200 lb +  +  5 years of rainwater +  5 years = 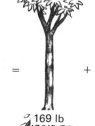 169 lb +  199 lb, 14 oz

ter—the basic substance in which other chemicals are dissolved. Nearly everything that moves into, within, between, and out of organisms moves in water.

What of the missing 2 oz of soil in van Helmont's experiment? Although this figure easily might represent experimental error, given the possibilities for gain or loss of soil in a long-term and rather crude experiment of this kind, it well may have represented a real loss of certain essential mineral elements.

Essential Elements

In the three centuries since van Helmont's experiment, botanists have clarified plant nutrition. By 1840 it was known that plants require not only water but also carbon dioxide (CO_2) from air and magnesium, potassium, and phosphorus from soil. In the 1860s Julius Sachs (1832–1897) discovered that plants could be grown without soil (Fig. 17-2), using a solution of water and five mineral salts: potassium phosphate (KH_2PO_4), calcium nitrate ($Ca[NO_3]_2$), potassium nitrate (KNO_3), magnesium sulfate ($MgSO_4$), and iron phosphate ($FePO_4$). The specific elements in these compounds that were considered essential for plant growth are calcium, iron, phosphorus, nitrogen, magnesium, and sulfur. In addition, hydrogen obtained from water, oxygen from water and the atmosphere, and carbon from atmospheric carbon dioxide were required.

For many years it was thought that the foregoing elements, called **major elements,** or **macronutrients,** were the only necessary ones. However, between about 1914 and 1930 it was discovered that other elements also are required if healthy plants are to be maintained. The reason for this tardy discovery is interesting. It seems that the best chemicals available to Sachs and other early workers in plant nutrition contained a number of impurities in very small (trace) amounts; in fact, some were present even in the glass from which culture vessels were made. Then, when purer chemicals and better glassware became available, an unexpected result was that plants grew less well than formerly. It soon was discovered that some of the impurities were essential for the growth of plants. Among these **trace elements,** or **micronutrients,** are copper, boron, cobalt, manganese, zinc, and, for some plants, silicon, iodine, selenium, and perhaps one or two others. Today, with ultrapure chemicals and ultraclean glassware available, plants and even isolated plant cells can be grown in solutions in which all the essential major and trace elements are clearly defined (Table 17-1).

How may a solution culture be used to further an understanding of plant growth? One way in which it has

TABLE 17-1
Composition of a medium for culture of plant cells

Component*	Concentration (milligrams/liter)
Macronutrients	
$CaCl_2 \cdot 2\ H_2O$	440.000
$FeSO_4 \cdot 7\ H_2O$	27.800
KH_2PO_4	170.000
KNO_3	1,900.000
$MgSO_4 \cdot 7\ H_2O$	370.000
NH_4NO_3	1,650.000
Micronutrients	
$CoC_2 \cdot 6\ H_2O$	0.025
$CuSO_4 \cdot 5\ H_2O$	0.025
H_3BO_3	6.200
KI	0.830
$MnSO_4 \cdot 4\ H_2O$	22.300
$Na_2MoO_4 \cdot 2\ H_2O$	0.250
$ZnSO_4 \cdot 7\ H_2O$	8.600
Other components	
Indoleacetic acid (plant hormone)	1.000
Kinetin (plant hormone)	1.000
Sucrose (carbohydrate)	30,000.000
Vitamins	
Inositol	100.000
Thiamine HCl (vitamin B_1)	0.100

Source: Data from *Physiologia Plantarum* 18:119.
*The H_2O is water of hydration absorbed by the mineral salt.

FIGURE 17-2
Sachs's mineral nutrition experiment. (Redrawn with permission from J. F. Bonner)

proved useful is in the determination of plant deficiency symptoms. Obviously, if plants do not grow well in certain kinds of soil, it would be helpful if their appearance could be associated with deficiencies of one or more of the essential elements. By growing plants in solutions lacking only one of the elements and noting the plant response, characteristics accompanying the deficiency can be determined (Fig. 17-3).

Normally, soil contains cations and anions in sufficient quantity to support plant growth. However, three important elements, called **fertilizer elements,** often become depleted in fields, gardens, and lawns and commonly are replaced by the application of fertilizers. The three fertilizer elements are **nitrogen, phosphorus, and potassium,** often abbreviated **NPK.** Bags of fertilizer are designated 6-6-6, 20-10-0, 12-6-4, or some other combination. The numbers are N-P-K, in that order, and refer to the percentage of each essential element.

Roots and Root Hairs

In cross section, a young **root** shows a central region of xylem (i.e., a protostele), usually in a three- to five-pointed star configuration. Between the arms of xylem are phloem cells. Outside the xylem and phloem is the **pericycle,** a cell layer from which new branch roots arise. The pericycle is enclosed by a layer of cells called the **endodermis.** Walls of most root cells are permeable to water and dissolved substances **(solutes),** but walls of endodermal cells are sealed by waterproof material in the form of a **Casperian strip** (Fig. 17-4; see also Fig. 17-6 for a diagram illustrating the endodermis and Casperian strip).

The endodermis, together with a rather thick layer of storage parenchyma, composes the **cortex** of the root. The cortex is surrounded by a layer of epidermal cells, some of which are specialized as **root hairs** (see Figs.

FIGURE 17-3
Symptoms of plants grown in solutions deficient in one or another element. (Redrawn with permission from J. F. Bonner)

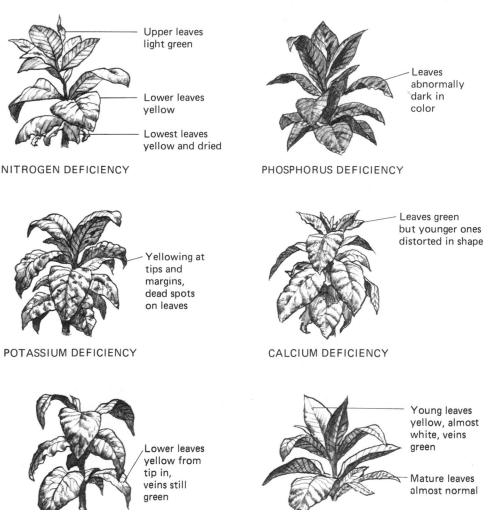

NITROGEN DEFICIENCY — Upper leaves light green; Lower leaves yellow; Lowest leaves yellow and dried

PHOSPHORUS DEFICIENCY — Leaves abnormally dark in color

POTASSIUM DEFICIENCY — Yellowing at tips and margins, dead spots on leaves

CALCIUM DEFICIENCY — Leaves green but younger ones distorted in shape

MAGNESIUM DEFICIENCY — Lower leaves yellow from tip in, veins still green

IRON DEFICIENCY — Young leaves yellow, almost white, veins green; Mature leaves almost normal

FIGURE 17-4
Cross sections of the root of
the buttercup. **A.** Cross section through the entire root.
B. Enlarged view of the vascular tissue.

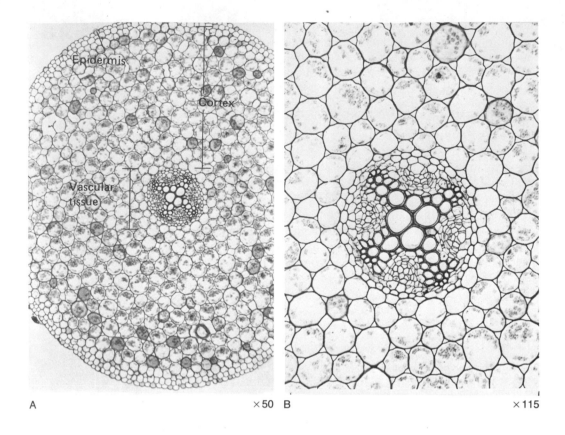

A   ×50 B   ×115

16-13 and 17-5). Within the cortex, the pericycle, xylem, and phloem constitute the inner root region called the **stele.**

About 90 percent of all water absorption takes place through root hairs. Root hairs are present only near the tips of young roots, and because they are delicate and usually shortlived, normally surviving only a few weeks or at most months, continuing root development is necessary to replenish the older, dying root hairs. Measurements and calculation of roots and root hairs of a rye plant have shown that over 14 million young roots and 14 billion root hairs can develop in just four months of growth.

Roots are continually growing and exploring the soil. As they grow, new root hairs are produced which wrap around the soil particles, from which they absorb water and nutrients (see Fig. 17-5). This solution of water and minerals does not remain in the root-hair cells for long. Instead, it is passed to the interior cells of the root, first from cell to cell in the cortex region and then through the cells of the endodermis and pericycle into the xylem.

The passage of the soil solution through the endodermis is especially important. Normally, the walls of living plant cells are porous, resembling blotting paper

in this respect, and much of the water and its solutes travels through the wicklike walls of root hairs and cortical cells, not through the cells themselves. However, when this solution reaches the wall of an endodermis cell, passage is blocked by the Casperian strip (Fig. 17-6B). Here, the solution must move by osmosis through the plasma membrane of the endodermal cell. In this way the endodermis controls movements of water and nutrients into the xylem. Because of this involvement of osmosis, the root is comparable to an osmometer: the cytoplasm and cell membranes of the endodermis and other living cells act like the semipermeable membrane of an osmometer bag (Fig. 17-6A).

The pressure developed by diffusion into the root is called **root pressure** and is sufficient to move water, dissolved minerals, and other molecules into the xylem (here the solution is known as **xylem sap**) and to support the upward movement of sap in the xylem of the stem and branches of the plant. Although root pressure in some cases is strong enough to move sap 30 m (100 ft) up the stem of a tree, the movement is slow, and on hot, dry days negative pressures sometimes occur in stems of large plants. For these reasons, other explanations for rapid transport of sap in plants have been developed.

FIGURE 17-5
Scanning electron micrographs of root hairs. **A.** Section of rice grass root with a covering of soil particles held in place by root hairs. **B.** Same section with soil removed. **C.** Cross section of root. The hollow cortex is typical of many water-loving plants. **D.** Enlarged view of soil particles and root hairs. (Photos from L. H. Wulstein and S. A. Pratt, 1981, Scanning electron microscopy of rhizosheaths occuring in Indian ryegrass, *Amer. J. Bot.* 68:408)

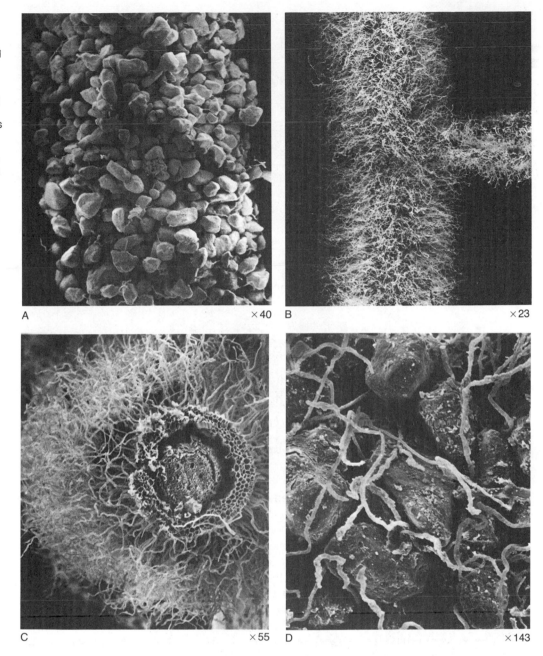

A   ×40   B   ×23

C   ×55   D   ×143

Once sap is present in the xylem, it becomes distributed throughout the tissues of the plant by passage through the branching system of xylem tracheids and vessels in the vascular bundles of the root, stem, and leaves. Finally, reaching the ultimate veins of the leaf, it passes into the intercellular spaces of the mesophyll and palisade cells and is available for photosynthesis. Not all the sap is taken up by leaf cells, however. Much of it passes as water vapor out through the stomata to the outside of the leaf, and this, it turns out, is very important in supplying energy for water transport in the plant.

TRANSPORT IN PLANTS

Transpiration and Water Transport

The movement of water vapor out of the leaf is called **transpiration**. Transpiration is believed to be responsible for pulling water up the stem of the plant from the roots by a process called **transpiration pull**. Water molecules are particularly cohesive in the liquid state due to hydrogen bonding, but with the application of heat energy from the sun or some other source, their kinetic energy

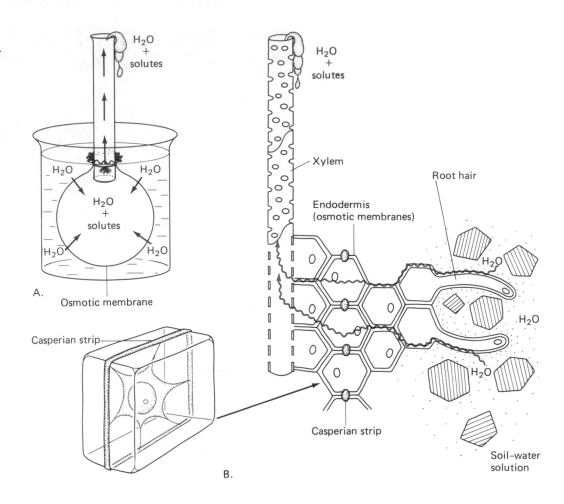

FIGURE 17-6
Root pressure. **A.** Osmometer.
B. Root as an osmometer.

H_2O + solutes

H_2O

H_2O

H_2O + solutes

H_2O

H_2O

A.

Osmotic membrane

Casperian strip

H_2O + solutes

Xylem

Endodermis (osmotic membranes)

Root hair

H_2O

H_2O

H_2O

Casperian strip

B.

Soil–water solution

eventually exceeds their cohesive forces and they form water vapor. As water molecules change from the liquid state to the vapor state, tension is exerted on molecules still in the liquid state. The many water molecules evaporating from within leaves by transpiration are capable of exerting great tension on the columns of sap water in the xylem of the stem; this tension, which can be measured, is great enough to move water to the tops of the tallest trees.

The physical principles of transpiration pull can be demonstrated by a simple laboratory experiment using a device called an **atmometer.** The atmometer consists of a porous ceramic bulb atop a glass tube. Bulb and tube are filled with water, and the open end of the tube is passed through a stopper and immersed in a flask of water. A water-filled inlet tube, also inserted through the stopper into the flask, is used as a measuring device (Fig. 17-7A). An air bubble injected at the bottom of the inlet tube will remain stationary when it reaches the horizontal bend of the tube if there is no water flowing into the atmometer. If, on the other hand, there is a flow of water as a result of evaporation from the surface of the

atmometer bulb, the bubble will move correspondingly. This apparatus shows the system capable of continuously moving water upward as it evaporates from the surface of the ceramic bulb and also reveals the effects of changes in air temperature, relative humidity, and air currents on the rate of water movement. A parallel experiment, in which the cut end of a leafy branch is substituted for the porous atmometer bulb and its glass tube, produces about the same results (Fig. 17-7B). In this case, the leaf substitutes for the ceramic bulb; the **stomata** of the leaf's surface are equivalent to the pores of the bulb. Similar experiments were done many years ago, using an entire tree hoisted with a derrick and placed with its cut end in a tank of water. The results were about the same and were interpreted to signify that transpiration pull could account for transport of water and solutes even in very large plants.

The plant walks a kind of tightrope in its water economy; on one side are requirements for photosynthesis and transport, which require open stomata, and on the other the danger of excessive water loss through the stomatal openings. If the latter occurs, the plant will

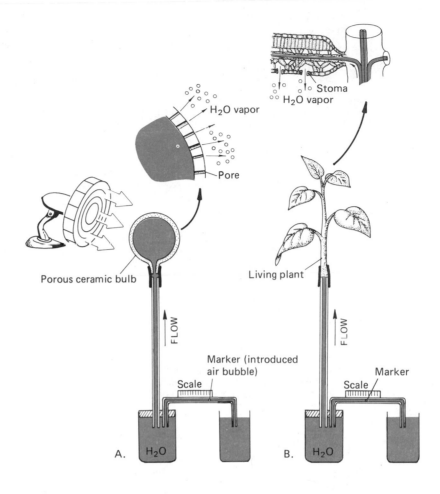

FIGURE 17-7
Transpiration. **A.** Atmometer. Water molecules evaporating from the porous ceramic bulb exert a pull on the liquid water in the glass column, and water moves upward, as indicated by a bubble marker. **B.** Living plant substituted for the ceramic bulb. The process in both systems is accelerated if the bulb or leaves are exposed to wind or higher temperatures.

wilt and many of its vital processes will come to a halt until the water balance is restored; if wilting is severe, the plant will die.

The opening and closing of stomata are controlled by guard cells (Fig. 17-8), which are therefore major devices for control of the movements of water vapor and other gases between the leaf and the environment. Although much remains to be learned about how guard cells work, it is apparent that they are involved in a feedback system in which photosynthesis plays an important part (see Chapter 2, Fig. 2–31).

Food Transport

Photosynthesis, occurring in leaf cells as well as in other green parts of plants, produces large quantities of sugar. The common plant sugars are glucose, fructose, and sucrose, but other sugars such as the 3-, 4-, 5-, and 7-carbon sugars of the Calvin-Benson and Hatch-Slack cycles (see Chapter 3) also are formed, as are fats and proteins. Fats and proteins are not produced directly by the Calvin-Benson or Hatch-Slack cycle, but the products of

those cycles (sugars) are used by the plant to make lipids and amino acids as well as other organic molecules, including starch and cellulose.

Many cells and tissues of higher plants are not photosynthetic. Some, such as the pith, are located in the interior of the plant and are not reached by light. Others, even though superficially located (e.g., root hairs), are in darkness continually. Nonphotosynthetic tissues require food transported from the photosynthetic tissues of the plant if they are to live and carry on their work.

Experiments in which portions of stems are cut away show that food is conducted in the bark of the stem and not the xylem. Other, more precise, experiments have localized the transport of food in the sieve tubes of the phloem. The cytoplasm of sieve tube elements is continuous from cell to cell, passing through pores in the end walls (the sieve plates) and also through smaller sieve areas in the lateral walls. Although much of the movement in sieve tubes is longitudinal, there also is lateral movement out of the tubes through the sieve areas and into surrounding living cells and tis-

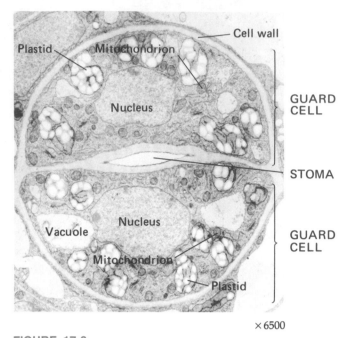

Plastid — Mitochondrion — Cell wall

Nucleus

GUARD CELL

STOMA

Vacuole

Nucleus

GUARD CELL

Mitochondrion

Plastid

×6500

FIGURE 17-8
Electron micrograph of guard cells of *Vigna sinensis*, a member of the pea family, in cross section. Note the greatly thickened walls of the guard cells in the region bordering the stomatal opening. (Photo courtesy H. Mitrakis and B. Galatis)

sues. In addition, some of the food of the sieve tube is passed on to ray cells and transported laterally into the interior of older stems and roots.

There has been a good deal of speculation about the way food is transported in the sieve tubes. One possibility is that the cytoplasm passes food from cell to cell by active transport. However, to equal measured rates of transport, each sieve tube element would have to be emptied in a second's time. This seems too rapid to be accounted for by the cytoplasm alone, which only occupies the periphery of the element. A more likely explanation is that most of the food in the liquid interior of the sieve tube elements flows through the tubes under pressure. How can this hypothesis be tested? Ingenious experiments use tiny insects called **aphids**, which feed on the sap of plants. The aphid, in some unknown way, can locate sieve tubes and then insert its long tubular mouth, called a **stylet**, through the surface cells of leaves and stems into a sieve tube. The pressure of the sugary juice in the sieve tube then force-feeds the aphid, often to the point where it exudes excess food through its anus (Fig. 17-9). Anyone who has parked a car under a large, shady maple tree knows the result: the car will be speckled with tiny spots of dried aphid juice, called honeydew.

FIGURE 17-9
Aphid sucking sap, and its use in studying food transport in plants.

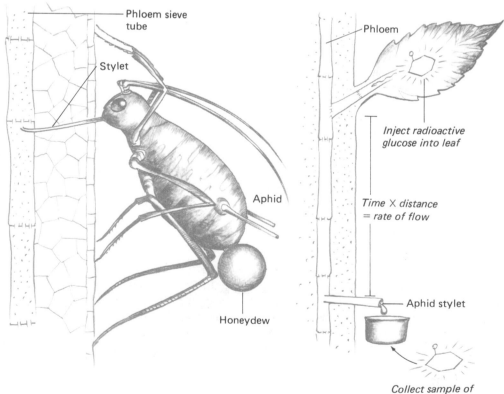

Phloem sieve tube

Stylet

Aphid

Honeydew

Phloem

Inject radioactive glucose into leaf

Time × distance = rate of flow

Aphid stylet

Collect sample of phloem exudate and test for radioactivity

The aphid has been used as a sampling device in studies of food transport in the sieve tubes. First, the aphid is detached from its stylet; that leaves the stylet as a tiny probe into the sieve tube. The stylet will continue to exude liquid under pressure for several days. Then, if radioactive glucose (glucose containing radioactive carbon atoms) is injected into a leaf some distance from the stylet, part of the radioactive glucose gets into the phloem and is transported in the sap; it can be collected as it drips out of the aphid stylet. In this way its rate of movement can be followed with a Geiger counter. Such studies and others have provided botanists with a theory on food transport called **mass flow.**

Mass flow can be demonstrated by connecting two osmometers together with a glass tube as shown in Fig. 17-10A. One osmometer bag contains a concentrated sugar and water solution and is analogous to a leaf cell. The other osmometer bag contains a sugar solution of low concentration and is comparable to a root cell. The glass tube between the bags is the equivalent of a sieve tube connecting the leaf cell with the root cell (see Fig. 17-10B). The whole apparatus is immersed in water. With this system, a mass movement of sugar molecules can be demonstrated from the high sugar concentration in one osmometer to the solution in the other osmometer. This flow is not simply diffusion but rather a flow of sugar (mass flow) carried along with water molecules,

which enter the osmotic bag containing a high solute concentration and exert enough pressure to force water molecules through the connecting tube and out of the second bag. The latter does not, at first, have a sugar concentration high enough to develop an equalizing pressure.

It has been proposed that movement of food in the phloem works about the same way as the mass flow of sugar in the connecting tube of the apparatus and that the equivalent of the paired osmometers is a leaf cell and a root cell. However, is the sieve tube really equivalent to an inert glass tube? As noted previously, xylem vessels and tracheids are nonliving, so in a real sense they are like glass tubes. Phloem cells, on the other hand, have to be living for transport to occur. If poisons are injected into the bark of a tree so that the sieve tubes remain intact but are killed, food transport stops. The reason for this is not known.

Although the mass-flow hypothesis does not explain the necessity of living phloem cells for food transport to occur, no other hypothesis, including that of active transport by cytoplasmic streaming or membrane actions, adequately explains the high rates of food transport in phloem. Many processes probably are involved.

REGULATION IN PLANTS

Many of the activities of plants are controlled by physical processes such as those just described. The opening and closing of stomata is an example in which photosynthesis, osmosis, and active transport are involved in the movements of gases in the leaf. Regulation of many vital processes in plants results from interplay between outside factors, including moisture, sunlight, and temperature, and the internal conditions of water pressure or tension, localized concentrations of nutrient molecules, plant hormones, and other regulatory chemicals.

Because plants lack muscles, plant movements usually are produced by changes in water pressure within cells, as in opening and closing of stomata, or by processes of differential cell growth. The integration of many plant movements and other functions is done chemically, often by plant hormones; nerves or nervelike structures have never been discovered in any plant.

Turgor Movements

Some movements of plants are brought about by changes in water pressure, or **turgor,** within cells. A classic example is the folding of leaflets of the sensitive plant, *Mimosa pudica*, in response to touch or temperature stimulation (Fig. 17-11). "Sleep movements," such as the folding of the leaflets of locust trees at night, also are

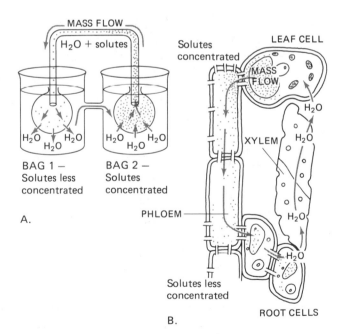

FIGURE 17-10
Mass flow. **A.** Mass-flow experiment. Water under greater osmotic pressure enters bag 2 and moves to bag 1, where pressure is less, forcing water out. Sugar is carried by mass flow of water from bag 2 to bag 1. **B.** Equivalent of part A as seen in the relationship of cells of leaf, root, and phloem elements.

FIGURE 17-11

The sensitive plant, *Mimosa pudica*. **A.** Ten seconds after all leaflets were touched. **B.** Recovery after 20 minutes.

A.

B.

turgor movements, as is the sudden snapping shut of the leaf of Venus's flytrap. These movements are produced by a loss of water from groups of strategically placed cells. In the sensitive plant and in locusts and other plants characterized by the repeated foldings and unfoldings of leaflets, the turgor-movement cells are located in swellings called **pulvini** (sing. **pulvinus**) at the bases of leaves and leaflets. The controlling principles of turgor changes are not understood completely, but the mechanisms appear to involve active transport of certain ions and to require hormone participation.

Growth Movements

Growth movements in plants are **nutations, nastic movements (nasties),** and **tropisms.** All involve changes in the orientation of plant organs, and all are brought about by the unequal growth of cells of the plant. If the growth movement is spontaneous and apparently self-controlled, it is a nutation; if it occurs in response to some external stimulus but is not unidirectional, it is classified as a nastic movement; if the response is unidirectional, it is called a tropism.

Nutations   The graceful, swirling growth movements of plants seen in time-lapse movies are produced by the alternate expansions of cells in a highly characteristic pattern of cell growth. The plant, in these movements, follows a spiral path of upward or outward growth. In some cases these movements, called nutations, appear not to serve any particular function, but in others, such as the twisting of tendrils, stems, and branches of climbing vines, the function of the nutation is attachment and anchorage of the plant (Fig. 17-12).

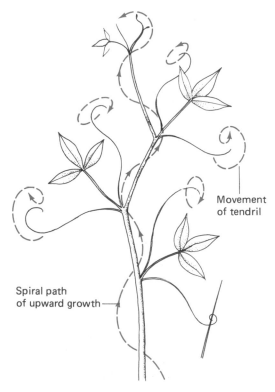

Movement of tendril

Spiral path of upward growth

FIGURE 17-12
Nutations.

Nasties   The opening of bud scales (highly modified leaves), young leaves, and flower petals, and in some cases the alternate opening and closing of flowers, are examples of nastic movements, or nasties. Nasties are produced by unequal growth of cells at the bases of these structures. When the bud is young, the bud scales or petals, as the case may be, grow more rapidly on the underside; when older, the bud scales or flower parts grow more rapidly on the upper side, and the bud or flower opens up. In some flowers these alternate growth movements are repeated every day, causing them to open in the morning and close at night. Nasties are responses to external, nondirectional stimuli such as light and dark or seasonal changes in temperature.

Tropisms Tropisms are curvatures of plant organs in response to directional stimuli. Included are **positive geotropism** (growth toward the pull of gravity), **negative geotropism** (growth away from the pull of gravity), **phototropism** (growth toward a light source), and **thigmotropism** (growth in response to touch) (Fig. 17-13). Sometimes included as a growth movement is the tropism of roots toward water, called **hydrotropism**, which probably is a general growth response rather than a tropism. Tropisms are among the most interesting plant phenomena known, because the role of hormones is fairly well understood and because these processes exemplify homeostasis so nicely.

How a plant "knows which side is up" is a question of great interest. As noted previously, balance organs in animals function by the movement of liquids and sometimes small particles within chambers equipped with sensitive hairs. There are no balance organs per se in plants; rather, the cells of the plant individually and collectively act as balance organs. When a plant is turned on its side, starch grains (or possibly other small particles) drift to the lower side of each cell (Fig. 17-14). Then, in some way not yet understood, the cells initiate

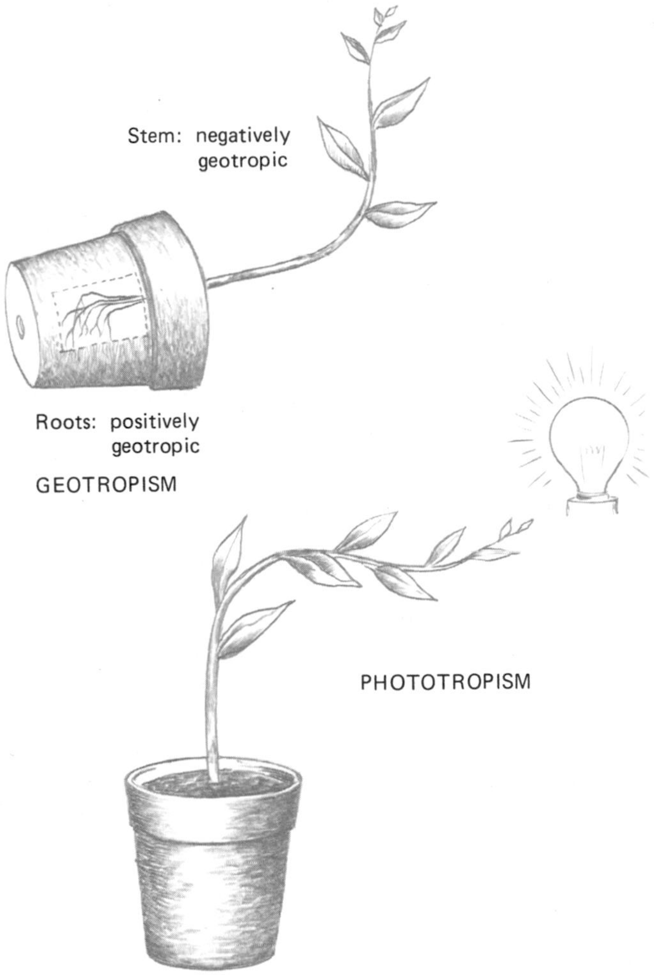

Stem: negatively geotropic

Roots: positively geotropic

GEOTROPISM

PHOTOTROPISM

FIGURE 17-13
Tropisms.

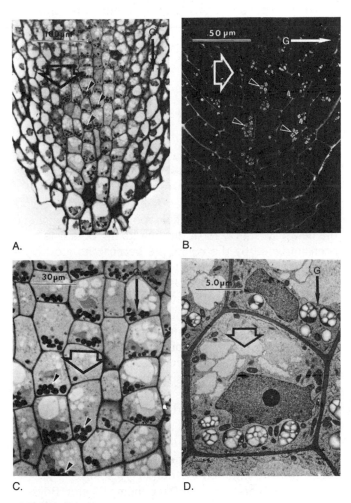

FIGURE 17-14
Geotropism in roots of the garden pea—role of starch grains. G indicates direction of gravity pull. **A.** Longitudinal section of a root tip in which placement was vertical (gravity parallel to the axis). Note position of starch grains, small arrows. **B.** Dark-field illumination of root-tip longitudinal section. Root was horizontal (ie., at right angles to gravity). Note starch grains tend to accumulate laterally (the "down side") in root cells. **C.** Enlargement of cells in A; starch grains again indicated by arrows. **D.** Electron micrograph of cells of root tip. (Photos courtesy J. S. Ransom and Randy Moore)

the tropism that results in the stem bending upward and the root downward.

How do plants tell light from darkness or "know" from which side the light is coming? Phototropic responses of cells involve light-sensitive pigments located in specific regions such as the stem tip or in young leaves. These pigments are present in very low concentrations but nevertheless are able to induce phototropisms. There is some uncertainty about the identity of the pigments of phototropism, but they are known to be sensitive only to blue light and to the blue wavelengths in daylight.

Both geotropisms and phototropisms involve action of plant hormones, which become locally concentrated in response to gravity or light and either promote or inhibit growth of cells so as to produce positive or negative curvatures (toward or away from the initial stimulus) in plant organs.

Hormones

Discovery of Hormones   When a grain of oats or some other grass is planted in moist soil, the first detectable sign of germination is the emergence of a colorless, conical structure called the **coleoptile**. The function of the coleoptile is the protection of the delicate shoot of the young plant as it emerges from the soil. Shortly after emerging, the coleoptile stops growing and the new leaves grow forth through its tip. Botanists have been intrigued by the coleoptile because its growth responses are very sensitive to light and gravity. The oat coleoptile has been adopted as an important experimental tool by plant physiologists, and it is used widely as a sensitive detector of plant hormones.

Charles Darwin became interested in the coleoptile, wondering how it was able to detect light and bend toward it. He did several simple experiments which had a far-reaching impact on science and agriculture (Fig. 17-15). First he placed little tinfoil hats over the tips of coleoptiles and discovered that when he did this the coleoptiles lost their power to bend toward the light. Then, by placing tinfoil tubes around the bases of coleoptiles, he showed that the tip of the coleoptile alone, when exposed to light, initiated the bending response.

Some years later, a Dane named Peter Boysen-Jensen repeated Darwin's experiments, adding a few embellishments of his own (Fig. 17-16). He found that if he cut the tip off the coleoptile, the stump would cease elongating, but when he replaced the tip on the stump, or when the tip was replaced with a thin slice of gelatin between it and the stump, the coleoptile regained its ability to grow upward and also its ability to bend toward light. This suggested that some substance capable

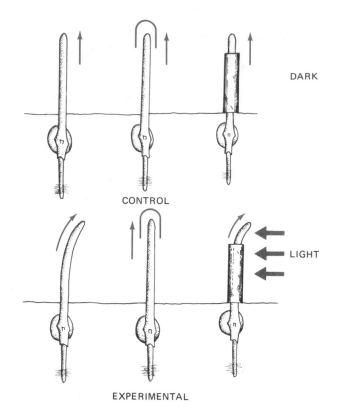

FIGURE 17-15
Darwin's experiments with coleoptiles.

of moving through gelatin was produced in the coleoptile tip and that this substance was responsible for elongation and phototropism. Later, experiments done by A. Paal showed that the coleoptile tip substance would not diffuse through cocoa butter and that the stimulus also was blocked by platinum foil and mica. Thus it could be said that the substance in question was water-soluble but not fat-soluble and was not electrical. Had it been electrical, it would have been transmitted across the platinum foil.

In the late 1920s a young plant physiologist named Frits Went became interested in the work of Darwin, Boysen-Jensen, and Paal and began some experiments to determine the nature of the coleoptile tip substance and how it produced tropisms. In a key experiment, he cut the tip from a coleoptile and placed it on a block of gelatin (Fig. 17-16). After an hour he discarded the tip and placed the block of gelatin on the cut surface of the coleoptile, but to one side. The experiment was carried on in a very weak red light, to which the coleoptile normally is insensitive. The cells on the side of the coleoptile beneath the block began to grow and produced a typical tropism. This experiment showed conclusively not only that a chemical was responsible for the elon-

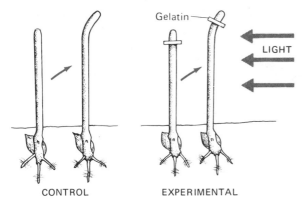

CONTROL   EXPERIMENTAL

BOYSEN–JENSEN 1910

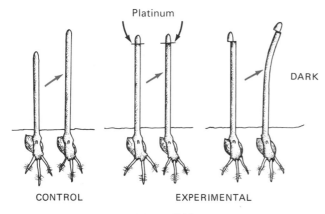

CONTROL   EXPERIMENTAL

PAAL 1919

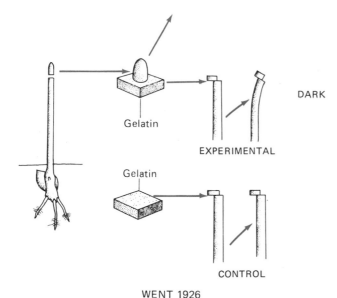

WENT 1926

FIGURE 17-16
Experiments of Boysen-Jensen, Paal, and Went.

gation of the coleoptile but also that localized concentrations of it could produce tropisms. Went named the unidentified substance **auxin,** a term now applied to any chemical capable of making a coleoptile bend. Auxin was the first plant hormone to be discovered.

Auxins

Auxins are considered to be plant hormones (**phytohormones**) because, like animal hormones, they are synthesized in one place and transported to another place where they produce a specific effect. About 10 years after Went discovered the nature of auxin, the substance was isolated and identified chemically by Kenneth Thimann as an organic molecule named **indoleacetic acid,** commonly called IAA (Fig. 17-17). Since then, IAA has been found to have a variety of actions, including stimulation of cell elongation, production of geotropisms as well as phototropisms, promotion of fruit setting and the development of unpollinated flowers, control of leaf-fall, or **abscission** (Fig. 17-18), inhibition of lateral bud growth, and stimulation of cell division.

In the decades since the discovery of IAA, other natural auxins have been isolated, but IAA appears to be one of the most widespread. Analysis of its chemical structure has led to the production in the laboratory of many synthetic auxins. Some of these are very powerful and have been used widely in agriculture and other applications. One of them, **2,4-dichlorophenoxyacetic acid (2,4-D)** is used as a herbicide because it selectively kills broad-leaved plants (dicotyledons) but will not harm narrow-leaved plants (monocotyledons) (Fig. 17-18). Thus it is not only a useful selective herbicide but has in fact made possible the high crop yields obtained from hybrid maize. Formerly, maize had to be planted in widely spaced rows to allow for cultivation, but the use of 2,4-D and even more powerful synthetic auxins as weed killers has eliminated the need for mechanical cultivation and permits planting in close rows for increased productivity.

The discovery of auxin, and the subsequent use of natural and synthetic auxins in agriculture, is worth billions of dollars in the world economy. Yet this important technology owes its existence to some simple experiments done over a century ago with grass coleoptiles and little tinfoil hats.

A simple chemical having many auxinlike properties is the hydrocarbon gas **ethylene** (Fig. 17-17). Its effects on plants were discovered when leaking city gas, produced from coal, resulted in misshapen stems and branches and in the rapid ripening of stored fruits. Years ago it was discovered that the active ingredient in coal gas that caused these effects was ethylene.

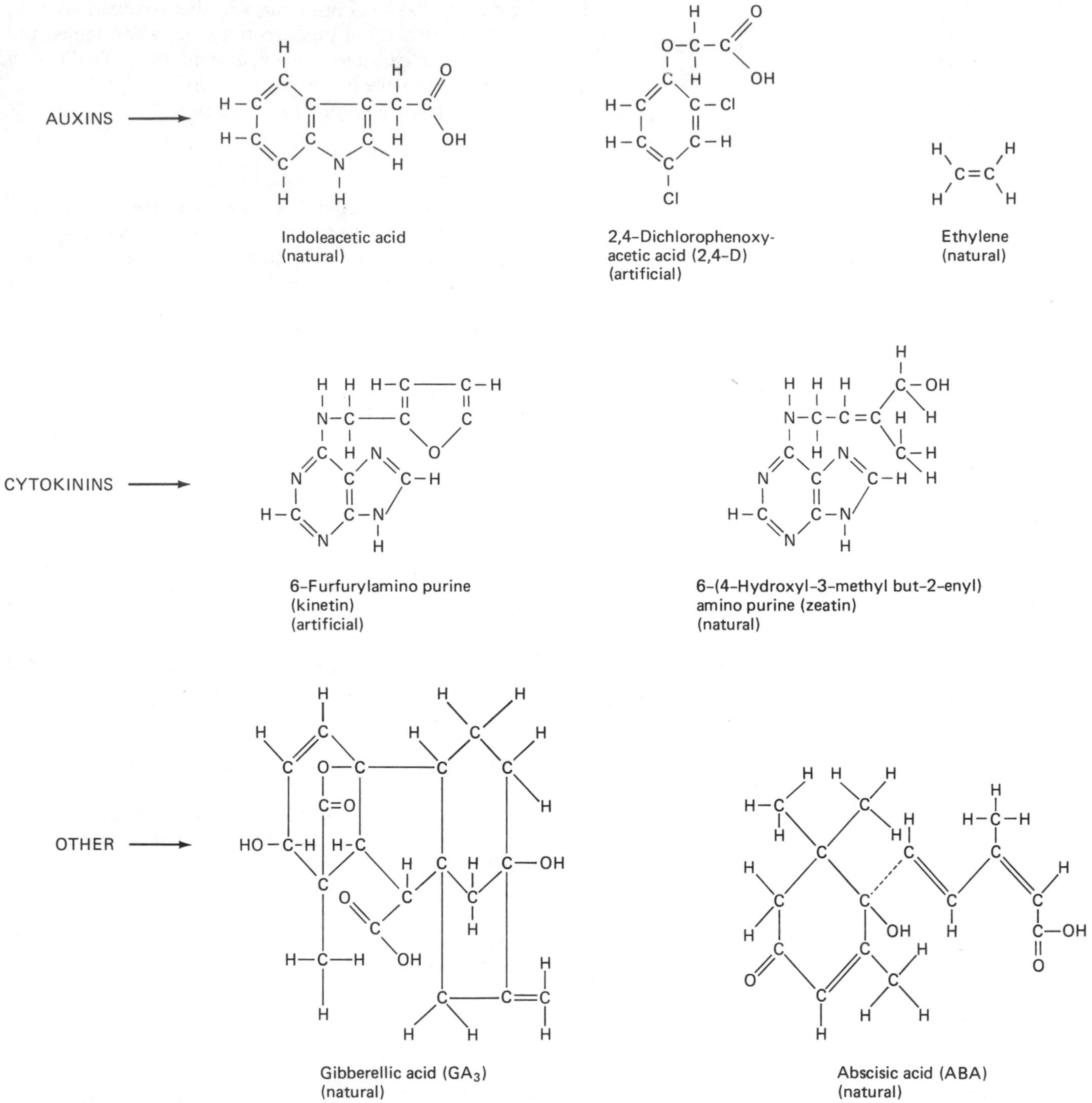

FIGURE 17-17
Structures of some important plant hormones and regulators.

Ethylene is a simple, 2-carbon molecule, and in recent years scientists have learned that it is produced by plants themselves. For example, fairly large amounts of ethylene are produced by rotten fruit; this is one reason for the proverb that a rotten apple will spoil the whole barrel—ethylene produced by a rotten apple will cause nearby apples to ripen rapidly and spoil. The principle of this role of ethylene is used in marketing fruits. Unripened fruit such as green bananas can be ripened rapidly by exposing them to ethylene gas; this technique

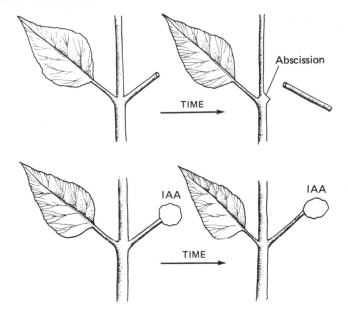

FIGURE 17-18
Leaf abscission experiment. In this simple experiment, a debladed leaf (upper drawings) will develop an abscission layer at the base of its remaining petiole and become abscised. However, if indoleacetic acid is applied in a paste to the debladed petiole stump, an abscission layer will not be formed and the petiole will remain firmly attached to the stem for an indefinite time. It is thought that a healthy leaf blade produces sufficient IAA to inhibit formation of an abscission zone. However, in autumn when the leaf becomes senescent, IAA no longer is present in sufficient concentration to prevent abscission.

allows green fruit to be shipped to market and then ripened quickly for sale.

Gibberellins For quite a few years auxins were thought to be the only natural plant hormones. Then Japanese botanists discovered another class of plant hormones in fungus-infected rice plants. Rice planters long had known of a disease of rice plants which they called *baka nae*, characterized by very tall and spindly growth. This disease is caused by a fungus named *Gibberella*, and in 1935 E. Kurasawa and T. Yabuta isolated and chemically analyzed from this fungus a plant hormone which they named **gibberellin** (Fig. 17-17). Because of World War II, the news of their discovery did not make much of an impact, and it was not until the 1950s that interest in gibberellin became widespread. By now at least 37 different kinds of natural gibberellins have been found and are known to control many different functions in plants.

Gibberellic acid, a common form of gibberellin, can make Mendelian dwarf pea plants grow tall, induce germination in some kinds of dormant seeds, initiate flowering in specific plants in which flowering has been inhibited by the wrong light or temperature regime, and

cause the formation of certain germination enzymes. It also is involved in fern reproduction (refer to Chapter 20.)

One of the most interesting and important examples of the action of gibberellic acid is in the germination of barley grains (Fig. 17-19). Barley is widely used in the brewing industry because the germinating grains contain enzymes that convert starch into glucose. This permits brewers to use starch as a source of sugar for their fermentations rather than depending on sugar or sugary fruits as does the wine industry. During germination of barley, the embryo produces gibberellic acid, which diffuses out into a layer of cells just inside the seed coat. This layer, called the **aleurone layer,** is activated by gibberellic acid and as a result produces and secretes the enzyme **alpha amylase,** which then breaks down the starch in the interior of the seed into glucose.

Because enzymes are catalysts, they are active in low concentrations and are capable of continuous action. Therefore the addition of partially germinated barley grains to other starch sources such as potatoes permits their conversion to glucose also. The processing of barley for brewing involves germinating the barley and then drying it at low heat to kill the embryo while not damaging the enzyme. This technique is called **malting.** Usually an extract (called **malt**) of the enzymes of the germinated and dried barley is prepared for use in brew-

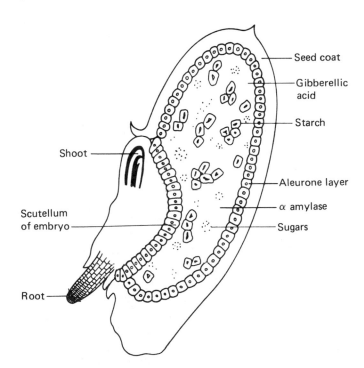

FIGURE 17-19
Gibberellic acid and the germination of barley. In this example gibberellic acid is produced by the embryo in cells of the scutellum (the cotyledon homolog) and diffuses out into the kernel to the aleurone cells, where it induces the production of alpha amylase.

ing. The action of gibberellic acid in the production of barley amylase is a typical hormone action.

Cytokinins A new class of plant hormones was discovered some years ago at the University of Wisconsin by F. Skoog and C. Miller. Oddly, Skoog and Miller did not find the new hormone in a living plant but in an old jar of dried herring sperm (a common source of DNA). To their surprise, an extract of this material produced some dramatic growth responses in cultured plant cells and tissues, including stimulation of cell division and the development of buds (refer to Chapters 3 and 16 for discussions of tissue culture). They were able to isolate and identify the active ingredient in the extract, naming it **kinetin** (Fig. 17-17). Although kinetin itself has never been found in plants, several natural kinetinlike substances have been discovered. They are called **cytokinins** and are considered as important as gibberellins and auxins in regulating vital processes in plants. Among the effects of cytokinins are promotion of bud development, stimulation of cell division (in combination with IAA), prevention of chlorophyll decomposition and protein degeneration, regulation of some aspects of dormancy of buds and seeds, and promotion of seed germination.

Brassins In preceding chapters, mention was made of a number of animal hormones called steroids (cholesterol-like molecules containing 5- and 6-membered rings of carbon atoms). Until recently, steroidal plant hormones were unknown. In the past decade, however, steroidal compounds with hormone properties have been isolated from pollen of members of the mustard family (Brassicaceae). These compounds, called **brassins,** have been found to accelerate plant growth when applied to such diverse plants as barley, elm, and bean. They accelerate the growth of crop plants apparently by increasing metabolic activity in meristems. In experiments with brassin-treated barley seedlings, leaf and stem growth was increased 23–27 percent over the control plants. The greatest effect of brassins is to accelerate the growth of relatively slow-growing plants in a population. The effect, therefore, in crop plants may be to reduce phenotypic variability, producing a uniform stand of plants.

Growth Inhibitors Dormancy of buds and seeds is an important plant response. Many physical and chemical factors interact to regulate such dormancy, but a plant hormone called **abscisic acid (ABA)** has been discovered to play a key role as a **growth inhibitor** (Fig. 17-17). Application of abscisic acid to rapidly growing shoots of trees causes them to stop growing, lose their leaves, and form winter buds protected by bud scales. These are the

same responses as occur naturally in winter dormancy, and it is believed that abscisic acid is a natural factor in the onset of such dormancy as well as in the control of seed dormancy and germination. Because many of the actions of abscisic acid are opposite to those produced by gibberellic acid, abscisic acid is considered to function in a hormonal feedback system as a gibberellin antagonist in many aspects of plant growth and development.

Recently it has been found that the downward growth of roots (positive geotropism) can be induced by abscisic acid. In response to the settling of starch grains on the lower sides of root cap cells in a horizontally extended root, abscisic acid moves to the lower side of the growth region of the root tip (Fig. 17-14). Here it inhibits cell elongation, whereas the cells in the upper side of the root tip elongate normally. The result is a downward growth of the root tip. Abscisic acid also is believed to be involved in stomatal opening and closing.

FIGURE 17-20
Chrysanthemums treated with a growth retardant, AMO-1618. Treated plant at left, untreated plant at right. (Photo courtesy U.S. Department of Agriculture)

Modes of Hormonal Action Although plant hormones are widespread and capable of producing dramatic effects, their mode of action is not well understood. In some of their actions, auxins, cytokinins, gibberellins, and abscisic acid are thought to act by derepressing genes or by activating mRNA translation and the synthesis of specific enzymes.

Stimulation of aleurone cells by gibberellic acid results in the production and secretion of amylase; this procedure is thought to involve derepression of genes that direct the synthesis of a group of enzymes, including alpha amylase. IAA may operate similarly in regulating cell enlargement. It has been suggested that IAA together with IAA-receptor proteins may act to derepress a gene directing the synthesis of the cellulose-digesting enzyme **cellulase**. This enzyme, diffusing out of the cell into the cell wall, may make the cell wall more plastic by partially digesting cellulose fibers of the wall. Then osmotic pressure could produce cell elongation. In addition, auxins and possibly other phytohormones may produce some of their effects by altering the permeability of cell membranes or by affecting the movement of substances by active transport through the plasma membrane.

Growth Retardants A number of synthetic molecules have been found to retard plant growth without otherwise affecting the health of plants. A widely used example is AMO-1618 (Fig. 17-20). Such substances are not, however, considered to be hormones because they do not seem to have any naturally occurring equivalent in plants. Their general action is to suppress the elongation of internodes, and they often are used in the production of shortstemmed flowering plants such as chrysanthemums and poinsettias.

Photoperiodism

In 1920 W. W. Garner and H. A. Allard, working in the U.S. Department of Agriculture, discovered that a robust variety of tobacco called Maryland mammoth, which differed from other varieties in its time of flowering, could be induced to flower by controlling the day length to which it was exposed. Most tobacco varieties require long summer days for flowering, but Maryland mammoth bloomed in winter when days were short. Now it is known that many plants require specific day lengths, or **photoperiods,** in order to flower. Some, such as Maryland mammoth and poinsettias, are **short-day plants** (Fig. 17-21); others, including many tobacco varieties and other summer-blooming flowers, are **long-day plants.** A few plants do not have photoperiod require-

ments and are called **day-neutral plants.** Some others, including sugarcane, fail to flower when photoperiods are too long or too short; these are termed **intermediate-day plants.**

Study of the photoperiodic response has shown that the length of the dark part of the photoperiod is of considerable importance in inducing or preventing flowering. This is particularly true in short-day plants, which require a long, uninterrupted period of darkness for flowering. Even though maintained under short-day conditions, a short-day plant will not flower if the night is interrupted by even a few moments of light. Long-day plants, on the other hand, have no requirement for an uninterrupted period of darkness and will flower even under a continuous light.

Phytochrome

How does a plant sense light and measure its duration? Research, much of it done in the U.S. Department of Agriculture by a chemist, Sterling Hendricks, and a botanist, Harry Borthwick, has shown that a sensitive blue pigment is active in many photoperiodic responses. This pigment, called **phytochrome,** exists in two forms in the plant. When exposed to daylight or to the **red** component of light (wavelength 660 nm), phytochrome changes into a light-green form capable of absorbing only a narrow band of infrared light called **far red** (wavelength 730 nm). This far-red-absorbing form of phytochrome is symbolized P_{fr}. When P_{fr} is kept in darkness, it reverts slowly

FIGURE 17-21
Effect of photoperiod on the flowering of poinsettia, a short-day plant. The plant on the left has flowered and was maintained on a short-day cycle of less than 11 hours of light per 24-hour day; the plant on the right, which has not flowered, was grown under long-day conditions (more than 11 hours of light per 24-hour day). (Photo courtesy Jane K. Glaser)

to the red-absorbing form, P_r (blue green in color). In a short-day plant, when the dark period is long enough for the P_{fr} to revert to the P_r form (usually more than 12 hours), the plant will begin to develop flowers. If, however, the night is interrupted even by a brief burst of light, the phytochrome that has been converted to P_r will revert instantly to P_{fr}. Because the remaining dark period will not be long enough for the complete conversion of P_{fr} to P_r, the plant will not flower (Fig. 17-22).

While P_{fr} normally reverts to P_r during darkness, the conversion can be accelerated by exposing the pigment to far-red light. A short-day plant will flower even if the dark period is interrupted by a flash of light, if that is followed by exposure for a few minutes to far-red light. Experiments of this kind show once again that short-day plants require phytochrome to be in the P_r form in order for flowering to be initiated. Long-day plants, on the other hand, will not flower if the phytochrome is kept in the P_r condition by darkness or by far-red light, but require the conversion of P_r to P_{fr} as normally occurs during long days and short nights.

Experiments have shown that the leaves are the site of the plant's phytochrome and that a plant will flower under the appropriate photoperiod even if all but one of the leaves is removed, but not if all its leaves are detached. However, phytochrome by itself is not capable of initiating flowering but acts on other plant molecules, which in turn induce flowering. The nature of these molecules has not been determined, but a hypothetical hormone, called **florigen,** has been postulated as the product of phytochrome action. Although florigen has not been isolated or characterized chemically, grafting exper-

iments suggest its presence. A short-day plant kept on a short-day photoperiod may be grafted to a plant of the same species which is shielded in such a way that it can be kept in continuous light. The plant kept in continuous light ordinarily will not develop flowers, but when grafted with the plant exposed to a short-day photoperiod, will produce flowers. This suggests that a chemical substance or substances induced by phytochrome can diffuse through the graft union and induce flowering in another plant which ordinarily would not produce flowers.

Other light-sensitive processes in the plant also seem to be controlled by phytochrome. An interesting example is Grand Rapids lettuce seed, which requires exposure to light to germinate. This light requirement may be satisfied by using red light, and the action may be reversed by far-red light. Thus it would be suspected that the response is mediated by phytochrome, and in fact, this is the case. Grand Rapids lettuce seeds are very sensitive to red and far-red light. The seeds may be "turned on and off" many times by alternate exposures to red and far-red light, and they either will or will not germinate, depending on the wavelength to which they were last exposed (Fig. 17-23).

The time of flowering is of immense ecological importance, for it must coincide with weather conditions if the plant is to succeed in producing flowers and viable seeds. Many of the plants of temperate zones, where the summer days are long and warm but the winter days are short and frigid, are long-day plants. Thus their time of flowering coincides with optimum conditions for plant growth, flowering, and seed production. The flowering of

FIGURE 17-22

Photoperiodism and flowering. At top a long-day plant (LDP) and a short-day plant (SDP) are grown under short-day (long-night) conditions—only the SDP flowers. At bottom, the conditions are reversed. Under long-day (short-night) conditions, only the LDP flowers. The form in which phytochrome occurs is indicated by the horizontal bands; PR = red-absorbing form; PFR = far-red-absorbing form; PFR/FR = incomplete conversion.

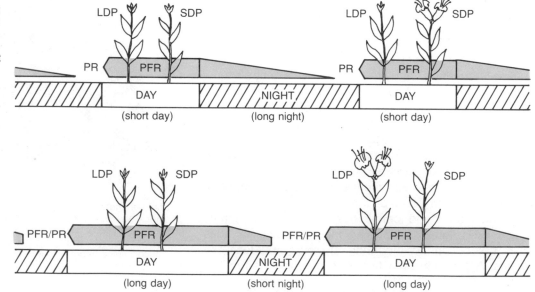

FIGURE 17-23
Light and dark responses of Grand Rapids lettuce seeds. Seeds in dish 1 were maintained in white light from a fluorescent tube. Those in dish 2 were kept in darkness. Seeds in dish 3 were kept in darkness except for 15 minutes of red light. Seeds in dish 4 were kept in darkness except for exposure to 15 minutes of red light immediately followed by 15 minutes of far-red light. The total length of the culture period in each case was 48 hours.

many important agricultural crops is similarly timed for maximal productivity during the growing season. The consequences of mistimed seed production by important grain crops, for example, would be food shortages and starvation for many human populations. Fortunately, flowering in many useful plants is very rigidly controlled by the phytochrome system and seldom goes wrong.

SUMMARY

The acquisition, transport, and utilization of nutrients is common to multicellular plants and multicellular animals. In plants the soil is the source of nutrients; the plant itself makes its food by photosynthesis. Soil is the principal source of water and essential minerals; the atmosphere is the source of oxygen. Soil water and its solutes enter the plant through the roots. Especially active in this uptake are root hairs, which are in intimate contact with the soil particles and the thin film of capillary water around those particles. Water and dissolved minerals pass into the interior of the root, mostly by absorption into the permeable cell walls. However, the endodermal cells have impermeable walls, and here the transport is by osmosis and active, energy-requiring transport (root pressure). Root pressure pushes water, dissolved molecules, and ions (sap) into the xylem and finally into the leaf interior. Here transpiration, the evaporation of water through the stomata, exerts a tension (transpiration pull) on water in the plant interior, and this contributes to water transport.

Foods, such as sugars, soluble protein, and so on, are transported in the phloem, most likely by a combination of cytoplasmic streaming and mass movement of water and solutes (mass flow) resulting from high osmotic pressures in the top of the plant and lower pressures below.

Coordination of plant functions appears to be controlled largely by interacting systems of plant hormones. Phenomena such as diffusion and osmosis, often coupled with growth by cell division, are regulated by hormones and produce plant movements. In tropisms, for example, cells elongate more rapidly in one region of a stem or root and cause bending. External stimuli such as light and gravity bring about differential distribution of growth-accelerating auxins, which are responsible for initiation of differential growth. Among major plant hormones are auxins (indoleacetic acid is the major natural auxin), cytokinins, gibberellins, and inhibitors such as abscisic acid.

Flowering in many plants is controlled by photoperiod through light influence on the structure of the blue pigment phytochrome. Changes in phytochrome are believed to regulate flowering through a hypothetical flowering hormone called florigen. The use of natural and synthetic plant hormones and of differing photoperiods has made it possible to control plant growth in new ways and has contributed to increased agricultural productivity.

major nutrient	transpiration	ethylene
minor (trace) element	transpiration pull	2,4-D
humus	stoma	gibberellin
gravel	photosynthesis	cytokinin
sand	food transport	kinetin
silt	mass flow	brassin
clay	turgor	abscisic acid
capillary water	turgor movement	photoperiod
ion exchange	growth movement	long-day plant
active transport	tropism	short-day plant
fertilizer element	nutation	day-neutral plant
root	nasty	far-red light
root hair	phytohormone	red light
Casperian strip	coleoptile	phytochrome
root pressure	auxin	florigen
xylem sap	indoleacetic acid	

QUESTIONS FOR REVIEW AND DISCUSSION

1 How does a plant absorb fertilizer elements and other nutrients from the soil?

2 How might you prove (or disprove) that transpiration pull effectively moves water upward in the stem of a plant?

3 Trace the movement of water molecules from the soil, into the roots, up the stem, into the leaf, and out of the stomata.

4 How has the rate of food transport in the phloem been demonstrated?

5 How do plants sense gravity, and what responses do they display with respect to it?

6 Describe a typically phototropic response in terms of light sensing, plant movements, and hormonal regulation.

7 Describe the role of auxin in the phototropic response of the coleoptile. How was this discovered?

8 Describe the action of gibberellic acid in the germination of barley grains. Of what possible economic significance is this action?

9 How do plants sense photoperiod? How do they differ in their photoperiodisms? How was photoperiodism in plants discovered? What is its ecological significance?

10 Why does a flash of light during the night often prevent the flowering of short-day plants?

SUGGESTED READING

BIDDULPH, S., and O. BIDDULPH. 1959. The circulatory system of plants. *Sci. Amer.* 200(2):44–49. (A nicely illustrated introduction to the subject.)

BIDWELL, R. G. S. 1979. *Plant physiology*, 2nd ed. New York: Macmillan Co. (The subjects of soils and water and solute transport are treated comprehensively in this excellent textbook.)

COHEN, I. B. 1976. Stephen Hales. *Sci. Amer.* 224(5):98–107. (Interesting story of the first scientist to measure human blood pressure and to show circulation of sap in plants.)

EPSTEIN, E. 1973. Roots. *Sci. Amer.* 228(5):48–59. (Review of how plants procure and transport nutrients through their roots.)

LAETSCH, W. M., and R. E. CLELAND, eds. 1967. *Papers on plant growth and development.* Boston: Little, Brown and Co. (Contains many milestones in the study of plant regulation, those of Went, Garner, and Allard included.)

LEOPOLD, A. C., and P. E. KRIEDEMANN. 1975. *Plant growth and development,* 2nd ed. New York: McGraw-Hill Book Co. (A comprehensive treatment of plant activities and their control.)

MULLER, W. H. 1979. *Botany: a functional approach,* 4th ed. New York: Macmillan Co. (Good chapter on regulation of plant development.)

STEWARD, F. C. 1964. *Plants at work.* Reading, Mass.: Addison-Wesley Publishing Co. (A modern classic.)

WAREING, P. F., and I. D. J. PHILLIPS. 1970. *The control of growth and differentiation in plants.* New York: Pergamon Press. (Emphasizes physiology of flowering processes.)

ZIMMERMAN, M. H. 1963. How sap moves in trees. *Sci. Amer.* 208(3):132–42. (How does sap get to the tops of the tallest trees?)

III

The Diversity of Life

18 Biological Diversity

I n 1977 the snail darter, a small fish thought to exist only in a 17-mile stretch of one of the tributaries of the Tennessee River, halted the multimillion-dollar Tellico Dam project. The decision to stop the dam construction was immediately controversial, in part because the darter's commercial value is nil, but mostly because hundreds of jobs and millions of dollars in contracts were at stake. The decision therefore was appealed, eventually reaching the Supreme Court of the United States, which decided in favor of preservation of the fish. (Congress subsequently acted to restore the project pending relocation of the fish population. Last word is that the fish is doing well in another tributary, and other populations have been found elsewhere.) Although this particular story seems to have a "happy" ending, the question remains: What values are to be placed on preservation of plant and animal species when such existence conflicts with human progress?

Approximately 1.5 million species of living organisms have received scientific names, but for the most part just brief descriptions and localities have been recorded. Only a handful of the world's organisms have been studied extensively or are really well known scientifically; even fewer have been investigated in great detail and those principally because of their direct impact on human life. Among plants, maize (corn) probably has been studied most extensively; among animals, fruitflies, laboratory rats, and humans have been and continue to be objects of intensive study. *Escherichia coli*, a bacterium, is the best known unicellular organism.

It is thought that several times as many unknown species exist as have been given Latin names. One estimate is that there are about eight million more species in the world today than have been described and named scientifically. Most of these unknowns probably live in tropical rainforests, which are particularly rich biological environments. Unfortunately, rainforests also are highly vulnerable to human exploitation; in cutover areas, the exposed soil is rapidly washed away by the frequent heavy rainfalls. In many of these regions, human populations are expanding at a dramatic rate, requiring more and more wood, fuel, and cropland. Moreover, the plants and animals are so diversified that eventual restoration of the original forest may never be possible. The Food and Agriculture Organization (FAO) of the United Nations estimates that about 40 percent of the tropical forests of the world have been destroyed during the past 150 years. Some forecasts limit survival of tropical rainforests to 20 years. Whether 20 years or 200 years, the chances are great that species of plants and animals will be destroyed before their existence has been recognized by science. The irretrievable loss of organisms, some perhaps of potentially great scientific and commercial value, must be a frequent occurrence.

Calaway Dodson, an eminent tropical botanist, has described the near loss of a valuable tree species, the Rio Palenque mahogany from Ecuador. About 10 years ago a tract of 170 acres of tropical forest near the Palenque River was set aside as a biological station. One kind of large tree on the station property and in nearby forests furnished a particularly durable wood, resistant to rot and termite attack, for local use. However, it was not until 1973 that the tree was recognized as a unique species and given a Latin name. Subsequently, forests surrounding the station were destroyed to make room for agriculture, with the result that the only known surviving specimens of the Rio Palenque mahogany are in the 170 acres of the station forest. Fortunately, the tree is readily propagated and its commercial potential will eventually be realized. Yet, had it not been for the chance establishment of a biological station, the species would have been lost without ever having been known to science and commerce.

It is difficult to establish criteria for preservation of species when they appear as obstacles to human progress. This is especially true if the species in question is a pest. No doubt fruit shippers would have been delighted had they been able to exterminate the fruitfly *Drosophila melanogaster*; yet it has been the source of much of our knowledge of the relationships between inheritance of characteristics and chromosomes. Suppose that the blue mold *Penicillium notatum* could have been eliminated forever as a destroyer of fruit and other foods. There would have been a sigh of relief that another human enemy had been defeated, and penicillin, the greatest drug in history, would never have been discovered. Much the same statement can be made of the pink bread mold *Neurospora crassa*, which has played a role in significant genetic studies and in the winning of Nobel prizes. How many drosophilas, penicilliums, and neurosporas remain to be discovered? How many have been exterminated before they ever became known?

BIOLOGICAL VARIETY

Life has many attributes that distinguish it from nonlife, some of them readily apparent in actions such as running, swimming, flying, song, and speech. Others are subtle, for example, the slow growth of a pine tree and the evolution of species. Perhaps the most dramatic characteristic of all is the overwhelming diversity of life. Primitive humans recognized a few hundred kinds of life;

ancient Greeks described a few thousand kinds of organisms; now organisms are numbered at more than a million. Yet, underlying life's diversity are unifying principles that enable organisms to be studied from a rational point of view.

Two centuries ago plants and animals were thought to have originated on earth in their contemporary form, each a distinct and immutable species. Individual differences within a familiar species were looked on as variations from the norm, not as evidence of long-term trends away from the original types. Nevertheless, even though species were not thought to be interrelated through evolution, it was found that they could be arranged in groups whose members exhibited common characteristics and that these groups could be arranged in larger groupings such as families and classes. Toward the end of the eighteenth century a few thoughtful individuals dared to suggest that species had not remained fixed in all their characteristics since creation, but that the similarities and dissimilarities among species that made classification possible were the result of evolution.

THE SPECIES CONCEPT

One of the more elusive concepts in biology is that of the **species**; an exact meaning of the term is difficult to pin down. Although some species such as lions and tigers are readily differentiated, others do not seem so clearcut. For example, not long ago an unknowing individual attempted to collect bounty on two beautiful Norwegian elkhounds he had shot under the assumption they were coyotes. Many persons, of course, would have no trouble distinguishing between dogs and coyotes, but to the uninitiated, species distinction often are not apparent. What, then, are major criteria used in defining and naming species?

Structural characteristics are important in defining species, but in the final analysis it is reproductive behavior that sets species apart. A species may be defined as a group of similar individuals which under natural conditions is separated genetically and reproductively from other groups. Dogs and coyotes, though similar in appearance, are separate species because they normally do not interbreed. Although from time to time crosses occur between related species such as dogs and coyotes, or lions and tigers, such matings are the exception rather than the rule and normally do not occur in nature. Moreover, in nature, should chance interspecific (interspecies) mating occur, survival of the offspring is unlikely. The greater the evolutionary distance between species, the less the likelihood of interbreeding and survival.

MAJOR GROUPS OF ORGANISMS

Many unifying characteristics lead biologists to believe in the interrelationships of all life on earth. The cellular nature of life suggests that life had a common origin, because there are only two basic cell types among organisms: **prokaryotic cells** and **eukaryotic cells**. Prokaryotes, which include bacteria, cyanobacteria (formerly blue-green algae), and a newly discovered green cell named *Prochloron*, are typically small cells of simple structure; their DNA is "naked," that is, not contained within chromosomes in a nucleus. Cell reproduction occurs by a simple splitting process, and sexual reproduction is unknown. Many prokaryotes are nonlocomotory, but if motion occurs it consists of gliding produced by minute movements of the cell boundary or of swimming by means of simple whiplike flagella composed of single fibers.

The manner in which the more complex eukaryotic cells evolved from prokaryotes is not well understood (refer to Chapter 2). Eukaryotic cells have their DNA incorporated into chromosomes, which in turn are contained in a nucleus. In addition, they possess a number of organelles that perform specific functions; in prokaryotes, such functions are not compartmentalized. Among the organelles of eukaryotes are mitochondria, which carry on biological oxidation; chloroplasts, in which photosynthesis occurs; and other membranous structures such as endoplasmic reticulum and Golgi bodies, both of which have transport and secretory functions. Although some motile eukaryotic cells move by creeping about on jellylike extrusions (amoeboid locomotion), most motile unicells possess whiplike flagella composed, in almost all cases, of 11 parallel fibers (the 9 + 2 pattern).

At the molecular level, evidence of unity of life and a common origin is abundant. All cells have the same kind of plasma membrane composed of two lipid layers. All cells store genetic information in DNA, and all transcribe the genetic code into messenger RNA, which is used by ribosomes to make various kinds of proteins. All cells use the same general kind of amino acids, sometimes called "**left-handed**" (**L**) amino acids, in making their proteins, and many other chemical similarities such as identical cytochromes exist, attesting to the belief that the different forms of life recognized today did not originate coincidentally but had a common origin.

During the more than three billion years since life's origin on earth, lifeforms have become more and more complex, but it would be a mistake to suppose that more complex forms completely superseded less highly evolved organisms. Simple organisms representing primitive stages in evolution have persisted and in some instances appear little changed from their remote ancestors. A curious example is a strange little fossil cell, *Kakabekia umbellata*, which looks like a miniature umbrella and occurs in Canadian rocks approximately 1.9 billion years old (see Fig. 18-1). At the time *Kakabekia* was dis-

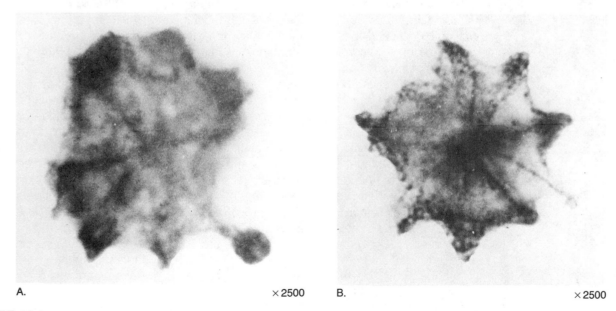

A.   ×2500   B.   ×2500

FIGURE 18-1
Kakabekia umbellata, an umbrella-shaped, fossilized microorganism from the 2-billion-year-old Gunflint Formation of Canada. The specimen at the left shows the bulbous "handle"; that at the right shows the "ribs" of the "umbrella." (Photos from E. S. Barghoorn and S. A. Tyler, 1965, *Science* 147:563–77. Copyright 1965 by the American Association for the Advancement of Science)

covered, no comparable living cells were known. Not long ago, however, living *Kakabekia* cells were discovered in ammonia-rich soil from Harlech Castle in the village of Harlech, Wales. *Kakabekia* has an absolute requirement for ammonia, which is supplied in the Harlech Castle environment by urine from livestock and humans. Although high ammonia concentrations are unusual environmental constituents, *Kakabekia* presumably evolved in ammonia-rich habitats and presently survives only where ammonia concentrations are high. This requirement is of great theoretical interest, because it is thought that the earth's primitive atmosphere was rich in ammonia gas.

Other examples of "living fossils" will be encountered in future chapters. They not only contribute to life's diversity but also are milestones along evolutionary routes.

Kingdoms of Life

When all organisms, living and fossil, are considered from the standpoint of their major characteristics and evolutionary relationships, it becomes possible to subdivide them into several large groups, or **kingdoms**. Biologists have different opinions regarding the number of kingdoms and the organisms to be included in each, but in most cases organisms with similar structural and functional characteristics are grouped together. Major characteristics used in constructing life kingdoms are cell structure, complexity of organization, mode of nutrition, and whether or not locomotion occurs.

Most older classifications divided organisms among two kingdoms and lumped prokaryotes and eukaryotes together. The **plant kingdom** included both photosynthetic organisms (algae and land plants) and

ARCHAEBACTERIA—A NEW KINGDOM

The newest group of organisms to be discovered may actually be the oldest. Called the **archaebacteria** and placed in a new kingdom, neither eukaryotic nor prokaryotic, the archaebacteria seem to their discoverers, Carl R. Woese and colleagues at the University of Illinois, to be particularly well suited to conditions that prevailed in the early evolution of life on earth. They are anaerobic and can live in atmospheres rich in carbon dioxide and methane but poor in oxygen.

The distinctions between prokaryotes and eukaryotes are well defined and clear, including presence or absence of a nucleus, chloroplasts, and mitochondria. These distinctions have led to speculation regarding origins of eukaryotes from prokaryotes and to the theory that eukaryotes of today arose from a simpler cell which, by incorporating bacteria and cyanobacteria (formerly blue-green algae) that became transformed into mitochondria and chloroplasts, developed capacities for aerobic respiration and photosynthesis (see Fig. 2-20). Although this hypothesis is an attractive one and explains a number of rather puzzling questions, such as why chloroplast and mitochondrial DNAs are more like DNAs of bacteria than DNAs found in chromosomes of eukaryotic cell nuclei, the identity of the original eukaryote has always been something of a problem. Presumably, eukaryotic ancestors evolved from prokaryotes or prokaryotelike organisms, but which ones?

When only superficial characteristics of bacteria are examined, it is difficult to discern evolutionary relationships among them. The common cell shapes—rods, spheres, and spirals—have arisen many times and do not provide much

organisms that obtained energy by food absorption (bacteria and fungi). A major diagnostic feature of organisms placed in this kingdom was the presence of enclosing cell walls, and the majority of members were nonmotile. To the **animal kingdom** were assigned motile organisms whose cells lacked walls and whose nutrition was **holozoic** (i.e., by ingestion of particulate food). Within each kingdom, organisms were further subdivided on the basis of structural complexity. For example, the simplest plants were the prokaryotic bacteria and cyanobacteria (also known as blue-green algae), followed by other algae and fungi. Next came simple land plants such as mosses, then ferns, gymnosperms (pines, spruces, etc.), and finally flowering plants. In the animal kingdom unicelled protozoans formed a base upon which was superimposed a succession of more and more complex forms, including sponges, worms, mollusks, insects and other arthropods, and finally vertebrates such as fish, amphibians, reptiles, birds, and mammals.

Although two-kingdom classifications have the advantage of at least superficial simplicity, they lump together some rather unlikely bedfellows: prokaryotes with eukaryotes, for instance, and nonphotosynthetic fungi with photosynthetic green plants. The modern trend in classification is to break up the old two-kingdom plan into several kingdoms. In such plans the subdivisions of the older classifications are not altered; rather, they are redistributed among additional kingdoms. This book assumes the five-kingdom arrangement proposed by R. H. Whittaker of Cornell University (Fig. 18-2). Whittaker uses the following criteria in characterizing kingdoms: complexity of cell structure, unicellularity versus multicellularity, and mode of nutrition. The kingdoms and their major characteristics are as follows:

information about the direction of evolution or about which forms are the most primitive. Study of genes, however, can be informative, because mutations have altered to greater or lesser degrees the original nucleotide sequences in different lineages of organisms. Ribosomal RNAs from almost 200 species of bacteria and eukaryotes have been analyzed, and their nucleotide sequences show, as might be expected, that eukaryotes are quite distinct from prokaryotes. Among prokaryotes, however, which seem to fall largely into one group, the eubacteria ("true bacteria"), one variant group was discovered. The members of this group are now called archaebacteria because they are thought to represent an evolutionary prokaryotic offshoot possibly predating the origin of eubacteria. Although having no nucleus and otherwise resembling eubacteria, at the molecular level archaebacteria are no more like eubacteria than they are like eukaryotes, and therefore the idea that archaebacteria constitute a separate kingdom of microorganisms, coequal with monerans and protistans, is now becoming accepted.

Because scientists now recognize two ancient lineages of microorganisms, the archaebacteria and the eubacteria, a better case can be made for an even more remote and simpler ancestral cell, which early in earth's history gave rise to these two divergent groups of prokaryotes. What about the origin of the eukaryotic cell? Because at the molecular level eukaryotes share some charcteristics of both eubacteria and archaebacteria, the possibility exists that the ancestral eukaryote arose by an earlier symbiosis between cells of the two groups. For example, the eukaryotic nucleus appears to contain genes of three kinds: some presumed to be of eubacterial origin, some apparently derived from archaebacteria, and others from some unidentified, third microorganism. A later symbiosis, then, would have involved incorporation of blue-green algal and bacterial cells to originate the chloroplasts and mitochondria of present-day eukaryotes.

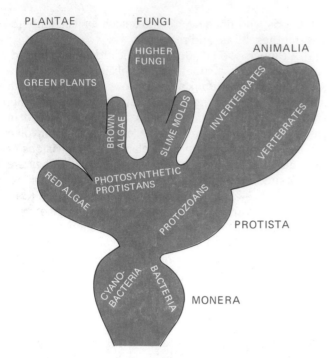

FIGURE 18-2
Broad relationships of the five kingdoms of life. (Based on R. H. Whittaker, 1969, *Science* 163:150–160. Copyright 1969 by the American Association for the Advancement of Science)

Monera Included are all prokaryotes (bacteria, cyanobacteria, and the recently discovered green organism *Prochloron*). All are unicellular organisms of simple cellular organization (i.e., walled cells; "naked" DNA; no mitochondria, chloroplasts, or other organelles). Their nutrition is by absorption of dissolved food or, if food is manufactured, by photosynthesis or **chemosynthesis**. The latter process uses energy derived from chemical reactions rather than sunlight.

Protista Protista includes a majority of unicellular eukaryotes (i.e., cells with nuclei, mitochondria, endoplasmic reticulum, plastids, Golgi bodies, 9 + 2 flagella) (Color plates 1A; 5A–C). Many protistans are holozoic (among them such familiar examples as amoebas and paramecia), but certain groups are photosynthetic.

Fungi Fungi are eukaryotic plantlike organisms. Nearly all are multicellular but of simple form. Many are microscopic (yeasts and certain molds), but large fungi (mushrooms, brackets, puffballs, etc.) also are common (Color plates 1B, C; 5D; 6A, C). Their mode of nutrition is absorption of foods made soluble by decay processes.

Plantae Plantae includes a majority of multicellular photosynthetic eukaryotes. (Color plates 7–9). Although

the range of types is not as great as in the animal kingdom, considerable diversity exists. Main subgroups are algae, mosses, ferns, and seed plants. (Some biologists prefer to include the algae with the protista. We do not choose to do so because of their general multicellular construction and because certain algae are clearly plant-like. Refer to Bold et al. in readings for Chapter 19.)

Animalia Members of Animalia are multicellular holozoic eukaryotes (color plates 10–12). The principal mode of life is pursuit and ingestion of food. Animals exceed all other kingdoms in numbers of species (an estimated 1.5 million, most of which are insects). Among them are such diverse forms as sponges, jellyfish, flatworms, roundworms, mollusks, insects, starfish, and vertebrates.

Principles of Classification

The broad principles of classifying organisms are illustrated by the characteristics of the five kingdoms. To a considerable degree cell structure and organization as well as mode of nutrition are also used in arranging subgroupings within the kingdoms. In the protistans, for example, presence or absence of flagella and type of nutrition differentiate major subgroups. In the plant kingdom, cell structure and pigmentation differentiate several major groups, and among animals as well as plants, increasing complexities of organ systems are important criteria. In addition, modes of reproduction are of great diagnostic importance in classification within kingdoms. External versus internal fertilization and **ovipary** ("egg birth") versus **vivipary** ("live birth") distinguish major categories of animals; in plants, presence or absence of seeds separates primitive and advanced groups.

History of Classification The naming of organisms is an essential aspect of human survival and has been so since before humans kept written records. At first, word of mouth transmitted information from generation to generation regarding the uses and perils of various plants and animals. With the invention of writing, books about plants (herbals) and animals (bestiaries) began to appear (see Fig. 18-3), listing both real and fictitious examples together with brief descriptions. A majority of these were rather rough attempts to group similar organisms with one another: oaks with pines, birds with bats, whales with fish. By the eighteenth century classification had become a more sophisticated science.

The great classifier of the eighteenth century was Carolus Linnaeus (1707–1778), who devised systems of classifying organisms according to structural features and gave organisms Latin names. Linnaeus sometimes

FIGURE 18-3
Lily plant from a fifteenth-century herbal.

is credited with the invention of the scientific **binomial**, that is, the latinized **genus** (pl. **genera**) and **species** name for each kind of organism. Thus the Virginia deer became *Odocoileus virginiana*, the red oak *Quercus rubra*, and the house cat *Felis catus*. The binomial concept was used before Linnaeus and continues to be employed unwittingly when someone refers to a "white-tailed deer" as distinct from a "mule deer" or to "red," "white," or "black oak." Because people in every region of the world have given plants and animals different common names, spoken in hundreds of languages and thousands of dialects, the necessity for a common scientific nomenclature is obvious. Moreover, many organisms such as fossils and microorganisms are known only by their scientific names. In other cases scientific and common names are the same: *Gorilla, Yucca, Dahlia, Lemur, Chrysanthemum, Eucalyptus, Hippopotamus, Loris,* and so forth.

Classification Guidelines   Any logical set of criteria may be used in classification. For example, in classifying plants all red flowers might be lumped into one **taxon** (pl. **taxa**, a general term for a unit of classification), all yellow flowers in another, and so on. Systems of this kind are said to be "artificial," in that they disregard evolutionary lines of descent. Roses and tulips, for instance, are not closely related but could be classed together if flower color were the sole characteristic considered. Pre-Darwinian classifications tended to be rather arbitrary,

because they often were based on artifical characteristics. After Darwin, classifications tended to be based on evolutionary criteria and hence are "natural" rather than artificial.

Systematists, as biologists who study classification are called, use a variety of principles in determining position of an organism in a natural system of classification. Essentially, classification is based on studies of similarities and differences in organisms. Many characteristics are examined, including **morphology** (form as a whole), **anatomy** (internal structure), **cytology** (cell structure), **ontogeny** (development of the individual), **reproduction**, and **biochemistry**.

A keystone in classification is the concept of **homology**. Homology is simply an expression of the evolutionary relationship of comparable structures in different organisms or, in some cases, different structures of one organism. Familiar examples of homology are the human arm, the foreleg of a horse, the wing of a bat, and the flipper of a porpoise. All are modified forelimbs, although each serves a distinctly different function. In plants the petals, stamens, and pistils of flowers are thought to be homologous with leaves, as are some types of thorns and tendrils.

To be considered homologous, structures must have both structural and developmental similarities. Structures with functional similarity, on the other hand, are not necessarily homologous but may instead be products of evolutionary parallelism. For example, wings of insects, birds, and bats are functionally similar but dissimilar in terms of homology. Insect wings develop from an exoskeleton (external skeleton), not found in vertebrates at all. Wings of birds and bats, although homologous as forelimbs, are not homologous in terms of wing architecture; rather, they are products of parallel evolution and developed quite differently (greatly elongated, skin-covered fingers in bats; feather-covered forelimbs with vestigial fingers in birds). Nonhomologous but functionally similar structures are said to be **analogous**. In plants analogous structures include flattened green stems that function as leaves; spines and thorns that are modified leaves and branches, respectively; underground storage organs that are modified stems (Irish potato) or roots (sweet potato); and many others (See Fig. 18-4 for examples of homologous and analogous structures.)

In addition to structural homologies, modern biology relies heavily on physiological and chemical characteristics in charting evolutionary pathways. For example, chromosomal analysis may involve special staining methods which cause certain banded regions of comparable chromosomes to fluoresce under ultraviolet light.

Similarities and differences in chromosome banding have been used to calculate evolutionary distances between species, genera, and groups of genera, called **families**. Comparisons between chemical constituents of plants and animals also give useful clues to their evolutionary position. Minor differences in classes of enzymes are used to differentiate closely related species. DNA comparisons have indicated the evolutionary positions of a variety of organisms. Blood typing reveals similarities and dissimilarities among human races. A group of pigments called **flavonoids** give similar information about plant interrelationships. Modern **taxonomy** (study of

classification) relies on many sophisticated techniques which were unknown to biologists a generation ago.

Higher Taxa

Organisms in different kingdoms are classified by various groups of specialists according to rules laid down by international associations. Botanists, for example, depend upon the International Botanical Congress meeting every four years to consider classification of plants and fungi under the International Code of Botanical Nomenclature. Similarly, rules for naming and classifying ani-

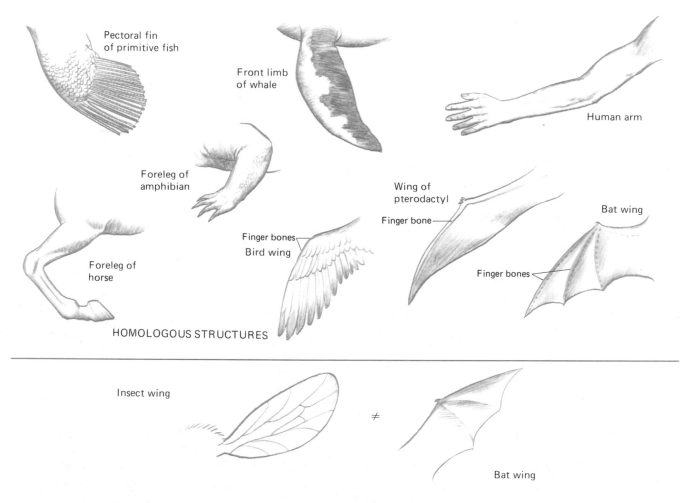

FIGURE 18-4
Homology and analogy. The upper group of figures features appendages that represent modifications of the vertebrate forelimb. They are homologous as forelimbs. The lower group features analogous but non-homologous appendages. Insect wings are not modified forelimbs. Bat and bird wings are homologous as forelimbs but not homologous in terms of modifications of bones and integuments for flight (i.e., finger bones and skin form bat wings; arm bones, hand bones, and feathers form bird wings).

mals and protistans are established when the world's zoologists meet at intervals in international zoological congresses. Even though rules of classification may at times seem arbitrary and even absurd, they must be followed, because the alternative is chaos.

In classification, the major groups are kingdoms. Each kingdom is subdivided in descending order into **phyla** (animals) or **divisions** (plants), and sometimes **subphyla** and **subdivisions**, then into **classes, orders,** families, genera, and species. For example, the oak tree is classified as follows:

Kingdom Plantae
    Division Magnoliophyta
        Class Magnoliopsida
            Order Fagales
                Family Fagaceae
                    Genus *Quercus*
                        Species *rubra*

and humans are:

Kingdom Animalia
    Phylum Chordata
        Subphylum Vertebrata
            Class Mammalia
                Order Primates
                  Family Hominidae
                    Genus *Homo*
                      Species *sapiens*

SUMMARY

In evolution theory as in biological classifications, the species is the basic unit of diversity, a species defined as a group of similar individuals sharing a common pool of genes. Members of a species are capable of interbreeding but normally do not interbreed with other species. The gene pool contains variant genes which are the result of mutations. In any given set of environmental circumstances, certain combinations of genes, including mutations, may be more favorable. Nature "selects" these combinations for survival.

Much information useful to classification comes from the fossil record. Based on past history and present diversity of life, classifications have been constructed in attempts to place organisms within an evolutionary framework. Other characteristics, including morphology, cytology, ontogeny, reproduction, and biochemistry, also are used. The concept of homology is often employed in determining taxonomic relationships. In its broadest aspect, a classification system distributes organisms among kingdoms. One such system is the five-kingdom classification that includes Monera (prokaryotes), Protista (unicellular eukaryotes), Fungi (multicellular plantlike eukaryotes), Plantae (multicellular photosynthetic eukaryotes), and Animalia (multicellular holozoic eukaryotes).

KEY WORDS

species concept	classification	ontogeny
prokaryotic cell	morphology	homology
eukaryotic cell	anatomy	analogy
kingdom of life	cytology	

QUESTIONS FOR REVIEW AND DISCUSSION

1   What are some reasons for the protection and preservation of endangered species?

2   What is the status of rainforest regions of the world? Can rainforests be easily restored? Explain.

3   Give a useful definition of the term "species."

4   State four characteristics of organisms that suggest a common evolutionary basis of life.

5   Name five kingdoms of life and name an organism representative of each kingdom.

6   List three examples of plant or animal structures that may be homologous, and explain their possible origins.

SUGGESTED READING

BATES, M. 1961. *Man and nature*. Englewood Cliffs, N.J.: Prentice-Hall. (An overview of biology with emphasis on humans in the environment.)

IRVINE, W. 1955. *Apes, angels, and Victorians*. New York: McGraw-Hill Book Co. (An interesting account of the development of Darwinian evolution.)

LACK, D. 1953. Darwin's finches. *Sci. Amer.* 188(4):66–72. (A nicely illustrated summary of adaptive radiation as exemplified by the Galápagos finches.)

SCIENTIFIC AMERICAN. 1978. Vol. 239, no. 1. (The entire issue is given over to discussion of various aspects of evolution.)

STEBBINS, G. L. 1966. *Processes of organic evolution*. Englewood Cliffs, N.J.: Prentice-Hall. (An uncomplicated and well-illustrated introduction to the processes of evolution.)

VIDAL, G. 1984. The oldest eukaryotic cells. *Sci. Amer.* 250(2):48–57. (Interesting electron micrographs of billion-plus-year-old cells.)

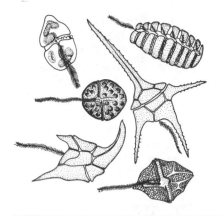

19

Monera, Protista, and Fungi

nvention of microscopes about 300 years ago led to the discovery of many hitherto unseen forms of life now recognized to be one-celled organisms. As microscopes improved, the components of larger cells became known, but until recently the structure of bacteria, which are the smallest cellular organisms, remained enigmatic. Then with the development of electron microscopes about 35 years ago, it was found that bacteria, together with closely related **cyanobacteria** (blue-green algae), differed from all other microorganisms, and for that matter from all other cells, in not containing nuclei. Because they lack nuclei and are simple cells, bacteria and cyanobacteria are referred to as prokaryotes and are classified in the kingdom **Monera**.

Many other unicellular organisms are placed in the kingdom **Protista**, an assemblage of eukaryotes encompassing a diversity of organisms. A third group of mixed unicellular and multicellular organisms often included in studies of microbiology (study of microscopic organisms) are the **Fungi**. Although some fungi are macroscopic (i.e., sufficiently large to be studied in some detail without a microscope), many are microscopic in size and therefore are considered to be true microorganisms.

MONERA

Monerans (*monos*, "single") include bacteria, cyanobacteria, and a recently discovered green prokaryote named *Prochloron*. Cyanobacteria and *Prochloron* are **photosynthetic** organisms and therefore **autotrophic** (self-feeding). A few members of the true bacteria (**Eubacteria**) also are autotrophic, deriving food by **chemosynthesis**, a process that derives energy for food manufacture by oxidizing certain mineral substances in the environment (ammonia, sulfur, etc.). Most of the bacteria, however, are dependent on external food supplies and are **heterotrophic** ("other-feeding"). Sexual reproduction is unknown among monerans, although gene transfer can occur in some bacteria by a process known as "**bacterial conjugation**" (refer to Chapter 5, pp. 111–14, and Fig. 5-15). Asexual reproduction occurs by a nonmitotic process called **fission**. Fission can take place either by a pinching-off process or by crosswall formation.

Bacteria

Bacteria, often associated in the public mind with "germs," **disease**, and death, are necessary in the biosphere. They are recycling agents of nature; by their digestive processes they keep nutrients in circulation and available for all forms of life. There are many examples of their ecological roles, among them the processing of nitrogen in the atmosphere and in organic and inorganic compounds.

Nitrogen-fixing bacteria change the nitrogen in air (N_2, or nitrogen gas), which is unusable in most organisms, into forms such as nitrates and ammonia, required by green plants for growth. Nitrogen-releasing bacteria break down animal excretions and dead plants and animals into simpler nitrogen compounds, making nitrogen available once again for other organisms to use. Bacteria also are active in many other biosynthetic processes that produce commercial and household substances such as yogurt, sour cream, butter, cheese, soy sauce, vinegar, antibiotics, and industrial solvents such as acetone and several kinds of alcohols.

In many of these actions, bacteria function by secreting enzymes into their immediate environment. Bacterial enzymes digest foods and other matter outside the cell, making it possible for the resultant nutrients to be absorbed.

Structure   Bacterial cells are relatively simple in structure. The principal forms are **rods** (**bacilli**; sing. **bacillus**), **spheres** (**cocci**; sing. **coccus**), **spirals** (**spirilla**; sing. **spirillum**), and **filaments** (**actinomycetes**) (Fig. 19-1). Some forms have delicate threadlike swimming structures called **flagella** which propel the bacteria through liquids (Fig. 19-2).

A typical bacterium of the bacillus type might measure about 1 μm in diameter by 4 μm in length. (A micrometer [μm] is one-millionth of a meter; a human hair is approximately 50 μm in diameter [see Fig. 2-7].)

Examination of bacteria by ordinary classroom microscopes shows very little internal structure, but in recent years highly accurate and precise cutting instruments have been developed to prepare sections of cells for examination by electron microscopy.

Some sectioned bacterial cells reveal the following structural details (see Fig. 19-2 for a diagram of a bacterial cell). Prokaryotic cells consist of a living part, the **protoplast**, enclosed within a nonliving **wall** of several layers, including an outermost, jellylike **sheath**. The outer part of the protoplast consists of a delicate plasma membrane, within which is a viscous cytoplasm containing numerous ribosomes. The latter give the cytoplasm a granular appearance when viewed with an electron microscope.

In the center of a bacterial cell is a tangled circular strand which unfolded is about 1 mm long. This strand, called a **genophore**, and sometimes a **bacterial "chromosome,"** is composed of DNA and is attached at one point to the plasma membrane. It is at this point of attachment that genophore replication takes place prior to cell reproduction. A structure called a mesosome is the site of genophore replication and the initiation of cell division.

Some kinds of bacteria produce internal bodies, or **endospores**, that function as survival cells. An endospore is a small cell, enclosed within a resistant wall,

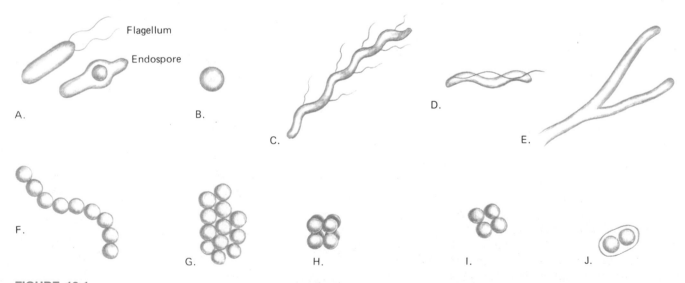

FIGURE 19-1
Some bacterial forms. **A.** *Bacillus*. **B.** Coccus. **C.** *Spirillum*. **D.** Spirochaete. **E.** Actinomycete. **F.** *Streptococcus*. **G.** *Micrococcus* (*Staphylococcus*). **H.** *Sarcina*. **I.** *Gaffkya*. **J.** *Diplococcus*. (*Note*: Latin names of specific organisms are always in italics)

PLATE 5

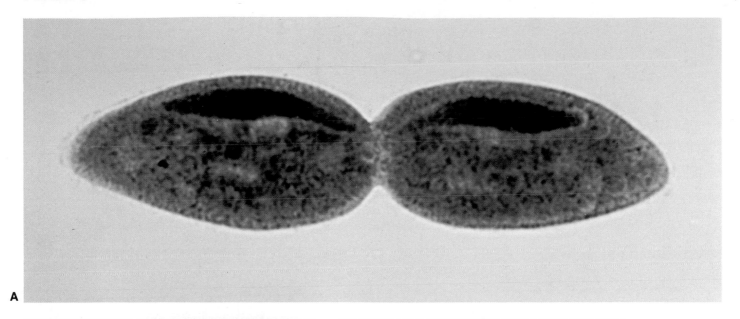

A

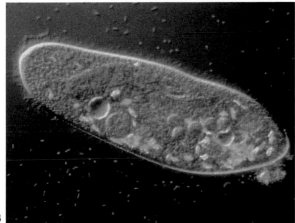

B

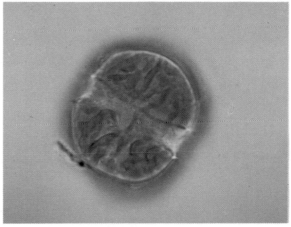

C

D

A. Asexual reproduction (binary fission) of a ciliated protozoan, *Paramecium caudatum.* Kingdom Protista, Phylum Protozoa, Class Ciliata. (Carolina Biological Supply Co.) **B.** *Paramecium caudatum.* Cilia are clearly visible as also are two contractile vacuoles. (Nova Scientific Corp.) **C.** Dinoflagellate, *Gonyaulax* sp., one of the red-tide organisms. Kingdom Protista, Phylum Pyrrhophyta. (J. Waaland/Biological Photo Service) **D.** Body (plasmodium) of a slime mold, *Physarum polycephalum.* Kingdom Fungi, Division Gymnomycota. (Carolina Biological Supply Co.)

PLATE 6

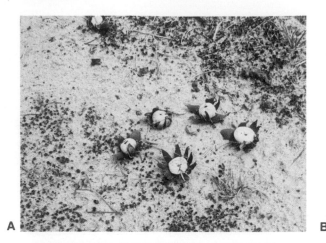

A. Earthstars, *Geaster* sp. Kingdom Fungi, Subdivision Basidiomycotina (club fungi). (Arthur Muller) **B.** Lichens. Lichens combine cells of prokaryotic cyanobacteria with those of fungi. Different body forms are seen here: flat foliose lichens, encrusted crustose lichens, and branched fruticose types. **C.** *Amanita pantherina,* a poisonous club fungus. Subdivision Basidiomycotina. (Turtox/Cambosco)

PLATE 7

A. Mixed green (Division Chlorophyta) and red (Division Rhodophyta) algae in the shallow water of Florida Bay. **B.** The sea palm, *Postelsia,* an intertidal brown alga. Division Phaeophyta. Vancouver Island, B.C. **C.** Two species of coenocytic green algae from same locale as (A): parasol-like *Acetabularia* and feather-form *Caulerpa.* **D.** A liverwort, *Conocephalum* sp. Division Bryophyta. (Nova Scientific Corp.) **E.** The hairy cap moss, *Polytrichum commune.* Thousands of individual leafy green gametophyte plants; the erect brown and gold structures are sporophyte plants of the same species. Division Bryophyta. **F.** Leafy green gametophytes of the bog moss, *Sphagnum* sp. Division Bryophyta. (Nova Scientific Corp.)

PLATE 8

A. An Australian tree fern, *Dicksonia antarctica.* Division Pteridophyta.
B. A club moss, *Lycopodium annotinum.* Division Lycopodiophyta. **C.** *Cycas circinalis,* a cycad growing in Castleton Gardens, Jamaica. Division Pinophyta. **D.** The maidenhair tree, *Ginkgo biloba,* a primitive seed plant (one of the gymnosperms) noted for having motile, flagellated sperm. Division Pinophyta. (Kenton E. Brooks) **E.** An ancient bristlecone pine (*Pinus anistata*) growing near the summit of Mt. Evans in Colorado. Division Pinophyta.

PLATE 9

A

B

C

D

A. Flowers of the cannonball tree, *Couroupita elegans,* produce two kinds of pollen, fertile and sterile, the latter as a bee attractant. Division Magnoliophyta. **B.** Seed dispersal by wind as seen in the milkweed. **C.** Pollination of flowers of a *Heliconia solomonensis* by a bat. Certain flowers of tropical plants are pollinated at night by nectar-feeding bats. (John Kress)
D. Emerald hummingbird feeding on (and probably pollinating) flowers of a bromeliad, Haita. (Walter Judd)

PLATE 10

A. Common tube sponge, a coral reef inhabitant. Phylum Porifera. (James A. Bohnsack) **B.** A common freshwater cnidarian, *Hydra* sp. Asexual reproduction by budding is illustrated; the lateral dome-shaped bulge is a sexual organ (gonad). (Carolina Biological Supply Co.) **C.** Polyps (feeding branches) of a soft Gorgonian coral. Phylum Cnidaria. (James A. Bohnsack) **D.** Sea anemone, *Metridium,* Phylum Cnidaria, on a hard coral (also Cnidaria). (Nova Scientific Corp.) **E.** Two marine flatworms grazing the surfaces of a colony of tunicates. Flatworms: Phylum Platyhelminthes; tunicates: Phylum Chordata (protochordates). (James A. Bohnsack)

PLATE 11

A. A marine nudibranch mollusk (Phylum Mollusca). (James A. Bohnsack) **B.** Squids, *Sepioteuthis sepioides,* Phylum Mollusca. (William Stephens/Tom Stack & Associates) **C.** Marine polychaetes, "fanworms." Phylum Annelida. (James A. Bohnsack) **D.** Hermit crab and its borrowed gastropod shell, exemplifying camouflage and escape behavior. (Nova Scientific Corp.) **E.** Planktonic water flea, *Daphnia.* Phylum Arthropoda. (Nova Scientific Corp.) **F.** Sea cucumber, *Cucumaria miniata,* an echinoderm. Phylum Echinodermata. (Carolina Biological Supply Co.)

PLATE 12

A. Male stickleback prodding female to induce her to lay eggs in the nest built by the male. (Dwight Kuhn) **B.** Male stickleback emitting sperm to fertilize eggs in nest. Note the red belly, coloration typical of breeding males. (Dwight Kuhn) **C.** Mimicry. Jordan's salamander, *Plethodon,* (right) is the model for *Desmognathus* (left), the mimic. Phylum Chordata, Subphylum Vertebrata, Class Amphibia. (Edmund D. Brodies, Jr./ Biological Photo Service) **D.** American alligator, *Alligator mississippiensis*. Subphylum Vertebrata, Class Reptilia. (James A. Bohnsack)

FIGURE 19-2

Composite of the structure of a bacterial cell based on electron-microscope studies.

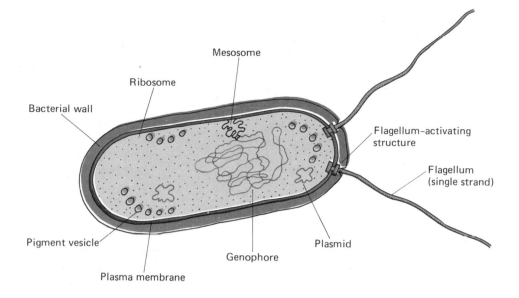

that is formed by the parent cell. When the parent cell dies, the endospore survives. Some endospores can withstand many hours of boiling temperature; their hardiness is a major factor in difficulties encountered in exterminating some bacteria. An anaerobic bacterium, *Clostridium botulinum*, the cause of an often-fatal form of food poisoning called **botulism**, produces endospores. Botulism, sometimes a serious problem in certain kinds of home-canned foods because the spores can survive boiling, can be avoided by careful food preparation. Boiling food for 15-20 minutes immediately before use destroys the botulism toxin.

Reproduction   Under ideal conditions (ample food, optimum temperature and moisture, etc.) bacterial cells may reproduce every 20 minutes (Fig. 19-3). Their reproduction is relatively simple, consisting of genophore duplication followed by transverse division of the entire cell. Genophore duplication is especially significant, because each new cell must necessarily receive a complete quota of genophore DNA.

Theoretically, descendants of a single bacterium, all reproducing at a maximum rate, would equal the mass of the earth in a short time. Such ideal conditions are not approached in nature, however; food supplies and space soon become exhausted, and poisonous byproducts of bacterial life processes accumulate in the surrounding environment and interfere with further growth and reproduction.

A parallel can be seen in the growth of bacterial populations and human populations (Fig. 19-4). At first there is a period of slow growth (**lag phase**), while organisms adjust to their environment. Then, if food and space are ample, reproduction is rapid, resulting in an

exponential growth phase during which the population doubles at each interval of time. Later, as food and space diminish, population growth slows and reaches a stationary **plateau phase**. Finally, because of food and

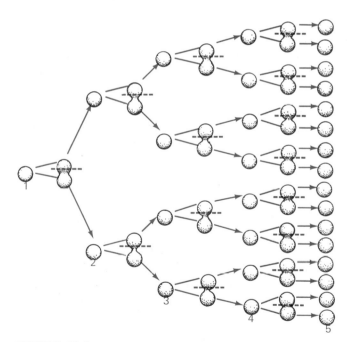

FIGURE 19-3

Bacterial reproduction showing maximum potential increase in numbers of cocci in five divisions (a doubling time of about 20 minutes). Without checks on such growth, the population would reach an astronomical number by the end of a day. (Redrawn from *Fundamentals of microbiology*, 9th ed., by Martin Frobisher et al. Copyright © 1974 by W. B. Saunders Company. Reprinted by permission of W. B. Saunders, CBS College Publishing)

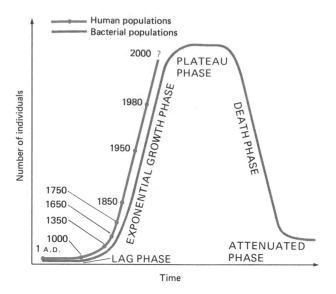

FIGURE 19-4
Population growth curves in bacteria and humans.

space shortages and accumulated wastes, the death rate exceeds the growth rate, and the population declines in an accelerated **death phase**, dwindling to a small number of organisms living in an impoverished environment (**attenuated phase**).

FIGURE 19-5
Cell structure of a cyano-bacterium.

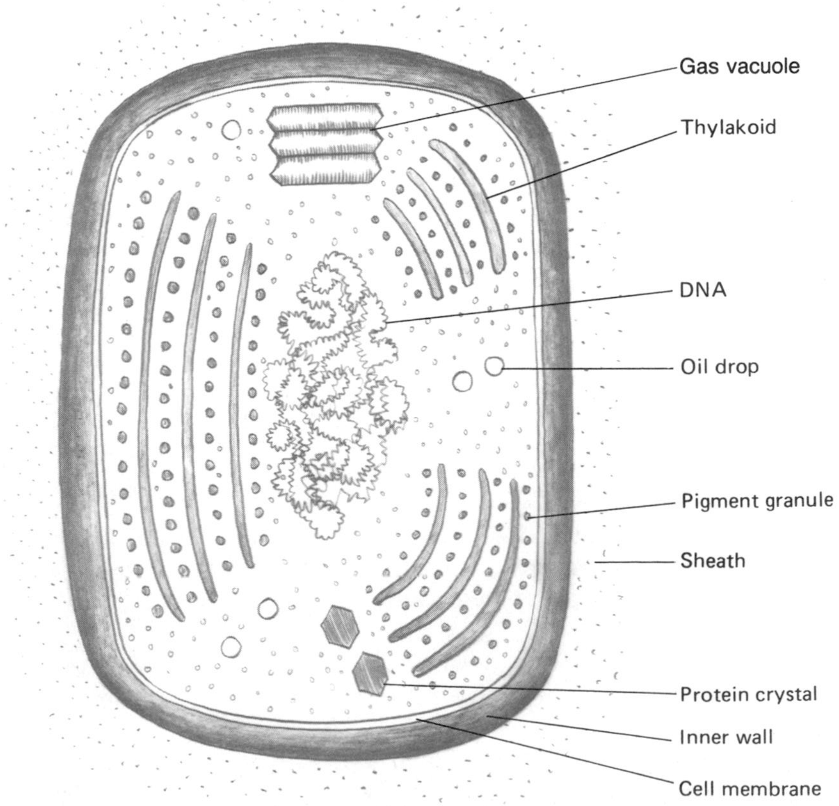

Cyanobacteria

The photosynthetic **cyanobacteria**, also called **blue-green algae**, comprise a second major group of prokaryotes. (**Algae** is a general term applied to many simple photosynthetic organisms, most of which are eukaryotes.) Cyanobacteria range in color from olive green to deep purple, but a bluish green is most common. The colors come from a combination of **chlorophyll-a** (green) and **phycoerythrin** (red) and **phycocyanin** (blue), both **phycobilin** pigments.

Cells of cyanobacteria are structurally similar to bacteria but usually larger (see Fig. 19-5); they resemble bacteria also in reproduction by simple cell division (**fission**), and in absence of a nucleus. However, they differ from bacteria in having **thylakoids**, parallel arrays of membranes that contain photosynthetic pigments. Cyanobacteria also have gas vacuoles, which account for their ability to regulate their buoyancy so as to rise or sink in water.

Many cyanobacteria are nonmotile, and none are flagellated. A few types are able to move by a kind of gliding action produced by rippling movements of the sheath region.

Ponds, lakes, and other bodies of water sometimes appear murky green because of the billions of algal cells

Gas vacuole

Thylakoid

DNA

Oil drop

Pigment granule

Sheath

Protein crystal

Inner wall

Cell membrane

present. Such population explosions of minute photo-synthetic cells are called **blooms**, and often consist of cyanobacteria cells, although eukaryotic algae also cause blooms. (In 1964 a bloom of cyanobacteria in Lake Erie covered approximately 2100 km² [800 mi²]). Blooms are produced by accumulation of nutrients in surface waters, providing the resource for rapid algal reproduction.

Some algal blooms are caused by upwelling of nutrients from bottom sediments as a result of changes in water temperature in spring and fall; others are caused by pollution by nutrient-rich sewage and fertilizers from adjacent fields. Algae die in great numbers in the aftermath of a bloom, and their decomposition produces oxygen depletion, because oxygen is consumed by decay processes. As a consequence of oxygen depletion, fish and other organisms die, adding to the decay and death in the aquatic environment. These processes contribute to ecological aging, called **eutrophication**, and are a leading cause of damage to natural pond, lake, and stream environments.

Prochloron

In 1975 a new form of prokaryotic organism was described by R. A. Lewin and N. C. Withers and named *Prochloron*. *Prochloron* resembles cyanobacteria in structure but has photosynthetic pigments of the type present in green land plants (chlorophyll-a and -b rather than chlorophyll-a and phycobilins). It generally is thought that early eukaryotes evolved from ancient prokaryotes; the discovery of *Prochloron* strengthens the hypothesis that green plants evolved from a photosynthetic prokaryote.

PROTISTA

Unicellular eukaryotes are classified in the kingdom Protista (*protistos*, "primary"), a loosely related assemblage of principally water-dwelling organisms. Many ways of life are represented by protistans, including **autotrophism, predation, parasitism,** and other forms of **symbiosis.** A considerable number of protistans are photosynthetic microorganisms constituting the so-called grass of the oceans, or **phytoplankton** (all phytoplankton are microscopic, photosynthetic, water-dwelling organisms).

A majority of protistans are motile: many of them move rapidly by beating of flagella or cilia; others move slowly by a type of crawling locomotion using pseudopodia or, in the case of diatoms, by laying down a slime track.

Predation is a way of life among one large group of protistans, the Protozoa, and here there are such interesting examples as the ciliate *Didynium* eating another ciliate larger than itself (*Paramecium*) (Fig. 19-6).

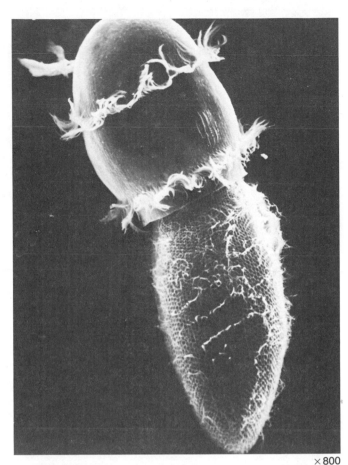

×800

FIGURE 19-6
Scanning electron micrograph of the capture of one protozoan (*Paramecium*) by another (*Didynium*). (Photo courtesy G. Antipa)

Cell Structure

Cells of protistans are typically eukaryotic and contain membranous **organelles,** each of which has unique functions in the life of the cell (Fig. 19-7). Among the more prominent of these are the nucleus with its chromosomes; chloroplasts (not present in all protistans), which convert light energy into food energy; mitochondria, which transform food energy into ATP energy; Golgi bodies, which secrete enzymes; plasma membrane; and flagella or cilia. The last have the basic eukaryotic 9 + 2 structure.

Reproduction

Protistans, like monerans, have an immense reproductive potential. In an optimal environment they may reproduce several times a day, so short-term population explosions occur periodically. These population increases almost invariably are the result of **asexual reproduction,** in which each parent cell divides by **mitosis**

FIGURE 19-7
Generalized protistan cell, illustrating some eukaryotic structures.

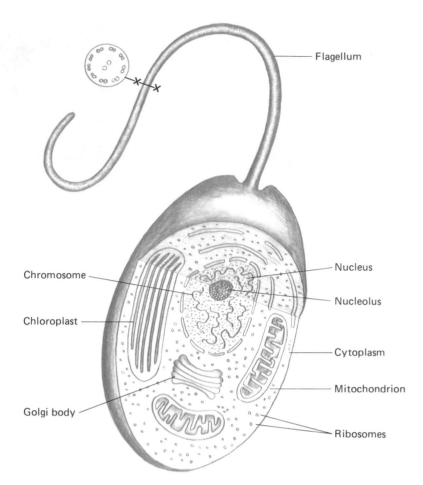

Flagellum

Chromosome

Nucleus

Nucleolus

Chloroplast

Cytoplasm

Mitochondrion

Golgi body

Ribosomes

to produce pairs of cells. Mitosis is the process of asexual cell reproduction common to all protistans (Color
plate 5B).

Most protistans also reproduce by **sexual reproduction**, in which two cells combine their nuclei, their
chromosomes, and usually their cytoplasm, to form a
new cell, the **zygote**. The nucleus of the zygote therefore
is **diploid**, or $2n$, whereas the original nuclei of the sex
cells prior to their combination had one set of chromosomes and were **haploid**, or $1n$.

In all organisms that reproduce sexually, **meiosis** is
present, reducing the $2n$ number of chromosomes to the
$1n$ number occurring in sex cells (in most cases called
gametes). Without meiosis, each time sexual reproduction occurred the chromosome sets would again double
and would soon reach an unsupportable number.
Meiosis maintains $1n$ and $2n$ chromosome numbers generation after generation (refer to Chapters 6 and 7).

Life Cycles

Protistans as well as other eukaryotic groups of organisms can be characterized by their manner of reproduction, specifically in terms of the point in their life at

which meiosis occurs (the reproductive history of a complete generation of an individual organism is termed its
life cycle).

Many protistans remain in the $1n$ condition for the
greater part of their lives, reproducing mitotically to generate more $1n$ individuals. Then some individuals become $1n$ gametes, or produce $1n$ gametes, and fuse in
pairs to produce a $2n$ zygote. The $2n$ zygote then divides
by meiosis to form $1n$ individuals. Commonly, four $1n$
offspring are produced by meiosis, because meiosis is a
two-stage process in which the first division (a reductional division) reduces the chromosome number by
one-half, and the second division (an equational, mito-
sislike division) maintains the $1n$ condition. This type of
life cycle may be called **zygotic meiosis** (Fig. 19-8).

A second common type of life cycle in protistans
involves $2n$ individuals that reproduce mitotically, thus
generating more $2n$ individuals. Just prior to sexual reproduction, $2n$ cells produce $1n$ gametes by meiotic divisions. Commonly, each $2n$ cell makes four $1n$ gametes.
These combine in pairs to form $2n$ zygotes, which in turn
divide mitotically, forming $2n$ individuals. This kind of
life cycle, in which the only $1n$ cells are the gametes, may
be termed **gametic meiosis** (Fig. 19-8).

FIGURE 19-8
Protistan life cycles. **A.** Zygotic meiosis follows fertilization; adult is haploid. **B.** Gametic meiosis precedes gamete formation; adult is diploid.

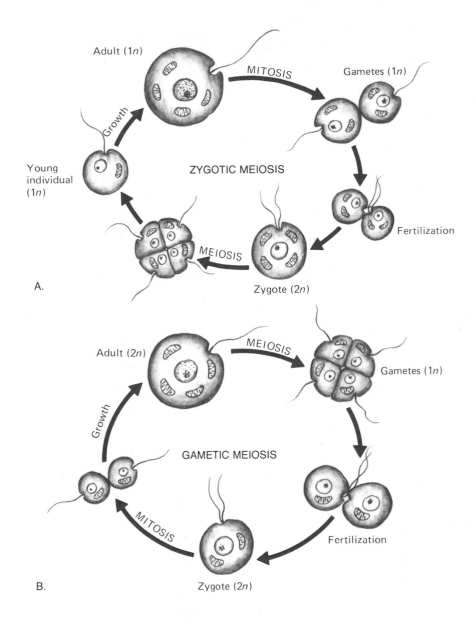

Adult (1*n*)

MITOSIS

Gametes (1*n*)

Growth

ZYGOTIC MEIOSIS

Young individual (1*n*)

Fertilization

MEIOSIS

Zygote (2*n*)

A.

Adult (2*n*)

MEIOSIS

Gametes (1*n*)

Growth

GAMETIC MEIOSIS

MITOSIS

Fertilization

Zygote (2*n*)

B.

Diversity

It is difficult to know with any degree of certainty how many species of protistans there are. Estimates vary because some classifications divide groups into more species than others do. Conservative estimates number protistan species at around 22,250, but some experts say there may be more than 100,000. Major groups of protistans are photosynthetic **dinoflagellates**, **diatoms**, and several classes of **protozoans**, which include both photosynthetic and nonphotosynthetic types.

The Algal Question

The classification of organisms is not an exact science. Few experts agree on the number of categories in which the organisms of our world should be distributed or on

which organisms should be placed together in one or another group. The reason is that many characteristics of a variety of life forms are overlapping. Photosynthesis is an example. It occurs in monerans, protistans, and plants. Should all photosynthetic organisms be classified in one phylum? What about unicellularity versus multicellularity? Protistans are mostly unicellular or colonial, so should green algae (see Chapter 20), many of which are unicellular, also be called protistans? What about those green algae that are multicellular and quite plantlike? Or brown algae, which usually are very large multicellular organisms? Although many biologists include photosynthetic organisms generally known as algae in the kingdom Protista, we prefer to include them in the plant kingdom. (You might wish to refer to the opinion of Dr. Harold Bold of the University of Texas, who groups the algae with green land plants—see the reference to his book [Bold et al.] in the list of readings at the end of this chapter—then decide for yourself.)

Dinoflagellates

Occasionally, ocean waters, especially in protected bays and lagoons, turn red with billions upon billions of small, red-colored protistans called **dinoflagellates** (Fig. 19-9 and Color plate 5c). This bloom produces the so-called **red tide** that kills fish and other forms of life. Red tides are not, as some persons believe, caused by modern-day pollution. In fact, they are not a new phenomenon; Lewis and Clark reported a fish kill by red tide when they arrived at the mouth of the Columbia River on their famous expedition of 1804–05. Today similar outbreaks are reported from time to time along the eastern and western seaboards; two dinoflagellates, *Gonyaulax* and *Gymnodinium*, are the cause, for they release a poisonous chemical into the water when they die and decay.

Dinoflagellates are a broad group of flagellated photosynthetic eukaryotes. Some of them are phosphorescent, and at night cause the wakes of ships and splashes of oars to glow in a lovely luminescence. Many have elaborately sculptured walls composed of several interlocking plates, not unlike armor, and are especially remarkable when viewed with a scanning electron microscope.

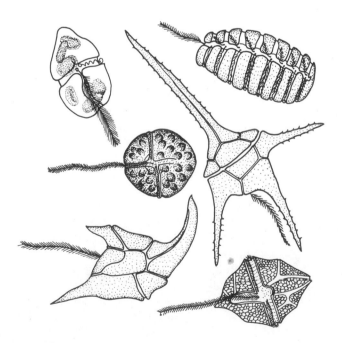

FIGURE 19-9
Dinoflagellates.

Most dinoflagellates are not toxic to animal life; in fact, they constitute an important food source. They are an abundant form of phytoplankton, furnishing food for small **zooplankton** (nonphotosynthetic eukaryotic microorganisms). Together, phytoplankton and zooplankton comprise **plankton**, upon which other aquatic animals feed. The smaller of these in turn are preyed upon by larger animals. Thus phytoplankton serves as the basic energy source for life in the sea, just as grasses and other green plants are the basis of life on land.

Dinoflagellates reproduce principally by asexual reproduction; sexual reproduction is almost unknown, having been reported in one species only.

Diatoms

The light from headlights is reflected by the center stripe on a highway at night because of the addition to paint of billions of glassy shells of microscopic marine organisms called **diatoms**, obtained from ancient deposits of dead diatoms in the form of **diatomaceous earth**. Diatoms secrete beautifully sculptured pairs of glasslike shells called **frustules** that surround the living cell and fit together like a box with its cover or like the lid and bottom of a petri dish (Fig. 19-10). The latter is probably a better comparison, because diatom walls contain **silicon**, the major component of glass.

Diatoms reproduce both asexually and sexually. In asexual reproduction a diatom divides by mitosis and each daughter cell receives one frustule, then manufactures a new, second frustule which fits inside the first. The result is that those diatoms inheriting the inner frustule keep getting smaller and smaller (Fig. 19-11). Eventually the frustules become restrictive and are discarded; then new, large frustules are manufactured.

Sexual reproduction in diatoms, when it occurs, usually involves gametic meiosis. The 2*n* diatom produces 1*n* gametes by meiosis. These unite in pairs, forming 2*n* zygotes which develop into mature diatoms.

Living diatoms are concentrated in upper zones of bodies of water, where they receive maximum sunlight and constitute an important phytoplankton component. When they die, their frustules drift down and slowly form a bottom sediment that over long intervals of time may solidify into diatomite rock. Ancient seas produced deposits of diatomite hundreds of feet thick. When these deposits were lifted by major geological upheavals, they formed surface deposits of diatomaceous earth, or **diatomite**, which now are mined on a large scale. Blocks of diatomite are used as building materials and in making filters and absorbents or are ground into powder for use in paints and polishing compounds.

FIGURE 19-10
Scanning electron micrograph of diatoms. **A.** Centric form. **B.** Pennate type. (Photos from G. A. Fryxell and G. R. Hasle, 1980, The marine diatom *Thalassiosira oestruii*, Amer. J. Bot. 67:904–14)

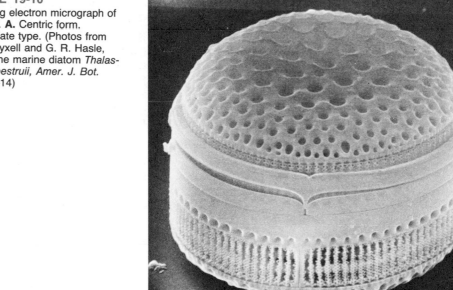

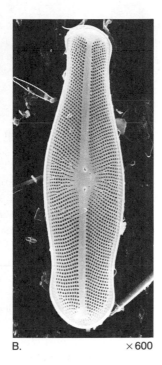

A.   ×5250   B.   ×600

Protozoans

Protozoans are a diversified group of protistans whose typical mode of nutrition is ingestion of particulate food, a type of feeding called **holozoic**. *Amoeba* and *Paramecium* are studied in biology classes as typical holozoic protozoans. However, some protozoans such as *Euglena* are photosynthesizing organisms sometimes classified as algae. Other protozoans live by absorbing soluble food, often occurring as internal parasites in other animals. Among these are the causal organisms of human ma-

laria, dysentery, and other diseases. Protozoans often are subdivided into four classes: **Flagellata, Ciliata, Sarcodina,** and **Sporozoa.**

Flagellata   One of the ten scourges of ancient Egypt written about in the Old Testament was a plague in which the rivers turned to blood. Although its true nature is unknown, some biologists think this scourge was caused by a bloom of a protozoan named *Euglena sanguina* (red *Euglena*). This euglena is rather common in livestock waterholes enriched by urine and droppings,

FIGURE 19-11
Asexual reproduction of diatoms. (Redrawn from *Plant biology* by Knut Norstog and Robert W. Long. Copyright © 1976 by W. B. Saunders Company. Reprinted by permission of W. B. Saunders Company, CBS College Publishing)

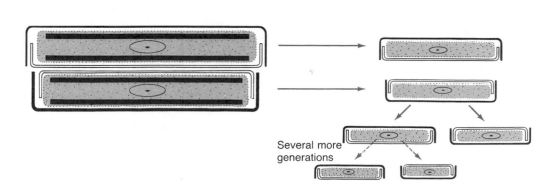

Several more generations

and its populations under such circumstances increase so greatly that the water turns red.

A common euglena often studied in biology laboratories is green rather than red and has conspicuous chloroplasts and a flexible skeleton called a **pellicle** (Fig. 19-12). Also visible is a bright orange-red **eyespot** associated with the shorter of two flagella. This species, named *Euglena viridis* (green *Euglena*), as well as all other euglenas, reproduces only asexually.

Flagellates occupy diverse habitats and niches. Many are free-living organisms; others are important symbionts, living in the guts of cattle, wood-eating cockroaches, and termites, where they digest cellulose and provide food for their hosts. Evidence of the importance of the protozoan parasite to its host is shown by a method of termite extermination used in the tropics. A paste of ground-up plant material known to be a laxative for termites is placed in termite tunnels, and when termites eat this paste they develop diarrhea and eliminate their symbionts. Thereafter, the termites are unable to digest cellulose and starve to death. More precise laboratory experiments in which the termite symbionts are killed without direct injury to the host have verified this relationship.

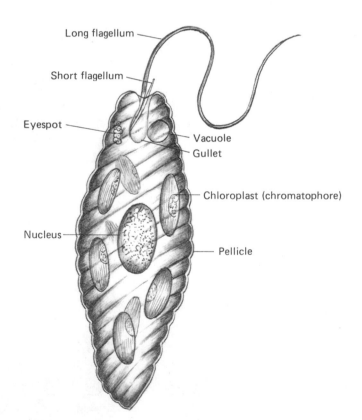

FIGURE 19-12
Structure of *Euglena*.

Other species of Flagellata are parasites dangerous to humans and other animals; among them are trypanosomes that cause African sleeping sickness and leishmanias that cause a severe liver infection called kalaazar. Both are transmitted to humans by biting insects: trypanosomes by tsetse flies, leishmanias by sandflies.

Ciliata   Ciliated protozoans are for the most part motile, active predators, but some are sedentary and feed on particles sucked into mouthlike cavities by water currents produced by circumoral rows of beating cilia. Others, such as *Stentor*, described in Chapter 6, can be sedentary at times as well as free-swimming.

If an organism of one cell could be said to resemble a multicellular animal in the things it does, *Paramecium* would be a good example. Although *Paramecium* is unicellular, its cytoplasm is complex and its organelles are arranged in such a way as to provide an amazingly complex set of functions (Figs. 19-13, 19-14, and Color plate 5A, B).

In addition to their capability for rapid asexual reproduction by mitotic division (Fig. 19-13), paramecia reproduce sexually by **conjugation**, a process in which two individuals become fused together laterally and exchange $1n$ nuclei formed by meiosis (a somewhat modified zygotic meiosis cycle; Fig. 19-8). After nuclear exchange and fusion of the recently exchanged nuclei to produce $2n$ nuclei, the conjugants separate and again reproduce asexually. In this case the nuclear exchange is the equivalent of gametic exchange and recombination of genetic material (genes), which is also the important function of sexual reproduction.

Feeding paramecia are impressive in their flexibility and powers of navigation. Numerous cilia located in pairs along ridges and grooves of semirigid pellicle (Fig. 19-14) beat rapidly and create a current that moves food into the **gullet**, where **food vacuoles** (cytoplasmic pockets containing food and digestive enzymes) are formed.

Between every pair of cilia is a highly modified cilium, called a **trichocyst**, which contains a stinging hair that can be ejected forcefully against an enemy. (Trichocysts are characteristic structures in ciliates but also occur in some dinoflagellates). Other obvious structures are a **macronucleus**, a large accessory nucleus which plays a principal role in regulating cellular processes, and a small **micronucleus**, which is involved specifically in mitosis and meiosis during asexual and sexual reproduction. Also remarkable are anterior and posterior **contractile vacuoles**, which eliminate excess water taken in during feeding.

Sarcodina   The class Sarcodina includes relatively formless **amoebas** as well as **foraminiferans** and **radi-**

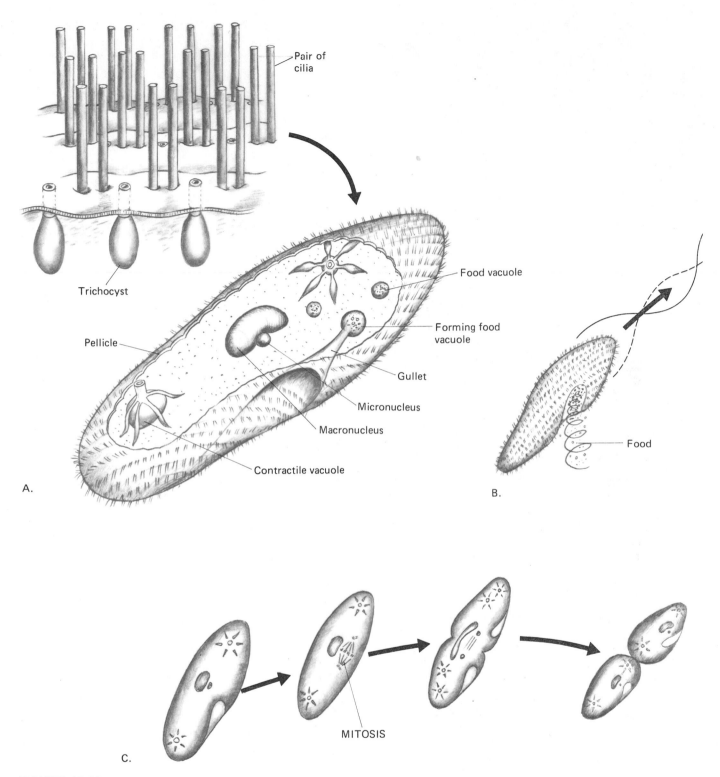

FIGURE 19-13
Paramecium structure and function. **A.** Cellular organization. **B.** Feeding action. **C.** Asexual reproduction.

Labels in figure:
Pair of cilia
Trichocyst
Pellicle
Food vacuole
Forming food vacuole
Gullet
Micronucleus
Macronucleus
Contractile vacuole
Food
MITOSIS
A.
B.
C.

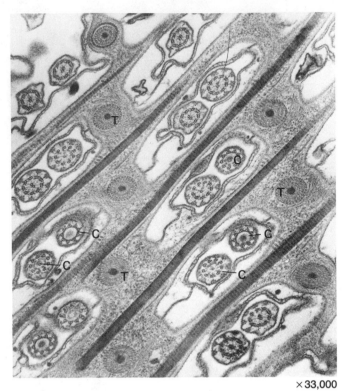

×33,000

FIGURE 19-14
Electron micrograph of a portion of the surface of a paramecium, showing the grooves and ridges of the outer coat, or pellicle. Visible are the stumps of cilia (C) (note their 9 + 2 structure) and adjacent pits within which are located the stinging hairs, or trichocysts (T). (Photo courtesy K. Hausmann, reproduced from *Jour. Cell Biology* 69:313–26, by copyright permission of the Rockefeller University Press, New York, N.Y.)

olarians, which are tiny, shelled protozoans of great structural beauty (Fig. 19-15, 19-16, and Color plate 1A). All have a characteristic creeping movement called **amoeboid locomotion,** in which shape changes of the protoplasm are produced by flowing movements of the cytoplasm. Cytoplasmic extensions, called **pseudopodia,** protrude in the direction of movement and are used both in locomotion and in capturing and ingesting food. This locomotion is easily observable in the *Amoeba proteus* of biology laboratory classes (Fig. 19-15). (Amoeboid locomotion also occurs in some cells of multicellular organisms. The white blood cells of humans and other animals move and feed by amoeboid movements.)

Many amoebas live within the digestive tracts of animals, and some cause diseases. For example, *Entamoeba histolytica* (Fig. 19-15) is the causal organism of amoebic dysentery, a common ailment in regions of poor sanitation. This is a serious disease and can be fatal if the amoebas invade the liver and brain.

Foraminiferans and radiolarians are characterized by elaborate shells or skeletons (see Fig. 19-16). Fora-

miniferans bear shells of calcium carbonate (limestone) that look like microscopic snails; radiolarian shells are composed of silicon. Shells of both accumulate in bottom sediments, eventually forming a type of chalk. The white cliffs of Dover, England, are composed of foraminiferan shells. Different species are characteristic of different geological formations and are used by petroleum geologists to identify oil-bearing deposits.

Sporozoa   Each group of protozoans has its disease-inducing members, but undoubtedly the most serious and widespread are malaria-causing members of the class Sporozoa. **Spores** are reproductive cells formed by many organisms; characteristically they are numerous and commonly are enclosed by resistant membranes so that they can survive drying or other traumas. Spores frequently function as infective cells and may be transmitted in food, water, or dust, or by biting insects such as mosquitoes and flies. The spores that propagate human **malaria** are transmitted by mosquitoes (Fig. 19-17).

About fifty species of malaria organisms infect warmblooded animals from birds to primates. Principal malaria species infecting humans are: *Plasmodium falciparum,* the cause of a severe malaria prevalent in tropical countries; P. *vivax,* the cause of tertian relapsing malaria in temperate regions (Fig. 19-17); and P. *malarieae,* which causes 72-hour relapsing malaria (quartan malaria).

All forms of human malaria are transmitted by *Anopheles* mosquitoes. Malaria once was widespread, but in the last hundred years it has nearly vanished in the United States and has greatly declined elsewhere, owing to better sanitation and mosquito control. However, it remains the leading infectious disease in the world, and even now in underdeveloped countries an estimated 1.5 million persons have malaria.

Origin of Animals   In their motility and holozoic nutrition, protozoans resemble **metazoans** (multicellular animals) and it is generally thought that metazoans evolved from a protozoan ancestry. There are many differences of structural organization and complexity among members of the animal kingdom, from simply organized sponges to complex and highly organized forms such as birds and mammals, but all are aggregates of multitudes of cells.

In theory, multicellular animals could have evolved in one of several ways. They might have begun as simple aggregates or **colonies** of identical cells and then, by functional specialization of groups of cells (**tissues**), become metazoans. Alternately, already complex protozoans, perhaps resembling some of the more elaborate ciliates, might have become multinucleate and subsequently partitioned into cells already specialized for certain functions (locomotion, digestion, reproduction,

FIGURE 19-15
Common amoeba, *Amoeba proteus*, and *Entamoeba histolytica*, the cause of amoebic dysentery.

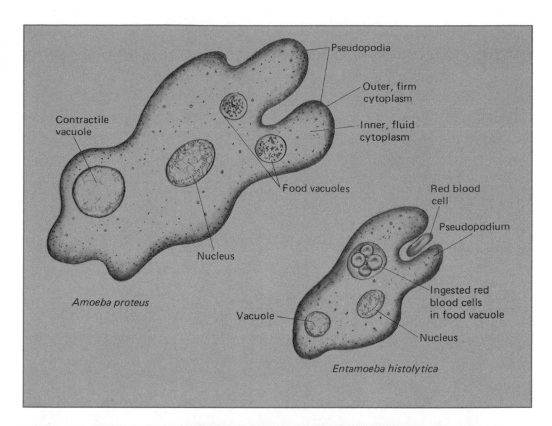

FIGURE 19-16
Scanning electron micrograph of foraminiferans and radiolarians. (Photo courtesy Scripps Institution of Oceanography, University of California, San Diego)

×125

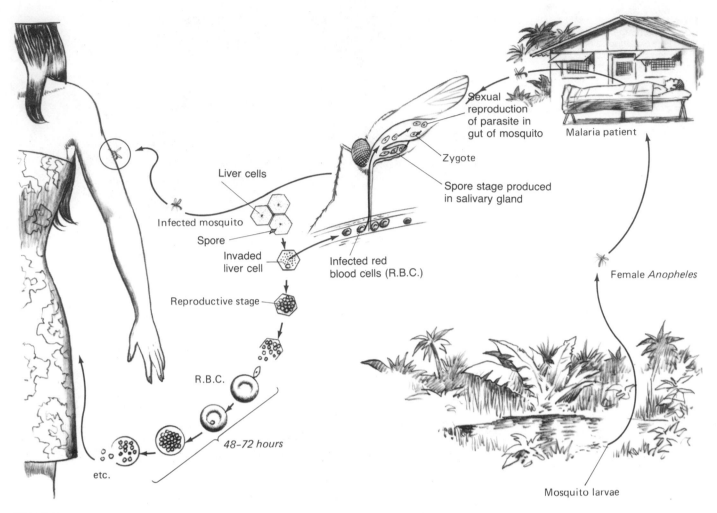

FIGURE 19-17
Life cycle of malaria *Plasmodium vivax*.

etc.). A third possibility is that some metazoans began as colonies (e.g., sponges) and others descended from multinucleate ciliates. Although each of these hypotheses is plausible, the supporting evidence is only circumstantial.

FUNGI

Fungi play much the same ecological role as do bacteria because they are, first and foremost, natural recycling agents, decomposing plants and animals and their products. Their characteristic mode of nutrition is called **saprobic**, and fungi often are referred to as **saprobes** ("decayers").

Although many fungi are saprobic, others are parasites infecting living organisms; some of these can be quite destructive. An example is the parasite responsible for the potato blight, which was the cause of the Ireland potato famine of the 1840s: a million persons died of starvation and related causes and another million were forced to migrate to other countries. The potato blight is still a problem in some parts of the world, as are the rusts and smuts that devastate wheat and corn fields. Fungal parasites of animals include those of athlete's foot, ringworm, and other skin diseases, as well as others which are annoying and sometimes dangerous human pathogens.

Fungi also may play a positive role in the human economy. Many are important to food production. The **yeasts**, a group of unicellular fungi, are of immense value to bread-making and brewing industries. Nor can the importance of fungi to the drug industry be overlooked; **penicillin** is of major importance, and other drugs, including hallucinogens, also are obtained from fungi. Not all fungi, of course, are microorganisms.

END TO LOCUST PLAGUES?

Since the days of the pharaohs, swarms of locusts have repeatedly devastated crops, leaving famine and death in their wake. In some of the poorer, underdeveloped countries of Africa and Asia, they are almost an annual occurrence. Swarms darkening the sky over hundreds of square miles have been reported.

Grasshopper swarms can be destroyed by expensive spraying with high levels of pesticides, but these chemicals also kill other insects, many of them beneficial, and ultimately find their way into the human food chain. Scientists have for some years been looking for effective biological controls, and now it appears that one has been found.

A natural protozoan parasite of grasshoppers, *Nosema locustae*, has been studied in the field and laboratory by John E. Henry and colleagues at the U.S. Department of Agriculture Experiment Station in Bozeman, Montana. The parasite infects the fat reserves of many grasshopper species, including those of greatest impact on agriculture. It is a spore producer, and feeding spore-infected lettuce to grasshopper nymphs results in production of adults carrying about 3.9 million spores each. The adult grasshoppers are frozen and kept in storage. When needed, the frozen grasshopper carcasses are crushed, releasing the spores, which are sprayed on bran, the commonly used bait. The spores of one grasshopper are sufficient to treat four acres of wheat and will reduce the density of grasshoppers by about 50 percent. The spores persist in the area for several years until the density of grasshoppers becomes too low to support the parasites.

Fleshy fungi such as **mushrooms, brackets, cups,** and **morels** are comparatively large and visible. Yet some classes of fungi are identifiable only with the aid of a microscope, and other classes with highly visible members contain microscopic genera and species.

There are three major groups of fungi: **slime molds (slime fungi), whip fungi,** and **higher fungi (bread molds, sac fungi,** and **club fungi).** Of these, the slime molds seem more like amoeboid protozoans than fungi and may in fact not be related very directly to the other three groups. Even their mode of nutrition differs, because they feed on bacteria and other microorganisms rather than by absorption of dissolved nutrients. Their inclusion with fungi is mostly traditional, but they are named and classified with other fungi under the rules of the international botanical code and are studied by **mycologists** (specialists in fungi; *myc-*, "fungus").

The other fungal groups typically have a threadlike filamentous organization, although one-celled types occur. The origin and relationships of whip fungi are con- troversial. It appears that some simple types with motile gametes and spores may have evolved from protozoans, but others seem to be more closely allied with algae. The bread mold group resembles the whip fungi in cytoplasmic structure, but its members are reproductively much more like sac and club fungi. Sac and club fungi have a number of basic similarities and probably are more closely related to each other than to other fungi.

Slime Fungi

Slime molds, or slime fungi, are a generally colorful group of organisms classified in the division **Gymnomycota** (*gymno-*, "naked," *myco-*, "fungus"). During the nonsexual phases of their lives, they exist as a **plasmodium,** composed of "naked" protoplasm, that slowly creeps over soil and rotting vegetation. Although often associated with decay, plasmodia do not actually feed on rotting matter but rather on microorganisms of decay. The colors of different species range from white or light gray

FIGURE 19-18
Sporangia of the common slime mold *Stemonitis*. (Photo by Carolina Biological Supply Company)

through yellow, orange, pink and red, to purple and blue black.

Slime molds are scientifically important experimental organisms. Because they are essentially large masses of protoplasm, they provide unique opportunities for the study of living matter.

At times when there are changes in available nutrient and light conditions, slime molds enter a sexual phase and form **sporangia** (sing. **sporangium**, "spore container"), which often are quite intricate and beautiful. Spores are produced in great numbers by meiotic divisions within sporangia, and hence are **meiospores** (Fig. 19-18). Meiospores often are transported great distances by air currents; those that land in moist places germi-

nate and form minute amoeboid cells called **myxamoebas**, which are capable of mitosis. They divide and redivide, forming additional myxamoebas. Eventually, myxamoebas function as gametes, fusing in pairs and producing 2*n* amoeboid zygotes. The zygotes develop into characteristic slime molds by repeated nuclear divisions and cytoplasmic growth.

Whip Fungi

Whip fungi, classified in the division **Mastigomycota** (*mastigo-*, "whip," *myco-*, "fungus") have flagellated cells at some stage. They generally are constructed of elongate tubes containing many nuclei and covered by a continuous, nonpartitioning wall. This condition is referred to as **coenocytic** (*coeno-*, "united," *cyte*, "cell") and, except for a few algae of similar structure, is a unique form of organization. The tubes are **hyphae** (a term also applied to cellular filaments found in other fungi), and the many hyphae (sing. **hypha**, "thread") of an individual fungus comprise a feltlike **mycelium** (pl. **mycelia**). Among the whip fungi are classes that live primarily in water and have flagellated spores (**zoospores**) and gametes, and transitional terrestrial forms that have both motile and nonmotile spores (Fig. 19-19). Several species of water molds (*Allomyces*) are microscopic whip fungi of great (microscopic) beauty and are found growing on dead vegetation (Fig. 19-19).

One of the aquatic whip fungi is the fish mold *Saprolegnia* (see Fig. 19-19), sometimes seen as a furlike mycelium covering the fins of living aquarium fish. The late-blight fungus, *Phytopthora infestans*, which caused the destructive potato famine in Ireland, has characteristics

FIGURE 19-19
Two representatives of the whip fungi, both aquatic. *Allomyces,* a water mold, grows on dead vegetable matter; hemp seeds are used as "bait" in laboratory cultures. *Saprolegnia,* often called "fish mold," is sometimes seen growing on the fins of aquarium fish.

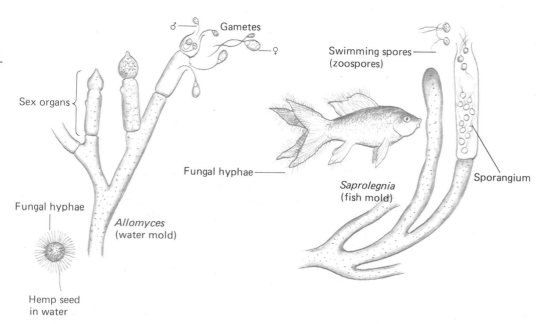

FIGURE 19-20
The black bread mold (*Rhizopus nigricans*). This common mold is shown growing on bread. The asexual phase is illustrated; sexual reproduction is by conjugation of + and − types of individuals (not shown).

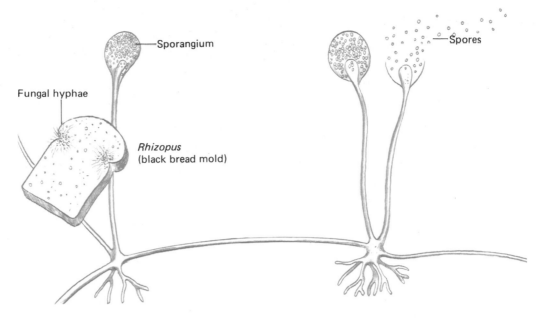

Sporangium

Fungal hyphae

Rhizopus
(black bread mold)

Spores

intermediate between aquatic and terrestrial forms and exemplifies the transitional type of whip.

Higher Fungi

In contrast with the fungi of the preceding groups, the "higher fungi" of the division **Amastigomycota** (*a-*, "without," *mastigo-*, "whip," *myco-*, "fungus") produce no flagel-

lated cells and are for the most part terrestrial. Many of them have large, fleshy bodies: "fruiting bodies" in the form of **morels, mushrooms, brackets**, etc. (Figs. 19-21 and 19-22).

Bread Molds One subdivision (**Zygomycotina**—*zygo-*, "joined," *myco-*, "fungus") of the higher fungi is characterized by tubular multinucleate hyphae not divided into

HOW PIGS FIND TRUFFLES

In the oak woods of Europe, where truffles have been harvested for centuries, pigs often are used to sniff out the buried treasures, sometimes locating them nearly a yard below the surface. The nature of the olfactory clue to the site of a buried truffle has long been a mystery. Recently, however, three German scientists, R. Claus, H. O. Hoppen, and A. Karg, discovered that truffles contain a steroid molecule characterized by a musklike scent. (Steroids are lipids that have certain carbon atoms joined in 6-membered rings, and they include cholesterol as well as several sex hormones.) The steroid responsible for the scent of truffles appears to be identical to a steroid produced in the testes of pigs and secreted by the salivary glands during premating activity, thus explaining a pig's interest in rooting in the ground. It also has been suggested that the truffle steroid may account for human interest in this fungus—the same steroid is found in the testes of human males and is secreted by armpit sweat glands. Although the suggestion that people like truffles because they smell like armpits might seem a bit farfetched, the results of several tests suggest that musklike scents are attractive to many individuals.

FIGURE 19-21
Sexual reproduction of an ascomycete.

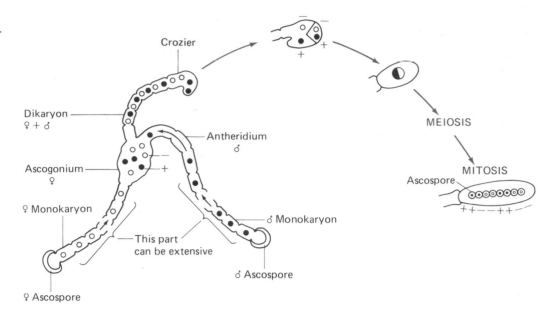

cells (i.e., coenocytic). In this respect this group, commonly known as **bread molds**, resembles the whip fungi.
Unlike them, the bread molds possess no flagellated reproductive cells. Instead, they reproduce asexually by the
production of spores and sexually by conjugation. In
their conjugation, hyphae of similar appearance, but
physiologically different and simply called **plus** (+) and
minus (−), approach, fuse together, and produce a zygote. The zygote undergoes meiosis and produces new,
haploid + or − individuals which reproduce by spores
as shown in Fig. 19-20. The common **black bread mold**,
Rhizopus nigricans, is the best-known example.

Sac Fungi Members of the subdivision **Ascomycotina**
(*ascus*, "sac," *myco-*, "fungus"), class **Ascomycetes**, are
probably the most useful fungi in the human economy;
products from this group would be sorely missed. Penicillin is the product of a sac fungus, *Penicillium notatum*,
as are Roquefort cheese (*P. roquefortii*) and Camembert
cheese (*P. camembertii*). A group of one-celled sac fungi
are the yeasts necessary for production of bread as well
as beer and other beverages. Truffles and morels, also
sac fungi, are sought after as gourmet food items, and
Neurospora crassa, a pink bread mold, is a sac fungus important in studies of the chemistry of genes (refer to
Fig. 5-2).

On the other hand, some parasitic sac fungi are
very destructive, disease-causing agents. Among these
are fungi causing chestnut blight and Dutch elm disease,
and others that cause ergot disease of barley, wheat, and
rye, human athlete's foot, respiratory ailments, and men-
ingitis. Saprobic sac fungi destroy great quantities of

food, clothing, and wooden objects; most mildews found
in clothing, leather goods, bedding, shoes, and so on,
are ascomycetes.

Sac fungi get their name from their distinctive sac-
like reproductive organ, the **ascus** (pl. **asci**), within which
ascospores are produced by meiotic divisions (Fig.
19-21). In addition to ascospores, nonsexual spores
known as **conidia** (Color plate 1C) are produced in great
numbers. Some ascomycetes produce only conidia and
are relegated to the division **Deuteromycota** (*deutero-*,
"secondary," *myco-*, "fungus"), which includes all fungi
for which no sexual stages have been discovered.

Although there is great variation in structure
among sac fungi, many produce a characteristic reproductive body called an **ascocarp**. The ascocarp is formed
in part from uninucleate hyphae originating in conidia of
opposite sexual or mating types (usually called plus and
minus), and in part from binucleate hyphae formed as a
result of the combining of protoplasm and nuclei from
uninucleate hyphae (see Fig. 19-21). As shown in Fig.
19-22, the female organ, an **ascogonium**, of an individual
composed of hyphae having uninucleate cells (a **monokaryon**) receives nuclei from the sex organ, the **antheridium**, of a male monokaryon. A new individual (a **dikaryon**) composed of hyphae with binucleate cells is formed.
The hyphae of the dikaryon produce hook-shaped structures, called **croziers**, at their tips. In a crozier, two nuclei
of female origin and two of male origin (pluses and mi-
nuses, respectively) migrate such that one female and
one male (or one + and one −) nucleus become isolated by formation of a wall at the tip of the crozier.
These two nuclei fuse to form the only true 2*n* nucleus

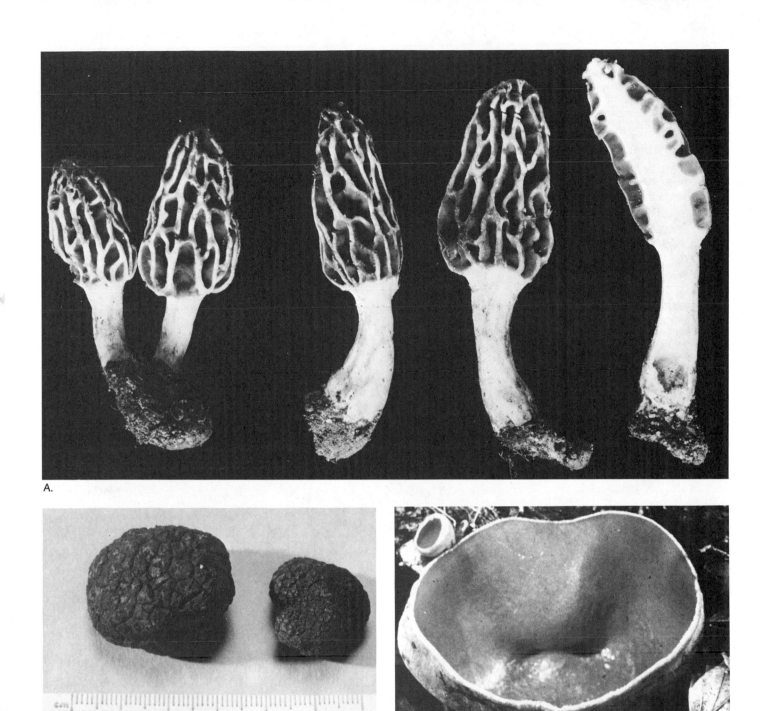

FIGURE 19-22
Ascocarps of some sac fungi. **A.** Morels. **B.** Truffles. **C.** Scarlet cup. (A, photo courtesy U. S. Department of Agriculture; B, reprinted with permission from W. D. Gray, 1970, *The use of fungi as a food*; copyright The Chemical Rubber Co., CRC Press, Inc.; C, photo courtesy Ward's Natural Science Establishment)

in the life cycle. Following fusion, the 2*n* nucleus undergoes meiosis, then mitosis, to produce four to eight 1*n* ascospores. Ascocarps of molds and mildews are microscopic, but those of "fleshy" ascomycetes such as morels, truffles, and scarlet cups are of mushroom size (Fig. 19-22).

Fleshy fruiting bodies of fungi differ in cellular organization from the cells of either animals or plants, which tend to be arranged in rather regular rows, tiers, or clumps; cells of fungal hyphae have a tangled, weblike relationship. One of the mysteries of biology is how such a seemingly irregular arrangement produces a body of such regularity as a mushroom or morel.

Club Fungi Members of the club fungi are members of the subdivision **Basidiomycotina** (*basi-*, "base," *myco-*, "fungus"), class **Basidiomycetes**, because of their distinctive club-shaped reproductive organ, the **basidium**. They are the most familiar fungi; their fruiting bodies, or **basidiocarps**, include such well-known structures as mushrooms, puffballs, stinkhorns, earth stars, coral fungi, and brackets (Fig. 19-23 and Color plates 1B, 6A, C).

FIGURE 19-23
Fruiting bodies of club fungi.
A. Bracket fungus (*Fomes*).
B. Puffball (*Calvatia*). **C.** Earth star (*Geaster*). **D.** Sacred mushroom (*Psilocybe*). **E.** Destroying angel (*Amanita verna*). **F.** Edible field mushroom (*Agaricus*).
G. Coral fungus (*Clavaria*).
H. Stinkhorn (*Phallus*).

Most club fungi are saprobes and have extensive hyphae that absorb nutrients from decomposing vegetation. Many also form root complexes, called **mycorrhizas,** with trees, shrubs, and other higher green plants. It has been estimated that as many as 90 percent of trees and shrubs have mycorrhizal associations from which both the fungi and the higher plant derive nutrients (fungi obtain carbohydrates; trees derive minerals).

In nature, old logs, dead roots, humus, and other plant debris are the niche of many kinds of club fungi. The ecological role of club fungi is absorption of nutrients from dead organisms, processes which result in the recycling of nutrients in the natural economy. For example, wood is composed largely of cellulose, which in turn consists of units (molecules) of sugar (glucose) linked together in long chains (polymers). These chains of glucose cannot be utilized as food by most animals, but fungi have the ability to break them down into individual glucose molecules which can be used by the fungus and indirectly by other organisms feeding upon them.

Reproduction of club fungi resembles that of sac fungi in overall aspect but differs in the nature of the meiotic portion of the life cycle. Whereas sac fungi produce meiospores (ascospores) within a saclike structure, club fungi form their meiospores (**basidiospores**) at the tips of basidia (Fig. 19-24). The basidia in turn are borne on or within a basidiocarp inside a puffball, in pores of bracket fungi, or on gills of mushrooms. (A mushroom life cycle is included in Fig. 19-24.)

There are not many recorded deaths in the U.S.A. from eating poisonous mushrooms, but collectors are well advised to use great caution in eating wild mushrooms. In Europe, where mushroom gathering is popular, hundreds become ill each year from eating poisonous mushrooms and many die.

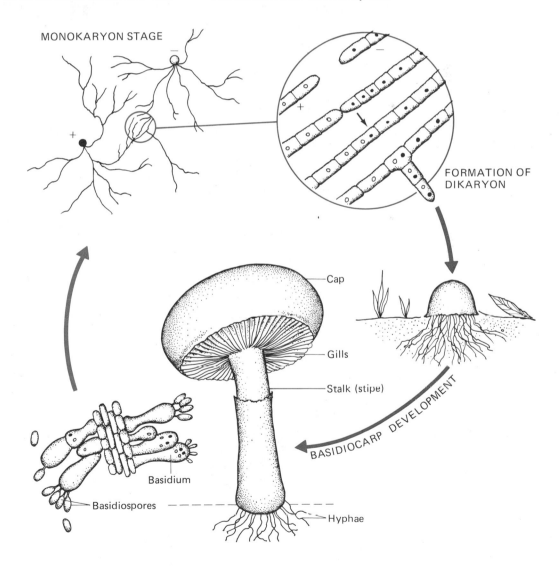

FIGURE 19-24
Mushroom reproduction and growth.

MONOKARYON STAGE

FORMATION OF DIKARYON

Cap

Gills

Stalk (stipe)

BASIDIOCARP DEVELOPMENT

Basidium

Basidiospores

Hyphae

Unfortunately, there are no simple tests to distinguish poisonous from nonpoisonous mushroom species. Security lies in recognizing edible mushrooms and avoiding all others. Some species are exceedingly toxic; the destroying angel, *Amanita verna*, and some of its close relatives are deadly even in small amounts (Color plate 6C).

Another poisonous mushroom, *A. muscaria*, with an attractive white stalk and orange red cap, can be quite toxic; it also is the source of a hallucinogen. A cult of Vikings, the berserkers, are thought to have attained their murderous frenzies by eating *A. muscaria* (Fig. 19-25). In more recent times, amanitas were used by Russian peasants. In Mexico hallucinogenic mushrooms are sold in rural markets and are eaten during rituals that have religious and psychic connotations. Lewis Carroll, writing *Alice in Wonderland*, seems to have had hallucinogenic mushrooms in mind when he had the caterpillar tell Alice, "One side will make you grow taller, and the other side will make you grow shorter."

LICHENS

Lichens are unusual in that they are symbiotic associations of two quite different forms of life, an alga and a fungus. Usually, the fungus partner is a sac fungus and the alga is a member of the green algae (**Chlorophyta**). Sometimes, however, the algal component is a blue-green alga, and the composite organism thus crosses the line between prokaryotes and eukaryotes.

Lichens exemplify **symbiotic mutualism**, an arrangement in which both partners achieve something from their relationship. Experiments have shown that fungus and lichen symbionts can be cultured separately and later reassembled into a functional lichen, but this occurs only if the culture medium will not support growth of either fungus or alga by itself. The combined alga and fungus can synthesize nutrients that neither can make alone.

FIGURE 19-25
Amanita muscaria, the fly agaric, a hallucinogenic mushroom. (Photo by J. R. Waaland, University of Washington/Biological Photo Service)

Lichens are found in nearly all environments, including tree bark, rocks, and soils. The reindeer lichens that grow in the Arctic and furnish food for caribou, reindeer, and musk-oxen look like tiny shrubs; other lichens are flat and leaflike, or look like encrusted paint (Fig. 19-26 and Color plate 6B).

SUMMARY

The kingdom Monera includes bacteria, cyanobacteria, and *Prochloron*. Bacteria are mostly heterotrophic prokaryotes, although a few are autotrophic; the cyanobacteria and *Prochloron* are autotrophic.

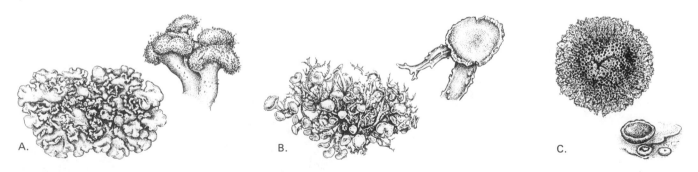

FIGURE 19-26
Some common lichen types **A.** Foliose (leaflike) type. **B.** Fruticose (branching) type. **C.** Crustose type. Enlarged reproductive structures (ascocarps) are also shown.

Bacterial cells are relatively simple in structure. The principal forms are rods (bacilli), spheres (cocci), spirals (spirilla), and filaments (actinomycetes). Some forms have flagella and are motile. Bacteria secrete enzymes that digest food and other matter outside the wall, making it possible for the resultant nutrients to be absorbed. Bacterial enzymes and other products are important in many ways. Some are used in the preparation of butter, cheese, and other foods. Other commercial bacterial products are solvents such as alcohols and acetone.

Cyanobacteria are photosynthetic organisms having blue and red accessory photosynthetic pigments which in combination with chlorophyll give them their characteristic colors. They are similar in structure to bacteria, but none are flagellated. Some move by a gliding or oscillating motion thought to be produced by rippling actions of their outer wall layers.

The predominantly unicellular protistans occupy an intermediate position between the prokaryotes and three great groups of fundamentally multicellular forms of life: fungi, plants, and animals. Although modern classifications are nearly always based on evolutionary relationships, the protistans are so varied that their differences often seem more pronounced than their similarities. Nearly all organisms classified as protistans are microscopic and were unknown before the invention of microscopes. Included are a heterogeneous group of photosynthetic microorganisms such as dinoflagellates, diatoms, and protozoans, including amoebas, paramecia, and malaria parasites.

Fungi are nature's recyclers, breaking down dead organisms and their waste products and recycling chemicals by decomposition. Among the fungi are the whip fungi (Mastigomycota), characterized by a tubular and multinucleate organization of protoplasm; the slime molds (Gymnomycota), whose body plan is that of an amoeboid mass of protoplasm; and the higher fungi classified as Zygomycotina (bread molds), Ascomycotina (sac fungi), and Basidiomycotina (club fungi). Familiar ascomycetes are the morel and truffle, considered delicacies by many gourmets, as well as yeasts and *Penicillium* (source of penicillin). Representative basidiomycetes are the mushrooms, puffballs, brackets, earth stars, rusts, and smuts.

Formal classification of the groups mentioned in this chapter can be found in the appendix.

KEY WORDS

Monera	gametic meiosis	myxamoeba
Protista	red tide	whip fungus
autotrophic	diatom	coenocytic
heterotrophic	phytoplankton	hypha
bacterium	zooplankton	mycelium
genophore	Protozoa	zoospore
disease	Flagellata	sac fungus
cyanobacteria	Ciliata	ascus
algal bloom	conjugation	monokaryon
Prochloron	Sarcodina	ascospore
organelle	pseudopodium	ascogonium
asexual reproduction	Sporozoa	antheridium
mitosis	malaria	dikaryon
sexual reproduction	fungus	crozier
zygote	saprobe	club fungus
diploid	slime mold	basidium
haploid	plasmodium	mycorrhiza
meiosis	sporangium	basidiospore
zygotic meiosis	meiospore	lichen

1 Briefly describe the bacterial cell and its different forms

2 What is nitrogen fixation? In what way are members of the kingdom Monera involved in nitrogen fixation?

3 Why do some bacteria in food survive boiling temperatures?

4 Discuss the term "biodegradable" in relation to bacterial action and ecology.

5 Discuss the action of antibiotics as well as some dangers connected with their use.

6 What is algal bloom? What are some underlying causes of algal blooms, and of what importance are they ecologically and to human societies?

7 Compare a representative of Protista with a typical member of Monera. Include similarities and differences in structure, reproduction, and mode of nutrition.

8 Based on your present understanding of biological principles, in what respects does *Paramecium*, a unicellular organism, resemble a multicellular animal? Discuss in terms of motility, ingestion of food, defense mechanisms, digestion, excretion, and reproduction.

9 What is the importance of phytoplankton? What are some phytoplanktonic organisms? What is the red tide?

10 Based on the material in this chapter, discuss the importance of monerans and protistans to human societies.

11 What is implied by the phrase "saprobic way of life?"

12 What reproductive structure do yeast cells and morels have in common?

13 Consider the form of a mushroom. How is it formed and what is its cellular nature?

14 On the basis of your present understanding of the structure of plants and animals, do their cells bear the same relationship to overall form as do those of a mushroom?

15 Are there any simple tests to determine whether or not a wild mushroom is poisonous?

16 When you observe a mushroom growing in the earth or a bracket attached to a tree, what generalization can you make regarding the nature of the unseen part of the fungus?

**SUGGESTED
READING**

AHMADJIAN, V. 1963. The fungi of lichens. *Sci. Amer.* 208(2):122–32. (The roles of the fungal and algal partners in the lichen association are discussed.)

ALEXOPOULOS, C. J., and C. W. MIMS. 1979. *Introductory mycology*, 3rd ed. New York: John Wiley and Sons. (An up-to-date introductory presentation of the structure and classification of fungi, with special emphasis on their importance to people.)

ALEXOPOULOS, C.J., and H. C. BOLD. 1967. *Algae and fungi*. New York: Macmillan Co. (A simple and readable review of the major groups of algae, protistans, and fungi.)

BARNES, R. D. 1980. *Invertebrate zoology*, 4th ed. Philadelphia: Saunders College Publishing. (Contains a lucid presentation of the major types of protozoans.)

BLOCK, T. D. 1951. *Milestones in microbiology*. Englewood Cliffs, N.J.: Prentice-Hall. (A collection of original reports of major advances in bacteriology.)

BOLD, H. C.; C. J. ALEXOPOULOS; and T. DELEVORYAS. 1980. *Morphology of plants and fungi*, 4th ed. New York: Harper and Row. (An excellent comprehensive textbook of morphology, emphasizing algal evolution.)

BRILL, W. J. 1977. Biological nitrogen fixation. *Sci. Amer.* 236(6):68–81. (Explains the roles of bacteria and cyanobacteria in nitrogen fixation and surveys the biochemistry of the process.)

ECHLIN, P. 1966. The blue-green algae. *Sci. Amer.* 214(6):74–81. (Surveys the principal kinds of blue-green algae, their ecology, and their importance to humans.)

GRAY, W. D. 1959. *The relation of fungi to human affairs.* New York: Henry Holt and Co. (Facts and fallacies regarding the use of fungi in human societies are related in a most interesting manner.)

HAWKING, F. 1970. The clock of the malaria parasite. *Sci. Amer.* 222(6):123–31. (A good explanation of the life cycle of this sporozoan in its human host. Emphasis is on the timing of the relapsing phase.)

HAYASHI, T. 1961. How cells move. *Sci. Amer.* 205(3):184–204. (Discusses amoeboid locomotion, ciliary action, as well as some other aspects of cell movement.)

LITTEN, W. 1975. The most poisonous mushroom. *Sci. Amer.* 232(3):90–101. (Mushroom collectors and fanciers will profit from this account of mushroom poisoning.)

McCOSKER, J. 1977. Flashlight fishes. *Sci. Amer.* 236(3):106–14. (Discusses how some marine fish use luminescent bacteria to operate their "flashlights.")

STANIER, R.Y.; M. DOUDOROFF; and A. E. ADELBERG. 1976. *Microbial world,* 8th ed. Englewood Cliffs, N.J.: Prentice-Hall. (A brief but nicely presented overview of the protistans is included, together with a statement about their possible evolutionary relationship.)

WOESE, C. R. 1981. Archaebacteria. *Sci. Amer.* 244(6):98–122. (These unusual organisms are neither prokaryotes or eukaryotes.)

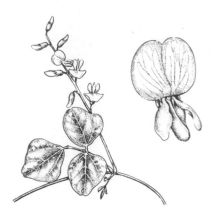

The Plant Kingdom

20

Most of us have rather well-defined images of the organisms we call "plants," and therefore it may come as a surprise to learn that a single microscopic green cell is as much a "plant" as is an oak tree or a petunia. As is often the case, the criteria by which science defines a group of organisms do not always fit the popular image.

Present evidence suggests that multicellular green plants invaded the land more than 400 million years ago, when the earth was far different from the green and pleasant place it is today. In that distant era, torrents of water ran down eroded hills unimpeded by vegetation, washing out sand, silt, and minerals and adding to the dissolved salts which the oceans contain to this day. It is likely that nothing resembling soil existed on the earth's exposed crust, because soil formation requires the presence of organic matter from decaying vegetation and protection of that material from erosion. The only agents then capable of forming potential soil particles were wind and rain, but as soon as rocks and stones were ground finely enough to constitute silt and dust, they were blown and washed away. Only on great mudflats at the mouths of rivers and in shallow lakes could multicellular life have flourished.

No doubt the first land vegetation consisted of pioneering forms of algae that grew on those mudflats, could withstand the alternate drying and wetting, and eventually could become semi-independent of the aquatic environment. Modern land plants and green algae have similarities of pigmentation, cell division, and sexual processes, supporting the concept of their evolutionary relationship. Possibly the transition from sea to land involved plants something like those pictured in Fig. 20-16.

Perhaps one and a half billion years ago the first organisms that we might, by stretching definition a bit, call plants made their appearance in the waters of the earth. Until then life was represented only by the prokaryotic bacteria and cyanobacteria. The evolutionary leap by which eukaryotic green cells arrived is by no means certain. This much we do know: eukaryotic cells of intricate shapes—little "stars" and "globes" not unlike the miniature "Christmas-tree ornaments" we call *diatoms* and *desmids*—were a part of a rich assemblage of unicellular life existing almost a billion years before the first multicellular organisms came into existence. (If you wish to explore this topic more fully, refer to the article by Gonzalo Vidal, referenced at the end of Chapter 2.) Given the variety of photosynthetic one-celled life presently known to us, it would be difficult to assign these first eukaryotes to a particular kingdom—they may have been protistans, and perhaps that would be the "safest" assumption, but they equally well might have been similar to one or another of the three major groups of eukaryotic algae. Possibly we shall never know.

ALGAE

In the five-kingdom classification used in this book, unicellular photosynthetic organisms are placed in Protista, whereas those that are predominantly multicellular are classed in Plantae. Among the latter are three groups of algae of which some are unicellular, some are multicellular colonial types, and a few are true multicellular plants. These three algal groups are classified and named on the basis of pigmentation as well as other features and are known as **green, red,** and **brown algae.**

Green Algae

Green algae, members of the division **Chlorophyta** (*chloro-,* "green," *phyta,* "plant"), constitute the largest group of algae, both in total numbers of species and in distribution. Most green algae are small plants; often a microscope must be used to identify structural details. Their characteristic pigments are **chlorophyll-a** (blue green), **chlorophyll-b** (yellow green), and several **carotenoid pigments,** which range from yellow to orange red. A blend of these pigments produces the grass-green color for which the group is named.

Green algae appear to have originated from motile unicellular ancestors, and to have evolved as three distinct lineages: **motile colony algae, filamentous algae,** and **coenocytic algae.** Filamentous algae are considered to have given rise to green land plants; the other two types probably are evolutionary dead ends (see Fig. 20-1).

Motile Colony Algae   Motile colony algae are classified according to cell number and reproductive complexity. The simplest are few-celled colonies composed of biflagellate cells. A unicellular form named *Chlamydomonas* represent the basic cellular unit in the series, because all motile colony algae are composed of chlamydomonaslike cells—ovoid cells having a large cup-shaped chloroplast partially surrounding the nucleus and two anterior flagella near which is a red-orange light receptor, called the **eyespot,** or **stigma** (Fig. 20-2; also see Fig. 2-9).

The most spectacular and complex motile colony alga is *Volvox,* a comparatively large ball of up to 65,000 chlamydomonaslike cells. *Volvox* is large enough to be seen without a microsope, although barely pinhead size. Under a dissecting microscope volvoxes are truly beautiful objects, rolling about in water like small emerald globes (Fig. 20-3).

FIGURE 20-1
Evolutionary lineages among green algae.

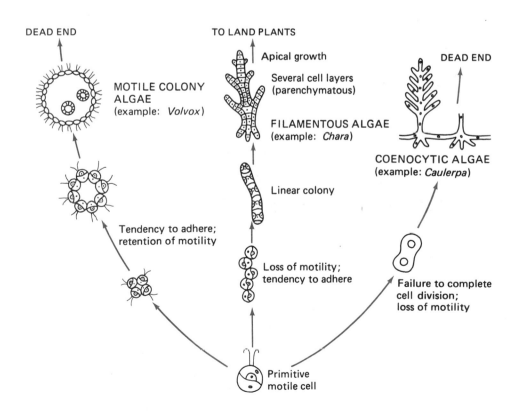

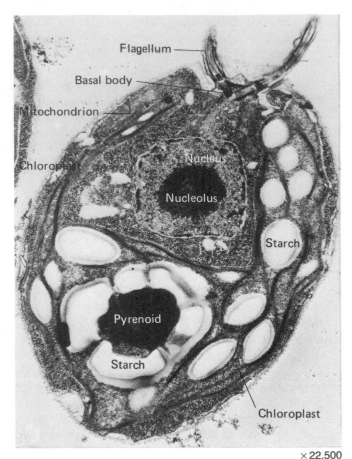

FIGURE 20-2

Electron micrograph of *Chlamydomonas*. (Photo reproduced from D. L. Ringo, 1967, Flagellar motion and fine structure of the flagellar apparatus of *Chlamydomonas, J. Cell Biol.* 33:543–71, by copyright permission of The Rockefeller University)

×22,500

Labels on figure: Flagellum, Basal body, Mitochondrion, Chloroplast, Nucleus, Nucleolus, Starch, Pyrenoid, Starch, Chloroplast

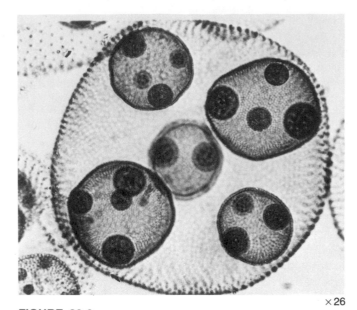

×26

FIGURE 20-3

Volvox. The smaller spheres inside are daughter colonies. (Photo by Carolina Biological Supply Company)

Filamentous Algae All higher green plants are composed of three-dimensional arrangements of many cells. Possible green algal ancestors of higher plants are algae characterized by similar cellular construction. Filamentous algae are thought to have evolved from single-celled greenalgae by formation of linear colonies, which by further specialization evolved into a variety of filamentous forms. One of these filamentous algae often studied is *Ulothrix*, a freshwater genus growing in tufts composed of individual filaments of cells. Its reproduction features zygotic meiosis, which is predominant in the majority of filamentous algae (Fig. 16-4; also see Fig. 20-4).

Some cells of *Ulothrix* divide internally by mitosis and produce zoospores (which, because they are created by mitosis, also are called **mitospores**), each looking like a tiny *Chlamydomonas* cell except for having four flagella instead of two. Zoospores swim about for a few hours, then settle down on a rock or other firm object, lose their flagella, and by repeated mitotic divisions form a new filament. This is the common mode of asexual reproduction.

Sometimes changes in light or in water temperature induce filaments to form gametes instead of mitospores. Gametes of *Ulothrix* differ from mitospores in having two rather than four flagella and in fusing in pairs to form zygotes. Zygotes divide by meiosis, each forming four flagellated, 1*n* meiospores which produce new filaments in the same manner as do zoospores.

Not all algae have flagellated reproductive cells. A beautiful filamentous freshwater alga named *Spirogyra* often is studied in biology classes. *Spirogyra* seems a rather specialized type, because its cells have helical chloroplasts studded at frequent intervals with pyrenoids (Fig. 20-5). Reproduction in *Spirogyra* occurs by **conjugation,** during which tubes develop between cells of adjacent filaments. One filament is a functional male plant, and the protoplasm of its cells migrates, in the form of amoeboid gametes, through the conjugation tubes into the cells of the female plant. The resultant zygotes develop thick walls and become **zygospores.** Meiosis takes place within the zygospore, and new 1*n* filaments develop when zygospores germinate.

One group of green algae, the class **Charophyceae,** is structurally advanced in comparison with most other green algal classes and consequently has been associated with mosses in some classifications. Although comparatively complex in cellular organization, members of Charophyceae have a long evolutionary history and are

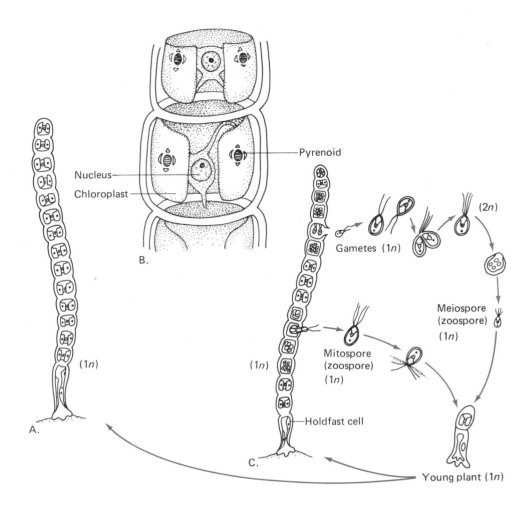

FIGURE 20-4
Ulothrix, a filamentous green alga. **A.** Form exhibited by a nonreproductive, 1*n* filament. **B.** Cell of *Ulothrix*. **C.** Reproduction in *Ulothrix*.

Nucleus

Pyrenoid

Chloroplast

B.

(1*n*)

A.

Gametes (1*n*)

(2*n*)

Meiospore (zoospore) (1*n*)

Mitospore (zoospore) (1*n*)

(1*n*)

Holdfast cell

C.

Young plant (1*n*)

among the oldest fossilized algae known. They constitute important evolutionary land marks, because of the assumption that simpler green algae must have predated them by many millions of years.

The **stonewort,** *Chara,* a rather common member of Charophyceae, lives in freshwater ponds (Fig. 20-6). *Chara* is a beautiful branching alga with some advanced plant features. Its stems are composed of a row of elongated inner cells surrounded by a layer of tubular sheathing cells, giving *Chara* a **parenchymatous** plant body. (The term "parenchymatous" refers to a simple plant tissue composed of several cell layers.) At frequent intervals along the stem of *Chara,* whorls of branches form; near these, multicellular male and female sex organs develop. The male organ produces tiny, biflagellate sperm cells, and the female organ produces a large, nonmotile egg. This differentiation of gametes into motile sperm and large nonmotile eggs is called **oogamy** (oh-ah-gamy) and is typical of reproduction in all higher green plants.

Chara also resembles land plants in its growth, which occurs at the tip of stems and branches, not by

cell divisions throughout the plant body. This style of growth is called **apical growth** and is found in all higher land plants. Another feature of *Chara* found in higher plants is its jointed appearance. The joints, called **nodes,** are places where branches occur. Regions between nodes are **internodes.** In higher land plants leaves or branches, or both, are attached to the stems at nodes.

Chara not only resembles land plants in appearance but it shares features of cell structure and cell division with green land plants (it is oogamous, its sperm are of the moss type, and a cell plate is formed in mitosis just as in land plants and unlike most algae). For these and other reasons, the group of green algae to which *Chara* belongs is thought to be the key to deciphering the evolution of the "green line" leading to all higher land plants. For further information on this subject, refer to the article by Linda Graham referenced at the end of the chapter.

Coenocytic Algae Some of the most beautiful marine green algae are single giant cells. The mermaid's wine-

FIGURE 20-5

Reproduction in *Spirogyra*, exemplifying zygotic meiosis.
A. Vegetative cell in which the nuclei, spiral chloroplasts, and pyrenoids are seen. **B.** Early stage in conjugation. **C.** Male gametes in the process of moving through the conjugation tubes into the female cells. **D.** Zygotes. (Photo by Carolina Biological Supply Company)

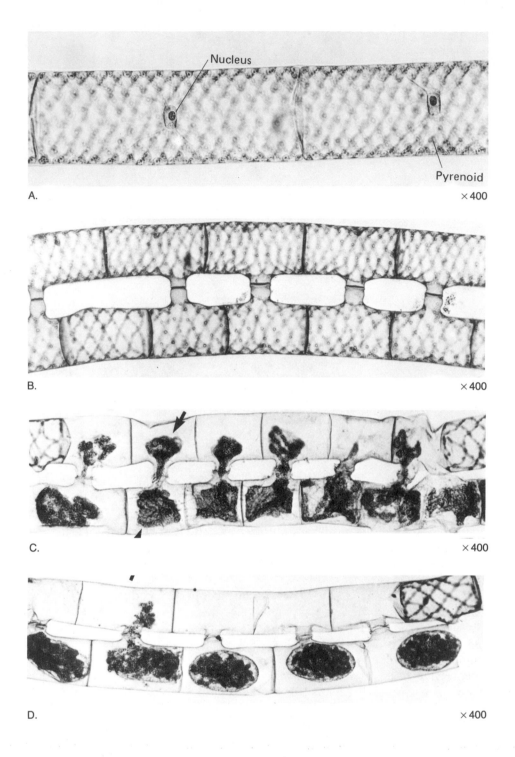

Nucleus

Pyrenoid

A.   ×400

B.   ×400

C.   ×400

D.   ×400

glass, *Acetabularia* (see Color plate 7A), is a coenocytic form, as are many-branched types such as *Codium*, which often attaches itself to scallops and other shellfish, and *Caulerpa*, a large and extensively branching plant, although technically a single cell (see Fig. 20-7). The term "coenocytic" was discussed in Chapter 19 and refers to a type of nonpartitioned multinucleate cell structure.

Evolutionary Importance Present-day green algae arc thought to have evolved from ancient and now extinct flagellated protistans. The ancestral forms may have originated by symbiosis between heterotrophic flagellated cells and photosynthetic prokaryotes (also discussed in Chapter 2). Until recently, as noted in the preceding chapter, this origin was questionable, because

414   PART THREE | THE DIVERSITY OF LIFE

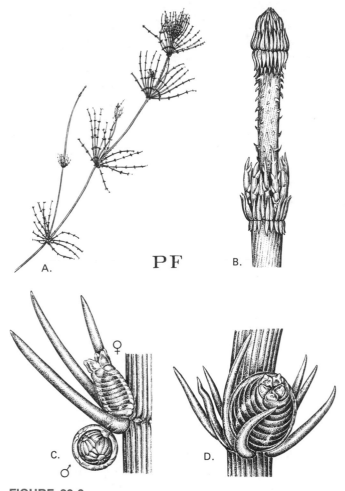

PF

the only prokaryotic algae known were blue green (cyanobacteria), with unique phycobilin pigments and lacking chlorophyll-b. The discovery of *Prochloron*, which has the photosynthetic pigments (chlorophyll-a and -b) and carotenoids characteristic of green algae and higher green plants, strengthens the hypothesis that green algae originated by symbiosis with green prokaryotes.

Red Algae

Almost all red algae are saltwater plants, growing mostly in warmer oceanic waters (Color plate 7A). They are important reef-building organisms. Coral reefs usually are considered to be an animal community laid down over millions of years by coral polyps, yet many coral reefs owe their existence to **coralline algae,** which secrete limestone and form a compact stony barrier to the forces of breaking waves.

Some red algae are important human foods (Fig. 20-8). In Japan a red alga (*Porphyra*) is cultured in shallow coastal waters, where it is grown on nets and mats of bamboo. When the crop is mature, algae are harvested from small boats and processed for food. **Agar-agar,** a gel once used in making jellies and now widely used in bacteriological work to solidify culture media, is a product of several species of Malaysian red algae of the genus *Gelidium.*

Red algae (division **Rhodophyta**—*rhodo-,* "red," *-phyta,* "plant") resemble other plants in possessing chlorophyll-a; in addition, they have **chlorophyll-d** and two accessory phycobilin pigments, **phycoerythrin** (red) and **phycocyanin** (blue). The latter two closely resemble red and blue phycobilin pigments of cyanobacteria (formerly blue-green algae), which, as in red algae, are lo-

FIGURE 20-6
Chara. **A.** Plant showing nodes, internodes, and branching pattern. **B.** Apex of *Chara.* **C.** Oogonium *(top)* and male gametangium *(bottom).* **D.** Enlarged oogonium.

FIGURE 20-7
Coenocytic green marine algae. **A.** *Codium.* **B.** *Caulerpa.* Enlargements of branches are also shown.

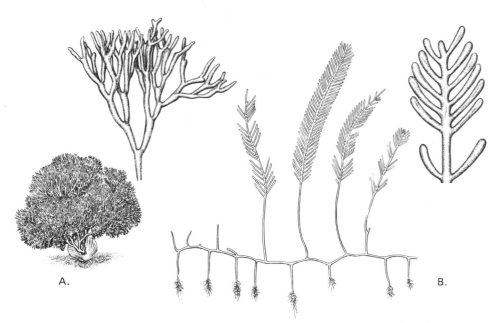

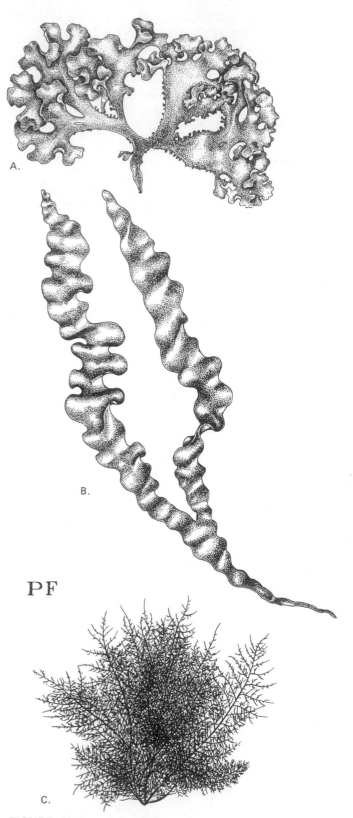

PF

FIGURE 20-8
Species of edible red algae. **A.** *Chondrus crispus,* or Irish moss, found in the North Atlantic and used in various foods. **B.** *Porphyra,* a food in Japan. **C.** *Gelidium,* source of agar-agar.

cated in particles called **phycobilisomes** attached to the thylakoids. This has led biologists to speculate that, of all plants, red algae are the most direct eukaryotic descendants of cyanobacteria (see Fig. 20-9). The phycobilins and chlorophylls give living red algae colors ranging from olive green to purple; most red algae are not distinctly red until they die.

Red algae are among the deepest growing algae in oceans, owing to their phycobilins, which are able to absorb green light (green light penetrates water more deeply than other wavelengths). This absorbed light energy is transferred from the phycobilins to chlorophyll, which initiates the chemical reactions of photosynthesis. It is because of their indirect role in photosynthesis that the phycobilins are called **accessory pigments**.

Most red algae are comparatively small plants, ranging from a few millimeters to about half a meter in length. A few forms are microscopic and unicellular. Multicellular species characteristically have cells connected by end-to-end constrictions that leave a pore, or **pit connection**, between adjacent cells (Fig. 20-10). This structure bears a resemblance to the arrangement of cells in fungi, particularly the sac fungi, which also have pit connections. For this and other reasons, some biologists suggest that red algae may be related to fungi.

Brown Algae

Like red algae, most brown algae (division **Phaeophyta**—*phaio-*, "dusky," *-phyta,* "plant") live in salt water, although they tend to be found in colder regions. Some are of microscopic size, but the majority range from an inch to many feet in length (Fig. 20-11). Two common types are **rockweeds** and **kelps**. The former include species of *Fucus,* which are found in intertidal zones, where they are alternately wet and dry as tides advance and recede, and the sea palm *Postelsia,* which grows on surf-washed rocks (Fig. 20-12 and Color plates 2C, 7C). Kelps grow in deeper water and are among the largest plants, sometimes attaining lengths of more than 50 m (160 ft). One group of brown algae, belonging to the genus *Sargassum,* is free-floating, giving its name to the region southeast of Florida called the Sargasso Sea (see box, p. 420).

Uses   Off the California coast, in cold waters brought down from the Bering Sea by the California Current, are great kelp beds. These are harvested each year by barges fitted with mowers that cut the plants a few feet below the surface. Kelps regrow rapidly, sometimes at a rate of 1 m (3 ft) in 24 hours, and are harvested repeatedly without apparent harm; much more damage is done to them by discharge of sewage into Pacific bays, and some formerly rich beds have been destroyed in this way. Be-

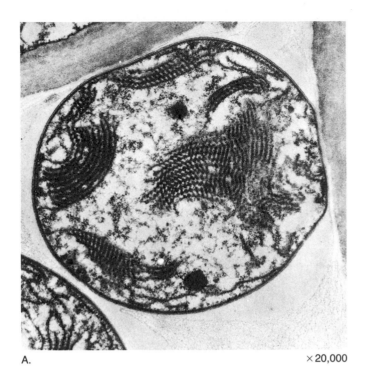

A.   ×20,000

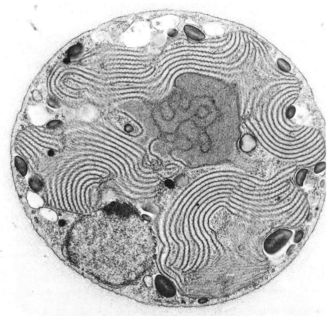

B.   ×17,000

FIGURE 20-9

A. Cell of the cyanobacterium *Nostoc*. B. Unicellular red alga *Porphyridium*. Note the similar arrangements of phycobilisomes.
(A, photo courtesy Charmian Aherne; B, photo courtesy Elizabeth Gantt)

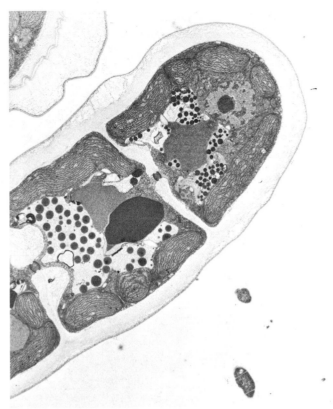

×4600

FIGURE 20-10

Cells of the red alga *Ptilota*. The membranes of the chloroplasts are parallel, resembling in this respect cyanobacterial membranes.
(Photo from J. L. Scott and P. S. Dixon, 1973, Ultrastructure of tetrasporogenesis in the marine red alga *Ptilota*, *J. Phycol.* 9:29–46)

cause kelp beds are a habitat of numerous sea creatures, including sea otters and seals, their loss has ecological as well as commercial significance.

Kelps and rockweeds once were a major source of iodine and were harvested intensively for that purpose, but currently they are more in demand as a source of complex substances called **phycocolloids**. Phycocolloids include **algin**, sodium alginate, and alginic acid and are used in ice creams and puddings (as thickening agents), pharmaceuticals, beer (to help maintain foam), paints (dripless types), soaps, and other products.

Structure and Reproduction   Cells of brown algae contain distinctive chloroplasts having parallel membranes (thylakoids) in groups of two or three (Fig. 20-13); in larger forms such as kelps and rockweeds some different types of cells are distinguishable. Although cell specialization is a feature more characteristic of land plants than of algae, some brown algae are composed of cells specialized as covering cells (epidermis), as photosyn-

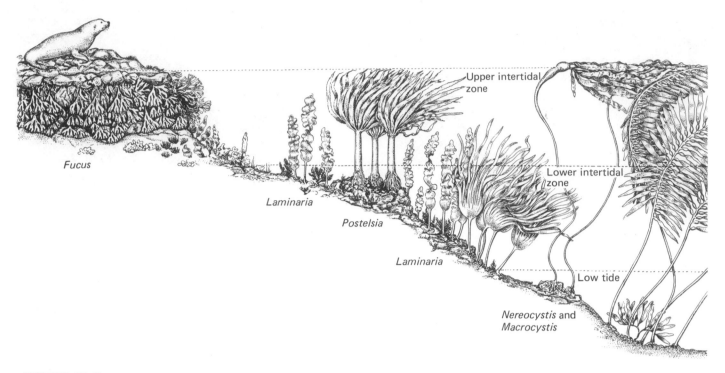

Upper intertidal zone

Lower intertidal zone

Low tide

Fucus

Laminaria

Postelsia

Laminaria

Nereocystis and *Macrocystis*

FIGURE 20-11
Marine brown algae association of the northern Pacific coast. Algae in the upper intertidal zone are bathed in seawater once every 24 hours; those in the lower intertidal zone are under water twice each day.

thetic or storage tissue, and as simple food-conducting elements.

Photosynthetic cells of brown algae contain characteristic chlorophyll-a and **chlorophyll-c,** and brown pigments named **fucoxanthins;** the latter give brown algae their characteristic coloration. Living brown algae

commonly are olive green, and dead algae are a distinct yellow brown.

Most brown algae exhibit **alternation of generations.** Some microscopic brown algae have **gameto-phytes** and **sporophytes** of about the same size and form. Among kelps, the spore-producing stage, or spo-

FIGURE 20-12
Postelsia, the sea palm, a brown alga of the upper intertidal zone of the North American Pacific coast.

FIGURE 20-13
Chloroplast of the brown alga *Padina vickersiae.* Chloroplasts of brown algae contain thylakoids consisting of two or three parallel membranes. (Photo courtesy Clinton J. Dawes)

×15,750

rophyte generation, is represented by the large kelp plant (up to 50 meters long), whereas the gametophyte stage is microscopic. Among rockweeds, the sexual phase is very simplified, and true gametophytes are non-existent.

LAND PLANTS

Some botanists, impressed with the size and complexity of many of the brown algae, have thought it possible that green land plants might have evolved from ancestors resembling sea palms (*Postelsia*), Devil's aprons (*Laminaria*), or other brown algae. These have organs resembling roots, stems, and leaves (see Fig. 20-12) and can survive alternate exposures to wetting and drying. Attractive though this concept is, the cellular biology of brown algae does not support it. Cellular organization, pigmentation, and the structure of gametes are not consistent with ancestry to the green land plants. Instead, a group of green algae, of which *Chara* is a member, seems now to exhibit the combination of characters most consistent with land-plant evolution.

As plants became more and more independent of water, they developed characteristic tissues and organs such as **stems, roots, leaves,** and **reproductive structures.** Epidermal cells of the first land plants probably developed a waterproof **cuticle** rather early in their evolution, thus cutting down on water loss by evaporation. The formation of the cuticle, however, posed a new problem: how to permit nutrients and gases to enter and leave the plant. Evolution of root systems, as distinct from stems, solved problems of nutrient uptake. Cells of root epidermis, which lacks a cuticle, can take up water and dissolved minerals; water loss due to evaporation is less of a problem because roots are underground. Aerial parts of plants, though equipped with cuticles, have evolved systems of pores that admit carbon dioxide and permit oxygen to leave. At first these probably were simply holes in the epidermis and therefore rather inefficient, allowing water to escape from the plant's interior. Later, guard cells were evolved that were capable of opening and closing the pores (stomata) and thus limiting water loss.

Most algae are limp and flexible plants which depend on surrounding water for support. Photosynthesis in such a plant can never be very efficient on land, with half its surface buried in mud. Consequently, as plants evolved on land, supporting tissues capable of holding a plant erect developed.

Eventually, certain supporting cells may have become modified to function as conduits for the transport of water and nutrients. These vascular tissues enabled plants to transport food, water, and minerals over considerable distances and so to grow much taller than simpler, nonvascular land plants. In this way competition between plants for space and sunlight took on a new dimension, and the first trees came into being.

As time passed, many types of land plants evolved and began to cover the earth and protect the newly formed soil, which they enriched by their own death and decay. Plants had conquered the land and made it possible for animal life also to exist away from the sea.

Bryophytes

The actual sequence of evolutionary events leading from green algae to land plants is not known. The simplest land plants are **bryophytes** (division **Bryophyta**—*bryo-*, moss, *-phyta*, "plant"), and some botanists think that these plants typify ancestral land plants. Bryophytes, which usually grow in moist habitats, are represented by three classes: **mosses, liverworts,** and **hornworts** (Color plate 7D–F). There are about 14,500 species of mosses, 9500 species of liverworts, and 320 species of hornworts.

Typically mosses are small plants, seldom attaining heights of more than a few centimeters (Fig. 20-14), although one New Zealand moss, *Dawsonia*, sometimes attains a stature of about 60 cm (2 ft). Mosses usually grow in very dense stands, giving the appearance of a lush green carpet, so much a part of Japanese gardens and landscaping. Bog mosses of the genus *Sphagnum* grow so densely in wet places that they eventually produce thick layers of **peat** which provide an exploitable resource of soil additives for gardening and in some countries (Ireland) a source of fuel.

THE SARGASSO SEA

O ne of the most interesting forms of free-floating brown algae is *Sargassum.* This group of brown algae makes up the vegetation of the Sargasso Sea in the Atlantic Ocean east of Florida and the Bahamas. In this region is the Bermuda triangle, a mysterious area where ships and planes are reputed to vanish without trace. Actually, the safety record of ships and planes here is no worse than anywhere else; but fear of this region has a long and persistent history. In ancient times, ships were thought to become becalmed forever in the Sargasso Sea, and fiction pictured it as a dead, still, seaweed-covered place where ancient sailing ships with skeletons still manning the rigging lay motionless (Fig. 1). Columbus's sailors dreaded the area, although their ships sailed through it with no trouble.

The Sargasso Sea is an area about 1600 by 3200 km (1000 by 2000 mi), with an axis more or less east to west. It is a sunny, open place with floating

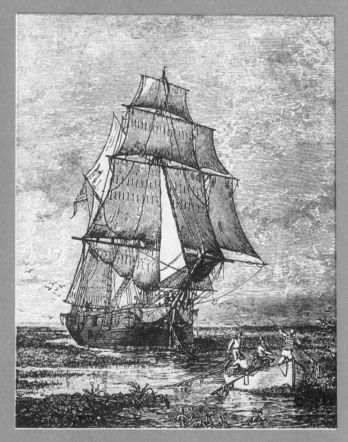

FIGURE 1
Sargasso Sea in fiction. Coleridge wrote *The Rime of the Ancient Mariner* about a ship becalmed here: "Water, water everywhere, nor any drop to drink." (From The Bettmann Archive, Inc.)

patches of *Sargassum* and expanses of open water. It is not known how *Sargassum* has become so concentrated here, but through evolution the plants have become adapted to open water, rather than being anchored to the bottom as is the case with most other brown algae. Small, berrylike float bladders contain gas and keep the plants at or near the surface, where they reproduce by budding and fragmentation. A rich life of sea animals lives in the *Sargassum,* and its members have evolved protective forms and colorations so as to be quite inconspicuous (Fig. 2).

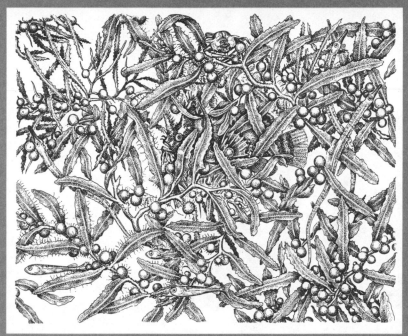

FIGURE 2
Sargassum association.

Bryophyte Reproduction Among green land plants, free-living gametophytes, as described in Chapter 16, are found only in the reproduction of mosses, ferns, and a group of vascular plants commonly known as "fern al- lies." The features that separate bryophytes from other land plants are their relatively simple nature and their emphasis on the gametophyte generation as a self- sufficient stage of the life cycle. Green, photosynthetic gametophytes occur in all classes of bryophytes, and in all cases sporophytes are attached to and nutritionally dependent on gametophytes. Some mosses have simple food- and water-conducting tissues, but the majority of bryophytes are nonvascular plants.

We are at this point interested mostly in knowing what mosses (and some of their close relatives called liverworts and hornworts; see Fig. 20-17) can tell us

about the evolution of reproduction in land plants. We have alluded to the dependency of their sporophytes on their gametophytes, something not found in more ad- vanced land plants, in which sporophytes are an inde- pendent generation, but it is entirely possible and plau- sible that independent sporophytes might have evolved from dependent ones by developing their own absorp- tive organs. The sporophytes of certain mosses have chloroplasts and can photosynthesize; theoretically, if they had rhizoids and could absorb water and minerals they might be able to function as free-living sporo- phytes. Experiments with detached moss sporophytes have shown them capable of completing their portion of the life cycle independently of the gametophyte. They do so, however, only when kept in a solution of water and minerals.

FIGURE 20-14
Gametophytes and sporophytes of the hairy-cap moss *(Polytrichum commune)*. A few sporophytes *(arrows)* are present among the more numerous gametophytes.

The reproductive organs of bryophyte gametophytes are the **antheridium** (male) and the **archegonium** (female) (Fig. 20-15). The antheridium is a "lollipop"-shaped structure containing thousands of biflagellate sperm. When the sperm are released, they are attracted to and swim to the archegonium, a vaselike organ containing a single egg. The archegonium (pl. archegonia) produces and secretes sperm attractants (sucrose is one), and when a film of water is present, the sperm swim to the archegonia and fertilize the eggs within them. Then follows a stage, the **embryo**, which is so characteristic of green land plants that in some classifications the entire assemblage of green land plants is called Embryophyta. The land-plant embryo is defined as a young, multicellular sporophyte which is dependent on its parent for its food supply.

The moss embryo atop the gametophyte plant rapidly develops into a mature sporophyte (Fig. 20-15) con-

sisting of a basal nutrient-absorbing organ, the **foot**, a **stalk** or **seta**, and a spore-producing organ, the **sporangium**—in mosses usually referred to as the **capsule**. Within the sporangium, **spore mother cells** undergo meiosis and produce quartets (**tetrads**) of meiospores. These are shed through a complicated saltshakerlike opening and germinate to form threadlike gametophytes if they happen to fall in a humid area. The threadlike gametophytes (called protonemata, "early threads") form buds, which grow into gametophytic moss plants, and the cycle is repeated. Other bryophytes have similar reproductive cycles, although the overall appearances of their gametophytes and sporophytes differ (Fig. 20-17).

Gametophytes of mosses are restricted to areas in which surface water is at least periodically present, because the sperm require water to effect fertilization. Naturally the moss sporophyte also is restricted to these same habitats because it remains attached to the gametophyte for the duration of its life. It is a "ball-and-chain" relationship that restricts the ability of mosses to colonize certain habitats effectively.

Some botanists think land plants may have evolved from green algae already possessing alternation of generations. Others favor an origin like that depicted in Fig. 20-16, in which sporophytes evolved as a result of zygotes becoming multicellular instead of undergoing meiosis. If land plants originated from algae having dependent sporophytic cells, as illustrated, we may seriously consider that ancestral mosses or mosslike plants were a step in the direction of higher land plants. Most botanists think more evidence is needed before a commitment is made in favor of this theory, but current studies seem to be tipping the scale in favor of the bryophytes.

Liverworts and Hornworts Liverworts and hornworts, like mosses, are plants of humid habitats. The gametophytes, as in mosses, are free-living and the sporophytes are dependent on them. Gametophytes of liverworts are prostrate (Color plate 7D) and either ribbonlike or leafy (leafy liverworts). Their sporophytes are either short stalked or globose (Fig. 20-17). Gametophytes of hornworts resemble small, flat, dark-green cushions with irregular margins; their sporophytes are cylindrical, long-lived, and have a basal meristem (Fig. 20-17). It has been suggested that sporophytes of this kind may represent an early stage in the evolution of independent sporophytes, characteristic of vascular plants.

Vascular Plants

Vascular plants, which outnumber the approximately 25,000 species of bryophytes about 10-1, include nearly

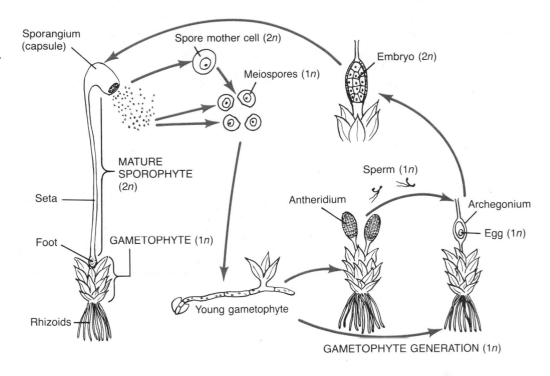

FIGURE 20-15
Key stages in the life cycle of a moss. The dimensions of sex organs, gametes, and spores are exaggerated.

all economically useful plants and are predominant members of the terrestrial environment. The sporophyte generation is comparatively large—sporophytes of California redwoods are the largest organisms in existence today—but their gametophytes are of near-microscopic dimensions. We generally divide present-day vascular plants into two groups: those in which the gametophytes are free-living (ferns and fern allies) and those in which the gametophytes are enclosed (**seed plants**).

Origin   The nature of the first plants to invade the land is not known for certain, but it is thought that mosses and liverworts might have been pioneers in land plant evolution, just as they are in ecological succession. However, although some fossils of liverworts have been discovered that are more than 300 million years old, fossils of even older vascular plants have been found. Thus the oldest true land plants may not have been mosses at all but vascular plants.

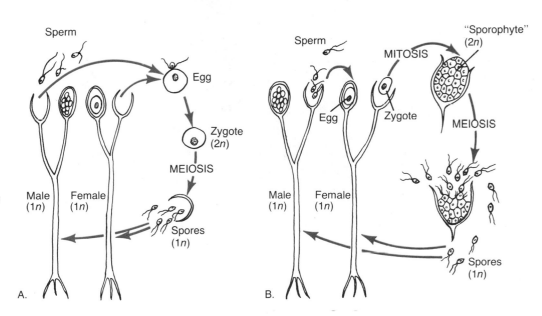

FIGURE 20-16
A hypothetical stage in the evolution of the sporophyte generation. A green alga having separate male and female gametophytes and zygotic meiosis as in **(A)**, retains its zygote, as in **(B)**, which then divides by mitosis instead of meiosis. A small, dependent "sporophyte" is formed and then undergoes meiosis to produce many haploid swimming spores. This condition is approached in the green alga *Coleochaete* (see the article by Graham listed in the readings at the end of this chapter).

ECOLOGY OF MOSSES

I n 1967 a new volcanic island appeared off the southern coast of Iceland. The new island, which was named Surtsey, grew rapidly and soon was more than a mile wide. Even as the cinders were cooling, Icelandic biologists made plans to monitor it for signs of life. Within a few months mosses began to appear among the ashes, forming a ground cover; later, other plants appeared. Mosses, despite their delicate and fragile appearance, are **pioneers** among plants. Wherever sufficient moisture is present, mosses colonize the bare rocks and disturbed soil. In Iceland, a land famous for its volcanoes and lava flows, hundreds of square miles of lava fields are carpeted with a thick blanket of moss. Eventually, sufficient humus accumulates beneath the moss plants to enable other plants to gain a foothold. In time, the mosses are replaced by grasses, shrubs, and low trees. (Initial stages in this kind of vegetational development are occurring on the lava produced by the 1980 eruptions of Mt. St. Helens in Washington State.) The vital ecological processes by which pioneer plants establish themselves in barren environments and eventually create conditions suitable for the growth of other plants is called **succession.**

One interesting succession in which a family of mosses act as pioneers is the transformation of a barren pond into a peat bog and finally into dry land. Species of the genus *Sphagnum* have leaves and stems composed in part of empty water-holding cells (Fig. 1 and Color plate 7F). These cells enable the *Sphagnum* plant to absorb quantities of water and to form a thick mat at the edges of a pond or lake. As *Sphagnum* continues to grow, it forms a floating mat, which finally may cover the surface water at the borders of the pond. Meanwhile, dead *Sphagnum* accumulates beneath the mat and forms a thick layer of peat. Eventually, peat and other vegetation complete the transition to dry land (Fig. 2).

 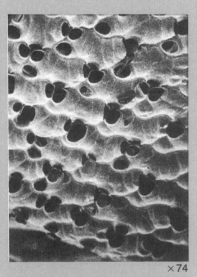

×74   ×74

FIGURE 1
Scanning electron micrographs of the leaves of the bog moss, *Sphagnum.* The leaves are composed of narrow green cells and larger, hollow, water-holding cells. (From H. N. Mozingo, 1969, *The Bryologist* 72:484–88; photos by Edwin R. Lewis)

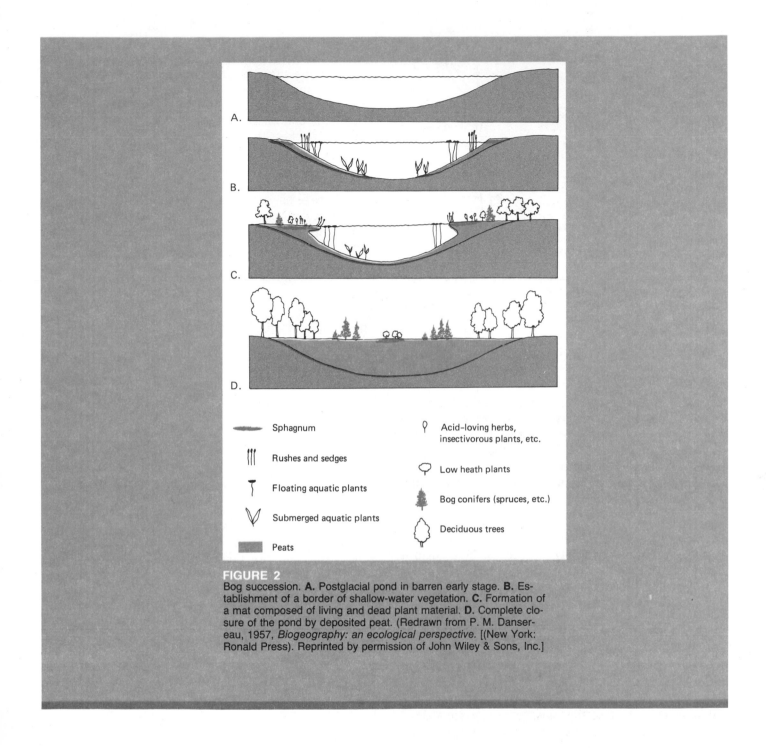

FIGURE 2
Bog succession. **A.** Postglacial pond in barren early stage. **B.** Establishment of a border of shallow-water vegetation. **C.** Formation of a mat composed of living and dead plant material. **D.** Complete closure of the pond by deposited peat. (Redrawn from P. M. Dansereau, 1957, *Biogeography: an ecological perspective.* [(New York: Ronald Press). Reprinted by permission of John Wiley & Sons, Inc.]

In 1859 a Canadian geologist named William Dawson (1820–1899) described a fossil plant, new to science, which he named **Psilophyton** (sy-lo-fyton) *princeps*. This ancient plant (more than 250 million years old) was vascular, with a solid core (protostele) composed of xylem tracheids (see Figs. 20-18 and 20-19).

The concept emerging from research of paleobotanists is that the earliest vascular plants were simple leafless and rootless plants with forking stems. Presumably these plants were photosynthetic; they bear guard cells and stomata, testifying to requirements for gas exchange consistent with photosynthesis. Some fossils are

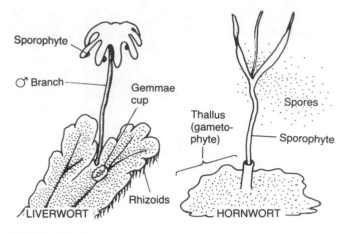

FIGURE 20-17
Liverworts and hornworts. The **gemmae** cup on the surface of the liverwort gametophyte is an asexual reproductive structure which produces small disk-shaped buds capable of growing into new gametophytes.

lowing the devastation of a forest fire, when the ground lies bare, bracken ferns (*Pteridium*) provide a ground cover, in the shade of which seedling trees grow. In the tropics ferns have proliferated abundantly in numbers and species, ranging from delicate, filmy ferns to robust tree ferns and enormously extended climbing varieties, one of which—*Gleichenia*—has leaves as long as 50 m (165 ft). *Gleichenia* produces almost impenetrable tangles on hillsides and in gardens and wherever else the forest has been disturbed or destroyed.

Tree ferns reach heights of 10 m (33 ft) or more and are particularly common in cut-over rainforests; together with other weedy trees, shrubs, and vines they produce jungles (Fig. 20-21 and Color plate 8A). Yet also in the tropics are found the most delicate of ferns, filmy ferns that shrivel up in a few moments after exposure to dry air.

An almost universal fern characteristic is the complex, multiveined, divided leaf, which as noted earlier can be very large. The botanical term for leaves of the fern type, whether large or small, is **megaphyll** ("big

recognizable as sporophytes, bearing terminal globose to ellipsoidal meiosporangia. (The spores within have three-cornered markings and occurred in tetrads. Thus they were products of meiosis; note that in living land plants, tetrads are direct products of meiosis.)

Little is known about gametophyte structure of the oldest vascular plants.

Fork Ferns   Do any modern-day plants look like *Psilophyton*? Dawson named *Psilophyton* because he thought it resembled **Psilotum,** which belongs in a very small group of two genera and perhaps a dozen species of present-day tropical and subtropical plants sometimes called **fork ferns.** A rare fork fern, T*mesipteris* (mes-ip-teris) *tennensis*, occurs in Australian rainforests; a more common one, *Psilotum nudum*, grows in Florida (Fig. 20-20). Both *Psilophyton* and T*mesipteris* are classified in the division **Psilophyta** (*psilo-,* "bare," *-phyta*, "plant").

Botanists would like to believe that *Psilotum* is a modern descendant of the most ancient vascular land plants, but unfortunately the links with the past are missing, as there are no connecting fossils to indicate any actual relationship with *Psilophyton.*

Ferns   Ferns (division **Pteridophyta**—*pteris*, "fern," *-phyta*, "plant") are the largest living group of primitive plants. They are widespread and grow in all environments, from the subarctic to the equatorial tropics. Fol-

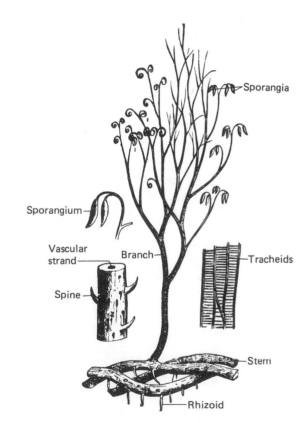

FIGURE 20-18
Dawson's reconstruction of *Psilophyton*. Note the tracheids.

A.   ×138

B.   ×368

FIGURE 20-19
Scanning electron micrographs of fossilized *Psilophyton*. **A.** Proto-
stele. **B.** Tracheids. Tracheids are more complex than shown by
Dawson. (Photos from C. Hartman and H. A. Banks, 1980, *Amer. J.
Bot.* 67:404–8)

leaf"). Although the name describes a large leaf, mega-
phylls technically are leaves having branching veins; they
are thought to have evolved by modification of branches
of primitive plants.

Some ferns have simple leaves with single blades,
but most have **pinnately compound leaves** in which the
blade is subdivided into leaflets. If the leaf has secondary divisions it is a **bipinnate leaf,** and if further divided
it is said to be **tripinnate** (Fig. 20-21 and 20-23). Many
persons recognize ferns from their compound leaves. Another characteristic of ferns is the coiling of young

leaves. Such coiled immature leaves are often called **fid-
dleheads** (Fig. 20-21).

Fern stems, with the exceptions of trunks of tree
ferns, are **rhizomes** extending horizontally, sometimes
for many feet. ("Rhizome" is the general term applied to
any elongate underground stem.) Internal structure of
fern stems varies considerably. Some stems are protostelic, but most are **siphonostelic,** which means that the
xylem and phloem are in the shape of a cylinder having
a central core of pith composed of parenchyma cells.
This basically is the arrangement found in stems of seed
plants (conifers, cycads, flowering plants, etc.).

Fern roots arise randomly from the stem surface
rather than as a prolongation at the base of the stem.
Any root arising randomly from a stem is an **adventitious root,** and in some ferns these roots are so numerous and closely spaced that they produce a mantle of
tough, intertwined fiber. This is especially obvious at the
base of tree-fern stems; blocks of this material are harvested and sold to be used as planters or supports for
growing orchids, bromeliads, and philodendrons.

Fern Reproduction   The basic parts of the fern reproductive cycle were outlined in Chapter 16, and the point
was stressed that in ferns and in all other vascular plants
the sporophyte generation is large and independent,
while the gametophyte is a small, nonvascularized plant.
We should now like to add a few particulars to that generalized account.

Persons who grow ferns as house plants or observe
them in nature sometimes are puzzled by the appearance of rusty brown spots or patches on the undersides
of some or all leaves. Such leaves are **sporophylls**
("spore leaves") and the brown spots are clusters of
sporangia, each containing relatively few spores (from 16
to 64 in most species) (Fig. 20-22). The spore clusters are
a characteristic of ferns, and their shape and arrangement can be used to identify and classify many species;
each cluster of sporangia is a **sorus** (pl. **sori,** "heap") and
it may be circular (as in Fig. 20-22), rectangular, or extend over much of the leaf surface or along its margins.
Spores commonly are produced in great numbers and
are dispersed far and wide by air currents. They germinate in moist, shady habitats, producing gametophytes
called **prothalli** (sing. **prothallus,** "early twig").

Mature fern gametophytes often are heart-shaped,
green, and photosynthetic, just 1 mm or so long, and
bear on their undersides hairlike absorbing cells called
rhizoids as well as male and female sex organs (antheridia and archegonia, respectively) (Fig. 20-23). Antheridia
produce multiflagellate sperm which swim in surface water to gametophytes having vase-shaped archegonia.

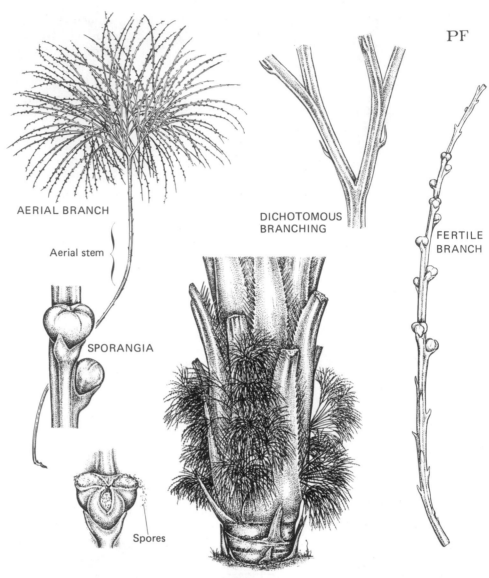

FIGURE 20-20
The fork fern *Psilotum nudum* growing as an epiphyte on a palm tree.

PF

AERIAL BRANCH

Aerial stem

SPORANGIA

Spores

DICHOTOMOUS BRANCHING

FERTILE BRANCH

FIGURE 20-21
Tree ferns. **A.** Australian tree fern, *Dicksonia antarctica*. **B.** Closeup view of a fiddlehead of a fern tree.

A.

B.

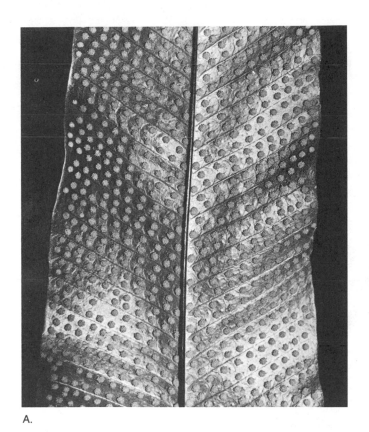
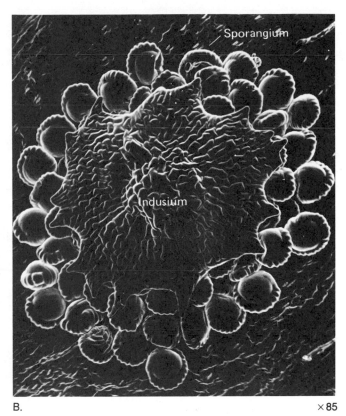

A.   B.   ×85

FIGURE 20-22
Fern sporophylls and sori. **A.** A part of the fertile leaf (sporophyll) of a strap fern showing its underside covered with sori, each composed of 20–30 sporangia. **B.** A single sorus of the Christmas fern. Note the umbrellalike covering, the indusium, protecting the sporangia underneath (sori of the strap fern lack indusia and are unprotected). (B, photo courtesy of P. Dayanandan)

Although both male and female sex organs are borne on a single gametophyte, they usually do not develop at the same time. It has been shown experimentally that plant hormones called **gibberellins** can induce antheridial formation on very young gametophytes. In nature, the first gametophyte to grow from spores of the same age secretes a type of gibberellin called **antheridogen,** which causes adjacent, younger gametophytes to form antheridia. At the same time, these earliest gametophytes develop archegonia, and, if unfertilized, they too eventually develop antheridia and self-fertilization occurs.

After fertilization, the zygote grows into a young sporophyte, or embryo, which remains attached to the gametophyte but shortly develops leaves and a root system and becomes independent. The life history of a bracken fern is illustrated in Fig. 20-23.

Horsetails If botanists were somehow to bridge the time barrier and go on a field trip in a Paleozoic forest,

they would find trees and shrubs quite unlike those of today; however, they would also see some that bear a superficial resemblance to today's plants (see Fig. 20-24). Among the latter would be some tall trees (named *Calamites* [cal-a-my-tees]) whose jointed trunks remind them of large bamboos. However, looking more closely at these Paleozoic "bamboos," they would discover that their structure and functions actually resemble those of a type of small present-day rush called a **horsetail** and that these trees are not so bamboolike after all.

Horsetails are the living descendants of giant Paleozoic rushes. All present-day species constitute a single genus, *Equisetum* (division **Equisetophyta**—*equus-*, "horse," *seta,* "hair," *-phyta,* "plant") The sporophytes of *Equisetum*, as well as those of their extinct relatives, bear cones (called **strobili;** sing. **strobilus**) composed of a central axis and scalelike appendages equipped with sporangia (Fig. 20-25). Strobili usually are thought to represent compact, often highly modified

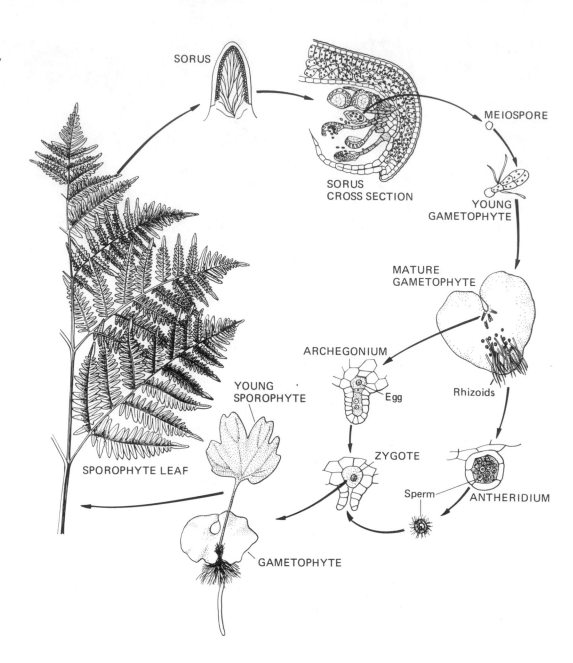

FIGURE 20-23
Life history of a bracken fern,
Pteridium aquilinum.

SORUS

SORUS
CROSS SECTION

MEIOSPORE

YOUNG
GAMETOPHYTE

MATURE
GAMETOPHYTE

ARCHEGONIUM

Rhizoids

Egg

YOUNG
SPOROPHYTE

SPOROPHYTE LEAF

ZYGOTE

Sperm

ANTHERIDIUM

GAMETOPHYTE

stem endings in which fertile leaves (sporophylls) bear one or more sporangia. This is an example of homology (see Chapter 18).

Sporangia of *Equisetum* produce remarkable spores bearing **elaters,** appendages that extend or retract in response to wetting and drying (Fig. 20-26). This action results in transport of the spores over short distances and no doubt aids them in becoming embedded in soil and humus. Not so long ago a paleobotanist treated 250-million-year-old spores of an extinct giant rush with acid to free them from surrounding rock. Surprisingly, these

ancient spores still had the ability to contract and expand in response to changes in humidity.

Meiospores of *Equisetum* develop into photosynthetic gametophytes somewhat resembling fern prothalli.

Horsetails are found in damp places along river banks and lake shores. In former years they were gathered and used as scouring pads to clean pots and pans. *Equisetum* stems and branches contain silicon and therefore have a harsh texture, making them suitable for scouring. This is reflected in a second common name,

FIGURE 20-24
Reconstruction of a Carboniferous forest. In addition to the giant horsetail *Calamites,* other primitive tree forms were present. (Photo courtesy Field Museum of Natural History, Chicago)

scouring rush. The tender, succulent, cone-bearing shoots of some species of *Equisetum* were gathered by Indians for food. Because of the developing spores with their rich store of food, the cones and shoots are probably quite nutritious.

Club Mosses   In the same Carboniferous forests where giant horsetails grew were trees with grasslike leaves belonging to a group of plants called **club mosses** (division **Lycopodiophyta**—*lyco-,* "wolf," *podo-,* "foot," *-phyta,* "plant"). One giant club moss, *Lepidodendron,* grew to a height of 30 m (100 ft) and a diameter of nearly 1 m (3 ft) (see Fig. 20-24). Club mosses are cone-bearing plants and, like the giant horsetails, are represented today only by much smaller living species (Fig. 20-27). Most species produce spores in clublike cones, hence the name.

The earliest club mosses were vascular plants of the Devonian period, nearly 400 million years ago. They gave rise not only to giant club mosses but the smaller plants represented today by *Lycopodium* (Fig. 20-27 and Color plate 8B) and a few other genera and species. Stems of living club mosses are typically protostelic. Often they are prostrate or climbing, but sometimes they form small, erect plants. Roots are adventitious and fibrous, like those of ferns, but less numerous. Leaves of club mosses are a type called **microphylls** ("little

leaves"). The term implies that such leaves are always small, but in actuality microphylls of *Lepidodendron* sometimes were several feet long. This misleading terminology can be clarified by defining a microphyll as a leaf with only one vein.

The life cycle of *Lycopodium* basically resembles that of ferns; however, some club mosses have a more elaborate reproduction approaching that of seed plants such as pines, spruces, and their relatives. An example is the fossil club moss, *Lepidodendron,* which produced two kinds of spores: large **megaspores** and very small **microspores.** Production of microspores and megaspores is called **heterospory.** (Species of *Lycopodium,* as well as bryophytes and most ferns, bear spores of one size and are said to be **homosporous.**) Heterosporous plants include a living club moss, *Selaginella* (Fig. 20-28), its relative the quillwort, *Isoetes* (iso-et-eez), a few aquatic ferns (including the nitrogen-fixing *Azolla* described in Chapter 3), and all seed plants.

Heterospory and the Evolution of Seeds

In the discussion of reproduction in flowering plants (Chapter 16), it was pointed out that two kinds of meiospores were involved: microspores and megaspores. These produce, respectively, male and female gametophytes. Unlike gametophytes of ferns and fern allies,

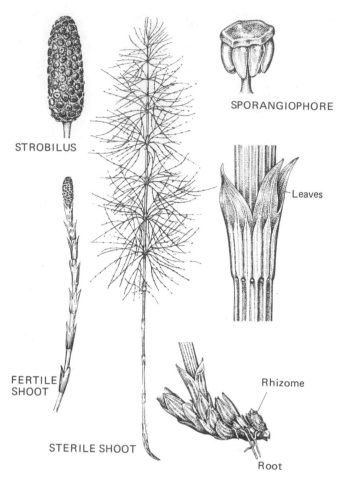

FIGURE 20-25

Equisetum sylvaticum, a species having both sterile, photosynthetic branches and fertile, cone-bearing branches produced by an underground rhizome.

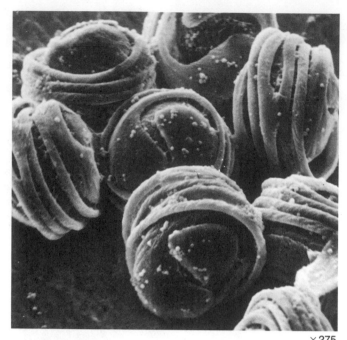

×275

FIGURE 20-26

Scanning electron micrograph of *Equisetum* spores showing locomotory appendages (elaters). (Photo courtesy P. Dayanandan)

FIGURE 20-27

Representative species of *Lycopodium, L. clavatum,* photographed in a Wisconsin woods.

flowering-plant gametophytes are not free living but remain enclosed within parental tissues. How did this condition come about in terms of plant evolution? Probably an early step was heterospory of the kind we find in some of the club mosses. We would like to be able to explain reproduction in flowering plants as having evolved by a series of stages starting with ferns or fern-like plants; examination of heterospory in the club moss *Selaginella* is helpful toward that end.

In species of *Selaginella* and the extinct *Lepidodendron,* club-shaped cones are composed of fertile leaves (sporophylls) which bear two kinds of sporangia: **microsporangia** and **megasporangia**. These produce, respectively, microspores and megaspores, which in turn develop into small male gametophytes and comparatively large female gametophytes (Fig. 20-28). Both male and female

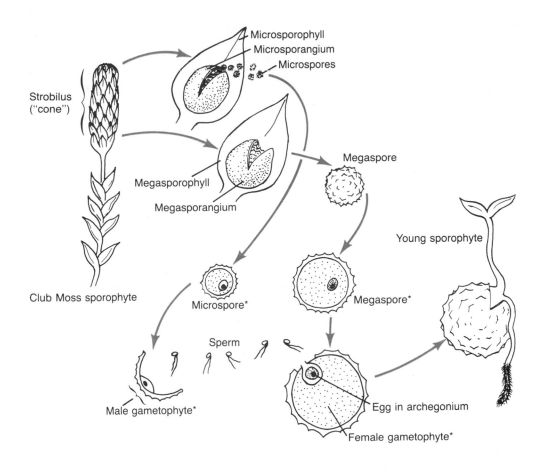

FIGURE 20-28
Reproduction of a heterosporous club moss. The two kinds of meiospores—microspores and megaspores—are produced in the same cones in some species or separately in cones of male and female sporophytes in other species (asterisk indicates exaggerated size). This reproductive cycle corresponds to that of *Selaginella*, a small heterosporous club moss of world-wide distribution.

Microsporophyll
Microsporangium
Microspores
Strobilus ("cone")
Megaspore
Megasporophyll
Megasporangium
Young sporophyte
Club Moss sporophyte
Microspore*
Megaspore*
Sperm
Egg in archegonium
Male gametophyte*
Female gametophyte*

gametophytes remain within their original spore walls, which open slightly at the time of fertilization. Thus we see, in these partially enclosed gametophytes, a possible intermediate stage in the evolution of completely enclosed gametophytes of seed plants. Fertilization occurs between male and female gametophytes of *Selaginella* when the spores from which they grow happen to have landed near each other. However, surface water still is required for fertilization; the sperm must swim from the male gametophyte to the female; once fertilized, the female gametophyte and its embryo are functionally like a small seed.

Seed Plants

Seed plants, reptiles, birds, and mammals all seem to have evolved in unison, as though the time were ripe for a break with their aquatic ancestors. All share a freedom from the requirement for water in their reproduction, reproducing by internal fertilization of egg cells. Although the processes differ in plants and animals, the ecological consequences have been the same: extensive colonization of uplands became possible.

Seed Ferns

At the time reptiles were just beginning their evolution from their amphibious ancestors, during a period referred to as the **Carboniferous** or "coal age," seed plants also were evolving. In fact some early seed plants were among the most abundant vegetation, and because their foliage looked rather fernlike (Fig. 20-29), the Carboniferous has erroneously been called the "age of ferns." It is more appropriately the "age of **seed ferns**." Seed ferns bore pollen organs and seeds directly on their foliage, and the seeds thus were truly naked or **gymnospermous**. All gymnosperms are classified in the division **Pinophyta** (from P*inus*, the genus of pine, *-phyta*, "plant"); see Appendix for subgroupings. It is uncertain what modern plants descended from seed ferns, but cycads probably are their most direct descendants, and it is thought that at least one group of seed ferns may have given rise to flowering plants. An even more intriguing possibility has been raised by S. V. Meyen, a Russian botanist, who suggests that *Ginkgo*, a well-known living tree of Chinese origin, may be a living seed fern (see Fig. 20-32 and Color plate 8C). Most botanists, however, believe *Ginkgo* is re-

FIGURE 20-29

A Paleozoic seed fern, *Medullosa noei*. The plant **(A)**, which stood about 5 m (16 ft) tall, had large divided leaves and a rather soft and weak stem. It probably grew like a large *Philodendron*, leaning against sturdier trees (perhaps such species as are shown in Fig. 20-24). Some of its leaves bore bell-shaped pollen organs **(B)** composed of numerous, fused microsporangia, others bore seeds **(C)**. (Reprinted by permission from *Bot. Rev.* 22:45–80. Copyright 1956 by W. N. Stewart and T. Delevoryas and The New York Botanical Garden)

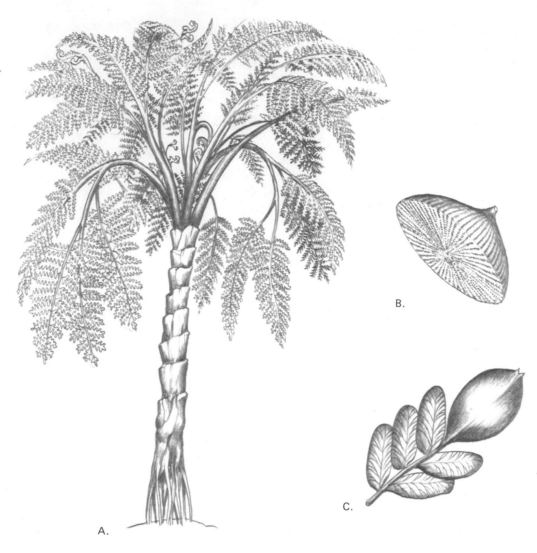

A.

B.

C.

lated to pines and other conifers (see later discussion) through an ancestry possibly as ancient as the seed ferns.

Living Gymnosperms

Seed ferns differed considerably from most present-day gymnosperms. The latter are for the most part trees, and the majority bear their pollen and seeds in cones. The familiar living **conifers** ("cone-bearers") of the Northern Hemisphere are redwoods, pines, spruces, junipers, and firs. In the Southern Hemisphere, different gymnosperms occur: the auracarias, typified by the Norfolk Island pines, now widely sold as house plants; *Podocarpus*, a broad-leaved ornamental introduced in warmer climates of the United States, and other less familiar species.

One well-known gymnosperm with a long history on earth is the maiden-hair tree, or *Ginkgo*, of which only one species now exists. Ginkgoes are known to occur in the wild state only in remote valleys of China. They are most familiar as ornamental plantings and are reputed to be unusually resistant to insect pests and diseases. An even more ancient group of gymnosperms are the **cycads** (Color plates 4B and 8D). Only 10 genera and about 140 species live today, but they once were much more diversified and widespread. Cycads are palmlike in appearance, but unlike palms they bear large and sometimes colorful cones (Fig. 21-30). Only one species, *Zamia pumila*, is found in the United States.

Ginkgo and the cycads are heterosporous plants; in both, flagellated, swimming spermatozoids are produced (see Fig. 20-31 for a photo of a cycad spermatozoid), but these spermatozoids never see the light of day. Microspores are produced as in club mosses, and these become tiny immature male gametophytes, commonly called **pollen grains** (Fig. 20-32). The pollen grains produce flagellated spermatozoids but only after they have been drawn into a part of the female sporophyte called

FIGURE 20-30
Mexican cycad *Dioon spinulosum.* Note the large seed cones at the center of the large compound leaves.

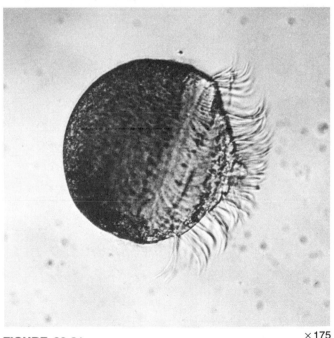

FIGURE 20-31   ×175
Living spermatozoid of the Florida cycad, *Zamia.*

the **ovule.** The ovule is a vaselike structure with an opening, the **micropyle,** through which pollen grains enter. Within the ovule, a female gametophyte develops from a megaspore—note that in seed plants megaspores never are shed. The pollen grain when securely enclosed within the **ovule** (the micropyle closes at this stage) grows into a tiny male gametophyte which produces multiflagellated, swimming spermatozoids (usually 2, but up to 16 in one species of cycad). The spermatozoids swim about in fluid within the ovule, somehow find their way to the archegonia of the female gametophyte, and fertilize the egg. An embryo develops from the egg, and the ovule and its contents become the **seed.** Because the seeds of *Ginkgo,* cycads, related plants known as **conifers** (pines, spruces, etc.), and some others do not have seeds enclosed in a fruit (as in an apple), they are called **gymnosperms** (*gymnos,* "naked," *sperm,* "seed").

The motile gametes of *Ginkgo* and cycads are important because they furnish an evolutionary link between seed plants and nonseed plants, all of which have flagellated gametes. However, in all seed plants other than *Ginkgo* and cycads, male gametes are not flagellated and are not actively motile, but instead are conveyed directly to the eggs by pollen tubes (refer to Fig. 16-4 to review fertilization in flowering plants).

Conifers For most persons, conifers are represented by the pine tree. Conifers are woody gymnosperms and are important ecologically, forming vast woodlands in many parts of the world (Fig. 20-33 and Color plate 13E). They also are important commercially, because much of the lumber trade of the Western world is based on them. When European settlers first arrived in the New World, they found great stretches of magnificent conifers extending from Florida northward into Nova Scotia and westward to the Great Lakes. In colonial times the finest of these trees were reserved for ship masts for the British Navy; they also furnished logs for the crude dwellings of pioneers and for later, more elegant constructions.

After the United States achieved independence from Britain, coniferous forests began to be exploited. However, it was not until the invention of steam-powered saws in the mid-1800s that destruction of the great forests really started, first in New England and then westward through the Great Lakes area. By the 1900s all virgin stands of pines were gone in the eastern half of the United States, and major lumbering operations shifted to the Pacific coast, where the great redwoods, Douglas firs, cedars, and pines presented new challenges and new opportunities.

The history of lumbering in the United States is one of unparalleled exploitation and waste. Even now, new inroads are being made in the remaining forests of

ARCHAEOPTERIS, CALLIXYLON, AND THE ORIGIN OF CONIFERS

For years botanists considered a Devonian fossil named *Archaeopteris* to typify the early ancestors of ferns. *Archaeopteris* was discovered, described, and named by the Canadian botanist William Dawson in 1871. The name means "ancient fern," and indeed the plant's leaves seem to be much divided and fernlike. Then, in the 1960s, Charles Beck, a botanist at the University of Michigan, found and described some stem and branch fragments of another ancient fossil named *Callixylon* ("beautiful wood") with attached leaves of *Archaeopteris* and realized that they represented one and the same plant (Fig. 1).

10 ft

B. Anthony

FIGURE 1
Reconstruction of *Archaeopteris* showing the entire tree and one leafy branch. Note that the overall aspect of the model resembles a living conifer. [From C. B. Beck, *Amer. J. Bot.* 49:373 (1962); 58:758 (1971)]

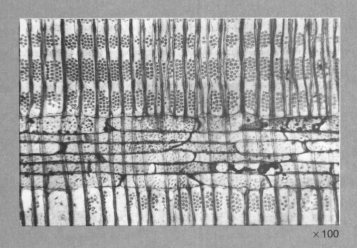

×100

FIGURE 2
Fossilized wood of *Callixylon (Archaeopteris)* showing the bordered
pits. (Photo courtesy C. B. Beck)

Botanists and paleontologists acclaimed this remarkable discovery because
Callixylon had long been a puzzle. It evidently had been a large tree with a trunk
1 m (3 ft) or more in diameter, and its wood was composed of tracheids bearing
bordered pits, much like those of modern conifers (Fig. 2). These characteristics
led botanists to think that *Callixylon* was a conifer, or at least related to conifers.
Unfortunately, no one had found either its leaves or reproductive organs, and so its
evolutionary relationships remained mysterious.

Beck's discovery of *Archaeopteris* leaves connected with *Callixylon* stems not
only clarified the question of the kind of leaves borne by *Callixylon* but shed light
on its reproduction. It had been known that *Archaeopteris* foliage bore
heterosporous sporangia but not seeds. Its reproduction was probably similar to
that of *Selaginella;* hence the newly recognized combined *Callixylon-Archaeopteris*
could not be a true gymnosperm. Further study of *Archaeopteris* leaves have
shown them to be much less fernlike than formerly thought, and they now are
considered to consist of simple, spirally arranged leaves on branches, rather than
much-divided compound leaves (Fig. 1). Although the name *Archaeopteris* ("old
fern") is not appropriate, because *Archaeopteris* is not a fern, it was given to the
entire plant for reasons of priority (*Archaeopteris* is an older name than *Callixylon*).

The significance of *Archaeopteris (Callixylon)* is that conifers are shown to
have heterosporous nonseed ancestors with vegetative characteristics much like
those of living forms. The origin of gymnosperms is placed earlier in the Paleozoic
than had been suspected, a link between certain plants of doubtful evolutionary
significance has been revealed, and a new class of plants known as
Progymnosperms has been brought to the attention of scientists.

the Pacific coast from northern California to Alaska.
Within the lifetime of the present generation of Americans, all the primeval forests will be gone; their place
will be taken by second- and third-growth woodlands of
inferior productivity and whose wood is fit mainly for paper pulp, particle board, plywood, and fence posts.

Conifers are the largest of the plant kingdom. A
giant cypress 25 km (15 mi) south of Oaxaca, Mexico, is
16 m (53 ft) in diameter, and the largest California trees
(*Sequoiadendron*) are 11 m (36 ft) in diameter and 100 m
(325 ft) tall. Conifers may also be the oldest organisms
on earth in terms of individual survival. The largest red-

FIGURE 20-32

Reproduction of *Ginkgo,* a primitive gymnosperm tree. *Ginkgo,* unlike most other gymnosperms, has male gametophytes that produce flagellated motile sperm. The male strobilus is composed of sporangia (microsporangia) in which microspores develop into three-celled pollen grains, at which stage they are shed (asterisk indicates exaggerated size). There is no female strobilus; pairs of ovules are present at the tips of stalks as shown above. Inside each, four megaspores are produced but only one is functional; the other three megaspores die.

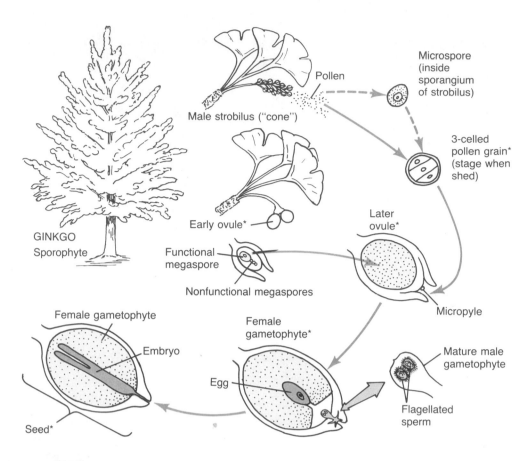

GINKGO
Sporophyte

Male strobilus ("cone")

Pollen

Microspore (inside sporangium of strobilus)

3-celled pollen grain* (stage when shed)

Early ovule*

Later ovule*

Functional megaspore

Nonfunctional megaspores

Micropyle

Female gametophyte

Embryo

Female gametophyte*

Egg

Mature male gametophyte

Flagellated sperm

Seed*

woods and big trees date from Roman times, and an even older tree, a bristlecone pine found in the U.S. Southwest, is calculated to be about 5000 years old (Color plate 8E.) A recent estimate of *El Tulé,* the Oaxaca cypress, places its age at 6000 years, making it the oldest living organism known.

Pine Structure and Reproduction   Pines are most often selected for study by biology classes because they are the most widely distributed of Northern Hemisphere conifers and are most familiar to the majority of people. The tall, columnar stems of pines are a major source of planks, beams, boards, and poles for telephone and power lines.

A cross section of a pine stem can be used to illustrate the general characteristics of gymnosperm wood. The wood is composed of a series of growth rings, each representing one season's growth (Fig. 20-34). Each such **annual ring** is produced by a vascular cambium, which by repeated divisions of its cells produces, to its inside,

FIGURE 20-33

Stand of virgin sugar pine in California. (Photo courtesy W. H. Hodge)

FIGURE 20-34

Pine stem. **A.** Cross section of three-year-old stem showing pith (P), annual xylem rings (X), phloem (Ph), and cork (Co). Cambium (C) position is indicated by the arrow. Ducts (D) are resin canals. **B.** Cross-sectional view of xylem showing small tracheids (T) of summerwood of one annual ring and springwood of the next annual ring. A ray (R) also is seen. **C.** Longitudinal section of xylem showing bordered pits (BP) of tracheids and a ray.

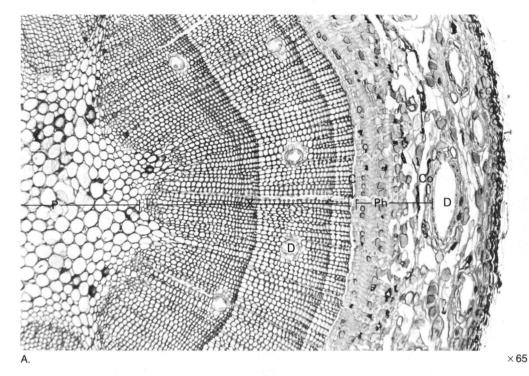

A. ×65

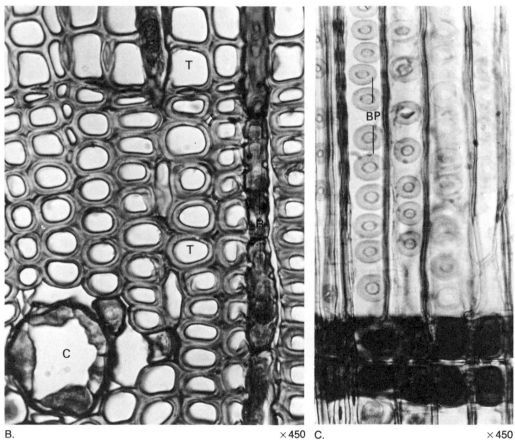

B. ×450   C. ×450

cells that will mature into **wood** (xylem) and, to the outside, cells that will mature into the **inner bark** of the tree (phloem). The **outer bark**, the layer called **cork**, is produced by discontinuous layers of **cork cambium** that develop in the older phloem tissues.

Microscopic examination of xylem shows it to be composed of tracheids bearing a type of pit partially overarched with wall material to form a border (see Fig. 20-34). A major factor in identification of fossilized wood as gymnospermous is presence of tracheids equipped with these **bordered pits;** they are characteristic of living gymnosperms as well.

Phloem in gymnosperms is composed of sieve cells arranged end to end so as to form sieve tubes. A cork cambium develops in older phloem tissue and most gymnosperm stems have a thick outer bark composed of cork.

At the center of the stem is a pith region, and radiating out from the pith are rows of transport cells called **rays.** Xylem and phloem transport water and nutrients up and down the stem; rays transport these materials laterally. Also seen in a cross section of the pine stem are channels in the wood and bark; these **resin ducts** secrete and transport resin produced in pine stems and other conifers.

The leaves of pines are its green needles, which occur singly in one species and in clusters of two to five in others. The needles are perennial, evergreen, and in cross section are seen to have a central vein surrounded by a layer of photosynthetic cells. The stomata are sunken in cuplike depressions, a characteristic of plants adapted to survival in dry regions. Sunken stomata are less likely to lose water by evaporation (Fig. 20-35).

Roots of conifers are generally similar to those of other vascular plants in that they lack a pith (see Figs. 16-13, 17-4, and 17-5). Externally, near the tip of very young roots, are a relatively small number of fragile root hairs which absorb water and dissolved nutrients from soil. Older roots lack root hairs and resemble older stems in having a thick xylem, annual rings, and bark.

In pines, separate pollen and seed cones are present on the same tree; pollen cones usually are quite small, whereas seed cones are considerably larger (Fig. 20-36). Reproduction in pines is a slow process but one not differing in essentials from that of *Ginkgo*, described earlier (Fig. 20-32), except (and it is an important exception) that the sperm are no longer flagellated. Instead, the male gametophyte forms a tube, the pollen tube, which grows right down into the female gametophyte and releases its two sperm in the vicinity of the egg (Fig. 20-37). One sperm fertilizes the egg, the other dies. Subsequently the zygote develops into an embryo and a seed is formed in which the female gametophyte functions as a source of food for the embryo and later, after germination, for the seedling plant. A year is required for pollination and fertilization to be completed and a second year for maturing of seeds (Figs. 20-36 and 20-37). Seeds of pines are winged and are borne in pairs on the

FIGURE 20-35
Cross sections of pine leaves.
(Photo by Carolina Biological
Supply Company)

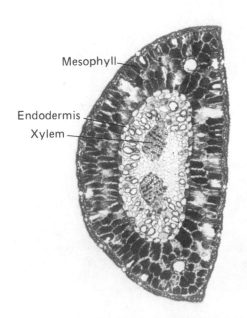
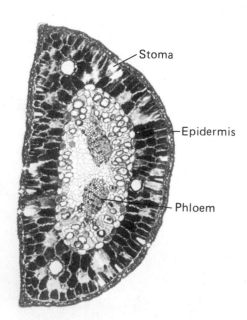

Mesophyll

Endodermis

Xylem

Stoma

Epidermis

Phloem

×33

FIGURE 20-36
Male (top) and female (bottom) cones of red pine (*Pinus resinosa*) in spring at the time of pollination. Male cones (pollen cones) occur in clusters; female cones occur singly. The male cones are shed after pollination, but the female cones develop for two years, becoming considerably larger.

upper surfaces of cone scales. Disseminated by gravity and wind, they germinate to form seedlings having many "seed leaves," or **cotyledons.**

Other Gymnosperms In addition to conifers and cycads, a few other living gymnosperm types occur in different regions of the world. They differ from the more familiar species in so many ways that one would not think them to be gymnosperms. Three genera—*Gnetum, Ephedra,* and *Welwitschia*—form a class of about 75 species of plants, collectively called **gnetopsids,** having unusual

growth forms (Fig. 20-38). *Gnetum* is a tropical shrub that looks quite a bit like a flowering plant; it has broad leaves not greatly different in appearance from those of a peach tree, for instance. Despite its overall appearance, *Gnetum* bears typical gymnosperm seeds colored bright red. *Ephedra* is a common plant of the U.S. Southwest. It has much-branched, green, photosynthetic stems and tiny scale leaves. A source of the drug ephedrine, *Ephedra* was used in pioneer times for the brewing of a medicinal tea and sometimes is called "Mormon tea." *Welwitschia,* a plant growing only in the deserts of southwestern Africa (Namibia), is probably the most bizarre plant on earth. During its lifetime, which may be several hundred years, it has only two leaves, which grow slowly but continuously from their bases. The leaves are leathery and tattered and prostrate. The flat-topped stem is mostly underground and bears cones at the edge of its exposed surface.

Flowering Plants—Angiosperms

No other group of plants is of greater importance to animal life today than flowering plants, or **angiosperms** (classified in the division **Magnoliophyta**—*Magnolia,* the genus of magnolia and related plants, *-phyta,* "plant"). Yet they are comparatively late arrivals on the earth's surface. If one were somehow able to travel back in time, one would find them to have evolved in comparatively recent times, along with birds and mammals. Indeed biologists believe that flowering plants may have made evolution of warm-blooded animals possible. Certainly the great herds of grazing animals, both wild and domestic, depend on grass (a flowering plant), and the ancestors of most herbivores of today evolved when the great grasslands of the world were formed around 65 million years ago.

Flowering plants are the most numerous plants on earth in terms of numbers of species (more than 275,000); in fact, there are more of them than of all other plants combined, and they rank first in economic importance.

Drugs of many kinds are products of flowering plants. Narcotics, some of which are so controversial in today's society, are obtained from flowering plants. Opium, the source also of heroin, is extracted from latex of the opium poppy (*Papaver somniferum*). Cocaine is an extract of the South American coca plant (*Erythroxylon coca*), and at one time the formulation for a popular soft drink contained coca extract and another ingredient prepared from the cola nut (*Cola nitida*). Tobacco and marijuana, as well as milder drugs in coffee, tea, and cocoa, are derived from flowering plants. Many other drugs—

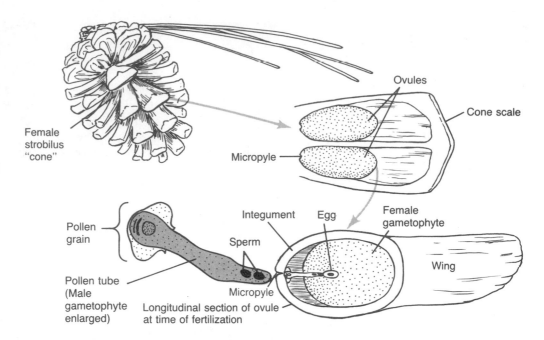

FIGURE 20-37
Several details of reproduction of a conifer (pine, spruce, etc.). The male strobilus ("cone") is not shown, but refer to Fig. 20-36. Note especially that the male gametophyte produces a pollen tube that grows directly toward the egg and that the sperm are not flagellated. Only one sperm fertilizes the egg. The seeds of conifers usually have wings and are dispersed by air currents and winds.

Female strobilus "cone"

Ovules

Cone scale

Micropyle

Pollen grain

Integument

Sperm

Egg

Female gametophyte

Wing

Pollen tube (Male gametophyte enlarged)

Micropyle

Longitudinal section of ovule at time of fertilization

FIGURE 20-38
Welwitschia, a member of the class Gnetopsida. (Photo courtesy Field Museum of Natural History, Chicago)

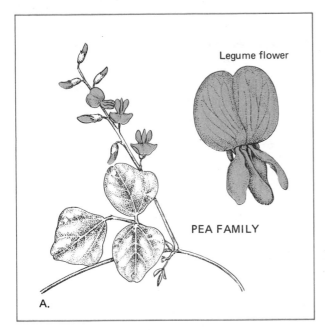

A.

Legume flower

PEA FAMILY

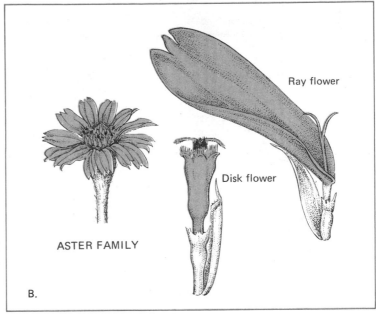

Ray flower

Disk flower

ASTER FAMILY

B.

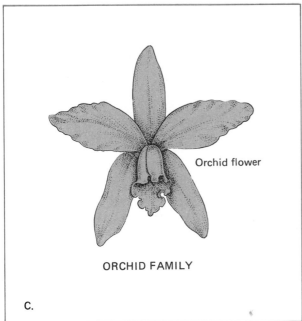

Orchid flower

ORCHID FAMILY

C.

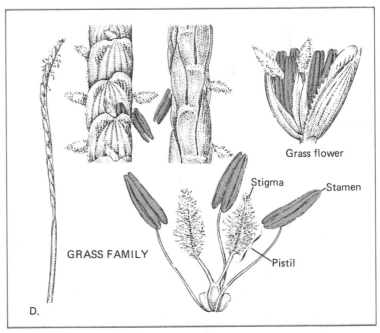

Grass flower

Stigma

Stamen

GRASS FAMILY

Pistil

D.

FIGURE 20-39
Some plant families. **A.** Pea flower (Fabaceae). **B.** Composite flower of *Wedelia* (Asteraceae). **C.** Vanilla orchid (Orchidaceae). **D.** Flower of a grass (Poaceae). Peas, asters, and orchids are insect pollinated; grasses are almost always wind pollinated.

about 40 percent of all medical prescriptions according to a recent tabulation—come from the angiosperms. Drugs such as ginseng, belladonna, atropine, quinine, curare, salicylic acid (aspirin), and digitalis have been known since ancient days. Others such as birth-control hormones and synthetic cortisone, both made from products of the Mexican yam, are more recent discoveries.

Our major food and fiber crops are flowering plants. Of great importance as food are the grasses, which include wheat, oats, rye, rice, maize (corn), and sugarcane. Potatoes, beets, turnips, beans, and peanuts also are important foods. Plants such as soybeans, maize, sunflower, safflower, and palms furnish much of the cooking oil in use, especially now that people are so

TABLE 20-1
Characteristics of monocots and dicots

	Monocots	Dicots
Examples	Lilies, orchids, palms, cereals, bamboos, etc.	Asters, roses, peas, potato, orange, walnut, etc.
Flowers	Flower parts in 3s and 6s (usually 3 sepals, 3 petals, 6 stamens, 3 carpels)	Flower parts in 4s and 5s (often 5 sepals, 5 petals, 5 stamens, 1 to many carpels)
Fruit	Fruit often dry; pod or kernel	Fruit often fleshy, pulpy
Embryo	Usually 1 cotyledon present	Usually 2 cotyledons present
Stems	Usually nonwoody; vascular bundles arranged evenly throughout	Often woody; vascular bundles form a ring
Leaves	Often linear, with parallel veins	Often broad, with a network of veins
Roots	Adventitious, fibrous	Usually a main root (taproot) and secondary branch roots

conscious of the health problems associated with eating animal fats. Cotton and flax (linen) are flowering plants, and rayon, a synthetic fiber, often is made from wood fibers of flowering trees such as aspens and poplars. Spices, including pepper, cinnamon, cloves, garlic, nutmeg, anise, sage, ginger, and many others, are obtained from flowering plants. Many other things that make life attractive and comfortable are similarly derived.

Although all flowers have essentially the same reproductive parts (Fig. 16-4 and 16-6), the accessory parts, petals and sepals, differ according to the pollination mechanism; those that are wind pollinated usually have inconspicuous petals and sepals (grasses, for example), those that are animal pollinated have showy flowers (Fig. 20-39 and Color plates 3C, 9, 14B).

Monocots and Dicots   Two main groups of flowering plants are recognized: **monocotyledons** and **dicotyledons**, abbreviated **monocots** and **dicots**. Monocots and dicots are readily distinguishable on the basis of several key characteristics (see Table 20-1 and Fig 20-40).

Flowering plants may be either short-lived, often fleshy **herbaceous** plants, such as common garden vegetables, or long-lived **woody** trees and shrubs, such as oak and elm trees. Most monocots are short-lived herbaceous plants, but a few, including palms, bamboos, and yuccas, are woody trees (Fig. 20-41). Dicots are considered to be fundamentally woody plants, even though many herbaceous genera and species are known. It is probable that the herbaceous growth habit is an adaptation to severe climatic conditions such as prolonged droughts or extreme cold, in which survival of woody plants would tend to be precarious.

Stems and Roots   The tissues and principal parts of flowering plants originate from meristems in very much

the same manner as previously described in Chapters 8 and 16 (refer to Figs. 8-13, 16-11, 16-12, and 16-13). Root and stem apical meristems produce the first or primary tissue systems of the growing plant; longitudinal growth is the result. Later on, principally in the stems and roots of dicots, lateral thickening occurs as a result of cell divisions in a vascular cambium (pl. cambia) producing secondary tissues, mostly of xylem and phloem.

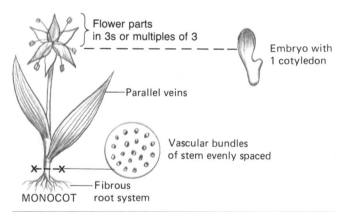

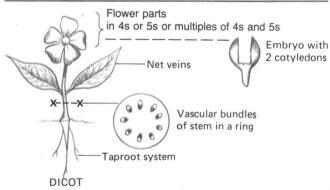

FIGURE 20-40
Structures of monocots and dicots contrasted.

FIGURE 20-41
Some woody monocots. **A.** Pony tail palm, *Beaucarnea,* a member of the *Agave* family. **B.** Dragon tree, *Dracaena.* **C.** Australian grass tree, *Xanthorrhoea,* member of the lily family.

A.

B.

C.

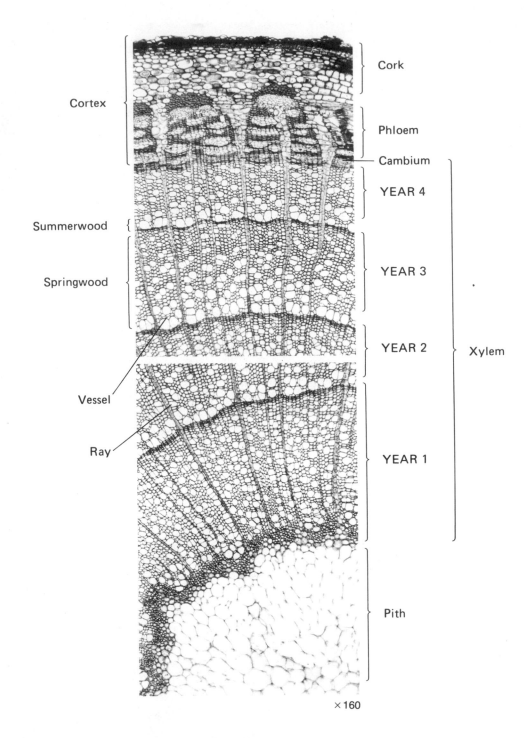

FIGURE 20-42
Cross-sectional view of a young stem of ash *(Fraxinus)* showing the relationship between cork, cortex, phloem, vascular cambium, xylem, and pith. The springwood and summerwood are visible. The large openings are xylem vessels.

Cork

Cortex

Phloem

Cambium

YEAR 4

Summerwood {

Springwood {

YEAR 3

YEAR 2

Xylem

Vessel

Ray

YEAR 1

Pith

×160

As a consequence of many years of cambial activity, a thick region of xylem is formed, constituting the wood of the stem. Cross sections of the wood of temperate-region dicot stems show yearly growth increments called annual rings. They result from the more rapid division of the cambial cells in the spring and early summer and also in the larger size of the xylem cells produced during this period (Fig. 20-42). Later, in summer, the cambium is less active and the cells smaller, making an annual ring composed of less dense springwood and denser summerwood. Tropical trees (see Color plate 4C) also may show xylem rings but these are usually not annual rings: they reflect wet and dry periods throughout the year. In some years there may be more than one ring; when rainfall and temperature are more or less uniform all year, there may be no growth rings at all.

Annual rings have been very useful in archeology and climatology, helping to place dates on wooden structures and also to reveal the history of past climates. A dry year, for example, would result in a narrow annual ring; a wet year would produce a wide ring. A sequence of wide and narrow rings in a tree, recently cut down and of known age, can often be matched sequence for sequence to the outer region of a much older and long-dead tree trunk. In that way a chronology can be constructed which not only can be used to put a firm date on archeologically significant remains of wooden structures, but also will be useful in reconstructing past climates.

The phloem does not accumulate in an older stem to the same extent as does the xylem, due in part to the greater production of xylem by the cambium, but also to the fact that phloem is on the outside of the cambium and is continually being compressed. Eventually, older phloem is converted into the cork of the outer bark. As the tree grows in diameter, older bark cracks, breaks, and is gradually sloughed off.

The situation in monocots is somewhat different than that just described. Not many monocot stems undergo any prolonged secondary thickening; they remain about the same diameter as long as the plant lives. The maize plant (*Zea mays*) is an example. You may have noticed, if you live in or have visited an area where palm trees are common, that the stem of a young palm is about the same diameter as an older one. There is no cambium, and very little secondary thickening growth occurs in palms. However, some monocot trees, including yuccas, pony tail palms, Australian grass trees, and a few others, do have cambia, and their growth somewhat resembles that of a woody dicot tree (Fig. 20-41).

Roots are basically quite similar in all groups of vascular plants. Like stems, they originate from an apical meristem of the embryo, but unlike most stems they remain protostelic (i.e., have no pith; see Fig. 17-4). Because roots lack leaves, they do not have nodes, internodes, or lateral buds. Branch roots develop from lateral growth points (new meristems) beneath the cortex. If a plant is a perennial dicot, its roots have a vascular cambium as does its stem; old roots exhibit annual rings and a corky outer layer.

In addition to the functions of absorption and anchorage, roots often have secondary functions and modification. Some fleshy roots serve as food-storage organs, as for example the taproots of carrots, beets, turnips, and sugar beets (Fig. 20-43), and roots of sweet potatoes have numbers of enlarged regions which serve as food-storage structures. (However, the Irish potato storage organ is a highly modified portion of a rhizome. The eyes of the Irish potato are its buds.) An important modification of the roots of a family of plants known as **legumes**

is a system of small nodules which harbor nitrogen-fixing bacteria. Some plants have prop roots that originate in above-ground parts of the stem and grow downward at an angle into the soil, where they both prop up the plant and absorb nutrients. Aerial roots, such as those of the ivies, are anchoring structures that aid the plant in its climbing growth habit. Other aerial roots such as those of orchids and philodendrons serve as both anchorage and absorptive organs.

Leaf Structure   The leaves of flowering plants display more variations than do those of any other group of green plants. They vary in size from a fraction of an inch, as in leaves of the chickweed, to nearly 10 m (30 ft) in the case of certain palms (Fig. 20-44A). The floating leaf of the Victoria water lily is large enough to function as a raft for a child (Fig. 20-44B).

Leaves of insectivorous plants are cunningly fashioned as traps for unwary insects. The Venus's flytrap (Fig. 20-44C) snaps shut on the fly that chances to touch its hairlike triggers; the sundew entangles its victims in a sticky fluid and then rolls them up in its leaf and di-

FIGURE 20-43
Sugar beet. (Photo courtesy U.S. Department of Agriculture)

FIGURE 20-44
Leaves of flowering plants. **A.** The leaf of the *Corypha* palm is the largest known, extending more than 10 m (30 ft) from the trunk. **B.** Leaves of the Victoria water lily measure more than 1 m (3 ft) across and can support the weight of a small child. **C.** Leaves of the Venus's flytrap are highly specialized, closing rapidly when "triggered" to catch insects. Many other leaf modifications are known among flowering plants. (C, photo by Carolina Biological Supply Company)

gests them; the pitcher plant is a miniature "tiger trap" into which an insect falls, drowns, and is digested.

Leaves of other plants may be so highly modified that they may not appear to be leaves at all. The spines of the locust tree are modified leaves, as are the tendrils of pea plants. The leaves of the common stonecrop are plump and succulent, functioning as water reservoirs as well as organs of photosynthesis. Despite these many variations, the structure of leaves has a common basis, and even in some highly modified ones a common organization can be discerned.

A typical leaf consists of a **blade** with a **median vein,** or **midrib,** and a **stalk,** or **petiole.** Leaves may be either **simple** or **compound,** and there are variations of each of these. Simple leaves have only one blade; compound leaves have several to many blades; both types have just one petiole. In both kinds there may be a variety of venation patterns (see Fig. 20-45).

The typical foliage leaf of a flowering plant has a **dorsiventral** organization. In other words, the upper portion differs from the lower. The petiole contains one or more vascular bundles, or **veins,** which branch out in characteristic patterns within the blade. Examination of these veins reveals that the xylem is uppermost and the phloem lowermost. This reflects the position of these two tissues in the stem, where the xylem is innermost and the phloem outermost; the veins simply reflect this arrangement of the stems when they depart into the leaf.

Origin of Flowers Darwin called the flowering plants an "abominable mystery" because he could not form a clear picture of their evolutionary history. Since Darwin's day, several theories have been proposed to explain the origin of flowering plants. Each theory has its adherents, and each has been abandoned by most botanists. *Gnetum*, for example, had been proposed as exemplifying the group of gymnosperms from which seed plants arose. The difficulty with this hypothesis is that *Gnetum* is a true gymnosperm, even though its stem and leaves look like those of flowering plants. No explanation of how a plant such as *Gnetum* might develop flowers and fruits has been forthcoming. Another hypothesis is that certain cycadlike plants of the Mesozoic era might have been ancestors of flowering plants, but this argument has even more obstacles to overcome than the *Gnetum* theory. Currently, most botanists believe that flowering plants evolved in the Mesozoic from seed ferns, but no one can say just which group of seed ferns might have been the true ancestors.

This failure to come to grips with the origin of the most important plant group has been explained by the theory that flowering plants evolved in dry uplands where fossil formation would have been unlikely. Led-

FIGURE 20-45
Some common variations in leaf structure. **A.** Simple leaf with **pinnate** ("feathered") venation. **B.** Simple leaf with **palmate** venation. **C.** Compound leaf with pinnate leaflets. **D.** Compound leaf with palmate leaflets.

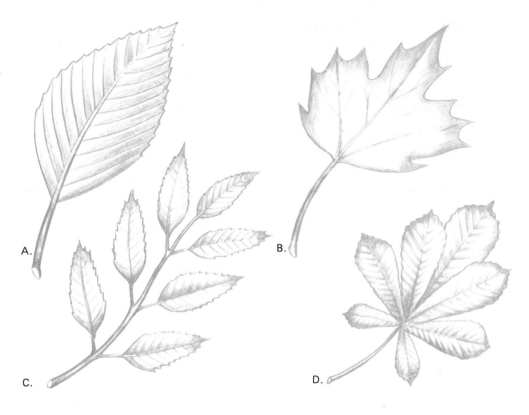

A.

B.

C.

D.

yard Stebbins, a theoretical botanist from the United States, has proposed that primitive flowering plants were small, shrubby, insect-pollinated plants that lived in hilly or mountainous regions where neither the plants nor their characteristic pollen would be preserved. In actuality, botanists still do not understand primitive flowering plants well enough to trace their ancestry. They are not even sure if all flowering plants are descended from the same ancestor or if major groups evolved separately by parallel evolution.

Although there are more than 275,000 species of flowering plants, comprising about 300 families, a comparative handful have been adopted by humans and domesticated for their use. Surprisingly, just a few families predominate in this listing, including grass (wheat, maize, etc.), pea (legumes), aster (sunflower, daisy, etc.), orchid, cabbage, tobacco (also tomato, potato), lily, palm, and apple (see Fig. 20-39).

SUMMARY

Eukaryotic algae are classified in three divisions: green, red, and brown algae. Green algae are the most numerous and widespread and are believed to have played a part in the evolution of higher plants. It is probable that green algae living in early Paleozoic time gave rise to the first land plants.

Red algae are producers of human foods and food additives. They characteristically are smaller than brown algae, often less than 1 m (3 ft) long, but most are above microscopic dimensions. Red algae contain red and blue accessory pigments called phycobilins that enable them to absorb more deeply penetrating green wavelengths of light in their photosynthesis.

Brown algae are mostly large marine plants and include giant kelps, smaller rockweeds, and floating *Sargassum*, for which the Sargasso Sea is named. Brown algae are ecologically important plants of intertidal zones as well as deeper waters. They are an important source of food additives, called alginates.

All algae are important ecologically as food-producing organisms. This importance is reflected in current interest in producing food for human and livestock use by growing some kinds of green algae in vats, tanks, and ponds.

The simplest land plants evolved from algae. Although much remains to be learned about these pioneers, it is known that the transition from water to land required plants to develop erect stems, a cuticle, wind-blown spores, and roots and a vascular system. Mosses and liverworts, which meet some of these specifications, may have been the first to invade the land, but the discovery of fossils of vascular plants more ancient than the oldest known fossil moss or liverwort casts some doubt on this theory.

Mosses and moss allies, and the related liverworts and hornworts, are considered the simplest green land plants. They differ from more complex plants such as ferns and club mosses in having sporophytes that are attached to and nutritionally dependent on their photosynthetic gametophytes. In other green land plants, often referred to as vascular plants, sporophytes are large and independent, whereas gametophytes are small.

Ferns are the more successful of modern spore-producing vascular plants, there being many more species of them than of horsetails and club mosses. Fertile fern leaves, called sporophylls, bear clusters (sori) of sporangia, and the spores when shed develop into small, often heart-shaped green gametophytes called prothalli. The prothalli bear both antheridia and archegonia.

An important feature from the standpoint of higher plant evolution is the production of two kinds of spores: microspores and megaspores. This condition, termed heterospory, is present in a few ferns and club mosses, but is common to all seed plants.

In their reproduction, seed plants are the equivalent of reptiles, birds, and mammals, because they also have internal fertilization. Thus they are independent of water in their sexual reproduction. The most primitive seed plants are seed ferns, an extinct group of gymnosperms that flourished in the late Paleozoic.

Present-day gymnosperms include conifers such as pines and spruces; *Ginkgo*, a primitive Chinese tree; and cycads. Conifers are the largest and oldest trees.

The largest division of the plant kingdom, in terms of number of species, is the Magnoliophyta, also known as angiosperms. The characteristic structure is, of course, the flower. The flower is composed of modified leaves called sepals, petals, stamens, and carpels. The stamens are microsporophylls producing microspores which develop into pollen grains (male gametophytes). The carpels are megasporophylls that bear ovules containing megasporangia and megaspores. One or more carpels make up a pistil. After pollination and fertilization the carpels develop into a fruit containing seeds (mature ovules).

There are two major groups of flowering plants: the monocotyledons and the dicotyledons. Monocotyledons have flowers commonly composed of units in threes (three sepals, three petals, three carpels, six stamens), parallel-veined leaves, fibrous roots, and soft stems with diffusely arranged vascular bundles. Dicotyledons commonly have flowers composed of units in fours or fives (four petals, four stamens, four carpels, etc.), net-veined leaves, taproots, and often woody stems. In young stems the vascular bundles are arranged in a ring.

Flowering plants are extremely important economically. They are food plants and also supply fibers (cotton, flax, hemp, etc.), drugs (digitalis, castor oil, etc.), and many other useful products.

The plants discussed in this chapter are included in the classification presented in the Appendix.

KEY WORDS

green alga	oogamy	homospory
motile colony	antheridium	heterospory
coenocytic alga	archegonium	gymnosperm
filamentous alga	gemma	integument
mitospore	sporangium	ovule
zygospore	spore mother cell	seed
alternation of generations	spore tetrad	pollen
sporophyte	*Psilophyton*	seed fern
gametophyte	protostele	*Ginkgo*
red alga	*Psilotum*	cycad
phycoerythrin	fern	spermatozoid
phycocyanin	megaphyll	conifer
brown alga	fiddlehead	bordered pit
fucoxanthin	siphonostele	ray
kelp	sorus	gnetopsid
algin	prothallus	angiosperm

bryophyte	horsetail	flower
moss	strobilus	monocotyledon
liverwort	club moss	dicotyledon
hornwort	microphyll	

QUESTIONS FOR REVIEW AND DISCUSSION

1 Differentiate between red, brown, and green algae on structural and chemical grounds.

2 Describe the ecological preferences of major marine brown algae.

3 What is the nature and usefulness of red and brown algal colloids?

4 What is a kelp? Of what importance are kelps ecologically and commercially?

5 What roles do bryophytes play in ecological succession?

6 Describe an archegonium. An antheridium. What basic requirement must be met for fertilization to occur? What is the significance of this from an evolutionary standpoint?

7 In what respects do bryophytes resemble green algae? In what respects do they differ?

8 Define "heterospory" in one sentence. What is its evolutionary significance?

9 What, if any, ecological importance can be attributed to ferns? They are known to colonize habitats long distances from parental plants. How is this possible?

10 What is a seed?

11 What is unique about fertilization in cycads and *Ginkgo*? What is the significance of this in plant evolution?

12 Compare in overall aspects (not in details) reproduction in conifers and flowering plants. What are the similarities, and the differences?

13 How did flowers evolve? How do flowers exemplify coevolution?

SUGGESTED READING

CRANDALL-STOTLER, B. 1980. Morphogenetic designs and a theory of bryophyte origins and divergence. *BioScience* 30:580–85. (Interesting speculation on the vascular nature of mosses and the origins and evolutionary relationships of bryophytes.)

DELEVORYAS, T. 1977. *Plant diversification*, 2nd ed. New York: Holt, Rinehart and Winston. (A brief evolutionary treatment of the plant kingdom.)

DOYLE, W. T. 1970. *The biology of higher cryptogams.* New York: Macmillan Co. (One of the best paperbacks dealing with the structure and functions of mosses, ferns, and fern allies.)

ESAU, K. 1977. *Anatomy of seed plants*, 2nd ed. New York: John Wiley and Sons. (Beautifully illustrated with light and electron micrographs and drawings. A good reference book for those who wish more than a superficial encounter with plant anatomy.)

EWART, B. 1975. Sex and the single *Volvox. Nat. Hist.* 84(4):50–54. (Discusses evolution of sex and the life of *Volvox.*)

FOSTER, A. S., and E. M. GIFFORD, JR. 1974. *Comparative morphology of vascular plants*, 2nd ed. San Francisco: W. H. Freeman and Co. (Beautifully illustrated. A clear and concise evolutionary study of vascular plants with emphasis on fossils as well as living forms.)

GRAHAM, L. E. 1984. *Coleochaete* and the origin of land plants. *Amer. J. Bot.* 71(4):603–08. (This significant paper details recent evidence pertaining to the evolution of green land plants from certain green algae.)

HEISER, C. B., JR. 1973. *Seed to civilization: the story of man's food.* San Francisco: W. H. Freeman and Co. (An interesting and informative source of information about economic botany.)

JACQUES, H. E. 1949. *Plant families: how to know them*, 2nd ed. Dubuque, Iowa: William C. Brown Co. (A useful guide to common plant families.)

JENSEN, W. A. 1973. Fertilization in flowering plants. *BioScience* 23:21–27 (Presents new insights into fertilization in flowers based on electron microscopy and histochemistry.)

MAJOR, R. T. 1967. *Ginkgo*: the most ancient living tree. *Science* 157:1270–73. (How has *Ginkgo* managed to survive over many millions of years?)

RAVEN, P. H.; R. F. EVERT; and H. CURTIS. 1981. *Biology of plants*, 3rd ed. New York: H. S. Worth Co. (A well-illustrated textbook of botany, especially strong in the areas of plant structure and function.)

SCHMID, R., and M. J. SCHMID. 1975. Living links with the past. *Nat. Hist.* 84(3):38–45. (The remarkable story of America's oldest living trees, bristlecone pines.)

SPURR, S. H. 1979. Sylviculture. *Sci. Amer.* 240(2):76–90. (Forest management, nutrient manipulation, and genetic selection can treble forest productivity.)

TAYLOR, T. N. 1981. *Paleobotany: an introduction to fossil plant biology*. New York: McGraw-Hill Book Co. (A modern and comprehensive textbook on fossil plants. Includes an interesting chapter on the origin of vascular plants.)

21

The Animal Kingdom

I f the impressive characteristics of animal life were to be enumerated, the almost overwhelming diversity would be near the top of the list. Every conceivable habitat, whether the sea, the land, or the air, has its animal occupant. Consider the coral reef (Fig. 21-1). Not only is the rocky face of the reef largely one vast colony of simple animals, but representatives of nearly all animal groups are present in blazing colors (Color plates 2A, B, 10A–E, 11A, C).

The diversity of animal life exceeds that of all other organisms: animals range from comparatively simple multicellular eukaryotes such as coral polyps, through worms, mollusks, and insects, to the most advanced types of vertebrates, such as the birds, the reptiles, and the mammals.

This diversity is so great that biologists are unable to agree on how many major categories of animal life should be recognized. Because the enumeration of animal phyla depends on whether certain small groups of animals are classified separately or combined with larger groups, the number of phyla ranges from 15 to 25 or more in different classifications; these phyla represent a vast spectrum of structural, developmental, behavioral, and ecological forms.

THE INVERTEBRATES

The majority of animal species are **invertebrates,** sometimes referred to as "animals without backbones" to distinguish them from members of the phylum Chordata to which animals having backbones belong. As we have seen, skeletal features are markedly different in the various animal phyla and therefore they are useful criteria in identification and classification. Also useful are the nature and state of development of reproductive, nervous, digestive, circulatory, and excretory systems, as noted in several of the preceding chapters. In addition, such features as relative size, kind of symmetry, mode of nutrition, and habitat preferences can be important in placing an animal in a family or in a phylum and in tracing its evolutionary relationships.

Poriferans—Sponges

Sponges, in the phylum **Porifera** (*porus,* "pore," *ferne,* "to bear"), are the most primitive invertebrate animals. They are common in all seas, usually in shallow water, but about 150 species of small sponges occur in fresh water. Sponges range in length from 1 cm (½ in) or so to nearly 1 m (3 ft); the smaller forms tend to be radially symmet-

FIGURE 21-1

Coral-reef scene in the Florida Keys, illustrating some common reef organisms. **A.** Brain coral, *Diplora labyrinthiformes*. **B.** Brain coral, *Colpophyllia natans*. **C.** Soft coral, *Eunicea* sp. **D.** Staghorn coral, *Acropora cervicornis*. **E.** Pillar coral, *Dendrogyra cylindrus*. **F.** Fan coral, *Gorgonia ventalina*. **G.** Fan coral, *Gorgonia flabellum*. **H.** Barrel sponge, *Xestospongia muta*. **I.** Tube sponge, *Callispongia plicata*. **J.** Parrot fish, *Sparisoma viride*. **K.** Blue angelfish, *Holocanthus isabelita*. **L.** Long-spined sea urchin, *Diadema antillarum*. **M.** Grunt, *Haemilon plumerieri*. **N.** Starfish, *Echinaster sentus*. **O.** Purple gorgonia, *Pseudopterogorgia americana*. **P.** Sea urchin, *Lytechnius variegatus*.

rical, the larger ones range from radial symmetry to assymmetry (Fig. 21-1H, I; Color plate 10A).

Based on cellular organization, sponges fall somewhere between typical multicellular organisms and colonies of protozoanlike cells. All adult sponges are sedentary and grow attached to objects in oceans, lakes, or streams. The living sponge is composed of several kinds of cells plus nonliving skeletal material. Irregular particles of silicon or calcium carbonate, called **spicules** (Fig. 21-2), are an important structural element of sponge skeletons, as is spongin, a fibrous proteinaceous substance. "Bath" and "auto" sponges are composed solely of spongin remaining after the cells have been allowed to rot away. As can be imagined, a sponge dock is notorious for its foul odors.

Unlike other animals, sponges lack a mouth and a digestive tract. Their bodies consist of a more or less rigid skeletal framework supporting a system of channels and pores. Flagellated cells propel water through this system and eventually to the outside through one or more comparatively large openings (Fig. 21-2). As water passes through the sponge, plankton and food particles are trapped by cells lining the internal passages.

There are no specialized sex organs in sponges; certain cells become transformed into eggs and sperm. After fertilization, the zygotes develop into flagellated larvae, which swim about for a short time, then settle down on some solid object and develop into sessile (nonmotile) adult sponges. Sponges can reproduce asexually also, sometimes by simple fragmentation—each piece growing into a complete sponge—and sometimes by the production of small buds which break free and become new individuals.

Because sponges lack an organized digestive system and seem more colonial than true multicellular organisms, they are considered by many biologists to be transitional between Protista and Animalia. Most zoologists consider sponges to be an early evolutionary offshoot, a "dead end," rather than part of the mainstream of animal evolution.

Cnidarians—Hydras, Corals, Jellyfish, and Anemones

Structure and Reproduction At the base of animal evolution is a group of radially symmetrical animals, members of the phylum **Cnidaria** (*cnidus*, "sting"), which are simple yet structurally and functionally advanced compared with Porifera. Cnidarians, also called **coelenterates,** exhibit one of two basic body forms. The **polyp** form (also called the **hydroid**) is exemplified by hydras, sea anemones, and corals (Color plate 10B–D); the **medusoid** form is represented by jellyfish (Fig. 21-3). Some cnidarians display both polyp and medusoid stages, which alternate in a life cycle. Their bodies are basically two-layered: an outer cell layer of ectoderm is separated from an inner endoderm by a jellylike **mesoglea** ("middle glue"), which is especially thick in jellyfish (Fig. 21-4).

There is no circulatory system. Food simply circulates in the gastrovascular cavity, and because no body cell is far removed from a food source, all receive nourishment. Undigested food particles pass to the outside through the mouth opening; there is no excretory system. Respiration is simple also: direct exchange of oxygen and carbon dioxide occurs between individual cells

FIGURE 21-2
Longitudinal section of a simple sponge. Larger and more complex sponges, such as "bath" sponges, are equipped with more channels and more intermediate flagellated chambers.

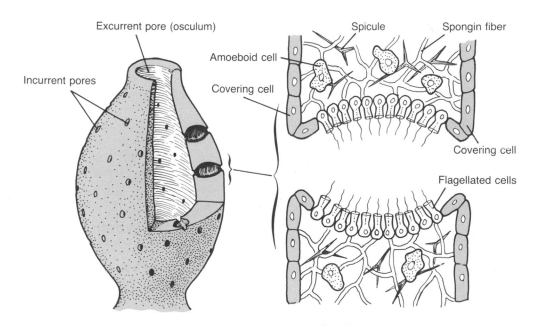

Excurrent pore (osculum)

Incurrent pores

Spicule

Amoeboid cell

Covering cell

Covering cell

Flagellated cells

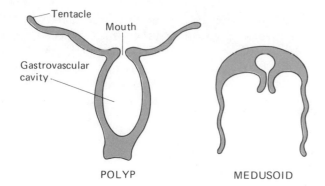

FIGURE 21-3
Polyp and medusoid forms of cnidarians. The polyp form at left corresponds to a small freshwater species called *Hydra;* the medusoid type at right is characteristic of jellyfish in general.

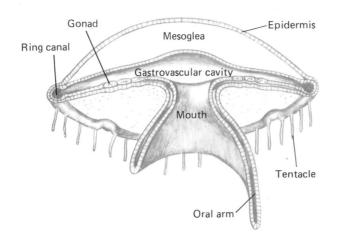

FIGURE 21-4
Medusa of *Aurelia,* an Atlantic jellyfish.

and the environment. So we see that in many ways the cells that make up the body of a *Hydra* are nearly as autonomous as are those of unicellular protistans.

Sex in cnidarians has a simple basis. Sperm and eggs may be shed into water, where fertilization occurs, or sperm may swim to eggs partially enclosed within ovaries. In most cases zygotes develop into ciliated larvae called **planulae** (sing. **planula**); in a few instances the planula stage is bypassed. Depending on the species, planulae may develop into adults resembling their parents or into forms quite unlike the parental type. The latter process is called **metagenesis** (Fig. 21-5). Although variation exists in metagenesis, medusoid individuals commonly are sexual and their zygotes develop into polyps. The polyps reproduce asexually, budding off small medusoid individuals.

Diversity For purposes of recognition and classification, cnidarians may be arranged in three classes. One class, **Hydrozoa,** includes solitary polyps, such as *Hydra,* and others, such as *Obelia,* that form small **medusas** (Fig. 21-5). One complex hydrozoan, the Portuguese man-of-war, combines polyps and medusoid individuals in a large floating colony supported by a gas-filled float (Fig. 21-6). The Portuguese man-of-war is armed with highly toxic nematocysts that can inflict painful injuries to humans.

A second class, **Scyphozoa,** is represented by **jellyfish,** in which the medusoid stage is dominant. Jellyfish are constructed basically like hydroids, although a thick layer of mesoglea in the umbrella-shaped **bell** tends to disguise the resemblance (Figs. 21-3 and 21-4). Jellyfish mesoglea is rather dense and resilient. Circular muscles in the bell contract rhythmically, at each pulse reducing the bell's diameter and forcing jets of water out of the bell's opening. The result is a rather slow but effective swimming action.

Sea anemones and **corals,** of the class **Anthozoa,** are characterized by the lack of a medusoid stage (Fig. 21-1 and Color plate 10). Corals are interconnected (colonial) polyps partially embedded in an extensive limestone skeleton of their own secretion. Their ability to form oceanic reefs is well known. Anemones are much more complicated versions of *Hydra* and are common in coral-reef environments. Many are large and colorful, adding the same dimension of beauty to the reef community as flowers do to fields and gardens. Like the other cnidarians, anemones and corals are armed with **nematocysts** (see box, p. 460).

Nematocysts are remarkably intricate structures, the more so when we consider that they are single cells. Each nematocyst is a tiny inverted harpoon coiled within the cell (Fig. 21-7) and held under tension. The coiled harpoon is discharged when a triggerlike projection is touched. There are two basic cell types: penetrants and volvents (sticky "lassos"). Nematocysts are abundant on the tentacles of cnidarians and as many as 25 percent can be fired at one time. They differ somewhat in structure and toxicity from group to group: those of certain corals (fire corals) are very poisonous as are those of the Portuguese man-of-war. Both can infect painful stings if brushed against by an unwary bather. Following discharge, new nematocysts are produced by differentiation of interstitial cells in the tentacles or body walls.

Platyhelminthians—Flatworms

Among the most beautiful of coral reef inhabitants are brightly colored **flatworms** (Color plate 10E), whose flattened bodies undulate gracefully as they swim about or graze along the coral. They are known as polyclads and

FIGURE 21-5
Metagenesis in *Obelia*,
a cnidarian.

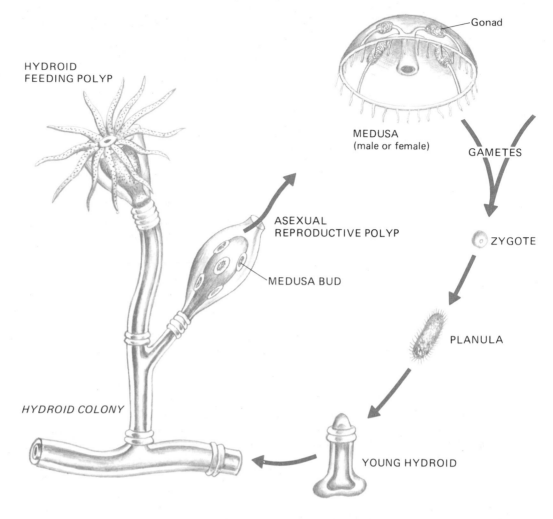

HYDROID
FEEDING POLYP

ASEXUAL
REPRODUCTIVE POLYP

MEDUSA BUD

HYDROID COLONY

Gonad

MEDUSA
(male or female)

GAMETES

ZYGOTE

PLANULA

YOUNG HYDROID

are related to smaller freshwater planarians described in Chapters 9 and 13 (Figs. 9-3 and 13-1). Although flatworms, as noted in those earlier chapters, are structurally more complex than cnidarians, they likewise lack a circulatory system, their excretory systems are rudimentary, their digestive system is a gastrovascular cavity with but one opening, and their bodies are solid (acoelomate, see Fig. 8-5).

Flatworms (in the phylum **Platyhelminthes**—*platys*, "flat," *helmines*, "worm") are monoecious, each individual being both male and female, but they are not usually self-fertilizing. Most free-living kinds copulate in pairs and exchange sperm, but some parasitic species such as tapeworms are first male, then female, thus avoiding self-fertilization. In some primitive flatworms a penis is present but there are no reproductive ducts. In these species the fertilization is hypodermic: the penis is pushed through the body wall of the partner and deposits its sperm in the vicinity of the ovaries.

During their evolution, many genera and species of flatworms have shifted from a free-living existence to

parasitism. Principal parasitic flatworms are tapeworms and flukes (Fig. 21-8). Flukes more closely resemble the planarian form than do tapeworms and have a mouth and gastrovascular tract. They inhabit internal ducts and passages of organs of higher animals, as well as their surfaces, feeding on body fluids and the products of digestion. Various kinds are found in the skin, eyes, liver ducts, blood vessels, lungs, and intestines, where they anchor themselves by suckers located near the mouth.

Tapeworms are composed of many sections arranged in series to form a long ribbon. A knoblike anterior section, the **scolex**, is equipped with hooks and suckers adapted for anchorage. The remaining sections, called **proglottids**, are identical reproductive units. A tapeworm therefore is not one animal but a colony of individuals. Tapeworms live in the intestines of vertebrates, commonly in the small intestine, where food undergoes digestion. They lack gastrovascular systems and absorb digested food directly into their bodies.

Tapeworm eggs do not directly grow into tapeworms but instead, when ingested, develop into burrow-

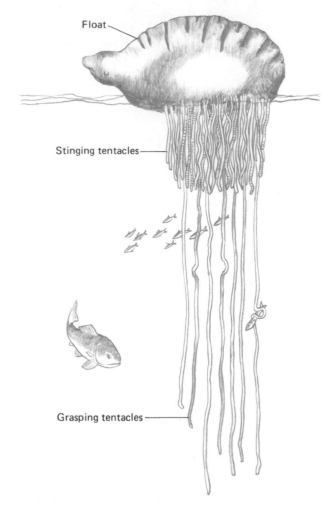

FIGURE 21-6
Portuguese man-of-war, a colonial cnidarian.

Nematodes—Roundworms

Roundworms, of the phylum **Nematoda** (*nema*, "thread") are free-living as well as parasitic animals in which for the first time hollow rather than solid bodies are encountered. Their cylindrical bodies contain a fluid-filled cavity, the **pseudocoelom** (see Fig. 21-9), and are covered by a thick but flexible **cuticle** secreted by underlying cells.

In roundworms we encounter, for the first time in our survey of major animal groups, a tubular digestive tract with two openings, a mouth and an anus. This, we will see, as we continue our progress through the animal kingdom, permits further specialization of the head vs. the tail region, for it concentrates food-capturing and food-grinding processes at one end of the tract, while regions specialized for food digestion, nutrient absorption, and waste deposition are arranged in an "assembly-line" sequence.

We also discover in roundworms an elementary circulatory system in the form of a fluid in the pseudocoelom. This fluid can collect and convey nutrients, gases, and other substances to and from cells. In addition, ciliated excretory tubules collect and transport metabolic wastes from the pseudocoelomic fluid to the outside.

A simple nervous system consisting of dorsal, ventral, and lateral nerve cords and branch fibers is present. Most roundworms move by a whiplike motion produced by contractions of longitudinal muscles in the body wall. Studies of one species of small roundworm have shown that only six neurons control touch sensitivity and locomotion. If touched near the tail, the worm moves forward; if touched near the head it moves backward. If either or both the anterior or posterior neurons are destroyed (with a laser beam), the animal will, respectively, move only forward, only backward, or show no response. The simplicity of this system has been exploited to study the relationships between genes and elements of the touch–response relationship, including synapses and neuromuscular junctions.

ing larvae that encyst in the organs of a new host. If the flesh of this alternate host is eaten, the cysts develop into tapeworms in the intestines of the consumer. Alternate hosts of human tapeworms are fish, livestock, and wild game species. Snails and fish serve as the alternate hosts for flukes.

FIGURE 21-7
Nematocyst structure. Undischarged nematocyst (**A**) is a highly modified cell containing an inverted and coiled harpoon (penetrant) or lasso (volvent). A ciliumlike trigger controls the discharge of the nematocyst by hydrostatic pressure. **B.** A penetrant is shown as having penetrated the body wall of a prey organism. Poison is injected through the hollow tube of the harpoon. **C.** Volvents are depicted as entangled about the antenna of a small crustacean.

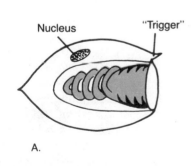

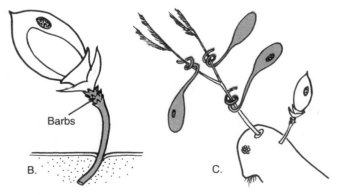

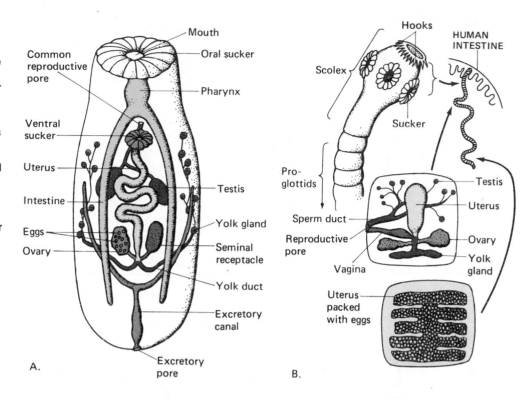

FIGURE 21-8
Parastic flatworms. **A.** The Chinese liver fluke infests humans. **B.** A tapeworm of the type found in the human small intestine. Flukes and tapeworms commonly infect alternate hosts, spending part of their lives in some animal other than man. Commonly the eggs of both pass to the outside in human excrement and, in the absence of proper sanitation, are transported in polluted water, infected dust, or food. The fish-human tapeworm, for example, grows to several meters length in the human intestine. Its eggs may enter a lake or stream and become attached to gills of fish, migrate as larvae into muscle fibers, and develop into resistant cysts. If a human eats poorly cooked or smoked fish the cysts survive and develop into adult worms in the host's intestines. Some flukes have complex life cycles involving as many as three alternate hosts.

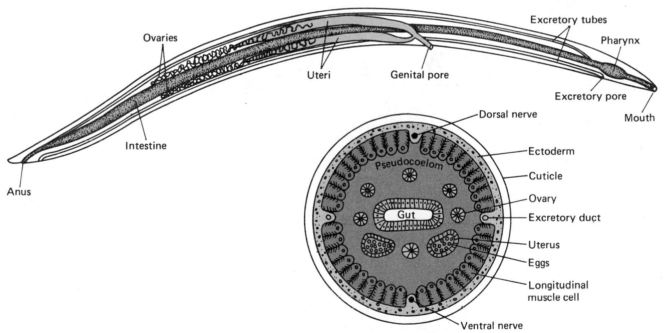

FIGURE 21-9
Ascaris, a roundworm parasite, in longitudinal and cross section.

It has been said that if every living plant and animal with the exception of parasitic roundworms were suddenly to vanish from the earth, it would be possible for an observer from another planet to reconstruct the diversity from the little heaps of species-specific roundworms left behind where a plant or animal had existed. This is perhaps an exaggeration but it emphasizes an important point: roundworms are exceptionally successful parasites, inhabiting a variety of plant and animal life, including humans. It should be noted here that

CLOWNFISH AND SEA ANEMONES

Sea anemones are armed with powerful offensive and defensive weapons found on their tentacles: batteries of nematocysts. An unwary crab or fish, brushing against the tentacles, triggers the discharge of numerous nematocysts and thereby causes its own death.

A group of fish, however, manages to live freely among the tentacles of its host anemones without discharging the dangerous nematocysts. Many researchers have tried to determine the nature of the fish's immunity ever since the relationship was first noted by C. Collingwood in 1868. The relationship between this fish—the clownfish—and the sea anemone is a **symbiotic** one, meaning that two totally unrelated organisms live together. The kind of symbiosis shown by the clownfish and their hosts is called **mutualism** because both partners in the symbiosis profit from the relationship.

Although many species of fish associate with sea anemones, the clownfish (also called anemone fish) have received the most research attention. About 27 species of clownfish have been found living with about 13 species of sea anemones. Clownfish are common in the tropical reef areas of the Indo-Pacific Ocean and the Red Sea. As adults they are boldly patterned with vertical white bars and stripes against a bright background, commonly orange (see Fig. 1). Some species of clownfish are restricted to one host species of sea anemone; other anemone fish may live among two or three different species. Whether "specialist" or "generalist," each individual anemone fish seems to be an **obligate** (complete or invariable) **symbiont,** except during its larval stage.

Some early researchers believed that the clownfish could live in their host anemones because the latter lacked nematocysts. Not so. Others felt that the nematocysts of the hosts were weak; that is, the anemones were incapable of

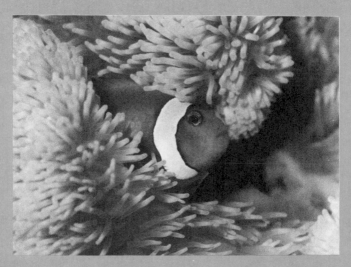

FIGURE 1
Sea anemone and clownfish. (Photo from Phil and Loretta Herman/
Tom Stack & Associates)

stinging prey as large as a fish. Not true. Serious experimental research began with Davenport and Norris in 1958. Their research and that of several others in the 1960s and 1970s indicated that there were two main hypotheses to be tested:

1 The symbiont fish, either by its behavior or by the physicochemical stimuli it provided the anemone, somehow altered the cnidarian's behavior and thereby inhibited nematocyst discharge.

2 The fish was able to alter its outer mucous covering and thereby avoid being stung.

All fish have a mucous or slimy covering, and this seemed to be the target area for further study. D. Schlichter has suggested recently that the clownfish gets a substance from the anemone, incorporates it into its own slimy covering, and thus "fools" the anemone into acting as if the fish were merely one of its tentacles. The mucous layer of the fish seems to be the research bull's-eye. The latest, from R. Lubbock of Cambridge University in England, indicates that the mucous layer of the fish is indeed the site of its immunity, but not because of anything it gets from the host anemone. Rather, the fish is immune because of what it lacks in its mucus. Detailed analyses of the mucus of clownfish and of nonsymbiotic fish clearly indicated a number of chemical differences. Nonsymbiotic fish have in their mucous layer chemicals that stimulate the anemone to discharge its nematocysts. These excitatory chemicals seem to be lacking or fewer in number in the mucous layer of the clownfish. Lubbock has reduced the problem to one of cellular recognition: the anemone can clearly recognize a number of compounds found in the mucus of nonsymbiotic fish, compounds that result in the discharge of nematocysts. The clownfish are like "stealth bombers": they cannot be recognized as prey by the anemones.

In the mutualistic relationship between the clownfish and their hosts, the major benefit to the fish seems clear: protection from their enemies. The anemone's nematocyst-armed tentacles enfold the clownfish in a protective "embrace." What of the benefits to the anemone? Many researchers have recorded how the attendant clownfish brings food to the anemone, shoving the "gifts" deep into the tentacles of their associate. Also, clownfish have been seen to clean debris from the oral discs surrounding the mouths of their hosts.

The case of the clownfish and the sea anemone clearly indicates how research often asks more questions than it answers, and this in turn leads to renewed observation and experimentation. Science is dynamic, not static.

many successful parasites do not measurably reduce the well-being of the host-organism; after all, the survival of the parasite depends on the survival of the host. However, some forms of roundworm infections are detrimental to the host and can result in death.

Roundworms include a number of human parasites, among them **hookworms, pinworms, eyeworms,** and **filaria worms** that live in lymph glands and cause **elephantiasis.** Others are **Guinea worms** more than a meter (3 ft) long that migrate in the human body and come to rest beneath the skin, where they are clearly

visible; *Trichinella* worms causing **trichinosis;** and *Ascaris lumbricoides*, a large (about 15 cm [6 in] long) intestinal parasite of humans and certain domesticated mammals, sometimes used as a study specimen in biology classes (Fig. 21-9).

Annelids—Segmented Worms

One of the great architectural and industrial advances of the present century has been the modular structure, based on similar units of construction which can be repeated in different ways to produce an integrated system. Modular construction, or **segmentation,** is an old story in animal evolution. It occurs in the segmented worms, called **annelids** (phylum **Annelida**—*annellis,* "ring"), which include **earthworms, polychaetes (marine worms),** and **leeches,** and also in modified form in other phyla. Another widely used term for segmentation is **metamerism.** Here also for the first time are organisms characterized by a coelom (see Figs. 8-5 and 21-10).

Bodies of annelid worms are constructed of basically similar segments, but some are modified depending on location and specialized function. Segmentation is principally an adaptation for movement. Annelids lack a rigid skeleton, but a fluid-filled coelomic chamber in each segment constitutes a flexible support, or **hydroskeleton,** for the body and for the action of circular and longitudinal muscles, which coordinate with each other and the hydroskeleton to produce an accordionlike movement.

Digestion, Circulation, and Excretion Annelids have a digestive tract consisting of a muscular **pharynx,** which acts as a pumping organ in food intake; an **esophagus;** a "stomach" where food is digested; and a long **intestine** where nutrients are absorbed and nondigested matter is compacted and finally expelled through the anus as castings. In most annelids nutrients are circulated through the body by a closed circulatory system (see Figs. 8-6 and 21-10) consisting of semicircular pairs of "hearts," major dorsal and ventral vessels, and branches extending to muscles, skin, intestine, and various organs (leeches have a much reduced circulatory system).

Metabolic wastes collect in the coelomic fluid in each segment, from which they are eliminated by the action of paired excretory organs called **nephridia** (sing. **nephridium).** The inner end of a nephridium is a ciliated funnel which drains the coelomic fluid; the other end empties to the outside, or in some cases into the intestine. Nephridia may be compared to kidneys of vertebrates in function, but unlike kidneys do not filter waste matter directly from the bloodstream.

Nervous System The generalized annelid nervous system consists of an anterior cluster of neurons, the **brain ganglion,** or **brain,** from which a double ventral nerve cord extends posteriorly (Fig. 21-10). Additional ganglia are present in the nerve cord, and lateral nerves extend from them to muscles, organs, and sensory receptors. Among the latter are light receptor cells capable of sensing changes in light intensity. In leeches and polychaetes, light receptors occur in clusters and form simple eyes, and in one marine worm a lens is present, so images probably are

formed. However, earthworms lack eyes and can sense and respond only to changes in light and darkness. Chemical (taste) receptors are found in the body wall and are most numerous in the anterior segment **(prostomium).** Touch receptors are widely distributed in the body wall. Polychaetes have, in addition, touch-sensitive antennae and balancing organs called **statocysts.** Statocysts are fluid-filled chambers containing small crystalline particles and lined with sensitive hair cells. Movements of the particles enable the animal to sense directional changes. Statocysts are found in representatives of other animal phyla as well as in annelids; their principle is repeated on a more elaborate scale in the balancing organs of vertebrates (the semicircular canals).

Communication The question of communication between animals is an interesting one, particularly in species such as earthworms in which visual and auditory communications seem ruled out. Earthworms, when disturbed or hurt, can generate alarm signals that cause other earthworms to retreat from the vicinity. The signal is in the form of a copious flow of mucus from the alarmed individual. This secretion contains a **pheromone** (communication chemical) to which other earthworms respond; the mucus may have the added effect of repelling some predators. Some polychaete annelids have their external fertilizations synchronized by the simultaneous production of sex pheromones from neighboring males and females. Reception of the sex pheromones brings about gamete expulsion into the surrounding seawater, where fertilization occurs.

Reproduction Reproduction in marine annelids is simple. The sexes are separate, and testes and ovaries shed sperm and eggs directly into the water, where fertilization occurs. The fertilized egg develops by spiral cleavage into a ciliated larva called a **trochophore** (see Fig. 21-11). Trochophores mature into small worms of a few segments, and these grow into adults by segment proliferation.

Earthworms, unlike polychaetes, are **hermaphroditic,** having ovaries and testes **(gonads)** located in different anterior segments of the same individual. At certain times of the year, earthworms copulate and exchange sperm. Afterward, each worm secretes a bandlike encircling cocoon into which is deposited a mixture of eggs and sperm. The cocoon slips off the worm and its two ends contract to seal off the interior, where fertilization of eggs and development of young worms then occurs. Annelids are protostomes and exhibit spiral cleavage in early stages of embryo development.

Diversity Earthworms and their near-relatives are called **oligochaetes** ("few bristles"), because their seg-

FIGURE 21-10
Structures of the earthworm, a
representative annelid. **A.** Longitudinal section of internal anatomy. **B.** Cross section in the region of the intestine.

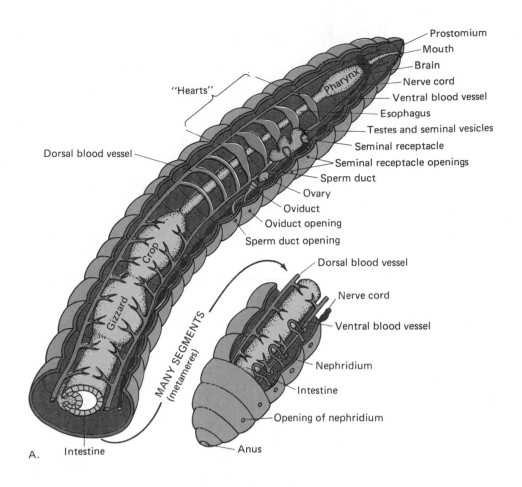

Prostomium
Mouth
Brain
Nerve cord
Ventral blood vessel
Esophagus
Testes and seminal vesicles
Seminal receptacle
Seminal receptacle openings
Sperm duct
Ovary
Oviduct
Oviduct opening
Sperm duct opening

Pharynx

"Hearts"

Dorsal blood vessel

Crop

Gizzard

MANY SEGMENTS
(metameres)

Dorsal blood vessel
Nerve cord
Ventral blood vessel
Nephridium
Intestine
Opening of nephridium
Anus

A. Intestine

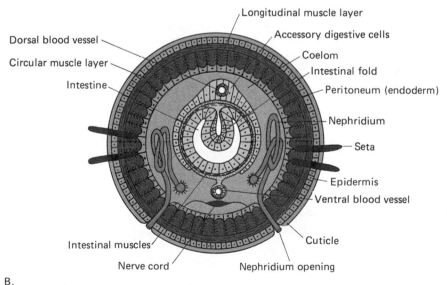

Longitudinal muscle layer
Accessory digestive cells
Coelom
Intestinal fold
Peritoneum (endoderm)
Nephridium
Seta
Epidermis
Ventral blood vessel
Cuticle

Dorsal blood vessel
Circular muscle layer
Intestine

Intestinal muscles
Nerve cord
Nephridium opening

B.

ments bear bristles, or **setae,** which are relatively few in
number. Setae aid in locomotion by serving as gripping
devices. Marine worms are known as polychaetes ("many
bristles") because they have many setae at the tips of
lateral appendages, the **parapodia.** Parapodia resemble

stumpy legs and are used in locomotion and as respiratory organs. Some sedentary polychaetes called **fanworms** dwell in burrows in the sea bottom and bear
upon their heads colorful featherlike feeding organs
which they extend into the water (Fig. 21-11 and Color

THE CASE OF THE DISAPPEARING FLAGSTONES

A narrow path running across part of my lawn was paved in 1843 with small flagstones, set edgeways; but the worms [earthworms] threw up many castings and weeds grew thickly between them. During several years the path was weeded and swept; but ultimately the weeds and worms prevailed, and the gardener ceased to sweep, merely mowing off the weeds, as often as the lawn was mowed. The path soon became almost covered up, and after several years no trace of it was left. On removing, in 1877, the thin overlying layer of turf, the small flagstones, all in their proper places, were found covered by an inch of fine mould [humus].

In his customary thorough manner, Charles Darwin went on to describe the way in which earthworms burrow under and remove soil from beneath objects and deposit the soil in small heaps or "castings" at the mouths of their tunnels. Although individually insignificant, collectively earthworms bury stones, coins, flagstones, even buildings beneath inches and sometimes feet of humus. Darwin, in *Formation of vegetable mould through the action of earthworms* (1881), related that Roman villas, erected more than 1,500 years previously and abandoned perhaps 200 or 300 years later, had actually sunk several feet as a result of the action of earthworms. In some cases, this lowering was so uniform that the tiles of elaborate floor mosaics still maintained their original patterns.

Earthworms feed on dead leaves and other vegetable matter which they draw down into their burrows and digest. They also ingest the soil they excavate in their burrows, extracting whatever nourishment can be derived, and deposit the residue at the surface. This is the "mould" or humus about which Darwin wrote. Because earthworm populations are generally high, even in relatively dry climates, the contributions of earthworms to the topsoil is cumulatively substantial—by some estimates as much as an inch and a half in 10 years time.

Some doubts have been expressed about whether earthworms swallow soil solely for the purpose of making burrows, but Darwin thought they did. He described castings of pure chalk and even bits of concrete and mortar. Apparently earthworms must eat their way through compact soil. Even so, earthworms are most abundant where they have access to leaves and other vegetable debris. In a hot and humid garden in Calcutta, India, earthworm castings 6 inches tall were deposited overnight; in temperate regions, such as Darwin's England, the castings seldom exceed an inch in height. The collective action of earthworms (an early estimate of 53,767 worms per acre is cited by Darwin) is such that the weight of castings ejected at the surface of an acre of land each year may reach 15 tons. Darwin concluded his remarkable essay on earthworms by saying that "an archaeologist ought to be grateful to worms, as they protect and preserve for an indefinitely long period every object, not liable to decay, which is dropped on the surface of the land," and "worms prepare the ground in an excellent manner for the growth of fibrous-rooted plants and for seedlings of all kinds." And again, "It may be doubted whether there are many other animals which have played so important a part in the history of the world, as have these lowly organised creatures."

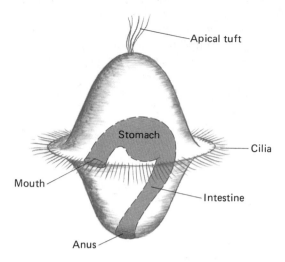

FIGURE 21-11
Trochophore larva of a marine annelid. The trochophores of mollusks are similar.

plate 11C). Recently discovered giant **tubeworms** also fall in this category; tubeworms have a symbiotic association with chemosynthetic bacteria, the bacteria serving as a food source.

Leeches (Fig. 21-12) have neither setae nor parapodia but use the suckers at either end of their bodies to pull themselves along, inchworm fashion. They commonly are aquatic, but many live on land, especially in moist tropical areas. Their bodies are flattened and bear a sucking mouth containing sharp teeth. Many leeches feed on blood, as is well known; once attached to a host, they may consume three to five times their own weights. Such feedings may be sufficient for weeks and months. In the past, large (up to 20 cm [8 in] long) "medicinal" leeches were used to reduce swellings and bruises and were part of the stock of drugstores, where they were kept in jars of water until needed.

FIGURE 21-12
Representative annelids.

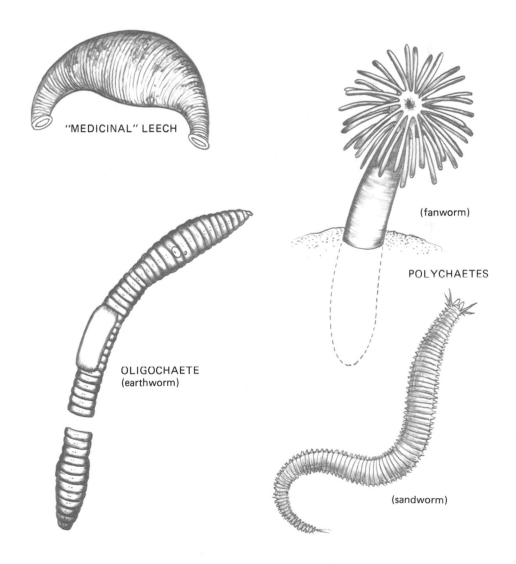

Mollusks—Snails, Clams, and Squids

Members of the phylum **Mollusca** (mollusks—*molluscus*, "soft, thin shelled") include ancient organisms that lived in Paleozoic seas as long ago as 500 million years. They constitute the second largest animal phylum, with some 128,000 species. Some (squids and octopods) are the largest invertebrates known.

Structure Mollusks are considered to be coelomates, although the body cavity is restricted to a relatively small region surrounding the heart and a portion of intestine. Their bilaterally symmetrical bodies are composed of three regions: a **head/foot, mantle** and **shell**, and **visceral mass** (Fig. 21-13 and Color plate 11A, B). The head/foot region is variously modified in the major groups of mollusks. In some it is primarily an organ of

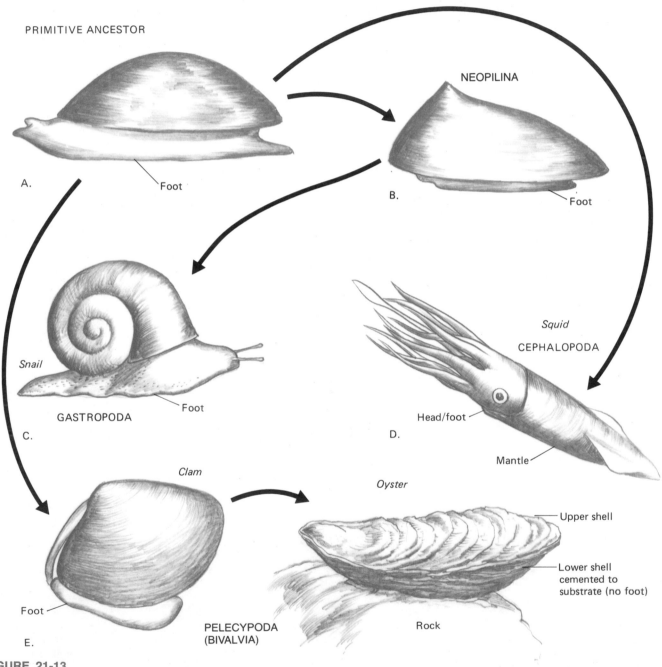

PRIMITIVE ANCESTOR

A. — Foot

NEOPILINA

B. — Foot

Snail

GASTROPODA

C. — Foot

Squid

CEPHALOPODA

D.

Head/foot

Mantle

Clam

Oyster

Foot

PELECYPODA
(BIVALVIA)

E.

Rock

Upper shell

Lower shell cemented to substrate (no foot)

FIGURE 21-13
Postulated evolutionary relationship among mollusks.

locomotion; in others it bears antennae, eyes, a mouth, and sometimes tentacles. The mantle is a specialized region of skin that secretes a shell and encloses a **mantle cavity,** in which respiratory and excretory structures are located. The visceral mass contains digestive, circulatory, and nervous systems.

Diversity   Mollusks are an exceptionally diverse group of animals (Fig. 21-13), second only in number of species to the **arthropods,** which include the myriad of insect species. Among the most primitive mollusks are the **chitons,** of the class **Amphineura.** They are all marine and inhabit shallow areas such as the zone between the tides. Chitons are bilaterally symmetrical, flat, and oval-shaped, and the dorsal side is covered by plates. A powerful foot musculature exerts great suction, enabling the organism to attach itself to intertidal rocks. A unique molluscan feeding organ, the **radula,** permits chitons to graze on algae by rasping this plant food from the rocks. Respiration is by gills, and a trochophore larval stage occurs in the life cycle.

The class **Monoplacophora** includes the famous "living fossil" *Neopilina,* dredged up in 1952 from the great depths of the Pacific Ocean off the coast of Costa Rica. *Neopilina* is clearly segmented, and to some zoologists this feature offers living proof of the annelid ancestry of mollusks (Fig. 21-13).

From these primitive groups evolved the more advanced and familiar mollusks. **Snails, slugs, limpets, nudibranches,** and the like comprise the largest group of mollusks, the class **Gastropoda.** Many gastropods—snails, for example—have spirally coiled shells, but others have vestigial shells or none at all. In a typical gastropod with a coiled shell, the early larva is bilateral, but coiling brings about a twisting of the digestive tract so that the anus, normally posterior in the body, winds up near the mouth, in the animal's anterior region. In addition, the whole visceral mass is rotated 180° and comes to lie above, or dorsal to, the head. The adult snail is asymmetrical because of the coiling and atrophy of certain organs on one side of the body.

Gastropods are broadly distributed in the seas, in fresh water, and on the land. They are among the few invertebrate groups that have evolved truly terrestrial species. In most land snails, the gills of their aquatic ancestors have been replaced by a vascularized mantle cavity that acts as a lung. Such fully terrestrial snails are said to be **pulmonate.**

The class **Pelecypoda** (*pelekys,* "hatchet," *podo-,* "foot"), sometimes called class **Bivalvia,** includes such familiar forms as **clams, mussels, scallops,** and **oysters** (Fig. 21-13). Pelecypods are sedentary filter feeders. Their modified gills act as strainers, filtering out their plank-

tonic food from water passing through the mantle cavity. Some giant bivalves of the Great Barrier Reef off the northeast coast of Australia have shells a meter (3 ft) or so in length and weigh as much as 250 kg (550 lb). In the Middle Ages these shells were especially treasured and were used as baptismal fonts in some of the great cathedrals.

In another evolutionary direction, the ancestral mollusks gave rise to a group of large tentacled forms that became rapidly moving carnivorous predators, the **cephalopods.** The class **Cephalopoda** (*cephalo-,* "head," *podo-,* "foot") includes **squids, cuttlefish, octopods,** and the shelled **nautiloids** (Fig. 12-13). All cephalopods are marine. These active predators locomote by a kind of jet propulsion. Large size in squids evolved through abandonment of the external shell of more primitive forms and the development of powerful muscles. The complex behavior of cephalopods is integrated by a highly advanced nervous system. The giant nerve fibers (**axons**) of

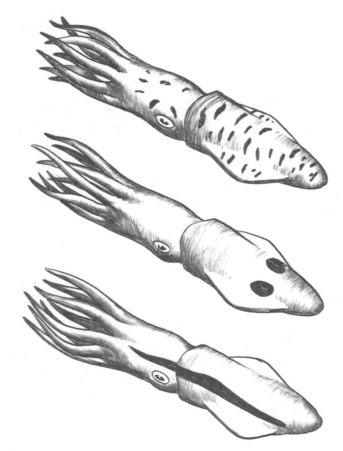

FIGURE 21-14
Variety of visual displays brought about by chromatophore changes in the skin of the active, predatory cephalopods. (Based on M. Moynihan)

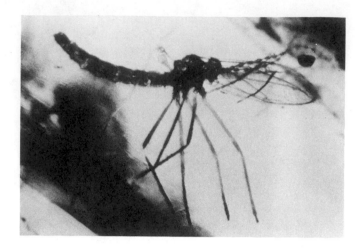

FIGURE 21-15
Insect in amber. (Photo by W. B. Saunders, Bryn Mawr College/Biological Photo Service)

squids are large enough to permit insertion of electrodes; study of these giant axons has added greatly to the knowledge of neurophysiology. The skin of squids and octopods is richly supplied with pigment-bearing cells called **chromatophores**. This feature enables these

active predators to show a diversity of visual displays (see Fig. 21-14), which can be changed rapidly and indicate a variety of moods or motivations.

Arthropods

Of the approximately 1.5 million named species of animals, nearly three-fourths are classified in the phylum **Arthropoda** (*arthron*, "joint," *podus*, "foot"). Arthropods are among the most rapidly evolving organisms known, yet among them are some of the oldest animal fossils. **Trilobites**, looking something like present-day pill bugs (compare Figs. 21-19A, D), were common in the ancient oceans of the early Paleozoic era half a billion years ago and may represent the ancestral stock of modern arthropods. Subsequently, during later Paleozoic periods, all the major arthropod groups evolved. Giant dragonflies with wing spans of nearly a meter (3 ft) flew about in the swamps of the Carboniferous period, and huge cockroaches as well as many other arthropods recognizable as related to present-day genera also were present. Arthropods continued to proliferate and diversify during the Mesozoic era. Some of these became the most perfectly preserved animal fossils, having become trapped in sticky gums and resins of ancient conifers which sub-

FIGURE 21-16
Generalized structure of arthropod segments and appendages **(A)** and major arthropod features **(B).** Typically, the arthropod digestive tract consists of a foregut, midgut (stomach), and hindgut (intestine). The circulatory system is said to be "open," meaning that blood circulates in cavities (sinuses) in the trunk and appendages. Blood vessels ending in the sinuses transport blood via a dorsal heart to all parts of the body. The main nerve cord of the body is a ventral double trunk having ganglia in each body segment, of which the brain is the anteriormost.

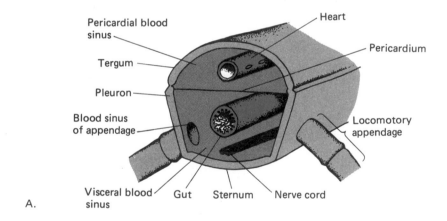

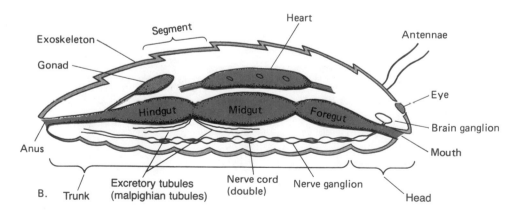

sequently fossilized into amber. Many amber insects have been studied and seem little different from those of today (Fig. 21-15). It would, however, be a mistake to think that the evolution of arthropods is largely past history. Important advances in the understanding of genetic mutation and rates of evolution have resulted from studies of such arthropods as fruitflies and moths.

Medieval knights rode forth to battle clad in plates of jointed metal, but long before the dawn of humans, arthropods enclosed within jointed armor ventured into new habitats and niches. Key to the proliferation and perpetuation of arthropod classes, orders, families, genera, and species is their evolution of body armor (exoskeleton) and jointed appendages. These first evolved in animals that lived in water and provided support, protection, and mobility after arthropod occupation of terrestrial habitats.

Structure Arthropods are characterized by a segmented body covered with a jointed exoskeleton. The generalized arthropod is a bilaterally symmetrical, multilegged animal having a body composed of a **head** and a **trunk** (Fig. 21-16B); the latter may be further differentiated into **thorax** and **abdomen**. The head is composed of several fused segments, the appendages of which are modified and serve as antennae, mouth parts, and grasping organs. Each trunk segment is composed of four plates (a dorsal **tergum**, two lateral **pleura** [sing. **pleuron**], and a ventral **sternum**), together with a pair of locomotory appendages (Fig. 21-16A). Variations from this general form can be interpreted in terms of two evolutionary trends: elimination of certain segments and fusion of others, and modification of appendages (antennae, legs, pincers, paddles, jaws, etc.).

The arthropod exoskeleton contains a complex substance called **chitin**, and may in addition be impregnated with calcium carbonate and calcium phosphate (components of limestone), so as to be tough and durable. Because the exoskeleton is not expansible, the growing arthropod must periodically shed the old exoskeleton (**molt**) and grow a new and larger one. The jointed exoskeleton serves not only as armor but also for attachment of muscles, and consequently it functions in locomotion. Arthropods are accomplished burrowers, walkers, runners, jumpers, swimmers, and flyers.

Major sensory structures are antennae (except in spiders and scorpions), eyes, **statocysts** (balance organs), sound receptors, and chemoreceptors (taste and odor receptors). Antennae function as both touch receptors and chemoreceptors (some arthropods have been shown to lose their ability to discriminate between chemical substances when their antennae are removed). Chemical secretions play an important role in commu-nication among arthropods, particularly among insects that use pheromones in pathfinding, sex attraction, and other communication. Specialized touch receptors attached to thin diaphragms in the exoskeleton function as sound receptors, but are present only in insects such as crickets and cicadas that employ sound in communication.

Arthropod eyes are of two types: **ocelli (simple eyes)** and **compound eyes**. Both types of eyes are capable of image formation, and both may be present in one individual. Insects may have three ocelli and a pair of compound eyes; spiders lack compound eyes but have as many as eight ocelli, some of which are capable of producing sharply defined images. A few arthropods lack eyes altogether.

The compound eye is composed of a variable number of rodlike units called **ommatidia** (sing. **ommatidium**) (Fig. 21-17). Each ommatidium is a simple eye equipped with its own lens system. Thus the image the crab or insect perceives is actually a mosaic of partial images formed by each eye unit. The compound eye is very sensitive to minor movements as the image of an object shifts from unit to unit.

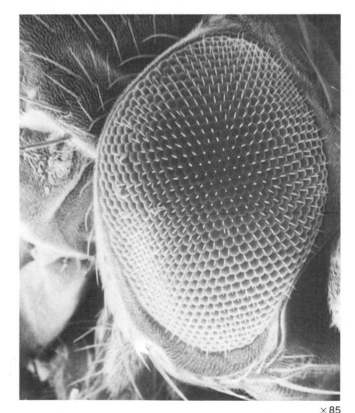

×85

FIGURE 21-17
Scanning electron micrograph of the compound eye of *Drosophila*. (Photo courtesy Dian Molsen)

Reproduction As would be expected in a large phylum whose members occupy diverse habitats, reproduction varies considerably in arthropods. Among aquatic species, fertilization may be external or internal; in land arthropods, fertilization is internal. Larval stages occur in many arthropods, and the young may pass through a series of developmental stages (collectively called **metamorphosis**), each differing in form and feeding habit (refer to Fig. 11-5). In other arthropods, development is direct, and young individuals are miniature versions of adults. All arthropods are protostomes and are characterized by spiral cleavage of embryonic cells.

Major Groups Arthropods may be subdivided on the basis of habitat as well as structural specializations (Table 21-1). The least specialized and most primitive arthropods were extinct bottom-dwelling marine animals called trilobites (Fig. 21-18A). Trilobites may have been the ancestral stock from which all other arthropods evolved, among them members of several groups of aquatic and land arthropods characterized by presence of mouth pincers (**chelicerae**; sing. **chelicera**). **Horseshoe crabs** (Fig. 21-18B) and **eurypterids** (a group of scorpionlike arthropods) are aquatic representatives. Horseshoe crabs are but little changed from ancestors which swam in Ordovician seas 450 million years ago; eurypterids have long been extinct, yet may be ancestral to present-day chelicerates (**spiders, scorpions, mites, and ticks**) (Fig. 21-18C and Table 21-1).

A second major group of arthropods includes **crustaceans, centipedes, millipedes,** and **insects,** all of which are characterized by the presence of **mandibles** (jawlike mouth parts). Crustaceans are diversified and include **crayfish, lobsters, crabs** (Color plate 11D), **barnacles** (Fig. 21-18E and Color plate 2C), **shrimp,** small, superabundant planktonic animals called **copepods** and **branchiopods** (Color plate 11E), and land-dwelling **isopods** such as wood lice and pill bugs. Crustaceans are primarily aquatic animals having numerous appendages (commonly 10 or more pairs) and a body composed of a fused head and thorax (**cephalothorax**) and a segmented abdomen. Centipedes and millipedes, as their names imply, are many-legged arthropods but differ in the number of appendages per segment (two per segment in centipedes, four per segment in millipedes). Centipedes are carnivorous and usually fast moving; millipedes are slow-moving vegetarians (Fig. 21-18F, G).

Insects Insects constitute the largest class of organisms on earth. Although about 750,000 species have been described and named, it is estimated that there are 10 million. In terms of numbers of species, numbers of

TABLE 21-1
Characteristics of major arthropod classes

Class	Representatives	Characteristics
Trilobita	Trilobites	Extinct, many-legged aquatic animals with compound eyes; 1 pair antennae
Merostomata	Horseshoe crab	Primitive 12-legged marine animals with large head-thorax (cephalothorax); no antennae; small abdomen; chelicerae; compound eyes; book gills*
Arachnida	Spiders, scorpions, mites, ticks	Eight-legged land animals; some poisonous, some parasites; chelicerae; simple eyes; book lungs*; no antennae
Chilopoda	Centipedes	Many-segmented land dwellers; 1 pair legs per segment; mandibles; 2 antennae; rapid moving; some poisonous; compound eyes
Diplopoda	Millipedes	Many-segmented land dwellers; segments fused in pairs, giving appearance of 4 legs per segment; mandibles; 2 antennae; slow moving; some poisonous; simple or no eyes
Crustacea	Crabs, lobsters, copepods, barnacles	Many appendages of different functions; mostly aquatic with gills; mandibles; compound eyes; 4 antennae
Insecta	Insects	Body of head, thorax, abdomen; land dwellers with tracheae; 6 legs; mandibles; 2 antennae; compound eyes; often 2 or 4 wings

*Modified gills consisting of many parallel folds of membranes and tissues and in which blood flows and past which water or air currents move.

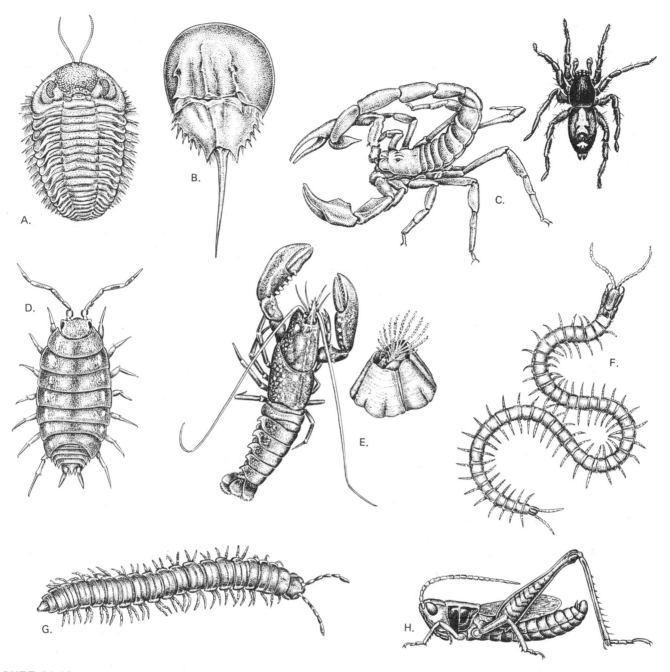

FIGURE 21-18
Major groups of arthropods. **A.** Trilobite. **B.** Horseshoe crab. **C.** Scorpion and spider. **D.** Wood louse.
E. Lobster and barnacle. **F.** Centipede. **G.** Millipede. **H.** Grasshopper.

individuals, and their ability to survive in habitats of all kinds, they are the most successful inhabitants of the earth.

Insects are characterized by bodies composed of a head bearing one pair of antennae, compound eyes and ocelli, and mouth parts adapted variously for biting, chewing, or sucking. Six legs are present on the thorax,

which is composed of three fused segments. The abdomen is elongate and segmented. Insects breathe by means of spiracles and tracheae; excretory organs are malpighian tubules (Fig. 21-19).

Insects display many fascinating structural, reproductive, and behavioral characteristics. Among these are flight, complete metamorphosis, and complex social be-

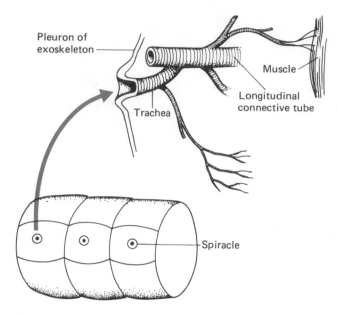

FIGURE 21-19
Spiracles and tracheae.

havior. Insects are the only arthropods capable of self-sustained flight. Although scientists do not know how insect wings have evolved, it has been suggested that they originated as flat projections of the thorax that enabled primitive insects to glide short distances and to land upright.

Wings of insects are folds of skin (epidermis) covered by cuticle, and commonly are four in number (two in *Diptera* [fly, mosquito, etc.]). Insect flight is based on contraction of powerful thoracic muscles that act in opposition to expand or contract the tergum to which the wings are pivoted (see Fig. 21-20).

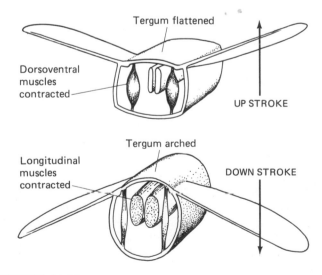

FIGURE 21-20
Wing structure of insects.

One important ecological function of insects is flower pollination. Fully two-thirds of flowering plants are insect pollinated. It is thought that flowers evolved in response to insect pollination and that certain groups of insects such as bees, wasps, moths, butterflies, and some flies in turn developed special structures that aided in finding, transporting, storing, and consuming nectar and pollen produced by flowers. Color patterns, fragrances, nectars, and shapes of flowers can be viewed as attractions for specific insect pollinators (e.g., the reddish-veined petals and putrid odor of **carrion flowers** are fly attractants).

Another well-known ecological function of insects is the consumption of carrion. This characteristic is used to advantage in natural history museums, where colonies of **carrion beetles (dermestids)** are kept and used to clean the skeletons of small animals. They do so rapidly and efficiently.

Insects directly or indirectly contribute to disease in many plants and animals and are known to transport spores of fungi from plant to plant. Their roles as vectors of malaria and sleeping sickness were described in an earlier chapter.

Echinoderms

One group of ancient invertebrate animals started out as bilaterally symmetrical organisms whose body walls contained hard calcareous plates forming rather rigid internal skeletons that did not easily decompose but were preserved as fossils. As a result, this group—**Echinodermata**—has become one of the best known phyla of ancient animals. Echinodermata (*echinus*, "spiny," *derm*, "skin") includes **starfish, brittle stars, crinoids (sea lilies), sea urchins, sand dollars,** and **sea cucumbers** (Fig. 21-21 and Color plates 2C, 11F). Not only were echinoderms present in considerable numbers and diversity during the early Paleozoic era, they also are abundant today, often in forms similar to those existing hundreds of millions of years ago. Therefore they have been and continue to be interesting and useful animals in the study of evolutionary mechanisms and trends, not only for their own sake but because they probably were the most direct evolutionary link between invertebrates and vertebrates.

Early in their evolution echinoderms lost their bilateral symmetry and evolved a unique, five-part radial symmetry. Even cylindrical and disklike forms such as sea cucumbers and sea urchins exhibit this symmetry, in that five rows of appendages extend along their bodies. Larval stages, however, have maintained ancestral bilateral symmetry.

A unique characteristic of echinoderms is a **water vascular system,** composed of a circular canal (**ring**

FIGURE 21-21
Representative echinoderms.
A. Sea star (starfish). **B.** Sea
urchin. **C.** Sand dollar. **D.** Sea
cucumber. **E.** Sea lily.

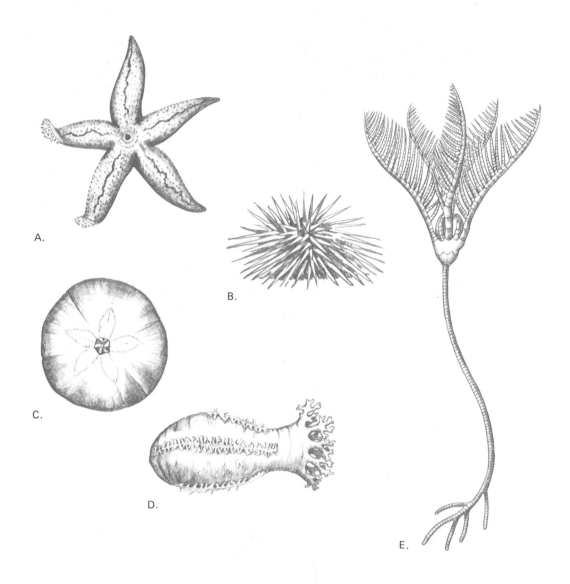

A.

B.

C.

D.

E.

canal) and five radial canals extending into the arms in starfish and brittle stars or along the body in sea urchins, sand dollars, and sea cucumbers. Along the radial canals are lateral canals to which are attached rows of appendages called **tube feet** (Fig. 21-21). Each tube foot is equipped with muscles that squeeze it, much as one would squeeze the bulb of an eyedropper. The squeezing causes the tube foot to be extended, its direction of extension under neuromuscular control.

The functions of the water vascular system are locomotion, food gathering, and gas exchange. In starfish, sea urchins, and sea cucumbers, tube feet are used in locomotion, producing a characteristic gliding movement. In other echinoderms the tube feet may be used in burrowing (brittle stars) or in feeding (crinoids).

At times echinoderms may have a dramatic impact on their environment. In recent years the crown-of-thorns starfish has devastated reef systems in the west-

ern Pacific Ocean, munching steadily on its diet of coral polyps. One possible reason for its abundance was that populations of a major starfish predator, the Pacific triton conch, were at a low ebb, Pacific triton shells being very beautiful and much sought after. Conditions now seem to be improving, however, and the coral seems well on its way to recovery.

Other Invertebrate Groups

In a survey such as this, it is inevitable that some groups of organisms will be omitted from consideration and discussion. A number of lesser phyla, some of only a dozen or so species, have not been included. There appears no remedy for this, but the reader should appreciate the fact that the diversity in the animal kingdom far exceeds the examples introduced here. Information about these excluded phyla may be found in comprehensive zoology textbooks.

The construction of an evolutionary tree of invertebrates relies on a number of key features as well as on lesser characteristics. The presence or absence of a coelom separates primitive and more advanced forms, and embryological development is an important consideration in evolutionary classification, as are symmetry and segmentation. Accordingly, the phyla studied herein can be arranged in an evolutionary sequence characterized by two diverging invertebrate lines (see Fig. 21-22). It is generally accepted that cnidarians occupy a basal position in the mainstream of animal evolution. In one direction have evolved flatworms, roundworms, mollusks, annelids, and arthropods. In the other direction, the echinoderm line appears to have led to **chordates**.

THE CHORDATES

Chordates, as their name implies, are members of the phylum **Chordata** (*chorda*, "cord"), animals with backbones—either rods (notochords of tough connective tissue) or vertebral columns. Other chordate features are: a dorsal hollow nerve cord (spinal cord); pairs of lateral gill openings, called pharyngeal gill slits, in the anterior of the gut (the pharynx); closed circulatory system; body segmentation, manifested as a series of muscle blocks; a body cavity, or coelom; and a postanal tail, which may or may not persist in the adult.

The origin of chordates is open to speculation, but it is likely that echinoderms are their nearest relatives. This may seem a strange relationship, because the starfish form is so unlike the typical chordate body plan; nevertheless, both chordates and echinoderms are deuterostomes and resemble each other in their early embryological development (refer to Chapter 11, Figs. 1 and 2, pp. 244–45 for a comparison of protostomes and deuterostomes).

It has been suggested that the remote ancestors of echinoderms and chordates were sessile filter feeders,

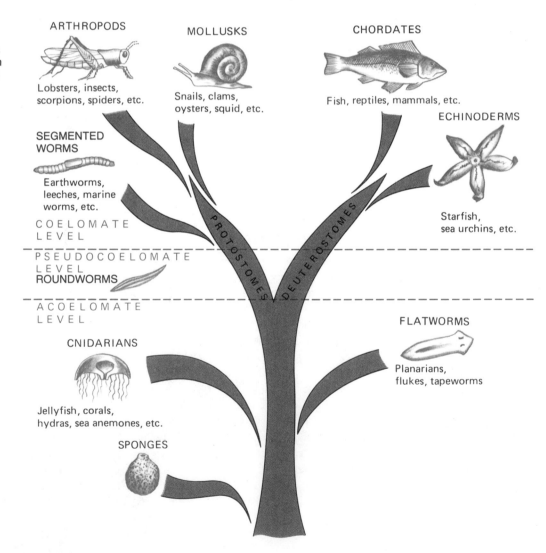

FIGURE 21-22
Relationships of major animal groups. Protostomes are animals in which the embryological mouth remains a mouth; deuterostomes are animals in which the embryological mouth becomes the anus and a new mouth forms.
(see Box p. 244)

ARTHROPODS
Lobsters, insects, scorpions, spiders, etc.

MOLLUSKS
Snails, clams, oysters, squid, etc.

CHORDATES
Fish, reptiles, mammals, etc.

ECHINODERMS
Starfish, sea urchins, etc.

SEGMENTED WORMS
Earthworms, leeches, marine worms, etc.

COELOMATE LEVEL

PSEUDOCOELOMATE LEVEL
ROUNDWORMS

ACOELOMATE LEVEL

PROTOSTOMES

DEUTEROSTOMES

FLATWORMS
Planarians, flukes, tapeworms

CNIDARIANS
Jellyfish, corals, hydras, sea anemones, etc.

SPONGES

superficially similar to some of the more elaborate cnidarians such as sea anemones. From these, starfish and crinoids (sea lilies) evolved, in one direction, by elaboration of tentacles into "arms." In another direction, the tentacles were not converted into arms; instead, the throat region, or **pharynx,** became a feeding and respiratory organ equipped with gill slits. These larval stages became the fishlike ancestors of present-day vertebrates (Fig. 21-23).

Chordates are coelomate animals with a "tube within a tube" body construction (Figs. 8-1, 8-3 to 8-6). The innermost tube is the gut; the outer tube is the body wall. Between the two lie visceral organs supported by folds of the coelomic lining called **mesenteries** (refer to Fig. 9-5).

Although coeloms of invertebrates and chordates are structurally similar, they have different embryological origins. The chordate coelom is formed by budding of the embryonic gut (i.e., it is of endodermal origin), whereas the invertebrate coelom originates by splitting of the mesoderm.

Protochordates

Whether or not the evolution of chordates followed the scheme just outlined, there are basic similarities between echinoderms and chordates that strongly suggest that they are more closely related to each other than, for example, to arthropods or mollusks. Illustrative of early stages in the evolution of chordates are several groups of living genera collectively termed **protochordates** ("first chordates"). One group, the **cephalochordates** ("head chords"), is represented by *Amphioxus*, a tiny fishlike filter feeder (Fig. 21-23). A second group, the **tunicates** (often called "sea squirts"), is composed of sedentary adults which have free-swimming, bilaterally symmetrical larval forms. Adult tunicates look more like sponges than they do vertebrates, but their larvae are more like the early evolutionary stages of all chordates (Color plate 10E).

If vertebrates are in fact descended from tunicate-like ancestors, it is the result of their retention and elaboration of the larval tunicate form and not their descent from the sessile adult. Whenever this kind of evolution of larvae occurs, it requires that the life cycle be telescoped in such a way that the ability to reproduce sexually is attained by the juvenile stage. This kind of evolution is called **neoteny.**

Cyclostomes

Cyclostomes (round-mouthed fish) are true "living fossils," the present-day descendants of ancient jawless fishes (class **Agnatha**—*a-*, "no," *gnathos*, "jaws"). One of

FIGURE 21-23
Hypothetical evolutionary family tree showing the ancestry of echinoderms and chordates. (Redrawn with modifications from A. S. Romer, *The vertebrate story*, Rev. ed., Chicago: The University of Chicago Press. © 1933, 1939, 1941, and 1959 by The University of Chicago Press)

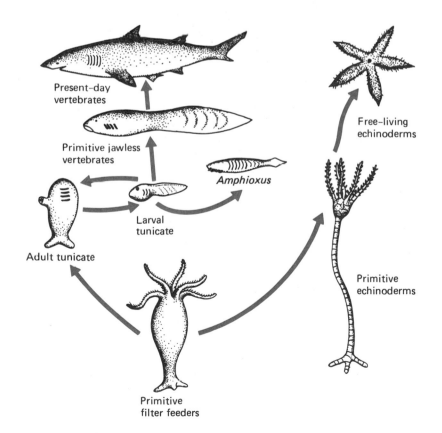

Present-day vertebrates

Primitive jawless vertebrates

Amphioxus

Free-living echinoderms

Larval tunicate

Adult tunicate

Primitive echinoderms

Primitive filter feeders

the cyclostomes, the **lamprey** (Fig. 21-24), has in recent decades increased in numbers in the Great Lakes, where it parasitizes salmon and trout, feeding on their blood. Opinions vary as to the reasons for this increase, but it may have been caused by ecological imbalances resulting from the impact of many human activities. In certain of its structural features the lamprey resembles protochordates, but its notochord, which persists throughout its life, is partially enclosed by cartilaginous vertebrae. Other structural elements are cartilaginous, including the supporting structures of its gills and a shield that partially supports and protects a relatively small brain.

Sharks and Rays

Sharks and rays (also basking sharks—huge plankton-eating creatures up to 15 meters long, skates, and a peculiar shark relative, the *Chimaera*) are living descendants of another group of ancient vertebrates, the placoderms. Placoderms were among the first jawed fishes.

Both ancient and present-day sharks as well as their relatives, **skates** and **rays,** are **elasmobranchs** (class **Elasmobranchii**—*elasmo-,* "plate," *branchia,* "gills") (Fig. 21-25). Unlike other fish, elasmobranchs have no **swim bladder** to aid the regulation of their buoyancy, so sharks often remain continually in motion to maintain their equilibrium and keep from sinking to the bottom of the sea. They are assisted in this by having very large buoyant livers well supplied with a hydrocarbon called squalene. Skates and rays, on the other hand, are bottom dwellers.

Bony Fishes

In terms of vertebrate evolution (Fig. 21-26) **bony fish** (class **Osteichthyes**—*osteon,* "bone," *ichthyo-,* "fish") occupy an important position. All advanced vertebrates, amphibians through mammals, are believed to be descended from ancient bony fishes. Bony fish (Figs. 21-27 and 21-28 and Color plates 2B, 12A, B) appear in the fossil record about the same time as sharks and placoderms. The earliest representatives had lungs, and some primitive **lungfish** still survive in Africa, South America, and Australia. Their ecological relationships probably are quite similar to those of the ancient lungfish; they live in lakes and streams subject to periodic drying, and the ability to breathe air is of obvious survival value.

The majority of ancient bony fish migrated into the sea, where lungs were not required, and their lungs became transformed into the swim bladders found in nearly all modern bony fish. Some primitive lungfish became more and more adapted to life on land by trans-

FIGURE 21-24
Lamprey, full body and mouth views.

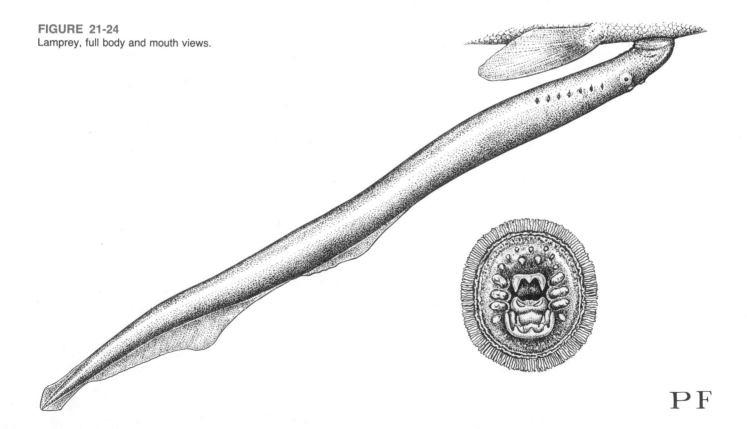

PF

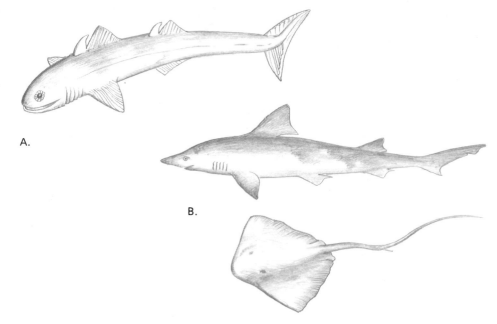

FIGURE 21-25
Primitive and present-day elasmobranchs. **A.** Devonian elasmobranch. **B.** Present-day shark and ray. (Redrawn from *The vertebrate body,* 5th ed., by A. S. Romer and T. S. Parsons. Copyright © 1977 by W. B. Saunders Company. Reprinted by permission of W. B. Saunders Company, CBS College Publishing)

A.

B.

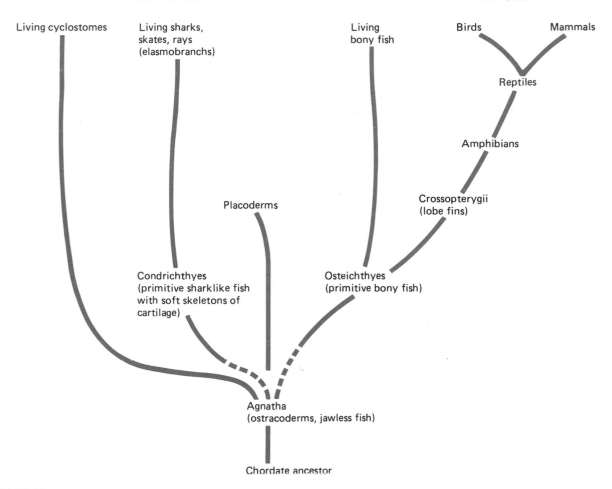

CARTILAGINOUS SKELETONS

BONY SKELETONS

Living cyclostomes

Living sharks, skates, rays (elasmobranchs)

Living bony fish

Birds

Mammals

Reptiles

Amphibians

Placoderms

Crossopterygii (lobe fins)

Condrichthyes (primitive sharklike fish with soft skeletons of cartilage)

Osteichthyes (primitive bony fish)

Agnatha (ostracoderms, jawless fish)

Chordate ancestor

FIGURE 21-26
Evolutionary family of chordates. Dashed lines indicate uncertainty.

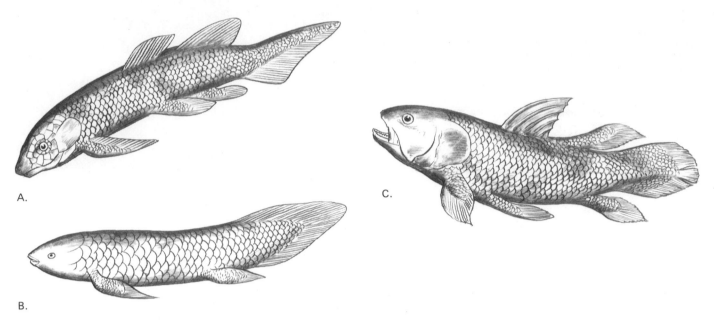

A.

B.

C.

FIGURE 21-27
Bony fish. **A.** Fossil lungfish, *Dipterus.* **B.** Australian lungfish, *Epiceratodus.* **C.** Coelacanth, *Latimeria.* (Redrawn from *The vertebrate body,* 5th ed., by A. S. Romer and T. S. Parsons. Copyright © 1977 by W. B. Saunders Company. Reprinted by permission of W. B. Saunders Company, CBS College Publishing)

formation of their pectoral and pelvic fins into front and back legs and development of devices that prevented excessive water loss.

One group of large and aggressive ancient lungfish named the **Crossopterygii** were common in Devonian times, 400 million years ago. They had stumpy lobed fins and probably were the stock from which amphibians evolved (Fig. 21-27). Crossopterygians disappeared from the fossil record during the late part of the Paleozoic era and were believed long extinct. Then in 1939 some African fishermen caught a large, strange fish off the coast of South Africa, which turned out to be a crossopterygian, specifically of a suborder whose members are called **coelacanths**. This survivor of an ancient group of fish was named *Latimeria;* since the first capture other specimens have been caught and studied in detail. (A magnificent full-scale model of *Latimeria* is on display in the Chicago Field Museum of Natural History.) *Latimeria* is not a functional lungfish but does possess a vestigial lung. Its greatest scientific value is that it confirms the accuracy of earlier studies done exclusively with fossil material. In addition, it extends knowledge of the soft

FIGURE 21-28
The yellow perch *(Perca flavescens),* a common representative bony fish. The lateral line organ consists of a series of pressure sensitive pits, possibly enabling a fish to detect motion in the surrounding water.

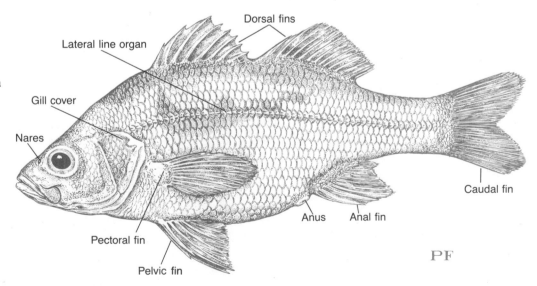

parts of these ancient vertebrates, which were not preserved as fossil specimens.

Today bony fish are found in all waters from the deep sea to tiny rivulets (Fig. 21-28). Although they differ greatly in size and shape, all basically are of similar structure. They, as well as sharks and rays, are important links in aquatic food relationships between different groups of organisms. Small fish feed on plankton and in turn constitute food for larger predatory fish, as well as for other invertebrate and vertebrate predators.

Amphibians

It is thought that a primitive fish with lungs gave rise to the **amphibians** (class **Amphibia**—*amphi-*, "on both sides"), an important evolutionary development that transpired during the Devonian period at about the same time that land invertebrates such as spiders, scorpions, and primitive insects evolved. Modern amphibians have smooth, scaleless skins, and water retention is rather inefficient. Consequently, those that live on land must maintain their water balance by inhabiting humid environments and by secreting mucus to keep the skin moist.

There are two major groups of amphibians, the tailed **urodeles** and the **anurans**, which are tail-less when adult (but tailed when larvae). The urodeles are represented by salamanders, newts, mudpuppies (Fig. 21-29A, Color plate 12C), Congo eels, and others. The largest living urodele, the Japanese river salamander, somewhat resembles a mudpuppy but is considerably longer (up to 1 meter). Frogs and toads are the only living anurans (Fig. 21-29B, Color plate 14A).

The organ systems of amphibians do not differ greatly from the generalized systems of vertebrates described in earlier chapters (see Figs. 9-5–9-9, 14-3–14-5). The nervous systems and sensory organs of fishes and amphibians are similar to each other in many ways. Both have a brain enclosed with a **cranium** (braincase) and a dorsal hollow spinal cord enclosed and protected by **vertebral arches**, general vertebrate characteristics (see Figs. 9-5 and 13-15). Brain and spinal cord bear pairs of

FIGURE 21-29
Amphibians. **A.** The mudpuppy, *Necturus*, a urodele having both gills and lungs. **B.** The common grass frog, *Rana pipiens*, a tail-less anuran.

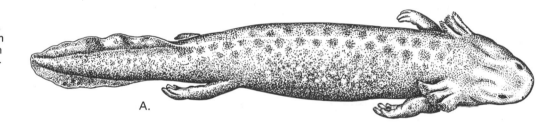

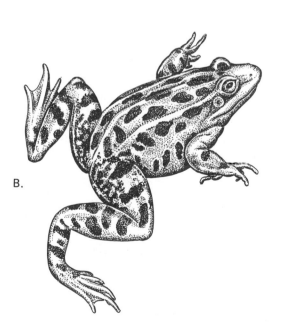

cranial and spinal nerves extending to various parts of the body (Fig. 13-14).

One new feature, an accommodation to life on land, is the ear. This consists of an outer membrane stretched over an auditory canal. The **eustachian tube,** a connection with the pharynx, regulates air pressure in the ear. An ear bone, or **columella,** conducts sound vibrations to the fluid-filled sacs of the **inner ear.** Here the sound waves produce movements in **sensory patches** of cilia and result in nerve impulses to the brain (see Fig. 13-11).

The ear is a remarkable device for conversion of atmospheric sound waves into mechanical energy (movement of **middle-ear** bones) and ultimately into nerve impulses (electrical energy). (The principle is similar to that of a telephone.)

Reptiles

Amphibians were predominant terrestrial animals for about 50 million years. Yet their occupation of the land environment was incomplete because of their requirement for water in which to reproduce. The evolution of **reptiles** from amphibians occurred about 310 million years ago. The ability of reptiles to complete reproduction on land, as well as other adaptations to terrestrial environments, permitted them to become dominant animals for the next 150 million years.

Reptiles (class **Reptilia**—*reptilis,* "creeping") such as **dinosaurs** flourished and became diversified in the Mesozoic era, and present-day reptiles also evolved dur-ing this time. In the Mesozoic nearly every major terrestrial niche was filled by a specialized kind of reptile.

There are four distinct groups of modern reptiles: (1) *Sphenodon,* (2) **crocodiles, gavials, caimans,** and **alligators,** (3) **turtles,** and (4) **lizards** and **snakes** (Color plate 12D and Figs. 21-30 and 21-31). *Sphenodon,* the tuatara of New Zealand, is the only surviving member of a group of extinct primitive reptiles. It is a large, lizardlike reptile with a third, middle eye located beneath the skin of its head. The eye can do little more than detect the presence or absence of light.

Crocodiles are the largest living reptiles. Nile crocodiles may exceed 5 m (16 ft) in length and weigh up to 1000 kg (2200 lb). Gavials are similar to crocodiles and occur primarily in India. Both have narrow jaws, and the fourth pair of lower jaw teeth remain exposed when their mouths are closed. Alligators and caimans have broader heads than crocodiles and gavials, and their teeth are not similarly exposed when their mouths are closed.

Turtles are an ancient race of reptiles, with close counterparts in the Mesozoic era. They have a body armor composed of bony epidermal plates fused with ribs and vertebrae. Their body is further protected by broad, horny scales formed by the outer epidermis. Before the plastic age these horny plates supplied the material called **tortoise shell,** used to make ladies' combs and other useful and decorative articles.

Lizards and snakes are classified in the order Squamata, a name meaning "scaly." Lizards are widespread in the tropics. American lizards include anoles (Fig. 21-30), iguanas, the horned toad, and poisonous Gila

FIGURE 21-30

A pair of Cuban anoles *(Anolis sagrei).* These common American reptiles are useful insectivores around gardens and patios in the southeastern United States and the Caribbean islands. An immature male is at the top of the drawings, a mature male in courtship posture is shown below. The courting male repeatedly flashes his bright red dewlap in the presence of another anole. Note the external ear opening behind the head.

PF

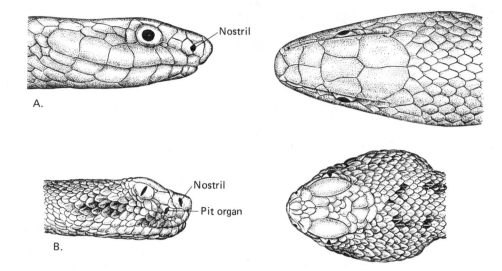

FIGURE 21-31
Venomous and nonvenomous snakes. **A.** Top and side views of the nonvenomous, common black snake, *Coluber constrictor.* **B.** The Mojave rattlesnake, *Crotalus scutulatus,* in top and side views. Rattlesnakes belong to the pit viper family, characterized by slitlike eye pupils, heat-sensitive pits, and a pair of hinged, tubular fangs that inject venom into prey.

monsters. The largest lizard is the Komodo "dragon," a monitor lizard of Indonesia which may attain a length of 4 m (13 ft).

Snakes are limbless reptiles with skeletal features that permit them to swallow whole animals larger in diameter than themselves. In addition to the expansible jaws, they lack breastbones, whose presence would impede the swallowing of such large prey, and their ribs terminate freely at the ventral side of the body. These free rib ends are associated with transverse, bandlike belly scales and are an aid in locomotion. However, snakes do not "walk with their ribs," as is popularly supposed. Primarily, locomotion is by an undulatory gliding in which body curvatures, as well as the belly scales, exert force against the ground and enable the snake to thrust itself forward.

Most poisonous snakes are **pit-vipers,** having a temperature-sensitive depression, the **pit,** located on either side of the head, just forward of the eyes (Fig. 21-31). Rattlesnakes, vipers, copperheads, and water moccasins are examples. The pit functions as a heat-sensing device, enabling the pit viper to detect the approach of a warm-bodied animal at distances of a meter or so.

Birds

In the middle of the Mesozoic era there lived a small dinosaur with well-developed hind legs, a long tail, a lizardlike head with toothed jawbones, and slender front limbs ending in clawed fingers. When the first fossils of this animal were discovered in 1861, they excited great interest in the scientific community, for preserved in rock were impressions of feathers (Figs. 23-11 and 21-32). The fossil was named *Archaeopteryx* ("ancient wing"), and

even then it was thought to represent an early stage in the evolution of birds, possibly from reptiles.

It is somewhat controversial whether or not *Archaeopteryx* could fly. Some researchers believe that this primitive bird (reptile?) probably could not fly but instead hopped and glided as it pursued its prey. However, recently the weight of evidence seems to be shifting to powered flight for *Archaeopteryx.* Several scientists have pointed out that the feather impressions of the fossils clearly indicate that they were fully aerodynamic, and were in fact flight feathers, which in modern flying birds

FIGURE 21-32
Archaeopteryx as imagined in pursuit of prey.

TWILIGHT OF THE GIANT REPTILES

When the subject of dinosaurs is brought up, the popular image is of huge, lumbering, brainless creatures almost preordained for extinction. True, many were indeed huge, brainless, and lumbering, but that obscures the fact that reptiles great and small continue to be an amazingly successful and diversified group of vertebrates. They were predominant for nearly 150 million years, whereas birds and mammals have been dominant lifeforms for about a third that time, and modern humans have existed for fewer than 100,000 years.

Although reptiles first appeared in the late Paleozoic, their period of greatest diversity was the Mesozoic. During the Jurassic and Cretaceous periods, from roughly 180 to 65 million years ago, reptiles occupied every major niche now inhabited by birds and mammals: the sea, the land, and the air.

Whereas many ancient reptiles were of modest proportions, as most are today, others were of awesome size. A nearly complete skeleton of a large herbivorous dinosaur named *Brachiosaurus,* which in the flesh probably weighed 80 tons, stands 12 m (40 ft) tall. An even larger dinosaur recently discovered and nicknamed "supersaurus" may have been the largest land vertebrate ever to live. It is estimated to have weighed 100 tons (more than 15 bull elephants) and to have stood 20 m (65 ft) high. At one time such large hervivorous dinosaurs were thought to have lived in lakes and streams, as do hippopotami today; it did not seem possible that their great weight could have been supported on land. More recent evaluations of their skeletal structure suggests that they were fully capable of land locomotion. However, other questions remain. Most puzzling is that of their feeding behavior. *Brachiosaurus* and similar species had rather small heads and weak dentition not in keeping with giant appetites. How did they manage to consume tons of vegetable matter every day? A bull elephant eats 100–250 kg (300–600 lb) of food each day, and "supersaurus" must have required several times as much. Another question is whether the large dinosaurs were warm-blooded or cold-blooded. If the former, they would have required much more food in order to maintain a uniform body temperature than if the latter. All present-day reptiles are cold-blooded. (Note, however, that their blood temperature is variable and approximately that of the prevailing air temperature—lower on a chilly day, higher when days are hot.) Although the largest dinosaurs probably were cold-blooded, their vast bodies could have stored heat, so in effect they were functionally warm-blooded much of the time. Smaller reptiles, now extinct, but including those groups that gave rise to birds and mammals, may have been truly warm-blooded.

Dominating the exhibition hall of the Field Museum of Natural History in Chicago is a full-scale model of the largest animal ever to fly. Its fossilized bones were discovered in Texas just a few years ago, hence the nickname "Texas pterosaur," and it was one of an order of flying reptiles that ranged from crow size to the size of a small airplane. The "Texas pterosaur," *Quetzalcoatlus northropi,* is estimated to have had a wingspan of at least 20 m (65 ft), which brings up the question of how such large animals were able to fly. They may have soared like hang-gliders, but their relatively weak legs and claws do not seem capable of climbing to the heights required for successful launchings; nor could they run to take off. Moreover, the regions in which they lived may have been relatively flat. They probably were capable of leisurely flight by a combination of wing flapping and soaring. A German biologist once made a flying model of one of the smaller pterosaurs and proved that the design was capable of powered flight. Other questions, about their feeding and reproductive habits, also are being pondered. Were they carrion eaters like vultures and condors, or did they soar over the shallow seas, capturing and feeding on fish? Were they egg layers like modern reptiles? If so, how were the wings of the young folded up within the eggs? The eggs must have been quite large to accommodate the wings, so how did the mother fly with the added weight? These questions and others have stimulated a resurgent interest in reptiles of the past and present.

The fossil record is interrupted by several great gaps, or **discontinuities.** The first of these occurred in Archeozoic times about 1.6 billion years ago; the second separated the Proterozoic era from the Paleozoic 600 million years ago; the third ended the Paleozoic some 230 million years ago; and the most recent occurred at the end of the Mesozoic era about 63 million years ago. Each of these so-called revolutions was more or less cataclysmic, and in each many fossil lines of plants and animals disappeared. By the end of the Cretaceous period (last period of the Mesozoic) all the giant reptiles of land, sea, and air were gone, and birds and mammals had become commonplace. Several theories for this massive extermination have been advanced. Perhaps the climate began to fluctuate widely, destroying the vegetation required by the larger herbivorous dinosaurs and in turn bringing on the extinction of the large carnivorous species. Possibly, smaller and faster mammals began to prey upon the eggs and young of those reptiles incapable of defending their nests against agile raiders.

One interesting theory of dinosaur extinction is that radiation from a supernova occurring in a nearby nebula may have destroyed many forms of life. The larger reptiles, unable to burrow or find other shelter, would have been especially vulnerable. There are other hypotheses. One is based on the discovery of a narrow band of clay containing the rare element iridium in deposits dated to the end of the Cretaceous. This deposit may have resulted from the impact of a meteor 10 km (6 mi) or so in diameter, since iridium is a component of meteors and meteorites but is not commonly present in sediments. Chance hits by meteors of this size are thought to happen at intervals of 100 million years or so, each time producing a crater about 160 km (100 mi) across and scattering into the atmosphere great clouds of dust, perhaps sufficient to block sunlight for two or three years. If this happened, many plants would have died or become dormant and food for the plant eaters would have become scarce. The largest reptiles would have been the most vulnerable, depending as they did on huge daily quotas of plant matter. In two years' time not only would large herbivores have become extinct but also many of the larger carnivores. This scenario, which calls for a sudden die-out of large reptiles, may actually have happened, but it is generally thought that most of the Mesozoic reptile species became extinct gradually over tens of thousands of years.

show asymmetric vanes (see Fig. 21-33), with the narrower vane providing a leading edge to the flow of air.

It also has been proposed, on the basis of circumstantial evidence, that *Archaeopteryx*, and possibly a group of four-footed reptiles called **therapsids**, were warm-blooded animals, more accurately called **endotherms** ("inner heat"). The ability of endotherms to maintain a constant body temperature at levels above or below surrounding air temperature is associated with the ability to live and move rapidly in many different environments, including colder regions. This so-called warm-bloodedness requires high rates of metabolism and, if animals are to survive in cold climates, some kind of insulating coat is necessary. Perhaps feathers may have been more important to *Archaeopteryx* as insulation than for flight; it has been suggested also that therapsids had some kind of insulating coat (fur or feathers). Present evidence supports the theory that birds evolved from small reptiles of the *Archaeopteryx* type, and that mammals evolved from therapsid reptiles.

Flight Birds (class **Aves**—*avis*, "bird") have mastered the aerial habitat to a truly remarkable degree. Although flight probably originated as an escape mechanism, it has other ramifications, including feeding and migration. Other animals, including insects, fish, bats, and the extinct pterosaurs, have evolved flight, but none are or were capable of the long, sustained flight of migratory birds, the speed attained by falcons, or the agility shown by swallows, swifts, and hummingbirds (Color plate 9C).

Forelimbs **(wings)** of birds have three regions of about equal length: the upper arm, or **humerus;** the forearm, composed of two bones, the **radius** and the **ulna;** and a long, highly modified **hand** in which the bones are reduced in number and fused together. The strong flight muscles of the breast are attached to the humerus, and feathers connected to the forearm function analogously to the wings of an airplane. Flight feathers of the hand region propel the bird forward, as do propellers of an airplane, and in addition control direction (see Fig. 21-33).

Muscles of most birds demonstrate special adaptations for flight. In particular, breast muscles are highly developed and massive, extending from the keel of the breastbone to the underside of the humerus (Fig. 21-34). Large **pectoralis major muscles** provide power for the downstroke of the wing. Smaller, **supracoracoideus muscles** extending from the breastbone through a hole in the shoulder bone to the upper side of the humerus move the wings upward in flight. In a strong flyer such as the pigeon, the breast muscles constitute more than one-third of the total body weight.

Feathers, a distinguishing characteristic of birds, are modified reptilian scales. True scales of the reptilian type are present on legs of birds, and the beak has a horny covering composed of modified scales. Feathers develop in skin pits called **follicles.** Follicles are periodically active in some birds, producing new sets of feathers once or twice a year following shedding, or **molting;** many birds, however, replace their feathers gradually.

There are three major types of feathers: large **flight feathers** of the wings and tail; **contour feathers** covering the body; and **down feathers,** which provide insulation against heat and cold (Figs. 21-35 and 21-36). Flight

FIGURE 21-33
Bird wing.

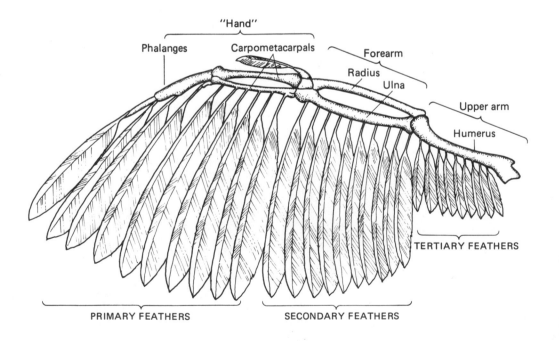

"Hand"

Phalanges   Carpometacarpals   Forearm
Radius
Ulna
Upper arm
Humerus

TERTIARY FEATHERS

PRIMARY FEATHERS   SECONDARY FEATHERS

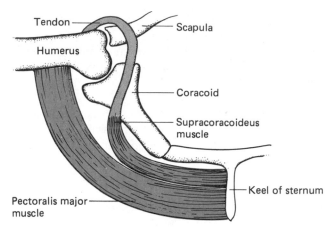

FIGURE 21-34
Arrangement of flight muscles in birds. Two sets of breast muscles act in opposition to each other to provide the powerful downstroke and the recovery stroke.

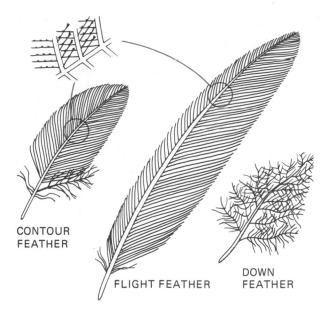

FIGURE 21-35
Major feather types of birds.

feathers and contour feathers are produced in localized areas called **feather tracts**; down feathers are more widely distributed over the body.

Reproduction Whenever behavioral characteristics of birds are mentioned, their construction of nests, laying and incubation of eggs, and care of their young often come first to our minds. Birds are not unique in this, of course; the Nile crocodile is well-known for its defense of its young, and even among fishes and invertebrates, examples of parental care of young are plentiful. Never-

theless, nesting and care of young are particularly well-developed behavioral traits in most birds.

The act of copulation in birds requires considerable cooperation between the male and the female. Because males of most species lack a penis, introduction of sperm requires rather precise appositioning of cloacal openings. Sperm pass from the male into the cloaca of the female, then into her oviduct, where they retain via-

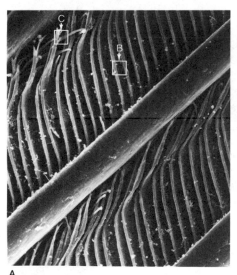

A.

B.

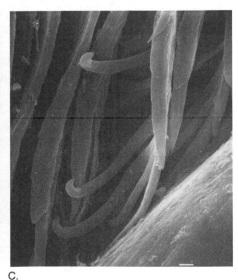

C.

FIGURE 21-36
Scanning electron micrographs of features of a flight feather of the green parrot. **A.** View of the outer surface, secondary and tertiary divisions. **B.** View of under surface of green parrot flight feather in area indicated in A, tertiary and quaternary divisions. **C.** An enlargement of a portion of A, as indicated, showing the barbs that maintain the integrity of the feather. (Photos courtesy of Bart Schutzman)

bility long enough to fertilize a clutch of eggs (as long as 30 or so days because normally one egg is produced per day) (Fig. 21-37). The actual egg of a bird is the yolk part, a single large cell packed with stored food. After fertilization, the white and shell are added by secretory glands of the oviduct. In the absence of copulation or viable sperm, many birds produce infertile eggs, like most hen's eggs bought at the market.

Bird's eggs usually are deposited in a nest, not all at once in reptilian fashion but commonly one per day over a period of several days, until a number characteristic of the species is attained. Then incubation begins. Even though some eggs in a clutch may be a week or more older than the most recently laid egg, hatching usually is synchronous. Embryo development is delayed until the start of incubation.

In most species the clutch of eggs is tended faithfully, and almost all birds incubate their eggs, keeping them near their normal body temperature. The exceptions are a few tropical species, which prepare communal nests in sand or rotting vegetation and depend on ground temperatures to maintain proper conditions, and nest parasites such as cowbirds and cuckoos, which lay their eggs in nests of other birds.

Sensory Organs Birds have comparatively large brains, much larger in proportion to their body weight than those of fish, amphibians, or reptiles. However, the large brain of birds may not indicate a comparatively high order of intelligence, for most of the brain mass consists of the **optic lobes,** which control vision. Birds have extremely keen sight, and high-flying species have telescopic vision capable of spotting very small objects on the ground.

The senses of hearing and balance are well developed in birds, balance being of particular importance in flight. Although some birds, such as owls, have feather tufts suggestive of external ear flaps, the external ear of birds is a simple opening of a passageway leading to the eardrum of the middle ear. A rodlike bone (the **columella**) in the middle ear transmits sound waves to a coiled chamber (the **cochlea**), which contains sensitive ciliated patches of cells (**hair cells**). Semicircular canals function as organs of balance as they do in fish, amphibians, and reptiles.

The many and varied songs of birds are produced by vocal cords as in mammals, but these are not located in a voice box (**larynx**) at the upper end of the trachea as in mammals, but are found instead in a chamber, the **syrinx,** at the lower end of the trachea (Fig. 9-9). Sound waves thus come from deep within the respiratory passages, which accounts for the resonant and penetrating qualities of bird song.

Navigation Many birds are capable of long, sustained flights. The tiny ruby-throated hummingbird migrates nonstop across the Gulf of Mexico, a distance of about 1000 km (600 mi). It should not be able to do so, given its high metabolism and almost continual need for food. The fact that hummingbirds do make these flights, apparently without mishap, is something of a biological mystery. Also astonishing is their ability to navigate over such distances. Some other birds, however, display even more amazing navigational skills. For instance, golden plovers regularly make intercontinental flights of more than 12,000 km (7500 mi) over land and sea, yet return to their place of origin, sometimes to the same nest site.

How birds navigate continues to be a subject of considerable speculation and research. It appears that during migration, birds rely on local landmarks and regional geographic features such as rivers, streams, and mountain ranges for precise orientation. They also can use celestial navigation by sun and stars in their long-distance flights, as shown by some intriguing experiments in which migrating birds have been trapped and placed in a planetarium under an artificial night sky. In such circumstances, birds face in the direction of their regular migration route as simulated by the artificial

FIGURE 21-37
Reproductive and excretory systems of the chicken.

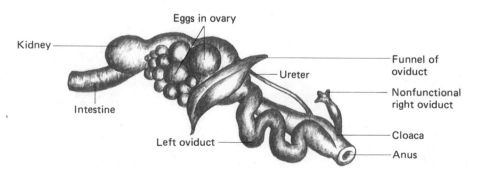

Kidney — Eggs in ovary — Ureter — Funnel of oviduct — Nonfunctional right oviduct — Cloaca — Anus — Intestine — Left oviduct

FEMALE REPRODUCTIVE STRUCTURES

stars in the planetarium ceiling. For more finely tuned localization of nesting and breeding sites, it has been postulated that birds use the earth's magnetism.

Diversity The class Aves is represented by 27 orders of living birds. The largest in number of species is **Passeriformes** (perching birds), which includes such familiar examples as crows, larks, jays, robins, starlings, finches, and sparrows (Fig. 21-38). Perching birds are fine flyers, but as their name indicates, spend much of their non-flying time perching in trees and shrubs. Most birds have a mechanism that allows them to sleep while perching. The Achilles tendon in the heel is connected with ligaments that curl around the inside of the toes (Fig. 21-38). As birds relax on the perch, legs and toes bend and the Achilles tendon tightens and pulls the toe ligaments. The more the bird squats, the tighter its grasp.

In the human economy, the most important group of birds is the predominantly ground-dwelling order **Galliformes,** which includes chickens, pheasants, peacocks, grouse, and quail. In these days of protein deprivation in many lands, the assembly line production of hens and eggs is a major protein source for peoples both in developed and undeveloped countries.

Some birds spend almost all their waking hours in endless flight; others do not fly at all. Hummingbirds, of the order **Apodiformes** ("without feet"), are not only remarkable for their endurance and high rates of metabolism but also because they are the smallest birds known (Color plate 9C). At the other extreme are the **ratites,** a group of large flightless birds with a keel-less sternum. Ratites include emus, ostriches, rheas, kiwis, and the largest of all birds, the extinct **elephant bird** (*Aepyornis*) of Madagascar and the **giant moa** (*Dinornis*) of New Zealand. The giant moa stood 3 m (9 ft) or more tall and became extinct only a thousand or so years ago, perhaps being hunted to death when Polynesians invaded its domain.

Many examples of niche specialization, evolutionary parallelism, adaptive radiation, and interesting behavior patterns are evident among the orders of birds, and their study is one of the most enjoyable hobbies the world over. Bird watching owes much of its appeal to the diversity of bird life observable in almost any habitat, from crowded cities to open countryside and wilderness.

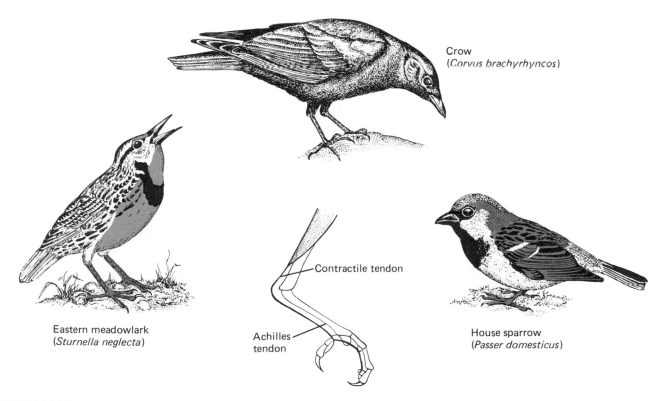

Crow
(*Corvus brachyrhyncos*)

Eastern meadowlark
(*Sturnella neglecta*)

Contractile tendon

Achilles tendon

House sparrow
(*Passer domesticus*)

FIGURE 21-38
Some members of Passeriformes, the largest order of birds both in total numbers and number of species. (Adapted from C. S. Robbins, B. Bruun, and H. S. Zim, 1966, *Birds of North America*, illustrated by Arthur Singer, © 1966 by Western Publishing Co., Inc., Racine, Wis. Reprinted by permission)

Mammals

The ascendancy of **mammals** to a dominant position in the biosphere is a comparatively recent event in earth history. Even though mammals originated in the age of reptiles, their evolution coincided with and probably resulted from the origin and spread of great grasslands that first appeared in the early Cenozoic era 60 million years ago after the extinction of dinosaurs.

Reproduction   Mammals are named for their milk-secretory **mammary glands.** All mammals nourish their young with **milk,** a nutriment composed of water (87 percent), sugar (5 percent), fat (4 percent), proteins (3 percent), minerals (1 percent), and vitamins. These approximate percentages are for cow's milk; values differ for different mammals.

All mammals except one small group, the **monotremes,** give live birth (**vivipary**) to their young. Monotremes, a group represented only by the Australian **duck-**billed platypus and the **spiny anteater (echidna)**, are egg layers, resembling in this respect birds and reptiles. Most other mammals are **placental animals,** because the embryo during its development is nourished by food exchanged between the blood of the mother and that of the embryo through an organ called the **placenta.** Marsupials (kangaroos, opossums, etc.) bear their young alive, but in a very embryonic state. Throughout their long development, the young are nourished by mammary glands in the mother's pouch rather than through a placenta.

The presence of hair is another mammalian characteristic, although, as noted earlier, at least one ancient group of reptiles, the therapsids, may also have had hair. Because both hair and feathers are thought to be modified scales of the reptilian type (Fig. 21-39), origin of both birds and mammals from warm-blooded, possibly furry or feathered reptilians is not improbable.

It has been suggested that ancestral mammals, like the platypus and echidna of today, laid eggs and cared

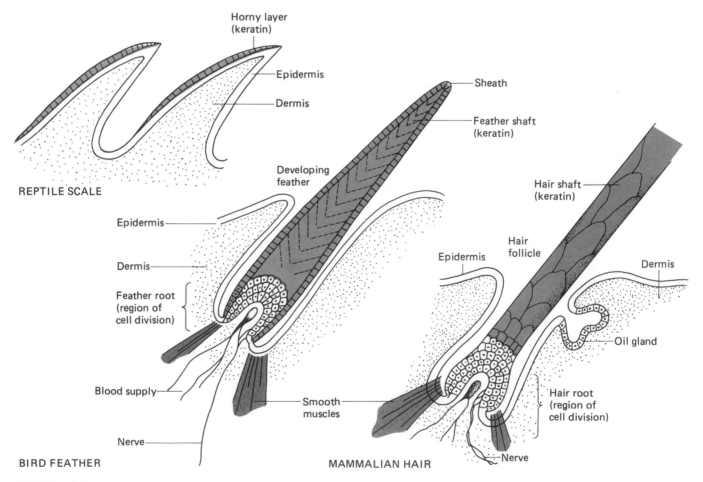

FIGURE 21-39
Structural similarities of reptilian scales, bird feathers, and mammalian hair.

for their young after hatching. Perhaps the mammary glands originated from sweat glands in the skin that became localized and specialized. The mammaries of monotremes are simply patches of secretory glands located on the abdomen, and the young receive nourishment by licking these patches; milk is produced by both males and females. Mammaries with nipples came later in more advanced mammals through further evolutionary steps.

A first step in evolution of mammalian vivipary was retention of eggs in the uterus. Subsequently, the placenta evolved as a nutritive link between the uterine wall and the embryo.

Viviparousness frees parents from nest tending and reduces the prenatal energy budget of the female parent because production of numerous eggs with large quan-

tities of yolk and shell is not required. In comparison with other vertebrate groups, fewer offspring usually are produced per breeding pair of mammals during their lifetimes. There is considerably greater parental time and energy invested in postnatal care of offspring, and the relative number of surviving young tends to be high.

Diversity The class **Mammalia** (*mammalis*, "breast") is subdivided into three subclasses: the primitive **Prototheria** ("first beast"), represented by two Australian egg-laying genera and species, the duckbilled platypus (see Fig. 21-40A) and the echidna; the **marsupials**, of the subclass **Metatheria** ("intermediate beast," Fig. 21-40B); and all other mammals, of the subclass **Eutheria** ("true beast"). The members of these subclasses include some 18 orders, a few of which will be singled out to show the

A.

B.

C.

D.

FIGURE 21-40
Representative mammals. **A.** Platypus. **B.** Koalas. **C.** Shrew. **D.** Beluga whale. (A, photo © MCMXXC Steve Martin/Tom Stack & Associates; B, photo by Warren Garst/Tom Stack & Associates; C, photo by Rod Planck/Tom Stack & Associates; D, photo by John R. Lewis/Tom Stack & Associates)

FIGURE 21-41
Adaptive radiation exemplified in the evolution of the major orders of mammals.

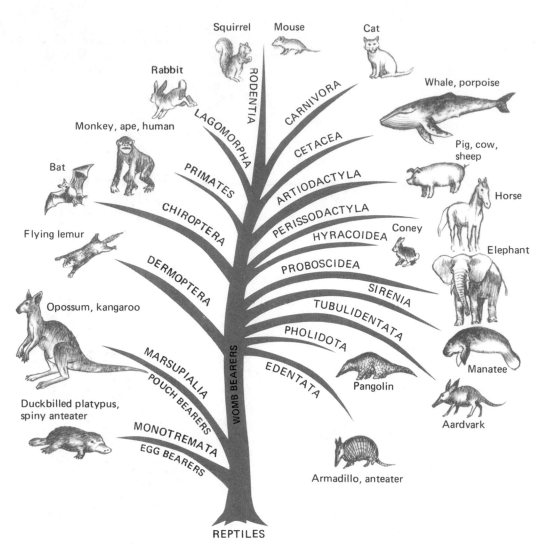

diversity of form and habit found among mammals. In addition to the monotremes and marsupials, mentioned earlier in this chapter, are mammals specialized for life in seawater, for flight, for feeding on vegetation, for predation, and for intellectual creativity.

Eutherian mammals may be arranged in four evolutionary series:

1 An **insectivorous** ("insect-eating") line including shrews, moles, bats, anteaters, and primates

2 Whales, dolphins, and **porpoises**

3 Rodents and **rabbits**

4 **Carnivores** and **ungulates** (hooved animals)

Each of these groups is presumed to have evolved by adaptive radiation from common ancestors (Fig. 21-41).

The most primitive group of eutherians are the shrews and moles. Shrews are amazingly active little mammals with rates of metabolism equaling those of hummingbirds (Fig. 21-40C). They will die if deprived of food for only a few hours, and although they feed primarily on insects, they are capable of attacking and killing other small animals. Water shrews, which are only about the size of a human thumb, regularly attack, kill, and eat frogs many times their size.* Moles, which are larger members of the order, can consume an incredible number of earthworms; they spend nearly all their time burrowing in pursuit of food. Hedgehogs, the largest members of the order, kill and eat snakes, from whose bites they are protected by an armor of spines. Bats are closely related to shrews but have evolved remarkable powers of flight through the development of leathery wings formed by webbing between their very long fin-

*For a fascinating account of the life of water shrews, see the chapter on them in *King Solomon's Ring* by Konrad Lorenz, listed in the reading section of Chapter 22.

gers. Primates, which include monkeys, apes, and humans, are thought to be distantly related to shrews.

Members of the whale order (**Cetacea**) have become objects of fascination the world over (Fig. 21-40D). They usually are considered highly intelligent mammals, some think second only to humans. All are aquatic mammals that bear flipperlike forelimbs and lack hind limbs. They are for the most part very social and exhibit a keen sense of responsibility for their fellows, protecting them when attacked and tending them when ill. They are noted for their agility and imitative talents. As an example of the latter, one dolphin was observed to imitate a scuba diver cleaning the viewing ports of the aquarium, even making sounds of air valves and mimicking the bubbles coming from the face mask. Despite their attractiveness, cetaceans have been mercilessly hunted through the centuries for their oil and sometimes for their meat (Color plate 6D). Now all but two major nations, the Soviet Union and Japan, have banned the hunting of whales and their relatives.

Rats, mice, squirrels, muskrats, beavers, and other rodents (**Rodentia**) are characterized by two pairs of long, chisel-like incisors (foreteeth). Rabbits and hares (**Lagomorpha**) also have two similar pairs of incisors but differ from rodents by having another small pair lying behind the larger front pair.

Dogs, cats, lions, bears, and many other mammals are members of **Carnivora,** an order of flesh-eating mammals. Although sharply differing in behavior and nutrition, carnivores are thought to share a common evolutionary ancestry with hooved mammals (**Ungulata).** Ungulates include even-toed pigs, cows, deer, giraffes, hippopotami, and so on, and odd-toed horses, tapirs, and rhinoceroses. Relationships of these and other mammals are illustrated in Fig. 21-41.

Primates

The term **primate** ("foremost in rank") is most often thought of in reference to the great apes: gorillas, chimpanzees, and orangutans. Primates, however, include mammals that many persons would scarcely conceive as related to apes and monkeys, not to mention humans.

Prosimians The most primitive living primates are small- to medium-sized unmonkeylike species such as lemurs, bush babies (galagos), and lorises, a group called **prosimians** (Fig. 21-42). The prosimian body shows some of the same adaptations to life in trees that are intensified in higher primates such as monkeys. These adaptations include large eyes and keen color vision, as well as hands and feet capable of grasping tree limbs, rather than simply clawing them as does a squir-

FIGURE 21-42
Bush baby, *Galago senegalensis.* Galagos are nocturnal prosimians about the size of a large squirrel, with large eyes and hands and feet well adapted for grasping. (Redrawn from D. Freeman, 1977, *The love of monkeys and apes,* London: Ardea London Ltd.)

rel. This is made possible by a truly fascinating modification of the mammalian limb: the **opposable first digit** (thumb and great toe). It is this feature that largely distinguishes primates from other mammals. Prosimians have flat nails rather than claws. Nails evolved in primates as a tool in connection with the adaptation of hands and feet for grasping and feeding. Fingernails are used as tools in many ways, particularly in picking up small particles.

The earliest known fossils of prosimian primates are those of small lemurlike mammals of the Paleocene epoch, roughly from 65 to 54 million years ago. During

FIGURE 21-43
Lemurs, modern prosimians. (Photo courtesy Richard Mueller)

that time the continents of Eurasia and North America were joined together in the supercontinent of **Laurasia.** The climate of Laurasia during the Paleocene was warmer than today; the vegetation was tropical, and the great regions of temperate grasslands and dry forests had not yet appeared.

Primitive prosimian primates resembling the lemurs of today (see Fig. 21-43) were present in Laurasia. Subsequently, the continents separated, and the equator shifted southward, as did the focus of primate evolution. The primitive primates of Europe moved into Asia and

Africa, and those of North America moved into Central and South America. From these two lines of primate evolution arose two distinct groups of advanced primates: **New World monkeys** and **Old World primates.** Meanwhile, prosimians survived and are represented in the present by the lorises, galagos, and lemurs of tropical Asia, Africa, and the Malagasy Republic, respectively.

New World Monkeys New World monkeys include such less familiar forms as the **marmosets** and **tamarins** as well as the more familiar **squirrel monkeys, howlers, spider monkeys,** and the **capuchin,** or "organ-grinder's" monkey, trained to do tricks and collect coins (see Fig. 21-44). Many New World monkeys have flat faces with broad noses and divergent nostrils. Some species have long, **prehensile** ("grasping") tails. The prehensile tail of the spider monkey is a marvelous organ, in reality a third hand. The underside of the tail tip is hairless and is richly supplied with sensitive nerve endings; in addition, the skin in this region is raised in fingerprintlike **dermatoglyphs,** which provide additional friction for grasping slippery objects.

FIGURE 21-45
Gibbons. Gibbons move swiftly through the trees by brachiation, swinging from limb to limb as shown in the upper photo. Gibbons also exhibit **bipedal** ("two-footed") **locomotion,** as shown in the lower photo. Compare this bipedalism with the **quadrapedal** ("four-footed") **locomotion** of the orangutan in Fig. 21-49.

FIGURE 21-44
Capuchin, or "organ-grinder's" monkey, a New World monkey. (Photo courtesy Tom Hutchinson)

Old World Primates   Old World primates include **gib-bons, baboons,** and **macaques.** The rhesus macaque of southeast Asia is used as an experimental primate in behavioral and medical laboratories around the world (the human Rh blood group system gets its name from the first two letters of this macaque's name). Less famil-iar Old World primates include **mangabeys, geladas, drills,** and **mandrills** (the last is one of the most bril-liantly colored of all mammals), and the widely distrib-uted African **vervet.** Old World primates do not have a prehensile tail and some species, such as the gibbons (Fig. 21-45) and the Barbary macaque, are tail-less.

Primate Evolution

Old World primates diversified about the same time as the New World monkeys and are represented by three descendant groups: **monkeys, apes,** and **hominids** (hu-mans and near-humans) (Fig. 21-46).

Some recent interpretations of the interrelation-ships among these three divergent families have led to the conclusion that the groups may have evolved from an Oligocene primate, *Parapithecus*, or a near relative, about 38 million years ago. *Parapithecus* was smaller than present-day monkeys and apes and was at home in the treetops, although perhaps not specialized for tree-swinging locomotion (**brachiation;** see Fig. 21-45) as are its modern descendants. Brachiation requires rotatable arms so that the body can be twisted as the individual

swings between branches. This rotatability of the arm is as important as the flexibility of the hand in primate evolution. A very important adaptation of the hand is the ability of most primates to move their fingers separately. This flexibility made possible the development and use of tools and weapons by early humans.

Apes   During a period of about 10 million years follow-ing the appearance of *Parapithecus*, one line of evolution diverged in the direction of the monkey family, a second line in the direction of the great apes and hominids.

There is some uncertainty about the ancestral pri-mate of present-day apes and humans, but a small, semi-erect ape named *Dryopithecus* ("oak ape") seems a likely candidate. *Dryopithecus* was discovered in the French Pyr-enees in 1856. Subsequently, fossils identified as *Dryopith-ecus* have been found in Pakistan, India, Asia Minor, Af-rica, and China. It evidently was a wide-ranging genus and inhabited tropical and semitropical forests, where it fed on leaves and fruit.

Several species of *Dryopithecus* are known, and these variously tend to resemble modern chimpanzees, orang-utans, and gorillas (see Fig. 21-47). A fossil ape, discov-ered in Tanganyika in 1948 by the well-known anthropol-ogists Louis and Mary Leakey, and named *Proconsul*, also appears to be a species of *Dryopithecus* (D. *proconsul*).

Present evidence suggests that between 10 and 15 million years ago *Dryopithecus* gave rise to three now-extinct genera of apes: *Sivapithecus*, *Gigantopithecus*, and

FIGURE 21-46
Phylogenetic tree depicting evo-lution of Old World primates and possible relationships of austral-opithecines and humans.

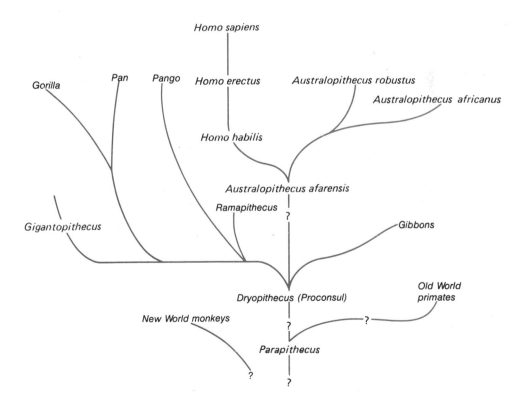

FIGURE 21-47
Skulls of apes. **A.** *Gorilla.*
B. *Dryopithecus.*
C. *Ramapithecus.* (After S. I.
Rosen, 1974, and E. L. Simons,
1977)

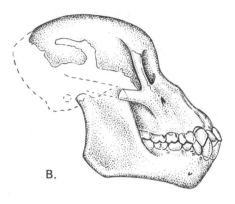

B.

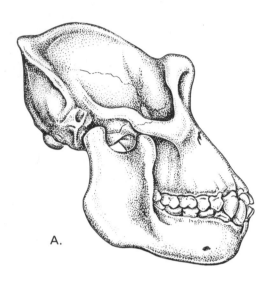

A.

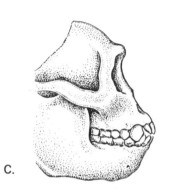

C.

Ramapithecus, from the last of which modern apes now are thought to have evolved. In addition, *Dryopithecus,* or one of its close relatives, may have given rise to the hominid line of primate evolution.

Hominids In 1925 a South African paleontologist named Raymond Dart discovered the skull of a fossil hominid which he named **Australopithecus africanus**. Since then, more complete skeletons of A. *africanus* have been found, and a good bit has been learned about its body structure and even about some of its habits.

A. *africanus* was a small hominid about 1.25 m (4 ft) tall. It had a pelvis much more like that of humans than that of apes and walked with its head erect. This is deduced from the forward position of the **foramen magnum,** the opening at the base of the skull through which the spinal cord passes. The back of the skull also lacked strong ridges. In apes, neck muscles are attached to ridges, supporting their heads in a forward-thrusting position characteristic of animals accustomed to walking on four limbs. Even so, A. *africanus* was apelike in certain features. Its brain was scarcely larger than a gorilla's. The lower jaw was massive, rounded, and lacked a chin point (Fig. 21-48).

Recent studies suggest that A. *africanus* may not have been directly in the main line of human evolution, but rather was an offshoot of an earlier species discovered in Africa's Rift Valley by Donald Johanson in 1974. This species was named A. *afarensis* after the Afar Triangle area of Ethiopia where the fossils were discovered. A partial skeleton, the most complete of any australopithecine yet found, was determined to be at least 3 million years old and is referred to by its discoverers as "Lucy," after the Beatles' song "Lucy in the Sky with Diamonds." Although their heads were apelike, Lucy and other australopithecines walked erect, and their hands were like those of humans. There is no evidence, however, that they made or used stone tools. Nevertheless, Lucy and her relatives may have been ancestral to true humans.

The manner in which hands evolved is uncertain. It is difficult to say whether they evolved in their present form because primates tended to have weak jaws and short muzzles, therefore benefitting by the extra help of hands, or whether hands evolved first and resulted in reductional evolution of facial features. Whatever the evolutionary order, hands—particularly in hominids— grew in dexterity as jaws and teeth lost their usefulness

FIGURE 21-48
Australopithecus africanus, a reconstruction. (Redrawn from J. E. Pfeiffer, 1969, *The emergence of man,* Philadelphia: Harper and Row, with permission of Phillip V. Tobias)

umes ranging from 800 to 1000 cc. The brain volume of modern humans averages around 1350 cc.

The Great Apes

Four species of great apes represent surviving lines of possible evolutionary descent from the Miocene ape *Dryopithecus:* the orangutan (*Pongo pygmaeus*), the gorilla (*Gorilla gorilla*), the chimpanzee (*Pan troglodytes*), and the pygmy chimpanzee (*P. paniscus*). All are found in Africa, except the orangutan, an endangered species restricted to the rainforests of Borneo and Sumatra.

Orangutans Orangutans are large-bodied apes with long arms and relatively short legs (Fig. 21-49). This species shows gross sexual **dimorphism,** with adult males being much larger than females; furthermore, fully adult males have large cheek flanges somewhat like facial blinders, as well as a huge throat pouch.

Orangs are adapted for life in the trees, but adult males regularly come to the ground either while foraging or when traveling long distances from one fruiting tree to the next. Ground travel is far less common among females. Fruit is the preferred food of these large apes, but they also eat leaves, bark, flowers, buds, and insects (especially ants and termites).

Gorillas Like orangutans, gorillas are primarily herbivorous. The mountain gorilla, main focus of observers' attention, lives in the mountains of eastern Africa. Its diet

as offensive weapons and devices for the killing and tearing of prey. Moreover, hominid evolution demonstrates an increased correlation between the growing inventiveness of the human mind and manual skills. For example, tools, even the most primitive of knives and spears, represent substitutes for powerful teeth and jaws. The discovery of the usefulness of fire both for cooking and for warmth also had an impact on human evolution. Cooking food makes it soft enough to be chewed and eaten by weak-jawed creatures.

Important as tools and fire were in human evolution, their use required intelligence. Manual dexterity and the use of tools and fire evolved together with the creative mind. Evidence of the dawning of the human intellect can be seen in the increasing brain capacity of prehuman apes and ape-men. *Dryopithecus,* the gibbonlike ancestor of apes and humans, probably had a brain capacity of about 300 cc. *Ramapithecus* had a brain of about the same volume as the gorilla (around 500 cc). The brain of australopithecines was a bit larger, on the order of 600 cc, and the earliest true human brains had vol-

FIGURE 21-49
Orangutan. (Photo courtesy Richard Mueller)

consists mainly of leaves (it would be classified as a **folivore**), with little emphasis on fruits. By contrast, the lowland gorilla of western Africa eats a greater proportion of fruit. Unlike orangs, mountain gorillas obtain most of their food on the ground, using a locomotory mode called **knuckle walking.**

In recent years, studies of gorilla ecology and social behavior have been conducted by a number of researchers, primarily by George Schaller and Dian Fossey and her associates. Schaller's study was the first in-depth look at gorilla biology in the field and was based on careful observations of apes that were habituated to the presence of the observer. Fossey continued use of this technique, and her long-term studies—over 10 years now—are still providing greater insight into the behavioral biology of this largest of primates.

Chimpanzees There are two species of great apes classified in the genus *Pan*: the common chimpanzee, P. *troglodytes,* and the pygmy chimpanzee, P. *paniscus.* The common chimpanzee has the largest distributional range of any great ape; it occurs in rainforests and wooded savannas from the west coast of Africa across the center of the continent to a point just west of the Indian Ocean in Tanzania. The pygmy chimpanzee has a very restricted range south of the Congo River.

The common chimpanzee is the best known of all great apes, due largely to the continuing research efforts of Jane Goodall and her associates at the Gombe Stream Research Centre in Gombe National Park, Tanzania, as well as to a Japanese research team working in the Mahali Mountains of Tanzania.

With the exception of humans, chimpanzees are unrivaled in their use of tools, which may be used in threat displays, in obtaining food, or in warding off insect swarms. The use of carefully prepared sticks and grass blades to capture termites and ants has been documented on film.

Chimpanzees learn easily, and can be skilled entertainers. They have been used also in scientific research, being sent into space in early U.S. efforts (Fig. 21-50). Recent studies in ape–human communication have shown them capable of learning the sign language of the deaf and of communicating their desires by recombining words into sentences.

The First Humans

Java Man and Peking Man In 1892, in a river bank in Java, a Dutch paleontologist named Eugene Dubois found the top of a skull and a thigh bone remarkably similar to the human femur. The remains were about 300,000 years old. He thought the bones represented a

FIGURE 21-50
"Astrochimp," an early U.S. astronaut, receives a reward after a successful space flight. (Photo courtesy NASA)

"missing link" between apes and humans, postulated in 1868 in the wake of the Darwinian revolution by the German anatomist and paleontologist Ernst Haeckel (1834–1919).

Dubois named his fossil **Pithecanthropus erectus** ("the ape-man that walked erect"); more often it is referred to as **Java man** (Fig. 21-51). Java man was immediately controversial; too little remained of the skull to enable scientists to say without qualification that it was human. In fact, Dubois later came to believe that his discovery was the remains of a giant gibbon. Then in 1929, in China, the anthropologist Davidson Black found portions of about 40 skeletons of an ape-man he called **Sinanthropus,** often referred to as **Peking man.** From Java man and Peking man come the most complete descriptions of primitive human beings. (One of the unfortunate consequences of World War II was the loss of the complete collection of skeletons of Peking man. They had been crated prior to the Japanese invasion of Peking, but had not been stored in a safe place; sometime during the Japanese occupation they disappeared. Fortunately, they had been completely measured, photographed, and cataloged prior to their disappearance.)

Finally, in 1964, in the Olduvai Gorge in Africa, Louis Leakey found a skull of a much more ancient hominid than either Java man or Peking man. This skull, which was given the name **Homo habilis,** and another H. *habilis* skull uncovered by Richard Leakey are the earliest human fossils known (Fig. 21-53). At first they were thought to be more than 2 million years old, possibly coexisting with all known species of *Australopithecus.* Now it is generally recognized that A. *afarensis* preceded H. *ha-*

FIGURE 21-51
Java man *(Homo erectus).*

bilis by up to a million years. However, A. *africanus* and the larger A. *robustus* possibly lived and persisted well into the era of early humans. Some researchers think that the australopithecines fit into the human family tree as shown in Fig. 21-48.

Today Java man and Peking man are considered members of a cosmopolitan race of early humans, sufficiently advanced to be classified as **Homo erectus**. In addition, recent datings of a newly discovered skull of H. *erectus* indicate that these direct ancestors of modern man, **H. sapiens**, roamed Africa at the same time as the later australopithecines, further proof that australopithecines are not the direct ancestors of modern humans.

The early ancestry of humans is an area of research that can be aptly described as fluid; controversies, newly discovered fossils, and reinterpretation of data all help to change ideas on human evolution almost overnight.

However, this is what science is all about. As data accumulate, old ideas give way to new, which in turn evolve into even newer discoveries and concepts.

Homo sapiens There is a quiet valley in Germany named after the great Reformation hymnist Joachim Neander. Neanderthal ("Neander's valley") is the site of a number of limestone caves in which human bones had been discovered from time to time, along with remains of extinct cave bears and other ancient mammals. One of the oddities of history is that many of these human skeletons were thought to have been recent victims of natural catastrophes and were therefore given Christian burial in local churchyards. In 1856, however, a skull from one of the caves was brought to the attention of scientists and excited a great deal of interest. At first, this skull was controversial. Because of its prominent brow ridges and low forehead, it was suggested by some to be only that of a village idiot. Nevertheless, it was soon accepted as representing an ancient and extinct race of humans and was named **Neanderthal man.**

Neanderthal man lived about 35,000 to 75,000 years ago and then rather quickly disappeared from the fossil record, replaced by humans with more advanced cultures. However, some anthropologists believe that Neanderthal man lives on in modern human beings as a result of early interbreeding with our most direct ancestors, H. *sapiens sapiens*. Since the first discovery of Neanderthal man, many other similar remains have been found in caves in France, Asia Minor, and elsewhere. Neanderthal humans were robust people with strong bones, heavy jaws, and deep brow ridges. They stood a bit shorter than modern humans—1.5 m (5 ft) or so—but had erect posture, not the slouch so often pictured. Their squat stature was mostly because of short shin bones and heavy musculature. Neanderthal men and women were both powerful creatures (Fig. 21-52).

Neanderthal humans usually are classified as a subspecies of H*omo sapiens* (H. *sapiens neanderthalensis*), but occasionally separated as H. *neanderthalensis*. There appears little justification for the latter. Their brain capacities are within the human range—1200 cc–1800 cc (see Fig. 21-53). They made chipped flint tools and crude bone carvings and demonstrated reverence for the dead, as shown by excavated burials in which the dead were accompanied by tools and ornaments and even flowers. Moreover, in spite of the conventional picture of Neanderthal man as a kind of ancient *yeti*, or "big foot," considerable individual diversity of form and stature is found in Neanderthal remains. Some were tall, some had chin points, some had low foreheads and some high, some had heavy brow ridges and some not. In short, there probably were as many distinct types among them

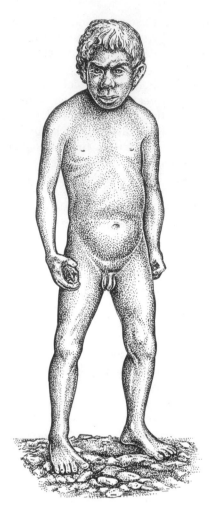

FIGURE 21-52
Neanderthal man, a reconstruction of a young boy. (Based on a reconstruction sculpture by M. M. Gerasimov)

as would be found in a crowd of humans in any society of today.

Modern Humans

Scientists do not know how long modern humans, *Homo sapiens sapiens*, have existed. Fossils as old as 30,000 years are known, placing modern humans as contemporaries of Neanderthal humans. Possibly these species were not so distinct after all but represented different cultures and variations at the extremes of a large and diverse group. The prototype of modern humans was discovered in a cave at Cro-Magnon in south central France in 1868. **Cro-Magnon man** was tall, erect, and in many aspects indistinguishable from *Homo sapiens* of today. Cro-Magnon humans had large, strong, and narrow skulls with strong jaws. The brain capacity of some Cro-Magnon skulls is 1800 cc, larger than skulls of contemporary humans (Fig. 21-53). Many other specimens of the Cro-Magnon type have since been found, and they are now known to be members of a culture that used sophisticated tools of stone and bone, including knives, spears, hammers, axes, awls, barbed fishhooks, and eyed needles. They had perfected the art of painting (see Fig. 21-54), and they domesticated the pig, sheep, goat, and cow (dogs had been domesticated earlier). It is believed that Cro-Magnon culture originated in the Middle East or perhaps in India.

In mystery dramas and fiction, the medical examiner deduces from a skull and a few bones that the victim was a middle-aged Chinese-American bricklayer and so establishes the scientific basis for the solution of the crime. Actually, as the British paleontologist W. E. LeGros-Clark points out, unless one examines skeletal remains exhibiting the extremes of racial variation, it is impossible to ascribe a racial origin to the skulls and bones of contemporary humans. Thus the features commonly associated with race are principally superficial

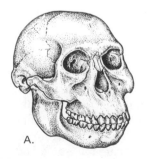
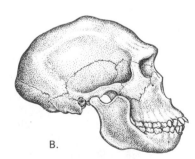
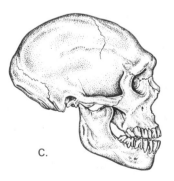
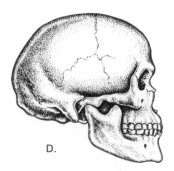

A.　　　　B.　　　　C.　　　　D.

FIGURE 21-53
Comparative aspects of hominid skulls. **A.** Skull of *Homo habilis.* **B.** Skull of *H. erectus.* **C.** Skull of Neanderthal human. **D.** Skull of modern human.

FIGURE 21-54
Charging bison, painted by a Cro-Magnon artist on the wall of a cave at Altamira, Spain. (Courtesy of the American Museum of Natural History, negative no. 317637)

characteristics such as skin pigmentation, hair color, and the like, and variation within any race tends to overlap with characteristics of other races. Although it has been suggested that the present races of humans represent early descent from primitive races of hominids such as the several subspecies of *Homo erectus*, there appears no basis for this suppostion in light of what is currently known about the early history of modern humans. Rather, the races of humans constitute populations of one widespread species.

SUMMARY

Invertebrates are a diverse group of animals without backbones. There are more genera and species of invertebrates—more than one million—than in all other groups of organisms combined. Of these, by far the most numerous are insects, classified in the phylum Arthropoda.

Sponges are the simplest multicellular animals; they lack digestive, excretory, and circulatory organs.

Corals, jellyfish, *Hydra*, and other similar organisms of the phylum Cnidaria are radially symmetrical and more complex than sponges.

Flatworms are a bit more advanced than cnidarians, having a branching gastrovascular cavity and bilateral symmetry. Roundworms are pseudocoelomates with slender cylindrical bodies tapered at both ends and a digestive system consisting of a tubular gut opening at both ends (a mouth and an anus).

Annelids show major evolutionary advances in their segmentation (metamerism) and possession of the first true coelom. Key annelid groups are the oligochaetes, polychaetes, and leeches.

Mollusks are bilaterally symmetrical animals exemplified by slugs, snails, clams, oysters, and squids. Their bodies are soft, but often protected by shells secreted by a fold of the body wall called the mantle.

The most diversified group of invertebrates belongs to the phylum Arthropoda. Arthropods are characterized by an exoskeleton containing chitin, and jointed appendages employed in walking, jumping, swimming, feeding, and flying.

Echinoderms, phylum Echinodermata, represent one end of an evolutionary line of invertebrates, arthropods another. Echinoderms are characterized by a radial, five-part symmetry, a water vascular system, an internal skeleton, and an exclusively marine habitat.

Members of the phylum Chordata are differentiated from other groups of animals by a solid elastic rod called the notochord, which in primitive forms serves as a "backbone." In early embryonic stages advanced chordates have a complete notochord, which is subsequently replaced by a jointed backbone composed of vertebrae.

The simplest chordates are the protochordates, jawless and often highly specialized animals such as sea squirts (tunicates), which bear a superficial resemblance to sponges.

The oldest known vertebrates are the jawless fish. They lived during the early Paleozoic era some 400 million years ago. Only a handful of jawless fish, the cyclostomes, exist today. The first jawed fish appeared in the early Paleozoic era; their descendants are elasmobranchs (sharks, rays, and skates) and Osteichthyes (bony fish).

Modern amphibians are the tailed urodeles (salamanders, newts, etc.) and tail-less anurans (frogs and toads). Their eggs are laid in the water and develop by the metamorphosis of a free-living larval stage into the adult form.

Reptiles are cold-blooded land animals totally free of the requirement for water in which to complete their reproduction. They produce shelled eggs in which the developing embryo is protected within a liquid-filled sac, the amnion. Present-day reptiles are alligators, crocodiles, turtles, lizards, snakes, and *Sphenodon*, the tuatara of New Zealand.

Birds and mammals differ from reptiles and amphibians in being warm-blooded animals with four-chambered hearts and high metabolic rates. Both further differ from reptiles in having an insulating body coat of feathers or fur.

Mammals nourish their young with a milk secreted by mammary glands, hence the name Mammalia. Among the major groups of mammals are the primates. Prosimians are the most primitive primates, and include lemurs, bush babies, and lorises. More advanced primates generally are classified as New World monkeys and Old World primates. The latter include Old World monkeys, apes, and humans.

The human race may have diverged from the ape line at least 10–15 million years ago. The first true human, classified as *Homo erectus*, inhabited Africa as long ago as 1.5 million years. *Homo erectus* and probably a 3-million-year-old African hominid named *Homo habilis* are the immediate ancestors of all the races of modern humans, classified as *Homo sapiens sapiens*.

A brief classification of the major animal groups is included in the Appendix.

KEY WORDS

sponge	cephalopod	anuran
coelenterate (cnidarian)	arthropod	reptile
polyp	exoskeleton	dinosaur
hydroid	trilobite	*Archaeopteryx*
mesoglea	trachea	Passeriformes
nematocyst	spiracle	therapsid
jellyfish	compound eye	Prototheria
coral	echinoderm	Metatheria
sea anemone	water vascular system	Eutheria
flatworm	chordate	mammal
roundworm	vertebrate	placenta
cuticle	pharyngeal gill slits	mammary gland
annelid	filter feeder	primate
polychaete	protochordate	prosimian
oligochaete	cyclostome	hominid
hydroskeleton	elasmobranch	bipedalism
statocyst	bony fish	*Dryopithecus*
trochophore	crossopterygian	*Australopithecus*
mollusk	lungfish	*Homo habilis*
chitin	amphibian	*Homo erectus*
gastropod	urodele	*Homo sapiens*

QUESTIONS FOR REVIEW AND DISCUSSION

1   How are flatworms and roundworms advanced structurally and physiologically compared to cnidarians?

2   Briefly describe the kinds of adaptive radiation observed in the phylum Mollusca. For example, what evolutionary modifications have occurred in the foot region?

3   Compare segmentation in an annelid worm, a trilobite, lobster, and an insect. What evolutionary trends are apparent?

4   Why do biologists think that echinoderms and chordates are evolutionarily related despite their striking differences in appearance?

5   What is the "larval hypothesis" of chordate evolution?

6 How did swim bladders originate? What group of animals possess swim bladders?

7 What features distinguish the mammals from other classes of vertebrates?

8 What extremes of adaptive radiation are identifiable among mammals?

9 What does brachiation imply with respect to primate evolution?

10 Relate bipedal locomotion to manipulative skills, the invention of tools, and the evolution of humans.

SUGGESTED READING

BAKKER, R. T. 1975. Dinosaur renaissance. *Sci. Amer.* 230(4):58–78. (Were dinosaurs warm-blooded?)

CALDER, W. A., III. 1978. The kiwi. *Sci. Amer.* 239(1):132–42. (Discusses evolution of kiwis by adaptive radiation from ratites isolated by continental drift.)

CARR, A. 1955. *The windward road.* New York: Alfred A. Knopf (Updated and reissued in 1984; a popular and fascinating account of the migration of the green sea turtle and descriptions of other tropical animals.)

CHALFIE, M. 1984. Genetic analysis of nematode nerve-cell differentiation. *BioScience* 34(5):295–99. (Interesting but brief article about use of mutant roundworms to study nerve-cell differentiation.)

COLE, C. J. 1984. Unisexual lizards. *Sci. Amer.* 250(1):94–100. (Parthenogenetic birth is the rule in some populations of whiptail lizards.)

EVANS, H. E. 1978. *Life on a little-known planet.* New York: E. P. Dutton. (The world of insects, written clearly and with wit.)

FINNELL, R. B. 1983. The joy of birds. *Nat. His.* 92(9):3–104. (The entire issue is devoted to various aspects of bird biology.)

GOREAU, T. F.; N. I. GOREAU; and T. J. GOREAU. 1979. Corals and coral reefs. *Sci. Amer.* 241(22):124–36. (An excellent description of limestone reefs and the coral polyps that build them.)

JOHANSON, D. C., and M. A. EDEY. 1981. Lucy. *Science 81* 2(2):48–55. (A popularized version of the discovery and significance of *Australopithecus afarensis*.)

KIRSCH, J. A. W. 1977. The six-percent solution: second thoughts on the adaptedness of the Marsupialia. *Amer. Scientist* 65:276–88. (A thought-provoking article on evolution and ecology of marsupials.)

LEWIN, R. 1983. How did vertebrates take to the air? *Science* 221:38–39. (Favors cursorial origin for flight.)

———. 1983. Were Lucy's feet made for walking? *Science* 220:700–02. (Despite some controversy "Lucy" [*Australopithicus afarensis*] seems to have walked upright.)

———. 1983. Is the orangutan a living fossil? *Science* 222:1222–23. (Probably.)

PILBEAM, D. 1984. The descent of hominoids and hominids. *Sci. Amer.* 250:84–96. (Recent evidence strengthens the relationship between African great apes and *Homo*.)

ROPER, C. F. E., and K. J. BOSS. 1982. The giant squid. *Sci. Amer.* 246(4):96–105. (Giant squids, virtually unknown in detail to science until recently, now are being studied. An account of some new discoveries about these largest of invertebrates.)

TRINKHAUS, E., and W. W. HOWELLS. 1979. The Neanderthals. *Sci. Amer.* 241(6):118–33. (Neanderthals were a unique race of humans who ranged far and wide prior to and during the past ice age. Interesting.)

WALKER, A., and R. E. F. LEAKEY. 1978. The hominids of East Turkana. *Sci. Amer.* 239(2):54–66. (An excellent description of discovery and dating procedures in search of early humans.)

WASSERHUG, R. 1984. Why tadpoles love fast food. *Nat. His.* 93(4):60–69. (Unlike many biologists who look on tadpoles as a ''phase'' that adult frogs and toads must go through, the author focuses his research emphasis on the larvae.)

YONGE, C. M. 1975. Giant clams. *Sci. Amer.* 232(4):96–105. (In addition to a discussion of the ecology of these giant reef dwellers, information on the structure of bivalves is presented.)

IV

Ecology and Evolution

22

Behavior

I n the opening pages of his delightful book *King Solomon's Ring*, the eminent **ethologist** (scientist who studies animal behavior) Konrad Lorenz describes the following incident: Two men wearing bathing trunks are walking along a river bank carrying on their heads a canoe, balanced between them. They are followed by 10 half-grown goslings and 13 tiny mallard ducklings cheeping anxiously as they scurry to keep up, and at the end of this "ridiculous parade" marches a half-grown hybrid of a duck and an Egyptian goose.

Why do goslings follow humans as if they were large mother geese? One explanation is that a newly hatched gosling quickly develops a bond with the first large animate object it sees, whether a mother goose, a dog, a human, or even a box with a ticking clock inside. (Mallard ducklings, on the other hand, will develop such a bond with a human only if the latter is squatting and emitting a quacking sound—see Fig. 22-1). This response by hatchlings is called **imprinting**, a concept originated years ago by Oskar Heinroth and Douglas Spalding. Understanding imprinting involves the studies of genetics, communication, and learning and is one of the many fascinating aspects of the important biological subject of **behavior**.

BEHAVIORAL COMPLEXITY

Simply defined, behavior is a movement in response to a stimulus. Plants exhibit behavior when they bend toward the light or close their leaves when touched, as in the case of the sensitive plant and Venus's fly-trap. Single-celled protistans exhibit behavior when they swim toward or away from light, when they recognize and single out prey, as distinct from nonprey, and when they conjugate with an individual of opposite mating type.

More complex forms of behavior are exhibited by simple multicellular animals with nervous systems capable of coordinating the action of sensory receptors and muscles. *Hydra*, although its nervous system is only a simple nerve net, demonstrates a fine bit of coordination in its feeding behavior. When it captures prey, its mouth opens in anticipation of the arrival of food. Study of this behavior has shown that the prey is first recognized only if it contains **glutathione,** a simple peptide found in many animal cells. This chemical stimulus causes the discharge of nematocysts into the prey and further leakage of glutathione from the wounded prey. *Hydra's* tentacles, in response to glutathione, retract toward its mouth; glutathione still oozing from the prey

505

FIGURE 22-1
Imprinting in ducks. Newly hatched ducklings will adopt and follow a moving, sound-producing object in place of a natural parent. The imprinting requires periodic repetition, a procedure known as **reinforcement.**

stimulates chemical receptors in the mouth region, causing the mouth to open. The automatic nature of these responses can be demonstrated by adding different concentrations of pure glutathione to the water surrounding a living *Hydra*.

As might be expected, more complicated behavior patterns are exhibited by organisms with nervous systems capable of storing information. Here, in addition to the automatic, genetically programmed behavior, which sometimes is called **instinct**, is behavior that is acquired through experience and is termed **learned behavior**, or simply **learning**.

Whether simple and apparently automatic, or complex and based on experience, behavior has a genetic basis which is inheritable. Although learning is acquired during the lifetime of the individual, it is dependent on the number and kind of neurons present and on their arrangement in the nervous system, characteristics that are inherited. Both simple and complicated patterns of behavior are subject to the forces of natural selection and survival, and therefore can become a part of the evolutionary history of the species.

The inheritance of stereotyped, seemingly automatic kinds of behavior is the easiest to demonstrate. Some of the most complex patterns of stereotyped behavior are those of the social insects, whose ritualistic behavior chains are intricate marvels. In a colony of social insects, such as ants or bees, there are morphological castes, each with its specific, genetically programmed behavior. The behavioral interrelationships of castes are often amazingly complex; examples are the cultivation of underground fungus gardens by some ant species and the transformation in other species of some worker ants into honey casks (Fig. 22-2). Another remarkable example of automatic behavior is seen in one species of tropical ant which makes nests of leaves held together with threads of silk. Because the adult workers of this species cannot spin silk, they hold the larvae, which have silk glands, and pass them back and forth over the nest as they spin out their threads.

Although most behavior patterns of invertebrates are inherited, so that individuals set about their simple or complicated chores from the moment of hatching, insects and even less complex invertebrates are capable of learning. Planarians can be trained to avoid an electric shock stimulus and to associate it with patterns of light and dark. Bees can be trained to associate food with signal objects of different shapes and colors, and ants can be trained to work their way through a maze. Nevertheless, acquired behavior plays only a minor role in the lives of most invertebrates. In many cases the life spans are much too short to benefit from a chain of experiences, and in social insects particularly, the individual is programmed from birth to death to function as part of a survival machine. Only among vertebrates does learning supplant stereotyped behavior to a large degree. The reason for this lies in the great complexity of the vertebrate brain; even the brains of fish are vastly more complicated than those of invertebrates.

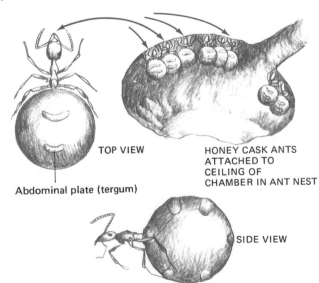

TOP VIEW

Abdominal plate (tergum)

HONEY CASK ANTS ATTACHED TO CEILING OF CHAMBER IN ANT NEST

SIDE VIEW

FIGURE 22-2
Honey cask ants. (Redrawn from W. Goetsch, 1957, *The ants*, West Berlin, Germany: Springer-Verlag)

Behavioral Responses

Patterns   Most behavioral actions are goal-oriented, not simply random actions that accomplish nothing. Some behaviors may appear random in that an organism may make a number of irregular and seemingly inconsistent movements. However, continued observation of these movements shows their total effect to be an adjustment to a stimulus. For example, seemingly random movements of microorganisms often result in the orientation of the organisms in specific regions relative to light, oxygen, or food—forms of behavior called **kineses** (sing. **kinesis**).

Somewhat more complex behavior patterns are patterns characterized by a directional movement in response to the stimulus. If the response involves just part of the body—an arm or leg, for instance—it is a **reflex**. If the response involves movement of the entire organism, it is called a **taxis** (pl. **taxes**). It is sometimes difficult to make hard and fast distinctions between reflexes and taxes; however, taxes usually involve several to many reflex actions, and in addition there may be a difference between the direction of the stimulus and the direction of the response. Reflex actions usually are directed toward or away from the point of stimulus (pulling hand from fire, slapping an insect bite, etc.). Taxes also may be oriented toward or away from a stimulus, but can be quite complex and involve more than one stimulus source. The escape reaction in some insects is an example of this. When a butterfly is startled, it often flees toward the sun. This orientation depends on two-eyed vision, with each eye receiving equal proportions of light. If one eye is blinded, the insect flies in circles.

Behavior patterns of the type just described were once called "instinctive" and were believed to be unvarying and unlearned. Yet even the simple escape of an insect toward light can be changed by experience. If, for example, an insect is given an electric shock each time it moves toward the light, it eventually will learn not to turn toward the light in an escape attempt. In the same way, higher animals can be trained, by exposing them to a succession of loud noises, not to leap whenever they hear a loud noise. This kind of learning is called **habituation**. Learning therefore is the capacity of an organism to change its behavior pattern as a result of experience. However, it is not always easy to say where automatic, genetically programmed behavior stops and learning begins.

Imprinting   Imprinting, though seemingly instinctive, is a result of rapid and early learning. Konrad Lorenz, who carefully studied imprinting, at first was reluctant to consider it a form of learning because it was so rapid, so stereotyped, and so permanent. When ducklings are incubated and reared in isolation, they may become imprinted by some moving object such as a football, a stuffed hen, or even a milk bottle, but the attachment sometimes is not permanent and a duckling later may adopt some other object as a "mother." On the other hand, wild ducklings removed from the nest shortly after hatching appear unimprintable to objects other than their natural mother.

Careful analysis of natural imprinting in ducklings has shown that it begins before the eggs hatch. The ducklings inside the eggs begin to cheep, and the mother duck clucks softly in response (Fig. 23-3). Part of this communication between mother and unhatched young serves to synchronize hatching. Eggs are laid over a period of several days. In the laboratory incubator, each egg hatches about three weeks after it is laid, whereas in nature the hatching date is synchronized so that all the young emerge at about the same time—obviously an important factor in their survival. The role

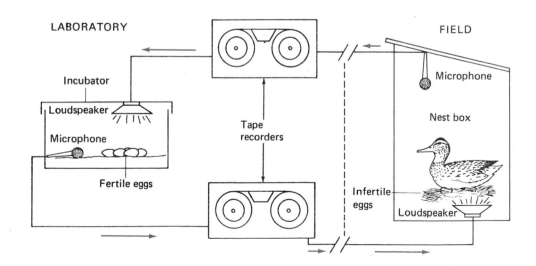

FIGURE 22-3
Imprinting experiment. In this experiment, sounds of unhatched ducklings in an incubator are transmitted to a female mallard in a remote nesting box. The female's responses are taped and transmitted to the incubating eggs. The ducklings in the incubator thus become imprinted. (Modified slightly from E. H. Hess, Imprinting in a natural laboratory. Copyright © by Scientific American, Inc. All rights reserved)

of vocal communication in synchronizing the hatch is shown when duck eggs are reared in an incubator and tape-recorded clucking sounds of a mother duck on her nest are played. Under such circumstances the ducklings hatch synchronously, just as they do in nature. Further study has shown that each mother duck has a distinct vocal pattern, some emitting single clucks, some clucking in triplets, and so on. It seems likely, therefore, that the imprinting behavior includes the ducklings' learning to recognize the sounds made by their mothers.

More complicated learning involves some form of reward or punishment. Human mothers reward a successful effort at learning with a sweet or a pat, and a failure with a scowl or even a whack on the rear. Some experiments with rats indicate that they learn some behaviors faster when punished for wrongs than when rewarded for rights, although most persons feel that animals respond better to reward than to punishment. Even though human learning seldom involves something as simple as an electric shock for each mistake or a piece of candy for each success, it does involve rewards and punishments. Often these are in the form of approval or disapproval of other individuals; sometimes they are the painful consequences of a foolish act.

Learning

Habituation, in which an animal gradually decreases its natural response toward some stimulus until the response disappears, is probably the simplest kind of learning.

A second simple form of learning is **conditioning**. In conditioning experiments, a specific signal is associated with a certain stimulus until it becomes capable of generating a response similar to that induced by the original stimulus. In the classic example, the Russian psychologist Pavlov harnessed a dog and then repeatedly blew powdered meat in its face, noting that the dog salivated each time this was done. Then Pavlov rang a bell every time he blew meat powder in the dog's face. Finally, he rang the bell without presenting food. Pavlov found that the dog salivated in response to the bell alone, a behavioral action he termed a **conditioned reflex** (see Fig. 22-4). Many other experiments of a similar nature and with similar results have been done.

A third form of learning is trial-and-error learning, the simplest kinds of which are very much like conditioning. A kitten learns by accident that pressing a lever rewards it with food, whereas a second lever of different

FIGURE 22-4
Conditioned reflex.

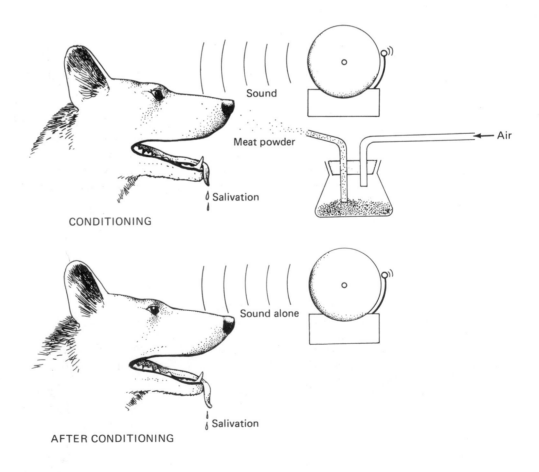

Sound

Meat powder

Air

Salivation

CONDITIONING

Sound alone

Salivation

AFTER CONDITIONING

shape punishes with an electric shock. Animals can be taught to solve rather complex problems using trial and error. A favorite tool in such studies is the maze, which ranges from a simple **Y** shape to a complex meandering course with many wrong turns and blind alleys (Fig. 22-5). Planarians successfuly learn **Y**-shaped and other simple mazes by trial and error. Social insects can learn more complex mazes; ants seem to be the most successful. Vertebrates, as expected, are superior to invertebrates in trial-and-error learning, and birds and mammals show some ability to apply experience gained from one trial-and-error situation to solving other, similar problems.

Reasoning

The different hierarchies of behavior can be associated with the hierarchies of animal classification. Like body structure, behavior is a product of evolution. Up to this point, behavior has been considered in terms of stereotyped actions and learned responses, in which a particular stimulus or pattern of stimulations calls forth a specific kind of behavior. As nervous systems became more and more complex, behavioral patterns also became more and more complex, involving highly organized and modifiable patterns and interactions with other individuals.

A final step in behavior, most apparent in the primates, has been the evolution of **reason**. Reason may be defined as the ability to solve problems by bringing various aspects of past experience to bear, without resorting to trial and error.

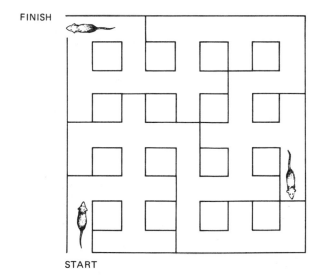

FINISH

START

FIGURE 22-5
Rats in a maze. A maze may have movable partitions so that more or less complex problems may be presented.

FIGURE 22-6
Simple detour problem soluble by reasoning.

There has been little success in experiments designed to show the reasoning abilities of invertebrates, lower vertebrates, and mammals other than primates.

A simple test of reasoning ability is the "detour" problem. In this test an animal is presented with food that it can see and smell but to which the pathway has been blocked; the animal must make a detour to reach the food (see Fig. 22-6). Most animals fail this test and reach the food only by trial and error; only monkeys and chimpanzees have shown themselves capable of solving the problem by the use of reason.

In a more complex detour problem than the one illustrated in Fig. 22-6, a chimpanzee was confronted by a bunch of bananas strung out of reach at the top of a room, and a scattered set of boxes. The chimpanzee demonstrated reasoning by stacking the boxes to make a platform from which it was able to gather the bananas (Fig. 22-7). (In *King Solomon's Ring*, Lorenz relates an amusing incident that occurred in the chimpanzee cage at the Berlin Zoo. A chimpanzee, confronted with a banana "problem" similar to the one depicted in Fig. 22-7, motioned to the keeper to move toward him, as he, the chimpanzee, moved away. Puzzled, the keeper followed until the chimpanzee clambered up on the keeper's shoulders and picked a banana from the stalk. Only then did the keeper realize he had been tricked.)

One of the difficulties with reasoning experiments of this kind is that one never is certain to what extent an animal draws upon past experiences, and how much is real innovation. That, of course, is also true of hu-

FIGURE 22-7
Chimpanzee demonstrating reasoning in solving a problem. (From R. F. Oram, P. J. Hummer, Jr., and R. C. Smoot, 1979, *Biology: living systems,* Columbus, Ohio: Charles E. Merrill Publishing Co.)

man problem solving and even of abstract conceptual thinking.

SOCIAL BEHAVIOR

In analyzing behavior so far, the principal concern has been with responses of individuals rather than with interactions between individuals. Even solitary animals must socialize at times so as to bring about mating and reproduction. Most social species have more complicated social organizations beyond the needs for reproduction. There are highly organized societies in which the roles of some individuals are subordinate to others and highly stereotyped, as for example the societies and castes of social insects. The social organization of other animals may be much less rigorously controlled by genetic programming, but behavioral interactions between society members are just as vital to the welfare of the individual and the group.

Why socialize? There are major advantages to social living, among which are improved detection of predators and common defense of such resources as food and space. A major disadvantage is increased competition for these same resources. However, in view of the thousands of highly social species known, the costs of sociality are clearly outweighed by the benefits of social living.

Communication

At the heart of any social organization is a system of **communication.** Figure 22-8 depicts how such a communication system operates. Signal generation is initiated by a **signaler,** also known as a **transmitter, sender,** or **actor.** The signaler uses any of a number of **channels** for signal generation, and the signal is carried by some **medium** such as air or water. The **receiver,** also known as the **recipient** or **reactor,** detects the signal by means of a **sensory system** and usually reacts to the signal. Thus, feedback stimulates the sender to generate a new signal, and so on.

Communications are usually classified on the basis of the sensory organs used to detect the signals. The commonly recognized channels are **visual, auditory, chemical,** and **tactile.**

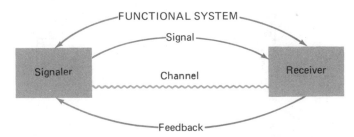

FIGURE 22-8
Diagram of a functional communication system.

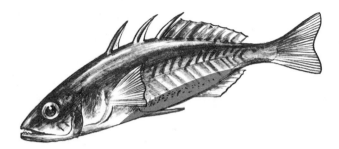

FIGURE 22-9
The red belly of the male three-spined stickleback functions in threat interactions between rival males.

Visual Communication Visual signals may involve color, form, movement, or any combination of these. The red belly of male three-spined sticklebacks is an effective visual signal associated with **agonistic** interactions between rival males (agonistic behavior is any conflict behavior, including threats, attacks, submission, etc.) (see Fig. 22-9 and Color plate 12A, B). Niko Tinbergen, a Nobel prize-winning behaviorist, showed in his experiments with sticklebacks that a male in reproductive condition would threaten and attack any object, animate or inanimate, as long as it was red.

The green heron demonstrates how form and movement may be combined in a visual display. As illustrated in Fig. 22-10, a threatening bird gradually increases the intensity of its display by erecting its feathers, flipping its tail, and bulging its eyes. At the height of the display, the heron appears to be much larger because of the raised plumage. In addition, the bird opens its mouth and gives a harsh call, further accentuating the visual signals. This threat display of the green heron is an excellent example of a **graded signal,** a signal that gradually increases in intensity until the full-blown form is exhibited.

By contrast, a **discrete signal** has an all-or-none form. There is no gradation; the signaler either signals or it does not. Again, the green heron provides an example—the snap display (Fig. 22-11). From a normal perched posture the displaying bird shows the full snap in one smooth motion, ending with a snapping together of the mandibles. There is no gradual buildup in intensity.

Firefly communication demonstrates how a signal system may be exploited. Male fireflies flash at various intensities and sequences which differ from species to species. Females respond to the flash patterns of males of their species by flashing their own patterns. The male-flash, female-flash-in-response sequence continues until the two individuals come into contact for mating.

Two beetle genera, *Photinus* and *Photuris,* exhibit an interesting and bizarre modification of typical flashing

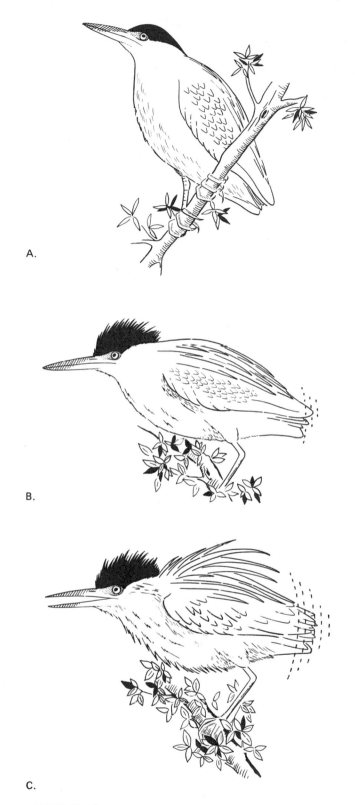

A.

B.

C.

FIGURE 22-10
Aggressive display of the green heron. **A.** Alert posture. **B.** Mild aggressive posture. **C.** Full aggressive display. (Based on A. J. Meyerriecks, 1972, *Man and birds: evolution and behavior,* New York: Bobbs-Merrill Co.)

FIGURE 22-11
Snap display of the green heron, a discrete (all-or-none) display.

behavior. The male of a certain *Photinus* species flashes his species-specific pattern until he detects the return flash pattern of a female of his species. Copulation soon takes place. Another male of this same *Photinus* species responds to the flashing "answer" of another female, only to be seized and devoured. What has happened? A

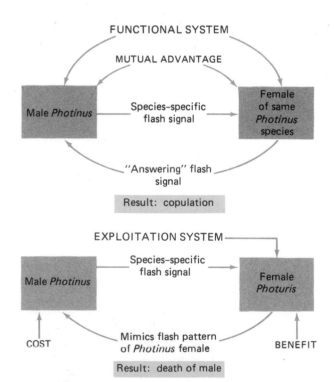

FIGURE 22-12
Diagram of how a functional communication system can be exploited by another species, in this case by "femmes fatales" of the genus *Photuris*.

FIGURE 22-13
Black-tailed prairie dog giving an alarm call, an example of a discrete vocalization. (Redrawn from J. A. King, The social behavior of prairie dogs. Copyright © 1959 by Scientific American, Inc. All rights reserved)

female of a *Photuris* species has mimicked the answering flash pattern of the *Photinus* species and lured the male to his death. James Lloyd, a leading student of firefly behavior, has aptly called such females "femme fatales." He also has observed that a single female *Photuris* may change the pattern of her answering flashes such that males of up to four different species are entrapped. This exploitation system is diagrammed in Fig. 22-12.

Auditory Communication   Communication by sound has obvious advantages when compared with visual signaling. Sounds can be produced regardless of light intensity; visual signals are maximally effective in bright light. Sounds can "go around" objects that lie between the signaler and the receiver; visual signals lose their efficiency rapidly depending on the nature of the intervening obstacles (trees, bushes, rocks, etc.). Sound signals travel far and fast; visual signals tend to fade rapidly unless the receiver is very close to the signaler. A major disadvantage of sound signaling is that it is essentially broadcast, and any receiver within range can detect the signal and react accordingly. This includes predators.

Like visual signals, sound signals may be discrete or graded. Alarm calls of prairie dogs are given in an all-or-none fashion (see Fig. 22-13). Vervet monkeys have discrete and distinctive alarm calls for different classes of predators.

A graded vocalization is exhibited by red deer (*Cervus elephas*) stags. Stags maintain harems, and they compete vigorously for additional females by engaging in roaring duels (Fig. 22-14). Two males roar back and forth until one of them increases the tempo of its roaring. Soon the other escalates its vocalization. The first stag retaliates in kind, and so on, until suddenly one gives up. The victor adds another female to his harem.

Males of many bird species use songs to defend their territories. To human ears many of these territorial advertising songs seem deceptively simple. However, the "ok-a-leee" song of the territorial male red-winged blackbird (Fig. 22-15) may contain all the following information for the receiver: (1) presence of the signaler, (2) species identity of the signaler, (3) sex of the singer, (4) motivation—will the signaler attack?, (5) identity of the individual, (6) exact location of the singer.

FIGURE 22-15
The territorial song of a male red-winged blackbird may contain detailed information for a receiver. (Photo by Charles G. Summers/Tom Stack & Associates)

Chemical Communication   Semiochemicals are any chemicals that act as signals in a communication system (Fig. 22-16); they may affect interactions between individuals of the same species (**intraspecific**) or between individuals of different species (**interspecific**). **Pheromones** are semiochemicals that signal between conspecifics; **allelochemics** mediate interspecific interactions.

There are three major categories of pheromones: **primer, releaser,** and **informer.** Upon reception, **primer pheromones** do not bring about an immediate behavioral change but instead affect the physiology of the receiving individual. For example, nymphs of a migratory locust species detect a primer pheromone called **locustol** in the feces of nearby conspecifics. Locustol affects the development of the nymphs such that they form wing buds, become darker in color, deposit fat in fat bodies, and eventually are transformed into devastating flying machines.

Releaser pheromones, when detected, bring about an immediate behavioral response. For example, when a harvester-ant worker comes upon a potential source of

FIGURE 22-14
Red deer stags engage in roaring duels over harems. (Sketch from a photo by T. H. Clutton-Brock/S. D. Albon, published in *New Scientist*, November 16, 1978)

ARTHROPOD CHEMICAL DEFENSES

Many arthropods have evolved structural features, colors, patterns, and behavioral immobility that render them all but invisible to predators. They closely resemble their immediate environment, whether a blade of green grass, brownish-gray bark, a twig, a leaf, or another background. Other arthropods flee or counterattack when under siege by a predator. Rather than grappling with their foe, numerous arthropods use chemical defenses for self-protection, many of which can be "fired" at long range. Specialized glands found in various parts of the body act as reservoirs for a battery of defensive toxins (see Fig. 1A).

Such defensive chemicals are called allomones; they may be injected directly into the body of the attacker whether by sting (honeybee), mandibles (centipedes), or chelicerae (spiders). Injectable venoms may be used for prey capture (recall how spiders incapacitate their prey) as well as for defense. The other major groups of chemical defenses, the noninjectable ones, are produced as oozes or as sprays. The toxin-producing glands and their associated storage reservoirs may be located anywhere in the arthropod's body.

A primitive chemical defense system used by many ant and beetle species employs formic acid as a deterrent; the acid is produced and stored in comparatively simple glands typically located at the end of the abdomen (Fig. 1B). More complex are the toxins produced by certain millipedes that ooze defensive benzoquinones from a series of glands found on both sides of the long, slender body. There are no glands on the heads of such millipedes, so when this vulnerable part of their anatomy is under attack, the millipede simply throws its body into a tight coil with its head tucked inside next to the protective glands. Other millipede species employ deadly hydrogen cyanide (HCN) as their key defensive poison. The glandular apparatus used here is much more complex and includes secretory lobes, conducting tubules, storage areas, "reactor" chambers (vestibules), and so forth (see Fig. 1C).

Another fascinating mode of chemical defense is used by bombardier beetles, so called because of the heat and noise they generate during active defense. The toxin-producing glands, like those of certain millipedes, are very complex (see Fig. 1D). Glandular tissue produces large quantities of hydroquinones and hydrogen peroxide, both of which are stored in a large reservoir. When under attack, the beetle opens the valve between the reservoir and the smaller vestibule. A battery of enzymes (catalases and peroxidases) are produced in the walls of the vestibule. These enzymes bring about an extremely rapid reaction—an explosion—that converts the reservoir contents to parabenzoquinone (the key allomone), plus oxygen and water, all at 100°C and accompanied by a very loud "pop" (rapid production of oxygen). The bombardier beetle also has the ability to rotate the nozzle tips of its paired abdominal glands so as to fire directly at an attacker. Furthermore, the defensive system is like an automatic weapon: the beetle can fire repeatedly until the glandular contents are exhausted.

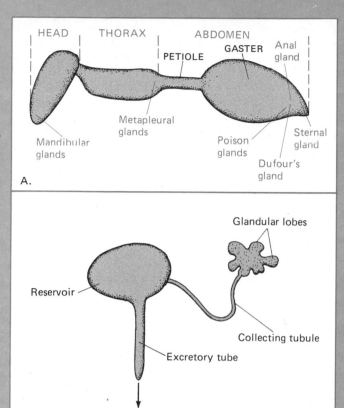

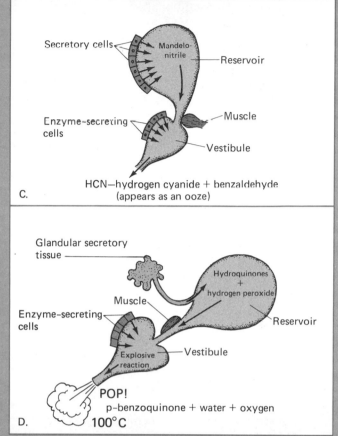

FIGURE 1

Arthropod chemical defenses. **A.** Defensive glands of ants. **B.** Simple defensive gland of the ground beetle. **C.** Complex defensive gland of some millipedes. **D.** Complex defensive gland of the bombardier beetle.

danger near its nest it gives off a tiny puff of **alarm pheromone** from glands in its head. Nearby workers detect the alarm pheromone, which "releases" alarm behavior on their part. In this way additional workers are recruited to deal with the danger.

A different kind of releaser pheromone is that found in leafcutter ants, which cut small pieces of leaves and carry them many feet back to their nest. How do these ants find their way back and forth? The workers follow a **trail-marking pheromone** initially laid down by the worker who discovered the food supply (Fig. 22-17 and Color plate 14E). Recruited workers reinforce the trail with their own trail-marking pheromones until the food supply is exhausted. The sterotyped nature of trail following in ants can be tested easily by placing a square of paper over the trail and after a time turning it at right angles, whereupon the ants will swerve unerringly to the end of the marked trail on the paper and will then halt in confusion until random wandering results in relocating the original trail, or in blazing a new one.

Most deer species provide examples of **informer pheromones**. These deer have glands between their hooves, and as they walk about they leave a personalized scent. Later, other deer come upon the scent and are thereby informed of the prior presence of the now-departed signaler. No behavioral changes are seen in the receiver, and no primer effects are known; presumably the receivers are simply informed of the passing of a certain conspecific.

As noted previously, allelochemics signal between species. Allelochemics are of two types: **allomones** and **kairomones**. The bombardier beetle's discharging of noxious chemicals at the head of an enemy is a classic example of the use of an **antagonistic allomone** (see Fig. 1D, pp. 514-515). **Methyl mercaptan**, the active ingredient of skunk smell, is a very effective deterrent to would-

be attackers; it too is a fine example of the effect of antagonistic allomones.

Most **mutualistic allomones** are scents given off by flowers so as to attract pollinators. While an insect is sipping nectar, it gets dusted with pollen, which is later transferred to a different flower that attracts this pollinator with its mutualistic allomones.

Kairomones are semiochemicals inadvertently given off by the signaler as a metabolic by-product. For example, mosquitoes are in part attracted to humans by the carbon dioxide content of their breath. The carbon dioxide is of course a normal by-product of human respiratory metabolism, but it acts as a kairomone because the mosquitoes have evolved the appropriate carbon dioxide detectors and are thereby attracted to the source of a potential blood meal. Predators that hunt by smell also avail themselves of kairomones unavoidably given off by their prey.

Tactile Communication Workers of some ant species, rather than laying down trail-marking pheromones after discovery of a food source, return to the nearest entrance of the nest, excitedly antennate a fellow worker, and stimulate it to follow. The new recruit keeps close contact with the recruiter and is thus led to the food source. This tactile behavior is called **tandem running**.

Grooming in primates and mutual preening in birds are other examples of tactile communication. The grooming baboon or chimpanzee removes bits of dead skin, tiny parasites, and so on, from the skin of an associate as a service, but the groomer may also be appeasing the aggressiveness of the individual being groomed. In fact, it is possible in some primate species to determine the rank of different individuals in a social group by who grooms whom, high-ranking individuals receiving much although sometimes giving little.

FIGURE 22-16
Semiochemicals.

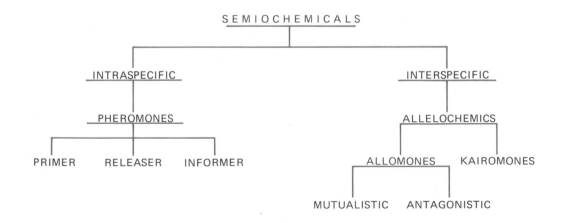

FIGURE 22-17
Leafcutter ants. The safarilike train follows a path first defined by chemicals secreted by the trailblazers. (From *The ways of the ant* by John Crompton. Copyright 1954 by John Crompton. Reprinted by permission of Houghton Mifflin Company)

Social Use of Space

Flocks of swallows or starlings precisely spaced as they perch on an overhead wire are exhibiting a simple use of space called **individual distance,** or **personal space.** It is as if each individual carried with it an invisible sphere which, if violated, would bring about an agonistic response such as a brief threat or quick peck. The behaviorist Peter Marler, working with a small European songbird called the chaffinch, showed that although this species has a more or less typical individual distance, the distance varies depending on such factors as sex, age, season of the year, and so forth. Male chaffinches have larger individual distances than do females, and the personal spaces of both sexes are larger in the breeding season, but again much larger in males, which are more aggressive.

Peck Orders   A more complex kind of spacing is found in **dominance–subordinance hierarchies,** or **rank orders,** first discovered in domestic hens. If a group of hens unfamiliar with one another are placed in an enclosure, they will quickly fight and establish a hierarchy of victors and vanquished. One hen, called the **alpha hen,** ranks at the top because she has defeated all her rivals. The **beta hen** has defeated all other hens except the alpha hen, and so on down the line. Such **peck orders** (so called because hens fight by pecking at each other with their bills) soon lead to stability in the flock. Each hen knows which other hen it has peck rights over and to whom it must defer when some resource such as water, food, roosting site, and so on, is in dispute.

The number 10 apparently is critical to hens. If a flock numbers 10 or less, a linear hierarchy forms ($1 > 2 > 3 > \ldots > 10$, where $>$ means "dominant to"); if more than 10 hens are in the flock, the hierarchy suffers some internal inconsistencies, such as $1 > 2 > 3 > \ldots > 15 > 2 > \ldots$, in which hens become confused about their peck rights over other hens in the flock. Apparently, hens cannot remember individual indentities if the flock gets too large.

Territoriality   Most animals have a characteristic **home range,** which can be defined as the area over which the individual or social group ranges in its daily activities (foraging, search for water, avoidance of predators, etc.). The size of the home range may be a matter of a few square yards in the case of a forest mouse or up to a thousand square miles, the home range of a pack of African hunting dogs.

An individual or social group tends to spend much time in some parts of the home range and very little in others. The most intensively used regions of a home range are called **core areas.** Typical core areas might be the site of a nest or burrow, a source of water, a group of trees in fruit, hiding places such as rock outcrops, and the like. The relationship between core areas and the home range is as follows:

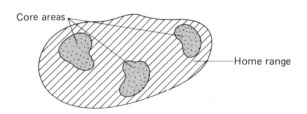

A **territory** is any defended area. Thus, if an individual or social group defends its entire home range (unlikely), the territory and the home range are one and the same:

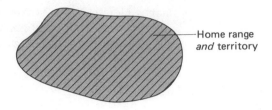

—Home range *and* territory

If the core areas are defended, but not the rest of the home range, the relationship is as follows:

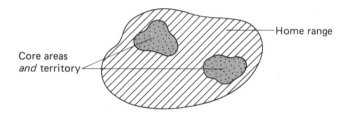

Core areas *and* territory

—Home range

Often there is an overlap between home ranges of neighboring individuals or social groups, but no overlap of their respective territories:

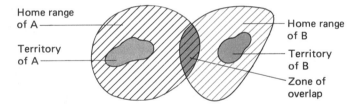

Home range of A

Territory of A

Home range of B

Territory of B

Zone of overlap

What is being defended is some vital resource such as a food supply, or a nest, burrow, or den site. Territorial defense usually is restricted to excluding conspecifics, but a number of animals also will exclude other species, especially if they are potential competitors for the vital resource. For example, red-winged and yellow-headed blackbirds, when they nest in the same cattail pond, engage in interspecific territoriality.

The concept of territorial behavior has been advanced greatly by studies of bird behavior. The German ornithologist Bernard Altum (1868) and the English ornithologist H. Eliot Howard (1920) independently came to the conclusion that the songs of male European songbirds function as long-range threats to rival males and also to attract a female to the singing male's territory. In other words, male birds adopt a plot of ground as their territory and then advertise their presence by singing. Rival males are warned not to trespass; if they do, the resident will threaten and attack. Females, however, are encouraged to come and examine the male as a prospective mate. Both Altum and Howard concluded that the main function of the territory was to provide food for the mated pair and their offspring.

The concept of territoriality has been extended and modified over time. There are species that show the classic type of territoriality, with the major function being defense of a food resource, but numerous other species gather no food whatsoever on their territories. Such defended areas may serve for courtship and mating, to provide a nest site, and so forth, but food is gathered someplace else. The green heron, for example, has separate nesting and feeding territories, each of which is defended against conspecifics.

The males of some bird species, such as grouse, move onto a traditional, communal display ground called a **lek** (from a Swedish word meaning "love play"). On the lek, males display and fight vigorously (see Fig. 22-18). Soon a territorial society is formed, with each male occupying and defending a small display site on the lek. The preferred sites are in the center of the lek, and these go to the most vigorous males. Females are now attracted by the hubbub on the lek but they ignore completely the displaying males on the edge of the region. Only the few males in the preferred sites get to

FIGURE 22-18
Male sage grouse assemble on a traditional, communal display ground called a lek, where they display and fight for the favored positions in the center of the area. (Redrawn from Scientific American. Copyright © 1978 Scientific American, Inc. All rights reserved)

FIGURE 22-19
Black-tailed prairie dogs live in social groups called coteries, which exhibit group defense of territory. ① Two prairie dogs approach and check if they are members of the same coterie. ② If they are, they engage in the recognition "kiss" ceremony. ③ If strangers meet, they elevate their tails, sniff each other, and attempt to nip. ④ The two prairie dogs return to their respective coterie territories after a brief interaction. (Redrawn from J. A. King, 1959, The social behavior of prairie dogs. Copyright © Scientific American, Inc. All rights reserved)

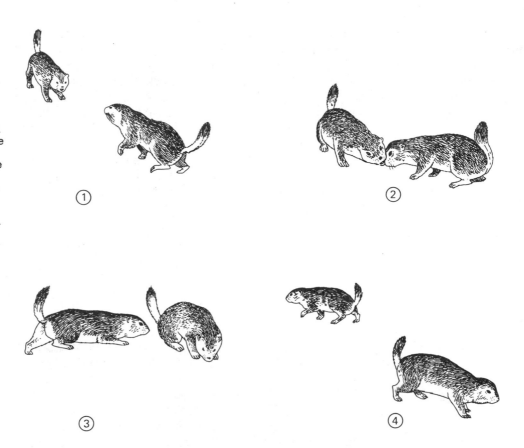

copulate; sometimes one male, the master cock, will effect most of the copulations in a season. The fierce fighting pays off in the genetic contributions to the next generation by just a handful of males each year. Thus the small territories of lek birds function as mating stations.

Black-tailed prairie dogs show an interesting example of social group territoriality: the defended area provides all food, and defense is conducted by an entire social group called a **coterie**. Fellow coterie members regularly groom each other, so individual recognition has a firm basis in many mutual interactions. If an individual is not recognized as a coterie member, it is threatened and driven from the collective territory (Fig. 22-19).

Social Organization

The charge of an African elephant is a spectacular and frightening display. Until recently it was believed that defense of the cows and calves was one of the main functions of the bulls, with one of the older bulls acting as leader of the herd, as well as its prime defender. When provoked, the leader bull would attack with an intimidating charge.

This myth was dispelled by a young English behaviorist named Iain Douglas-Hamilton. Painstaking study of known individuals over several years convinced Doug-

las-Hamilton that African elephants actually live in very stable family groups and larger kinship groups. A family group consists of an older female (the true leader of the society), several daughters, and their calves. The society of African elephants is a true **matriarchy**—a society dominated by a single female—because the dominant older female both leads and protects the family unit and larger kinship group.

The average size of an African elephant family unit is 10. A kinship group—an association of several or more family units related to one another—may number 40 or more. What about the males? Young bull elephants either leave their natal family unit voluntarily or are prodded into leaving by female aggressive behavior. These young bulls soon form temporary all-bull groups characterized by a complete lack of stability. Bulls are free to wander from group to group or to lead a solitary existence. When a bull is seen in association with a cow–calf unit, it is usually for reproductive purposes. The male checks to see if any females are in heat, mates with them if they are, and then leaves the family unit.

When foraging, a larger family unit may split into several smaller subunits. However when danger threatens, all family unit members gather around their leader, the matriarch, and if the danger escalates it is then that the spectacular charge of the dominant female is seen.

A society dominated by a single male—called a pa-triarch—is exemplified by the mountain gorilla. Long-range studies by George Schaller, Dian Fossey, and others have shown that mountain gorillas live in small- to medium-sized groups (average 17) comprised of several adult males and females, subadults, juveniles, and infants. Although it can be characterized as a multimale society, one of the older males is clearly the leader, or patriarch.

When fully reproductively mature, male mountain gorillas grow silver gray hairs on their backs. Some-times there are two or more of these silverback males in a gorilla social group; if so, one is always dominant to the other. This male is the true leader of the society, determining feeding and sleeping rhythms, routes of travel, and so forth. If danger threatens, he exhibits the spectacular chest-slapping display, followed by short, intimidating rushes accompanied by loud roars.

Both male and female mountain gorillas may leave their natal group. The females join another group or form the nucleus of a new group very quickly. Males,

MOLE-RATS

A tiny rodent weighing an average 35 gms, with a snout-to-tail length of about 100 mm, has excited the world of behavioral biology. Observations of its social behavior are being touted as one of the century's outstanding mammal discoveries.

The family Bathyergidae, African mole-rats or sand rats, is small when compared with other rodent groups. There are eight species, all confined to sandy soils from south of the Sahara to the Cape of Good Hope. The species that has brought attention to the group is the naked mole-rat, *Heterocephalus glaber*, of East Africa. This buck-toothed, burrowing rodent is nearly hairless, hence its common name. Its virtual lack of pelage may not be so strange for an underground burrower, but its unusual social structure is decidedly so!

Naked mole-rats live in large, complex burrow systems that comprise a central communal nest chamber with a series of radiating, branched foraging tunnels. Made up of some 20–30 individuals, the mole-rat society has a caste system much like that of such social insects as ants, honeybees, and the like. A single, dominant female leads the mole-rat society and does all the breeding during her tenure as "queen" of the social structure.

Spending her time mainly in the communal nest chamber with two or three other nonworkers (both males and females), the queen is tended by the worker caste (males and females) that makes up the bulk of the naked mole-rat society. The parallels with social insects are striking. Although they are not sterile, the worker females' ovaries remain quiescent by nature of some as yet unknown "suppressive factor" from the queen. If the queen is removed by an experimenter, then one of the larger worker females gradually becomes reproductively potent and takes over leadership of the colony.

The worker caste, like that of an ant society, cooperatively excavates and maintains the tunnels, forages for bulbs and roots, and tends the queen and her nonworking associates in the central chamber. This social structure is a far cry from the typical image of a single large dominant male mammal herding a harem of smaller intimidated and submissive females and their young. For obvious reasons, the little sand rats of Ethiopia, Somalia, and Kenya have helped to blur our usual "clear-cut" images of reality in the natural world.

however, may lead a solitary life for several years before they make their move to take over an established group. If they succeed in taking over a group, the genetic rewards are great, because they get to mate with the females of their new social unit.

The studies of Hans and Ute Klingel on the plains zebra, also known as Burchell's zebra, have provided much information about the **harem** type of social organization. The Klingels not only immobilized with drugs and marked captured zebras in various ways but also pioneered the now commonly used technique of photographing natural markings of zebras and later using a "mugbook" to recognize hundreds of individuals.

Plains zebra society is composed of family groups, bachelor groups, and solitary stallions. There are no solitary mares. Family groups consist of a single stallion and his harem of 1–6 mares and their foals. A family group may consist of as few as 2 zebras (a stallion and the first mare he has acquired) or as many as 16. The average size of such a group is 7. Bachelor groups range from 2 to 10 members, with a mean of 3.

Plains zebra family groups are characterized by their great stability, and their coherent nature is continually reinforced by strong personal bonds between mares and foals, stallion and mares, and stallion and foals. Stability is not disrupted when a harem loses its stallion; instead of breaking up and dispersing, the mares and their foals remain together until the group is taken over by a new harem master.

The mares in a harem maintain a strict rank order. This is most evident when the harem is on the march from one grazing area to another or to a water hole. The alpha mare takes the lead, followed by the beta mare, and so on down the line to the lowest-ranking mare at the very end of the progression. Foals rank just below their mother and follow directly behind her during the march. Lower ranking mares accord a foal standing near her mother the same rank as the mare. A mare new to the harem assumes the lowest rank.

The stallion is dominant to all harem members. Even though the alpha mare leads the progression during a march, the stallion controls group movements by moving from his position at the rear or to the side up to the lead mare and driving her in the desired direction. During the birth process the stallion stands guard over the delivering mare; other harem mares pay no attention. If danger threatens, the stallion and his harem form a semicircle and stare at the predator; if the predator draws closer, the mares and foals gallop away, with the stallion taking up the rear guard. The stallion threatens and even attacks the enemy by biting and kicking and is often successful in driving enemies from the vicinity of his harem.

When female zebras are about 15 months old they come into their first estrus. They indicate this not only by chemical means (pheromones) but also by a striking visual display: they stand spraddle-legged with their tails held high. They maintain this posture for long periods, and the display is irresistible to stallions in the nearby bachelor herds. The stallions immediately approach such a young mare and attempt to "cut her out" of the harem and start one of their own. It is then that prolonged and vigorous fights are seen between the bachelor stallions and the harem master. The harem master is usually overwhelmed by numbers and gives up from sheer exhaustion.

Young stallions are not initially driven from the harem by the lead stallion but rather leave peaceably of their own volition. The age at which they join a nearby bachelor herd is quite variable, but is typically before they are four years old. Although there is no rank order in the bachelor herd, an adult stallion does act as leader.

The diversity of social organizational types found within a group of closely related species is clearly shown by the primates. The most primitive social mode is called the **mother–infant family** and is exhibited by such distantly related species as mouse lemurs of Madagascar and orangutans of Borneo. This mode is characterized by a single adult male who has a large home range. This large area encompasses the smaller home ranges of several adult females and their offspring:

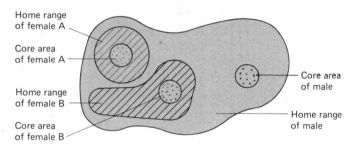

The closest social bonds occur between adult females and their offsping. The adult male has nothing to do with parental care, leadership, or defense. The male regularly patrols his large home range and checks the females to see if any are ready for mating. Young subadult males leave their mother's home range and occupy very small home ranges between the neighboring home ranges of adult males. Sometimes these young adult males lead a vagabond existence until they settle on the periphery of an established male's home range.

Although **monogamy** ("one mate") is not common among primates, there are several outstanding examples of this life style. Elegant white-handed gibbons, slender-bodied lesser apes that inhabit the canopy of the Malayan rainforest, live in small monogamous social units

The gibbon society is also a territorial one, with both parents active in vocal advertisement of their area. The male, however, is more active in actual defense of the territory when the social group engages in agonistic interactions with gibbons from neighboring groups. The marvelous locomotion of gibbons (brachiation) is nicely shown as the rival males swing through the canopy in their attempts to intimidate each other.

Another monogamous species is the titi monkey of northern South America. All-black in coloration, titis form gibbonlike parental family units made up of two permanently mated adults plus one or two of their offspring. Titis are territorial, and as with gibbons it is the male who is most active in area defense. The strong bonds between family members are shown whenever two titis sit side by side: they immediately entwine their tails (Fig. 22-20).

Multimale societies are common among many Old World monkey species (baboons, macaques, etc.). The society of mountain gorillas, with their patriarchal dominance, is one example. Olive baboons, common in the savannas of East Africa, live in a multimale society characterized by a different kind of male control.

Usually three fully adult males in an olive baboon troop form a **central coalition**. When a coalition member is threatened by another male in the troop, he solicits support from the other two coalition members by rapidly

FIGURE 22-21
Canine display, a common threat display of the olive baboon. (Redrawn from E. O. Wilson, 1975, *Sociobiology: the new synthesis*, Cambridge, Mass.: Belknap Press)

FIGURE 22-20
The strong bonds between titi monkey family members are shown here by the entwined tails. (Adapted with permission from J. Napier and P. Napier, 1967, *Handbook of living primates*. Copyright by Academic Press Inc. [London] Ltd.)

called **parental families.** A parental family consists of an adult male mated permanently with an adult female, plus one or two of their offspring, typically an infant and a juvenile. The adult male gibbon becomes intolerant of his older male offspring, and the latter normally leave the natal unit and establish their own small society at some distance.

glancing from his antagonist to his supporters. The rival male usually backs off when faced by a trio of threatening adults. A common threat display of this species is the canine display, a most intimidating visual signal (Fig. 22-21).

Young adult males seeking higher status attempt to single out a coalition member and harass him before aid is solicited. A vigorous young adult may be able to dominate coalition members on this one-to-one basis, but he will be forced to retreat in the face of the combined threats of the coalition. A very old coalition member, one whose canines have been worn down by years of a harsh grass diet, fails to support his fellow coalition members and eventually is replaced by one of the younger, vigorous status seekers.

When danger threatens the troop, coalition members rapidly move to the scene of the danger; their combined threats have been known to intimidate such powerful predators as cheetahs and leopards.

SUMMARY

Behavior is a movement in response to a stimulus. All behavior has a genetic basis and has evolved by natural selection. It ranges from simple reflex responses to highly original motions requiring reasoning. Many behavior patterns in lower animals appear completely stereotyped; the roles of castes of ants and bees are examples. These patterns were formerly considered instinctive, but that term has lost favor because learning has been observed even in very primitive organisms. Learning is behavior modified by experience.

Central to any kind of social organization is a system of communication. The commonly recognized communication modes are based on the sensory receptors of the receiving individual; visual, auditory, chemical, and tactile are standard modes.

Animals use space in a variety of ways, ranging from the simple individual distance through more complex systems such as dominance hierarchies, home ranges, and core areas. Much research has been conducted to study territorial behavior. The functions of territoriality are extremely diverse.

Equally diverse are the kinds of social organization. Among the many types recognized are single-male-dominated societies (patriarchies), single-female-dominated societies (matriarchies), harems, mother-infant families, and multimale societies. The primates, although closely related, show a broad spectrum of social organization types.

KEY WORDS

ethology	communication	rank order
behavior	visual communication	home range
instinct	agonistic behavior	core area
kinesis	graded signal	territory
taxis	vocalization	matriarchy
learning	semiochemical	patriarchy
conditioning	pheromone	parental family
imprinting	kairomone	coalition
trial-and-error learning	reasoning	
social behavior	tactile communication	

QUESTIONS FOR REVIEW AND DISCUSSION

1 Describe the conditioned reflex. Is it an example of learning?

2 Compare trial-and-error learning with reasoning. What is their common denominator?

3 How do predators exploit a functional communication system?

4 What are the commonly recognized modes of communication?

5 What are core areas? What relationship do they bear to home ranges? To territories?

6 What are some of the major functions of territoriality?

7 What is a matriarchy? Where are the adult males in such a society?

8 How does the society of black-tailed prairie dogs compare with that of mole-rats?

9 How does a young stallion plains zebra form a new harem?

10 What are some advantages and disadvantages of living in groups?

SUGGESTED READING

BUCK, J. and E. BUCK. 1976. Synchronous fireflies. *Sci. Amer.* 234(5):74–85. (Jungle fireflies flash in synchrony by means of a triggering mechanism in the brain that responds to flashes of neighboring fireflies and aids courtship in dense vegetation.)

CLUTTON-BROCK, T. 1982. The red deer of Rhum. *Nat. Hist.* 91(11):42–47. (Long-range studies on the Scottish island of Rhum lead to insight on the workings of natural selection.)

HESS, E. H. 1972. "Imprinting" in a natural laboratory. *Sci. Amer.* 227(2):24–31. (Interesting experimental approach to the study of imprinting.)

LORE, R., and K. FLANNELLY. 1977. Rat societies. *Sci. Amer.* 236(5):106–16. (The success of the Norway rat depends on a complex social structure and on its ability to live harmoniously without aggression and to communicate learning about poisoned food.)

LORENZ, K. A. 1952. *King Solomon's ring.* New York: Thomas Y. Crowell Co. (The most entrancing book about animal behavior ever written.)

MANNING, A. 1979. *An introduction to animal behavior*, 3rd ed. New York: Addison-Wesley Publishing Co. (Contains chapters on learning and motivation.)

McCOSKER, J. E. 1977. Flashlight fishes. *Sci. Amer.* 236(3):106–14. (Light organs of certain fish use symbiotic association with luminescent bacteria as an aid in communication.)

MENZEL, R., and J. ERBER. 1978. Learning and memory in bees. *Sci. Amer.* 239(1):102–10. (Bees learn and remember clues to food locations.)

MEYERRIECKS, A. J. 1972. *Man and birds: evolution and behavior.* New York: Bobbs-Merrill Co. (Exceptionally clear account of territoriality.)

PRESTWICH, G. D. 1983. The chemical defenses of termites. *Sci. Amer.* 249(2):78–87. (Highly specialized termite soldiers do their thing with an amazing array of defensive chemicals.)

RIBBANDS, R. 1955. The honeybee. *Sci. Amer.* 193(2):52–60. (Complexities of the social structure of the hive are discussed.)

SMITH, D. G. 1972. The red badge of rivalry. *Nat. Hist.* 86(3):44–51. (How a key male visual signal functions in territoriality.)

TINBERGEN, N. 1968. *The herring gull's world: a study of the social behavior of birds.* Garden City, N.Y.: Doubleday and Co. (A fascinating account of an experimental approach in the analysis of behavior.)

WILEY, R. H., JR. 1978. Lek mating system of the sage grouse. *Sci. Amer.* 238(5):114–25. (The lek mating behavior of America's largest grouse is discussed in terms of social behavior and evolution.)

WILSON, E. O. 1972. Animal communication. *Sci. Amer.* 227(3):53–60. (An array of communication modes is discussed.)

23

Evolution and Population Biology

Ernst Mayr, the well-known Harvard University professor of zoology, has called evolutionary theory the "organizing principle of biology." In the minds of most persons, evolution is synonymous with the name Charles Darwin, although Darwin (1809–1882) was not the first to conceive of biological evolution. In fact, his grandfather, Erasmus Darwin (1731–1802), believed in evolution and discussed it in lengthy and rather cryptic poems. Jean Baptiste de Lamarck (1744–1829), a French naturalist, proposed a theory of evolution in 1809 in which plants and animals were depicted as advancing from simple to more complex forms. Lamarck thought that characteristics of organisms were modified in response to use or disuse and that traits acquired in this way could be inherited. For example, the long necks of giraffes evolved because of constant neck stretching to browse in treetops (Fig. 23-2). In the Lamarckian view, necessity was the force of evolution. A more current example is the suggestion that humans eventually will lose their power of locomotion because of constant use of automobiles and other vehicles.

Currently, most biologists accept the modifications of Darwin's theory of evolution by **natural selection** as the most satisfactory explanation of organismal change and diversity. This theory, first proposed jointly by Charles Darwin and Alfred Wallace (1823–1913) (see Fig. 23-1) in 1858, states that environmental stresses do not mold organisms into new variations; rather, the environment selects from among variant individuals those best adapted for survival. (See Fig. 23-2 for a comparison of the **Darwinian** and **Lamarckian** theories of evolution.)

The **Darwin-Wallace theory of evolution** by natural selection was based on four major premises:

1 Organisms produce many offspring, and competition for food and space among them is so keen that only a few live long enough to mate and reproduce. Of these, the best adapted survive.

2 Among individuals of a species, **variations** in structure and function arise from time to time as random occurrences.

3 Such random variations sometimes may be beneficial, giving individuals with them a competitive advantage in life. The environment, in effect, selects such individuals for survival.

4 The result of natural selection is the evolution of new species.

In his book On the origin of species by natural selection, Darwin gave evidence for each of these premises, using exam-

FIGURE 23-1
A. Charles Darwin. **B.** Alfred Russel Wallace. (Drawing by James H. Hubbard)

ples both from nature and from plants and animals under domestication. The clarity with which he explained these main points and, perhaps more importantly, the painstaking way in which evidence was marshalled for each premise led to rapid and widespread acceptance of the theory.

Darwin was familiar with the work of the eighteenth century Scottish economist Robert Malthus, who observed that populations of organisms, including humans, have an almost infinite capacity for reproduction. Such a potential for "runaway" or **exponential** growth is illustrated in Fig. 23-3A. However, the finite nature of environmental resources acts as a check on excessive reproduction; rapidly expanding populations soon run out of such limiting factors as food, space, and so on. Exponential growth is rarely realized in nature; a more typical **growth curve** is that shown in Fig. 23-3B. In addition, Darwin knew that natural checks and balances restricted the growth of populations of organisms; even though an organism produces hundreds or thousands of progeny per generation, only a few live long enough to reproduce.

Some of the checks on population growth are indicated in Fig. 23-3C.

GENETIC VARIABILITY AND EVOLUTION OF SPECIES

Individual variation has a genetic basis, and the genes of each individual differ in some degree from those of every other individual. Although the members of a species have many genes in common, all possess various alleles that have been altered by past mutations. Alleles are responsible for some of the variability associated with individuality; other sources of genetic variability are the rearrangement of genes on chromosomes and the restructuring of chromosomes themselves. It is recognized, however, that not all variability in members of human populations or of other species is specifically genetic in nature and inherited. Even before birth the environment molds the individual in many ways. After birth, accidents produce disfigurements, diseases leave their residues of change. In addition, interactions between the individual and the physical environment and between individuals contribute to the making of a personality.

In the United States, records of average height of soldiers have been kept in every war since the Civil War. These records clearly indicate a height increase from generation to generation. Does this constitute evolution? Studies suggest that human height is 50–80 percent hereditary; the rest is the result of the environment. Nutrition, of course, is very important in determining human body size, and the gradual but significant increase in the average height of American men to a great degree is due to improved nutrition. The increase also reflects better medical care and better protection from disease through immunization and drugs. Nevertheless, relative height is controlled genetically. Environment can alter the phenotype of the individual but not the genotype.

Genetic variability in populations of humans or other organisms can be measured fairly accurately by using an analytical tool called **gel electrophoresis** (Fig. 23-4). In gel electrophoresis, proteins are placed on the surface of a column of an absorbing colloid called **polyacrylamide gel**. When an electric current is applied to the gel column and its absorbed proteins, individual proteins migrate at varying rates according to the net electrical charge of the particular protein. After a predetermined interval, the column is removed and stained so that each protein may be seen as a distinct band in the gel column.

The significance of gel electrophoresis in analyzing genetic variability depends on the nature of the gene. It

The inheritance of stretched necks gradually brought about the long neck of the modern giraffe

LAMARCK:
Use and disuse

SHORT-NECKED ANCESTORS

DARWIN:
Natural selection

ANCESTORS OF VARYING NECK LENGTHS

Natural selection of individuals better adapted to feeding among the treetops

PF

FIGURE 23-2
Lamarck's evolutionary theory compared with Darwin's.

will be remembered that each gene contains specifications for a unique protein, often an enzyme. A mutant gene will code for a protein that is different to a greater or lesser extent from the nonmutant protein. For example, the gene for sickle-cell anemia codes for a protein that differs from the normal protein by only one amino acid.

Comparatively minor changes in a protein can be detected by gel electrophoresis. By applying this method to a wide range of organisms, including fruitflies, humans, and plants, surprisingly consistent results have been found. A population of organisms has a common stock of genes, showing the same banding patterns by gel electrophoresis; in addition, they have variant genes

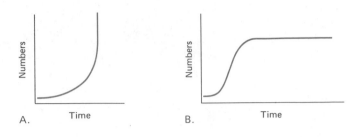

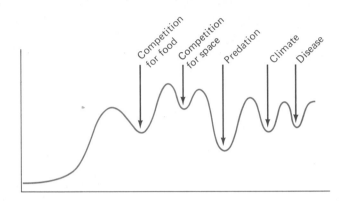

C.

FIGURE 23-3
A. Exponential growth curve. **B.** Logistic growth curve. **C.** Checks on population growth.

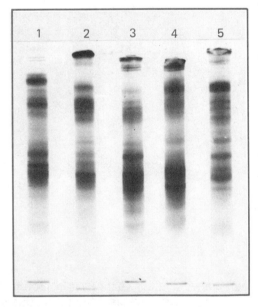

FIGURE 23-4
Separation of proteins in species and varieties of wheat by gel electrophoresis. Note the similarities and differences in the banding profiles of the five samples. (Photo from J. M. J. DeWet, D. H. Timothy, K. W. Hilu, and G. B. Fletcher, 1981, *Amer. J. Bot.* 68:269–76)

for 30–50 percent of the total. These variant genes within a population constitute some of the genetic variability upon which natural selection operates and, together with the nonvariant genes, are commonly called the **gene pool.**

Random Variation and Selection

Spontaneously occurring random variation is a keystone of evolution by natural selection. In his O*rigin of species*, Darwin drew heavily on his knowledge of variation in domestic animals and plants. He explained how human cultures had artificially selected individual plants and animals having certain desirable traits and through selective breeding had perpetuated them, eventually producing the many remarkably different strains and breeds of domesticated species (see Fig. 23-5). In a similar fashion, but over a much greater span of time, nature, acting on variation within species, has selected some forms for survival (e.g., lions and tigers) and others for extinction (e.g., saber-tooth cats).

Evolution as a Continuous Process

Darwin and Wallace thought that species in nature evolved by accumulation of many small changes which over thousands and millions of years added up to great differences among individuals of a common ancestry. New species arose when intermediate forms became extinct, leaving markedly different groups of individuals at the extremes of populations. Thus representatives of present-day species, genera, and families came to exist as distinct entitites. The Darwin-Wallace theory offers an explanation of the origin of closely related species such as lions and tigers, or coyotes and wolves, and also of broader relationships between, for example, birds and reptiles, or mosses and ferns. The present classification of organisms is based on an understanding of evolutionary relationships.

Cause of Variation

Darwin knew nothing of the sources of variation in species. Paradoxically, at about the same time as Darwin published O*rigin of species*, Gregor Mendel (1822–1884) published his theory, which ascribed variation to units of heredity (now termed **genes**) transmitted in definite ratios from parent to progeny. Mendel's theory, however, was not accepted during Darwin's lifetime, and there is no evidence that Darwin ever heard of it. In the early 1900s Mendel's contributions to science were "rediscovered," and there followed a reappraisal of evolution by natural selection.

A.

B.

C.

D.

FIGURE 23-5
Variation in pigeons resulting from human selection. **A.** The ancestral rock pigeon, the bird from which all domestic pigeons are descended. **B.** English carrier pigeon. **C.** Fantail pigeon. **D.** English pouter pigeon. (From C. R. Darwin, 1898, *The variation of animals and plants under domestication,* vol. 1, New York: Appleton and Co.)

For a time it appeared that Mendelian genetics were in conflict with Darwin's principle of random variation and evolution. Genes were not known to be mutable, and species seemed as fixed and invariable as before Darwin's day. Mendelian genetics explained the transmission of hereditary factors but did not explain the origin of genetic variations. This mystery was unraveled in the early decades of the twentieth century, principally by Thomas Hunt Morgan and his students at Columbia University. In studies of inheritance in the fruitfly, *Drosophila melanogaster*, it was found that gene changes, or **mutations,** occurred spontaneously from time to time. Although the frequency with which mutations occurred was very low, it was sufficiently high to account for genetic variation in fruitflies and other species.

The recognition that random origin of genetic variation could be explained by genetic mutations resulted in the wedding of Darwinian evolution and genetics. Modern evolution envisions natural selection as acting on variations that originate as gene mutations or rearrangements of genes and chromosomes. In addition, biologists have come to realize that variation is much more common in populations of species than Darwin had thought possible. All the genes of individuals in a species constitute collectively the gene pool upon which natural selection can act, and as a result·a species has come to be understood in terms of a group of organisms sharing a common pool of genes.

Competition for Energy

The Role of Natural Selection   Cells and organisms are energy-converting and energy-utilizing machines. The initial source of energy for the living world is sunlight (radiant energy). Radiant energy is converted by photosynthetic organisms into chemical energy (food), and nonphotosynthetic organisms obtain energy by consuming photosynthetic organisms directly or indirectly. Competition for energy exists at all levels of life and results in evolution of ecological strategies; no organism escapes the necessity of competing for life-giving energy. Hence an explanation for the prolific reproduction in all species is that genetic variation, coupled with the competition for energy, tends to ensure fitness and survival. Today evolutionary biologists measure fitness in terms of gene contributions to the next generation. Those members of a generation bearing optimal gene combinations survive because in a given situation they are the better competitors for energy.

Because all energy used by living organisms initially comes from sunlight, the supply would seem to be limitless and competition therefore unnecessary. However, utilization of solar energy depends on space and other factors. Green plants, the primary converters of solar energy into food, require exposure to light of sufficient intensity to drive the reactions of photosynthesis; thus plants compete with each other for optimal light exposure. In addition, there is competition for water and soil minerals. In essence, this means that those plants most effective in competing for space and nutrients on land or in water survive, whereas those that fall into shady places or are crowded onto infertile ground may not. Consumers of plants are in competition for food energy stored in plants. Here again, there is competition for energy among various groups of **herbivores** ("plant eaters"), and further competition among **carnivores** ("meat eaters"), which feed upon the herbivores.

Life Strategies and the Niche Concept   Imagine two groups of plants occupying an area in which the essentials for life are in optimal supply. All members are of uniform height and have about the same number and size of leaves. Suppose that in one group, genetic changes occur that produce plants with taller stems and broader leaves. Under this circumstance, the new plants might become more efficient competitors for light energy and crowd out the older types. In that case, they can be said to have evolved a new **life strategy** in the competition for environmental resources.* Many examples of life strategies among plants, animals, and other organisms are apparent, for instance, evolution of flight in insects, bats, and birds; complex social systems of ants and bees; plant–animal interdependencies exemplified by birds and fruits, and bees and flowers; and so on. The process of evolution, therefore, may be visualized in terms of evolution of new life strategies.

Organisms are readily associated with certain kinds of environments, that is, with characteristic **habitats.** If one wishes to study beavers, one looks for them in ponds and streams where adequate supplies of aspens and other trees are nearby to serve as food. Otters may be found in the same habitat as beavers. Nevertheless, although beavers and otters may share a common habitat, their life strategies differ. One is a herbivore, the other a carnivore, and they are not in direct competition for the same energy resource. The ecological concept of **niche** is applied to differentiate between habitats and life strategies. Eugene Odum, a well-known American ecologist has said that "the habitat is the organism's address; its niche is its profession."

Evolution can be viewed in terms of niche diversification. Random variation and natural selection produce

---

*Although the term "life strategy" is often used in biology, care must be taken not to imply purposeful behavior on the part of most organisms. Their ecological responses are the result of natural selection.

structural and functional specializations that culminate in exclusive survival strategies known as niches.

EVIDENCE FOR EVOLUTION

Many kinds of evidence led Lamarck, Darwin, Wallace, and succeeding generations of evolutionists to think in terms of adaptability and change in living organisms. Today evolution has become an even more compelling philosophy because of numerous modern developments, especially in genetics, ecology, and behavioral biology.

Divergent and Convergent Evolution

In Patagonia, the southernmost state of Argentina, Darwin observed in the fossil record extinct mammals which were unlike present-day mammals. This gave him an appreciation of the role of time in the record of past life. He later visited the Galápagos islands, off the coast of Ecuador, and was confronted with evidence of evolution in living species of tortoises, sea iguanas, and birds. Each of the Galápagos islands held distinctive species of animal life, yet all these species seemed to him to bear similarities that suggested a common origin.

WOODPECKERLIKE FINCH

CACTUS-EATING GROUND FINCH

WARBLERLIKE INSECT EATER

SEED-EATING GROUND FINCHES

FIGURE 23-6
Adaptive radiation exhibited by Darwin's finches. These birds are believed to have evolved from a common mainland ancestor.

In the Galápagos, Darwin studied a group of birds (now called Darwin's finches) that he thought demonstrated **divergent evolution** from a common ancestry particularly well (see Fig. 23-6). This type of diversification now is called **adaptive radiation** (evolution from a common origin by adaptations to specific niches). Although all of Darwin's finches resemble one another, differences especially in overall size, beak development, habitat preferences, and feeding behavior are clearly observable. Certain ground finches bear massive beaks adapted to feeding on large seeds; others with smaller beaks feed upon small seeds or insects. Several species have evolved the interesting behavior pattern of using thorns to probe insect tunnels in cacti for prey. Adaptive radiation is most often observed in isolated land areas such as an island or in the islands of an archipelago.

In another form of niche diversification, natural selection in unrelated species may result in "look-alikes" as well as "act-alikes," trends referred to as **convergent evolution**. For example, harsh desert environments have selected plants adapted for survival in arid environments; such plants, called **xerophytes**, commonly have thick, fleshy stems with spines, reduced or nonexistent leaves, and other features associated with the cactus form. In Fig. 23-7 convergent evolution is illustrated by three desert plants, each representing a different plant family but all exhibiting a similar xerophytic form.

Other examples of convergent evolution are seen in species that mimic the structure, coloration, or behavior of other species (Color plate 12C). The viceroy butterfly mimics the wing coloration of the monarch butterfly, a strategy that enables it to escape predation by

FIGURE 23-7
Convergent evolution exhibited by an African euphorb (right), an American cactus (left), and an African *Stapelia* (foreground). Each plant represents a different family, but all display characteristics thought to be adaptations to desert conditions.

NO GAP IN THIS FOSSIL RECORD!

The first documented account in the fossil record of evolution of a species has been announced by Peter Williamson of Harvard University. Examination of fossil mollusks in some 400 m (1300 ft) of sediments near Lake Turkana, Kenya, Africa, has shown changes in a number of species over a period of several million years. Some species changed very little and appear identical to living species. However, other species, subjected to past environmental stresses caused by changing lake levels, apparently underwent rather rapid evolution. The evidence points to periods of comparatively quick evolution (over 50,000 or so years) interspersed among long periods of evolutionary stasis. The question therefore is not whether evolution of species can occur, but rather whether it occurs continuously or in spurts.

birds. The monarch is distasteful to birds, and the viceroy benefits from this. The European cuckoo, a parasite that lays its eggs in nests of other birds, produces eggs of the same color and markings as the host species. Many other examples of mimicry could be cited.

Evolutionary Reduction

Among the many indications of evolution is a trend for **reduction,** and sometimes disappearance, of structures. Lamarck thought this to be due to disuse, but it more properly reflects the effect of natural selection on variant structures. Most organisms live by a tight energy budget, and "excess baggage" tends to be eliminated by variation and natural selection. In many species, evolutionary reduction is apparent, as in the vestigial leaves of cacti and other desert plants or in the loss of toes in the evolution of horses. The latter trend often is cited because the fossil lineage of modern horses so nicely illustrates evolutionary progression (23-8).

Age of the Earth

In pre-Darwinian times the earth was believed old only in human terms. James Ussher (1581–1656), a protestant archbishop in Ireland, basing his estimates on biblical chronologies, reckoned the earth was created in 4004 B.C. In the mid-1700s scientists extended the estimate to between 70,000 and 500,000 years. Darwin believed the earth to be considerably older, possibly some millions of years, despite the fact that his scientific contemporaries, including the eminent Victorian physicist Lord Kelvin (1824–1907), estimated its age in terms of hundreds of thousands of years. The latter figures were based on faulty calculations of the sun's age; no fuel supply then was known that would have sustained the sun's energy for hundreds of millions, to say nothing of billions, of years.

The current estimate of the earth's age is based on rates of decay of radioactive elements. A number of natural radioactive elements such as radium and uranium break down into other elements at a slow and steady

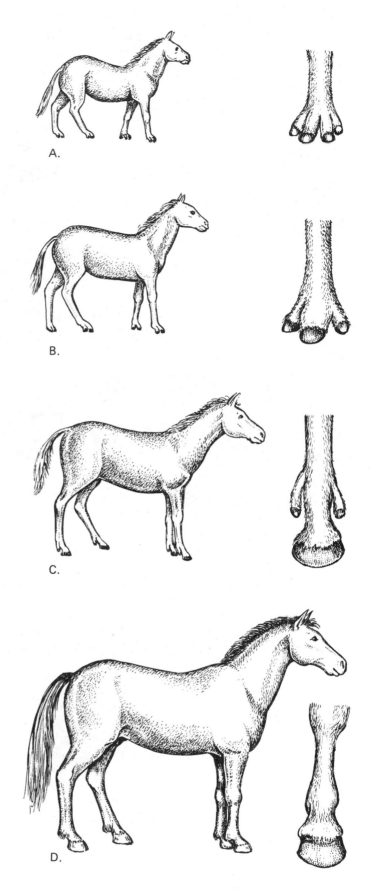

A.

B.

C.

D.

FIGURE 23-8
Evolution of horses. Depicted here are four of the many types that evolved in radiating lineages during Eocene and later times. Forefeet of each type are at the right. **A.** Eocene ancestral horse, *Hyracotherium* (also called *Eohippus*), a small mammal, was approximately 60 cm (2 ft) tall, having forefeet with four toes and hindfeet with three toes. **B.** *Miohippus,* an Oligocene horse only a bit larger than *Hyracotherium,* had three functional toes on each foot. **C.** *Merychippus,* a Miocene horse, was pony sized and three-toed, with the two lateral toes nonfunctional. **D.** Modern horse, *Equus* (Pleistocene to present), is one-toed but its front feet retain vestiges of lateral toes in the form of pairs of splinterlike bones not visible externally.

FIGURE 23-9
Grand Canyon. The rock layers exposed by the cutting action of the Colorado River are seen. The oldest rocks, near the bottom of the canyon, are over a billion years old. (Photo courtesy U.S. Geological Survey)

rate and hence can be used as "clocks." Uranium 238 ultimately is converted to lead 206, and the measurement of the rate of this radioactive decay and the ratio of uranium 238 to lead 206 in rocks today allows the age of the earth to be calculated. The figure derived from these data is about 4.5 billion years.

Precise information about the age of fossils is important in arranging evolutionary sequences. In Darwin's day such information could be obtained only from **stratigraphy,** in which sequences in rock strata were used to date geological deposits. Rock layering is particularly evident in chasms and canyons, where the thickness of each layer can be correlated with known deposition rates of sand and mud (Fig. 23-9). This permits calculations that give a chronology of the earth's history. Today radioactive decay rates are used to verify data obtained by calculations of sedimentation rates and stratigraphy. The resultant estimated ages of fossils are generally accurate.

The Fossil Record

A fossil may be defined as any remnant of past life, and may include nearly intact organisms; parts of organisms such as bones, teeth, and tree trunks; or impressions of organisms or parts of organisms such as scales and leaves or even footprints. Petrified excrement also is considered an important fossil, because it may reveal something of the feeding habits of animals.

Some of the more spectacular fossils are the few completely preserved plants and animals found from time to time. Ears of corn more than 5000 years old have been preserved in Mexican caves and have been used in tracing the ancestry of modern maize. Perfectly preserved ants have remained for 50 million years embedded in hardened resins (amber) of conifers. Perhaps the most unusual preservations are those of Pleistocene mammoths frozen in arctic tundra, their flesh so well preserved it has been eaten by hungry Siberians (see Fig. 23-10).

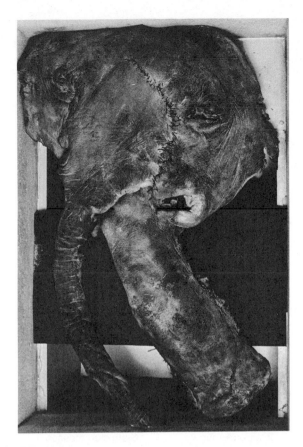

FIGURE 23-10
Remains of a young mammoth recovered from the Alaskan permafrost. Approximately 10,000 years old. (Photo from The Bettmann Archive, Inc.)

The most familiar fossils are petrified bones and tree trunks. Such petrifications are not actual bone or plant parts; rather, they represent a cell-by-cell replacement of the original material, usually with limestone or sandstone. Other fossils, called **compressions**, represent components of actual organisms compressed between layers of rock (Fig. 23-11).

The fossil record is far from complete, and many gaps remain. Yet it constitutes excellent evidence of the ongoing nature of evolution and the progression from simple forms of life to more complex ones. The most ancient rocks, on the order of 3 billion years old, contain only fossils of one-celled organisms. Next came simple worms and algae, followed by invertebrate animals and simple land plants, then reptiles, seed plants, and finally mammals and flowering plants (see Fig. 23-12 and Table 23-1, pp. 536-537).

NEO-DARWINISM

As noted earlier, Charles Darwin and Alfred Wallace recognized the importance of individual variation in evolution by natural selection (see for example Fig. 23-2). However, at that time nothing was known about the causes of individual variation, which now is known to be the result of gene mutation together with segregation and assortment of chromosomes and their genes during sexual reproduction. Application of genetic theory to Darwinian evolution sometimes is called **neo-Darwinism.**

Individual Gene Selection

Darwin believed that most individuals in natural populations were well adjusted to their environment, exemplifying the evolutionary concept of **fitness,** and that only an occasional individual was significantly different. He thought it was the occasional variant individual that natural selection acted upon. It is now known that all individuals in a population carry variant genes occurring in the gene pool of the population. Some of these genes are beneficial in a given situation, some may be of neutral value, and some are decidedly detrimental. Nevertheless, the lesson here is that all members of a population carry a sizable proportion of alleles, usually in the heterozygous condition.

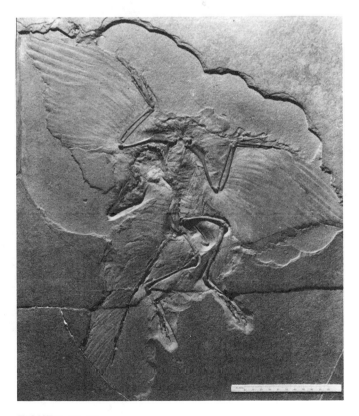

FIGURE 23-11
Compression of *Archaeopteryx*, an ancient bird, progenitor of modern birds. (Photo courtesy John Ostrom)

TABLE 23-1
Geological time scale and evolutionary sequence

Era	Years B.P.* (× million)	Period	Life Present

CENOZOIC	2.5	Quaternary
		  Recent
		  Pleistocene
			
			Human
			Mammoth

		Tertiary
		  Pliocene
		  Miocene
		  Oligocene
		  Eocene
		  Paleocene
			Hominids
			Hominoids
			Grazing mammals
			Birds
			Grasslands
			Insects

MESOZOIC	65 136 190	Cretaceous Jurassic Triassic	Flowers, insects Last dinosaurs Conifers, cycads Last seed ferns Dinosaurs, first birds First mammals First dinosaurs Seed ferns
PALEOZOIC	225 280 345 395 430 500	Permian Carboniferous Devonian Silurian Ordovician Cambrian	Reptiles expand Amphibians decline Last trilobites Seed ferns, ferns Giant club mosses, horsetails Seed ferns, tree ferns First reptiles Early vascular plants First amphibians First insects Bony fishes First land arthropods First land plants Algae Jawless fishes First vertebrates Trilobites, corals, sponges, echinoderms Algae
PROTEROZOIC	570 1,500 2,500	Precambrian	Primitive marine worms First eukaryotes Prokaryotes First prokaryotes

*B.P. = before present.

FIGURE 23-12
Evolutionary sequence of plants and animals.

Key to Organisms

A. Algae
B. Sponge
C. Cnidarians (jellyfish, anemones)
D. Annelid worms
E. Echinoderms
F. Flatworm
G. Bryozoan (upper) and brachiopods (lower)
H. Mollusks
I. Arthropods, trilobites, eurypterid, etc.
J. Jawless fish (left), early shark (right)
K. Crossopterygian fish
L. Lungfish
M. First vascular plants—psilophytes
N. Giant club mosses and horsetails
O. Amphibians
P. First reptiles
Q. Primitive mammal-like reptiles
R. Dinosaurs, pterosaurs, turtles, crododilians, first birds
S. Seed plants: seed ferns, cycads, conifers
T. Early mammals
U. Mammals
V. Australopithecines
W. *Homo erectus*
X. Neanderthal humans
Y. Modern humans
Z. Flowering plants

Darwin stressed natural selection at the level of the individual; today's evolutionary biologist emphasizes the effects of selection on an individual's genes and on the distribution of genes in a population. For example, the size of an animal is determined by a number of genes (polygenes), and individuals in a population often cluster about a mean size, with larger and smaller extremes occurring less frequently. Under some circumstances, selection may tend to accentuate the average size by eliminating the extremes, or selection can skew the size distribution toward one or another extreme if either contributes to individual fitness and survival.

Kin Selection

Natural selection operates at the level of the individual, or, in the view of many evolutionary biologists, at the level of the genes. In short, the reproductive strategy in evolutionary terms is "pass your genes on to the next generation." Paradoxically, there are a number of situations in which individuals do not pass their genes directly to the next generation but rather assist others in doing so. This seemingly altruistic behavior actually turns out to be quite selfish. The case of social insects with sterile worker castes is an outstanding example, one that long perplexed Charles Darwin. The situation was resolved with the concept of kin selection in the 1960s.

By assisting your close, reproductive relatives, with whom you share many genes, you are actually acting in your own, or your own genes', behalf. The closer the kin, the higher the probability that the related individuals share a number of genes in common. The concept of kin selection has helped to explain some seemingly suicidal behavior on the part of social animals. Why, for example, should individuals utter alarm calls that clearly bring the attention of a would-be predator to themselves? If the altruist succeeds in saving the lives of several close relatives, then it has in fact been successful in ensuring the passage of a high frequency of its own genes via the reproductive contribution of the survivors to the next generation.

Heterozygosity and Survival

One of the questions often asked about mutant genes, especially deleterious ones, is why they are not eliminated from the population by selection. There are several answers to this question. One is that although a mutant gene may be decidedly disadvantageous in the homozygous state, under some circumstances it may be advantageous if it exists in the heterozygous state. The sickle-cell anemia gene, which often is lethal in homozygotes but not usually debilitating in heterozygotes,

tends to be perpetuated in regions where malaria occurs, because the sickle-cell gene confers malaria resistance in the heterozygous state. Similarly, selection in favor of heterozygotes is thought to account for the perpetuation of apparently disadvantageous color genes of the English peppered moth, discussed later in the chapter.

Another reason that even deleterious genes persist is that genes tend to be perpetuated in the population by the random assortment inherent in the sexual processes. This is particularly evident in the case of recessive genes, which in some populations may be carried by a significant number of individuals (see box, p. 541). Finally, new mutations of genes occur from time to time and thus add mutant alleles to the gene pool.

Microevolution and Macroevolution

Could microevolution—relatively minor changes of genotypes within a species, resulting from gene mutation, sexual recombination, and natural selection—have produced the great differences, or macroevolution, observable in organisms of the past and present? Most evolutionists subscribe to the Darwinian viewpoint that microevolution, operating over immense spans of time, has resulted in macroevolution. However, some biologists propose that macroevolution may be saltational ("jumping"), proceeding by a series of relatively great steps rather than being a continual but gradual process.

The hypothesis of evolution by major jumps is based on the "sudden" appearance in the fossil record of markedly different species of plants and animals. Those biologists who support the Darwinian viewpoint suggest that the fossil record is spotty because intermediate forms have been lost through nonpreservation, or they remain undiscovered. According to other evolutionists, most species are very stable and persist on the average of 5–10 million years, then apparently give rise rather rapidly to substantially different, new forms. This kind of evolution, if indeed it occurs, would be a process in which species remain relatively unchanged for extended periods (thousands of years) during which new species emerge. This theory is called the punctuated equilibrium hypothesis. (see box, p. 532).

A genetic explanation of saltational evolution is that groups of genes may exist as complexes governing major characteristics of species. It has been suggested that such gene complexes, known as transposable elements, can shift positions on chromosomes, thus changing the assemblage of characteristics upon which species phenotypes are based. Complexes of genes also might be influenced by master or regulatory genes whose mutations could modify functions of an entire gene complex, thus producing relatively great changes in an organism. At this time, there is insufficient evidence to permit a clear choice between gradual and saltational evolution. If and when such evidence accumulates, one or the other theory may be accepted or modified, or possibly a synthesis of the two theories might be adopted.

POPULATION GENETICS

Hardy-Weinberg Law

Inherent in a Mendelian 3 : 1 phenotype ratio is a hidden ratio of genotypes consisting of one homozygous dominant to two heterozygous to one homozygous recessive. This is expressed as 1AA : 2Aa : 1aa. It should be noted, however, that the Mendelian ratio is the result of controlled matings between two individuals and not a summary of matings in a population, except in rare cases where homozygous parental genotypes occur in equal numbers. Usually, one parental genotype will predominate, producing genotypic and phenotypic ratios that differ from classical Mendelian ratios. However, if the ratios of parental genotypes in a population are known, the genotypes of succeeding generations as well as the frequencies of occurrence of specific genes in that population can be calculated.

In 1908 an English mathematician named G. H. Hardy and a German physician named W. Weinberg independently arrived at equivalent conclusions regarding gene frequencies in nature. Their concepts are embodied in the **Hardy-Weinberg law,** which states that in a large natural population, after one round of random mating, gene frequencies tend to remain constant from generation to generation, provided the effects of mutation, selection, and inbreeding are not taken into account. This means that in a population-genetics experiment in which the population is loaded heavily in favor of one genotype or another, population ratios of genotypes will stabilize at a specific ratio after one round of mating. For example, suppose that a large cage is stocked with 900 homozygous dominants (AA) and 100 homozygous recessives (aa). It can be shown mathematically that random mating in the population would produce an F_2 ratio of 81AA : 18Aa : 1aa. Theoretically, in all successive generations the genotype ratio would be stabilized in this same ratio. It should not change in the direction of either dominant or recessive homozygosity nor should it trend toward a Mendelian 1 : 2 : 1 ratio.

The significance of the Hardy-Weinberg law is that it provides a yardstick against which the forces of selection and other factors acting on genetic variability may be measured. In other words, it permits the calculation of population trends in nature. If it is known what the

genetic ratios would be if the genes were unaffected by natural selection, than any deviation from the expected ratio can be laid at the door of selection, mutation, migration, or other factors.

Hybrid Vigor and Survival

In Chapter 7 the subject of **hybrid vigor** was introduced in connection with plant and animal breeding. In the examples of hybrid maize and the mule, it was stated that heterozygosity conferred extra vigor in the hybrid organism. Hybrid vigor is operative also in the genetics of natural populations and in maintaining variability in the gene pool.

The relationship of sickle-cell anemia to malaria in human populations is a good example of the benefits of heterozygosity. As stated previously, the gene for sickle-cell anemia differs from the normal hemoglobin gene by just one nucleotide, and the proteins produced by these two genes differ from each other by only one amino acid. Yet this seemingly insignificant difference produces defective hemoglobin and misshapen red blood cells, leading to death in persons homozygous for the sickle-cell anemia allele. On the other hand, this deleterious gene confers immunity to malaria when, in the heterozygous condition, its action is partially masked by the normal allele.

In regions where the incidence of malaria is high, natural selection tends to favor individuals who are heterozygous for the sickle-cell anemia allele, while tending to screen out those who are homozygous either for normal hemoglobin (malaria-susceptible) or for sickle-cell hemoglobin (lethal condition). As a consequence, the sickle-cell gene is maintained in the gene pool at a relatively high level, simply because in the hybrid state it has survival value.

Human Blood Types

Human blood is classified according to a number of cross-reaction types, including A, B, AB, and O, as well as several others. Genes governing the expression of these various types are multiple alleles; that is, a series of blood-type genes exist but only a pair of them can be present in any one human individual. These are generally symbolized as I^A, I^B, and i, and in the series of types, A is $I^A I^A$ or I^Ai, B is $I^B I^B$ or I^Bi, AB is $I^A I^B$, and O is ii.

Interestingly, the frequencies of blood-type alleles are not the same in different human populations (Fig. 23-13). Among white Americans the frequency of the A (I^A) allele is 28 percent; that of the B (I^B) allele is 8 percent; and for O (i) the frequency is 64 percent. Among American Indians the frequencies are A (I^A) = 10 percent, B (I^B) = 0 percent, and O (i) = 90 percent. In pop-

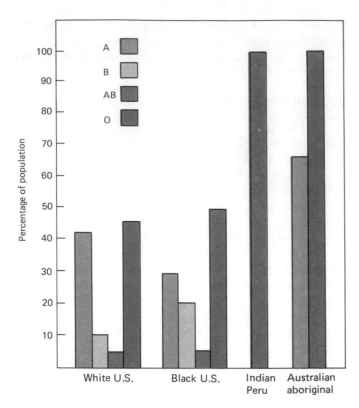

FIGURE 23-13
Distribution of A, B, AB, and O blood types in several human races.

ulations of Central American Indians, only type O is present. Other human races also show characteristic ratios of blood-type alleles, and it has been possible to trace both racial origins and racial migrations by comparing frequencies of these alleles.

It might seem from day-to-day experience that whether an individual has A, B, AB, or O blood type would make little difference, unless a mismatching of blood occurred during a transfusion. However, studies have shown that under some conditions, blood-type alleles may be related to survival. There is a slightly higher incidence of duodenal ulcer among persons of O-type blood, as well as increased susceptibility to the viruses of Asian flu. On the other hand, it has been suggested that the O-type allele confers resistance to syphilis, which originated in Central and South America. Resistance to prevalent diseases no doubt is responsible for some of the variations in blood types observed among human populations.

Speciation

About two million species of organisms have been described and named, and biologists estimate that about as many remain to be discovered. In fact, some biologists believe the number of undescribed species to be

two or three times larger. How did all these species come into being, and for that matter, what is a species?

A species is a group of individual organisms that share a great number of common features; the dog is an example. Even after thousands of years of domestication and human selection, dogs remain recognizable as a species. Not only do they look like one another, but even more importantly, they recognize themselves as belonging to the same species. They signify this awareness of their common identity by interbreeding with each other. The test of the validity of a species is that its members freely interbreed and produce offspring which themselves are fertile and freely interbreeding. A species may also be defined as a group whose members share a common gene pool.

Genetic Drift and the Founder Principle

Populations of animals and plants on certain remote islands are thought to have originated from chance migrants—a fertile female bird blown far astray by hurricane winds, similarly transported insects, plant spores and seeds, even refugee organisms clinging to pieces of driftwood. In some circumstances such occasional migrants may establish new populations of species in regions never before occupied by them. Isolated populations, descended from one or a few such **founder** individuals, demonstrate the **founder principle** of evolutionary biology (the finches of the Galápagos Islands are thought to exemplify this concept. Fig. 23-6).

In large populations, gene distribution ratios (gene frequencies) tend to remain relatively constant (refer to "The Hardy-Weinberg Law," this chapter, and "Calculating gene frequency," p. 541). Gene mutations and recombinations will not greatly alter gene frequencies and evolution will be slow. Conversely, in small populations mutations and gene recombinations may have more profound effects. Consider the hypothetical case of an albino among a handful of shipwreck survivors who populate an island. As time passes and numbers grow, the frequency of albinism will be markedly higher than in the original population on the mainland. When frequency distributions of all the genes in the new population are summed up, it is quite possible that they also will differ considerably from the original.

Evolution in small populations may in some cases occur more rapidly than in long-established large populations. One reason is that selection pressures in a new environment may be more relaxed (fewer predators, more open niches, lessened competition), so that individuals with novel gene combinations may survive. Additionally, a phenomenon known as **genetic drift** may alter gene frequencies more dramatically in small

CALCULATING GENE FREQUENCY

Mendel's laboratory experiments, and those of other researchers, proceeded on the basis of matings between equal numbers of parental types. For example, Mendel crossed equal numbers of tall (TT) peas and dwarf (tt) peas to obtain his hybrid (Tt) peas, and he was careful to ensure that in the next step only Tt peas were crossed with Tt peas. In this way he was able to obtain his F_2 ratio of 1TT : 2Tt : 1tt. It must be recognized, however, that the small world of the genetics laboratory does not reflect the situation in the real world. For example, the ratio of normally pigmented humans to those who are albino is not the 3 : 1 ratio of the genetics laboratory (1AA : 2Aa : 1aa), but is much greater. In human populations only about one in 20,000 individuals is an albino. The allele for albinism is therefore uncommon and is said to have a low **allele frequency.**

How many individuals carry the allele for albinism in the heterozygous state; that is, how many have the genotype Aa? The application of some simple arithmetic will provide the answer to this question. If, as seems to be the case, albino humans occur in a ratio of only one in 20,000 ($\frac{1}{20,000} = 0.00005$), then it can be assumed that normal pigmentation occurs in a ratio of 19,999 in 20,000 ($\frac{19,999}{20,000} = 0.99995$). What proportion of these 19,999 persons are heterozygous (Aa)? Substituting p for A and q for a, the Mendelian ratio 1AA : 2Aa : 1aa can be expressed algebraically as $p^2 + 2pq + q^2$. If the frequencies of the phenotypes in question differ from the Mendelian 1 : 2 : 1 ratio, they can be calculated from the known frequency of aa (q^2), which is 0.00005, as follows: a = $q = \sqrt{0.00005} = 0.007$. Then, because $p + q = 1$, A $(p) = 1 - 0.007 = 0.993$, and Aa $(2pq) = 2(0.993 \times 0.007) = 0.014$, or 280 per 19,999. This reduces to a frequency for Aa of about one in 70 persons. In the present population of the United States, there are about three million persons heterozygous with respect to albinism. This works out to about one in 70 Americans. If you are an unsuspecting Aa, your chance of meeting another Aa is about $\frac{1}{70} \times \frac{1}{70}$ or $\frac{1}{4900}$; the chance that a child of two Aa parents would be aa is $\frac{1}{4}$.

populations than in large ones. (Genetic drift is defined as the alteration of gene frequencies by chance rather than by selection.) Alleles disseminate more rapidly in a small population, but also may more readily be lost (by death of individuals carrying certain alleles or by their failure to breed). Genetic drift, which is seemingly unrelated to natural selection, is offered as an explanation for certain evolutionary trends that do not appear explicable on the basis of natural selection; for example, several puzzling discrepancies in human blood groupings, such as the high frequency of O and the absence of the B allele in Central American Indians.

One can see in such phenomena as the founder principle and genetic drift cause for concern over the fu-

ture of many threatened species of wild plants and animals, such as whooping cranes (now about 100 in number) and California condors (about 20 survivors). They must certainly have lost much of their diversity and, now that they are reared in captivity, genetic drift may in the long run profoundly change the nature of the species.

Isolating Mechanisms

We have pointed out that many biologists think that species ordinarily do not evolve by large evolutionary jumps. Rather, the process is a gradual one in which segments of a population develop characteristics that at first set them apart in minor ways, perhaps by slight differences in protective coloration or by food and habitat preferences. That being the case, probably an early step in the evolution of the species, also called **speciation,** is the formation of subpopulations, known as **races** (or sometimes **subspecies**), within a larger population. In order for these races to develop further into genuine species, according to the definitions of species given earlier, they must be isolated in such a way that their genes are not continually intermingled with those of the gene pool of the parent population. In nature this occurs through two kinds of **isolating mechanisms: geographic barriers** and **reproductive barriers.**

Geographic Barriers   The role of the geographic barrier as a species-isolating mechanism does not require much explanation. The barrier may be a geological change in a region that prevents the movement of individuals from one portion of a formerly cohesive region to another. The upthrusting of mountain ranges, the development of a river system, the flooding of a valley, and the rise of the oceans to flood intercontinental land bridges are examples. Geographic barriers may also be existing geographically isolated environments that become colonized by chance migrants from other populations. The Galápagos islands are an example, as are the Hawaiian Islands and a few other similarly isolated land masses. It is now generally held that many species evolved as the result of geographic separation, as exemplified by Darwin's finches of the Galápagos islands.

Geographic isolation on a large scale is illustrated by Australian marsupials. Primitive marsupials were isolated on the continent of Australia, probably by continental drift, in the early stages of mammalian evolution. In the course of some 50 million years they have evolved by adaptive radiation in such a way that they occupy many of the niches (carnivore, insectivore, browser, grazer, burrower, etc.) occupied in other continents by more advanced mammals. As a result, there are marsupial equivalents of such grazing and browsing animals as deer and antelope (kangaroos), large carnivores such as wolves (Tasmanian devil and Tasmanian wolf), rabbits (wallabies), monkeys and squirrels (phalangers and gliders), rodents (wombats), and others. Although now extinct, marsupial bears, saber-tooth cats, and lions once existed and may have preyed on 10-ft-tall kangaroos and wombatlike herbivores as large as rhinoceroses, also now extinct.

Geographic barriers physically prevent races from interbreeding with members of the original population. Further evolution of the isolated populations then may occur and in time result in the formation of new, distinct species. Among these evolutionary changes are differences in color, form, and behavior, as well as chromo-

somal differences that prevent successful interbreeding, should the isolated subpopulations be reunited with other remnants of the original population. Such reproductive barriers maintain the distinction of the new species and result in the reproductive isolation of species occupying the same area (see Fig. 23-14).

Reproductive Barriers   Differences in the timing of sexual activity may constitute a reproductive barrier. Flowers that open at night will tend to be isolated reproductively from those that bloom during the day, as will animals that are **nocturnal** (night active) from those that are **diurnal** (day active). Such species normally will not interbreed in nature, even though they are interfertile and may be induced to interbreed in the laboratory. Other behavioral isolating mechanisms acting to suppress interspecific breeding are known, as well as anatomical, physiological, and genetic barriers. Anatomical barriers range from size differences that make mating difficult to changes in reproductive organs that preclude copulation. Among physiological barriers are differences

1

Single population in a homogeneous environment

2

Differentiation of the environment and migration into new environments results in evolution of races (stippling)

3

Further migration produces geographic isolation of some races

4

Further changes in geographically isolated races produce reproductive differences that prevent interbreeding

5

If changes in the environment occur so that formerly isolated populations can exist together in the same region, the populations may remain distinct because of reproductive barriers and therefore can be considered different species

FIGURE 23-14
Stages in the evolution of a new species. (From G. L. Stebbins, *Processes of organic evolution*, 3rd ed., p. 132. © 1977. Adapted by permission of Prentice-Hall, Inc., Englewood Cliffs, N.J.)

in metabolic rates, in timing of reproductive cycles, and in phases of embryological growth. Genetic barriers include lethal genes, which reduce the reproductive potential, and chromosomal changes that interfere with meiosis and render offspring sterile.

Reproductive barriers often are related quite specifically to reproductive behaviors and characteristics. Some animals carry out elaborate and distinctive courtship rituals which are species-specific and tend to prevent interbreeding between species. Plumage differences in birds are a barrier to interbreeding, and differing songs also have an inhibiting effect. The eastern meadowlark and western meadowlark are very similar in appearance and occupy overlapping geographical ranges, but they are maintained as separate species by behavioral differences, including those of their songs. (Fig. 23-15).

In some cases the barriers may reflect the use of space. For example, if some species inhabit the tops of trees in a forest while similar and closely related species

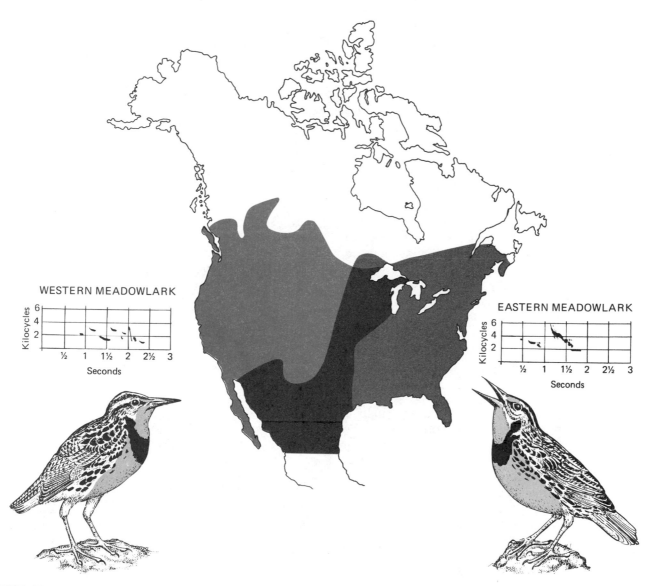

FIGURE 23-15
Meadowlarks. The eastern meadowlark differs slightly in plumage from the western meadowlark. Their songs differ significantly, that of the western meadowlark having an extra trill. These birds do not interbreed, although their populations overlap. (Redrawn from C. S. Robbins, B. Bruun, and H. S. Zim, 1966, *Birds of North America,* illustrated by Arthur Singer, © 1966 by Western Publishing Co., Inc., Racine, Wis. Reprinted by permission)

inhabit the lower shrub layers in the same region, it is unlikely that interbreeding will occur, simply because opportunities for sexual interaction are rare.

Although not a common occurrence, interbreeding between species can take place within the genus boundary, both in nature and in domestication. Natural hybrids between species of animals sometimes are observed (Color plate 2B), as are interspecific hybrids in plants. Oak trees are notorious for their production of hybrids, but interspecific hybridization in most genera is the exception rather than the rule. Barriers to interbreeding of species usually are present and keep the gene pools separated.

In many cases there may be no barrier per se to sexual reproduction between species or even genera, but the progeny turn out to be defective. This in itself constitutes an effective barrier to intermingling of genes between different species, because the hybrids cannot perpetuate themselves. The well-known cross between horses and donkeys is an example. The horse has a $2n$ chromosome number of 64; the donkey has 62 chromosomes per cell. Their offspring, the mule, with one $1n$ set of horse chromosomes and $1n$ set of donkey chromosomes, has 63 chromosomes per cell. The dissimilar sets of horse and donkey chromosomes cannot pair up prior to meiosis. As a consequence, chromosomes are misaligned in the first metaphase of meiosis and segregation of chromosomes is incomplete. Some gametes receive both members of a pair of chromosomes; other gametes receive neither. As a result, the sex cells are nonfunctional and the hybrid is sterile. This is an artificial relationship, because both parental species are domesticated animals. In nature, barriers of various kinds tend to prevent interbreeding, but when interbreeding does occur, genetic incompatibility often results.

Reproductive Isolation in Plants Flowers are reproductive organs of the most highly evolved group of plants, the angiosperms. Although most flowers have both male and female parts, crossing normally occurs between flowers of different plants of the same species, not between male and female parts of the same flower (refer to Chapters 16 and 20 for a review of flower structure).

Interspecific and intergeneric pollination and fertilization are prevented by a variety of mechanisms. In some species, flowers are open only during the morning hours; others are open in the afternoon or only at night or they may not be attractive to the same insect pollinator. Pollen and stigma interactions also may be involved, as in cases where foreign pollen is incompatible with the stigma of a flower. Walls of pollen grains contain proteins which act as recognition signals when coming in contact with the cells of the stigma of a flower of another species.

Within minutes of a pollen grain landing on a stigma, the proteins of the pollen grain wall begin to diffuse onto the surfaces of stigmatic cells. These surfaces have protein "recognition" sites and are capable of setting into motion rejection or acceptance reactions. If the reaction is one of rejection, blockage of the penetration of the pollen tube results, usually by secretion of a layer of polysaccharide (called **callose**) in the path of the pollen tube (see Fig. 23-16). That a recognition system is at work is demonstrated by experiments in which killed "friendly," same-species pollen is mixed with living "unfriendly," other-species pollen. The result is that the recognition system is "fooled" into accepting the other-species pollen.

Recognition factors in plant cell walls go by the general name **lectins,** and appear to have several functions in addition to pollen recognition. Lectins in root cell walls of legumes are thought to be responsible for establishing symbiotic relationships with nitrogen-fixing bacteria; others may be important in the formation of root grafts between trees of the same species, thus setting up water and nutrient transport linkages. Some extracted lectins are used in stimulating animal cells to divide in tissue culture. Bean seed extracts, called **phytohemoglutinens,** are used in cultures of human lymphocytes to induce mitotic divisions and are the basis of human chromosomal karyotyping (see Fig. 7-20).

Probably the most common barrier to crossing between plant species is a chromosomal difference that tends to prevent pairing in meiosis. In this kind of sys-

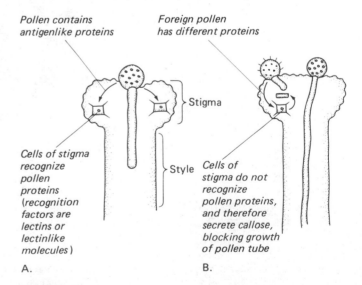

FIGURE 23-16
Pollen–stigma interactions. **A.** Pollination between members of the same species. **B.** Pollination between different species.

tem, pollination and fertilization occur and result in the formation of a new, hybrid plant. As in the case of the mule, the hybrid will have two unlike sets of chromosomes. Infertile pollen and abortive egg cells will be formed, and the plant will be sterile.

Plants, unlike animals, can propagate asexually by budding, by specialized underground stems and roots, or by special mechanisms in the flower that produce seeds asexually. Some races of hybrid plants are perpetuated indefinitely by asexual reproduction. When Karl Nageli, an eminent Austrian botanist, was approached by Gregor Mendel, seeking help with his genetics experiments using peas, the great Nageli suggested to Mendel that instead of working with peas, he turn his attention to de-

termining inheritance patterns in hawkweeds, a flowering plant in which Nageli himself was interested. It is now known that hawkweed reproduces only by asexual means, forming flowers and seeds without meiosis or genetic recombination. It would have been impossible for Mendel or anyone else to work out the genetics of this plant.

Polyploidy Rarely among animals but rather often in plants, multiples of the 1n chromosome number higher than 2n occur. This condition, in which three or more complete sets of chromosomes are present, is called **polyploidy,** and the condition may, in some circumstances, interfere with reproduction between individuals

FIGURE 23-17
Polyploidy in the artifical cross between the radish *(Raphanus)* and the cabbage *(Brassica).* Unfortunately, the new plant, *Raphanobrassica,* had none of the useful characteristics of the cabbage or the radish. A genome is the entire set of chromosomes inherited from one parent.

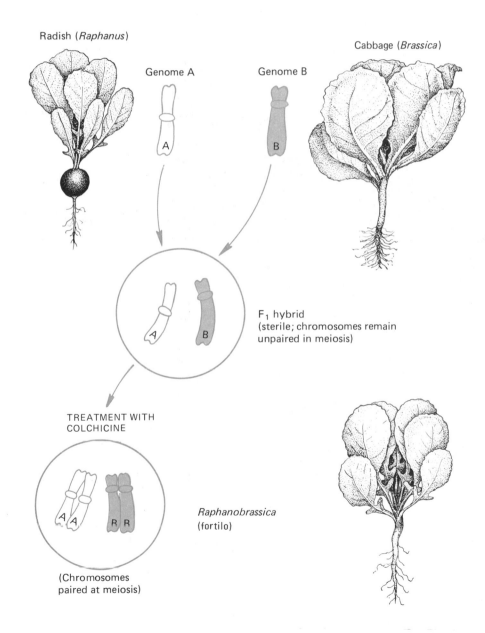

SOCIOBIOLOGY

In 1975 E. O. Wilson's book *Sociobiology* was published and became the foundation of a new biological subdiscipline—the study of the biological basis of animal behavior (including humans). Sociobiology profits from research on invertebrates, vertebrates, evolutionary theory, population biology, and classical ethology.

Since publication of Wilson's epochal book, scores of studies have appeared in the technical literature on such sociobiological research favorites as altruism, kin selection, inclusive fitness, reciprocal altruism, parental investment, parent–offspring conflict, selfish genes, and the like. Today evolutionary biologists believe that the name of the game is passing one's genes on to the next and subsequent generations. Adaptations have evolved to assist in promoting an organism's **fitness,** that is, the measure of that organism's genetic contribution to subsequent generations relative to the contributions of other genotypes in the population.

Altruism may be defined as self-destructive behavior performed for the benefit of others, and at first glance altruistic behavior seems to be maladaptive, promoting another's fitness over your own. The problem of explaining this apparent violation of the evolutionary rules has been largely solved by W. D. Hamilton's concept of **kin selection.** This theory states that if the altruist's behavior helps the survival of **close relatives,** who would have many genes in common with the altruist, then the behavior actually enhances the altruist's own **inclusive fitness,** which is its own genetic contribution **(personal** or **individual fitness)** to subsequent generations, along with the success of its relatives in passing on their genes (kin selection). Viewed in this light, the seeming altruist is in reality behaving in a selfish manner.

An example of altruistic behavior is shown by the wild turkeys of the Welder Wildlife Preserve in eastern Texas. Male siblings, when about six months old, leave their brood flock and join other juveniles during the winter. Next, the critical "decision" as to which males will mate with females is made, based on vigorous fighting. First, brothers fight until one becomes dominant (for life). These brotherhoods then fight until one pair of males becomes dominant over all other pairs in the flock. Finally, when flocks meet, interflock dominance is settled by displays and fighting. As a result, one male is dominant in any local population.

of a species. For example, if a polyploid group of plants arises in a population, a reproductive barrier may separate the polyploid group from the remainder of the population. Probably the most frequently encountered polyploid number is the **tetraploid** or $4n$ organism. Tetraploid plants commonly are larger than $2n$ plants; some well-known garden varieties, such as a large-flowered variety of snapdragon known in trade circles as "tetrasnaps," are $4n$. Tetraploid plants may be produced in the laboratory by treating $2n$ plants with the chemical **colchicine,** extracted from the autumn crocus. Colchicine breaks down the spindle in metaphase of mitosis so that chromosomes divide but remain in the same nucleus. When the plants are removed from the colchicine solution and rinsed off, normal mitosis resumes but cells remain $4n$.

Tetraploid plants, though large and productive, often are sterile and must be propagated vegetatively by cuttings and grafting. Sterility seems to result because the four pairs of chromosomes do not synapse normally in meiosis I and hence do not form normal gametes. Some naturally originated $4n$ plants are known, including

In February the males and females gather on the mating grounds. The different brotherhoods strut before the females in competitive displays. Brothers synchronize their behavior as they display, but when a female is ready to mate, the subordinate brother defers to his sib, the subordinate brotherhood defers to the more dominant one, and so on. In this way, mating is effected by very few males, but by altruistic behavior and kin selection, the nonmating males increase their own inclusive fitness.

Reciprocal altruism may also play a role in animal behavior. Here, the altruist's behavior enhances the fitness of a nonrelative, but the altruist has the expectation of reciprocal behavior at a later time. This is the "You scratch my back, I'll scratch yours" phenomenon. Of course, a high degree of sociality and the ability to tell individuals apart are necessary before such behavior can evolve.

Animal communication is an area in which ethologists have made great strides. The traditional ethological view is that the sender of a signal is conveying specific information to a receiver. For example, when a male bird sings on its territory at the start of the breeding season, it is believed that he is conveying precise information as to his species, sex, age, motivation, precise location, individual identity, and so forth. On the receipt of this information, the recipient would then react accordingly. If a male, it might flee from the area or be drawn into an agonistic encounter. If a female, it might draw nearer the signaler and engage in courtship activities leading to pair formation and mating. Recently, sociobiologists interested in behavioral ecology have contested this view and believe that much of animal communication is concerned with manipulation, coercion, deceit, and the like. They argue that it is to the advantage of the signaler to alter the behavior of receivers to its own benefit. This would make sense in view of the current emphasis in evolutionary theory on personal fitness.

Although sociobiology has made important contributions to the understanding of animal behavior, the field has not been without controversy. Wilson's book was heralded by most reviewers as a solid foundation for a new approach to attacking problems in behavior—except for the last chapter. Here, Wilson applies sociobiological thinking to human behavior. One result is that the "nature–nurture" controversy seems to have been resurrected. This age-old problem revolves around the question, What is more important in determining human behavior, genetic programming or cultural influence? Each side cites studies in support of its own views, and the controversy continues unabated.

wheat, cotton, and tobacco. These are fertile and are thought to be the result of crosses between different species, so that F_1 hybrids contain two distinct sets of chromosomes that do not synapse with each other. In such cases the original hybrids were sterile, but chromosomal doubling occurred in some plants so that two of each set of chromosomes were present. These chromosomes, then, were able to undergo synapsis. This chromosomal doubling effect was tested in a cross between the radish and the cabbage plant. When the chromosomes of the radish-cabbage hybrid were doubled

with colchicine, a true-breeding, fertile, new plant was produced (Fig. 23-17). In nature, chilling of the plant seems to have about the same effect as the laboratory use of colchicine, although it is less efficient.

EVOLUTION TODAY

Many of the examples of evolution of species cited in the foregoing paragraphs are case histories re-created by careful analysis of the characteristics of populations of organisms and their environments. In most instances, however, they are the culmination of natural evolutionary events. Therefore, biologists are always eager to find examples that provide clear evidence of ongoing speciation.

Few biologists are fortunate enough to observe processes of selection and evolution occurring so rapidly as to be apparent within one's lifetime. During the 1950s H. B. D. Kettlewell, an English biologist, examined collections of a tree moth named the peppered moth, and found that light-colored moths were much more numerous in collections made in the early 1800s than in more recent collections, in which dark-gray moths predominated. Why should this be? Kettlewell observed that the peppered moth hid during the day on tree trunks. In preindustrial England, tree trunks were lichen-covered and light gray in color. As the industrial age advanced, soot and other air pollutants killed the lichens and the tree trunks became dark. Kettlewell reasoned that light-phase moths were adapted for concealment on lichen-covered trees, whereas darker colored moths were adapted to hiding on the darker trees. He and his students did many experiments in which they proved that light-gray moths were highly subject to bird predation when released into woods near industrial areas. Likewise, dark-gray moths were subject to predation when released in woods in rural areas (see Fig. 23-18).

This natural selection of a race of moths adapted to life in polluted air is termed **industrial melanization.** Similar color shifts have been observed in other animals subjected to a change in the background coloration of their environment. Although industrial melanization of moths is evidence of natural selection in action, it is not considered to have resulted in creation of new moth

species during the immediate past. Rather, it is evidence of the shift of gene frequencies in gene pools of moth populations as a result of selection. In the decades since the end of World War II, England has made a conscientious effort to eliminate air pollution. As a result, lichens once again grow on forest trees, and an increase in light-gray moths has been noticed. Thus the trend toward an overwhelming preponderance of dark individuals seems to be reversing.

If the effects of air pollution were to be overcome completely, would the dark races of moths disappear? Probably not. Recent studies suggest an explanation for the presence of dark moths in natural environments where they tend to be more vulnerable to predation. Because the gene for melanin in the peppered moth is dominant, dark moths are of two types: homozygous (CC) individuals and heterozygotes (Cc). It has been suggested that hybrid vigor associated with the Cc genotype, in some as yet unknown way, may permit those moths to overcome the detrimental aspects of their lack of protective coloration.

An important lesson to be learned from industrial melanization is that the gene pool of any population contains alleles which permit exploitation of a changing environment. Even though in natural conditions alleles producing dark coloration render some moths vulnerable to predation, the presence of those alleles has enabled moths to adapt to the industrially altered environment.

SUMMARY

In his theory of evolution, Darwin emphasized the survival of individuals. The strongest male capable of defending a harem of females would transmit his characteristics to his offspring, whereas weaker males would be driven off, or perhaps killed, and would leave no progeny. Biologists still believe in survival of the individual, but the emphasis has shifted to the importance of gene survival. To study the fate of genes in subsequent generations, scientists must use evidence from population genetics.

An example often cited to illustrate the genetics of populations is the peppered moth, which has light and dark forms. In industrialized areas the dark form is more numerous because it is well camouflaged against the dark trunks of trees from which lichens have been removed by air pollution. In unpolluted regions the light form is less noticeable against the background of lichen-covered trunks and is the more abundant form. Such instances of natural selection lead scientists to believe that many similar changes occur in populations of plants and animals over time. If such populations of a species

FIGURE 23-18
Light and dark forms of the peppered moth, *Biston betularia,* as they might appear against the background of a lichen-covered tree trunk in an unpolluted environment.

become isolated by barriers to reproduction from other populations of the same species, so that interbreeding cannot occur, the genes of each population will diverge, and entirely new species eventually will evolve.

Barriers to interbreeding between populations and fragments of populations may be reproductive or geographic. In the one case, changes in the reproductive behavior of organisms may prevent their sexual reproduction with other, diverging populations and species. Geographical barriers also may isolate populations. Sub-

sequently, such isolated populations may go their own evolutionary way, and if reunited by removal of a barrier, or through migration or other means, the differences may be so great as to preclude interbreeding.

Modern evolutionary biologists stress not only inclusive fitness, which takes into account the genetic contributions of close relatives, but also a variety of kinds of selection, including stabilizing, directional, and disruptive selection.

KEY WORDS

Lamarckian evolution

Darwin-Wallace theory of evolution

natural selection

population growth curve

variation

genetic variability

neo-Darwinism

microevolution

macroevolution

punctuated equilibrium

Hardy-Weinberg law

hybrid vigor

allele frequency

speciation

isolating mechanism

genetic drift

founder principle

geographic barrier

reproductive barrier

reproductive isolation

polyploidy

industrial melanism

fitness

kin selection

QUESTIONS FOR REVIEW AND DISCUSSION

1 How has the concept of the gene pool modified Darwin's theory of evolution by natural selection?

2 What does gel electrophoresis of proteins show about the variability in the gene pools of organisms?

3 Explain why some genes which are deleterious in the homozygous state persist in the gene pools of populations.

4 The gene for human albinism occurs in the homozygous state (aa) in only about one in 20,000 individuals. Because it is of negative value to human survival, why has it not been eliminated completely?

5 Give a working definition of the term "species."

6 Outline, in brief, several stages in the evolution of new species out of a relatively homogeneous original population.

7 In what way may polyploidy in plants lead to new species?

8 Briefly discuss industrial melanization in relation to speciation.

9 What is meant by "inclusive fitness"?

10 Describe several different forms of natural selection, and discuss how each acts.

SUGGESTED READING

AYALA, F. J. 1978. The mechanisms of evolution. *Sci. Amer.* 239(3):56–69. (Discusses the molecular basis of variation, electrophoretic methods, and speciation.)

BISHOP, J. A., and L. M. COOK. 1975. Moths, melanism, and clean air. *Sci. Amer.* 232(1):90–99. (In this sequel to Kettlewell's studies, the effects of pollution abatement on moth populations are described.)

CALDER, N. 1974. *The life game.* New York: Viking Press. (Readable account of the exciting research issues in contemporary evolutionary biology.)

CLARK, B. 1975. The causes of biological diversity. *Sci. Amer.* 232(2):50–60. (The role of natural selection in maintaining diversity is discussed.)

DAWSON, T. J. 1977. Kangaroos. *Sci. Amer.* 237(2):78–89. (Discusses adaptation and radiation of kangaroos in Australia since their isolation by continental drift.)

KETTLEWELL, H. B. D. 1959. Darwin's missing evidence. *Sci. Amer.* 200(3):48–53. (Presents industrial melanism and its significance in the origin of species.)

Scientific American. 1978. Vol. 239, no. 1. (The entire issue is given over to discussion of various aspects of evolution.)

UNDERWOOD, J. H. 1979. *Human variation and human microevolution.* Englewood Cliffs, N.J.: Prentice-Hall. (Evolutionary concepts as they apply to humans.)

WILLS, C. 1970. Genetic load. *Sci. Amer.* 223(3):98–107. (The accumulation of mutations in the gene pool and the advantages and disadvantages of mutations are discussed.)

24

Life Strategies

For plants and animals, the name of the game is survival. The inherent driving force of evolution is the persistence with which organisms reproduce themselves. For all but humans, there appears to be no self-consciousness or awareness in this effort; it is the natural outcome of the almost limitless ability of DNA to store information and to produce new arrangements of itself (new genes) and for the cells to translate this information into new versions of organisms capable of meeting the challenges in nature. Humans tend to think in goal-oriented terms, and for many, these goals are visualized in a rather narrow, short-range manner. Success means a home, an expensive automobile, a good education for the children, a feeling of accomplishment, and the like. In the short range, this may be an accurate assessment of human values. In the long range, however, human goals must be judged in terms of their survival value—in terms of the survival of genes. Fundamentally, the survival of the human species requires competition among its members for a place in the sun, as is true of all other species. This is more than just a figure of speech, because directly or indirectly, the sun is the source of energy for all life, and survival and evolution of species can be boiled down to competition for energy. In analyzing evolutionary and ecological relationships, the energy relationship must always be kept in mind.

The total life situation is expressed by the concept of the **ecosystem**: interrelationships between populations of organisms and their environment. An ecosystem consists of food producers, food consumers, and the physical environment in which they live. The concept is a broad one, and an ecosystem may be as small as a laboratory culture or as large as the earth (sometimes called the "ecosphere"). It may be a natural habitat or an artificial model system such as a balanced aquarium. All ecosystems have the following in common:

1   An **energy flow**, including an initial input of energy, and food-web relationships, or nutrient cycling, among food-producer and food-consumer organisms.

2   A physical structure, which includes **essential, abiotic** (nonliving) components such as chemical elements, water, light, temperature, and climate.

3   **Biotic** (living) factors, including communities of interdependent living organisms and the organic substances produced by them (Fig. 24-1).

It is important to remember the conceptual nature of the ecosystem. The world is not composed of a specific number of ecosystems with sharp borders; the only requirement is that they satisfy the specifications just listed. Their dimensions depend on the choice of com-

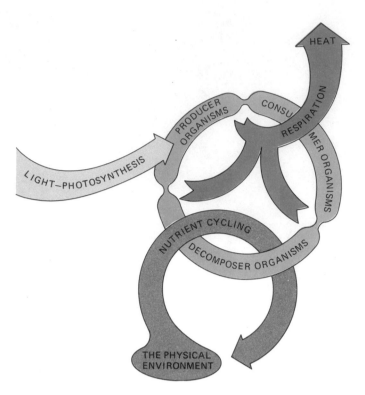

FIGURE 24-1
Ecosystem relationships.

ponents by which they are defined. For example, one biologist might choose to define a bog in northern Minnesota as an ecosystem, whereas another designates as an ecosystem the much larger coniferous forest of which the bog is but a part. In "real life," ecosystems may be **nested** (boxes within boxes), or **overlapping,** depending on which populations or organisms and which environments are being defined and measured. Nevertheless, despite the broadness of the ecosystem concept, there is general agreement about the dimensions, characteristics, and geographical boundaries of many natural ecosystems.

ENERGY FLOW IN ECOSYSTEMS

Various terms are used to describe the energy relationships between food producers and food consumers. A **food chain** is a sequence of organisms, beginning with plants (the food producers) and extending through various levels of food consumption. Each of these levels, called **trophic** ("feeding") **levels,** represents energy transfer steps. Usually, these energy transactions are not linear, but rather interwoven. Therefore, the term **food web** is a more appropriate description of relationships between trophic levels of the ecosystem (Fig. 24-2).

FIGURE 24-2
Food web.

We have learned in Chapter 1 that energy transfers are never accomplished without some energy loss—the first law of thermodynamics. The amount of energy available to organisms at each trophic level of an ecosystem, therefore, steadily diminishes as the number of energy exchanges increases. Much of the energy (80–90 percent) of the food available to the organisms constituting a trophic level is used up in their metabolism. Thus comparatively little of the input of food energy in one trophic level remains as food available to consumer organisms of the next trophic level. The result of this reduction in energy is a reduction in the numbers of organisms supportable in each trophic level, an outcome symbolized by a **pyramid of numbers** (Fig. 24-3). In this model of an ecosystem, producers are most numerous; consumers of the first level, or **primary consumers,** are less numerous, and predators, which may be called **secondary consumers,** are still less numerous.

There is another way of stating the pyramidal relationship of organisms in a food web. If the weights of all of the individuals in each trophic level are totaled, a **pyramid of biomass** is obtained (Fig. 24-4). This is a better measure of energy flow than numbers, which may conceal true trophic relationships. For example, two primary consumers with the same biomass, say rabbits and cattle, would be unequal in numbers. Many more rabbits than cattle could be supported by the same amount of vegetation.

Ecologists often characterize ecosystems by their productivity. One way to do this is to weigh all the or-

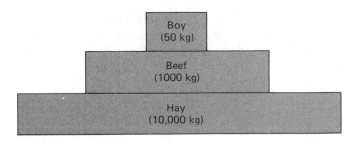

FIGURE 24-4
Pyramid of biomasses.

ganisms living within a representative area of the ecosystem to find the total biomas of the ecosystem. This figure, called the **standing biomass,** is useful in describing an ecosystem but does not always reveal energy flow or productivity of the system.

By analogy, measurement of the total capital of a human society might not always provide a true understanding of the productivity of the society. In one instance, the wealth could be tied up in bank accounts and be rather nonproductive; in another, wealth could be cycling rapidly in business transactions and be highly productive. In a forest community, energy could be tied up in old trees and productivity could be low, even though standing biomass is high. Therefore, ecologists prefer to measure energy relationships in ecosystems by yearly

FIGURE 24-3
Pyramid of numbers.

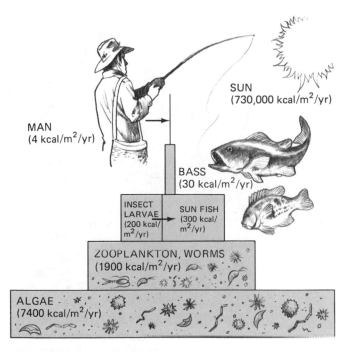

FIGURE 24-5
Energy pyramid for a Georgia fish pond. (Adapted from *Fundamentals of ecology*, 3rd ed., by E. P. Odum. Copyright © 1971 by W. B. Saunders Company. Reprinted by permission of W. B. Saunders, CBS College Publishing)

productivity. One way to do this is to calculate the amount of solar energy available to the system and then see how effectively this energy is used. In Fig. 24-5, the energy flow in the food web of a Georgia fish pond is illustrated. On a larger scale, such energy calculations give a comparative picture of the productivity in major world ecosystems.

ABIOTIC COMPONENTS OF ECOSYSTEMS

Essential Elements and Their Recycling

As energy flows through a food web, it is involved in chemical reactions at each trophic level, whereby atoms are combined, separated, and recombined. Only about 20 of the more than 100 elements present in the earth are used by organisms in these processes, and they are used over and over again as organisms are born, live, die, and decompose. Thus it is possible to follow the fates of the essential elements through biogeochemical cycles involving food webs and energy-flow relation-

ships. Some of the more important cycles are covered here.

The Water Cycle   A major characteristic of environmental water is its transitory nature. It is very mobile, as is illustrated in the **water** and carbon **cycle** shown in Fig. 24-6. In the environment, the existence and character of ecosystems depend to a great degree on the amount and availability of water.

Water makes up most (60–95 percent) of the mass of the living cell and, together with carbon dioxide, is the raw material of photosynthesis. When foods are oxidized in cellular respiration, water is given off as a by-product, along with carbon dioxide. Water also is the solvent in which nutrients are dissolved and the stream in which all molecular traffic moves into, within, and out of cells.

The Carbon Cycle   The **carbon cycle** connects the two fundamental life processes of photosynthesis and respiration. Carbon dioxide factors into both processes—as a raw material for photosynthesis and as a by-product of respiration.

AUTOTROPHIC ANIMALS?

In 1977 the submarine *Alvin* explored the Galápagos rift, a deep-ocean trench characterized by volcanic fissures and lava flows. At depths of over 2500 m (8000 ft), *Alvin's* oceanographers were astonished to discover large clams, mussels, and giant annelid tubeworms, hitherto unknown. Sunlight does not penetrate to such depths, and food therefore should be very scarce. What energy sources, then, do these organisms tap in order to live and reproduce?

Samples of warm water flowing out of cracks and vents in the lava were found to be rich in hydrogen sulfide, a compound certain chemosynthetic bacteria utilize to drive reactions leading to the conversion of carbon dioxide into carbohydrates. These bacteria are the primary food supply of filter-feeding animals such as clams and mussels. A puzzle, however, was the nature of energy conversion processes in the giant tubeworms, which live near warm-water vents and can grow to nearly 3 m (10 ft) in length. These organisms have no feeding organs, nor do they appear to absorb soluble food from their surroundings. Instead, they contain large numbers of symbiotic, carbon-fixing bacteria. These bacteria use hydrogen sulfide to fix carbon dioxide, and in doing so provide nutrients for their hosts. Thus, although the giant tubeworms are themselves not autotrophic organisms, the combination of bacterium and worm functions as if it were completely autotrophic.

FIGURE 24-6
Water and carbon cycle. Not shown are the underground water table and deep storage deposits, which are not directly available to plants.

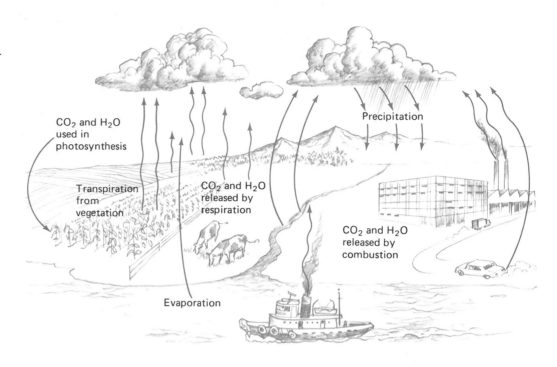

Carbon dioxide constitutes only about 0.03 percent of the present atmosphere, although the burning of fuels in automobiles and industry is gradually increasing the concentration. This increase is not detrimental to plant life, because, as it has been learned, photosynthesis becomes more efficient as carbon dioxide concentrations become greater. However, there is concern among environmentalists that increasing concentrations of atmospheric carbon dioxide may change climates by the **greenhouse effect,** in which excess carbon dioxide acts as a blanket of insulation, preventing the radiation of heat into space. Thus far, the greenhouse effect appears to have had little effect, but it may be a factor to be reckoned with if consumption of fuels continues at the present rate.

The Nitrogen Cycle The **nitrogen cycle** involves organisms of all trophic levels. Plants take up nitrates and ammonia from the soil and use them to make amino groups, which in turn are used to make amino acids and proteins. Plants, unfortunately, are not able to use nitrogen gas, which constitutes 90 percent of the atmosphere, but bacteria and cyanobacteria, living in the soil and sometimes in specialized root nodules of certain plants, can fix nitrogen by taking up gaseous nitrogen and converting it to ammonia. Other bacteria active in the nitrogen cycle convert ammonia into nitrite, nitrate, nitrous oxide, and nitrogen gas, thus completing the cycling of nitrogen in the environment (Fig. 24-7).

Some atmospheric nitrogen is converted into nitrate by lightning and by industrial processes. Other sources of nitrogen for plants are urea and other nitrogenous compounds resulting from the decomposition of animal wastes and plant and animal proteins. The nitrogen cycle is extremely important in the maintenance of soil fertility and plant productivity, which tend to diminish under conditions of intensive agriculture.

The Phosphorus Cycle Phosphorus atoms form the backbone of DNA, RNA, and important nucleotides such as ATP. Phosphorus is one of the three important fertilizer elements often in short supply in agricultural soils. Although frequently deficient in fields and gardens, phosphorus, together with nitrogen, is a major contributor to pollution of lakes and streams. Sewage adds large amounts of nitrogen and phosphorus to the environment, much of the latter coming from the degradation of phosphorus-containing detergents. Water running off fields and lawns fertilized with phosphates also is a source of pollution. The problem is a serious one, because excesses of such fertilizers in water bodies lead to uncontrolled growth of algae and bacteria. This causes **eutrophication,** an accelerated aging of lakes, in which the water becomes charged with organic and inorganic molecules, the bottom fills in with sediments, and oxygen levels decline due to the greatly increased bacterial action. It is estimated that Lake Erie, a glacial lake about 10,000 years old, has been aged about 15,000 years because of pollution.

Phosphorus becomes available in soils by the decomposition of phosphate rocks and by the breakdown of organic material. The dung of bats and seabirds,

FIGURE 24-7
Nitrogen cycle.

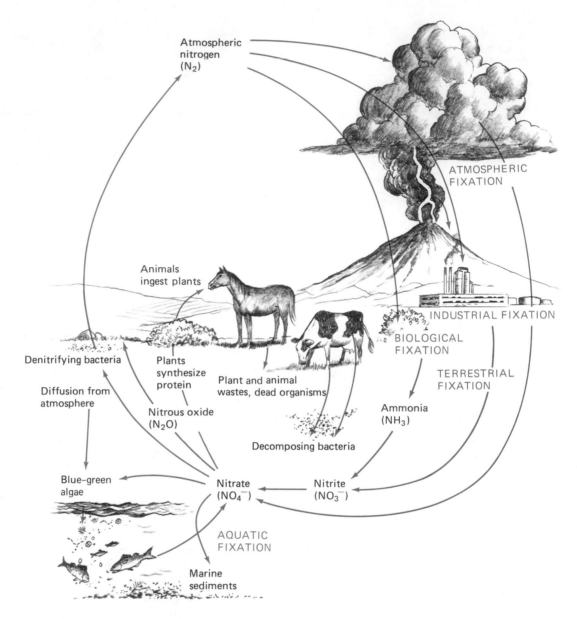

known as **guano,** is a particularly rich source of phosphate, but it is only locally abundant. Much of the phosphorus currently used in fertilizers the world over is obtained from phosphate rock mines. Florida is one of the leading producers of this important commodity. The world supply of available phosphorus is constantly diminishing because of the loss of phosphorus in sewage pollutants, which eventually find their way to the depths of the sea. There, the phosphorus remains unavailable. The conservation of phosphorus is a matter of urgent importance.

Light and Temperature

Light and temperature directly determine the nature of ecosystems. Not only are there specific optimal light and

temperature ranges within which plants and animals flourish, but short- and long-term fluctuations in light and temperature also regulate the lives of organisms. Both plants and animals respond to periodic changes in light and temperature, including day and night lengths and daily and seasonal variation in daylight and temperature. The flowering of plants and the reproduction of animals may be dependent on photoperiod, and every species of life has rather rigidly defined temperature requirements.

Climate

A particular combination of light intensities, photoperiods, temperatures, atmospheric moisture, and winds is referred to as a **climate.** Within broad regions of prevailing world climates are many local subclimates produced by geographical features such as high mountains and large bodies of water. For example, the climate atop a mountain may be severe enough to produce an ecosystem resembling that of the Arctic. This altitudinal effect is also latitude dependent. In the tropics a zone of arcticlike vegetation is found only on top of the highest mountains (above 4000 m [13,000 ft]), whereas in temperature regions of the United States a similar zone is found at lower elevations (about 1500 m [5000 ft]).

Regions of very limited extent often have highly localized climates and, as a result, may have their own distinctive ecosystems. Such a climate, called a **microclimate**, may be found on the side of a cliff, in a damp ravine, or even on the shady side of a large tree.

BIOTIC FACTORS IN ECOSYSTEMS

Ecosystems can be described in terms of ecological **communities**. A community is composed of the interacting populations of the various species of organisms occupying a common habitat. Accordingly, one can speak of a desert community, a forest community, or a coral reef community. Usually, an ecological community is more localized than the major ecosystems called **biomes** (which are described in Chapter 25). Biomes are widespread, a few to a continent, and communities are functional subunits of biomes, sometimes quite limited in scope (e.g., a pond or a mountaintop).

Two important characteristics of communities are a developmental history and a tendency to be stable at maturity. The developmental history is generally known as **succession**, and the stable state is usually referred to as the **climax community**. Both of these terms represent conceptual ideals, but are useful yardsticks by which community structure can be measured. The stable community tends to remain in a condition called the **steady state** or **homeostasis**. In previous discussions, the term "homeostasis" was used to describe steady-state processes maintained by feedback systems operating within organisms. For example, blood-sugar levels tend to remain steady because of feedback regulation by the hormones glucagon and insulin. In the ecological community, steady-state conditions are maintained by interactions between plant, animal, protistan, fungus, and moneran populations. Among these are producers at all levels of life, from green plants through various algae and photosynthetic protistans to cyanobacteria and other autotrophic bacteria. The consumers range through all levels of animal life (macroconsumers) and through fungi, heterotrophic protisans, and bacteria (microconsumers). Prey–predator interactions are well-known feedback systems in which numbers of both prey and predators are controlled by the relative abundance of each.

Population Phenomena

One of the better known population feedback interactions is that occurring in the Arctic between mouselike rodents called lemmings, their food supplies, and predators. In the American Arctic, lemming populations increase every three or four years, only to decline dramatically within a single year. As lemming populations increase, so do those of predators such as snowy owls (Fig. 24-8). The rhythmic declines of lemming populations may be due, in part, to increased predation, but stress and declining food supplies probably are the major factors. In northern Europe (but not in North Amer-

FIGURE 24-8
Population cycles of arctic lemmings and snowy owls.

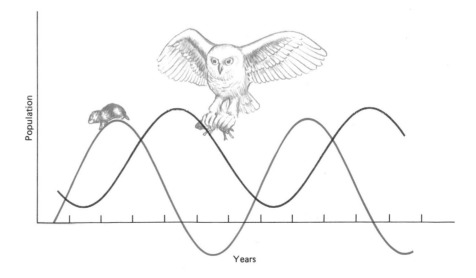

Population

Years

ica) the lemmings become so numerous that they migrate in hordes, moving to different parts of the Arctic, sometimes pushing southward, and even swimming out into the sea to their deaths. When lemming populations decline, there follows a southward migration of starving snowy owls. Such migrations occur every three or four years, and in the United States these arctic creatures are occasionally seen as far south as Georgia.

Limiting Factors Populations of many animal species undergo periodic fluctuations. Most of them, unlike lemming populations, increase in 7- to 10-year cycles. Attempts have been made to correlate these population cycles with sunspot occurrence, but cause-and-effect relationships of this kind have proven elusive. Population cycles having extreme highs and lows are, for the most part, characteristic of simple ecosystems of deserts and the Arctic, where there are only a few interacting species. Complex ecosystems with many interacting species seldom show major population fluctuations unless they are disturbed by some natural or man-made catastrophe.

The population controls in communities include space, food, rainfall, temperature, and other organisms, and are called **limiting factors.** Modification or elimination of even one limiting factor may cause a population to increase so dramatically that it outstrips its space requirements and food supply, thereby creating a **population explosion.**

Population explosions in nature are common. Usually, they occur when a limiting factor is changed or removed. Years ago, in a well-known bird sanctuary in the Midwest, the manager decided to trap skunks, because he believed they were consuming the eggs of nesting ducks and thus limiting the duck population. For a time this program succeeded, and the duck population increased, but later an even higher mortality of juvenile ducks was noted. The culprits proved to be snapping turtles, who seized and ate the ducklings as they swam behind their mothers. The demise of the skunk population had led to localized population explosions of both ducks and snapping turtles, of the latter because the skunks also prey on snapping turtle eggs. The resultant increase in turtle numbers then imposed a new limiting factor on the duck population (Fig. 24-9).

Exotic Species An introduced species of plant or animal is called an **exotic species.** The number of exotic species in North America is huge. Most of these species have been introduced by humans; a few have drifted in by quirks of nature (such as storms) or by chance migrations. Some of the more common exotic species are the English sparrow, the starling, the ring-necked pheasant, the wild boar, the walking catfish, the giant land snail, the carp, the Japanese beetle, the fire ant, the Brazilian bee, the water hyacinth (Fig. 24-10), and the Australian paperbark tree. There are many others. It should be noted that most of the species just named are now considered to be pests, although nearly all were believed to be desirable when first introduced.

The usual result of a plant or animal introduction is that the animal either fails to become established or it does very well indeed. A case in point is the English or house sparrow, which was introduced in the United States in the middle of the nineteenth century and quickly became abundant. A reason given for this population expansion is that the birds were accustomed to following horses, eating grain spilled from feed bags and undigested seeds in horse droppings. In the horse-and-buggy days, such food was plentiful. Since then, the sparrows have found a place in the environment, and their numbers have become stabilized. A more technical explanation of the success of the English sparrow is that it filled a **niche,** that is, a position in the energy flow pattern of the ecosystem where competition was minimal.

FIGURE 24-9
Relationship between skunks and turtles in limiting populations of ducks in a sanctuary.

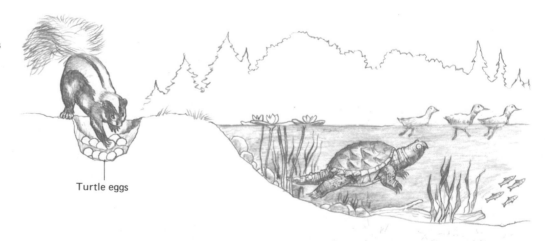

Turtle eggs

FIGURE 24-10
Water hyacinth, a major pest in tropical and subtropical waterways the world over, was introduced into Florida about the turn of the century. Millions of dollars are spent yearly on control measures.

Competition and the Niche Concept

Competition governs the structure of communities. It is most keen between individuals of the same species, who make the same demands on the elements of the ecosystem. Young seedlings compete against older trees for space and light, usually unsuccessfully. Only when a mature tree falls or is cut down do the young trees flourish. Competition also may exist between species for the same position in a food web, but usually in a stable community the roles of each species are different, and their requirements tend not to overlap. In other words, it may be said that each species has found its own niche. "Niche" is a useful term for the way of life of a particular species. In ecosystems, the niches tend to be compatible with one another, not competitive. Thus, in a human community, doctor may compete against doctor in the "doctor niche," but not against dentists or druggists in their niches. A grocer competes with another in the "grocery niche," but not with clothiers in their niche.

An interesting example of niches in a nonhuman community is found on some barren, rocky outcroppings in the Atlantic called the Rocks of St. Paul. The rocks are inhabited principally by seabirds. However, in this barren habitat exists a species of moth that lives by feeding on the shed feathers of the seabirds and a spider that lives on the moth (Fig. 24-11).

In an ecosystem—human or otherwise—the niche is the "profession" of the species. In the discussion on the genetics of populations presented in Chapter 23, it was noted that competition within a population tends to support evolutionary diversification. If the environment

changes, competition coupled with genetic diversity (heterozygosity of the gene pool) results in the filling of new niches. In this way, new subspecies—and eventually species—are created. One point must be made clear, however. A niche does not exist until an organism evolves a new role for survival. It is the organism and its life strategy that defines the niche, not vice versa. As an example, did the niche for automobile mechanics always exist, or did it only come into existence when competition forced a buggy manufacturer to invent the automobile? The answer should be obvious.

LIFE-STRATEGY ADAPTATIONS

The concept of niche involves another ecological concept, that of **life strategy**. As communities increase in complexity, niches increase in number, and the survival strategies of organisms increase correspondingly.

Life strategies can be fairly simple or very complex. A few of the more interesting life strategies of plants and animals are considered in the sections that follow.

r and K Selection

So-called ecological strategies have a genetic basis and are, therefore, subject to the pressures of natural selection. We can identify two broad strategies for the exploitation of the environment by organisms: rapid opportunism and long-term stability. These two strategies sometimes are referred to, respectively, as **r-strategies** and **K-strategies** (after equations symbolizing interactions between organisms and various kinds of environmental factors—**r-selection** and **K-selection**).

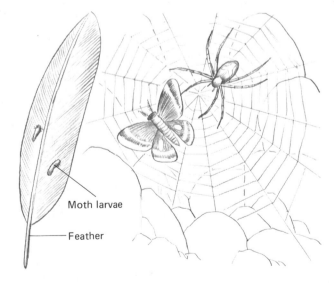

Moth larvae

Feather

FIGURE 24-11
Food web of birds, moths, and spiders on the Rocks of St. Paul.

CARNIVOROUS PLANTS

Contrary to science fiction, no plants on earth are capable of capturing and devouring large animals. However, a number of unrelated species of flowering plants independently have evolved methods of capturing and digesting small insects and crustaceans. The more spectacular of these use devices that actively trap prey. The Venus's flytrap, *Dionaea muscipula,* found in certain habitats on the coastal plains of North and South Carolina, is one of these (see Fig. 20-44C). A fly or spider touching two of the three hairs on the surface of the flytrap's leaf causes the "jaws" of the trap to snap shut. Then digestive enzymes secreted by glands in the leaf's surface are released, and these digest proteins in the body of the prey (Fig. 1). The resulting solution of amino acids is absorbed by the plant as a nutritional supplement. In the case of one insectivorous plant, the sundew, *Drosera rotundifolia,* up to 40 percent of the requirement for the amino acid arginine is obtained from trapped prey.

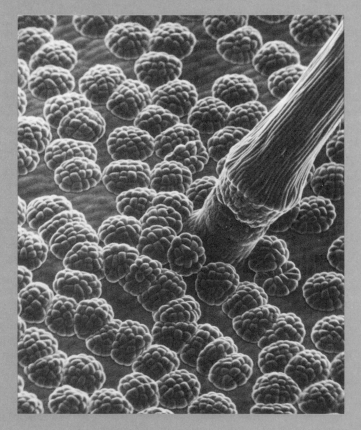

FIGURE 1
Scanning electron micrograph of the surface of a leaf of *Dionaea muscipula,* the Venus's flytrap, showing enzyme-secreting glands (multicellular rosettes) and the base of one of the three sensitive trigger hairs. (Photo courtesy P. Dayanandan)

The bladderwort, *Utricularia inflata,* traps zooplankton organisms, such as the small crustacean *Daphnia* (color plate 11E), in its submerged bladders. The bladders are constructed along the lines of fish traps, consisting of transparent spheres having a small opening at one end equipped with a trapdoor and surrounded by hairs. When a tiny victim touches one of these hairs, the bladder suddenly expands, the trapdoor opens, and the prey is sucked inside. Digestive enzymes within the bladders break down the proteins, which then are absorbed by the plant.

Other carnivorous plants, such as the sundew and butterwort, are the equivalent of flypaper, trapping insects in adhesives secreted by their highly specialized leaves. The butterwort produces several enzymes capable of breaking down lipids, proteins, carbohydrates, and nucleic acids; most other carnivorous plants have fewer enzymes, but all produce protein-digesting **proteases.**

A number of carnivorous plants have vase-shaped leaves and capture prey by simply drowning them in a mixture of enzymes and water at the bottom of the vase. The pitcher plant is an example. Its red-veined, pitcherlike leaves are lined with downward-pointing hairs. Once an insect falls in, climbing out is very difficult.

What advantages are conferred by the habits of carnivorous plants? Charles Darwin and his son, Francis, showed by experiments that sundews which were fed insects grew larger and produced more flowers and seeds than those on an insect-free diet. In addition, the ability to obtain amino acids from prey enables carnivorous plants to occupy nitrogen-poor habitats such as sand barrens and sphagnum bogs. The nutrients are absorbed from the prey quite rapidly, moving into the leaf within two or three hours, as shown by studies using proteins containing radioactive isotopes.

Organisms identifiable as r-strategists tend to colonize habitats controlled by variable and often disruptive forces (unstable climate, storms, fires, etc.); r-strategists are short-lived, often small in size, and rapidly reproducing. They are quick to invade a habitat, to expand in population, to decline, and to disperse. Such species often are known as "pioneers." K-strategists are long-lived "seize-and-hold" organisms (shrubs and trees vs. weeds). They tend to be large in size and to be long-term reproducers. Among animals, juvenile mortality of k-strategists tends to be relatively low and the investment in parental care is often high. K-strategists tend to occupy richer habitats in regions of stable climates and to be specialists in their life-history strategies. Some examples follow. Finally, r-strategists and K-strategists often are interlocking (see "Succession" in Chapter 25).

Mimicry

Among the most interesting life strategies are those that involve **mimicry** (color plate 12C). Mimicry is a form of interspecific communication in which an animal or a plant mimics its surroundings or another organism as a defensive or an offensive strategy. The organism may mimic its background so as to be relatively undetectable when motionless or it may mimic another organism, usually one that is ferocious or poisonous, but even, in some cases, one that is dead. Mimicry of another organism involves model and mimic populations, with the model population having some characteristic, such as toxicity, sting, or bite that gives its members some selective advantage.

There are many examples of mimicry. The monarch butterfly is avoided by birds because it is distasteful. The viceroy butterfly attains protection by mimicking the color and pattern of the monarch. The walking-stick insect closely resembles a twig and escapes detection in that way. The chameleon, as well as many other animals, including the octopus and flatfish such as the sole, are able to mimic the color, and in some cases the pattern, of their background by regulating the size of epidermal pigment cells. The Io moth has a pair of "eyes" on its upper wing surfaces and flashes them when attacked. Some caterpillars have fake head and eye markings on their posteriors and rear up in a threat posture when attacked or threatened with attack. The hognose snake plays dead when attacked or threatened and sometimes is a "ham actor": when turned over, it will give itself away by immediately turning belly-up again.

Probably the most amazing feats of mimicry are those of Old World species of cuckoos, nest parasites that lay their eggs among those of other birds. The different species, after many generations of nest parasitism, come to mimic the coloration of the eggs of the birds whose nests they parasitize.

Flowers also are notable mimics. The wasp orchid mimics the shape and scent of a female wasp and is pollinated when the male wasp attempts to copulate with it (Fig. 24-12). Carrion flowers mimic the veined pattern of dead animals and emit the odor of decay; they therefore are pollinated by carrion flies.

Symbiosis

Symbiosis is a life strategy involving two species of organisms in a close relationship. Biologists classify symbiotic relationships on the basis of the degree of support each symbiont derives from the other.

Predation may be considered a form of symbiosis not involving a close physical relationship between predator and prey. Predation in which the predator is physically dependent, living on or in the body of the prey (host) is **parasitism.** Parasitism, a form of symbiosis, is very much one-sided, as has been noted in previous chapters.

Two other forms of symbiosis occur in which hosts are not damaged. The first is **commensalism,** a close association of two organisms in which one partner derives some benefit while the other does not profit. The second, **mutualism,** is perhaps the most intricate and interesting symbiotic system. It is an intimate interrelationship of two organisms of different species to their mutual benefit. Mutualism can be facultative: one or both partners can exist independently of the other, as the algal and fungal partners in some lichens or *Anabaena* and *Azolla* in nitrogen fixation (discussed in Chapter 3). If the partners cannot live without each other the mutualism is obligate; an example is the termite and the cellulose-digesting microorganisms which live in its gut. Mutualism is more common than might be expected.

Finally, at the other end of the spectrum, is **amensalism,** in which organisms of one population repel other organisms. An example of amensalism is seen in the production of antibiotics by fungi and some other organisms.

Parasitism Parasitism is a modified kind of predation. The strategy of the predator is not to destroy its prey, but to enslave it. The more successful the parasite, the less the parasitized organism feels the presence of the parasite. It is in the best interest of the parasite that its host be healthy and vigorous, for if the host dies, so will the parasite. There are many variations on this theme, and parasites are more or less successful in their avoidance of actions that might result in the early death of their hosts.

One of the most interesting plant parasites is an orange-colored vine called **dodder** (Fig. 24-13). Dodder is a flowering plant and is completely dependent on its host because it has no chlorophyll and hence is nonphotosynthetic. It is therefore a **complete** or **obligate parasite** (that is, it is completely obligated to a life of parasitism). (Parasites that are less than completely dependent are called **facultative parasites.**)

By any measure, dodder is a successful parasite, for it does not usually kill its host. However, because dodder grows on annual plants, it dies when its host dies at the end of the growing season. But annual plants do not die until they flower and produce seeds. Therefore, dodder, if it is to reproduce itself, must flower and produce seeds in synchrony with its host.

Dodder is not naturally photoperiodic, but when it grows on a short-day plant, it acts like a short-day plant, and when it grows upon a long-day plant, it behaves like a long-day plant. This strategy is thought to demonstrate the existence of florigen, the hypothetical flowering hor-

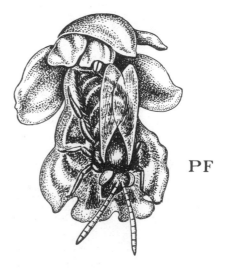

FIGURE 24-12
The wasp orchid, *Ophrys*. The color patterns of this flower mimic those of the female wasp.

mone. It will be recalled that when a photoinduced plant is grafted to a nonphotoinduced plant, both plants will flower because the flower-inducing substance has diffused through the graft union. Dodder, which develops the equivalent of graft unions by sending suckers into its host, probably receives a flower-inducing substance in about the same way and flowers accordingly. The result is that dodder and host live and reproduce together.

Commensalism Commensal associations between species are one-sided arrangements in terms of actual benefit. Small fish that live among the tentacles of sea anemones are not attacked by the nematocysts of the anemone. The fish probably derive protection, and perhaps food, by this association. Barnacles live on whales and probably do them no harm. Orchids belong to a class of plants known as **epiphytes,** which includes all species that grow on other plants but do not parasitize them. The orchid does no harm to the tree on which it grows, but does benefit from its orientation with respect to light and space.

Mutualism Mutualism is the association of two or more organisms with some benefit to all involved. Humans, for instance, are inhabited externally and internally by many microorganisms. Some of these are para-

sites, some are commensals, but some also are beneficial and therefore demonstrate mutualism; for example, intestinal bacteria that live on fecal matter in the colon. These bacteria seem to repel harmful bacteria and prevent infection.

More intimate and mutually beneficial relationships between animals and gut-inhabiting microorganisms are the mutualistic associations found in **ruminants** (e.g., goats, cows, deer, sheep) and termites, in which cellulose-digesting protozoans and bacteria provide glucose for their hosts. A cow or sheep would not be able to live very well on a diet of hay if it were not for the cellulose-digesting bacteria living in the rumen of its stomach. With the aid of their bacteria, ruminants can even digest paper. In one experiment, a ram on a diet of 150 g of filter paper daily was found to have digested up to 90 percent of this nearly pure form of cellulose.

Plants also benefit from bacterial symbionts. The nitrogen-fixing bacteria that live within the root cells of legumes are an excellent example (Figs. 3-14 and 3-15). They convert nitrogen gas into ammonia, which is required by the host plant for the manufacture of amino acids. In return, they receive certain nutrients and shelter. The association between a fungus and an alga to form the composite plant called a lichen also exemplifies mutualism (see Chapter 19).

FIGURE 24-13
Dodder, a parasitic plant, growing on alfalfa. Its flowering is synchronized with that of its host, presumably by transmittal of a flowering hormone from the host to the parasite. (Photo courtesy U.S. Department of Agriculture)

FLOWER ECOLOGY AND PLANT–ANIMAL COEVOLUTION

Magnolias are among the showiest of flowering plants. Their large and attractive flowers are considered to have evolved by **coevolution** with pollinating insects; that is, the two different organisms have evolved in a close ecological relationship characterized by compatible structures in both. For instance, the first insect pollinators probably were beetles that fed on petals and carpels of magnolialike flowers—some magnolias still produce sugary nectar as a surface secretion of petals, stamens, and carpels which serve as a food source for beetles. While feeding, beetles brush against the stamen and thus become pollen carriers.

Coevolution is a mutualistic relationship between two different organisms which, as a result of natural selection, become increasingly interdependent. This evolutionary trend is beautifully illustrated by complex flower-pollinator relationships. Many kinds of flowers have colorful petals and attractive fragrances and nectars capable of attracting insect pollinators or even birds and bats. In this relationship, flower form, color, and smell serve as a recognition signal to the pollinator, in effect saying, "here is a source of food." Food for the pollinators comes in two forms:

1 Nectar, a sweet fluid usually produced in specialized structures called nectaries within the flower, or sometimes in nectaries located on stems and leaves adjacent to flowers

2 Pollen, which is a source of food for bees and other pollinating insects

Nectar is, of course, mostly sugar but contains some amino acids and other nutrients; pollen contains lipids and proteins. Nectaries usually are strategically located so that the pollinator must brush past the pistil; there, pollen from other flowers previously visited is transferred to the sticky stigma of the pistil. In retreating from the flower, the insect rubs against the anthers and is dusted again with pollen. The importance of insects in pollination has been known for centuries; beekeeping based on this knowledge is older than recorded history. However, it has been only recently that the complexities of coevolutionary relationships have begun to receive intensive study.

Flower–pollinator mechanisms may be quite complex. The anther of the milkweed is horseshoe-shaped and catches on the foot of a bee as it visits the flower; pollen is carried away to pollinate flower after flower. The alfalfa flower provides a landing platform in the form of a modified petal. When a bee lands on it, the flower opens to admit it to the nectaries, and at the same time the stamens are depressed so that pollen is rubbed off onto the back of the bee.

Extrafloral nectaries located near flowers and attractive to ants, in particular, are one of the most intriguing aspects of nectar–insect coevolution. The ants protect flowers (and sometimes fruit) against nectar robbers (usually wasps) that cut their way into floral nectaries, robbing them of nectar but not pollinating the stigmas.

Recently, another example of coevolution has been described for several tropical flowering trees that produce two kinds of pollen. One kind of pollen is fertile and is produced by anthers located so as to brush against the backs of bees; the second kind is infertile and is produced by numerous anthers located on a "landing platform" part of the flower, where it is "offered" as food to the pollinator. The conspicuous flower of the cannonball tree, *Couroupita guianensis,* exhibits this coevolutionary reproductive strategy (Fig. 1). In effect, the flower sacrifices part of its energy to facilitate pollination by insects.

FIGURE 1
View of a flower of the cannonball tree, *Couroupita guianensis,* showing sterile anthers (S) and fertile anthers (F). Sterile anthers serve as a food attractant to bees and also as a "landing platform." (Photo courtesy Fairchild Tropical Garden)

Although plants have evolved structures and secretions attractive to pollinating insects and even birds and bats, they also have developed protective devices of a number of types. Among these are stinging gland cells that resemble tiny hypodermic needles (e.g., stinging nettles), noxious and even fatal poisons, and substances having hormonelike properties. Nicotine is a well-known plant poison, evolved as a guard against insect damage. Most persons think of nicotine as a drug present in smoking and chewing tobacco, but it is a powerful insecticide, still used as a greenhouse fumigant and spray. One of the more remarkable groups of protective compounds evolved by plants are **juvocimines,** which mimic juvenile insect hormones. Insect larvae feeding on leaves containing juvocimines are prevented from undergoing metamorphosis. Thus they never become breeding adults and their numbers are kept in check.

An interesting relationship between plants and ants has been described by Daniel Janzen. In an area of Malaysia noted for its poor, sandy soil and scrubby vegetation are found epiphytes with an exceptional nutritional strategy. The short, ball-shaped stems of these epiphytes are inhabited by ant colonies. The ants actually grow the epiphytes by harvesting the seeds and planting them in bits of soil strategically placed along the stem of the host tree. In addition, the ants fertilize the epiphytes with parts from dead insects deposited in special cavities within the plants. These deposits of fertilizer provide nitrogen for the growth of the epiphytes (Fig. 24-14).

Flower–Pollinator Coevolution

Flowers appear to be evolutionary adaptations guaranteeing pollination, fertilization, and reproductive success (Color plate 9A–D). Flowers and insect pollination probably evolved in tropical rainforests. Plants growing in more open situations often are wind pollinated. So are plants that grow together in immense stands, as do conifers. However, in the tropical rainforest, there is such a diversity of plant species that although plants are very numerous, no single species predominates. Moreover, the air currents there are often stilled. The diversity of plants and the absence of air currents lessens the chance for successful wind pollination. Because insects are the most abundant form of animal life in rainforests, it seems probable that strategies based on insect transport of pollen have originated there.

Plants and insect pollinator mechanisms are among the most fascinating life strategies. The flowers of the attractive mountain laurel of the Appalachian Mountains have stamens wedged in small pockets of the petals. When an insect applies slight pressure to the petals, the stamens are suddenly released, and pollen is explosively ejected onto the back of the insect.

Nectar, the sugary liquid produced by **nectaries,** is sought and collected by insects and other animals. It is produced by many kinds of flowers, usually in nectaries at the base of the flowers. In *Thunbergia*, an Indian vine, nectaries are present both within and outside the flower.

Those outside (see Fig. 24-15) attract ants. The ants, in turn, protect the flowers from robber bees that drill holes through the bases of the flowers into the internal nectaries and, if successful, escape without pollinating the flower.

Flower–pollinator interactions reflect the climate. Bee flowers such as snapdragons and clover generally grow in dry and open regions of the temperate zone. Flies are the common pollinators in high mountains and arctic areas, but serve as pollinators in the tropics also. Beetle-pollinated plants such as magnolias are common in the warm temperate and tropic zones, and bat and moth pollination is common in the tropics (Fig. 24-16). Wind pollination is common in temperate woodland and grassland where winds are prevalent. In temperate woodland, trees tend to flower in early spring, when air flow is maximal and before new foliage develops.

Dispersal Mechanisms

Most young plants have a greater probability of survival if they grow at some distance from their parent plants. Having the same requirements and the same life strategy as the parents, their chances of developing in close competition with the older parental plants are minimal. Therefore, mechanisms for fruit and seed dispersal are extremely important.

Various kinds of fruit and seed dispersal mechanisms have been evolved by plants, including winged

A.

B.

C.

D.

FIGURE 24-14
Myrmecophytes ("ant plants") growing epiphytically on trees in Malaysia. **A.** Several epiphytes in a tree. **B.** Closeup of a myrmecophyte stem. The entrances to ant tunnels are visible. Note the orchid on the top of the myrmecophyte (an epiphyte upon an epiphyte!). **C.** and **D.** Smaller myrmecophytes before and after sectioning, showing the inner tunnels and chambers inhabited by ants. (From D. H. Janzen, 1974, *Biotropica* 6:237)

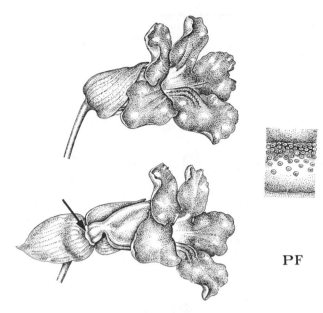

FIGURE 24-15
Extrafloral nectaries (arrow and inset at right) of *Thunbergia*. Note also that the flowers have obvious honey guides—the dark lines along the petals.

PF

fruits for wind dispersal, explosive mechanisms that eject seeds from fruit, and devices that attach fruits or seeds to an animal (Fig. 24-17).

Fruits, and in some cases seeds, are often highly colored and attractive. This physical appearance signals their edibility (directed more to birds than to mammals, as many mammals are color blind). Typically, the whole fruit, rather than the seeds, is transported by an agent.

Many fleshy fruits contain indigestible seeds, which pass through the alimentary tracts of animals. Cherry trees often grow profusely along fences, where birds perch after feeding. Mistletoe, a plant parasitic on certain trees, has a most interesting dispersal system. Its seeds have a sticky covering, and these seeds become attached to the beaks of birds. When the birds wipe their beaks on tree branches, the mistletoe seeds stick to the host tree and germinate.

Biological Clocks

The leaves of a bean seedling fold up every night and unfold each morning. It might be supposed that this is solely a light/dark response. However, if the seedling is kept in continuous light, the same once-a-day opening and closing behavior will continue for a week or more, although gradually becoming more and more erratic. It is as if every bean cell contained a tiny metronome

which ticked off the minutes and hours of the day, irrespective of light or dark.

Although there are no simple explanations of these timing mechanisms, called **biological clocks,** it has been suggested that they involve the repeated, periodic transcription of certain lengths of DNA called **chronons.** Depending on the speed of these repeated transcriptions, the cells would periodically make the specific enzymes governing the rhythmic behavior patterns. Rhythms are so common that they must be considered one of the characteristics of life, along with irritability, cell division, and growth. Cell division cycles, reproductive cycles, and migrations are some of the many activities governed by biological clocks.

Circadian Rhythms   In many plant species, the onset of flowering is regulated by the relative lengths of day and night. However, studies of this phenomenon, called **photoperiodism,** indicate that the specific day and night lengths must correlate with the internal rhythms of the plant if flowering is to occur. Thus a "short day" and a "long night" that total 36 hours instead of the normal 24 may not trigger flowering of a short-day plant, whereas a "short day" and a "long night" that add up to 48 hours or another multiple of 24 will induce flowering. These 24-hour cycles are called **circadian rhythms** (*circa*, "about"; *dia*, "day"). Plants as well as other organisms have 24-hour clocks. This should not be surprising, because all organisms on earth have evolved in an environment of 24-hour days.

Biological clocks have to be "set" by some stimulus. Usually, this stimulus is daylight, but other stimuli such as lunar cycles or seasonal temperatures also are known. Once set, the clock will continue to "run," even when the organism is kept in a constant environment. Fiddler crabs, for example, are dark in color at night and pale by day. Their 24-hour clock is set by the stimulus of sunrise, but if they are isolated in an experimental darkroom without clues to day or night, they will continue to darken and lighten in their characteristic day/night circadian rhythm. Fiddler crabs also are active at low tides and inactive at high tides, but this, unlike color change, is timed to the tides, which are based on lunar cycles. The tide-stimulated cycle of action and inaction continues when the crabs are removed from their natural environment.

Human patterns of sleep and wakefulness are basically circadian in nature, as demonstrated by the discomfort of jet lag after long cross-country or intercontinental flights. In addition, there seem to be pronounced biorhythms (periods of monthly activity alternating with depressed periods).

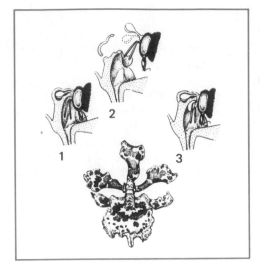

A.

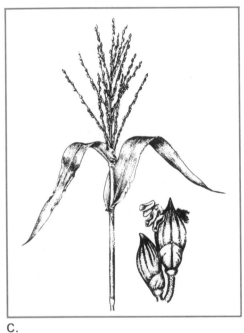

C.

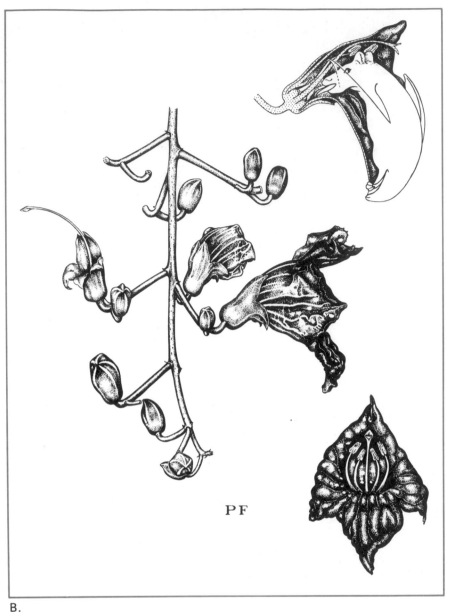

PF

B.

FIGURE 24-16
Pollination and pollinators. **A.** Bee pollination of an orchid. The bee gathering nectar touches the adhesive pollen sac (1), which becomes attached to its head (2). The bee then visits another flower, transfers pollen to the stigma, and picks up a second pollen sac. **B.** Bat-pollinated flowers of the African sausage tree, *Kigelia*. The structure of the flower provides a landing platform for the bat. **C.** Wind-pollinated flowers of panic grass. In the enlarged drawing of the flowers, the stigma is feathery, and the stamens have slender filaments. These are adaptations to wind pollination. Wind-pollinated flowers usually lack petals and are inconspicuous.

Two other behavior patterns of organisms regulated by biological clocks are dormancy and hibernation, which are set by such external stimuli as the onset of winter cold and the changing day length.

Dormancy Dormancy, the condition in which metabolism slows almost to a halt, is a survival mechanism of organisms at every level of biological classification. Bacteria, fungi, protistans, algae, mosses, and ferns form resistant spores that enable the species to survive periods of hostile environmental conditions. Dormant seeds of gymnosperms and flowering plants represent the same survival strategy. These are examples of survival mechanisms based on the formation of resistant reproductive

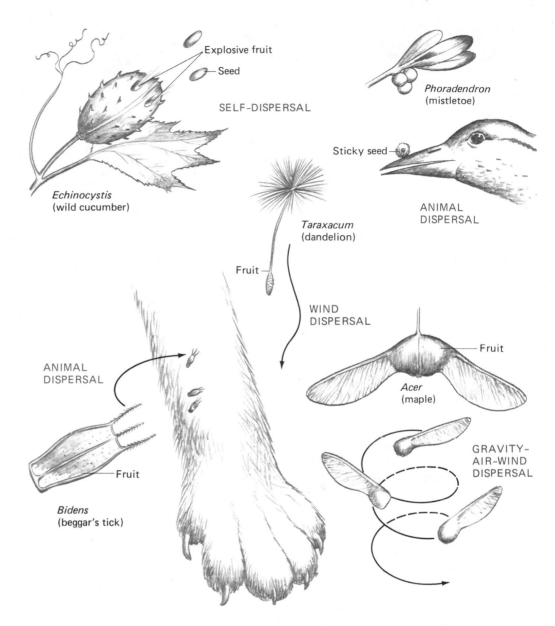

FIGURE 24-17
Some agents of fruit and seed dispersal.

Explosive fruit

Seed

SELF-DISPERSAL

Echinocystis
(wild cucumber)

Phoradendron
(mistletoe)

Sticky seed

ANIMAL
DISPERSAL

Taraxacum
(dandelion)

Fruit

WIND
DISPERSAL

ANIMAL
DISPERSAL

Fruit

Acer
(maple)

Fruit

GRAVITY-
AIR-WIND
DISPERSAL

Bidens
(beggar's tick)

units. In addition, adult organisms at every level of biological complexity use various versions of dormancy as survival strategies.

Deciduous trees and shrubs are a good example of this adult strategy. They, by and large, are plants of warm summers and cold winters. During the summer they produce a flush of new, photosynthetically efficient leaves, which in the fall they shed. They then remain dormant until the next summer. This would seem to be a wasteful process, and in comparison with strategies of tropical, evergreen trees, no doubt it is. However, the summer efficiency compensates for winter leaf loss and dormancy, and thus the deciduous plants survive. If competition

and selection were to "invent" some more efficient means of tree survival—say, a warm sap and good insulation—so that the plant were capable of functioning the year around, then no doubt this new strategy would supercede the presently successful one.

Hibernation Many animals **hibernate** during periods of inclement winter weather. Amphibians bury themselves in the mud at the bottom of ponds or burrow in soil below the frost line, remaining there until spring comes with milder temperatures. During this period of hibernation, their respiration slows to an almost imperceptible level, and they use only the food stored within

FIGURE 24-18
Scanning electron micrograph of a water bear, *Echiniscus spiniger.* (Photo courtesy Diane R. Nelson)

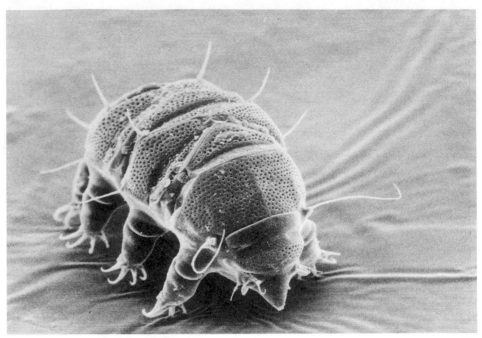

×500

their bodies. Some amphibians of the tropics behave similarly during the hottest and driest part of the year, by a strategy called **aestivation.** Some mammals also go into hibernation during the winter, but their metabolism is kept at a higher rate to maintain their body temperatures, and their food requirements place a higher demand on stored food, mostly body fat.

The champion hibernator is a tiny invertebrate known as the **water bear,** a member of a minor phylum, **Tardigrada** (Fig. 24-18). The name was coined by the Italian naturalist Spallanzani, who referred to the organism as *il tardigrado* ("slow stepper"). Tardigrades inhabit many niches from the Arctic to the Antarctic; terrestrial forms live in mosses, lichens, liverworts, and leaf litter. A 120-year-old specimen of dried moss has yielded living water bears. Their amazing capacity to withstand drying for long periods of time was first observed by Linnaeus, who called it "cryptobiosis" ("hidden life"). When dried, the water bear can lose up to 95 percent of its body water and remain alive in a complete vacuum. Obviously, this ability to become completely dormant is an important survival strategy enabling tardigrades to survive long periods in harsh environments. It also appears to be an important dispersal mechanism; dry and dormant water bears are dispersed widely by wind, water, and other animals.

SUMMARY

The ecosystem is the unifying concept of ecology. It includes the abiotic factors of the physical environment, the biotic factors (organisms and their products) that live within it, and the energy relationships that sustain them.

Energy relationships are often expressed in terms of food webs, in which the initial input of solar energy passes through trophic levels of producers (plants) and consumers (herbivores and carnivores). At each trophic level, the available energy diminishes drastically. The pyramid of diminishing energy that results may be expressed in numbers of organisms, in their total weight (biomass), or in their productivity (expressed in calories of energy). The physical environment may be viewed in terms of light, temperature, and other factors of climate, and also in terms of the availability of essential elements, which are maintained by continual recycling processes, for example, the water, carbon, phosphorus, and nitrogen cycles.

The biotic aspect of the ecosystem consists of its populations or organisms, which are interlocked with each other in the food web. Each species exhibits a different survival strategy, or niche. Among the most common life strategies are predation, parasitism, commensalism, and mutualism. Under natural circumstances, ecosystems and their communities of organisms maintain a steady-state relationship between producers and consumers and the spatial and nutritional factors of the environment. However, if by accident or design these relationships become upset, some populations may increase greatly. Exotic species, which generally are introduced without their limiting factors, often increase and undergo population explosions until they come under the control of new limiting factors.

KEY WORDS

ecosystem	water cycle	niche
energy flow	carbon cycle	life strategy
essential element	nitrogen cycle	mimicry
biotic factor	phosphorus cycle	symbiosis
food chain	eutrophication	parasitism
trophic level	community	commensalism
producer	succession	mutualism
primary consumer	climax community	coevolution
secondary consumer	limiting factor	biological clock
biomass	competition	dormancy

QUESTIONS FOR REVIEW AND DISCUSSION

1 What three factors do all ecosystems have in common? What are the physical dimensions of ecosystems? What can be said about their geographical sites?

2 Why is the term "food web" preferable to "food chain"? Why does a pyramid of biomass reveal more about energy flow relationships than a pyramid of numbers?

3 In what form is nitrogen available to plants? To nitrogen-fixing microorganisms? What is one survival strategy of nitrogen-fixing bacteria and higher plants?

4 Explain the term "microclimate," and give several examples.

5 Describe a feedback system regulating the populations of two animal species.

6 What is an exotic species? Why are they often troublesome?

7 Define the term "niche," and give an example. Do you have a niche? Explain.

8 Compare parasitism, commensalism, and mutualism. Give examples.

SUGGESTED READING

DAUBENMIRE, R. 1968. *Plant communities*. New York: Harper and Row. (A basic introduction to plant ecology.)

GILBERT, L. E. 1982. The coevolution of a butterfly and a vine. *Sci. Amer.* 247(2):110–21. (A plant defends itself against predation by using several interesting strategies.)

GOSZ, J. R.; R. T. HOLMES; G. E. LIKENS; and F. H. BORMANN. 1978. The flow of energy in a forest ecosystem. *Sci. Amer.* 238(3):92–102. (Food-web relationships and forest ecology are explored.)

LIKENS, G. E.; R. F. WRIGHT; J. N. GALLOWAY; and T. J. BUTLER. 1979. Acid rain. *Sci. Amer.* 241(4):43–51. (The nature of acid rainfall is examined historically; its chemistry is described, and corrective measures are discussed.)

NELSON, D. R. 1975. The hundred-year hibernation of the water bear. *Nat. Hist.* 84(7):62–65. (A fascinating account of cryptobiosis as a survival strategy.)

ODUM, E. P. 1971. *Fundamentals of ecology*, 3rd ed. Philadelphia: W. B. Saunders Co. (A standard textbook in general ecology.)

PALMER, J. D. 1975. Biological clocks of the intertidal zone. *Sci. Amer.* 232(2):70–79. (Experiments on the biological clocks of crabs are discussed.)

REVELLE, R. 1982. Carbon dioxide and world climate. *Sci. Amer.* 247(2):35–43. (Is the greenhouse effect real?)

SCHMIDT-NIELSEN, K., and B. SCHMIDT-NIELSEN. 1953. The desert rat. *Sci. Amer.* 189(1):73–78. (Describes how the desert rat conserves water.)

WICKLER, W. 1968. *Mimicry in plants and animals*. New York: McGraw-Hill Book Co. (An excellent, beautifully illustrated book on mimicry.)

25

Ecosystems

I n their natural habitats, plants and animals are generally considered to live in associations called **ecological communities.** In such communities, each individual interacts with individuals of other species in such a way that a relatively stable social structure is achieved. According to the community model, community structure is a reflection of both the **physical environment** (geological and climatic factors) and the **biotic environment** (numbers and kinds of organisms). Different kinds of communities are found in the Arctic than at the equator, in water than on land, and in arid regions than in areas characterized by high annual rainfall. Many natural communities have been studied and described—the temperate oak–hickory forest, the coral reef, the freshwater pond, the *Acacia* savanna, and so on. In each of these there is a characteristic association of plant, animal, protistan, fungal, and moneran species. The community model has predictive value: certain kinds of organisms can be expected in each specific community.

Ecologists define communities in various and very divergent ways. Some consider them to be well-organized, discrete associations of characteristic organisms. Others think of them as loosely arranged, rather variable associations of species, possibly without distinct physical boundaries. Both viewpoints seem to be valid, depending on the example chosen as well as on the different philosophies of the ecologists who interpret them.

Ecological communities, also referred to as **biotic communities,** are parts of **biomes,** which are continental in scope and contain several to many different kinds of communities. Biomes, in turn, constitute the world ecosystem, or **biosphere,** and usually are considered within the realm of **biogeography.**

COMMUNITIES

Definition of Community

A biotic community may be defined as any assemblage of populations of species living within a specific area or habitat. It is the living part of the ecosystem and is characterized by the interdependence of its members. Communities may be large or small, sharply delimited by boundaries or somewhat diffuse and overlapping. They do, however, have common features, including the following:

1 Dominant species of organisms

2 Physical habitat

3 Functional interrelationships of the component populations of species

In Chapter 24, the concept of niche as the "profession" of a population of organisms within a community was mentioned. The niche concept, which carries with it the connotation of survival strategies of species, also indicates interdependence among species and between species and their physical environment. The community therefore is not just a zoo or botanical garden in which different species are set out for display, but rather a dynamic ecological system composed of interrelated units.

Functionally, a community can be described in terms of its food-web relationships, but this often is quite cumbersome because of the many producer–consumer interdependencies. Therefore, the general practice is to name the community after the most obvious common species. Typically, these are plants, because plants do not move in and out of the community as animals often do. Thus ecologists speak of the California redwood community, in which the dominant plant is the coastal redwood (*Sequoia sempervirens*), and recognize that black bears, mule deer, ferns, mosses, insects, fungi, and many other organisms are present also.

Community Structure

The structure of a community is commonly assessed by the **quadrat method.** In this sampling procedure, a defined area, often a 10-m square of the community, is literally taken apart and all the organisms enumerated. When several quadrats are completed, the information is used to construct a model representing the structure of the entire community. Such ecological models may consist simply of statistical tables, charts, and graphs, or they may be diagrammatic profiles in which the forms of at least some components are visually presented (Fig. 25-1).

Succession

In addition to its characteristic structure, every community has a historical past, in some cases a very long one. For example, the great coniferous forests that once spanned the continents of North America and Eurasia are thought to have existed continuously for millions of years. No matter how long a community has existed, it will have undergone a series of developmental changes leading to its present form. These stages comprise **ecological succession,** or simply **succession,** and often are described in terms of the dominant plants. However, successional stages also have characteristic animal species, and in some cases one or more of these may control succession in the community. Grazing mammals

FIGURE 25-1
Profile of a deciduous forest in the St. Lawrence Valley. The top underground layer represents humus. Note that relatively few species are indicated. The layer of large trees represents the dominant beeches and maples of this community. Animals are not shown because of their mobility, but might be represented by a tabulation of species and their spatial distribution. (Redrawn from P. M. Dansereau, 1957, *Biogeography: an ecological perspective*, New York: Ronald Press Co.)

such as bison were important in maintaining the grassland stage of succession before they were wiped out and replaced by cattle. Kelp beds in the Aleutian Islands are controlled by sea otters which feed on and control the numbers of sea urchins. Without this control, sea urchins tend to overgraze the kelp and markedly change the structure of the community. **Keystone species** such as bison and sea otters appear vital to the stability of many communities.

There are two basic types of succession. The first is **primary succession,** the starting point of which is a barren site without soil, vegetation, or organic matter (i.e., rock or open water). The second is **secondary succession,** in which a previous community has been wholly or partially destroyed by some accident, such as fire, storm, flooding, or severe drought, or by human action such as clearing of a forest or the plowing of a prairie. Secondary succession usually begins on an already established soil.

Primary Succession Primary succession on rock surfaces is very slow and begins with reduction of the solid rock to fragments by various erosive forces such as freezing and thawing, mechanical action of wind and flowing water, and chemical action of water in dissolving minerals and holding them in solution. Lichens and mosses are often the first plants to become established. Eventually, these **pioneer** plants break up the rock surface and produce a cover that acts as a trap for blowing sand and dust and for waterborne particles. As these first plants die and decay, they add humus to the developing soil. The development of soil leads to the establishment of larger plants; usually, these are grasses and sedges (grasslike monocotyledons) and herbaceous dicotyledons. Later, shrubs and trees may take over as the dominant species. As succession progresses, microclimates develop. The shaded soil provides a moist, sheltered habitat for many small organisms, including bacteria, fungi, worms, and insect larvae. Small rodents find a home in the shelter of grasses and bushes, and the seeds of grasses and other flowering plants provide food. In some places, rain and snow water may accumulate to form small, temporary pools of water, which aquatic insects and amphibian larvae inhabit. Predators move in and feed on rodents and other primary consumers. Gradually, a balanced, interdependent community of plants and animals comes into existence.

A succession similar to that just described occurs on sand dunes (Fig. 25–2), except that the process begins where the fragmentation of rock has already taken place. Succession, therefore, occurs more rapidly than when the starting point is bare rock.

Primary successions have been observed also on lava flows in Hawaii, in Iceland, on the island of Krakatoa in the Indian Ocean, and recently, following the eruption of Mt. St. Helens in Washington state (Color plate 13A, B). In each case, communities characteristic of the regional climate gradually evolved. They are capable of maintaining themselves unless changes in the climate occur. In Iceland, the community prevalent at the end of succession is likely to be subarctic tundra; in Hawaii, a tropical rainforest. In all instances, the mature community is called the **climatic climax community.**

When primary succession begins in a freshwater pond or lake, the community at first is rather barren. Only a few organisms are present and the food chain is simple, consisting of phytoplankton (algae), zooplankton (protozoans and small invertebrates), and perhaps a few fish and larger invertebrates (Fig. 25-3). At this stage, it may be called an **oligotrophic** ("little food") body of water. Slowly, a rooted vegetation becomes established, first of emergent plants such as rushes and cattails along the shores and later of submerged bottom plants such as pond weeds. As plant and animal life increase, the action of decomposers adds nutrients to the water, and debris continues to accumulate at the bottom. The aquatic community is now described as **eutrophic** ("good food"). Emergent aquatic plants such as water lilies then begin to fill in the center of the pond. Later stages of succession are characterized by low oxygen levels, accumulated sediments, and reduced fish populations. These stages eventually lead to further eutrophication and finally to the death of the pond, which becomes transformed first into a marsh, then a prairie, and finally a woodland, if that is the climatic climax vegetation of the region.

Another kind of primary aquatic succession is seen in some marine environments. Coastal mudflats and river deltas may be inhabited by a distinctive vegetational community called the **mangrove association.** Mangroves are shallow-water trees equipped with stilt roots or specialized emergent roots called **pneumatophores,** which have air passages and enable the underwater parts of the plants to live. A common species in the Florida Gulf Coast and the Caribbean islands is the red mangrove, *Rhizophora mangle*, pictured in Fig. 25-4. The red mangrove has an interesting germination strategy. The embryo grows into an arrow-shaped seedling while the fruit is still attached to the tree. When the seedlings are released, they plummet into the mud or float off to be implanted elsewhere. By this process, mangroves create sheltered backwaters in which sediments accumulate

FIGURE 25-2
Primary succession on the sand dunes of Lake Michigan. Note the establishment of beach grass on the fore dune, shrubs on older dunes in the immediate background, and established forest on ancient dunes in the far background. New dunes and beaches arise in the aftermath of storms or as a result of changing water levels.

A.

B.

FIGURE 25-3
Primary pond succession. **A.** Postglacial pond in Wisconsin. Eutrophication is well advanced, with submerged and emergent vegetation in the pond, and sphagnum moss, cattails, pitcher plants, and shrubs at the pond margins. **B.** Eutrophication in a Florida pond. In this case, the pond has been invaded by water hyacinth, an African import, and the submerged vegetation is primarily *Hydrilla,* a serious pest in southern waterways and also an introduced species.

and form land. Much of the southwest coastal area of Florida, as well as many small islands along the Florida Keys, has been built by mangroves.

The mangrove community is a rich and productive one, harboring many kinds of marine organisms. It is especially important in furnishing shelter for young fish of many species. When the mangrove community is destroyed to make room for beach-front condominiums, hotels, marinas, and other developments, local fisheries often suffer a serious decline.

Secondary Succession Assume a garden is being started in a vacant lot. Possibly the lot is woodland and will need to be cleared of vegetation before plowing. More than likely this clearing was done long ago, and the consequences inherited in the form of a plot of weedy and scrubby vegetation. In any case, plowing the garden will turn back the clock of ecological succession to the barren soil stage.

The seeds of tomatoes, carrots, radishes, and lettuce are planted, the growth carefully tended, and, assuming consumer organisms have stayed away, the first crop harvested. The vegetable plants have grown very well on the exposed and sun-baked soil of the garden. Why were they able to do so? The answer to this question is that these common garden vegetables are really weeds in disguise. Humans selected them ages ago because of their ability to grow in the disturbed soil around the village dump. Gradually they were improved for food production by conscious or unconscious selection and breeding, while they retained their hardy characteristics.

If this hypothetical garden patch had gone unattended, the early stages of secondary succession would have become evident. First would have come a ground cover of wild weeds. Weeds have a most important place in nature; their role is that of the pioneer. They blaze the trail of community succession and furnish the first ground cover beneath which a longer lived perennial vegetation can become established.

As the garden grows unattended year after year, the weeds, most of which are annual plants, are supplanted by perennial grasses, and these are followed by shrubs and finally trees. Perhaps the first trees in the garden are cottonwoods and box elders—common short-lived, weedy species. Later these will be replaced by oaks and maples. These latter, together with associated organisms, form the climatic climax community, for they are the dominant species of the most complex community the climate is capable of supporting. In other regions where rainfall is infrequent, succession may scarcely progress beyond the barren annual weed stage. The climatic climax there would be a desert community, with the survival strategy of plants one of rapid germination, growth, and flowering.

Animals also succeed other animals in the development of a climax community. In one sand dune succession, the following progression of invertebrates was observed: first, white tiger beetles, sand spiders, and digger wasps; later, snails, wireworms, and ants; and, in the final stages, when microclimates were well established, earthworms, sowbugs, wood roaches, millipedes, and wood snails. In a grassland-to-forest succession in

FIGURE 25-4
The red mangrove in Florida. **A.** Thicket of red mangrove showing the stilt roots upon which oysters, barnacles, and algae grow. **B.** Emerging fruit and seedlings of the mangrove; note the elongated, arrowlike hypocotyls. **C.** Seedling implanted in mud.

Georgia, the grasslands were inhabited by meadowlarks and grasshopper sparrows; the later shrub-stage vegetation was inhabited by cardinals, yellow-breasted chats, and towhees; and the climax forest was populated with woodpeckers, warblers, thrushes, and flycatchers. Much the same story could be told of mammalian succession, with forest-dwelling types such as squirrels, lynx, and foxes replacing grassland forms such as field mice, badgers, and coyotes.

Many other kinds of succession are well known. A rotting log, for instance, furnishes an interesting example of small-scale succession—from early stages in which wood-boring beetles riddle the log with tunnels to later stages in which the wood is digested by fungi and bacteria until only a bit of humus remains. Each stage in this succession has a unique community of organisms associated in food-web relationships.

MAJOR WORLD BIOMES

Many of the characteristics of communities are controlled by climate. Climate also controls major ecosystems such as oceans, deserts, rainforests, and prairies. These major ecosystems are called **biomes**. In addition to the **oceanic biome,** there are approximately eleven large terrestrial ecosystems distributed around the world. Each has a distinct vegetational aspect related to

the prevailing climate (Fig. 25-5). Basically, these biomes are oriented along the earth's latitudes and tend to be shifted by ocean currents northward or southward along the continental coasts. The warm waters of the Gulf Stream, for example, give England, which has about the same latitude as Alaska, a climate and vegetation like that of South Carolina. Other factors such as mountain ranges, large bodies of water, and prevailing wind direction also introduce climatic variations and influence the pattern of regional vegetation.

Oceanic Biome

The oceans constitute more than two-thirds of the earth's surface, and for this reason alone, their total productivity is very great. However, the oceanic system is not particularly productive in purely relative terms. In general, concentrations of essential nutrients in the open ocean are low, and food production is considerably less than in the majority of terrestrial biomes. Only the top layer of the open ocean, penetrated by the sun's rays, is the scene of primary productivity, exclusively by phytoplankton (unicellular algae and photosynthetic protistans). Also, because the ocean is so deep, there is little recycling of nutrients, which tend to sink to the bottom and remain there.

Productivity in an ecosystem can be defined in several ways, including the total standing biomass, or the

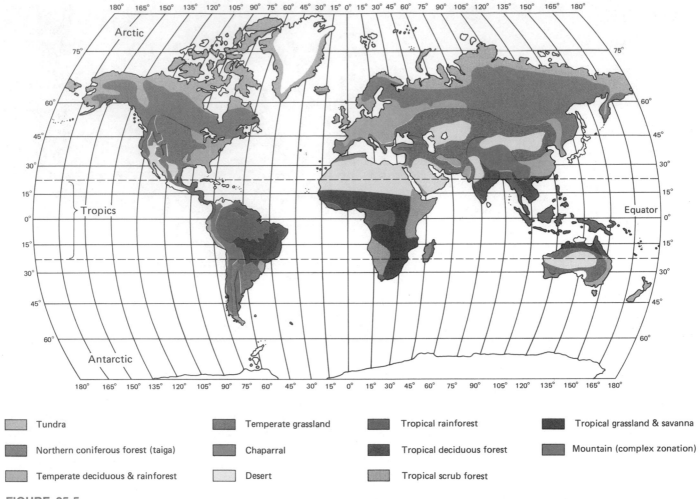

180° 165° 150° 135° 120° 105° 90° 75° 60° 45° 30° 15° 0° 15° 30° 45° 60° 75° 90° 105° 120° 135° 150° 165° 180°

Arctic

75°   75°
60°   60°
45°   45°
30°   30°
15°   15°
Tropics   Equator
0°   0°
15°   15°
30°   30°
45°   45°
60°   60°

Antarctic

180° 165° 150° 135° 120° 105° 90° 75° 60° 45° 30° 15° 0° 15° 30° 45° 60° 75° 90° 105° 120° 135° 150° 165° 180°

- Tundra
- Northern coniferous forest (taiga)
- Temperate deciduous & rainforest

- Temperate grassland
- Chaparral
- Desert

- Tropical rainforest
- Tropical deciduous forest
- Tropical scrub forest

- Tropical grassland & savanna
- Mountain (complex zonation)

FIGURE 25-5
Major biomes of the world.

biomass produced per unit of time per unit of area. The latter is the more accurate measure, because it gives the rate at which solar energy is converted into food. This productivity measure is commonly expressed in kilocalories of energy produced per square meter per year (kcal/m^2/yr). The productivity of the oceanic biome is on the order of 1000 kcal/m^2/yr.

The richest oceanic regions are the continental shelves, where there is a constant upwelling of deeper water, bringing with it nutrients from the bottom sediments. Nutrients also wash in from rivers and streams. The producers are phytoplankton, as well as red, brown, and green algae, and in some cases submerged flowering plants. (An abundant flowering plant is turtle grass, which harbors great populations of shrimp as well as fish that feed upon shrimp.) The primary consumers are zooplankton organisms (protozoans and copepods). The secondary consumers are larger animals, including coel-

enterates, worms, mollusks, arthropods, echinoderms, and vertebrates (Fig. 25-6). Productivities on the continental shelf range from 2000 to 6000 kcal/m^2/yr. These equal or exceed those of nearly all terrestrial biomes.

Coral reefs are amazingly complex shallow-water ecosystems in which coralline red algae as well as coral polyps play a major role (see Fig. 21-1, also Color plates 2A, B;10A, C–E). Virtually every animal phylum is represented in the coral-reef community, and highly colored crustaceans, echinoderms, mollusks, corals, sponges, and fish make for a beautiful scene.

Tundra

The **tundra** corresponds roughly in latitude to the Arctic Circle and is characterized by a low vegetation composed of a few species of grasses and sedges, together with mosses and lichens, perennial herbaceous dicoty-

PLATE 13

A. Mt. St. Helens, aftermath of eruption. Blasted ridge had no vegetation in 1980. (Roger Del Moral) **B.** Mt. St. Helens, same area as in (A) but in 1983. Limited recovery of a few survivors. (Roger Del Moral) **C.** Tundra, North America. Musk oxen, Devon Island, N.W.T., Canada. Subphylum Vertebrata, Class Mammalia. (Lawrence Bliss) **D.** Kittiwake gulls nesting on seaside cliff in western Iceland near the Artic Circle. Phylum Vertebrata, Class Aves. **E.** Boreal forest near Goose Bay, Labrador, Canada. The trees are principally black spruce, *Picea mariana,* interspersed with swampy, open areas called muskeg.

PLATE 14

A. Tropical rainforest, Malaysia. Wallace's flying tree frog can extend membranes between its legs and glide from branch to branch. Phylum Vertebrata, Class Amphibia. (Kiew Bong Heang) **B.** Lobster claw plant, *Heliconia* sp., is a common New World rainforest flower. Division Magnoliophyta. **C.** Rainforest in the New World, French Guiana. The variegated pattern of tree colors is indicative of great species diversity. (David Lee) **D.** Hut of Colombian rainforest dweller, made of palm thatch and surrounded by cultivated garden patch with bananas and taro. Such small-scale usage of the rainforest is not nearly as destructive as large-scale slash and burn. **E.** Leaf cutter ants are common rainforest insects. Phylum Arthropoda, Class Insecta. (David Lee)

PLATE 15

A. Mixed grass prairie, Wind Cave National Park, South Dakota, U.S.A. **B.** Subtropical savanna, Hunter Valley, Australia. **C.** Subtropical mist forest of tree ferns, Victoria, Australia. **D.** Coastal desert vegetation, Colombia, South America. **E.** Temperate deciduous forest, Ohio, U.S.A.

PLATE 16

A. Burning palm–palmetto scrubland, Florida, for land clearing. **B.** Lumbering and clear cutting tropical rainforest, Malaysia. (David Lee) **C.** Wind erosion after cultivation of prairie in American Midwest. (U.S. Department of Agriculture) **D.** Whale slaughter, Iceland, 1960. Iceland is a signator to the International Whaling Treaty and no longer engages in such whaling activity. **E.** Spraying citrus groves with fungicide. (U.S. Department of Agriculture) **F.** Industrial pollution, U.S.A. (Tom Stack)

FIGURE 25-6
One of many oceanic food webs.

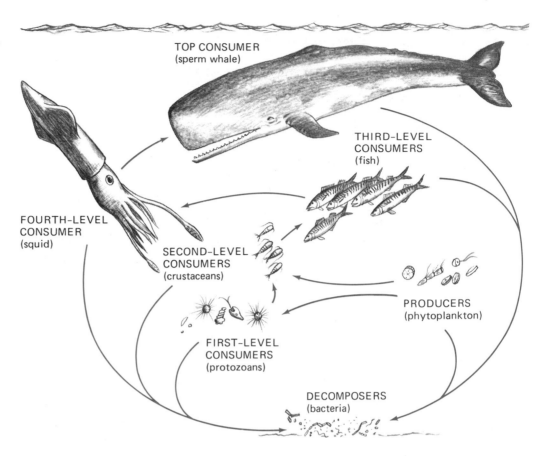

TOP CONSUMER
(sperm whale)

THIRD-LEVEL
CONSUMERS
(fish)

FOURTH-LEVEL
CONSUMER
(squid)

SECOND-LEVEL
CONSUMERS
(crustaceans)

PRODUCERS
(phytoplankton)

FIRST-LEVEL
CONSUMERS
(protozoans)

DECOMPOSERS
(bacteria)

ledons such as the arctic rose, and here and there dwarf birches and willows. In wetter areas, *Sphagnum* moss occurs, as does cotton grass (actually a sedge) and a few small orchids. Animal life may be abundant during the summer season, for the tundra is a favorite nesting site for such migratory birds as snow geese, swans, plovers, and other migratory birds (Color plate 13D). Herds of caribou and musk oxen also are present (Color plate 13C). Lemmings and other small rodents are prey for snowy owls, arctic foxes, and weasels. Other elements in the food web are swarms of flies and mosquitoes, arctic hares, and ptarmigan. The largest predators are wolves and polar bears, although the latter are more often found on ice floes.

Even though covered with ice and snow for much of the year, the tundra is not a region of high precipitation and in fact may be considered a very cold desert. Much of the ground moisture is tied up by **permafrost,** the permanently frozen layer of earth hundreds of feet deep. Consequently, productivity of the tundra is very low (only about 200 kcal/m²/yr). Even though the subsoil is permanently frozen, the surface thaws each summer,

supports the tundra vegetation, and can be cultivated. However, the surface thawing produces upthrust blocks of permafrost called **polygons,** and these make farming difficult.

No corresponding biome occurs in Antarctica, because the southern oceans occupy the same relative region as the tundra does in the Northern Hemisphere. However, on the Antarctic peninsula there is a sparse coastal vegetation composed of lichens, mosses, and two species of flowering plants. At lower latitudes in both the Northern and the Southern Hemispheres, a tundralike vegetation, the **alpine tundra,** is found on high mountains, where the prevailing climate is similar to that of the Arctic.

The lowest temperature ever recorded on earth, −106°C (−127°F), occurred in the Siberian tundra in 1960 near Vostok, where temperatures around −73°C (−100°F) are common. Permafrost covers fully 43 percent of the Soviet Union. It is reported that a Siberian pioneer named Fyodor Shargin attempted to dig through the permafrost to reach water, so he would not have to continually thaw ice and snow. He dug through the frozen, rock-hard soil for ten years, finally reaching a depth of 117 m (380 ft), but had in that time penetrated only about halfway through.

Taiga

Extending across the continents of Eurasia and North America is the **taiga,** commonly called the northern coniferous forest (Color plate 13E). The northern border of the taiga is the tundra, and the southern border is the deciduous forest and prairie. Most of the North American taiga lies in Canada, but elements extend into New England, the Lake states, and the Rocky Mountains. Long, dark, and cold winters and short, warm summers with many hours of daylight are major factors controlling the taiga. Permafrost is present in the northern portions of the biome. The dominant vegetation is the white spruce and birch woodland. Wet areas interrupt the woodlands and are characteristically *Sphagnum* bogs or cotton grass marshes, with later successional stages grown up into tamarack and black spruce. Moose, wolves, otters, ravens, beavers, wolverines, deer, grouse, lynx, and various species of hares are among the animal life in this ecosystem. The productivity of the taiga is higher than that of the tundra, but considerably lower than those of more temperate biomes, ranging from a low of 400 kcal/m^2/yr, to around 2000 kcal/m^2/yr at its southern extremities.

Temperate Deciduous Forest

Extensive **temperate forests** are found in central and eastern North America, western Europe, eastern Asia, and the southern tip of South America (Color plate 15E). These regions are characterized by warm summers and moderately cold winters, but no permafrost. The annual precipitation ranges from 80 to 160 cm (30 to 60 in) per year, and the productivity is moderately high, on the order of 3000 kcal/m^2/yr. The dominant vegetation in these forests is deciduous hardwoods, with softwoods intermingled. (Hardwoods are tree forms of flowering plants, not simply trees with hard wood. Conventionally, conifers are considered softwoods even though the wood of some species may be harder than that of some of the so-called hardwoods.)

At the center of the deciduous forest biome in North America (the Appalachian plateau) is an association of dominant trees called the **mixed mesophytic** ("humid but not wet") **forest,** the predominating tree species of which are beeches, maples, tulip poplars, oaks, and hickories. Hemlock, basswood, ashes, and magnolias also occur there. To the north and northeast of this central region, beech and maple tend to predominate; to the west (where it is drier), oaks and hickories

tend to be dominant; and in the east (in Pennsylvania and the Virginias), an oak and chestnut forest prevailed until the chestnut blight destroyed the chestnuts in the first part of this century.

Where the deciduous forest borders on the taiga, a transitional zone of conifers and hardwoods occurs. Vast tracts of land in New England and the Lake states once were covered with magnificent stands of white and red pine, but these regions had been logged barren by the beginning of the century. There now grows a weedy forest composed of soft maples and mixed softwoods and hardwoods such as jack pines, red pines, and poplars. This zone (mixed conifers and hardwoods) extends into the southeastern United States to the coastal plain, where pinelands predominate.

A wide variety of animal life occurs in the deciduous forests of North America, although their numbers have been considerably reduced because of lumbering and clearing for agriculture. In some central states, such as Ohio, up to 80 percent of the original forest is gone, and with it the vast flocks of passenger pigeons and wild turkeys that once fed on the plentiful acorns and other forest fruits. Now the common fauna are white-tail deer; red, gray, and fox squirrels; cottontail rabbits; foxes; raccoons; opossums; and diverse species of thrushes, warblers, owls, and sparrows.

The temperate deciduous forest region is economically important. Many of the less hilly portions have been cleared and now are used for a diversified agriculture. However, the soil is easily eroded, in part because of the hilly contours of the land and in part because of the relatively high rainfall (in comparison with the prairies to the west). The hilly portions where farming is not economical are used for lumber production and recreation.

Grassland

The great **grasslands** of the world are the **steppes** of Eurasia, the **prairies** of North America (Color plate 15A), the **pampas** of Argentina, and the **veld** of South Africa (Fig. 25-7). These occur where rainfall ranges from 25 to 75 cm (10 to 30 in) annually and where winters generally are cool to cold and summers are hot and dry. Such climates often are called **continental**. Productivity of the grassland is lower than that of the deciduous forest, averaging about 2500 kcal/m^2/yr. However, the generally flat to rolling topography allows intensive agriculture, and the agricultural productivity is high.

Several different communities make up the North American prairie, and each is characterized by the grass type that predominates. Eastward along and across the Mississippi River is the **tall grass prairie,** which in pi-

FIGURE 25-7
Prairie grasses and wildlife in Wind Cave National Park, South Dakota. The bison, *Bison bison*, once a prominent member of the North American grassland fauna, now is restricted to parks and preserves.

oneer days stood belly deep on horses. Westward is the **mixed grass prairie,** and still farther west is the **short grass prairie.** The tall grass prairie is now mostly "corn country," occupied by vast fields of maize in Indiana, Illinois, Iowa, and Missouri. The mixed grass prairie is the "wheat country" of the Dakotas, Nebraska, Kansas, and Oklahoma. The short grass prairie, once the home of great herds of bison, now is the "cattle country" of Montana, Wyoming, Colorado, and parts of Utah, New Mexico, and west Texas. However, rainfall patterns are variable, and periodic droughts produce "dust bowl" conditions and crop failures, particularly in the westernmost parts of the region.

Once abundant on the prairies of North America were bison, elk, mule deer, bighorn sheep, wolves, grizzly bears, sage grouse, and prairie dogs. Pronghorn antelope (not a true antelope), whitetail and mule deer, sharp-tailed grouse, prairie chickens, rabbits, hares, and many other animal species still occur in considerable numbers. The prairie marshes were the breeding grounds of many species of ducks and other water birds whose numbers have been severely depleted due to drainage of marshes.

Tropical Savanna

True **tropical savannas** (grasslands with scattered trees and shrubs, including numerous *Acacia* species) are found in central Africa, South America, and Australia (Color plate 15B). Rainfall in this ecosystem is moderately high (75–150 cm [30–60 in] per year), but highly seasonal, and the productivity is about 3000 kcal/m²/yr.

The biome is controlled in part by a yearly long dry period, during which fires sweep over vast areas. These fires maintain the grasslands by preventing the more extensive development of woodlands.

It is believed that the African savanna is the cradle of human evolution, and the anthropological investigations near Lake Rudolph in Kenya and in the Olduvai Gorge by the late Louis Leakey, his wife Mary, and son Richard, and by Raymond Dart and others in southeast Africa, have contributed considerably to the present understanding of hominid evolution. The savanna in Kenya, Tanzania, and Zimbabwe is famous for its large and diverse populations of animal life, well known from television and travel films. Lions, elephants, rhinoceroses, giraffes, gazelles, and zebras are among the familiar wildlife inhabitants.

In the western United States a dry scrubland vegetation called **chaparral** is maintained by frequent fires and may be considered a modified savanna. In Australia a similar community, the **mallee scrub,** is found. It resembles chaparral in overall appearance, although the species are completely different.

Deserts

Deserts occur in coastal areas bordered by cold oceanic currents, on the leeward side of mountain ranges, and in other areas where prevailing winds have already lost their moisture. In North America, South America, and Africa, deserts commonly occur on the seaward side of the western coastal regions (Color plate 15D). Deserts are also found in the "rain shadow" of the Andes and in their northern counterpart, the Rocky Mountains. Here the prevailing oceanic winds have already given up their moisture as they pass over the mountain ridges.

The controlling factor in the desert environment is the characteristically low annual rainfall, normally less than 25 cm (10 in) per year. Temperatures, however, vary greatly. There are cold deserts, such as the Gobi in Asia, and hot deserts, such as the Mojave in the United States and the Sahara and Kalahari in Africa.

Although their productivity is as low as 200 kcal/m²/yr, deserts are by no means lifeless regions. The plants are typical thorn-scrub vegetation with reduced leaf surfaces and extensive water storage tissues (cacti, euphorbs, etc.; see Fig. 7-6) or rapidly growing annuals that germinate, flower, and die within a couple of weeks. Animal life in the desert of the American Southwest includes peccaries, quail, coyotes, foxes, skunks, ring-tailed cats, kangaroo rats, hawks, and owls. Many of these animals are adapted behaviorally for a life of daytime burrowing to escape the heat and of nocturnal feeding. Some, such as the desert rat, conserve water so effi-

WINGED MITE OF THE JACK PINES

My good friend Roger Tyslan pulled his car to the side of the road and said "This is it. Jack-pine warbler country." I glanced at the nearby pines and confirmed Roger's statement when I read a sign nailed to one of the trees:

ENDANGERED SPECIES AREA
KIRTLAND'S WARBLER NESTING HABITAT
UNLAWFUL TO ENTER
WITHOUT WRITTEN PERMISSION
BETWEEN MAY 1 AND AUGUST 15

MICHIGAN DEPARTMENT OF NATURAL RESOURCES
IN COOPERATION WITH
U.S. FOREST SERVICE AND U.S. FISH AND WILDLIFE
SERVICE

Named in 1852 in honor of a Cleveland physician, Dr. Jared P. Kirtland, Kirtland's warbler, *Dendroica kirtlandii* (also known as the jack-pine warbler), is a member of North America's growing list of endangered species, a sad enumeration that includes the whooping crane and the California condor. Although not as spectacular as its more illustrious fellows, this mite of the wood warbler family is just as imperiled.

About 200 pairs remain, all confined during the breeding season to just three counties in the north-central part of lower Michigan. Only a few strays have turned up elsewhere during the spring and summer. As if a low population size were not enough, Kirtland's warbler has several other strikes going against it. The bird's habitat requirements are strict: stands of jackpine *(Pinus banksiana)* about the size of a Christmas tree, growing in dry, sandy soil with a moderate, low-growing ground cover. Forest fires, natural or man-made (and controlled), are necessary to

ciently that they excrete dry urine and need consume no liquid water.

Rainforest

There is no true tropical **rainforest** within the continental United States, but rainforests of limited extent are present in parts of Hawaii and the Virgin Islands, as well as in Puerto Rico and other Caribbean islands. More extensive rainforests are present in eastern Mexico and in Central America. A temperate rainforest is found in the Olympic Mountains of Washington and has some of the attributes of the true rainforest, including high rainfall and an abundance of epiphytic vegetation. However, it lacks the diversity of species characteristic of the tropical rainforest. Localized humid forests with some of the attributes of rainforests are present in a few places in the subtropics (Color plate 15C).

The great rainforests of the world are located in Brazil, the valley of the Congo River in Africa, Malaysia, the East Indies, and New Guinea (Color plate 14). All these areas are equatorial or subequatorial and are con-

insure proper nesting requirements. The warblers avoid stands of the pines when the trees begin to exceed about 13 feet in height.

The birds, strangely enough, do not nest in the trees but rather on the ground in the low cover. They forage for insects in the jack pines and the males sing while perched in the branches, but this species of pine can be replaced in special plantings by red pine; the warblers will nest there as long as the man-made habitat closely approximates the natural one. Strike two against the birds was delivered by cowbirds.

The brown-headed cowbird is one of a number of species called brood parasites; that is, females of such species lay their eggs in the nests of other birds, and all parental care devolves on the host "parents." As you may have guessed, the host's young often die for want of proper care usurped by their foreign nest mates. By the 1960s, cowbird parasitism had almost done in the warblers. Ornithologists, alarmed by the plunging population size of the remaining warblers, moved rapidly and succeeded in trapping and removing most of the cowbirds. Happily, warbler population size seemed to stabilize, but still at an alarmingly low 200 pairs. Suitable jack-pine habitat in north-central Michigan could support more Kirtland's warblers but the birds don't seem to be colonizing such areas.

The wintering grounds of this species may hold a clue to the dilemma. Kirtland's warblers have been found to winter only in the Bahamas: you can imagine the losses for a small songbird migrating from Michigan to a tiny spot in the Atlantic. However, other species of wood warblers face the same perils and also winter in the Bahamas. What's the problem with the Kirtland's warbler? A small percentage of yearling birds of many species manage to return to the exact place where they hatched the year before, but even if they are a bit off course they may very well find fellow members of their species with which to breed. But the Kirtland's warblers nest in a very tiny part of north-central Michigan; it takes pinpoint navigation to move back and forth from the Bahamas to suitable jack-pine country. Get a little bit off course and a male can sing his heart out but there will be no response from a female. Those strays mentioned earlier did not breed.

So what if the warblers of the jack pines join the ivory-billed woodpecker, the passenger pigeon, and hundreds of other extinct species? We doubt that saving the Kirtland's warbler will help us find a cure for cancer, but loss of this species will cheapen mankind. We are all part and parcel of the world ecosystem, and as the pace of extinction increases we may ponder when our time will come.

trolled climatically by high temperatures and very high annual rainfall (up to 1000 cm [400 in] per year), with little seasonal variation in either. The productivity of this biome is very high, up to 12,000 kcal/m²/yr. An explanation for this productivity, in addition to abundant rainfall and uniform temperature, is the multistoried structure of the rainforest vegetation.

The tropical rainforest is commonly composed of five layers, including a top layer of very tall emergent trees, a dense canopy of tall trees, an intermediate layer of smaller trees, a shrub layer, and a ground carpet of ferns, leafy liverworts, mosses, club mosses, parasitic flowering plants, and fungi (Fig. 25-8). The upper layers are the most productive, for they resemble a hanging garden, with every space between and on the trees filled with vines and epiphytes. Strangler figs encircle some of the trees, at times completely encasing and killing them. Birds and insects abound, and the rainforest is a particularly favorable environment for reptiles and amphibians. Mammalian life is not especially diverse; it is represented by large cats (ocelots and jaguars), tapirs, monkeys, sloths, marsupials, and rodents.

FIGURE 25-8
Profile of a tropical rainforest.

Strata
I
II
III
IV
V

A conspicuous feature of the majority of rainforest trees is their smooth, lichen-covered bark and flaring, buttressed trunks. Many trees bear flowers and fruit on their stems and branches, rather than on the smaller twigs. Almost all are evergreen and have large, leathery leaves equipped with drip tips like the spout of a pitcher. This leaf type is typified by the "rubber plant," much used as an ornamental house plant, although in nature it is a large rainforest tree. (See Fig. 25-9 for several common forms of rainforest vegetation.)

An interesting although localized ecosystem called the **cloud forest** is found in the perpetual mists of the high tropics. Cloud forests have dense stands of mosses, ferns, club mosses, and lichens, giving the effect of miniaturized versions of the forests found at lower levels. For that reason, cloud forests are sometimes called **elfin woodlands.**

BIOGEOGRAPHY

Communities and biomes, although presently fairly well defined, are not permanent ecosystems when regarded in terms of millions of years. For example, scientists know that Illinois, which just before the age of agriculture was occupied in part by the tall grass prairie biome and in part by the deciduous forest biome, many millions of years ago was a coal-age swamp forest composed of seed ferns and giant club mosses. So, too, every biome has an evolutionary past in which its appearance was quite different than it is today. It is also known that shifting climates have rendered uninhabitable some biotic environments that once were rich. Ant-

arctica, now mostly snow and ice, at one time was forested and had a diversified flora and fauna, and palm trees once grew in Greenland. Camels at one time were abundant in North America, and horses evolved there and later became extinct. Humans evolved in Africa and eastern Asia and have migrated only recently into North and South America. What is the explanation for all these changes in environments, shifts in populations, and origins and extinctions of species and ecosystems?

The science of **biogeography** is concerned with just such questions. Careful analysis of present distributions of plants and animals, and of their fossil ancestors, tells a story of the history of the earth's biomes and of the species that inhabited them. Probing studies of the crust of the earth, the oceanic deeps, the mountain elevations, the magnetic orientation of iron crystals in rocks, and the distribution of fossils present a fascinating picture of shifting poles, rising and sinking oceans, and slowly migrating continents.

CONTINENTAL DRIFT

Until the 1960s, most biologists and geologists had assumed that the continents had always existed in about the same relationship to one another. Intercontinental migrations of plants and animals were believed to have occurred via land bridges that were produced by land upheavals or that existed briefly as a result of lowered sea levels. These theories explained many "recent" migrations of populations occurring in the past 75 million years or so. Thus, humans invaded North America about 10,000 years ago via a land bridge that once connected

A.

B.

C.

D.

FIGURE 25-9
Rainforest vegetational types. **A.** Buttresses of a kapok tree. **B.** Epiphytes (bromeliads and vines).
C. Strangler fig growing on a dead palm trunk. **D.** Candle tree bearing flowers and fruit on its mature trunk.
(A, photo courtesy W. H. Hodge)

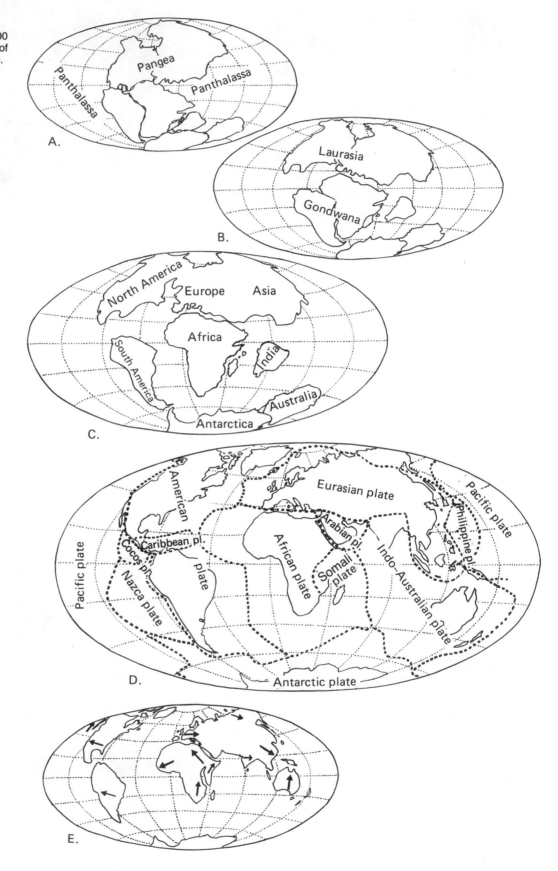

FIGURE 25-10
Continental drift. **A.** Pangea (200 million years ago). **B.** Breakup of Pangea (135 million years ago). **C.** Breakup of Gondwana (65 million years ago). **D.** Today. **E.** The future (50 million years from now).

Siberia and Alaska across the Bering Strait. The Central America bridge between North and South America has been made and broken more than once, but the present one, produced by the same upheaval that created the Andes, is fairly recent. For a long time the two continents were isolated from each other, so much so that each evolved its own unique collection of plants and animals. However, many biologists have felt uncomfortable with land bridges and chance migrations as the sole explanations for the past and present distributions of plants and animals. Only recently, with the broad acceptance of the geological theory of **continental drift** (Fig. 25-10), have satisfying answers to some of the most perplexing questions of biogeography been obtained.

Francis Bacon (1561–1626) was the first scientist to point out the correspondence between the outlines of the Atlantic coastlines of Europe, Africa, North America, and South America. However, it was not until about 1910 that an American geologist, E. B. Taylor, and a German geologist, Alfred Wegener, introduced the theory of continental drift, in which the continents were described as rafts of lighter rock floating on a sea of heavier, semifluid rock. Wegener proposed that there once existed only one major landmass on earth, a supercontinent that has been named **Pangea** (Fig. 25-10A). Pangea was composed of all the present continents and existed during the Paleozoic era and well into the Mesozoic. Thus, dinosaurs probably were able to migrate freely between the present Eurasia and North America. Distributions of seed fern fossils and cycads also show evidence of similar continuities. Cycads, for example, which are tropical and subtropical plants, have been found as fossils in Greenland, Alaska, and Siberia, as well as in Europe and Africa. Also, Antarctica once had forests of *Nothofagus*, trees very much like those now growing in South America.

In the early Mesozoic era, Pangea began to break apart into a northern continent (**Laurasia**) and a southern continent (**Gondwana**) (Fig. 25-10B). Then, about 100 million years ago, Africa, South America, Australia, and Antarctica became dissociated from one another; Eurasia and North America, however, remained united for another 35 million years until separating early in the Cenozoic era (Fig. 25-10C), when the great grasslands of the world were coming into existence.

Continental drift offers an explanation for the past distribution of many extinct plants and animals, as well as the present occurrence of certain groups of reptiles and primitive plants and mammals. However, land bridges, changing sea levels, and chance migrations offer the best explanations for the structure of present-day biomes and dominant orders of modern plants and animals.

SUMMARY

Communities, which are the functional unit of the larger, continental ecosystems called biomes, come into existence by ecological succession, the starting point of which is bare rock, sand, or water, and in the case of secondary succession, barren soil. Gradually, a soil is built, upon which pioneer species of plants and animals are established. These species in turn create habitats for other populations, until a stable community characteristic of the climate comes into existence (the climatic climax community).

About twelve major world ecosystems, or biomes, are recognized by ecologists. These are geographically oriented in relation to prevailing climatic conditions. The oceanic ecosystem is the most widespread, because oceans compose more than two-thirds of the earth's surface. Around the Arctic Circle is the tundra, a biome of very low productivity controlled by extremely long, cold winters and very short summers. Just below the tundra, and somewhat more productive, is the northern coniferous forest, or taiga, which covers large areas of Canada and Siberia. Southward is the productive, deciduous forest biome, found on all the continents in temperate regions of moderate to fairly high rainfall. The tropical rainforest, with its high rainfall, is the richest biome in terms of numbers of plant species and is by far the most productive. Two somewhat similar biomes of about the same productivity are the great grasslands of the world and the savannas. The African savanna is called the cradle of civilization because it is believed to have been the scene of human evolution. Often bordering grasslands and savannas are the deserts, which are dry biomes of low productivity and scanty rainfall.

The study of the distribution of plant and animal populations, called biogeography, attempts explanations not only of the present nature of the biomes but also of their evolution. Many theories to account for origins, extinctions, and migrations of plants and animals have been offered. These include changing sea levels, creation of land bridges, and continental drift.

KEY WORDS

ecological community	oceanic biome	savanna
primary succession	tundra	desert
secondary succession	taiga	rainforest
climatic climax community	temperate deciduous forest	biogeography
biome	grasslands	continental drift

QUESTIONS FOR REVIEW AND DISCUSSION

1 How is the structure of a community analyzed and its population enumerated?

2 Briefly describe and compare primary succession and secondary succession.

3 Describe an example of succession in your own human, home community. Who are the pioneers and the dominants?

4 Define "climatic climax community." What is the climatic climax community in your part of the country?

5 Why is the open ocean not as productive as the continental shelf?

6 What is permafrost? In what biomes does it exist? What nation is largely underlain with permafrost?

7 What is the nature of the savanna? Why may it have been the cradle of humankind?

8 What factors and forces have been responsible for the past and present distributions of plants and animals in world biomes?

SUGGESTED READING

BELL, R. H. V. 1971. A grazing ecosystem in the Serengeti. *Sci. Amer.* 225(1):86–89. (Describes interactions of producers and consumers in a dry habitat.)

COX, G. W., and M. D. ATKINS. 1979. *Agricultural ecology: an analysis of world food production systems.* San Francisco: W. H. Freeman and Co. (A comprehensive discussion of how people manipulate ecosystems in their production of food.)

HORN, H. S. 1975. Forest succession. *Sci. Amer.* 232(5):90–98. (Discusses the effect of leaf arrangement on succession.)

ISAACS, J. D. 1969. The nature of oceanic life. *Sci. Amer.* 221(3):146–62. (Discusses relationships among organisms of the open ocean.)

KURTEN, B. 1969. Continental drift and evolution. *Sci. Amer.* 220(3):54–64. (Continental drift explains puzzling discontinuities in populations of reptiles and primitive mammals.)

POWERS, C. F., and A. ROBERTSON. 1966. The aging Great Lakes. *Sci. Amer.* 215(5):94–104. (A dramatic account of man-made eutrophication.)

RICHARDS, P. W. 1973. The tropical rain forest. *Sci. Amer.* 229(6):58–67. (An excellent description of the structure of an important ecosystem.)

RICHARDSON, J. L. 1980. The organismic community: resilience of an embattled ecological concept. *BioScience* 30(7):465–71. (Discusses the discrete "organismic" ecological community and the "coincidental community" viewpoint.)

WILSON, E. O.; T. EISNER; W. R. BRIGGS; R. E. DICHERSON; R. L. METZENBERG; R. D. O'BRIEN; M. SUSMAN; and W. E. BOGGS. 1973. *Life on earth.* Sunderland, Mass.: Sinauer Associates. (Includes an excellent chapter on biogeography.)

26 Human Ecology

In 1780, a census found the population of Ireland to be about 2.5 million. The figure perhaps should have been higher, for census taking was not then the refined process it is today. The few roads that existed were poor, and even when people were located, they were not always truthful. Most of the Irish at that time were tenant farmers on small plots owned by landlords living in England. The typical farm was about three acres, half in meadow, with perhaps a cow, several pigs, and a few chickens for livestock. The farmer and his family lived in a one-room hut of stone or peat. Fields were tilled with a horse and plow shared with several neighbors; the potato garden was worked with a spade. The cash crops were wheat and oats, all of which went to pay the rent, along with what meat, eggs, butter, and milk could be spared. Potatoes were the main food of the family (Fig. 26-1), and the staple diet was boiled potatoes and milk. It was the food from the cradle to the grave, for the toothless very young to the toothless very old, and it was very easily produced. Cut-up seed potatoes were laid on the sod in burned over meadows and bogs, then covered with soil from shallow trenches to make the so-called "lazy beds." Conditions for growing the potato were ideal, and the yields were high.

Even though far from prosperous, life in Ireland for the rural poor was apparently not unhappy. The climate was pleasant, the food was ample, and people found time for singing and dancing and arguing the political problems of the day, chief among which was the autocratic and repressive rule exercised by England.

THE IRISH FAMINE OF 1845–47

Eighteenth century Irish men and women married young and had large families. Typically, a young man and his teen-aged wife obtained the lease of a couple of acres, erected a hut, and started a family, which in time might number between eight and ten children. Consequently, the population increased dramatically. In 1845 the population of Ireland stood at about 8.5 million, a rate of increase from 1780 unparalleled anywhere else in Europe and which even today is barely surpassed by the wildly expanding population of Central America and Africa. As the Irish population grew, the land became crowded. By the third decade of the 1800s, when food shortages began to occur, the average population density was 215 people per square mile. The potato crop

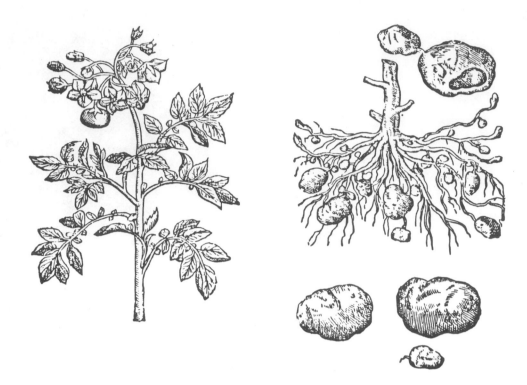

FIGURE 26-1
Cultivated or Irish potato *(Solanum tuberosum).* The potato was brought to Spain in about 1570 from the Andes of western South America. It reached England in about 1586 and was introduced to Ireland about 1586 by Sir Walter Raleigh. (From J. Gerard, 1663, *The Herbal, or General History of Plants)*

barely lasted from one season to the next. Meanwhile, in America, the **late blight,** a hitherto unknown disease of potatoes, appeared and then spread to Europe, possibly from infected potatoes thrown overboard by some ship's cook in a European port. Soon, the blight made its way to Ireland. It was first recognized there as a serious threat in the fall of 1845, when it greatly reduced the stock of seed potatoes needed for the following year. The spring and summer of 1846 in Ireland were unusually wet, a condition that favored the spread of the blight. In August the disease struck everywhere, wiping out the entire potato crop.

There was little government assistance in 1846. A few workhouses existed where the poor might find food and shelter in return for their labor, but these were not geared to cope with a national disaster (Fig. 26-2). There were no emergency loans, no food stamps, no Medicare, no CARE packages, no United Nations Relief Fund or International Red Cross. When Sir Robert Peel, the prime minister of Britain, realized the crisis at hand in Ireland, he attempted to import maize and wheat from America to feed the starving people. He could not do so at first because of the "Corn Laws," which prohibited importing grain into Great Britain in competition with English farmers. Finally, after a desperate political struggle, Peel got the Corn Laws repealed, but even then the food was not given freely to those in need, but rather was sold to them. The government at first charged a 30 percent profit, so as not to undercut the local merchants; later

they reduced the profit to 15 percent. The poor Irish could afford neither amount. The starving and destitute people benefited little from the imported food. Only the Quakers furnished free relief, setting up soup kitchens throughout the land. Meanwhile, people starved and died by the hundreds of thousands.

Horror stories were common. In County Cork a woman and two children were found dead and half-eaten by dogs; in a nearby cottage five more corpses were

FIGURE 26-2
Starving Irish women and children gather before the gates of a workhouse. (From M. O'Brien and C. C. O'Brien, 1972, *The story of Ireland,* New York: Viking Press. Reprinted by permission from the Mansell Collection Ltd., London)

found. In still another hut a dying man lay in a bed with his dead wife and two dead children, while in an adjacent crib a starving cat devoured a dead infant. No one knows how many starved to death or died of starvation-associated diseases. The best estimate is that more than a million persons died. Another million and a half emigrated, principally to the United States but also to Canada and Australia.

What caused the blight? The British government appointed a team of scientists to investigate. These scientists spent a few weeks in the field and then retired to their offices. After deliberating for some months, they issued 70,000 copies of a set of complicated and unworkable instructions for peeling, grating, and straining the rotting potatoes in order to salvage a bit of starch. Later, they published four huge volumes of their reports, and with that, ended their "investigation."

Various causes of the blight were proposed, among them static electricity, smoke and steam from locomotives, vapor from the center of the earth, and guano from seabirds. Regarding the last, it was suggested that muslin be draped over the potato fields to keep the gulls from defecating on the plants.

In 1845 a country clergyman, the Reverend M. J. Berkeley, noticed that whenever potato plants were attacked by the disease, a tiny growth of fungus appeared on the blighted leaves and tubers. He thought the fungus caused the blight and was seconded in this by a French scientist named Martagne. Together they announced their discovery of the new fungus, but were loudly rebuffed by the scientific community, who believed that fungi were always the product of putrefaction, never the cause of it. Only in the early years of the present century was the life cycle of the potato fungus, now known as *Phytophthora infestans*, worked out, and its association with the late blight disease of potatoes made clear.

HUMAN IMPACT ON ECOSYSTEMS

Today the great Irish Famine of 1845–47 is largely forgotten. Nevertheless, it stands as an example of the dangers of unchecked population increases and the overdependence on a single food source. Is there a similar danger now? Much of the world depends on cereals, principally rice, maize, and wheat. Only the United States, Canada, and Australia consistently produce surpluses for export. What if a blight were to strike the grain fields of these three countries? If losses to rusts, smuts, or other blights were sufficiently great to prohibit the export of cereals to other nations, famines of much greater proportions than that in Ireland would certainly occur. Even without future blights, it is doubtful that these three nations can continue indefinitely to supply enough food to support the ever-increasing populations of Asia and Africa.

Food-web systems in nature are usually intricate and self-sustaining. Population explosions most often occur when the checks and balances of the community are interfered with and the food web becomes simplified. In simple communities with only a few interacting species, populations fluctuate a great deal more than in complex communities where feedback regulation is based on many limiting factors. In a natural grassland or forest ecosystem, many plant species support the populations of primary and secondary consumers, and loss of one plant species or one consumer is usually not disastrous. The chestnut blight, for example, did not destroy the deciduous forest community in the eastern United

LOST AND FOUND?

Found, along one mile of two-lane highway in Kansas: 770 paper cups, 730 cigarette packages, 590 beer cans, 130 pop bottles, 120 beer bottles, 110 whiskey bottles, 90 beer cartons, 90 oil cans, 50 paper livestock feed bags, 30 paper cartons, 26 magazines, 20 highway maps, 16 coffee cans, 10 shirts, 10 tires, 10 burlap bags, 4 bumpers, 4 shoes (no pairs), 2 undershirts, 2 comic books, 2 bedsprings, 270 miscellaneous items. (From *Audubon*, May 1970, with permission)

States. However, when a simple ecosystem such as the tundra loses a species, the result may be catastrophic for the species that remain. This explains the concern environmentalists have for the safety of the arctic tundra and their insistence several years ago that great care be taken in laying the Alaskan pipeline.

Although a grassland community is naturally stable, when it is altered to make room for a potato or wheat field, trophic levels are eliminated, and the food web becomes simpler and more direct (Fig. 26-3). The ecosystem tends to become unstable. In the conversion of a grassland or forest community to agriculture, primary producers are reduced to only one or a few species.

Extensive plantings of one plant species such as potatoes, wheat, or maize results in a form of single-crop agriculture called a **monoculture**. In monocultures, the number of consumers is drastically reduced by insecticides, traps, scarecrows, fences, and so on. As a result, nearly all the food produced by plants is eaten by livestock and directly or indirectly consumed by humans. In Ireland in the 1840s, the energy flow was very direct—from sun to potato plants to human consumers. The instability of such food webs is obvious, and disasters are predictable. Monoculture patterns of agriculture are efficient, but they are very vulnerable, and can be ruined by widespread disease or inclement weather.

Conversion of the natural ecosystem to the agricultural or industrial community also greatly accelerates the effect of the erosive forces of nature (Color plate 16A–C). Consider for a moment that in a stable natural

FIGURE 26-3
Human impact on the ecosystem. (Redrawn with modifications from P. Dansereau, 1957, *Biogeography: an ecological perspective*, New York: Ronald Press Co.)

community, nutrients are continually cycled, and the fertility of the soil and the productivity of the community are maintained, perhaps even increased.

In contrast, in the United States, where agriculture is on the average less than 200 years old, thousands of square miles of fertile land have been ruined through human use. In some easily eroded areas in the southeastern states, and in the wheat-growing regions of Washington State, the topsoil has been washed away to a depth of 1 m (3 ft) and more (Fig. 26-4). This leaves the much less fertile subsoil for growing food for the future. It is possible, of course, to grow healthy plants even in sterile sand, provided they are supplied with a solution composed of the essential elements, but when the topsoil is gone, agriculture comes to resemble a solution culture dependent on fertilizers rather than natural nutrients to grow a crop. This need for fertilizers in turn exerts a much greater demand on the energy resources of the human community to supply the chemicals required.

Many experiments have shown that small, carefully tended farms in which nutrients are carefully husbanded and recycled can remain highly productive for many generations, with little loss in soil fertility. Unfortunately, agriculture the world over is trending toward expanding monocultures and the accompanying extensive use of herbicides, insecticides, fertilizers, and heavy equipment powered by fossil fuels (Color plate 16E, F). This increased mechanization requires large acreage and heavy capital investment. It forces small operators off the land and into cities. Displaced individuals who become wards of society require even greater agricultural productivity

FIGURE 26-4
Erosion in abandoned fields. (Photo courtesy U.S. Department of Agriculture)

and mechanization to support them. Thus there seems to be no escape from the continued mechanization of agriculture to produce food for an increasing world population and no escape from the accompanying spread of hunger.

THE GREEN REVOLUTION

The development of hybrid maize was but an early example of an agricultural development now known as the **green revolution**. The green revolution began more than a century ago, when it was found that certain minerals added to the soil resulted in increased crop production. It continues today in the work of agricultural scientists world-wide. Of particular relevance are the efforts of agronomists to develop high-yielding crops, the so-called "miracle" strains of cereals, for the "third world."

Mechanization

Mechanical devices came into wide use in agriculture in the Western world about a century ago and made it possible for fewer persons to produce more products. One of the pioneers in the green revolution was Eli Whitney, an American inventor from Massachusetts who, in 1793, was visiting friends at a cotton planation in Georgia. He was appalled at the tremendous amount of hand labor involved in pulling cotton fibers loose from the cotton seeds to which they are attached in the boll. This work was done, of course, largely by slave labor and contributed to a considerable degree to the need for maintaining large slave crews on the great plantations. It occurred to Whitney that the job could be done better and faster by machines, and in a few days he had invented a practical machine that cleanly separated cotton fibers. This machine is still called the cotton gin (a contraction of "cotton engine").

The cotton gin revolutionized the cotton industry. Hundreds of thousands of new acres had to be planted in cotton to keep up with the machines. The resulting rise of a wealthy southern aristocracy, the necessity of the continued employment of thousands of slaves, the erosion of the soft red soil which left millions of acres depleted, and the politics of it all, are well known.

Other agricultural inventions have had similar impacts. Cyrus McCormick, a poor Virginia farmer, founded a great industry with his invention about 1840 of the reaper, a machine that cut wheat and bound it into sheaves. Before the reaper, men, and often women, toiled long hours cutting grain by hand using scythes and sickles. Later, other inventors found ways to incorporate all the operations of cutting and threshing into one large machine, the combine. This machine and many others contributed to the industrialization of farming in the great grain-growing regions of Canada, Australia, Argentina, the United States, and the Soviet Union.

Agricultural Chemicals ("Agrochemicals")

In the early 1930s, plant hormones were discovered, and from these were developed products that could function in many ways to control plant growth. It became possible to kill weeds with chemicals and at the same time increase the yields of crop plants. The use of weed killers on hybrid maize has led to crop yields as much as 300 percent larger than the best yields of 50 years ago. Discovered about the same time as these weed killers were deadly insecticides such as DDT.

One problem with all these modern practices is that they favor large-scale use. A farm family living on a small plot of land finds it difficult to compete in a marketplace that relies on large-scale farming, where machines cost hundreds of thousands of dollars and the smallest farmland worked may be several hundred acres.

Another problem of growing concern is contamination of the environment with pesticides, herbicides, and their breakdown products. This problem was brought to the attention of the general public by Rachel Carson's book *Silent Spring*, which related dire effects of certain pesticides (notably DDT) on wildlife. The banning of DDT in the United States was an early result of *Silent Spring*. However, a number of problems associated with the use of synthetic chemicals in agriculture remain.

"Miracle" Plants

About three decades ago, largely through the efforts of the United Nations Food and Agriculture Organization (FAO), research was begun to develop more productive agriculture methods for use in less developed countries (LDCs). Because many LDCs are subject to unfavorable growing conditions, in particular periodic droughts and sometimes terrible flooding, and because their agriculture is primitive, food shortages are common among them (Fig. 26-5). Add to this the further complication that in many LDCs human populations are expanding at very high rates, and the dimensions of the food problems become even greater. A partial solution to these problems has been attained through the use of improved agricultural techniques and new, high-yielding varieties of grain crops, mainly rice, wheat, and maize.

FIGURE 26-5
The green revolution. Rice tends to be a labor-intensive crop in LDCs, as shown in this photo showing rice being transplanted into a paddy. Obviously, a high-yielding variety would be important in increasing the amount of food produced per work unit. (Photo courtesy U.S. Department of Agriculture.)

Much of this work has been carried out at two institutions: the International Rice Research Institute, in the Philippine Islands, and the International Center on Maize and Wheat, in Mexico. Realizing that some of the food shortages in LDCs were attributable to poor methods of farming, FAO placed emphasis on improving methods of tillage, on water use and conservation, and also on harvesting efficiency. A limited increase in mechanization was recommended, although not to the extent that large holdings and expensive machinery would be necessary. Small tractors, cultivators, and harvesting machines, all suitable for use on small farms, were developed and produced. More importantly, a number of high-yielding varieties of rice, wheat, and maize were developed. They are not hybrids, which require fresh seed-purchases year after year; rather they are inbreds (as described in Chapter 7). The productivity of these new varieties is more than triple that of the types formerly in wide use, which for the most part were traditional types of no known genetic documentation. These native varieties generally were highly variable, often not high producers, and usually subject to lodging (i.e., being broken down by wind and losing their kernels). The most successful of the new inbred types are so-called "dwarf" and "semidwarf" wheats and rices—often referred to as "miracle" wheats and rices. There is, however, nothing miraculous about either the methods by which they were developed or the manner in which they grow. They are the products of modern genetics. An American botanist, Norman Borlaug, has been instrumental in developing a number of the new varieties; he was awarded the Nobel Prize for his work, which has been carried on principally at the International Center on Maize and Wheat in Mexico.

The "miracle" varieties of wheat and rice are genetically uniform, short-stemmed strains bred for resistance to lodging and disease as well as for high yields. They do, however, require careful cultivation and use of chemical fertilizers. In addition they often require careful attention to their water requirements—the wheats particularly. Because droughts are not uncommon in India, Mexico, and Africa, the dwarf wheats often need irrigating, in addition to fertilizers. In spite of these problems, the wheat yields in India, to select a large-scale example, have increased 4 to 7 times since the green revolution began there.

The picture is not entirely rosy, however. The new cereal varieties require "educated" farmers—not all small farmers in LDCs have the "know-how" to take advantage of modern farming methods. Another problem is that fertilizers, irrigation, and machines are expensive and the small landholder often cannot afford to modernize. Consequently, the new methods tend to favor larger operators, and small farmers have been forced out. To quote a 1981 Indian Government report, "The new technology brought with it the expulsion of tenant farmers

AGRICULTURE AND HUMAN SURVIVAL

Why did our ancestors switch from hunting and food gathering to agriculture? The answer seems to be that population pressure forced them into a more highly productive method of ensuring a food supply, but evidence has been lacking. Archeological data from central Illinois show that here this shift occurred about 3000 years ago, and Dr. Mark Cohen of the State University of New York at Plattsburgh cites evidence of a general decline in human health at that time among inhabitants of the lower Illinois River valley. The diet changed from one relatively rich in protein to one high in carbohydrates—evidence of a decline in hunting success and of dependence on vegetation for food. The decline in health and increased mortality persisted for several centuries, until more productive agricultural methods were developed. The shift to agriculture seems not to have resulted from adoption of an immediately superior means of obtaining food, but rather was a matter of gradually discovering methods of raising plants to stave off starvation.

and sharecroppers, the enlargement of cultivated areas by their proprietors, the increase in the numbers of the landless, and a growth of mechanization and productivity." In addition, irrigation of dry lands in India, Africa, and Mexico has brought about increased soil-salt deposition in the topsoil (so-called "salt pans") that is ruinous. Generally, however, the green revolution's influence in LDCs has been highly visible and positive.

Although some persons think that the green revolution may in fact be much less of a blessing than originally envisioned, the ever-increasing world population requires huge quantities of cheap food—amounts that most countries are presently unable to produce. What is the answer? Obviously, the green revolution is not a complete solution, especially if each increase in the food supply finds many additional hungry mouths to consume it.

HUMAN POPULATIONS

Growth

The earliest humans were roaming gatherers of food and spent most of their time hunting, fishing, and collecting edible wild plants. They raised no crops and kept no livestock, but were solely dependent on nature for all of life's necessities. Slowly, however, this changed, and through cultivation of plants and domestication of animals, people were able to produce enough food to free themselves from the constant search for their next meal. About 10,000 years ago, in the Middle East, a few advanced tribes had changed from gathering food from the wild to producing food in fields, gardens, and meadows. Having a dependable food source allowed people to live together; villages and towns came into existence, and civilization followed.

Excavations of a Stone Age village at Jarmo, near the Tigris River in Iraq, tells something of life about 9000 years ago (Fig. 26-6). The villagers, numbering perhaps 100, lived in two dozen or so clay-walled huts. Examination of fossilized human feces (coprolites) shows that the inhabitants enjoyed a varied and balanced diet. They had domesticated dogs, goats, and sheep, and raised barley and wheat. Apparently, the villagers had some leisure time, as indicated by numerous clay effigies of animals and female fertility goddesses.

Although no one knows how domestication of animals and plants actually started, the first domestic animals were probably very young wild animals kept for pets. Perhaps, if taken early enough, they became imprinted to humans and thus domesticated. Plants such as wheat and barley very likely originated from wild harvests. Humans would have harvested the largest grains and therefore unconsciously selected those plants most suitable for cultivation. The earliest gardens and fields probably grew in the refuse piles outside the village.

Through the centuries, until the Industrial Revolution some 200 years ago, the life of villagers—whether African, Asian, or European—differed little from that at Jarmo. However, some villages more strategically located than others became centers of trade and industry. From these, the great cities of antiquity arose, reaching a zenith first in Egypt about 4000 years ago and then in Athens, Alexandria, Carthage, and Rome.

The human population of the world at the time of Jarmo probably stood at about 5 million, slightly less than the present population of Chicago. Then, following the invention of agriculture and the resulting availability of new sources of food, it increased, reaching around 130 million at the time of Christ. Here it stabilized, increasing very slowly through the Middle Ages, until the invention of steam power in England introduced the Industrial Revolution (Fig. 26-7). The world population doubled between 1750 and 1850 and since then has been doubling about every 50 years. By the year 2000, it is expected to be around 6.3 billion; the end of this explosion is not in sight.

Causes of Overpopulation

It is generally agreed that the cause of the present human overpopulation has been the removal of many of the limiting factors that formerly checked reproductive excesses. In addition to ample food and shelter, advances in sanitation and medicine are the factors most responsible for these changes. Average human longevity has increased markedly because of the decrease in infant mortality and the extension of the life span of adults. Neanderthal man had a life expectancy of about 29 years; in classical Greece it was around 35 years and in Elizabethan England about 45 years. At present, life expectancy in the United States is about 75 years. Further increases are expected, although not as dramatic as those in the past.

Reproductive Potential

Every species has a maximum reproductive potential. A pair of frogs may produce 300 fertile eggs during the year and may continue breeding for three years. If all 900 eggs produced during their life span grew into breeding adults (450 pairs) equally successful in rearing their young, in three years nearly 7,000,000 frogs would be produced by the original pair. At this rate of reproduction, in just a few more years the world would be literally covered with frogs. However, long before this happened,

FIGURE 26-6
Village in the New Stone Age in
northern Europe. The dead were
sometimes cut into pieces and
spread over the fields in fertility
rites performed by the women of
the village.

diminishing food and space would have begun to limit
the growing frog population.

Populations never attain their maximum reproductive potential. For example, in a stable population of
frogs, the long-run average of reproductive success is
such that, of all the progeny produced by a breeding pair
during their lifetime, only 2 survive to become breeding
adults. Therefore, of the possible 7,000,000 successful
frog offspring, 6,999,998 are lost through death of potential parents by predation, disease, and in some cases migration. The sum of these population limitations is

called the **environmental resistance**. In equation form,
reproductive potential − environmental resistance =
population increase (P − R = I).

In calculating the growth of human populations,
the terms **birth rate** and **death rate** are used, rather than
reproductive potential and environmental resistance,
and these generally are expressed in terms of annual
births and deaths per thousand persons. Thus, birth rate
− death rate = growth rate. If the total population of a
nation is multiplied by the growth rate, then the annual
population increase can be determined. The follow-

FIGURE 26-7
Early steam locomotives and trains in England, 1830s. The increasing ability to transport agricultural and industrial products was an important factor in the expansion of the population during the Industrial Revolution. (Prints from Historical Pictures Service, Inc., Chicago, Ill.)

ing are the current data for North America and Latin America:

Births per 1000—North America, 15 (1.5%); Latin America, 38 (3.8%)
Deaths per 1000—North America, 9 (0.9%); Latin America 10 (1.0%)
Population growth rate—North America, 0.6%; Latin America, 2.8%

The total population increase per year may be calculated by multiplying the present population by the growth rate, as follows:

North America—233 million × 0.6% = 1.398 million
Latin America—308 million × 2.8% = 8.624 million

These figures can be used to calculate the population doubling times for these regions: North America, 117 years; Latin America, 25 years. For the world, the doubling time has been calculated at about 35 years.

Currently, Mexico has the highest growth rate of any nation (3.6 percent).* Its population can be projected as doubling every 20 years and in 50 years to exceed that of the United States. Mexico City then will be the largest city in the world, with an estimated 50 million people. Other Latin American countries with high population growth rates are El Salvador (second highest

in the world—3.5 percent) and Nicaragua (third highest—3.4 percent). The dangers of these high rates of growth for all of Central and North America cannot be overemphasized. The situation is similarly distressing elsewhere in the world. Nearly all the nations of Africa have population-doubling times of 20–30 years, as do those of Asia. Algeria has the highest growth rate in Africa (3.3 percent), followed closely by Zimbabwe, Morocco, and Nigeria. In Asia, population growth in Jordan is highest (3.3 percent), then Saudi Arabia (3.0 percent), and the United Arab Emirates (3.0 percent). With the exception of Albania (2.5 percent), all the nations of Europe have growth rates under 1.0 percent, and three—Austria, West Germany, and Luxembourg—have negative growth rates.

Presently, the worst effects of overpopulation are felt in the LDCs of the world. It is sometimes difficult to say whether such lands are overpopulated because they are underdeveloped, or underdeveloped because they are overpopulated. The problem is not a new one. In 500 B.C., the Chinese sage Han Fei-Tzu wrote:

In the old days, there were fewer but wealthier people. People now think that five children are not too many and, if each child has five more children, there will be 25 before the death of the grandfather. Therefore, there are more people and less wealth, and people work harder for less. The life of a nation depends on the people having enough food, and not on the number of people.

*This and other population estimates are based on *World population estimates, 1980* (Washington, D.C.: The Environmental Fund).

Carrying Capacity

In stable natural communities, populations are maintained at levels somewhat lower than those supportable by the renewable food and other resources available. These resources are called the **carrying capacity** of the environment. Obviously, if a population of lemmings, say, increases to the point where they eat more seeds and leaves than the plants produce as surplus each year, then the reproductive capacity of the vegetation and its productivity become reduced. The inevitable result is that the lemmings, having "outgrown" their food resources, will have to migrate or die. In this case vegetation might be thought of as the capital investment of the lemming population and the yearly surplus of food as the interest paid on that capital. If the population grows such that it can no longer live on the interest paid by its capital, then it must live by cashing in its capital. The consequences of this in human life are well known, and in nature they are no different.

In northern Arizona there is an isolated highland forest known as the Kaibab Plateau, a region of about 2000 km^2 (1200 mi^2 or 700,000 acres). In the early years of this century, it carried a population of about 4000 deer and a few mountain lions and wolves, which, together with hunters, kept the deer population well below the carrying capacity of the environment. Then in 1907 a well-intentioned government program of predator control was instituted. The lions and wolves were wiped out, and the results were spectacular—by 1924 the deer population reached 100,000. Unfortunately, the deer had by this time greatly exceeded the carrying capacity of the Kaibab forest, which probably could have sustained a deer population of around 30,000. The deer ate all the vegetation and then began to die of starvation. In two years 60,000 deer died, and by 1940 the population was around 10,000, where it has remained since (Fig. 26-8). This is about the present carrying capacity of the Kaibab forest, which has been reduced by grazing and browsing. The lesson is clear: excess populations deplete their

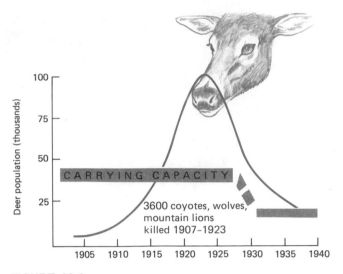

FIGURE 26-8
Kaibab deer population.

100

75

50 CARRYING CAPACITY

25

3600 coyotes, wolves,
mountain lions
killed 1907–1923

Deer population (thousands)

1905 1910 1915 1920 1925 1930 1935 1940

"capital investment" so that from then on it returns a lowered "income."

Trophic Levels

Cereal grains (wheat, rice, maize, oats, and barley) are the basic foods of the world. In Africa and Asia, the average poor person consumes about 175 kg (350 lb) of grain per year. In North America the average person consumes about 800 kg (1600 lb) per year. However, only around 20 kg (40 lb) of grain is eaten directly by North Americans in their bread, breakfast cereals, and pastries; the rest is consumed in the form of meat and poultry raised on grain. Thus, human populations exist at different trophic levels. The poor Asian or African is a primary consumer (herbivore), whereas the average American or Canadian is a secondary consumer, and makes a greater demand on the ecosystem than does a person living in an underdeveloped nation.

The extent to which food is a limiting factor can be seen in India, which in 1965–67 suffered a bad famine because of a rainfall shortage for two consecutive years. Each of those years, the United States shipped one-fifth of its wheat crop to aid India, but even though aid of this kind prevented mass starvation, malnutrition was widespread. When India, the second most populous nation in the world, held its Olympic trials in 1968, not one of its athletes met the minimum qualifications for participation.

Environmental Resistance

Food sets the upper limits on population growth; lower limits are set by predation, disease, and other restraints.

In studies of crowded populations of caged rats and mice, many animals have shown signs of stress that tend to limit population increases. In particular, crowding leads to overdevelopment of the adrenal cortex and decreases in the development of gonads and of fertility. It is not known if similar overcrowding will result in decreased human fertility, but it is believed that human reproduction also tends to be self-limiting. The maximum lifetime reproductive potential of a human female is probably close to 30 offspring, but that number is never attained and scarcely ever approached. In present-day Ireland the population is self-limiting because women tend to delay marriage until 30 or 35 years of age, and thereafter to practice continence to a greater degree than is common elsewhere. Even in countries with wildly expanding populations, birth restraints are practiced.

The destruction of the environment, as in the case of the Kaibab deer, acts as a check on population increases through reduction of food production. The tropical rainforest occurs in the most overpopulated and underdeveloped regions of the world. It is being converted to agriculture at such a rate that some ecologists have estimated it will be gone by the end of this century. Ironically, although it is the most productive ecosystem in the world, it grows on the most easily destroyed soil known—**laterite,** a type of red clay. When exposed to air, laterite turns into brick and in fact was used by ancient civilizations as a paving for roads and for building temples. When vegetation of the rainforest is destroyed to make way for agriculture, the usual result is that in a few years the surface layers of laterite become hard and bricklike. In Dahomey, in tropical west Africa, the rainforest was completely removed some 60 years ago and the land planted in crops. Now there is nothing left but a rocky desert incapable of supporting human life at more than a bare subsistence level. Much the same can be said about Haiti, an impoverished Caribbean nation with a high population growth rate, a deplorable health record, and illiteracy in excess of 80 percent. Like Dahomey, Haiti is on its way to becoming a desert.

There is good evidence that many desert regions of the world were once fertile and supported relatively large human populations. Carthage, in North Africa, was at one time the center of a fertile agricultural region. Bad agricultural practices have reduced it to a desert capable of supporting only a few nomadic tribes. Similar tales can be told about once-fertile regions in the Middle East, Central America, and elsewhere.

Environmental pollution also acts to limit populations. A bacterial culture will grow unchecked for a while, but will decline even before its food is exhausted. Analysis of the culture medium shows that it has become

FIGURE 26-9
Deformation of a young tern as a consequence of pesticides in the environment. (Photo courtesy The American Museum of Natural History)

poisoned with toxic bacterial waste products. Human populations are not very different, as shown by case histories of many pollutants of human origin. Hydrocarbons and other wastes pollute the air in the vicinity of large cities, and studies have shown that as a consequence the risk of cancer has greatly increased. Pesticides and herbicides also add to the risk of human disease. The cumulative effects are not always recognized until considerable damage has been done. A case history is DDT, a chlorinated hydrocarbon discovered in Switzerland about 40 years ago. DDT became a widely used insecticide, and its discoverer, Paul Müller, was awarded the Nobel Prize in 1948. Since then, it has accumulated as a nonbiodegradable environmental contaminant and has caused such destruction to wildlife that its use has been banned in all advanced nations. The ban in the United States apparently has not been completely effective. Some pesticides still contain residues of DDT, not listed as "ingredients." DDT residues are eaten by herbivores and concentrated at each trophic level of consumer organisms, killing smaller animals and interfering with reproduction in larger ones (Fig. 26-9). DDT has been found in penguins in the Antarctic and in mothers' milk in the United States.

Recently, another chlorinated hydrocarbon, polychlorinated biphenyl (PCB), was fed by mistake to some cows in Michigan. Before the error was discovered, PCB killed several thousand cows and made a number of farm families seriously ill. Traces of PCB have since been found in the milk of nursing mothers in many communities in Michigan. However, toxic products of civilization are by no means new problems. The pipes that carried water in ancient Rome were made of lead, and it is believed that lead poisoning was common among Romans. It has even been suggested that lead poisoning was the cause of the decline of Roman power and the demise of the Roman Empire.

Photosynthesis and the Human Race

There is a great contrast in living standards between the two-thirds of the world's population living in the LDCs and the one-third who live in the developed nations. The latter enjoy a per capita income 3–10 times that of the former, and the gulf between them is widening. Although this difference in wealth can be translated into the presence or absence of fine automobiles, color televisions, well-constructed homes, and many other conveniences and comforts, good nutrition is the greatest differential factor. There are two causes for the poor nutrition of people in the underdeveloped Third World: (1) their populations are too great to be supported by their agriculture and industry, and (2) their agriculture is inefficient.

People who live in industrialized societies enjoy food high in proteins. Proteins are essential to the proper functioning of cells, and health suffers when the right kinds are not consumed in the right amounts (Fig. 26-10). In contrast, poverty-stricken people tend to exist mainly on diets high in carbohydrates and low in proteins. Unfortunately, animal cells cannot make their essential proteins from carbohydrates alone; they require amino acids. Because animals are not able to synthesize amino groups, they must get them from plants they eat. However, once they have obtained amino groups from their foods, they can transfer them from molecule to molecule and make most of the amino acids found in proteins (there are 20 different amino acids in proteins, although any one protein may contain only some of them). Even so, there are a few amino acids—the essential amino acids—that animals cannot make at all. These must be obtained directly or indirectly from plants (i.e., directly from plants or from animals that have been fed on plants). It would seem logical, therefore, that plants

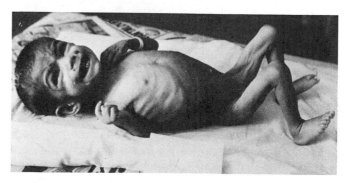

FIGURE 26-10
The devastation of malnutrition. (Photo courtesy Public Health Department, Iran; issued by the Food and Agriculture Organization of the United Nations)

could supply all the proteins necessary for life. To a degree this is true, but plant proteins are usually deficient in some essential amino acid (often lysine). For this reason, a diet containing animal protein is more apt to satisfy the protein amino acid needs of an individual human or other animal. Botanists currently are working to develop new strains of plants that are not only high in proteins but also have a balanced supply of essential amino acids.

The reason that people in less developed lands do not adopt a diet containing greater amounts of animal protein is simple: animals are very inefficient in converting plant proteins into animal proteins. Thus, if plant foods are already in short supply in a country, the situation would become much worse if plants were fed to livestock rather than to humans. For this reason, scientists and governments are looking for improvements in plant protein content and quality, as well as increases in the productivity of agriculture, to help people in the Third World.

Crop productivity of industrialized nations differs markedly from that of LDCs. In Cuba, the sugarcane yield per hectare (2.5 acres) is less than one-third that in Hawaii; soybean yields in Indonesia are only about one-third those achieved in Canada; wheat yields in India are about one-fifth as great as they are in the Netherlands; rice crops in Japan are three times as productive as those in Brazil. What makes the difference? Higher productivity is not the result of intrinsically better soil, greater rainfall, or any other set of natural conditions. The great difference is in the **energy subsidy** of agriculture.

The developed countries use much more energy in the form of fuel to run tractors, manufacture and apply fertilizers, and irrigate and harvest crops. Without petroleum, now increasingly imported from the Middle East and Venezuela, the economies of such completely oil-dependent countries as the Netherlands would immediately deteriorate. The United States, which is rapidly approaching a state of oil dependence, may soon lose its place as the world's "breadbasket." This is why it is so important that methods of tapping other sources of energy are found.

The Energy Crisis

It is no coincidence that as the population crisis deepens and the prospects for feeding a hungry world continue to diminish, a crisis in energy also occurs. The increasing comfort of the life of the average North American sets the standard for the world, but is maintained by energy supplies increasingly obtained from outside sources. This dependence on external sources is even more evident in other major industrialized societies, few of which at present are as self-sufficient in energy as North America.

Estimates of the total world energy reserves vary, but the oil reservoirs of the Middle East are expected to run out sometimes in the first or second decade of the next century. Meanwhile, other energy sources will be developed, but none will be cheap, and all carry with them dangers to the environment. Conversion to coal as a major energy source in the United States will require strip mining of large areas. Reclamation of coal-stripped land is never very successful, despite the advertisements of coal producers. At best, stripped and restored coal fields have recreational value; seldom can they be restored to their former agricultural productivity.

At present, atomic energy is a dangerous source of power. Reactors can be operated in comparative safety, but a safe disposal method for radioactive wastes, which retain their deadly contamination for thousands of years, has not been found. One solution is to bury them in the ocean depths, but even this gives no assurance against eventual contamination of surface waters.

Fission reaction (hydrogen fission), which produces very little radioactive waste matter, will probably become an important energy source, but thus far the technology has been elusive.

One of the greatest nonpolluting forms of energy is photosynthesis. Even now, when no special efforts have been made to adapt photosynthetic processes toward freeing the world from dependence on fossil fuels, photosynthesis produces over 100 times as much energy as do all human activities, including coal mining, oil and gas wells, nuclear power, and wind and water power. In addition, if coal, oil, and gas are considered, since they are the products of ancient plants, photosynthesis far eclipses any other energy source. It is the most direct and efficient means of tapping solar energy, and new and better ways of using it need to be found.

Zero Population Growth

When the number of births equals the number of deaths in a population, the result is a stable population, and the growth rate is zero (**zero population growth,** or **ZPG**). Currently Austria, West Germany, and Luxembourg have attained ZPG, and Sweden, Denmark, East Germany, Norway, and France are near it. Japan approached ZPG for a while through rigid birth control and legal abortions, but its growth rate seems to be increasing. The United States, with an indigenous growth rate of 0.7 percent, might have achieved near ZPG, but illegal immigration has raised its true growth rate to the highest of any nation with an advanced economy. Zero population growth is a world ideal, but currently unattainable because the present rate of growth (2.0 percent) promises a doubling of the world population (presently about 4 billion) in 35 years and 30 billion people struggling for a bare existence a hundred years from now.

The "good news" is that calculations give the carrying capacity of the world ecosystem at about 30 billion people. The "bad news" is that a population that high will require extermination of all animal life other than human and conversion of all arable land to crops. In addition, the human population will have to live on the polar ice caps or on rafts in the sea. Long before then, environmental resistance will probably reduce the rate of human population increase. In fact, recent estimates of population growth in certain emerging Third World nations, including China and India, indicate their population growth rates may be starting to diminish. This is a hopeful sign for the future. Nevertheless, populations in these and many other parts of the world continue to increase, and it is clear that great troubles lie ahead of us; all our mental resources will be taxed to find solutions.

Conclusion

There is little doubt that the present human population explosion is a temporary phenomenon and will be self-limiting in the future, but how it becomes self-limiting is a matter of grave concern. Meanwhile, populations continue to grow, and the energy crisis is a reality. As enlightened individuals who understand something of the nature of biological processes, it is up to us to begin applying some of the principles we have learned. As informed individuals and as advanced nations, we must extend every aid to less developed nations to help them limit birth rates and improve the quality of their life. Population reduction will be an important means of attaining world peace and prosperity.

SUMMARY

A famous case history of the death of an ecosystem is the great Irish potato famine of 1845–47. Destruction of the potato crop caused the death of a million Irish people and the migration of a million and a half. What often is forgotten is that the famine was preceded by an unprecedented population explosion.

All organisms including humans have the capacity to crowd the earth in just a few generations. This reproductive potential is held in check by environmental resistance—the sum of limiting factors, including food, space, disease, and predation, in the ecosystem. However, when limiting factors are removed, as is now true of the human population, populations increase beyond the carrying capacity of the environment. This point has already been reached in some nations and will be reached in others in just a few more decades. It is a biological problem, and its biological solutions are known. Whether biological principles will be applied in time to salvage the world ecosystem remains to be seen.

KEY WORDS

monoculture
green revolution
"miracle" rice

"miracle" wheat
reproductive potential
environmental resistance

carrying capacity
energy crisis

QUESTIONS FOR REVIEW AND DISCUSSION

1 What were the principal factors responsible for the Irish potato famine of 1845–47?

2 What is the long-term effect of human population on the ecosystem? Give an example based on your knowledge of history.

3 How is the growth rate of a human population calculated? What is the current growth rate in the United States? In Mexico?

4 Compare carrying capacity and environmental resistance. Are they essentially the same? Explain.

5 What effect does stress due to overcrowding have on caged animals? What might this portend for human populations?

6 What would you say to a remark to the effect that tropical rainforest regions should supply ample food for future human populations? To a statement that coal will meet future enegy needs? To an advocate of atomic energy?

SUGGESTED READING

COALE, A. J. 1974. The history of the human population. *Sci. Amer.* 231(3):40–51. (A nice history of human populations from the Stone Age to the modern era.)

DEMENEY, P. 1974. The populations of underdeveloped countries. *Sci. Amer.* 231(3):148–59. (Illiteracy, income maldistribution, the status of women—all contribute to overpopulation.)

FREEDMAN, R., and B. BERELSON. 1974. The human population. *Sci. Amer.* 231(3):30–39. (A rather pessimistic view of the future of human populations. Stabilization may not occur before the 15 billion mark is reached.)

FREJKA, T. 1973. The prospects for a stationary world population. *Sci. Amer.* 228(3):15–23. (A somewhat more optimistic view of world populations than usually expressed by demographers.)

GWATKIN, R., and S. K. BRANDEL. 1982. Life expectancy and population growth in the third world. *Sci. Amer.* 246(5):57–65. (Third World populations, projected to triple in the next century [from three billion to nine billion], can be checked only by reducing birth rates drastically.)

HARLAN, J. R. 1976. The plants and animals that nourish man. *Sci. Amer.* 235(5):88–97. (An interesting account of the dawn of agriculture and the domestication of plants and animals.)

NATIONAL ACADEMY OF SCIENCES, USA. 1984. *Genetic engineering of plants—agricultural research opportunities and policy concerns.* Washington, D.C.: National Academy Press. (Presents information on the green revolution in LDCs.)

NIEDERHAUSER, J. S., and W. C. COBB. 1959. The late blight of potatoes. *Sci. Amer.* 200(5):100–112. (Describes the Irish potato famine briefly and examines the present status of the late blight disease.)

REVELLE, R. 1974. Food and population. *Sci. Amer.* 231(3):160–70. (Mass starvation seems to be around the corner for peoples in many underdeveloped lands.)

Appendix:
Classification of Life

Kingdom Monera—prokaryotes
    Division Bacteria (about 1500 species)—*Bacillus*
    Division Cyanobacteria (about 1500 species)—*Anabaena, Nostoc, Oscillatoria*
    Division Prochlorophyta (1 known species)—green prokaryotes (*Prochloron*)

Kingdom Archaebacteria—a recently discovered group of primitive bacterialike
    microorganisms.

Kingdom Protista—unicellular eukaryotes
    Phylum Pyrrophyta (about 1100 species)—dinoflagellates (*Gymnodinium, Gonyaulax*)
    Phylum Chrysophyta (5000+ species)—golden algae (diatoms)
    Phylum Protozoa (15,000+ species)
        Class Flagellata—*Euglena, Giardia,* trypanosomes, leishmanias
        Class Sarcodina—*Amoeba, Entamoeba,* infusorians, radiolarians
        Class Ciliata—*Paramecium, Didinium, Vorticella, Stentor*
        Class Sporozoa—*Plasmodium*

Kingdom Fungi
    Division Gymnomycota (165+ species)—slime molds
    Division Mastigomycota (1800+ species)—whip fungi (*Saprolegnia, Rhizopus,*
        Phytopthora)
    Division Amastigomycota—higher fungi
        Subdivision Ascomycotina (25,000+ species)—sac fungi (*Neurospora,* yeasts, ergots,
            morels, truffles)
        Subdivision Basidiomycotina (25,000+ species)—club fungi (*Fomes, Geaster, Psilocybe,*
            Agaricus, Amanita, Clavaria, Phallus, Marasmius, corn smut, wheat rust)
        Subdivision Deuteromycotina (25,000+ species)—imperfect fungi (*Penicillium*)
    Division Lichenes (25,000+ species)—lichens (crustose, foliose, and fruticose
        lichens)

Kingdom Plantae
  Division Chlorophyta (7000+ species)—green algae (*Chlamydomonas*, *Volvox*, *Ulothrix*, *Spirogyra*, *Ulva*, *Chara*, *Acetabularia*, *Codium*, *Caulerpa*)
  Division Rhodophyta (4000+ species)—red algae (*Porphyridium*, *Ptilota*, *Nemalion*, *Gelidium*, *Porphyra*, *Chondrus*)
  Division Phaeophyta (1500+ species)—brown algae (kelps, rockweeds, *Postelsia*, *Padina*, *Fucus*, *Sargassum*, *Laminaria*)
  Division Bryophyta (24,500+ species)—mosses (*Dawsonia*, *Bryum*, *Sphagnum*), liverworts (*Marchantia*), hornworts (*Anthoceros*)
  Division Rhyniophyta—extinct leafless, rootless plants (*Psilophyton*, *Rhynia*)
  Division Psilophyta (about 12 species)—fork ferns (*Psilotum*, *Tmesipteris*)
  Division Lycopodiophyta (800+ species)—club mosses (extinct *Lepidodendron* and other similar trees, *Lycopodium*, *Selaginella*, *Isoetes*)
  Division Equisetophyta (about 20 species)—extinct *Calamites*, *Equisetum*
  Division Polypodiophyta (9000+ species)—*Pteridium*, *Gleichenia*, *Cyathea*, other ferns including Christmas fern and water ferns
  Division Pinophyta—gymnosperms
    Class Lyginopteridopsida—seed ferns (all extinct)
    Class Cycadopsida (about 120 species)—cycads (*Zamia*)
    Class Ginkgoopsida (1 species)—*Ginkgo biloba*
    Class Pinopsida (about 575 species)—pines, spruces, firs, cedars, cypresses, *Araucaria*, *Sequoiadendron*, *Podocarpus*
    Class Gnetopsida (about 75 species)—*Gnetum*, *Ephedra*, *Welwitschia*
  Division Magnoliophyta—flowering plants
    Class Liliopsida (about 75,000 species)—monocotyledons (monocots) (lilies, orchids, maize, wheat, bamboo, palms, others)
    Class Magnoliopsida (about 200,000 species)—dicotyledons (dicots) (beans, peas, oaks, elms, cotton, flax, roses, apples, peaches, many others)

Kingdom Animalia
  Subkingdom Parazoa
    Phylum Porifera (5000+ species)—sponges
  Subkingdom Metazoa
    Phylum Cnidaria (Coelenterata) (10,000+ species)
      Class Hydrozoa—*Hydra*, *Obelia*
      Class Scyphozoa—jellyfish
      Class Anthozoa—corals, sea anemones
    Phylum Platyhelminthes (10,000+ species)
      Class Turbellaria—*Dugesia*, other planarians
      Class Trematoda—flukes
      Class Cestoda—tapeworms
    Phylum Nematoda (12,000+ species)—*Ascaris*, *Trichinella*, filaria worms, hookworms, pinworms, other parasitic and free-living nematodes
    Phylum Rotifera (1500+ species)—rotifers
    Phylum Annelida (7500+ species)
      Class Polychaeta—polychaetes, fanworms
      Class Oligochaeta—earthworms
      Class Hirudinea—leeches
    Phylum Mollusca (128,000+ species)
      Class Amphineura—*Neopilina*
      Class Gastropoda—snails, slugs, conches
      Class Pelecypoda—clams, oysters
      Class Scaphopoda—tooth shells
      Class Cephalopoda—nautiloids, squids, octopods

Phylum Arthropoda (900,000+ species)
   Class Trilobita—trilobites (extinct)
   Class Merostomata—horseshoe crabs
   Class Arachnida—scorpions, spiders, mites, ticks
   Class Crustacea—shrimps, lobsters, crabs, barnacles, sow bugs
   Class Diplopoda—millipedes
   Class Chilopoda—centipedes
   Class Insecta—insects
Phylum Onychophora (65 species)—*Peripatus*
Phylum Echinodermata (5000+ species)
   Class Echinoidea—sea urchins, sand dollars
   Class Ophiuroidea—brittle stars
   Class Crinoidea—sea lilies
   Class Asteroidea—starfish
   Class Holothuroidea—sea cucumbers
Phylum Chordata (40,000+ species)—animals with backbones
   Subphylum Urochordata (1600+ species)—tunicates
   Subphylum Cephalochordata (30 species)—*Amphioxus*
   Subphylum Vertebrata (38,000 + species)—animals with vertebral columns
      Class Agnatha (50+ species)—extinct jawless Paleozoic fish, living cyclostomes
      Class Elasmobranchii (600+ species)—extinct jawed fish with a cartilaginous skeleton; living sharks and their relatives, rays and skates
      Class Osteichthyes (20,000+ species)—bony fish (lungfish, crossopterygians, bony fish with swim bladders)
      Class Amphibia (3000+ species)—amphibians (salamanders, frogs, toads)
      Class Reptilia (6000+ species)—reptiles (dinosaurs, lizards, crocodiles, alligators, turtles, snakes)
      Class Aves (8600+ species)—birds
      Class Mammalia (4000+ species)—mammals
         Subclass Prototheria—egg layers (duckbilled platypus, echidna)
         Subclass Metatheria—pouch bearers (kangaroo, oppossum)
         Subclass Eutheria—advanced, placental mammals (mouse, whale, human)

Glossary

abdomen Posterior region of an animal body containing viscera.

aberration Abnormal chromosome structure or complement.

abiogenesis Hypothetical development of living organisms from nonliving matter.

abortion Induced termination of pregnancy.

abscission Leaf or fruit fall by formation of a separation layer of cells at the base of the organ.

absorption spectrum Pattern resulting from the absorption of various wavelengths of radiant energy by a pigment.

accessory pigment Nonchlorophyll pigment that augments absorption of radiant energy in photosystems.

acid Substance that upon dissociation releases hydrogen ions, producing a pH of less than 7.

acoelomate Lacking a coelom.

acrosome Enzyme-containing cap of a spermatozoid.

actin Fibrous cytoplasmic protein associated with internal cellular movements.

action potential Change in the electrical charge distribution in outer membranes of an active cell.

action spectrum Conversion of absorbed radiant energy of different wavelengths.

activation energy Energy required to initiate a chemical reaction.

active site Region of an enzyme molecule to which a substrate molecule is bound temporarily.

active transport Movement of molecules in cells as a result of energy-expending actions.

adenine One of the two purine bases of nucleic acids.

Note: For specific chemical compounds and organisms, refer to the index.

adenosine Component of nucleic acids composed of adenine and a pentose sugar.

adenosine diphosphate (ADP) Adenosine coupled with two phosphoric acids, forming one pyrophosphate bond. See adenosine.

adenosine monophosphate (AMP) Adenosine coupled with one phosphoric acid. See adenosine and cyclic-AMP.

adenosine triphosphate (ATP) Adenosine coupled with three phosphoric acids, forming two pyrophosphate bonds; major usable source of chemical energy in cells.

adipose tissue Fat-storing tissue in animals.

ADP See adenosine diphosphate.

adrenal cortex Outer region of the adrenal glands.

adrenal gland One of a pair of endocrine glands adjacent to the mammalian kidneys; source of adrenaline and other hormones.

adrenal medulla Inner region of the adrenal glands.

adventitious root Root originating laterally from a stem.

aestivation Summer "sleep"; summer dormancy.

afferent neuron See sensory neuron.

agar-agar Colloidal component of cell walls of red algae; extract is used as a solidifier in bacteriological media.

agglutination Clumping of red blood cells.

agonistic behavior Conflict behavior.

alchemist Philosopher of the late Dark Ages to the Middle Ages (A.D. 1100–1500); studied the nature of matter; recognized seven metallic elements.

alcohol Organic molecule containing a hydroxyl group joined to a nonaromatic carbon atom to which no other oxygen atoms are bonded.

aldehyde Organic compound possessing a CHO group.

aleurone layer Outer cell layer of a plant endosperm.

alga One of a diverse group of simple unicellular, colonial, or multicellular plants, usually aquatic.

algal bloom Population increase of an alga.

algin Colloidal, highly hygroscopic polysaccharide occurring in cell walls of brown algae.

alginate *See* algin.

alginic acid *See* algin.

alkali Substance that releases hydroxyl ions in water, producing a pH greater than 7.

allantois Waste storage sac of the embryo of reptiles and birds; a transport organ in mammalian embryos.

alleles Genes on homologous chromosomes occupying similar loci but carrying contrasting inheritance factors.

allele frequency Percent occurrence of an allele in a population.

allergen Any substance, but usually proteinaceous, capable of causing allergy and of inducing production of antibodies.

alpha ray Positively charged radioactive particle, equivalent to the nucleus of an atom.

alternation of generations Plant life cycle of alternating haploid, gamete-producing and diploid, meiospore-producing generations.

alveolus Sac; in vertebrates, one of the air sacs of the lungs.

amine Derivative of ammonia with one, two, or all three of the hydrogen atoms of ammonia replaced by hydrocarbon radicals.

amino acid Nitrogen-containing organic molecule that is the basic structural unit of proteins.

amniocentesis Analysis of fetal defects by withdrawing amniotic fluid.

amnion Sac enclosing the embryo of higher vertebrates.

AMP *See* adenosine monophosphate.

amphibian Member of a class of vertebrate tetrapods having both gills and lungs.

anabolism Any phase of metabolism in which molecular synthesis occurs.

anaerobic respiration High-energy-yielding respiration occurring in the absence of oxygen.

analogous Functionally but not developmentally similar.

anaphase Stage in cell division in which chromosome sets migrate toward the spindle poles.

anatomy Study of the internal structure of organisms; structure of an organism.

androgen Testosterone or testosteronelike hormone.

aneuploidy Having one or several extra or fewer chromosomes than the normal complement, but not plus or minus a complete set.

angiosperm Flowering plant; enclosed seeds.

animal pole Hemisphere of an animal egg where yolk density is least.

anion Ion with a negative charge.

annelid Any segmented worm of the phylum Annelida.

annual Flowering plant completing its life cycle within one year.

anode Positively charged electrode.

Note: For specific chemical compounds and organisms, refer to the index.

anther Microsporangium of the stamen; male pollen-bearing structure of a flower.

antheridium Sperm-producing organ of land plants.

antheridogen Gibberellinlike hormone secreted by a fern gametophyte which induces development of antheridia on adjacent gametophytes.

antibiotic Substance produced by one organism capable of inhibiting growth of another organism (commonly a fungal or bacterial product).

antibody Protein formed in the animal body capable of interacting with an antigen.

anticodon Group of three nucleotides that acts as "recognition site" in binding tRNA to mRNA.

antigen Foreign proteinaceous substance capable of producing an immune reaction.

antipodal cells Cells, usually occurring in threes, in the embryo sac of flowering plants; have no known function.

antitoxin Antibody in serum capable of neutralizing a specific toxin.

anuran Tail-less amphibian (e.g., frog).

anus Excretory aperture of the gut.

aorta Major artery in vertebrates.

ape Any of the large tail-less primates (orangutans, chimpanzees, gorillas) exclusive of humans.

apical dome In plants, the shoot tip including an apical meristem.

apical growth In plants, growth by cell divisions at apices or tips of shoots and roots (if present).

apical meristem Meristem of a shoot or root tip.

apogamy Vegetative development of a sporophyte plant from a gametophyte; i.e., without gametic fusion.

apospory Development of a gametophyte plant from a sporophyte by vegetative reproduction, i.e., without spores (without meiosis).

appendage Outgrowth of a cell or organism having a specific function, usually sensory or locomotory.

appendicular skeleton Part of the vertebrate skeleton composed of bones of the appendages.

archegonium Vaselike, egg-containing organ of land plants.

archenteron Primitive gut with only one opening, as in hydras or the gastrula embryo stage.

artery Blood vessel that carries blood away from the heart.

arthropod Animal having an exoskeleton and jointed appendages; member of the phylum Arthropoda.

ascocarp Fruiting body of an ascomycete.

ascogonium Female sex organ of ascomycetes.

ascus Sac-like, meiospore-producing organ of sac fungi.

association neuron *See* connector neuron.

asymmetrical, asymmetry Lacking planes of symmetry.

atom Smallest unit of an element; comprised of electrons, protons, and neutrons.

ATP *See* adenosine triphosphate.

atrium Chamber; heart chamber that receives venous blood.

attenuation (medical) Production of a weakened form of pathogen by culturing *in vitro* or in a laboratory animal.

autolysis Digestion of a cell or its components from within.

autonomic nervous system Part of the vertebrate peripheral nervous system associated with involuntary actions.

auxin Plant growth hormone capable of inducing coleoptile curvature.

axial gradient Pole-to-pole concentration gradient of any metabolic substance occurring in an organism or an appendage.

axial skeleton Skull and spinal column of vertebrates.

axillary bud Lateral bud in the axis of a leaf.

axon An extension of a neuron that usually carries impulses away from the cell.

bacteriology Study of all aspects of bacteria.

bacterium Microscopic prokaryotic cell; may be pathogenic.

bark Outer covering of woody plant stems and older roots; includes phloem, cortex, and cork.

Barr body Condensed, dark-staining inactive X chromosome in human female cells.

base Substance that upon dissociation releases hydroxyl ions, producing a pH greater than 7.

base pair Association in DNA of a complementary purine and pyrimidine.

base pairing *See* base pair.

basidiocarp Fruiting body of a basidiomycete.

basidiospore Haploid meiospore of a club fungus (basidiomycete).

basidium Club-shaped cell of basidiomycetes, usually forming four meiospores (basidiospores).

behavior Motor responses to stimuli that result in some kind of spatial adjustment.

β-galactosidase Enzyme that hydrolyzes lactose.

beta (β) ray Negatively charged particle (electron) of radiation.

biceps Large flexor muscle of the upper arm of tetrapods.

biennial Flowering plant completing its life cycle within two years.

bilateral symmetry Divisible into like halves in one plane only.

bile Mixture of degradation products of blood and the bile pigments, secreted by the liver.

bile duct Duct leading from the gallbladder to the intestine of vertebrates. Also called the *common bile duct.*

biochemical mutant Gene mutation producing an altered protein, or inability to synthesize a particular protein.

biodegradable Capable of being decomposed by organisms.

biogenetic law Theory that embryos pass through stages reflecting evolutionary ancestry.

biogeography Study of the distribution of species in the world's ecosystems.

biological clock Endogenous (inner) time-measuring component of cells and organisms.

biological oxidation Cellular respiration; oxidation of food.

biomass Total mass of all organisms and their products in an ecosystem.

biome Major ecosystem.

biosynthesis Synthesis of organic compounds by living cells.

Note: For specific chemical compounds and organisms, refer to the index.

biotic factor Organisms in the ecosystem and their products.

bipedalism Going about on two legs.

bipolar neuron *See* connector neuron.

blastopore Opening of the primitive gut in a developing embryo.

blastula Hollow, ball stage of early animal embryology.

blood Gas-transporting liquid of animal bodies.

blood group Set of blood-serum antibodies cross-reacting with blood antigens.

blood type *See* blood group.

blue-green alga Photosynthetic prokaryote containing chlorophyll-a and phycobilins; now cyanobacterium.

B-lymphocyte White-cell-producing antibodies in the blood.

bond *See* chemical bond.

bone Calcified connective tissue of vertebrate skeletons.

bony fish Member of the class Osteichthyes, having skeletons of bone rather than cartilage.

book lung Gill-like respiratory organ of arachnids.

bordered pit Xylem pit having circular, overarching borders.

botany Scientific study of plant life.

Bowman's capsule Cuplike membrane surrounding a kidney glomerulus.

brachiation Travel by swinging from arms.

brain Major aggregation of neurons, usually at the anterior end of a nerve cord.

brassin Steroid growth hormone found in plants.

brown alga Fucoxanthin-containing alga, of the division Phaeophyta.

bryophyte Simple land plant, generally considered nonvascular (mosses, liverworts, and hornworts).

buffer Salt of a weak acid which tends to suppress extremes of pH.

C-3 pathway *See* Calvin-Benson cycle.

C-4 pathway *See* Hatch-Slack pathway.

C-3 plant Plant photosynthesizing exclusively by the Calvin-Benson cycle.

C-4 plant Plant having the Hatch-Slack pathway of photosynthesis.

calcitonin Animal hormone that prevents bone resorption.

calorie Unit of heat energy required to raise 1 g of water 1°C.

Calvin-Benson cycle Pathway of carbon dioxide fixation in which 3-carbon sugars are produced. Also called the C-3 *pathway.*

calyx In plants, a whorl of sepals.

CAM Crassulacean acid metabolism; a form of photosynthesis.

cambium Meristematic cells beneath the bark of angiosperms and gymnosperms from which new wood and bark originate.

CAM plant Plant having the CAM pathway.

capillary Smallest blood vessel of vertebrates.

capillary water Water adhering to soil particles in a thin film.

carbohydrate Any of a group of organic molecules including or composed of simple sugars.

carbon cycle Cycling of carbon from carbon dioxide to organic compounds in the ecosystem.

cardiac Pertaining to the heart.

cardiac muscle Striated, branching muscle cells of the heart.

carnivore Flesh-eating mammal characterized by dentition specialized for tearing.

carotenoid Red or orange accessory photosynthetic pigment of plants.

carpel Organ of which the pistil of a flower is composed; equivalent to a megasporophyll.

carrying capacity Number of consumer organisms supportable on a sustained basis by producer organisms of an ecosystem.

cartilage Elastic, flexible connective tissue component of the vertebrate skeleton.

Casperian strip Suberized band encircling the tangential walls of the endodermal cells in roots and sometimes stems.

catabolism Metabolic degradation of complex molecules into simple ones.

catalysis Process that employs a catalyst to accelerate a chemical reaction.

catalyst Substance that accelerates chemical reactions but that is not used up in them.

cathode Negatively charged electrode.

cation Ion having a positive electrical charge.

cell Discrete mass of living matter surrounded by a membrane.

cell cycle Phases of cell reproduction.

cell membrane Lipid–protein membrane enclosing the cytoplasm of a cell.

cell plate Initial partition between two sister plant cells, formed in the final stages of mitosis.

cell theory Statement that all life has a cellular basis.

cellulose Carbohydrate (glucose polymer) that is the major component of plant cell walls.

cell wall Nonliving enclosing layer outside the cell membrane, common to all plants, fungi, and bacteria.

central cell In plants, a cell of the embryo sac of flowering plants that unites with a male gamete to produce the endosperm.

central nervous system. Vertebrate brain and spinal cord.

centriole Cytoplasmic organelle forming the spindle pole during mitosis and meiosis; rarely present in plant cells.

centromere Region of the chromosome to which spindle fibers are attached.

cephalization Evolutionary trend toward specialization of a head region in animal bodies.

cephalopod Class of mollusks including squids and octopods.

cerebellum Midbrain of vertebrates.

cerebrum Large forebrain of vertebrates.

cetacean Member of the order of whales and porpoises (Cetacea).

chelicerate Having pincerlike mouth parts; one of two subdivisions of arthropods, including scorpions and spiders.

Note: For specific chemical compounds and organisms, refer to the index.

chemical bond Force that binds atoms together in molecules.

chemical energy Energy involved in the assembly of atoms into compounds or resulting from the breakdown of compounds.

chemical evolution Theory of development of macromolecules in an atmosphere in which free oxygen was absent.

chemical reaction Interaction between molecules characterized by a change in the nature of the molecules.

chemiosmotic theory Theory that membrane separation of anions and hydrogen ions occurring during cellular oxidations generates sufficient free energy to produce ATP synthesis.

chemoreceptor Sensory neuron stimulated by certain aromatic gaseous molecules.

chemotaxis Movement of an organism in response to a chemical stimulus.

chiasma Structure formed by intertwining chromatids during prophase of meiosis I, when an exchange of chromatic material may take place.

chitin Mucopolysaccharide substance composing invertebrate exoskeletons and cuticles.

chlorenchyma Thin-walled, chloroplast-containing living cells of plants.

chlorophyll One of several forms of green pigments; photoactive.

chloroplast Any chlorophyll-containing photosynthetic organelle.

cholesterol Lipid steroid component of cell membranes, but also associated with coronary artery disease.

chondrin Gel-like secretion of cartilage cells, of which the matrix of cartilage is composed.

chordate Animal having an internal axial notochord.

chorion Outer embryonic membrane of reptiles, birds, and mammals; contributes to the placenta in the last.

chromatid One of two longitudinal strands composing a metaphase chromosome.

chromatophore Pigment granule of a pigment cell.

chromosomal aberration Any atypical structural or numerical modification of a chromosome or chromosomes.

chromosomal puff Diffuse area of a chromosome associated with transcription.

chromosome Threadlike, gene-bearing body in the nucleus; becomes conspicuous during meiosis and mitosis.

chronon Putative timing unit of organisms; possibly DNA sequences of certain lengths.

chyme Mixture of partially digested food, enzymes, and stomach fluids.

ciliate Ciliated protozoan, of the class Ciliata.

cilium Hairlike locomotory appendage of a cell, similar to a flagellum but shorter.

circadian rhythm Periodic cycle of biological clocks based on a 24-hour cycle.

circulatory system Any system of blood vessels, sinuses, and hearts capable of circulating a gas- and food-transporting fluid (blood).

circumcision In man, surgical removal of the foreskin of the penis.

citric acid cycle See Krebs cycle.

classification System of arranging organisms on the basis of common characteristics.

clay Fine-grained mineral component of soil, capable of forming a plastic colloid when wet.

cleavage Partitioning divisions of zygotes and early embryos.

cleavage division See cleavage.

climatic climax community Most enduring association of organisms comprising an ecosystem in a particular environment.

climax Culmination of copulation.

climax community See climatic climax community.

clitoris Small organ at the anterior or ventral part of the vulva; homolog of the penis.

cloaca Posteriormost chamber of the intestine in lower vertebrates.

clone Individuals descended by asexual propagation from a single parent and therefore genetically alike.

cloning Production of like individuals by asexual reproduction; replication of genes by introduction into bacterial cells.

closed circulation Having blood contained wholly within the heart and blood vessels.

club fungus Group (Basidiomycotina) of fungi producing meiospores borne on club-shaped reproductive cells.

club moss Coniferous, non-seed-bearing vascular plant.

cnidarian Any member of the phylum Cnidaria (jellyfish, corals, etc.).

cnidocyte Stinging cell of Cnidarians.

coalition Temporary alliance to thwart a dominant in many primates.

cochlea Snail-like chamber of the vertebrate inner ear.

codon Sequence of three nucleotides specifying for the synthesis of one amino acid.

coelenterate See cnidarian.

coelom Body cavity lined with epithelium.

coenocytic Nonpartitioning of cells, as in multinucleate tube cells of some fungi and algae.

coenzyme Small organic molecule required for many enzymatic reactions in cells.

coevolution Interdependent evolution of two different organisms in a symbiotic or other interdependent relationship.

cofactor Small molecule or ion required in addition to an enzyme in catalysis.

colchicine Alkaloid from the bulb of the autumn crocus; used to induce artificial polyploidy.

coleoptile Sheathing first leaf of grass embryos.

collar cell Flagellated cell of sponges.

collecting tubule Distal tubule of the kidney nephron that discharges urine into the renal pelvis.

collenchyma First-formed mechanical or strengthening tissue of stems.

colloid Permanent suspension of fine particles.

colon Posterior part of a mammalian large intestine, where water is resorbed.

Note: For specific chemical compounds and organisms, refer to the index.

colonial theory Hypothesis that metazoans evolved from unspecialized cell aggregates.

commensal Pertaining to a symbiotic relationship benefitting one member and doing no harm to the other member.

community In ecology, any group of several kinds of organisms living together in a definable relationship.

companion cell Auxiliary cell of phloem in angiosperms, adjacent to sieve tube elements.

competition Competition for energy by the organisms of an ecosystem.

competitive inhibition Inhibition of action of an enzyme or other molecule receptor by another molecule structurally similar to the reactant.

complement Blood proteins induced by infection to react with foreign bodies such as bacteria.

compound Substance composed of two or more different elements united in a definite numerical ratio by chemical bonds.

compound eye Eye of certain invertebrates; composed of many simple eyes.

compound leaf Leaf divided into leaflets.

condensation Process by which a gas is reduced in volume to a liquid or in which small molecules are joined to make a large molecule.

conditioning Involuntary response to a signal experimentally substituted for an original stimulus.

cone Color-sensitive, conical photoreceptor of the vertebrate eye.

conifer Woody, cone-bearing gymnosperm (pines, spruces, etc.).

conjugation Sexual mating of unicells.

connective tissue Fibrous and cellular structural tissue of animal bodies.

connector neuron Neuron serving as a link between two other neurons, commonly a motor and a sensory neuron. Also called an *association neuron* or a *polar neuron*.

consumer Organism that eats other organisms or their products to obtain energy.

continental drift Long-term migrations of landmasses due to sea-floor spreading.

contour feather Nonflight feather of birds forming a cover over the down feathers.

contraception Use of birth-control measures.

control Normal or natural state in an experiment.

controlled experiment Experiment conducted in duplicate with one experiment, the control, representing the natural state, the other the variable.

convergent evolution Evolution of nonrelated species into phenotypically similar forms.

copulation Mating behavior of dioecious animals; gamete exchange.

coral Colonial limestone-forming hydroids.

core area Central area of the range of an animal; that which may be defended.

cork Outer bark of older stems and roots produced by a cork cambium; composed of dead, suberized cells.

cork cambium Lateral meristem producing cork in woody and some herbaceous plants.

corpora cavernosa   Spongy erectile tissue of the mammalian penis

corpus callosum   Dense band of neurons between the cerebral hemispheres of the vertebrate brain.

corpuscle   Any small, discrete body, such as a blood cell.

corpus luteum   Yellow endocrine tissue that develops in a ruptured ovarian follicle after ovulation.

corpus spongeosum   Spongy erectile tissue in the mammalian penis and clitoris.

cortex   Outer region of an organ (e.g., of the kidney or the adrenal glands); outer portion of a plant stem or root.

cortisol   Hormone formed in the adrenal cortex.

cotyledon   Seed leaf of a plant embryo.

covalent bond   Chemical bond composed of a pair of electrons shared by two atoms.

cranial nerve   Nerve trunk arising from the vertebrate brain in pairs.

crinoid   Any stalked or sessile echinoderm (e.g., sea lily).

cristae   Shelflike partitions in the interior of a mitochondrion.

crop   Esophageal storage sac of birds.

crossing over   Interchange of genes, usually linked in inheritance.

crossopterygian   Primitive bony fish with stumpy, leglike fins.

crozier   Hook-shaped sexual cell of ascomycetes; juvenile fern leaf (fiddlehead).

crystalline   Pertaining to a substance in which the arrangements of atoms is regular.

cuticle   Membranous, usually chitinous layer covering the bodies of invertebrates; lipid layer coating plant epidermal cells.

cycad   Primitive palmlike seed plant.

cyclic-AMP   Cyclic adenosine monophosphate; regulatory molecule associated with many hormonal functions. *See* adenosine monophosphate.

cyclostome   Jawless primitive fish (e.g., hagfish, lamprey).

cytochrome   Any of several iron-containing respiratory pigments.

cytokinin   In plants, any of a group of bud-stimulating hormones.

cytokinesis   *See* cytoplasmic division.

cytology   Study of cell structure.

cytoplasm   Cell interior exclusive of the nucleus and nonliving inclusions.

cytoplasmic division   Stage of cell division in which the cytoplasm is divided between the two sister cells.

cytosine   One of three pyrimidine bases of nucleic acids.

cytoskeleton   Supporting structure of a cell's interior; composed of microtubules.

dark reactions   Carbon-dioxide-fixing reactions of photosynthesis.

daughter chromosome   One of a complete set of chromosomes produced in mitosis and meiosis.

day-neutral plant   Plant whose flowering is not induced by any specific photoperiod.

Note: For specific chemical compounds and organisms, refer to the index.

deamination   Chemical reaction in which an amino group is removed from an amino acid.

decarboxylation   Chemical reaction in which an acid radical (—COOH) is removed from an organic acid.

dehydration   Loss of water from tissues, cells, or organisms; loss of water between molecules, resulting in bond formation.

dendrite   Branched, afferent extension of a neuron.

deoxyribonucleic acid (DNA)   Macromolecule composed of a double strand of deoxyribose, containing nucleotide pairs that serve as the genetic code in cells.

derepression   Inactivation of a gene-repressing molecule for DNA transcription.

dermatoglyph   Fingerprint or similar epidermal ridges.

dermis   Skin of vertebrates; epidermis plus underlying connective tissue.

desert   Biome characterized by a very low annual precipitation.

determinate growth   Having a finite growth, ending in an adult phase with predictable size.

deuterium   Hydrogen atom, called "heavy hydrogen," composed of one proton, one neutron, and one electron.

deuterostome   Animal in which, during embryo development, the blastopore becomes the anus and a new mouth opening is formed.

development   Progression through the stages of maturation.

dialysis   Diffusion of small molecules of a solution through pores of a containing membrane with retention of large molecules.

diaphragm   Sheet of muscle in mammals separating the chest cavity from the abdominal cavity; cuplike contraceptive device.

diatom   Protistan cell enclosed by siliceous walls.

dichotomous   Even forking of a structure; pertaining to alternative but similar pathways.

dicotyledon   Flowering plant having embryos with two cotyledons.

dictyostele   Siphonostele having overlapping leaf gaps.

diencephalon   Posteriormost part of the vertebrate forebrain (cerebrum).

differentiation   In development, the process of cell specialization.

diffusion   Movement of particles from an area of high concentration to one of lower concentration.

digestion   Breaking down and solubilizing of food; hydrolysis of macromolecules.

digestive tract   Gut or alimentary canal in which digestion occurs.

dihybrid cross   Genetic cross involving two pairs of alleles.

dikaryon   Body of a fungus composed of binucleate cells.

dinoflagellate   Autotrophic flagellated protistan; member of the phylum Pyrrophyta.

dinosaur   One of a group of extinct, largely terrestrial reptiles.

dioecious   Literally, "two houses"; condition of separation of the sexes; male and female individuals.

diploid   Number of chromosomes characterized by chromosome pairing; the 2*n* number.

disease   Any atypical or unusual physiological or anatomical condition in an organism; any infective debilitating condition.

displacement reaction Chemical reaction in which ion exchange occurs between different reacting molecules.

dissociate Separate into two parts.

disulfide bridge Sulfur-to-sulfur bond between sulfhydral groups of two amino acids of a protein.

divergent evolution Radiant evolution; evolution of species from a common progenitor.

DNA *See* deoxyribonucleic acid.

dominant gene Gene whose expression prevails over that of its allele, which thus is termed recessive.

dormancy Long-term inactive or resting stage in the life of an organism.

dorsal Pertaining to the upper or back side.

dorsal hollow nerve cord Vertebrate spinal cord.

dorsal lip Upper rim of the blastopore having an organizing function in embryo development.

dorsiventral Having distinct dorsal and ventral surfaces.

double bond Two pairs of shared electrons that hold two atoms together.

double covalent bond *See* double bond.

double helix Three-dimensional, helical arrangement of the two strands of DNA.

down Fluffy inner layer of vaneless feathers of birds.

Down's syndrome Genetic defect, usually characterized by subnormal intelligence.

duodenum Relatively short length of the mammalian small intestine, adjacent to the stomach.

dyad Chromosome composed of two chromatids in meiosis.

ecdysone Molting hormone of arthropods.

echinoderm Any member of the phylum Echinodermata (starfish, sea urchins, etc.); invertebrates having radially symmetrical bodies.

ecological community *See* community.

ecology Study of organisms in relation to one another and to environmental factors.

ecosystem Any ecological association of organisms and a physical environment.

ectoderm Outer embryonic germ layer of animals.

ectotherm Animal deriving body heat principally from external sources (cold-blooded).

effector Cell or organ which when stimulated by a nerve impulse initiates a specific action.

efferent neuron *See* motor neuron.

egg Female gamete; shelled, reproductive unit of animals.

egg apparatus In flowering plants, the egg and two synergids.

elasmobranch Primitive jawed fish with a cartilaginous skeleton (sharks, rays, etc.).

elater In plants, a hygroscopic device that aids dispersal of spores of equisetums, liverworts, and hornworts.

electrical energy Flow of electrons along a conductor.

Note: For specific chemical compounds and organisms, refer to the index.

electron Negatively charged component of the atom.

electron carrier Cytochrome or other organic molecule functioning in electron transport in cells.

electron transfer Oxidation-reduction; transfer of electrons from donor to acceptor molecules.

electron transport Passage of electrons via a series of oxidation-reductions of electron carrier molecules.

electron transport system Linked set of cytochromes and associated enzymes and coenzymes functioning in biological oxidation.

electrophoresis Separation of polar molecules by use of an electrical field.

element Substance composed of only one kind of atom.

elementary particle One of certain structural subunits of mitochondria and chloroplasts thought to be sites of electron transport.

embryo Multicellular yet relatively undifferentiated juvenile individual; early stage in developing young.

embryogenesis Developmental stages of an embryo.

embryoid Plant embryo produced from somatic tissues; nonzygotic embryo.

embryo sac Female gametophyte of flowering plants.

emulsion Suspension (mixture) of lipid molecules, water, and a lipid-dispersing agent.

endergonic Pertaining to chemical reactions that consume energy and result in products having an increased energy content.

endocrine Pertaining to the hormonal products of ductless glands.

endoderm Inner embryonic germ layer of animals.

endodermis In plants, a layer of cells separating the cortex from the stele in stems and roots

endometrium Lining (mucosa) of the mammalian uterus.

endoplasmic reticulum Cell organelle composed of an extensive system of cytoplasmic tubules or flattened membranous sacs; functions in intracellular transport.

endorphin Neurosecretion having opiate properties.

endoskeleton Bony, internal skeleton.

endosperm Food storage tissue of a seed.

endotherm Animal deriving body heat from catabolism.

energy flow Transfer of energy from a source through a series of intermediates such as electron carriers in a cell or producer and consumer organisms in an ecosystem.

energy of activation Amount of energy necessary to initiate a chemical reaction.

enkephalin Member of a group of natural opiates found in mammalian brain tissues.

enucleated Pertaining to a cell that has lost its nucleus.

environmental resistance All the factors in the environment that limit population expansion.

enzyme Protein catalyst in living cells that in low concentration speeds the rate of chemical reactions.

epicotyl Part of the seedling axis above the cotyledons.

epidermis Single (in plants) or multiple (in animals) layer of cells that covers a tissue.

epididymis Sperm-collecting tubule of the mammalian testis.

epigenesis Theory that a stepwise development of embryos occurs, as opposed to preformation.

epiphyte   Literally, "plant on plant"; nonparasitic association of one plant growing upon another.

episome   Gene residing outside the chromosome; cytoplasmic gene.

epithelium   Outer layer of animal epidermis.

equational division   Second division in meiosis.

erythrocyte   Hemoglobin-containing cell; red blood cell.

esophagus   Muscular tube between the pharynx and stomach.

essential amino acid   One of several amino acids that animals cannot synthesize but must obtain by consuming plants.

essential element   *See* macronutrient, micronutrient.

essential mineral   *See* macronutrient, micronutrient.

ester linkage   Bonds between alcohols and fatty acids in lipids.

estrous cycle   Ovarian cycle; cycle of events from one ovulation to the next.

estrus   Period of sexual excitement at the time of ovulation.

ethology   Study of animal behavior.

eukaryotic cell, eukaryote   General higher cell type containing a nucleus and other organelles (as opposed to prokaryotic cell).

eurypterid   Extinct marine arthropod, ancestral to present-day chelicerates (scorpions, etc.).

eustachian tube   Tube between the pharynx and middle ear of vertebrates.

eutrophic   Pertaining to a mature ecosystem.

eutrophication   Enrichment process occurring in bodies of water; accumulation of nutrients.

evolution   Theory that organisms change over time so that descendants are morphologically and physiologically distinct from their ancestors.

evolutionary reduction   Evolutionary loss of structure or function.

excretion   Elimination of cellular wastes.

exergonic   Pertaining to chemical reactions that release energy and yield products having a decreased energy content.

exocrine   Pertaining to hormonal products of ducted glands.

exoskeleton   Outer covering of articulated, chitinous plates.

experimentation   Process of testing hypotheses.

explant   Initial piece of tissue placed in a nutrient culture.

exponential growth   Population growth in which numbers increased by a fixed proportion at equal time intervals.

external auditory meatus   Vertebrate ear canal terminating inwardly at the eardrum.

eyespot   Light receptor structure of certain protistans. Also called the *stigma*.

F$_1$ generation   *See* first filial generation.
F$_2$ generation   *See* second filial generation.
FAD   *See* flavin adenine dinucleotide.
Fallopian tube   Oviduct.
far-red light   Red light with wavelength between approximately 700 nm and 780 nm.

Note: For specific chemical compounds and organisms, refer to the index.

fat   *See* lipid.

fatty acid   Organic molecule composed of a hydrocarbon chain and terminal acid group.

feather tract   Patterned distribution of feathers.

feces   Excrement; waste product composed of undigested debris.

feedback, feedback control   System employing product output to regulate substrate input.

fenestra ovalis   Opening between the middle and inner ear through which sound waves are transmitted. Also called the *oval window*.

fenestra rotunda   Opening between the middle and inner ear. Also called the *round window*.

fermentation   Low-energy anaerobic oxidation reactions.

fern   Any broad-leaved, non-seed-bearing vascular plant of the division Pteridophyta.

fertilization   Union of sperm and egg.

fertilizer element   Nitrogen-, phosphorus-, or potassium-containing mineral salt.

fetus   Unborn mammalian juvenile in later stages of development.

fibrin   Insoluble protein that promotes blood clotting by forming a network of fibers.

fibrinogen   Globulin protein that associates with thrombin to form fibrin.

fibrous tissue   Structural network of fibers and cells in connective tissue.

fiddlehead   Coiled, immature fern leaf (crozier).

filiform apparatus   In plants, wall thickening in synergids that acts in secreting a pollen tube attractant.

filter feeder   Organism that feeds by straining plankton organisms from water.

first filial generation   Generation of offspring in a cross between homozygous dominant and recessive parents. Also called the F$_1$ *generation*.

first law of thermodynamics   Law stating that energy is neither created nor destroyed, merely changed from one form into another.

fitness   State of genotypic adaptation resulting from natural selection.

flame bulb   Ciliated cell associated with excretory tubules in some flatworms.

flame cell   Flagellated excretory cell of lower invertebrates.

flatus   Anal discharge of intestinal gases.

flatworm   Simple acoelomate worm; any member of the phylum Platyhelminthes (flukes, tapeworms, etc.).

flavin adenine dinucleotide (FAD)   Catabolic coenzyme functioning as an electron acceptor or donor in plant and animal oxidations.

flight feather   Large, vaned feather of bird wings.

florigen   Postulated flower-inducing hormone.

flower   Reproductive structure of angiosperms; fertile branch composed of highly modified sporophylls.

fluke   Leaflike parasitic flatworm of the phylum Platyhelminthes.

fluorescence   Emission of photons by a compound.

follicle   Small hollow sac, as in a hair follicle.

food Any organic substrate used by an organism in biological oxidation.

food chain Energy flow relationship between producer and consumer organisms in an ecosystem.

food vacuole Membrane-bounded cytoplasmic sac containing food and hydrolytic enzymes.

food web *See* food chain.

foramen magnum Large opening into the braincase through which the spinal cord extends into the vertebral column.

forebrain Anteriormost of three major divisions of the vertebrate brain.

fossil Any preserved part or product of a dead organism.

fovea As in *fovea centralis*; central, discriminatory region of the mammalian retina.

frontal lobe Anterior part of the cerebrum of the brain; site of thought and motor functions.

fucoxanthin Yellow-brown accessory photosynthetic pigment of brown algae.

fungus Any of a large group of heterotrophic plantlike organisms.

galactose Six-carbon sugar component of milk sugar (lactose).

gamete Sex cell; egg or sperm.

gametic fusion Fusion of gametes ($1n$) to form a zygote ($2n$).

gametophyte Gamete-bearing generation of a plant life cycle.

gamma (γ) ray Radiation emitted by radioactive decay, with energy greater than several hundred thousand electronvolts.

ganglion Cluster of nerve-cell bodies.

gastrolith Smooth-surfaced pebbles found in conjunction with dinosaur remains. Thought to have functioned as grinding stones in the stomachs of extinct reptiles.

gastropod Any of a large class of mollusks, including snails and slugs.

gastrovascular cavity Simple gut having only one opening.

gastrula Early stage in embryology characterized by a two-layered, vase-shaped body.

gemma Budlike unit of vegetative reproduction produced by some liverworts and mosses.

gene DNA coding for one polypeptide.

generative cell Cell of a pollen grain that produces two sperm.

gene splicing Incorporation of DNA into chromosomal DNA by enzymatic means or by transduction.

genetic cloning Propagation of selected genes by recombinant DNA techniques.

genetic code Three-symbol system of base pair sequences in DNA.

genetic engineering Any experimental or applied manipulation of genes.

genetic recombination Segregation of genes into gametes and their association in new combinations in zygotes.

genetics Study of the biology of heredity.

Note: For specific chemical compounds and organisms, refer to the index.

genophore Gene-bearing component of cells (a chromosome); specifically, the DNA of a prokaryote.

genotype Genetic makeup of an individual with respect to a trait.

geographic barrier Geographic feature isolating one part of a population from another.

geotropism Plant growth curvature induced by gravity.

germination Outgrowth and development of a seed embryo, pollen grain, or spore.

germ layer Embryonic cell layer (ectoderm, mesoderm, or endoderm).

gestation Period during which an embryo remains in the uterus.

giant chromosome Large, multistranded chromosome found in larval insects.

gibberellin One of a group of plant hormones influencing seed germination and cell expansion.

gill slit Opening between the pharynx and the animal's exterior.

gizzard Modified stomach of reptiles and birds functioning as a food-grinding organ.

gland Multicellular secretory body, such as the salivary glands, testis, etc.

glomerulus Knot of capillaries within the Bowman's capsule of a kidney nephron.

glottis Opening in the larynx between the vocal cords.

glycine Simplest amino acid.

glycogen Starchlike animal polysaccharide.

glycolysis Anaerobic oxidation of glucose yielding pyruvic acid, ATP, and $NADH_2$.

gnetopsid Any advanced gymnosperm of the class Gnetopsida (*Gnetum*, *Ephedra*, *Welwitschia*).

goiter Pathological condition of the thyroid gland.

Golgi body Cell organelle composed of stacked cisternae.

gonad Any gamete-forming body; testes and ovaries.

Graafian follicle A matured, fluid-filled ovarian follicle.

graded signal Visual communications of increasing intensity.

granulocyte One or another of three white-blood-cell types: basophil, neutrophil, eosinophil.

grasslands Major world ecosystem.

gravel Coarse mineral soil particles.

gray crescent Region of zygote acting as an organizer; site of sperm penetration.

green revolution Developments in plant breeding, soil treatments, and pesticides leading to higher crop yields.

ground meristem Embryological plant tissue giving rise to the pith and cortex.

ground state Stationary state of least energy in a physical system.

growth All phases of cell and organismal development, especially increases in size and number.

growth movement Plant movement effected by a growth curvature.

guanine One of two purines of DNA and RNA.

guard cell One of a pair of elliptical cells surrounding a stoma.

gut Alimentary canal.

gymnosperm Literally, "naked seed"; nonflowering seed plant (cycads, conifers, etc.).

habituation Accentuation or diminishment of a reflex or tactic response as a result of repeated stimulation.

hair cell Ciliated sound receptor cell of the vertebrate inner ear.

half-life Amount of time required for half the atoms of a radioactive substance to decay to a more stable form.

haploid Half-number of chromosomes in which all chromosomes lack homologs; having one set of unpaired chromosomes; the $1n$ number.

Hardy-Weinberg law Theory that, in a population, genotype ratios tend to remain constant over time.

Hatch-Slack pathway Set of photosynthetic reactions in which a 4-carbon intermediate molecule (malic acid) is produced. Also called the C-4 *pathway*.

Haversian canal System comprised of a central, blood-vessel-containing duct and radially disposed bone layers and bone cells.

HDL *See* high-density lipoprotein.

head/foot Anterior region of certain mollusks (cephalopods).

heat Form of energy; energy of motion of atoms and molecules.

heat engine Device utilizing change in heat (kinetic energy of molecules) to do work.

HeLa Human cancer cell line used in tissue culture.

helix Coil.

hemocoel Blood-containing body cavity of an open circulatory system.

hemocyanin Copper-containing blood pigment of certain invertebrates.

hemoglobin Oxygen-transporting molecule of blood composed of the nitrogen-containing compound heme, iron, and the protein globin.

hemolysis Breakdown of red blood cells.

hemophilia Inherited blood disease in which the absence of blood-clotting factors delays blood coagulation.

Henle's loop Tubule of a vertebrate kidney nephron.

heptulose Seven-carbon sugar.

herbaceous Pertaining to succulent, nonwoody vegetation.

hermaphrodite Individual having sex characteristics of both male and female.

heterospory Production of microspores and megaspores.

heterotrophy Requiring food from source other than self.

heterozygous With respect to genes, having unlike alleles, i.e., dominant and recessive.

hibernation Long-term, winter-passive phase of animal life during which the metabolic rate is comparatively low.

high-density lipoprotein Blood-serum component associated with consumption of unsaturated fats, and other factors.

hindbrain Posteriormost of the three main subdivisions of the vertebrate brain.

histamine Substance released during an infection or exposure to an allergen; contributes to allergic symptoms.

histone Type of protein associated with activation or inactivation of transcription by chromosomal DNA.

homeostasis State of physiological equilibrium in an organism.

home range Area over which an animal moves in pursuit of food.

hominid Human or humanlike primate.

homologs Structures that are developmentally alike but have different functions; chromosome pairs.

homology Being developmentally alike though possibly functionally dissimilar.

homospory In plants, production of one kind of meiospore.

homozygous Having only dominant or recessive alleles.

hormone Chemical "messenger" molecule produced in one tissue, transported in the blood, and capable of inducing a specific response in another tissue.

hornwort Thallose bryophyte bearing hornlike sporophytes.

horsetail *Equisetum*; member of the division Equisetophyta.

humerus Upper arm bone in tetrapod vertebrates.

humus Organic matter of soil, leaf litter, etc.

hybrid Having dominant and recessive genes of allele pairs.

hybrid vigor Hardiness expressed by heterozygous individuals.

hydrocarbon Organic compound containing only carbon and hydrogen.

hydrogen bond Attractive force between a hydrogen atom of a molecule and a strongly electronegative atom, such as oxygen, of an adjacent molecule.

hydrogen ion (H^+) positively charged ion of hydrogen formed upon removal of an electron from atomic hydrogen.

hydroid In plants, water-conducting cell in some mosses; in animals, sessile, hydralike life stage of coelenterates.

hydrolysis Digestion of fats, carbohydrates, and proteins by addition of water to split bonds.

hydrophobic Literally, "water-hating"; tending not to dissolve in water.

hydrophyllic Literally, "water-loving"; tending to dissolve in water.

hydroskeleton Rigidity produced by hydrostatic pressure within a coelom.

hydroxyl ion OH^- ion produced by dissociation of a base.

hypertension Condition of higher than the normal range of blood pressure in humans.

hyperthyroidism Condition characterized by an overactive thyroid gland.

hypertonic Having a higher solute concentration than a reference solution.

hypha One of many filamentous multicellular or coenocytic components of fungi.

hypocotyl Part of a seedling axis below the cotyledons.

hypoglycemia Condition of low blood sugar.

hypothalamus Region of the midbrain, below the thalamus, having a regulatory function, principally via the pituitary and the autonomic nervous system.

hypothesis Tentative assumption subject to verification or proof by experimentation.

Note: For specific chemical compounds and organisms, refer to the index.

hypotonic   Having a lower solute concentration than a reference solution.

immunity (medical)   Having defenses against infective agents (pathogens, toxins), usually by action of specific antibodies.

implantation   Attachment of an embryo to the wall of the uterus.

imprinting   Formation of a parental bond between the young of a species and (usually) a parent.

incisor   Foretooth.

indeterminate growth   Nonfinite growth pattern with no predictable end point.

induction   Inducement of differentiation in one region of an organism by another region of the organism or by some external force.

indusium   Shield-shaped or flaplike appendage covering a fern sorus.

industrial melanism   Selection and survival of dark-pigmented organisms in habitats colored by industrial soots, where organisms are inconspicuous.

infection thread   Invasion pathway of nitrogen-fixing bacteria into the cortex of a legume root.

inflorescence   Characteristic flower cluster.

information storage and retrieval   Any systematic method of storing and retrieving coded information.

ingestion   Taking in of food particles by a cell.

inguinal canal   Canal between the coelom and scrotum in male mammals.

inheritance   Transmission of genes from parents to offspring.

innate behavior   Unlearned, apparently inherited behavior. Also called *instinct*.

insect   Six-legged terrestrial arthropod.

instinct   *See* innate behavior

insulin   Hormone produced by the islets of Langerhans in the pancreas; promotes the conversion of blood glucose into tissue glycogen.

integument   In animals, the outer covering of the body, including hair, feathers, and fur; in plants, the outer layer(s) of the ovule and seed.

interferon   Body protein capable of inhibiting virus infections.

interphase   Phase of mitosis between cell divisions during which duplication of genetic material takes place.

interstitial fluid   Extracellular fluid in animal tissue.

intestine   Digestive–absorptive part of an alimentary canal.

intrinsic factor   Protein that facilitates the uptake of vitamin B_{12} by intestinal cells.

invagination   Inpocketing of a surface.

invertebrate   Member of a large subkingdom of animals lacking internal skeletons.

iodopsin   Photoreceptive pigment of vertebrate cone cells.

ionic bond   Force of attraction between oppositely charged ions (anions and cations).

ionic compound   Crystalline substance that loses its identitiy (i.e., dissociates) when dissolved in water.

irritability   Ability to respond to stimuli.

Note: For specific chemical compounds and organisms, refer to the index.

islets of Langerhans   Insulin- and glucagon-secreting cells of the pancreas.

isolating mechanism   *See* reproductive barrier.

isotonic   Having the same solute concentration as a reference solution.

isotope   Atom differing from the majority of atoms of an element by having additional neutrons; may or may not be radioactive.

jejunum   Region of the mammalian small intestine.

jellyfish   Any medusoid stage in members of the phylum Cnidaria; any scyphozoan (class Scyphozoa).

junction   Point of union between adjacent structures.

juvenile   Young of an organism.

kairomone   Attractant or recognition signal secreted by an organism inadvertently (metabolic by-product).

karyotype   Diagram of an organism's metaphase chromosomes.

kelp   Any of various large marine brown algae.

keratin   Protein composing hair, claws, fingernails, feathers, and the epidermis.

kidney   Urine-secreting organ of vertebrates.

kinesis   Direct behavioral response to a simple stimulus such as unidirectional light.

kinetic energy   Energy of motion.

kinetosome   Granule at the base of the flagellum that presumably controls motion.

kingdom   Major taxon or life category (i.e., plant, animal, fungus, protistan, moneran).

Klinefelter's syndrome   Disorder in human males resulting from an extra X chromosome in each cell.

Krebs cycle   Aerobic oxidative reactions in which acetyl is converted to carbon dioxide and energy is liberated. Also called the *citric acid cycle*.

labia majora   Outer fleshy folds of mammalian female genitalia.

labia minora   Inner fleshy folds of mammalian female genitalia.

lacteal   Lymph vessel in an intestinal villus.

lagomorph   Member of the rabbit and hare family.

Lamarckianism   Use–disuse theory of evolution.

large intestine   Posterior portion of the vertebrate intestine.

larva   Motile, free-living, subadult feeding stage in an animal life cycle.

Laurasia   Northernmost of two subcontinents derived from the world continent, Pangea.

LDL   *See* low-density lipoprotein.

leaf   Photosynthetic stem appendage of plants.

leaf gap   Parenchymatous region in the stem vasculature associated with a leaf trace.

leaf primordium   Mound of undifferentiated meristematic cells that develops into a leaf.

learning   Acquisition of a behavioral response as a result of repetitive stimuli.

lectin   Mucopolysaccharide plant product that acts as a "recognition" factor in pollinator–stigma and some other cell-to-cell interactions.

lenticel   Pore in the cork of the stem of a woody plant permitting gas exchange.

leptoid   Food-conducting cell in some mosses.

lethal gene   Gene that when expressed can cause the death of the organism carrying it.

leukocyte   White blood cell.

lichen   Symbiosis between an alga and a fungus.

life   Sum of all cellular, physiological, and developmental processes.

ligament   Strand of fibrous connective tissue binding bone to bone.

light reactions   Light-absorbing, hydrolytic reactions of photosynthesis.

light trap   Array of chlorophyll and accessory pigment molecules in the thylakoid; functions in photon absorption and electron transfer.

lignin   Component of the cell walls of fibers and xylem elements of wood.

limbic system   Part of the forebrain associated with such basic behavior as eating, drinking, and sexuality.

limiting factor   Environmental factor that prevails over others in limiting population growth.

linkage   *See* linked genes.

linked genes   Two or more nonallelic genes located on the same chromsome.

lipase   Lipid-digesting enzyme.

lipid   Organic compound composed of an alcohol and one or more linked fatty acids.

liver   Complex glandular and digestive organ of animals.

liverwort   Thallose bryophyte.

locus   Position that a gene occupies on a chromosome.

long-day plant   Plant requiring a light-exposure period exceeding a critical length in order to flower.

low-density lipoprotein   Blood-serum component associated with hypertension, saturated fats, and cholesterol.

lung   Gas-exchange organ of animals.

lymph   Fluid in the lymph vessels of vertebrates.

lymphokinine   Chemical produced by t-lymphocytes, capable of lysing foreign cells.

lysis   Dissolving, usually with water; dissolution.

lysosome   Membranous cytoplasmic vesicle containing hydrolytic enzymes.

lysozyme   Enzyme produced by animal tissues and having the capacity to dissolve (lyse) bacteria.

macroevolution   Theory that evolution proceeds by relatively large, abrupt changes.

macromolecule   Large organic molecule composed of numerous subunits.

macronucleus   Large type of nucleus found in ciliated protozoans.

macronutrient   Mineral element required in large amounts for the growth and development of an organism.

macrophage   Large white cell of blood and tissues, active in phagocytosis following an infection.

Note: For specific chemical compounds and organisms, refer to the index.

malaria   Relapsing fever of the blood spread by certain mosquitoes.

malpighian tubule   Tubular excretory organ of arthropods.

mammal   Vertebrate possessing mammary glands.

mammary gland   Milk-secreting gland of mammals.

mandible   Jaw.

mandibulate   Literally, "having jaws"; one of two subgroups of arthropods, including crustaceans, insects, millipedes, and centipedes.

marsupial   Pouch-bearing primitive mammal (e.g., kangaroo, opossum).

mass flow   Theory of transport of solutes in the phloem of plants.

mastax   Jawlike grinding organ of rotifers.

mast cell   Cells in animal tissues which produce histamine during an infection.

matriarchy   Family structure centered on a dominant female parent.

medulla oblongata   Posteriormost part of the vertebrate hindbrain.

medusa   Jellyfish stage of the coelenterate life cycle.

megaphyll   Multiveined leaf associated with a leaf gap in the vascular tissue of the stem.

megasporangium   Sporangium that produces megaspores.

megaspore   Large meiospore that produces a female gametophyte.

meiosis   Cell divisions in which the chromosome number is reduced from diploid to haploid.

meiospore   Any spore produced directly by meiosis; a sexual spore.

melanization   Selection of dark-colored races as a consequence of the darkening of background habitat coloration, as in industrial melanization.

memory   Recollection of past events.

Mendelian factor   Allele; gene.

Mendelian ratio   Ratio of F_2 offspring in a cross of F_1 hybrid parents.

Mendel's laws   Statements regarding segregation, assortment, and recombination of Mendelian factors.

meninges   Fluid-filled membranes surrounding the brain and spinal cord.

menopause   Permanent cessation of the menstrual cycle in an older human female.

menstruation   Periodic shedding of the endometrial lining layer of the uterus.

meristematic tissue   Group of undifferentiated dividing cells giving rise to specialized cells that make up the plant body.

mesentery   Epithelial sheet supporting the viscera of vertebrates.

mesoderm   Germ layer between the ectoderm and endoderm.

mesoglea   Jellylike layer between the ectoderm and endoderm in coelenterates.

mesophyll   Green, photosynthetic tissue in a leaf.

messenger RNA (mRNA)   Product of the transcription of a gene to a corresponding sequence of RNA nucleotides.

metabolism   Sum of all chemical processes occurring in the cells of an organism.

metagenesis Alternation of sexual and asexual stages in a life cycle.

metamere One segment of the body of a segmented animal.

metamorphosis In development, a marked change from a juvenile to an adult form.

metaphase Stage in cell division during which chromosomes are aligned in a plane at right angles to the spindle axis.

metazoan Multicellular animal higher in complexity than a sponge.

microbiology Study of microscopic organisms, especially bacteria.

microbody Small cytoplasmic organelle containing catalase and other hydrolytic enzymes.

microclimate Climate prevailing in a small segment of the total environment.

microevolution Theory that evolution proceeds by successive minor genetic changes.

microfilament Minute proteinaceous cytoplasmic fiber composed of actin.

micronutrient Element required by cells only in small amounts.

microphyll One-veined, primitive leaf type, not associated with a leaf gap.

micropyle Opening in the integument of the ovule of seed plants; opening at one end of the egg of insects.

microsporangium Sporangium producing microspores.

microspore Small meiospore that produces a male gametophyte.

microtubule Cytoplasmic tubule composed of protein subunits.

microtubule organizing center (MTOC) Site of microtubule synthesis in cells.

microvillus Microscopic projection of surface epithelium

midbrain Middle of the three major brain regions of vertebrates.

middle ear Region of the ear between the eardrum and the inner ear.

mimicry Convergent evolution of unlike species resulting in behavioral or morphological similarities; evolution of one species to resemble another; evolution of a species to resemble its background.

mitochondrion Organelle that is the site of respiratory enzymes and reactions of the Krebs cycle and the electron transport system.

mitosis Cell division in eukaryotes in which chromosomal numbers remain constant.

mitospore Asexual spore produced by a mitotic division.

mixture Composition of two or more substances not chemically bound to one another.

molar Rear, grinding teeth of vertebrates.

mole, molar solution One molecular weight in grams (gram molecular weight); one mole of solute per liter.

molecule Least particle of a substance; composed of two or more like or unlike atoms.

Note: For specific chemical compounds and organisms, refer to the index.

mollusk Shelled invertebrate of the phylum Mollusca (snails, clams, etc.).

monad One of anything, e.g., a single chromosome.

moneran Any prokaryotic cell.

monocotyledon Flowering plant having embryos with one cotyledon.

monoculture Practice of growing one crop over extensive areas.

monoecious Literally, "one house"; having both male and female structures.

monohybrid Genetic cross involving one pair of alleles.

monokaryon Body of a fungus composed of uninucleate cells of hyphae with unpaired nuclei.

monomer One of the repeating units of the same kind of molecule in a polymer.

monotreme Primitive, egg-laying mammal (e.g., duckbilled platypus).

morphology Study of the form of an organism; structure of an organism.

morula Early, "mulberry" stage in embryology, characterized by a solid ball of relatively few cells.

moss Leafy bryophyte; primitive leafy land plant, usually considered nonvascular.

motile colony Permanent aggregation of identical flagellated cells.

motor neuron Neuron that transmits a nerve impulse away from the central nervous system and to an effector cell or organ. Also called an *efferent neuron*.

mRNA *See* messenger RNA.

multicellular Composed of many cells.

multiple alleles Series of more than two alleles, any two of which may be present on a pair of chromosomes at a particular locus.

multiple genes Several pairs of alleles at different loci that determine a particular genetic trait.

muscle Tissue composed of fibrous cells capable of contraction and relaxation.

mutation Induced or spontaneous change in a gene.

mutualism Symbiotic relationship of mutual benefit to all symbionts.

mycelium Thallus or body of a fungus; composed of threadlike hyphae.

mycoplasma Primitive prokaryotic cell; obligate parasite lacking a cell wall.

mycorrhiza Association of plant roots with fungal hyphae in a mutualistic (symbiotic) relationship.

myelin sheath Many-layered Schwann-cell wrapping that covers an axon.

myomere Muscle segment of the vertebrate body.

myosin Protein fiber involved with actin in muscle contractions.

myxamoeba Amoeboid, unicellular phase of the slime mold life cycle.

nastic movement Growth movement of plants not directed toward or away from a stimulus.

nasty *See* nastic movement

natural selection Survival of variant types of organisms as a result of adaptability to environmental stresses.

nectary Nectar-secreting body commonly found in flowers but sometimes on leaves or stems near flowers.

nematocyst Stinging hair cell of coelenterates.

neo-Darwinism Revision of Darwinian evolution incorporating population genetics.

neoteny Becoming sexually mature in the larval state.

nephridium Ciliated excretory tubule of invertebrates.

nephron Tubular filtration unit of the vertebrate kidney.

nerve Fiber composed of nerve-cell axons.

nerve cord Major, elongated bundle of neurons.

nerve impulse Wave of electrochemical changes in plasma membrane permeability to sodium or potassium ions; travels the length of a neuron.

nerve net Primitive netlike system of neurons.

nerve trunk Major nerve branch composed of many nerve fibers; spinal and cranial nerves of vertebrates.

nervous system Any organized set of neurons constituting the entire neuronal system of an animal.

neural fold Dorsal longitudinal ectodermal fold that gives rise to the neural tube of an animal embryo.

neuron Nerve cell.

neurotransmitter Chemical capable of inducing an impulse at a synapse of nerve endings, between a nerve ending and an effector, or between a sensory structure and a neuron.

neutralization Reaction between an acid and a base yielding a salt and water.

neutron Uncharged unit of mass in an atomic nucleus.

neutrophil Neutral-staining (having no affinity for either basic or acidic dyes) white cell (i.e., leucocyte) of blood.

niche In ecology, the role of an organism in an ecosystem.

nitrogen cycle Cycle of oxidation and reduction of nitrogen and nitrogen-containing compounds in the environment.

node of Ranvier Gap between adjacent Schwann cells.

notochord Cartilaginous, rodlike backbone of lower chordates.

nucleic acid Molecule composed of a sequence of nucleotides (e.g., DNA and RNA).

nucleoplasm Protoplasm contained within a nuclear membrane.

nucleotide Molecule consisting of phosphate, a 5-carbon sugar, and a purine or pyrimidine.

nucleotide triplet Sequence of three nucleotides coding for synthesis of one amino acid.

nucleus Organelle in eukaryotic cells bounded by a double membrane and containing chromosomes.

nutation Circular or spiral growth movement of plants.

nutrition Process of absorbing nutrients.

occipital lobe Posterior region of the cerebrum of the brain; center for visual perception.

oceanic biome Major ecosystem, comprised of the oceans.

ocellus Simple eye of arthropods.

oligochaete Annelid worm having relatively few bristles (e.g., earthworms).

oligotrophic Pertaining to an immature ecosystem.

ommatidium Visual unit of the arthropod compound eye.

oncogene Cancer-causing gene.

ontogeny Development of an individual.

oocyte Cell that produces an ovum by meiosis.

oogamy In plants, reproduction by egg and sperm.

open circulation Circulatory system of combined blood vessels and blood sinuses.

operator locus Binding site on DNA that, when bound to a repressor protein, can inhibit transcription.

operon Functional inducible or repressible system of gene or genes plus inhibitor and operator loci.

opiate Sedative drug.

opiate receptor Receptor site of neurons of the central nervous system capable of binding a specific opiate.

opson Antibody capable of opsonization, e.g., forming coating on pathogens, making them subject to phagocytosis.

optic lobe One of a pair of lobes of the forebrain; site of the visual senses.

orbital Path of electron pairs around the atomic nucleus.

organelle Specialized, membrane-bounded structure that serves a specific function within a cell.

organic compound One of the carbon-containing molecules that form the basis of life.

organism Unicellular or multicellular individual living entity.

organizer Region in a cell or an organism capable of initiating differentiation and organization.

osmosis Tendency of water to diffuse through a differentially permeable membrane separating regions of low and high solute concentrations.

ossification Natural process of bone formation.

oval window *See* fenestra ovalis.

ovarian follicle Nest of mammalian ovary cells containing a single oocyte.

ovary Ovum-producing gonad in animals; site of ovules in flowers.

oviduct Tube conveying ova from the ovary to the uterus.

ovulary In plants, structure composed of one or more carpels enclosing one or more ovules. Also called *ovary*.

ovule Female gametophyte and enclosing integumentary layer(s) of seed plants.

ovum Unfertilized egg.

oxidation Energy-releasing process in which electrons are transferred from one atom or molecule to another.

oxidation-reduction reaction Reaction in which electrons are exchanged between molecules.

oxyhemoglobin Oxygenated hemoglobin.

palate Roof of the mouth cavity in mammals.

palisade layer Layer of columnar mesophyll cells in a leaf; commonly found in leaves of dicotyledons beneath the upper epidermis.

pancreas Large visceral gland producing and secreting digestive enzymes as well as hormones (e.g., insulin, glucagon).

Pangea Paleozoic supercontinent from which the present continents were derived.

Note: For specific chemical compounds and organisms, refer to the index.

parapodium Stumpy, unjointed leg of a polychaete worm.

parasexuality Direct exchange of DNA without gametic fusion.

parasite Symbiotic organism deriving benefit from the host, with no profit to the host.

parasympathetic nervous system Part of the vertebrate autonomic nervous system associated with inhibition of involuntary actions.

parathyroids Four small endocrine glands adjacent to the thyroid; regulate blood-calcium levels.

parenchyma Unspecialized tissue.

parental family Social group in which parental care is invested.

parthenogenesis Literally, "virgin birth"; development of an animal ovum without fertilization.

passive transport Transport of molecules into and out of cells by diffusion; not energy requiring.

pathogen Disease-causing organism.

pavement epithelium Animal epithelium composed of one or more layers of flat cells. Also called *squamous epithelium*.

pectoral Pertaining to the breast or chest region.

pellagra Vitamin deficiency disease resulting from lack of niacin (vitamin B$_3$).

pellicle Thin skin, such as a chitinous membrane, surrounding a protistan cell.

pelvic Pertaining to the pelvic girdle; supporting skeleton of the posterior appendages of tetrapods.

penicillin Antibiotic drug (*Penicillium notatum*) derived from ascomycetes.

penis Male organ of sperm intromission; homolog of the clitoris.

peptide Short chain of amino acids.

peptide bond Bond linking two amino acids in a peptide or polypeptide.

perennial Plant having a life cycle encompassing more than two years.

pericardium Sac surrounding the heart.

pericycle Outer cell layer of the stele of stems and roots.

peripheral nervous system Spinal and cranial nerves of the vertebrate nervous system.

peristalsis Wave of muscular contractions.

petal Member of the inner whorl of vegetative flower parts; considered to be a sterile sporophyll.

petiole Stalk of a leaf to which the blade is attached.

phagocytosis Ingestion of bacteria or other cells by an amoeboid cell, such as a white blood cell.

pharyngeal gill slit Opening between the pharynx and outside in chordates; functional gills in lower chordates.

pharynx Region of the alimentary canal immediately posterior to the mouth cavity; region of gill openings in fish.

phenotype Visible products of gene action.

pheromone Aromatic substance serving as a chemical communicative signal in animals, such as a sex attractant.

phloem Conduction tissue of plants through which foods that are synthesized move.

Note: For specific chemical compounds and organisms, refer to the index.

phospholipid Molecule composed of a fat molecule and a phosphate group; an important part of cell membranes.

phosphorus cycle Cycle of conversions of inorganic and organic phosphorous compounds between organisms and the environment.

phosphorylation Addition of a phosphate to a compound; addition of phosphate to ADP.

photon Quantum of light energy.

photon trap Photon-absorbing array of chlorophylls and accessory pigments.

photoperiod Length of time that an organism is exposed to a light/dark cycle and its effect on development and physiological processes.

photophosphorylation Production of ATP in the light reactions of photosynthesis.

photorespiration Site of auxiliary reactions whereby oxygen is processed as hydrogen peroxide in microbodies adjacent to chloroplasts of C-3 plants.

photosynthesis Process by which light energy is converted to chemical energy by chlorophyll, with the attendant production of carbohydrates from carbon dioxide and water.

photosystem I, photosystem II Sets of light reactions in which photophosphorylation, photolysis of water, and reduction of hydrogen carriers occur.

phototropism Growth movement of plants in response to unidirectional light.

phycobilin One of several red and blue accessory pigments occurring in cells of red algae and cyanobacteria.

phycocolloid Any of a number of colloidal, hygroscopic, polysaccharide, cell-wall components of red and brown algae (agar-agar, algin, etc.).

phycocyanin Blue accessory photosynthetic pigment found in red algae and cyanobacteria.

phycoerythrin Red accessory photosynthetic pigment occurring in red algae and cyanobacteria.

physiology Study of the physical and chemical processes occurring in living organisms.

phytochrome Blue to blue-green pigment associated with photoperiodism.

phytohormone Any of several plant hormones.

phytoplankton Microscopic, autotrophic component of plankton.

pineal body Small body arising from the forebrain of vertebrates, light receptive in some fish, reptiles, and birds and having hormonal functions.

pinocytosis Uptake of water and solutes by invagination of the plasma membrane.

pinocytotic vesicle Vesicle formed from pinocytotic invagination of the plasma membrane.

pioneer In ecology, a plant or animal occupying an early successional niche.

pistil Female organ of a flower; composed of one or more carpels.

pit In plants, a thin area in walls of adjacent cells.

pit connection Pore in the end walls of adjacent cells of fungi and red algae.

pith Central core of parenchyma in stems and sometimes roots.

pituitary Small gland associated with the hypothalamus of the vertebrate brain.

placebo Inert dosage of a drug or medicine given as an experimental control.

placenta Nutritive organ of the mammalian embryo, composed of embryonic and uterine tissues; ovule-bearing carpellary tissue in flowering plants.

planula Ciliated larval coelenterate.

plaque (medical) Deposit of lipid and connective tissue partially or wholly blocking blood flow in coronary artery disease.

plasma cell Antibody-synthesizing white blood cell.

plasma membrane Sheet of lipoprotein surrounding and containing the living matter of a cell.

plasmid Small loop or circle of cytoplasmic DNA.

plasmodium Vegetative phase of a slime mold; malarial parasite (*Plasmodium*).

plasmolysis Shrinking of a plant protoplast as a result of water diffusion from the vacuolar sap to the exterior of the cell.

platelet Cytoplasmic body of cellular origin in blood which contains blood-clotting factors.

pleuron Lateral body plate of an arthropod exoskeleton.

pluteus Ciliated larva of an echinoderm.

polar body Small degenerative cell produced together with the ovum by meiosis of an oocyte.

polarity Exhibiting linear diversity or pole-to-pole differentiation.

polar molecule Molecule having asymmetrical concentrations of electrical charges.

pollen Immature male gametophyte of seed plants.

pollination Transport of pollen by some agent to the female part of a cone or flower.

polychaete Many-bristled annelid (clamworm, sandworm, etc.).

polygenes Group of genes controlling the expression of one Mendelian trait.

polymer Molecule consisting of many repeating units of the same kind of molecule (monomer).

polyp Tentacled, sessile form of a colonial coelenterate.

polypeptide Long-chain amino acid polymer such as a protein or large protein fragment.

polyploidy Multiples of the base chromosome number higher than 2*n* (3*n*, 4*n*, 5*n*, . . .).

polyribosome See polysome.

polysaccharide Polymer composed of a repetitive sequence of monosaccharides.

polysome Chain of ribosomes; ribosomes aggregated on a strand of mRNA.

pons Enlargement in the vertebrate hindbrain connecting neurons of the cerebrum and cerebellum.

population Any geographically circumscribed aggregation of individuals of a species.

population growth curve Bell-shaped curve of population growth. See exponential growth.

posterior Hind or rear end of a structure or an organism.

Note: For specific chemical compounds and organisms, refer to the index.

potential energy Energy in a form capable of doing work when released.

prairie Major grassland region.

preformation Theory that the egg, sperm, and zygote contained a miniature of the adult organism.

prehensile Having a capacity for grasping.

primary consumer First-level consumer organism in a food chain; consumes producer organisms or their products; herbivore.

primary succession First stage of ecological succession, beginning with a barren substrate.

primate Advanced mammal having opposable thumbs, fingernails, a large brain, and binocular vision (monkeys, apes, hominids).

"primordial soup" Matter out of which life is said to have arisen.

primordium Early stage in development of a plant organ such as a root or leaf.

probability Likelihood of an event occurring; statistical prediction.

procambium Primary plant tissue giving rise to xylem, phloem, and vascular cambium.

producer Organism that manufactures food used by consumers; autotroph.

proglottid Segment of a tapeworm, equivalent to one individual of a colony.

progymnosperm Ancient non-seed-bearing extinct plant showing features of vegetative structure also present in modern gymnosperms (e.g., bordered pits).

prokaryotic cell, prokaryote Primitive, one-celled organism lacking a nucleus, mitochondria, and plastids (bacteria, cyanobacteria).

prophase First visible phase of mitosis, during which condensation of chromosomes occurs and the nuclear membrane disappears.

proprioceptor Sensory nerve ending associated with orientation and perception of position, tension, etc.

prosimian Primate antecedent of apes and monkeys; early primate.

prostaglandin Group of fatty acid compounds exhibiting a diversity of hormonal effects.

protein Complex organic molecule composed of amino acids linked by peptide bonds.

protenoid Cell-like aggregate of protein molecules.

prothallus Gametophyte generation of ferns.

protochordate Primitive chordate animal (tunicates, amphioxus, acorn worms, etc.).

protoderm Basic plant tissue giving rise to the epidermis.

proton Positively charged particle of matter in the nucleus of an atom.

protoplasm Living substance of cells.

protoplast Plant cell from which the wall has been enzymatically removed; any "naked" or wall-less cell in isolation.

protostele Simple central core of vascular plant tissue in a stem or root lacking a pith.

protostome Animal in which the blastopore becomes a mouth.

proximal tubule Kidney tubule nearest Bowman's capsule.

pseudocoelom Body cavity unlined by epithelium, as found in ascarine worms.

pseudopodium Leafless stalk of the gametophyte of sphagnum moss; footlike protoplasmic extension of a mobile cell.

pulmonary Relating to the lungs.

punctuated equilibrium Theory that evolution proceeds in spurts or jumps.

purine Basic compound composed of two rings of carbon and nitrogen atoms (adenine and guanine).

pupa Nonmotile stage in metamorphosis of an insect.

Purkinje fiber Specialized impulse-transmitting muscle cells in the vertebrate heart.

pyrimidine Compound comprised of one ring of carbon and nitrogen atoms (cytosine, thymine, and uracil).

quadrat Rectangular plot of prescribed dimensions laid out in a habitat for the purpose of enumerating species of organisms.

quantasome Elementary particle of chloroplast thylakoids; possibly the site of photosystems I and II.

quantum Fundamental unit of energy.

radial cleavage Cleavage pattern in which cleavage planes are offset radially.

radial symmetry Having radiating planes of symmetry.

radiant energy Energy consisting of electromagnetic waves.

radiation Radiant energy; energy produced by the decay of a radioactive element.

radicle First root of embryo plants.

radius One of two bones in the forearm.

radula Grinding organ in the mouth of gastropods (snails).

rainforest Tropical or subtropical evergreen forest characterized by an annual rainfall in excess of 250 cm (100 in).

rank order Position in a hierarchy of individuals; place in a peck order.

rationalism Concept that life can be explained and understood.

ratite Large flightless bird (ostrich, emu, kiwi, etc.).

ray Beam of radiant energy; outer flower of members of the aster family; radially oriented band of cells in the stele of a vascular plant.

reactant Substance participating in a chemical reaction.

recessive gene Gene whose expression can be masked by its dominant allele.

recombination To recombine; in genetics, to recombine genes via gametes or to splice genes together via recombinant DNA techniques.

rectum Posterior region of the mammalian colon.

red alga Member of the division Rhodophyta; eukaryotic, mostly marine alga having phycobilin pigments.

red blood cell *See* erythrocyte.

red cell *See* erythrocyte.

reduction Addition of a hydrogen atom to another atom or molecule, or the acceptance of an electron by an atom or molecule.

Note: For specific chemical compounds and organisms, refer to the index.

reflex Behavioral response in which part of an animal body responds to a stimulus.

reflex arc Chain of sensory and motor neurons functioning in reflexes.

regeneration Regrowth or restoration of missing parts of an organism.

renal pelvis Collecting chamber of the mammalian kidney.

replication Reproduction; reproduction of genes (DNA), chromosomes, cells, and so on.

repression Action of a repressor molecule; *see* repressor.

repressor Protein that binds to an operator locus and suppresses gene transcription.

reproduction Process of production of offspring (new cells, new organisms).

reproductive barrier Evolutionary process occurring in geographically or reproductively isolated populations.

reproductive potential Maximum theoretical rate of reproduction for a species.

reptile Member of the class Reptilia; any egg-laying, scaled tetrapod (snakes, lizards, turtles, etc.).

respiration Gas exchange in the lungs; biological oxidative processes including anaerobic and aerobic reactions.

resting potential Distribution of electrical charges in the membrane of a neuron at rest (i.e., when not transmitting a nerve impulse).

retina Light-receptive layer of the rear inner surface of the vertebrate eye.

rhizoid Unicellular or multicellular hairlike structure functioning in anchorage and uptake of water and, possibly, of nutrients.

rhizome Horizontal, usually below-ground stem of a vascular plant.

rhodopsin Visual pigment of the retina.

ribonucleic acid (RNA) Complex, single-stranded molecule composed of repeating nucleotides.

ribosomal RNA (rRNA) RNA that binds the subunits of a ribosome together.

ribosome Small cytoplasmic body upon which mRNA is translated into a polypeptide structure.

ring canal Circulatory vessel of echinoderms.

RNA *See* ribonucleic acid.

rod Photoreceptor in the retina of the vertebrate eye.

rodent Herbivorous mammal having chisel-like incisors.

root Nutrient-absorbing portion of the axis of vascular plants.

root cap Layer of cells covering a root tip meristem.

root hair Hairlike epidermal cell of roots having an absorptive function.

root pressure Hydrostatic pressure developed by water uptake of roots.

rotifer Any member of the phylum Rotifera; simple acoelomate microorganism having ciliated mouth parts.

round window *See* fenestra rotunda.

rumen Stomach chamber of a group of herbivorous mammals (ruminants), in which bacterial degradation of cellulose occurs.

ruminant Mammal possessing a rumen (e.g., cow, sheep, goat).

sac fungus Member of the division Ascomycota, producing ascospores in a saclike organ, the ascus.

sagittal Median dorsal-to-ventral, anterior-to-posterior section.

sand Coarse grains of soil minerals.

saprobe Organism, commonly a fungus or bacterium, capable of digesting and absorbing other, dead organisms.

sarcoplasmic reticulum Endoplasmic reticulum of the striated muscle cell.

savanna Tropical and subtropical scrubland.

schizophrenia Mental aberration or psychosis characterized by pronounced mood alterations, illusions, and delusions.

Schwann cell Thin sheetlike cell covering part of the axon of a neuron.

science Methodology of generating information about forces or objects that can be measured; based on observation, experimentation, and interpretation; the scientific method.

scientific method *See* science.

scientific model Conceptual construction thought to approximate or to explain the natural state of some phenomenon.

sclereid Sclerenchyma cell having thick, lignified secondary walls and numerous pits.

sclerenchyma Lignified supporting tissue of plants (fibers, sclereids, etc.).

scolex Anterior, anchoring proglottid of tapeworms.

scrotum Fleshy sac enclosing the mammalian testes.

scurvy Disease resulting from a deficiency of vitamin C.

sea anemone Sessile coelenterate hydrozoan.

secondary consumer Organism that feeds on primary consumers in a food chain.

secondary succession Stage in ecological succession starting with a disturbed habitat (after fires, etc.).

second filial generation Offspring of hybrid (heterozygous) parents. Also called the F_2 *generation.*

second law of thermodynamics Law stating that energy always tends to become dispersed and disorganized in the universe.

seed Ripened ovule containing an embryo.

seed fern Any of an order of primitive fernlike seed plants, now extinct.

self-assembly Assembly of macromolecules into cell components or virus particles, apparently by random juxtapositioning.

semen Secretion containing sperm.

semicircular canal One of three fluid-filled canals of the vertebrate inner ear.

semiochemical Any chemical that can act as a signal between organisms (e.g., pheromones).

senescence Aging or process of becoming aged.

sensory neuron Neuron that initiates a nerve impulse, transmitting it toward the central nervous system. Also called *afferent neuron.*

sepal Outermost, leaflike appendage of a flower.

serum Liquid portion of blood plasma after coagulation of blood cells has occurred.

Note: For specific chemical compounds and organisms, refer to the index.

seta Hairlike projection; stalk of a moss sporophyte.

sex Nature of the reproducing individual (i.e., male or female); *see* sexual reproduction.

sex chromosome One of a pair of chromosomes, usually of unlike morphology, associated with sex determination.

sex-influenced genes Genes whose expression depends on the presence or absence of sex hormones (e.g., baldness).

sex-linked genes Linked genes occurring on sex chromosomes.

sexual Pertaining to meiotic reduction of chromosome numbers, gamete production, and gamete fusion, as well as related processes and structures.

sexual reproduction Reproduction by gametes and fertilization.

shell Any deposited or secreted coating of a cell or organism, usually chitinous or calcareous.

short-day plant Plant requiring a photoperiod of shorter duration than a critical period of time.

sieve cell Conduction cell of phloem, usually a nucleated cell having pores (sieve areas) in side and end walls.

sieve tube In plant phloem, a multicellular tube composed of sieve cells or sieve tube members.

sieve tube member Cell composing sieve tubes of angiosperm phloem.

silt Fine grains of soil minerals.

simple goiter Enlargement of the thyroid gland as a result of iodine deprivation.

sinus Space within a tissue or organ.

siphonostele In plants, the vascular system of a stem (and sometimes a root) having a pith.

skeletal muscle Muscle fibers composed of multinucleated, striated cells. Also called *voluntary muscle.*

skeleton Internal or external supporting structure of an animal body.

slime mold "Naked" plasmodial fungus; member of Myxomycota.

small intestine Digestive and absorptive part of the vertebrate gut, between the stomach and large intestine.

smooth muscle Muscle fibers made up of elongated, nonstriated cells. Also called *involuntary muscle.*

social behavior Behavior of social groups composed of individuals of a species.

sodium/potassium-ion pump System of energy-utilizing ion carriers that transport sodium and potassium ions through the plasma membrane of the cell.

solar plexus Large abdominal ganglion of the autonomic nervous system.

solute Dissolved substance in a solution.

solution Usually liquid, with the molecules of the solvent and solute interspersed.

solvent Dissolving medium of a solution.

somatic Pertaining to the body; vegetative tissue as opposed to reproductive tissue.

somatic cell Any nonsexual cell of a multicellular plant or animal.

somatic nervous system. Part of the peripheral nervous system associated with voluntary responses.

sorus Cluster of sporangia in ferns.

speciation Evolution of a species.

species Ultimate category of taxonomic classification based on breeding potential or gene flow, morphology, anatomy, and distribution; collection of similar individuals.

species concept *See* species.

spermatid Postmeiotic male cell prior to transformation into a sperm.

spermatium Nonmobile male gamete of red algae and certain fungi.

spermatocyte Meiotic, sperm-forming cells.

spermatogonium Proliferating cell of the testes that gives rise to spermatocytes.

spermatozoid Flagellated, motile male gamete.

spherical symmetry Having an infinite number of planes of symmetry (e.g., a sphere). Also called *universal symmetry*.

sphincter Drawstring type of valve or closure in a tubular organ such as the intestine.

spicule Any crystalline, needlelike or trapezoidal component of a cell or tissue.

spinal cord Dorsal nerve cord of vertebrates.

spinal nerve One of a series of paired nerve trunks arising from the vertebrate spinal cord.

spindle Microtubular structure formed during mitosis associated with poleward transport of the chromosomes.

spiracle Respiratory opening.

spiral cleavage Cleavage pattern of many invertebrate classes in which each cleavage plane is offset slightly to that of adjacent cells.

sponge Any primitive multicellular animal lacking true tissues and organs; member of the phylum Porifera.

spongin Proteinaceous, fibrous component of the sponge skeleton.

spongy mesophyll Spongelike layer of photosynthetic cells within a leaf.

spontaneous generation *See* abiogenesis.

sporangium Unicellular or multicellular spore-bearing organ of fungi, algae, and green plants.

spore Commonly thick-walled reproductive cell, usually capable of withstanding adverse environmental conditions.

spore mother cell Spore in algae and land plants giving rise to a tetrad of spores by meiosis.

spore tetrad Four haploid cells (spores) produced by meiosis of a spore mother cell.

sporophyll Fertile, sporangium-bearing leaf.

sporophyte Spore-producing generation of a plant life cycle.

squamous epithelium *See* pavement epithelium.

stamen Male pollen organ of flowers.

starch Complex insoluble polysaccharide composed of glucose subunits.

statocyst Organ of equilibrium; a fluid-filled cavity having touch-sensing nerve endings and statoliths.

statolith Granule within a statocyst.

stele Vascularized core of the tissue of a plant stem or root.

stem Axis of a vascular plant to which leaves (if present) are attached.

Note: For specific chemical compounds and organisms, refer to the index.

sternum Vertebrate breastbone.

steroid Fat-soluble molecule composed of four hydrocarbon rings.

stigma Pollen-receptive part of the pistil of a flower.

stoma Epidermal pore in plants.

stomach Digestive organ of an alimentary tract.

strobilus Cone composed of sporangium-bearing, modified leaves (sporophylls).

substrate Medium upon which an organism grows; substance acted upon by an enzyme.

succession In ecology, the progression of organisms during the developmental history of an ecosystem.

supermale XYY individual.

symbiosis Intimate, often interdependent relationship between two unlike organisms.

symbiotic theory Theory that mitochondria and chloroplasts are derived from captive prokaryotes in the cytoplasm of primitive eukaryotes.

symmetrical, symmetry Regular in appearance; capable of bisection into like halves.

sympathetic nervous system Part of the vertebrate autonomic nervous system associated with stimulation of involuntary actions.

synapse To combine, as in chromosomal synapsis; junction between adjacent neurons or their extensions (axons and dendrites).

synapsis Pairing of homologous chromosomes during meiosis.

synaptic cleft Space between the synaptic knob of an axon and the membrane of the adjacent dendrite.

synaptic knob Ending of an axon at a nerve synapse.

synaptic vesicle Small sac containing neurotransmitters found in endings of efferent nerve fibers at a synapse.

syncytium Cell containing two or more nuclei and lacking partitioning membranes or crosswalls.

synergid One of (usually) two cells associated with the egg in the embryo sac of flowering plants.

tactile Relating to touch.

taiga Northern coniferous forest (subarctic forest).

target cell Cell in which a hormone induces a specific response.

taxis In animals, a behavioral reorientation to a stimulus.

telencephalon Posteriormost part of the vertebrate forebrain.

telophase Final phase in mitosis and meiosis, in which the cytoplasm is cleaved and two cells are formed.

temperate deciduous forest Broad-leaved deciduous forest biome of temperate regions.

tentacle Boneless feeding or sensory arm of an animal.

tergum Upper thoracic or abdominal plate of arthropod exoskeletons.

territory Area that an animal will defend against same-species intruders.

testis Sperm-producing male gonad.

testosterone Vertebrate male sex hormone.

tetany Condition in which muscles contract violently and remain contracted.

tetrad Four-part structure resulting from the duplication of each pair of homologous chromosomes in meiosis; group of four sister meiospores.

tetraspore Meiospore of red alga; any meiospore formed in cubical arrays of four spores.

thalamus Relay center beneath the vertebrate midbrain; composed of neurons in the lateral walls of the third ventricle of the brain.

therapsid Late Paleozoic reptile from which mammals evolved.

thermoperiod Cycle of daily temperatures with respect to growth and development of organisms.

thymine Pyrimidine base of nucleic acids.

thymus Endocrine gland, most active in juveniles, that produces hormones having immune functions.

thyroid Endocrine gland in the neck of tetrapods that produces the hormones thyroxin and calcitonin.

tissue In any multicellular organism, an association of cells having like structure and function.

tissue culture Process of removing tissue from an organism and growing it on a nutrient medium in sterile culture.

t-lymphocyte White blood cell capable of inactivation and destruction of other cells (e.g., microorganisms, cancer cells).

tonicity State of water balance (solute concentration) between the internal environment of a cell and the external environment.

totipotency Expression of the full developmental potential in a cell or tissue.

toxicity Having the ability to disrupt cells or interfere with cellular processes; poisonous.

trace element See micronutrient.

trachea Rigid, tubular air passage in the body of an animal.

tracheid Elongate, empty, pitted xylem cell having pointed ends that lack openings.

transcription Synthesis of DNA by copying the genetic coding of a parent DNA strand.

transduction Virus-mediated transfer of prokaryotic or eukaryotic DNA from cell to cell.

transfer RNA (tRNA) Amino-acid-transporting RNA molecule.

transformation Direct incorporation of foreign DNA by a bacterium.

transpiration Evaporation of water from within leaves and movement of the resulting water vapor through the stomata.

transpiration pull Tension exerted on water in xylem vessels and tracheids as a result of transpiration.

transposable element. See transposon.

transposon Segment of DNA capable of movement from plasmid to plasmid, from plasmid to chromosome, or chromosome to chromosome.

trichome Single-celled or multicellular hair growing from the epidermis of a plant.

triglyceride Lipid composed of three fatty acids and glycerol.

trilobite Primitive, extinct, many-legged marine arthropod.

triose Any 3-carbon sugar.

tripinnate Three-times divided (leaf).

Note: For specific chemical compounds and organisms, refer to the index.

triplet See nucleotide triplet.

trisomy Condition characterized by one additional chromosome.

trochophore Ciliated larva of annelids and mollusks.

trophic level Any stage or link in the energy transfer transactions in a food chain.

tropism Bending response of plants resulting from phytohormone-regulated, asymmetrical growth.

tropomyosin Component with troponin of muscle filaments.

troponin See tropomyosin.

trunk Stem of a tree; thorax and abdomen of a vertebrate body.

tube cell Cell of a pollen grain that elongates to form a pollen tube.

tubulin Subunit of microtubules.

tundra Alpine, arctic, or antarctic ecosystem characterized by permanently frozen subsoil (permafrost).

turgid Distended, usually with water.

turgor movement Plant movement brought about by changes in hydrostatic pressure in cells.

Turner's syndrome Female human disorder resulting from the presence of only one X chromosome.

ulna One of two bones in the forearm.

ultraviolet Light energy in wavelengths ranging between X rays and visible light.

umbilical cord Tubular connective of the body of a fetus and its placenta; contains blood vessels and connective tissue.

unicellular Composed of one cell.

uniformitarianism Concept that existing forces or actions have operated uniformly from the origin of the earth to the present.

universal symmetry See spherical symmetry.

uracil Pyrimidine base in RNA.

ureter Duct leading from the kidney to the urinary bladder.

urethra Duct leading from the urinary bladder to outside the animal body.

urinary bladder Storage receptacle for urine in vertebrates.

urine Waste product of the kidneys.

urogenital system Combined system of vertebrate kidneys and urinary ducts, together with gonads, gametic ducts, and sex organs.

uterus Enlarged chamber, connected with an oviduct, in which eggs are stored or embryos develop.

vaccination Immunization by introducing an antigen into the body.

vagina Tubular canal between the genital aperture and the uterus.

valence Measure of the bonding capacity of an atom.

variable Deviating from an established type; experimental subject as opposed to a control in the application of the scientific method.

vascular cambium Meristem in stems and roots giving rise laterally to xylem and phloem.

vascular tissue In plants, the tissue through which water and nutrients move; in animals, the blood and other components of a blood vascular system.

vas deferens   Sperm duct extending from the testes to a seminal vesicle or, in some cases, to the exterior of an animal body.

vasodilation   Expansion of blood vessels, increasing blood flow.

vegetal pole   Hemisphere of an animal egg where yolk density is greatest.

vein   Vessel that returns blood to the heart.

vena cava   Major vein by which blood returns to the heart of a land vertebrate.

ventral   Lower surface of an organ or organism.

ventricle   Heart chamber; cavity of the vertebrate brain.

vertebra   Bony or cartilaginous element of the backbone of vertebrates (higher chordates).

vertebrate   Any animal having a backbone composed of vertebrae (i.e., a jointed backbone).

vesicle   Small membranous sac.

villus   Fingerlike projection of the vertebrate intestinal lining.

virus   Noncellular, microscopic, infective particle composed of protein and nucleic acid (RNA or DNA).

visceral   Pertaining to the viscera; organs within the abdominal cavity.

visceral mass   Body region of a mollusk containing intestines and other visceral organs.

visceral muscle   *See* smooth muscle.

visible spectrum   Visible wavelengths of light.

vitamin   Food component required by plants and animals for replenishment of respiratory coenzymes.

vivipary   Bearing of young in the womb (uterus) until birth (as opposed to production of shelled eggs that require incubation).

vulva   External genitalia of female mammals.

wastes   By-products of metabolism.

water cycle   Continuous circulation of water between the atmosphere, oceans, and land.

water vascular system   Circulatory and locomotory system of echinoderms.

wheel organ   Ciliated, funnel-like structure in or near the mouth of a rotifer.

whip fungus   Any member of the division Mastigomycota; group having hyphae without crosswalls (i.e., coenocytic hyphae).

white cell   Any of several kinds of leukocytes or lymphocytes.

white light   Entire visible spectrum as perceived by the human eye.

wild type   Phenotype of a free-living organism; allele common in a wild population.

work   Application of energy resulting in an increase in orderliness; expenditure of free energy.

X chromosome   Chromosome associated with sex determination in many organisms, usually the only sex chromosome(s) present in females.

X ray   Radiation of short wavelength, having high penetrative capacity and capable of causing gene mutations.

xylem   Water-conducting tissue of plant stems, roots, leaves, etc.

Y chromosome   Sex-associated chromosome in many organisms, usually related to male sex determination.

yolk sac   Food storage sac of embryos of reptiles and birds; fluid-filled sac in mammalian embryos functioning as an initial source of blood cells.

z-line   Periodic transverse meshwork to which actin fibrils are attached in a striated muscle cell.

zoology   Scientific study of animal life.

zooplankton   Minute heterotrophic planktonic organisms, including protozoans, crustaceans, and various animal larvae.

zoospore   Motile spore.

zygospore   Zygote having a resistant shell or wall.

zygote   Cellular product of gametic fusion.

Note: For specific chemical compounds and organisms, refer to the index.

Index

The Authors

KNUT NORSTOG is presently serving as Research Associate at Fairchild Tropical Garden, Miami, working principally on the reproductive biology of cycads. He has had teaching duties at Luther College, Wittenberg University, the University of South Florida, and Northern Illinois University, covering general botany, introductory biology, psychology, electron microscopy, and plant development. His research has been concerned with ultrastructure of plant reproduction, plant tissue, and embryoculture. Recent research took Dr. Norstog to Colombia where he was instrumental in the recovery of a rare cycad lost from scientific collections since the late 1800s. Dr. Norstog has had approximately 50 research papers published, as well as several book chapters, and he is the coauthor of the book *Plant Biology*. Through the end of 1984, he served as Editor-in-Chief of the *American Journal of Botany*.

ANDREW J. MEYERRIECKS, currently Professor of Biology at the University of South Florida in Tampa, has also served on the faculties of Boston College and Harvard University. His specialty courses are animal social behavior and primate social behavior; he has also taught general zoology, organic evolution, genetics, and general biology. The focus of Dr. Meyerriecks's research is on the breeding and related behavior of herons, egrets, and bitterns. In addition to descriptive studies of these birds, he is interested in the origin and evolution of their various displays. Field studies have taken him to the marshes, swamps, and mangroves of Florida, Mexico, New York, and Massachusetts. His numerous research publications include papers in technical journals, a book (*Comparative Breeding Behavior of Four Species of North American Herons*), and major contributions to the *Handbook of North American Birds*.